Photonic Switching

Progress in Lasers and Electro-Optics

Peter W. E. Smith, *Series Editor*

Integrated Optics: Devices and Applications, edited by Joseph T. Boyd
Coherent Lightwave Communications, edited by Paul S. Henry and Stewart D. Personick
Photonic Switching, edited by H. Scott Hinton and John E. Midwinter
Semiconductor Diode Lasers, Vol. 1, edited by William Streifer and Michael Ettenberg
Microlenses: Coupling Light to Optical Fibers, edited by Huey-Daw Wu and Frank S. Barns

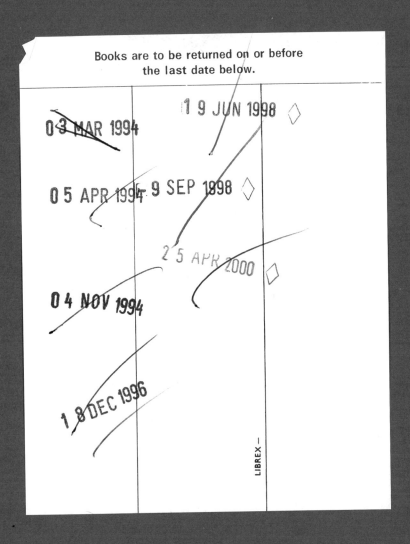

PROGRESS IN LASERS AND ELECTRO-OPTICS

Photonic Switching

EDITED BY

H. Scott Hinton

Department Head, Photonic Switching Department
AT&T Bell Laboratories

John E. Midwinter

British Telecom Professor of Optoelectronics
and
Department Head, Electrical and Electronic
Engineering Department
University College London

**A Series published for the
IEEE LASERS AND ELECTRO-OPTICS SOCIETY**
Peter W. E. Smith, *Series Editor*

This volume published in cooperation with the IEEE Communications Society

The Institute of Electrical and Electronics Engineers, Inc., New York

IEEE PRESS
1990 Editorial Board
Leonard Shaw, *Editor in Chief*
Peter Dorato, *Editor, Selected Reprint Series*

F. S. Barnes	W. K. Jenkins	M. I. Skolnik
J. E. Brittain	A. E. Joel, Jr.	G. S. Smith
S. H. Charap	R. G. Meyer	P. W. Smith
D. G. Childers	Seinosuke Narita	M. E. Van Valkenburg
R. C. Dorf	W. E. Proebster	Omar Wing
L. J. Greenstein	J. D. Ryder	J. W. Woods
J. F. Hayes	G. N. Saridis	

Dudley R. Kay, *Managing Editor*
Carrie Briggs, *Administrative Assistant*

Bernice Pettinato, *Associate Editor*

Copyright © 1990 by
THE INSTITUTE OF ELECTRICAL AND ELECTRONICS ENGINEERS, INC.
345 East 47th Street, New York, NY 10017-2394
All rights reserved.

PRINTED IN THE UNITED STATES OF AMERICA

IEEE Order Number: PC0253-5

Library of Congress Cataloging-in-Publication Data
Photonic switching / edited by H. Scott Hinton, John E. Midwinter.
    p.   cm. — (Progress in lasers and electro-optics)
  ISBN 0-87942-260-2
    1. Telecommunication—Switching systems—Equipment and
supplies. 2. Optical communications. 3. Optical data processing.
I. Hinton, H. Scott. II. Midwinter, John E. III. Series.
TK5103.8.P49 1989
621.381′045—dc20   90-4083
                                                                           CIP

Contents

Foreword ix

Introduction 1

Part 1: Overviews 3
On the Physical Limits of Digital Optical Switching and Logic Elements, *P. W. Smith* (*Bell System Technical Journal*, October 1982) 5
Digital Optics, Smart Interconnect or Optical Logic, *J. E. Midwinter* (*Physics in Technology*, May and July 1988) 24
Architectural Considerations for Photonic Switching Networks, *H. S. Hinton* (*IEEE Journal on Selected Areas of Communications*, August 1988) 38

Part 2: Devices and Components 57
Section 2.1: Electro-Optic Devices 59
Electrically Switched Optical Directional Coupler: Cobra, *M. Papuchon, Y. Combemale, X. Mathieu, D. B. Ostrowsky, L. Reiber, A. M. Roy, B. Sejourne, and M. Werner* (*Applied Physics Letters*, September 1, 1975) 59
Switched Directional Couplers with Alternating $\Delta\beta$, *H. Kogelnik and R. V. Schmidt* (*IEEE Journal of Quantum Electronics*, July 1976) 62
Polarization-Independent Optical Directional Coupler Switch Using Weighted Coupling, *R. C. Alferness* (*Applied Physics Letters*, November 15, 1979) 68
Directional Coupler Switches, Modulators, and Filters Using Alternating $\Delta\beta$ Techniques, *R. V. Schmidt and R. C. Alferness* (*IEEE Transactions on Circuits and Systems*, December 1979) 71
Low-Loss Polarization-Independent Electrooptical Switches at $\lambda = 1.3$ μm, *L. McCaughan* (*IEEE Journal of Lightwave Technology*, February 1984) 81
Strictly Nonblocking 8 × 8 Integrated Optical Switch Matrix, *P. Granestrand, B. Stoltz, L. Thylen, K. Bergvall, W. Döldissen, H. Heinrich, and D. Hoffmann* (*Electronics Letters*, July 17, 1986) 86
Performance of Two 4 × 4 Guided-Wave Photonic Switching Systems, *J. R. Erickson, G. A. Bogart, R. F. Huisman, and R. A. Spanke* (*IEEE Journal on Selected Areas in Communications*, August 1988) 89
Balanced Bridge Modulator Switch Using Ti-Diffused $LiNbO_3$ Strip Waveguides, *V. Ramaswamy, M. D. Divino, and R. D. Standley* (*Applied Physics Letters*, May 15, 1978) 96
Electrically Active Optical Bifurcation: BOA, *M. Papuchon and A. M. Roy* (*Applied Physics Letters*, August 15, 1977) 99
Electro-Optic X-Switch Using Single-Mode Ti:$LiNbO_3$ Channel Waveguides, *A. Neyer* (*Electronics Letters*, July 1983) 101
Optical Channel Waveguide Switch and Coupler Using Total Internal Reflection, *C. S. Tsai, B. Kim, and F. R. El-Akkari* (*IEEE Journal of Quantum Electronics*, July 1978) 103
Digital Optical Switch, *Y. Silberberg, P. Perlmutter and J. E. Baran* (*Applied Physics Letters*, July 1978) 108

Section 2.2: Logical Switching Elements 111
Optical Bistability, Photonic Logic, and Optical Computation, *S. D. Smith* (*Applied Optics*, May 15, 1986) 111
Use of a Single Nonlinear Fabry-Perot Etalon as Optical Logic Gates, *J. L. Jewell, M. C. Rushford, and H. M. Gibbs* (*Applied Physics Letters*, January 15, 1984) 126
GaAs-ALAs Monolithic Microresonator Arrays *J. L. Jewell, A. Scherer, S. L. McCall, A. C. Gossard, and J. H. English* (*Applied Physics Letters*, July 13, 1987) 129
Novel Hybrid Optically Bistable Switch: The Quantum Well Self-Electro-Optic Effect Device, *D. A. B. Miller, D. S. Chemla, T. C. Damen, A. C. Gossard, W. Wiegmann, T. H. Wood, and C. A. Burrus* (*Applied Physics Letters*, July 1, 1984) 132
The Quantum Well Self-Electroptic Effect Device: Optoelectronic Bistability and Oscillation, and Self-Linearized Modulation, *D. A. B. Miller, D. S. Chemla, T. C. Damen, T. H. Wood, C. A. Burrus, Jr., A. C. Gossard, and W. Wiegmann* (*IEEE Journal of Quantum Electronics*, September 1985) 135

Integrated Quantum Well Self-Electro-Optic Device: 2 × 2 array of Optically Bistable Switches, *D. A. B. Miller, J. E. Henry, A. C. Gossard, and J. H. English* (*Applied Physics Letters,* September 29, 1986) 149

Symmetric Self-Electro-Optic Device: Optical Set-Reset Latch, *A. L. Lentine, H. S. Hinton, D. A. B. Miller, J. E. Henry, J. E. Cunningham, and L. M. F. Chirovsky* (*Applied Physics Letters,* April 25, 1988) 152

Hard Limiting Opto-Electronic Logic Devices, *P. Wheatley, M. Whitehead, P. J. Bradley, G. Parry, J. E. Midwinter, P. Mistry, M. A. Pate, and J. S. Roberts* (*Photonic Switching: Proceedings of the First Topical Meeting,* 1987) 155

Section 2.3: Optical Amplifiers 159

Semiconductor Laser Optical Amplifiers for Use in Future Fiber Systems, *M. J. O'Mahony* (*Journal of Lightwave Technology,* April 1988) 159

All-Optical Repeater, *Y. Silberberg* (*Optics Letters,* June 1986) 173

Guided-Wave Optical Gate Matrix Switch, *A. Himeno, H. Terui, and M. Kobayashi* (*Journal of Lightwave Technology,* January 1988) 176

Section 2.4: Hybrid Switching Devices 181

Hybrid Optoelectronic Integrated Circuit, *R. I. MacDonald, D. K. W. Lam and B. A. Syrett* (*Applied Optics,* March 1, 1987) 181

4 × 4 OEIC Switch Module Using GaAs Substrate, *T. Iwama, T. Horimatsu, Y. Oikawa, K. Yamaguchi, M. Sasaki, T. Touge, M. Makiuchi, H. Hamaguchi, and O. Wada* (*Journal of Lightwave Technology,* June 1988) 184

Section 2.5: Spatial Light Modulators 191

Two-Dimensional Magneto-Optic Spatial Light Modulator for Signal Processing, *W. E. Ross, D. Psaltis, and R. H. Anderson* (*Optical Engineering,* July/August 1983) 191

Characteristics of the Deformable Mirror Device for Optical Information Processing, *D. R. Pape and L. J. Hornbeck* (*Optical Engineering,* November/December 1983) 197

The Current Status of Two-Dimensional Spatial Light Modulator Technology, *A. D. Fisher and J. N. Lee* (*Optical and Hybrid Computing,* 1987) 204

Section 2.6: Ultrafast Devices (Subpicosecond) 225

Femtosecond Switching in a Dual-Core-Fiber Nonlinear Coupler, *S. R. Friberg, A. M. Weiner, Y. Silberberg, B. G. Sfez, and P. S. Smith* (*Optics Letters,* October 1988) 225

Section 2.7: Optical Interconnects 229

Comparison Between Optical and Electrical Interconnects Based on Power and Speed Considerations, *M. R. Feldman, S. C. Esener, C. C. Guest, and S. H. Lee* (*Applied Optics,* May 1, 1988) 229

Optical Perfect Shuffle, *A. W. Lohmann, W. Stork, and G. Stucke* (*Applied Optics,* May 15, 1986) 239

Compact Optical Generalized Perfect Shuffle, *G. Eichmann and Y. Li* (*Applied Optics,* April 1, 1987) 241

Optical Implementations of the Perfect Shuffle Interconnection, *K.-H. Brenner and A. Huang* (*Applied Optics,* January 1, 1988) 244

Crossover Networks and Their Optical Implementation, *J. Jahns and M. J. Murdocca* (*Applied Optics,* August 1, 1988) 247

Hierarchic and Combinatorial Star Couplers, *M. E. Marhic* (*Optics Letters,* August 1984) 253

Combinatorial Star Couplers for Single-Mode Optical Fibers, *M. E. Marhic* (*Proceedings: Papers Presented at the Eighth International Fiber Optics Communications and Local Area Network Exposition [FOC/Lan 84],* 1984) 256

Small Loss-Deviation Tapered Fiber Star Coupler for LAN, *S. Oshima, T. Ito, K.-I. Donuma, H. Sugiyama, and Y. Fujii* (*Journal of Lightwave Technology,* June 1985) 261

Reflective Single-Mode Fiber-Optic Passive Star Couplers, *A. A. M. Saleh and H. Kogelnik* (*Journal of Lightwave Technology,* March 1988) 265

Efficient $N \times N$ Star Couplers Using Fourier Optics, *C. Dragone* (*Journal of Lightwave Technology,* March 1989) 271

Section 2.8: Beam-Combination 283

Design of an Optical Digital Computer, *M. E. Prise, M. M. Downs, F. B. McCormick, S. J. Walker, and N. Streibl* (*Optical Bistability-IV,* 1988) 283

Section 2.9: Spot-Array-Generation 287

Array Illuminator Based on Phase Contrast, *A. W. Lohmann, J. Schwider, N. Streibl, and J. Thomas* (*Applied Optics,* July 15, 1988) 287

Part 3: Networks and Systems ... 295

Section 3.1: Switching Networks ... 297

Ultrafast All-Optical Synchronous Multiple Access Fiber Networks, *P. R. Prucnal, M. A. Santoro, and S. K. Sehgal* (*Journal of Selected Areas in Communications,* December 1986) ... 297

Spread Spectrum Fiber-Optic Local Area Network Using Optical Processing, *P. R. Prucnal, M. A. Santoro, and T. R. Fan* (*Journal of Lightwave Technology,* May 1986) ... 307

Encoding and Decoding of Femtosecond Pulses, *A. M. Weiner, J. P. Heritage, and J. A. Salehi* (*Optics Letters,* April 1988) ... 315

Demonstration of High Capacity in the LAMBDANET Architecture: A Multiwavelength Optical Network, *H. Kobrinski, R. M. Bulley, M. S. Goodman, M. P. Vecchi, C. A. Brackett, L. Curtis, and J. L. Gimlett* (*Electronics Letters,* July 30, 1988) ... 318

WDM Coherent Optical Star Network, *B. S. Glance, K. Pollock, C. A. Burrus, B. L. Kasper, G. Eisenstein, and L. W. Stulz* (*Journal of Lightwave Technology,* January 1988) ... 321

Multiwavelength Optical Crossconnect for Parallel-Processing Computers, *E. Arthurs, J. M. Cooper, M. S. Goodman, H. Kobrinski, M. Tur, and M. P. Vecchi* (*Electronics Letters,* January 21, 1988) ... 326

Demonstration of Fast Wavelength Tuning for a High Performance Packet Switch, *M. S. Goodman, J. M. Cooper, H. Kobrinski, and M. P. Vecchi* (*Fourteenth European Conference on Optical Communications [ECOC 88], Conference Publication Number 292—Part 1,* 1988) ... 328

FDMA-FSK Star Network with a Tunable Optical Filter Demultiplexer, *I. P. Kaminow, P. P. Iannone, J. Stone, and L. W. Stulz* (*Journal of Lightwave Technology,* September 1988) ... 332

Section 3.2: Switches ... 341

Architectures for Large Nonblocking Optical Space Switches, *R. A. Spanke* (*IEEE Journal of Quantum Electronics,* June 1986) ... 341

Dilated Networks for Photonic Switching, *K. Padmanabhan and A. N. Netravali* (*IEEE Transactions on Communications,* December 1987) ... 345

Photonic Switching Modules Designed with Laser Diode Amplifiers, *J. D. Evankow, Jr., and R. A. Thompson* (*IEEE Journal on Selected Areas in Communications,* August 1988) ... 354

An Experiment on High-Speed Optical Time-Division Switching, *S. Suzuki, T. Terakado, K. Komatsu, K. Nagashima, A. Suzuki, and M. Kondo* (*Journal of Lightwave Technology,* July 1986) ... 363

An Experimental Photonic Time-Slot Interchanger Using Optical Fibers as Reentrant Delay-Line Memories, *R. A. Thompson and P. P. Giordano* (*Journal of Lightwave Technology,* January 1987) ... 369

An Optical Switching and Routing System Using Frequency Tunable Cleaved-Coupled-Cavity Semiconductor Lasers, *N. A. Olsson and W. T. Tsang* (*IEEE Journal of Quantum Electronics,* April 1984) ... 378

A Coherent Photonic Wavelength-Division Switching System for Broadband Networks, *M. Fujiwara, N. Shimosaka, M. Nishio, S. Suzuki, S. Yamazaki, S. Murata, and K. Kaede* (*Fourteenth European Conference on Optical Communication [ECOC 88], Conference Publication Number 292—Part 2,* 1988) ... 381

Eight-Channel Wavelength-Division Switching Experiment Using Wide-Tuning-Range DFB LD Filters, *M. Nishio, T. Numai, S. Suzuki, M. Fujiwara, M. Itoh, and S. Murata* (*Fourteenth European Conference on Optical Communication [ECOC 88], Conference Publication Number 292—Part 2,* 1988) ... 385

Section 3.3: Packet Switching ... 389

Deterministic and Statistic Circuit Assignment Architectures for Optical Switching Systems, *A. de Bosio, C. De Bernardi, and F. Melindo* (*Topical Meeting on Photonic Switching,* 1987) ... 389

An Optoelectronic Packet Switch Utilizing Fast Wavelength Tuning, *H. Kobrinski, E. Arthurs, R. M. Bulley, J. M. Cooper, E. L. Goldstein, M. S. Goodman, and M. P. Vecchi* (*GLOBECOM '88 IEEE Global Telecommunications Conference & Exhibition, Conference Record Volume II,* 1988) ... 392

Section 3.4: Systems Issues ... 397

Photonic Switching Technology: Component Characteristics versus Network Requirements, *J. E. Midwinter* (*Journal of Lightwave Technology,* October 1988) ... 397

Design of Lithium Niobate Based Photonic Switching Systems, *W. A. Payne and H. S. Hinton* (*IEEE Communications Magazine,* May 1987) ... 405

Optical Considerations in the Design of Digital Optical Computers, *M. E. Prise, N. Streibl, and M. M. Downs* (*Optical and Quantum Electronics,* January 1988) ... 410

Author Index ... 439

Subject Index ... 441

Editors' Biographies ... 447

Foreword

This book is one of the first to appear in the new series "Progress in Lasers and Electro-Optics" (PLEOS). The IEEE Lasers and Electro-Optics Society is sponsoring this series to provide background and up-to-date information in rapidly evolving areas of laser technology, laser applications, optical and opto-electronic devices, and optical signal processing.

Because the field has evolved so rapidly, there are very few contemporary books on lasers and electro-optics. In many cases, the topics of interest are still subjects of current research. This series has been designed to provide researchers, optical engineers, and professors teaching courses in these new areas of technology with access to the latest advances and understanding.

Other reprint volumes in the PLEOS series include:

Integrated Optics: Devices and Applications, edited by Joseph T. Boyd

Coherent Lightwave Communications, edited by Paul S. Henry and Stewart D. Personick

Semiconductor Diode Lasers, Volume I, edited by William Streifer and Michael Ettenberg

Microlenses: Coupling Light to Optical Fibers, edited by Huey-Daw Wu and Frank S. Barnes

The editors of each reprint volume are experts in their field and have written explanatory material to guide the reader through a carefully selected collection of reprints. I hope you will find this volume useful both for increasing your current knowledge and as a reference for years to come.

Peter W. E. Smith
Series Editor

Introduction

The term "photonic switching," which has come into common use only in the last few years, suggests a new high-performance technology that has the capability of revolutionizing the telecommunications and computing industries. By collecting in this volume a cross-section of what has been published on the subject, we hope to give the reader a balanced and reasonably up-to-date overview of what has been achieved, and in this way point him to what may be expected.

The understanding and utilization of this new technology has been investigated from the perspective of three different engineering disciplines. The first group of researchers are from an electronic-switching background. They perceive problems with electronic switches, often in the area of control rather than data throughput, and, observing the immense success of optical transmission and the claims of optical device researchers, conclude that optics can assist them.

The second group, optical device researchers, have strong backgrounds in optical fibers, planar "integrated optics," or III-V opto-electronics. For these researchers, it has been clear for many years that there are devices that can switch or route optical signals, and it is natural to seek applications for these devices at the nodes of an optical fiber transmission network. Many of these devices are virtually transparent in both the optical and data sense, so that they offer vast transmission bandwidths.

Finally, we find a group who come from an optical computing or nonlinear optics background, a field that has been in existence since the early 1960s and saw a massive resurgence of interest following the observation of bistable optical device action that seemed to imply that "optical logic" might be realistic for digital computing. Given the hope that such techniques might make possible massive parallel computers with huge data throughputs, it seems reasonable to assume that the same technology might lead to major progress in the design of digital switching machines.

The papers are organized by technology groupings, starting with the various types of discrete devices and moving toward the different types of "systems" that they might be used to construct. Here, it is important to note some very fundamental distinctions that occur in the design of systems and the consequent choices of components. A transmission system is simple in the sense that it is merely required to transport information at a certain data rate from point A to point B between well-defined interfaces and at a maximum defined error rate. A switching system, on the other hand, usually receives data from many sources and is required to reorder it so that it is redistributed to many sources. It is unusual for there to be a requirement to redirect a single 10-Gbit/s data stream from one destination to another. It is much more likely to receive an aggregate 10 Gbit/s of data from many (hundreds of) sources and to have to break it into its constituent messages and deliver it to many (hundreds of) different destinations. There are therefore problems at a number of different levels. Each individual data flow must be identified by source and destination. Thus there is associated with each message some control or routing information that must be used to establish the desired route. Key to the design of the switch is the relationship between the messages and their associated control or routing data.

To highlight the significance of this, we list below some well-known examples that occur in switching systems.

Block Switching. The need to reconfigure a telecommunications network to take into account the differing loads during the day or week, or in response to a special function such as a political convention! The reconfiguration rate of these types of switches is relatively slow compared to the bit rate, in a microsecond or longer.

Circuit Switching. The normal telephone switch sets up a "circuit" once the destination is dialed. Such a circuit is typically at 64 kbit/s for telephony but for future wideband services would be at a higher rate (150 Mb/s for broadband ISDN services). Once established, the circuit remains in operation for the duration of the call, which is likely to be a long time compared to the setting-up time. The control data are handled as a separate message that "precedes" the call through the system to establish a route. Notice that while many such voice channels, once multiplexed together, rapidly generate data rates that can fill a fiber, the individual channels are at relatively low data rates. The complexity of the system arises from the need to handle very large numbers of voice calls on an individual basis, unlike the transmission system, where there is little knowledge of how the traffic being transmitted is constructed.

Packet Switching. Whereas circuit switching is historically linked with holding a conversation, packet switching is linked with posting a letter. A packet of data, say 1000 bits in length, is "posted" into the network with an attached header that contains the "address" information. On arrival at the switching node, the header must be read, a suitable connection must be made, and the packet must again be posted to the next node or to its destination. At no time does a true circuit exist between sender and receiver.

Asynchronous Time Multiplexing. This is a variant of packet transmission and switching that is widely predicted to form the basis for future integrated multiservice networks where "bandwidth on demand" is offered to a customer rather than a series of separate networks, each specialized for a discrete function, packet, telex, telephone, etc. The user accesses more or less bandwidth by posting more or fewer packets/second. As a result, all services are carried in a packetlike format.

From these examples, we see that a number of themes emerge as items of key interest to any discussion of "photonic switching," of which transmission bandwidth is only one and perhaps the least significant. For example, we may list the following:

- ability to optically read and react to a routing header
- speed of setting up a path in a complex switching matrix
- transparency once route established
- scope for making messages "self-routing"
- extent to which electronic control negates any potential performance gain.

On every one of these issues, optics has something to offer, as will be demonstrated in the selection of papers that we have made. For example, data transparency is readily achieved in many optical switches but usually only at the penalty of electrical routing control. Optical "wiring" in intimate association with optical or electronic logic may overcome some of the problems of synchronization and control in very fast synchronous switches. And the colossal bandwidth offered by the optical transmission medium may open up some entirely new approaches to "self-routing," such as dense wavelength-division multiplexing and code division multiple access (CDMA).

Serious appraisal of the opportunities of this new technology by many telecommunications R&D laboratories is leading to a better understanding of its capabilities and future market value. We believe that it is rapidly becoming clear that optics will find a number of niche markets in switching quite soon and that these can be expected to grow and spread as wideband services become more widespread.

Part 1
Overviews

In this section we have gathered three papers that attempt to set some physical bounds to our subject. Since the competition is from electronics, we must appreciate what it can and cannot do. This is a moving target, but some limitations do emerge that are reasonably fundamental. In the same vein, optics also has some fundamental limitations, and these limitations are reviewed here. Finally, and perhaps most important, if we wish to design a new switch for a telecommunications network, then we must understand what is required within the network, and this imposes very significant constraints on our thinking. One might protest that this is tantamount to "hobbling" the new technology before it has ever learned to walk, let alone run. The inverse of this argument is that if you wish to be taken seriously by a switching manufacturer, then first understand his problems!

On the Physical Limits of Digital Optical Switching and Logic Elements

By P. W. SMITH

(Manuscript received February 5, 1982)

In this paper we identify and discuss some fundamental physical mechanisms that will provide limits on the speed, power dissipation, and size of optical switching elements. Illustrative examples are drawn primarily from the field of bistable optical devices. We compares the limits for optical switching elements with those for other switching technologies, and present a discussion of some potential applications of optical switching devices. Although thermal effects will preclude their wide application in general-purpose computers, the potential speed and bandwidth capability of optical devices, and their capability for parallel processing of information, should lead to a number of significant applications for specific operations in communication and computing fields.

I. INTRODUCTION

A number of recent developments have increased the interest in digital optical signal-processing devices and techniques. Laser technology has now advanced to the point that lasers are being used in consumer electronics. Optical fiber communication systems are being widely installed. Integrated-optics spectrum analyzers have been developed.

In the research stage it has been shown that optical fibers can be used to transmit information at rates approaching 1 THz.[1,2] This rate is much beyond the capabilities of any presently known electronic light detector. Thus, to utilize this information-handling capacity, some form of optical signal processing will have to be performed before the light signals are converted to electronic ones.

Low-power integrated-optics light switches[3] and low-energy integrated-optical bistable devices[4] capable of performing optical logic have been demonstrated. It is tempting to propose that such digital optical switching elements be used to construct high-speed computers as well as repeaters and terminal equipment for optical communications systems. To examine these possibilities we need to understand: (i) What are realistic possibilities for speed, power dissipation, and size for optical switching elements? and (ii) What are the fundamental limits imposed by the physics of the nonlinear interactions, and by the available optical materials?

Previous studies of these optical device limits have been made by several authors. The pioneering work of Keyes[5] examined several nonlinear optical processes and concluded that thermal considerations imposed severe limits on the use of optical logic elements. A similar assessment was made by Landauer.[6] A more optimistic conclusion was reached by Fork,[7] who suggested that by using suitable interactions and resonant structures, competitive optical elements could be realized. Recently Kogelnik[8] has examined the role of integrated-optics devices and has concluded that they may well be most useful in performing functions that cannot be provided by other technologies.

In this paper we will attempt to provide a perspective on the ultimate limits of optical switching elements, and the areas in which one might expect optical signal processing to offer a significant advantage over other technologies.

II. BISTABLE OPTICAL DEVICES

In this paper we will draw examples from the field of bistable optical devices. This is to some extent due to a bias of the author, as he has worked on these devices for several years. However, in many respects, bistable optical devices are the most basic binary optical systems, and they have a demonstrated capability for low-energy latching operation. In addition, these devices are extremely versatile and can function as optical limiters, differential amplifiers, and optical logic elements. Their transmission can also be controlled by another optical beam creating an "optical triode."

A generic bistable optical device is shown in Fig. 1. It consists of a Fabry-Perot resonator containing a nonlinear optical material. This nonlinearity can be either a saturable absorption (an absorption that decreases with increasing light intensity), or a nonlinear refractive index (a refractive index that increases or decreases with increasing light intensity). The bistability arises from the simultaneous requirements that the intensity of light inside the resonator, and thus the transmitted light intensity, depends on the resonator tuning (or the loss in the resonator), and the resonator tuning (or the loss in the resonator) depends on the intensity of light in the resonator. The resonator transmission exhibits optical hysteresis, as shown in Fig. 1b.

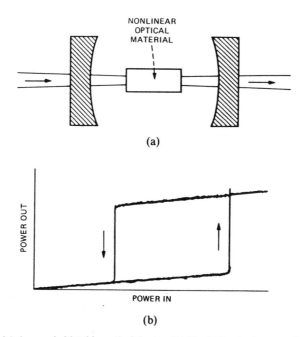

Fig. 1—(a) A generic bistable optical device. (b) Typical output power characteristic for a bistable optical device.

Optical bistability has been demonstrated using both intrinsic devices, which use materials with an intrinsic optical nonlinearity, and hybrid devices, which use a detector and an electrooptic modulator to create an artificial nonlinear medium.

Many types of bistable optical devices have been developed and studied. Small, integrated-optics hybrid devices have been operated with less than a picojoule of optical energy.[9] Two-dimensional arrays of bistable elements using a liquid-crystal light valve have been demonstrated for image-processing applications.[10] Nonresonant devices have been developed that use no Fabry-Perot resonator and thus have a broad frequency response and can switch very rapidly with a suitably fast-responding nonlinear material.[11,12]

III. SPEED AND POWER LIMITATIONS

In this section we will discuss the fundamental physical mechanisms that limit the performance of these devices. Although we will specifically consider bistable optical devices, the results will be generally applicable to any passive digital optical switching elements.

The switching speed of a bistable optical device is limited by the buildup time of the resonator, and by the response time of the nonlinear medium. In principle, the resonator response time can be made negligible by reducing the length of the resonator, or by using a nonresonant configuration. The ultimate limit is set by the response time of the nonlinearity. Materials exhibiting strong electronic nonlinearities are known with response times of $<10^{-14}$ seconds. To operate a device with such a short response time, however, requires high light powers.

The switching power and switching speed of a bistable optical device are not independent. For example, if the response time of a given device is dominated by the resonator buildup time, the response time can be reduced by a factor of two by halving the length of the resonator. However, as only half the length of nonlinear material is now available, twice the switching power must be used to reach the switching threshold.

R. W. Keyes[5] has discussed several physical processes that limit the switching power and speed of optical devices. For any (nonreversible) switching operation, it can be shown that a minimum energy of the order of kT must be dissipated (k is Boltzman's constant and T is the absolute temperature). Quantum mechanical considerations lead to the assertion that a switching operation must dissipate at least h/τ of energy (h is Planck's constant and τ is the switching time). These limits are shown in Fig. 2, which is a plot of the power required for a switching operation as a function of switching time, i.e., the time for which this power must be applied. The frequency label on the horizontal axis is appropriate if switching is being done repetitively so that a switching time limit implies a limit on the data rate.

Keyes has discussed the limitations imposed by the heat dissipated in a switching element. For continuous operation, this heat sets an upper limit on the achievable switching rate (a higher rate would result in an unacceptable temperature rise in the device). The region affected by such thermal considerations is also shown in Fig. 2, assuming a value of heat transfer coefficient (100 W/cm²) that is appropriate for liquid-cooled elements and a maximum acceptable temperature rise of 20°C. In many cases a lower temperature rise may be required because of the rapid refractive index change with temperature exhibited by most optical materials. It is important to note that a switching device can operate in this "thermal transfer" region provided that it is operated at less than the maximum repetition rate, or that not all of the switching power is dissipated in the device.

Keyes considered the case of an optical switching device that operates by absorbing light that saturates an atomic transition and changes the optical properties of the material. To achieve appreciable saturation, the condition

$$\sigma I > hc/\lambda \tau_D \tag{1}$$

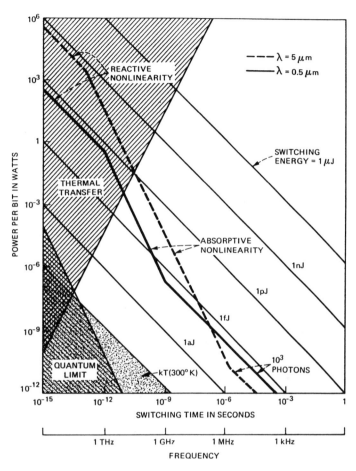

Fig. 2—Limitations on optical switching devices. The frequency scale (at bottom) applies for repetitive switching. The heavy lines indicate limits on optical switching devices for the case of $\lambda = 0.5$ μm (solid lines) and $\lambda = 5$ μm (dashed lines).

must be satisfied. Here, σ is the peak absorption cross section of the transition, I is the light intensity, h is Planck's constant, c is the velocity of light, λ is the wavelength of the light, and τ_D is the decay time of the transition. From the expression for the Einstein stimulated emission coefficient, we can write

$$\sigma = \frac{4\pi^2 |\mu|^2 T_2}{3\epsilon_0 h \lambda}, \qquad (2)$$

where $|\mu|$ is the dipole moment of the transition, T_2 is the inverse line width of the transition, and ϵ_0 is the vacuum permittivity.

The maximum intensity for a given input power is obtained in a waveguide geometry for which the input power, P, is given by

$$P \sim I\lambda^2 \tag{3}$$

(see Section IV). Combining eqs. 1, 2 and 3 we obtain:

$$P > \frac{3\epsilon_0 h^2 \lambda^2 c}{4\pi^2 |\mu|^2 T_2 \tau_D}. \tag{4}$$

To evaluate eq. 4 we will assume a large dipole moment for an atom given by

$$|\mu| = ea_0, \tag{5}$$

where e is the electronic charge and a_0 is the Bohr radius. We will further take the response time, τ, of the switching element to be

$$\tau = \tau_D \sim T_2. \tag{6}$$

(T_2 cannot be less than τ_D, and a larger T_2 implies a larger switching energy.) With these assumptions we find

$$P > 1.2 \times 10^{-24} \times \frac{\lambda^2}{\tau^2} \text{ W}, \tag{7}$$

where λ is in μm and τ is in seconds. This limit is plotted on Fig. 2 for $\lambda = 0.5$ μm and $\lambda = 5$ μm and labeled "Absorptive Nonlinearity." Two points should be noted. First, Keyes shows that a similar limit should also apply for the case of a second-order nonlinear effect [$\chi^{(2)}$ process] or for the case of a nonlinearity based on self-induced transparency. Second, it has recently been shown that it is possible to find systems in which one can use excitonic resonances[13] to obtain effective dipole moments appreciably greater than ea_0. An example using such a system is described in Section VI.

A different limit is found for the case of a third-order nonlinearity [$\chi^{(3)}$ process]. Let us consider a material exhibiting an optical Kerr effect, i.e., the refractive index has a term proportional to the light intensity. To obtain an appreciable effect, we require a change in phase shift through the medium

$$\Delta\phi > \pi. \tag{8}$$

If the refractive index change is $n_2 I$, where n_2 is the optical Kerr coefficient, then eq. 8 becomes

$$\frac{2\pi n_2 I \ell}{\lambda} > \pi, \tag{9}$$

where ℓ is the length of the element. The delay time is governed by the length

$$\tau = n_0 \ell / c, \tag{10}$$

where n_0 is the (linear) refractive index of the material. Combining eqs. 9, 10, and 3, we obtain

$$P > \frac{n_0 \lambda^3}{2n_2 c\tau}. \qquad (11)$$

Note that the dependence of P on τ is different from that in eq. 7.

How can we estimate n_2? We can argue that the nonlinear refractive-index term should be of the order of unity for light fields of the order of the atomic fields. Thus, we could write

$$n_2 I = n_0 \qquad (12)$$

for light fields of the order of e/a_0^2. Now

$$I = \frac{\epsilon_0 n_0 c E^2}{2}, \qquad (13)$$

where E is the electric field strength of the light. We combine eqs. 12 and 13 to obtain

$$n_2 = \frac{32\pi^2 \epsilon_0 a_0^4}{ce^2}. \qquad (14)$$

From eq. 14 we obtain $n_2 = 2.9 \times 10^{-17} [\text{W/cm}^2]^{-1}$. Much larger electronic nonlinearities are known, however. The polydiacetylene PTS has the largest-known value;[12] it is $n_2 \sim 6 \times 10^{-12} [\text{W/cm}^2]^{-1}$. The reason for this large value is that the electrons are relatively unconfined along the chain axis of the PTS molecules. Thus, effective distances much larger than a_0 are encountered, and as $n_2 \propto a^4$, nonlinearities much larger than our estimate are found. The limits represented by eq. 11 are shown in Fig. 2 for $\lambda = 0.5$ μm and $\lambda = 5$ μm and labeled "Reactive Nonlinearity." They are evaluated using the value of n_2 for PTS.

A third limit shown in Fig. 2 is derived from statistical considerations. A number of photons large compared with unity is necessary to define a switching state. We have somewhat arbitrarily taken 10^3 photons as this statistical limit, i.e.,

$$P\tau = 10^3 hc/\lambda. \qquad (15)$$

This limit is also plotted in Fig. 2 for $\lambda = 0.5$ μm and $\lambda = 5$ μm.

A word of caution is in order here. These limits that we have identified are "fuzzy" in that it may not be possible to do as well as these limits, or it may be possible to do somewhat better than these limits would indicate. In general, many kT of energy will be required for stable switching devices. On the other hand, the use of a high-finesse optical resonator will lower the required switching energy. These limits are intended to be used as a guide and an indication of the underlying physical mechanisms.

An optical resonator decreases the required switching power at the expense of a reduction in the bandwidth. A bistable optical device utilizing a lossless material with a refractive nonlinearity has a switching power that varies as

$$P \propto 1/F^2, \tag{16}$$

where F is the finesse of the resonator. In the region where the switching time is limited by the resonator,

$$\tau \propto F \tag{17}$$

so that the switching energy, defined as the product of the switching power and switching time, is given by

$$P\tau \propto 1/F. \tag{18}$$

If the switching time is limited by the response time of nonlinear material, τ will not depend on F, and

$$P\tau \propto 1/F^2. \tag{19}$$

The limits shown in Fig. 2 were computed assuming no resonator, i.e., for $F \sim 1$. We see that with high-finesse resonators, switching energies appreciably below the limits shown in Fig. 2 should be possible. The 10^3 photon limit will still apply, however.

IV. SIZE LIMITATIONS

To obtain the largest light intensity for a given input power, it is usually desirable to focus the input light. The light can be focussed to a cross-sectional area of $\sim\lambda^2$, but will diffract rapidly if not confined by some waveguiding structure. For this reason the lowest-power switching devices are likely to be those in which the light is guided in an optical dielectric waveguide with cross-sectional dimensions of $\sim\lambda$. (Kogelnik[8] has pointed out that with a smaller dielectric waveguide, the guided mode will extend beyond the waveguide walls so that the minimum light-beam cross section will always be of the order of λ. Light can be confined to smaller cross sections using metallic waveguides, but in this case large absorption losses will result.)

The minimum length of the waveguide (i.e., the device) will depend on the strength of the optical nonlinearity and on the finesse of the optical resonator. In many cases the linear absorption loss of the nonlinear medium will determine the resonator dimensions. Miller[14] has shown that for a high-finesse waveguide resonator containing a material with a nonlinear refractive index and a linear absorption loss, the lowest switching power will occur for a length of resonator such that

$$1 - R = A, \tag{20}$$

where R is the reflectivity of each of the resonator mirrors, and A is the absorption loss per pass through the nonlinear medium. If we write $A = \alpha\ell$, where ℓ is the length of the medium and α is the absorption coefficient, this condition requires

$$\alpha\ell = \pi/2F \qquad (21)$$

or

$$\ell = \pi/2\alpha F. \qquad (22)$$

Thus, the length of a device that is optimized for minimum switching power will be determined by the absorption coefficient and the finesse of the resonator. If the nonlinearity is caused by the absorption of light, then compact, fast, and efficient elements will require a large value of α. Under optimized conditions the switching power, P, varies as

$$P \propto 1/F. \qquad (23)$$

The optimum length of the device varies with R so that the resonator response time, τ, is independent of F. Thus, in this case we find the switching energy

$$P\tau \propto 1/F. \qquad (24)$$

In Table I, we show recent results from the literature on bistable optical devices. We have taken the switching time to be the "recovery time" of the device, i.e., the time constant for the return to equilibrium in the absence of a driving signal. It is important to note that in many cases a device can be switched on (or off) much more rapidly by the application of a short, intense driving pulse. The data in Table I shows a wide range of switching powers and speeds. As might be expected, the fastest devices require the highest switching powers. The lowest

Table I—Experiments: recent results

	Switching Power (watts)	Switching Time (seconds)	Switching Energy (joules)
Bistable Fabry-Perot resonators			
CS_2	3×10^5	5×10^{-10}	1.5×10^{-4}
Na vapor	10^{-2}	10^{-5}	10^{-7}
GaAs	2×10^{-1}	4×10^{-8}	8×10^{-9}
InSb	10^{-2}	$<5 \times 10^{-7}$	$<5 \times 10^{-9}$
Hybrid bistable Fabry-Perot resonator			
$LiNbO_3$	10^{-5}	5×10^{-8}	5×10^{-13}
Bistable liquid-crystal matrix			
Hughes liquid crystal light valve	5×10^{-7}	4×10^{-2}	2×10^{-8}
Nonlinear interface			
Glass—CS_2	2×10^5	2×10^{-12}	4×10^{-7}

reported switching energy is for a hybrid, integrated-optical bistable device. This low energy is possible because of the very large effective nonlinearity created by the electrooptic modulator with electrical feedback. Although for some applications long switching times or large switching powers are acceptable, in general one wishes to minimize each of these parameters. How much could we improve present bistable devices by shrinking device dimensions to provide shorter transit times and higher light intensities for a given input power?

We have extrapolated current experimental results to λ^2/n^2 cross-section waveguide devices with a length adjusted for minimum switching energy. We have assumed a finesse of 30 for the Fabry-Perot resonator and have considered the response time to be the undriven recovery time of the device. The results are shown in Fig. 3. It is interesting to note that with known devices and materials it should be possible to approach rather closely the fundamental limits for optical

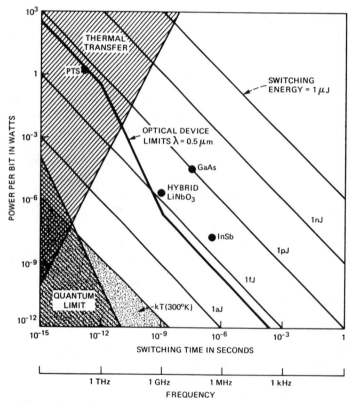

Fig. 3—The points represent performance limits extrapolated from current experimental values for waveguide Fabry-Perot resonators containing the polydiacetylene PTS, the semiconductor GaAs, the semiconductor InSb, and a hybrid device using the electrooptic crystal LiNbO$_3$.

devices; however, as we see in Table I, present laboratory devices are not yet developed to the point where they are close to these limits.

V. COMPARISON WITH OTHER SWITCHING TECHNOLOGIES

How do these projected results and limits that we have derived compare with those for other switching technologies? In Fig. 4 we show how the limits for bistable optical devices compare with the best reported values for two well-established switching technologies—semiconductor electronic devices, and Josephson devices. It is also interesting to see how these devices compare with a biological switching device—a neuron.

It is clear that in the 10^{-6} through 10^{-11} second region, one cannot hope to switch with substantially less power than that required for semiconductor electronic devices, and appreciably lower switching powers are possible with Josephson technology. In the 10^{-12} through 10^{-14} second region, however, optical devices appear to have no com-

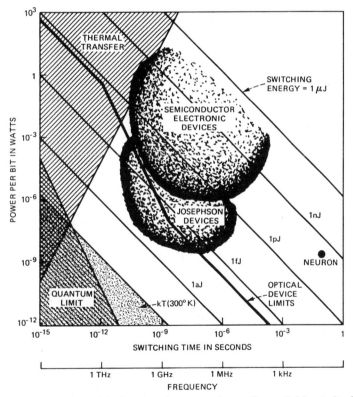

Fig. 4—A comparison with the operating range of two other switching technologies: semiconductor electronic devices, and Josephson devices. The operating point for a biological "switch"—a neuron—is also shown for comparison.

petition. This unique capability for sub-picosecond switching is one of the most exciting aspects of optical switching technology.

The switching power required in this short-time region puts the operating point well within the "thermal transfer" region discussed earlier. For this reason, it does not appear feasible to design a high-speed, general-purpose digital optical computer. However, for many applications these thermal limits may not present severe problems. Two points are worth noting. First, devices using a reactive nonlinearity do not depend on the absorption of the incident light. Thus, most of the switching power is transmitted by the device and the power dissipated is much less than the power required for switching. Second, for some applications, fast switching operations are required at relatively low duty cycles. In both cases the temperature rise in the switching elements will be much lower than the maximum value used in computing the "thermal transfer" region boundary.

There are many other factors that relate to the choice of a switching technology that cannot be shown on a power-time plot. In many cases it is desirable to perform some signal-processing operation on a light signal, either because the incoming signal is in the form of light or because freedom from electromagnetic interference is desired. Optical switching devices typically operate at room temperature. In many cases, they have extremely large bandwidths and can be adapted for many special functions such as rapid parallel processing of information. For these reasons, there will be cases where optical switching systems will be used, even in an area of Fig. 4 in which other technologies show a switching-energy advantage.

VI. ILLUSTRATIVE EXAMPLES

We have shown that because of thermal problems associated with the high packing densities required for rapid operation, optical switching elements are unlikely to be used as building blocks for a general-purpose computer. For certain specific applications, such as the integrated-optical spectrum analyzer recently developed for microwave signal processing, and optical computers for picture processing and pattern recognition,[15-17] special-purpose optical computers have already demonstrated their usefulness. We should also point out that fast optical switching devices are opening up a new time region for scientific studies, and picosecond spectroscopy is rapidly becoming an important field of research.[18]

In this section we will consider a few specific applications of optical switching elements, and see where an extrapolation of current technology might lead. In many cases nonlinear materials with a suitable combination of properties are not currently available. The materials we have chosen for these examples illustrate the wide range of properties that can be obtained.

6.1 Low-energy optical switch

The lowest "switch-off" energy currently demonstrated is 0.5 pJ for a hybrid bistable device (see Table I). Extrapolation of this figure with a LiNbO$_3$ device with minimum waveguide dimensions and assuming current detector technology, we find 1 fJ operation should be possible. A similar limit is found by extrapolating current figures for InSb devices at 5 µm wavelength. These figures might be reduced still further by using optical resonant structures related to those currently being employed for surface-enhanced Raman studies. (See, for example, Ref. 19.)

6.2 High-speed optical switch

The highest-speed operation will be obtained with a device utilizing a nonlinear material with an electronic nonlinearity. Such nonlinearities are believed to have response times in the range of 10^{-14} seconds. For minimum device response time, a nonresonant configuration should be used. A suitable configuration might be the self-focussing, bistable optical switch described in Ref. 20. By focusing the input to a spot size on the order of the wavelength of the light at the nonlinear material, adequate discrimination between "on" and "off" states could be obtained with a device length of 20 wavelengths. For a device using as nonlinear material the polydiacetylene PTS[12] and light of 1 µm wavelength, the response time would be 0.1 ps and the peak pulse switching power would be 100 W. This power is low enough that it might be reached with a mode-locked semiconductor laser diode.

6.3 4 x 4 optical switching network

If picosecond speed is not required, an optical distribution network could be formed with integrated optics technology on a LiNbO$_3$ substrate, as demonstrated by Schmidt and Buhl.[21] This example is somewhat different from the others we have given, in that electrical signals are used to control the distribution of the optical signals. Kogelnik[8] has addressed the question of the limits for the stepped $\Delta\beta$ couplers that comprise the switching units. He shows that the limiting electrical energy, E_{SW}, needed for one switch of the optical path is given by

$$E_{SW} \cdot \tau = 100 \text{ pJ} \times \text{ps}, \tag{25}$$

where τ is the transit time through the device. A directional coupler switch with a switching time of 110 ps and a transit time of 3 ps has already been demonstrated.[22] Switching times of 30 ps should be possible with 1 μm electrode gaps.[22] LiNbO$_3$ waveguides suffer severe problems at optical power levels higher than ~100 μW in the visible range. Much higher power levels are possible, however, if near infrared light is used.

6.4 Image amplifier

An optical image amplifier could be made using an array of bistable elements, as shown in Fig. 5a. Each element in the array is a self-focussing bistable element similar to that described in Ref. 20. A typical output characteristic for each element is shown in Fig. 5b. It can be seen that a weak input signal (I_S) produces a strong modulated signal at the output when the input light level corresponds to the

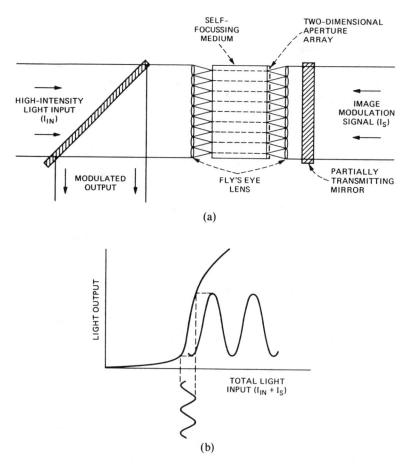

Fig. 5—An optical image amplifier. (a) Schematic diagram. (b) Operating characteristic of each element.

"knee" in the characteristic curve. Thus, each spatial element behaves as an "optical triode."

One possible nonlinear medium for this application would be a liquid suspension of sub-micron dielectric particles. As shown in Ref. 23, this medium exhibits a large nonlinear coefficient, although it has a slow response time on the order of a second. Such times may be acceptable for certain image-processing operations. If each element of the input image is focussed by the composite lens to a spot size of $\sim\lambda^2$, with an aqueous suspension of quartz particles one would require an input power of 10 mW/resolution element at $\lambda = 0.5\ \mu$m and 1 mW/resolution element at $\lambda = 5\ \mu$m. Recent calculations[24] indicate that input powers of $\sim 100\ \mu$W/resolution element and a response time of $\sim 1\ \mu$sec might be possible using suitably doped GaAs as the nonlinear medium. The verification of these ideas must await further experiments.

6.5 Optical time-division multiplexer and demultiplexer

As our final example, let us consider a high-speed optical time-division multiplexer (or demultiplexer) that might be used to multiplex picosecond optical pulses into a high-capacity optical fiber, and demultiplex the signals at the receiver to obtain low enough bit rates to allow handling by optical detectors and subsequent electronic systems.

A multiplexer can be made from a number of triggerable switching elements, as shown in Fig. 6a. A trigger pulse with the proper time synchronization is required to multiplex pulses as shown. Each element could be made from a properly designed bistable optical device, as shown in Fig. 6b. This bistable device consists of a suitable nonlinear optical material in a ring resonator. The ring geometry allows separation of the inputs and outputs. However, some polarization selectivity may have to be employed to avoid interference effects between the two input beams $I_{IN}^{(1)}$ and $I_{IN}^{(2)}$. Let us assume here that pulses in these two beams are never present simultaneously in the element. The output intensity depends on the total input intensity $I_{IN}^{(1)} + I_{IN}^{(2)} + I_{TRIG}$, as shown in Fig. 6c. If the input pulses are of intensity slightly less than the critical intensity corresponding to the "knee" of the curves, the output in the absence of a trigger input will consist solely of $I_{IN}^{(1)}$. However, during the time that a small trigger signal I_{TRIG} is present, the output will consist solely of $I_{IN}^{(2)}$. Because of the sharp "knee" in the characteristic curves, only a small I_{TRIG} is required to accomplish this switching.

In a similar way, one can perform demultiplexing by using an optical "tap," as shown in Fig. 7a. A similar bistable ring resonator serves as a triggerable "tap," as shown in Fig. 7b; the output characteristics are shown in Fig. 7c.

An appropriate nonlinear material for use in these devices might be

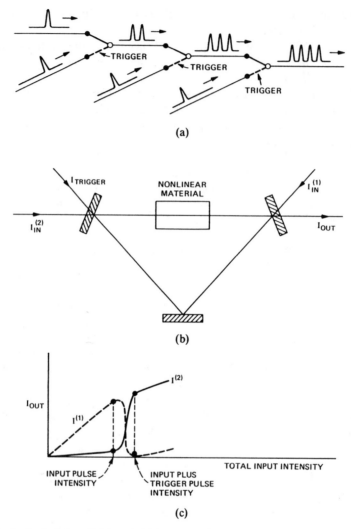

Fig. 6—Optical time-division multiplexer. (a) Overall schematic diagram. (b) Ring triggerable bistable element. (c) Output characteristic of ring bistable element.

the semiconductor CdS. It has recently been shown[13] that large nonlinear effects are found near the biexitonic resonance line. By using light at $\lambda \sim 4836$Å, we will obtain a sufficiently large nonlinearity to allow the device to operate at a pulse power level of about 1 mW with a time response of ~ 1 ps. This material requires cooling to liquid helium temperatures, however, and the wavelength range over which this operation can be obtained is small (~ 5Å). A material such as PTS will avoid these difficulties, but will increase the peak power required

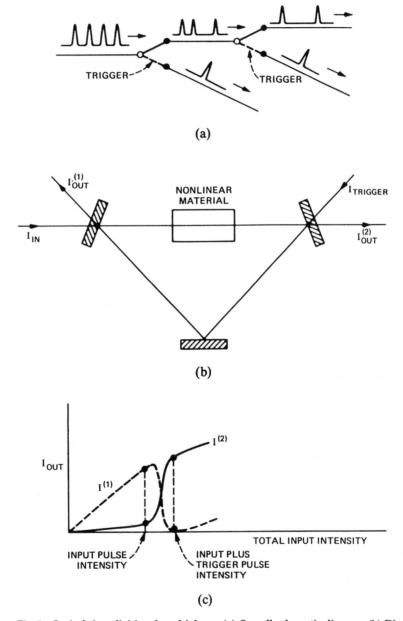

Fig. 7—Optical time-division demultiplexer. (a) Overall schematic diagram. (b) Ring triggerable bistable element. (c) Output characteristic of ring bistable element.

to the 1 W level. As this power is only required for 1 ps, however, the operating energy would still be a very reasonable 1 pJ.

VII. CONCLUSIONS

Having identified the physical limits for optical switching devices and discussed some specific examples, let us try to draw some general conclusions with regard to their future applications. The strong points of optical switching devices are:

(*i*) Speed: With an electronic nonlinearity or free-carrier generation in semiconductors, sub-picosecond switching times are possible.

(*ii*) Bandwidth: With a nonresonant bistable optical device or a nonlinear interface, a large fraction of the visible light bandwidth can be used.

(*iii*) Ability to treat directly signals already in the form of light.

(*iv*) Capability for parallel processing: With a liquid crystal bistable array, image processing has already been demonstrated.[15]

The weak points of optical switching devices are:

(*i*) High power is required for fast switching. This will tend to create thermal problems unless highly transparent materials are used.

(*ii*) Materials do not yet exist that have the ideal combination of properties for these devices.

(*iii*) Theoretical and practical problems involved in waveguide and microresonator formation in λ^3 volumes have yet to be overcome.

(*iv*) The minimum size of an optical switching element cannot be reduced below a volume of about λ^3 unless lossy metallic structures are used.

The field of digital optical switching is a dynamic one in which rapid progress is being made. New materials and devices are being proposed and studied. Let us close by proposing some areas where future work should be directed.

There is clearly a need for good optical quality, high-nonlinearity materials with low absorption coefficients. The development of such materials will have a great impact on future applications. Techniques must be developed for fabricating optical waveguides and optical resonant structures with dimensions on the order of optical wavelengths. For many applications, the problem of regeneration of signals associated with 'fan out' is important.[7] Some types of bistable lasers may well find use here. Finally, the general problem of the optics—electronics interface (i.e., making optics compatible with electronics)—is one that will require a good deal of attention if the most effective use is to be made of the tremendous potential of digital optical switching.

REFERENCES

1. L. F. Mollenauer, R. H. Stolen, and J. P. Gordon, "Experimental Observation of Picosecond Pulse Narrowing and Solitons in Optical Fibers," Phys. Rev. Lett., *45* (September 1980), pp. 1095-7.
2. A. Hasegawa and Y. Kodama, unpublished work.
3. R. C. Alferness, "Guided Wave Devices for Optical Communication," IEEE J. Quantum Elec., *QE17* (June 1981), pp. 946-58.
4. P. W. Smith and W. J. Tomlinson, "Bistable Optical Devices Promise Subpicosecond Switching," IEEE SPECTRUM, *18* (June 1981), pp. 26-33.
5. R. W. Keyes, "Power Dissipation in Information Processing," Science *168* (May 1970), pp. 796-801, and R. W. Keyes, "Physical Limits in Digital Electronics," Proc. IEEE, *63* (May 1975), pp. 740-67.
6. R. Landauer, "Optical Logic and Optically Accessed Digital Storage," *Optical Information Processing*, Nesterikhin, Stroke, and Kock, Eds., New York: Plenum, 1976, pp 219-54.
7. R. L. Fork, unpublished work.
8. H. Kogelnik, "Limits in Integrated Optics," Proc. IEEE, *69* (February 1981), pp. 232-8.
9. P. W. Smith, I. P. Kaminow, P. J. Maloney, and L. W. Stulz, "Self-Contained Integrated Bistable Optical Devices," Appl. Phys. Lett., *34* (January 1979), pp. 62-4.
10. U. H. Gerlach, U. K. Sengupta, and S. A. Collins, Jr., "Single-Spatial Light Modulator Bistable Optical Matrix Device," Opt. Eng., *19* (July/August 1980), pp. 452-5.
11. E. Garmire, J. H. Marburger, and S. D. Allen, "Incoherent Mirrorless Bistable Optical Devices," Appl. Phys. Lett., *32* (March 1978), pp. 302-22; P. W. Smith, W. J. Tomlinson, P. J. Maloney, and J.-P. Hermann, "Experimental Studies of a Nonlinear Interface," IEEE J. Quantum Elec., *QE-17* (March 1978), pp. 340-8; and J. E. Bjorkholm, P. W. Smith, W. J. Tomlinson, and A. E. Kaplan, "Optical Bistability Based on Self-Focusing," Opt. Lett., *6* (July 1981), pp. 345-7.
12. J.-P. Hermann and P. W. Smith, "Nonlinear Fabry-Perot Containing the Polydiacetylene PTS," Proc. of the XI Int. Quantum Elec. Conf., Paper T6, (June 1980), pp. 656-7.
13. A. Maruani and D. S. Chemla, "Active Nonlinear Spectroscopy of Biexcitons in Semiconductors: Propagation Effects and Fano Interferences," Phys. Rev. B, *23* (January 1981), pp. 841-60.
14. D. A. B. Miller, "Refractive Fabry-Perot Bistability with Linear Absorption: Theory of Operation and Cavity Optimization," IEEE J. Quantum Elec., *QE-17* (March 1981), pp. 306-11.
15. M. T. Fatehi, S. A. Collins Jr., and K. C. Wasmundt, "The Optical Computer Goes Digital," Optical Spectra, *15*, No. 1 (January 1981), pp. 39-44.
16. N. G. Basov et al., "Methods of Realization of an Optical Processor with Variable Operators," Sov. J. Quantum Elec, *8* (March 1978), pp. 307-12.
17. A. Huang, Y. Tsunoda, J. W. Goodman, and S. Ishihara, "Optical Computation Using Residue Arithmetic," Appl. Opt., *18* (January 1979), pp. 149-62.
18. *Picosecond Phenomena II*, R. M. Hochstrasser, W. Kaiser, and C. V. Shank, Eds., Berlin: Springer Verlag, 1980.
19. R. L. Fork and J. P. Gordon, unpublished work.
20. J. E. Bjorkholm, P. W. Smith, W. J. Tomlinson, and A. E. Kaplan, "Optical Bistability Based on Self-Focusing," Optics Lett., *6* (July 1981), pp. 345-7.
21. R. V. Schmidt and L. L. Buhl, "Experimental 4 x 4 Optical Switching Network," Elec. Lett., *12* (October 1976), pp. 575-7.
22. R. C. Alferness, N. P. Economou, and L. L. Buhl, "Fast Compact Optical Waveguide Switch Modulator," Appl. Phys. Lett., *38* (February 1981), pp. 214-7.
23. P. W. Smith, A. Ashkin, and W. J. Tomlinson, "Four-Wave Mixing in an Artificial Kerr Medium," Optics Lett., *6* (June 1981), pp. 284-6.
24. D. A. B. Miller, private communication.

DIGITAL OPTICS, SMART INTERCONNECT OR OPTICAL LOGIC?

Part 1

J E Midwinter

This two-part article examines the implications of using light for the wiring functions within electronics processors, starting from the simple point-to-point fibre interconnect and following through to the 'all optical' computer and optical 'neural network'. In so doing, the strengths and weaknesses of optics are highlighted and some of the problems and opportunities established.

Optical technology has come to dominate telecommunications cable transmission in the last five years through the medium of optical fibre. The cost/performance advantage is already so great and the potential for further development so enormous that it is hard to see how any other technology could rival optical for many decades to come, since it is clear that the transmission performance of single-mode fibre systems is primarily limited by the terminal and repeater equipment, not by the fibre cable itself (Midwinter 1986, Chown 1987).

Over the last decade, the astonishing capability of silicon integrated circuits (ICs) has been equally vividly demonstrated, although here the competing technology of III–V GaAs ICs remains the preferred solution for some specialist applications. Despite the power and speed of modern ICs, it is becoming apparent that straight-line performance projections on logarithmic plots of packing density, speed and particularly processing power are ceasing to be valid as the technology approaches some of its more fundamental limits (Keyes 1981). The practical limitations of communicating data along the metallised tracks on large chips, for both input/output (I/O) and internal communication, are starting to limit what can be achieved.

Both lines of reasoning point to the possibility that there may be a role for optical technology within processing systems and not merely for communications between systems. The field of digital optics is intimately concerned with examining this but, as we shall see in the ensuing sections, simple questions about the viability of optical interconnects within processing systems rapidly lead to deeper issues in which it ceases to be clear where the 'interconnect' stops and the 'processing' starts. In short, the distinction between communication and processing ceases to be clear cut.

Basic optical connections versus electronics

We must first address the question of when it is advantageous to replace a metal conductor with an optical pathway. The typical metal conductor used for communication at chip level presents itself to the drive circuit either as a capacitance to be charged to the logic level and subsequently discharged after its state has been tested at the remote end or as a transmission line with a characteristic impedance. The power consumed by the former increases with both modulation frequency and track length.

For an optical link using an electro-absorption modulator (EAM) and a remote optical power source as postulated below, the 'transmitter' is likely to appear primarily as capacitance at high bit rates, although photocurrent will give rise to a 'resistive' component of loss. Thus as a first approximation, we might equate the capacitances of the EAM and metallic strip conductor of length L to find a breakpoint in terms of electrical communication power. For a PIN multiple-quantum-well EAM of 30 μm diameter, we estimate a capacitance of 0.05 to 0.1 pF. Using the areal capacitance for 3 μm MOS circuits of 3000 pF cm^{-2}, we find that for a 3 μm wide track, a distance in the region of 600 μm leads to a similar capacitance. Longer than this will require more drive power than that required to drive the EAM for an optical link.

In lossy conductors, signals diffuse along their length and this gives rise to delay. Again for delay diffusion in 3 μm MOS metal tracks, we find typical figures of 0.1 ns cm^{-1}, which is expected to scale to 10 ns cm^{-1} with 0.3 μm geometry. Since diffusion delay scales with the square of distance, this projects to 0.1 ns mm^{-1}.

In a lossless conductor (a superconductor, perhaps?), the capacitance remains but the link looks more like a stripline and can be expected to show reflections unless matched; the drive power is that needed to drive the characteristic line impedance. In either case, the signal cannot propagate faster than its associated electromagnetic wave and power must be dissipated. For a metal on polysilicon stripline, we might expect a delay of the order of 100–150 ps cm^{-1}. Light could be expected to travel a little faster in free space (33 ps cm^{-1}).

In each case the message is clear for high-bit-rate operation (well in excess of 100 Mbit/s). Over short distances, 10–100 μm, metal tracks provide low-energy, wideband communication between electronic components but for distances in the region of 1 mm and above, communication by metal requires greater drive power and is characterised by undesirable delays. However, we also note that whilst optics might tackle the power problem favourably, as well as providing additional bandwidth, it will at best only have a secondary impact on the propagation delay given low-resistivity tracks. Indeed, optics might well increase the delay since the optical path could be longer if it is in free space above the chip as seems likely in many proposed situations.

Finally we note that the areal capacitance of above 3000 pF cm^{-2}, when driven at 300 Mbit/s with 3 V and 50% duty cycle, corresponds to a power of 1 W cm^{-2}, assuming 50% of the area to be metallised. Scaling this 3 μm geometry down by a factor f is expected to increase the capacitance by f but decrease the voltage by $1/f$, so that the power decreases as $1/f$. Hence, at 3 Gbit/s with the same parameters as above scaled, we would again expect 1 W cm^{-2}, already within a factor of ten of the desirable heat sink dissipation level. Such fundamental limits clearly set overall limits to the processing power that can be packed onto an electronic chip. We will examine later whether optics can improve on this performance.

These observations highlight several factors in high-speed complex processor design. Small, isolated all-electronic islands can both operate very rapidly and communicate freely within themselves. However, such electronic islands must be able to communicate with one another as well. Thus we suspect that communication between distant islands will be difficult, slow and power consuming if done electronically and that optics might be of assistance. Perhaps here is a first role for optics, one that is almost indistinguishable from chip-to-chip optical interconnect other than for the fact that the optical interconnects and associated logic chips might be constructed monolithically on a single substrate or wafer.

Optical interconnects

In the discussion above, we have implicitly considered replacing a wire or metal track with an optical path. This necessarily involves interfaces to carry out electronic to optical (E–O) and optical to electronic (O–E) conversions as well as an optical transmission path which might be glass fibre, planar dielectric waveguide or free space. Conventional optical links embrace a semiconductor laser or LED as source and a PIN or avalanche detector in the receiver. They are covered in detail in numerous papers on optical communications (see Haugen *et al* 1986 for example). We merely note here that generating light consumes electrical power and normally, the electrical to optical power efficiency is not good. For example, a semiconductor laser driven from a 3 V rail at 100 mA consumes 300 mW electrical power yet probably only delivers to the fibre 1 mW of optical power, an efficiency of −25 dB.

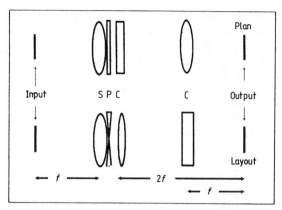

Figure 1. A simple optical system that performs the perfect shuffle wiring operation. The symbols S and C label spherical and cylindrical lenses, P signifies a prism.

Sophisticated lasers do substantially better but to integrate arrays of such devices monolithically with complex electronics is no mean task. LED sources are much worse, with the same electrical drive power typically delivering 10–100 µW of optical power into a fibre, −45 to −35 dB efficiency. We will propose a plausible solution to this problem later. At the receiver, the situation is better, since photodetectors typically operate at sensitivities of about $1\,\text{A}\,\text{W}^{-1}$. Consequently, with a few volts bias rail, the power efficiency in detection is close to unity. We conclude that the power efficiency of the E–O interfaces is likely to be a problem area if more than a few are required and efficiency will be a paramount consideration.

Following the success of optical fibre links for long-haul communication, it is natural to consider them for the present task. In the case of rack-to-rack and board-to-board connections, they are the most likely choice and are already being used in many systems. At the chip-to-chip level, the problem of interfacing the fibre onto the chip becomes major, since the typical optical connector is probably as large as the chip itself and multiple-fibre connectors seem entirely inappropriate. The use of fibre arrays has been proposed but from an engineering point of view, it is not clear that the problem of handling them has yet been solved entirely satisfactorily.

If we consider optical connection within the chip itself, then fibre seems entirely inappropriate and free-space propagation looks increasingly attractive. The most striking thing about a simple lens used to image a two-dimensional array of sources to a 2D array of detectors is that a single element provides many parallel wideband communication pathways. Interestingly, each pathway has an identical time delay, accurate to fractions of a picosecond so that it is also a 'zero time skew' interconnect. However, one may reasonably object that this does not seem to be a useful wiring scheme.

Using a similar layout but placing a single source at the lens focal point then produces a parallel beam impinging on the detector array. Pulsing a single laser will now deliver precisely synchronised 'clock' signals to any detector in the array. A complete optical system would require a microlens array over the detector array, with one microlens per detector and the spaces between lenses blanked off to prevent unwanted interference with the electronic circuit. This combination of large lens and microlens array would most probably be designed as a single hologram. Thus we are led to think of packages in which a hologram placed over the chip couples light from a single source to multiple detectors to provide precision timing and control signals to a large chip.

If we return to the imaging interconnect, we note that any wiring pattern that can be described using the language of imaging is potentially implementable in optics. Operations such as magnify, shear, reflect, shift, rotate and invert come to mind and can have great relevance for data processing, especially when applied to matrix processing. For example, overlaying two matrices with one rotated by ninety degrees from the other juxtaposes the appropriate elements to perform matrix multiplication. The optical system shown in figure 1 carries out the operations of one-dimensional twofold magnification, image shearing and overlaying to give the final operation of 'perfect shuffle', a key ingredient in many logical processors for switching or computing (Midwinter 1985). The steps are shown schematically in figure 2.

Space wiring schemes such as the perfect shuffle are simultaneously both very powerful and very inflexible. A few simple optical elements offer the possibility of wiring many independent lines in parallel but only in very simple and regular patterns. Printed circuits or IC metallisation tracks are also

Figure 2. A schematic drawing indicating the stages involved in carrying out a one-dimensional shuffle operation.

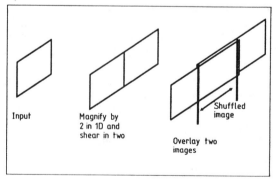

severely but differently constrained but by means of two or more layers can achieve very complex interconnection patterns. Hence we will see that the systems proposed to exploit the optical space wiring technology incur a penalty since they must exhibit extreme regularity and great simplicity. However, in return, they gain a high degree of logical modularity that is not unlike that of the electronic processors of the systolic array type.

Thick holograms

More generalised optical connections are possible in principle using thick holograms. In the most general case, one would like to connect an array of $N \times M$ source points to an array of $P \times Q$ receiver points with a connection pattern of arbitrary complexity. This would imply, in the limit, $N \times M \times P \times Q$ degrees of freedom for the wiring. If we consider the simpler case in figure 3 of connecting $N \times N$ inputs to $N \times N$ outputs via an intervening thick hologram, we note that as N increases, the degrees of freedom required increase as the fourth power of N whilst the volume of holographic material at best only increases as the cube of N. We conclude that complete generality of interconnection is not possible. In addition, the practical constraints of multi-exposure holograms in terms of scattering efficiency and random and crosstalk noise or scattering probably set very finite limits to the number of discrete interconnection patterns that can be achieved within a single multi-exposure element.

Stray light appears to be a particularly serious problem in such wiring schemes when one recognises that laser sources are generally obligatory from the point of view of power efficiency and that one is thus operating with coherent light. For wiring patterns where any single receiver point only receives power from a single source point, then to ensure that the received power is stable to within ±10%

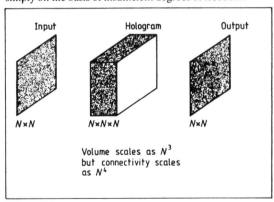

Figure 3. A schematic diagram of a holographic interconnect from $N \times N$ input pixels to $N \times N$ output pixels illustrating that entirely free interconnection is not possible simply on the basis of insufficient degrees of freedom.

(0.4 dB), the stray light level must be 26 dB down assuming that they are both phase coherent but randomly phased.

In the case that the receiving element is an optically activated threshold logic element receiving two signals from different sources, the problem is potentially much more severe, since not only stray light can affect the detected power: the relative optical phase of the two (or more) signals, each interacting with the reference or pump beam can also be significant. Multi-axis addressing may reduce the problems but they remain formidable. Thus any real system will clearly need to take care to minimise such effects.

A further issue to consider is the likely packing density achievable for all optical elements. Evidently, if an array is to operate in a digital logic mode, each device must be addressed uniquely, so that optical crosstalk between elements must be at a very low level, particularly if they are of the two-terminal type. This implies well separated elements. To quantify this statement, we use the formula for a focused Gaussian beam to relate far-field angle to spot size.

A Gaussian spot of diameter $2w$ at its $1/e$ points gives rise to a far-field 'half' beam angle, also at the $1/e$ points, of $W/\pi w n$ where W is the wavelength and n the refractive index of the medium. Noting that the optical system for such a processor is clearly expected to provide very high-quality 'images', we postulate the use of $f/2$ optics, implying a maximum half-field angle of approximately 0.25 rad. If we arbitrarily allow the diffraction-limited (half) beam to fill half of this, then we have the result that

$$w = W/0.25\pi n.$$

Taking the wavelength as 1 μm and n as unity (air), we have $w = 1.3$ μm. The $1/e$ spot diameter is 2.6 μm and we should probably double this again to 5.2 μm to enclose the outer fringes of the spot. We also note that non-Gaussian beams always focus to larger spot sizes, so this may well err on the optimistic. Moreover, it allows for no alignment errors or aberrations. The density of packing allowable will be technology dependent. Studies of thermal crosstalk have suggested that a 1:10 spot-to-space ratio is desirable. Using that figure, we would have active devices on 52 μm centres, leading to 200 per centimetre or 40 000 per square centimetre. Such figures are small compared with the device packing densities already achieved in electronics.

Optically activated logic

The terminology 'optically activated logic' embraces two broad categories: electronic logic with optical 'wiring' for I/O and clock, and 'all-optical' logic in

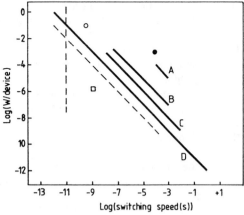

Figure 4. Plot of switching power against switching time for a variety of optical bistable elements and electronic logic gates. The GaAs and ZnSe devices are 'all-optional'. Since power generally scales with area, hybrid devices show much lower switching power for comparable areas (approximate linear dimensions given). ○ 2.5 μm GaAs; ●, 5 μm ZnSe; □, bistable laser; A, 250×750 μm optoelectronic logic device (OELD); B, 100 μm self electro-optic device (SEED); C, 10×30 μm OELD (projected); D, 1 pJ switching energy. The broken lines are speed and power limits for electrical devices.

which any electronic involvement is purely secondary and probably covert. We include both since we highlighted above some of the potential advantages of optical wiring and must now address the question of how to interface it to logic elements.

The case for all-electronic logic is well established, as it is both mature and extremely powerful, yet it has its limitations as noted above. We merely note that typical discrete electronic gates are characterised by power–speed products (switching energies) in the range 10 fJ to 1 pJ and, depending upon technology, switch at speeds measured in between nanoseconds and low picoseconds.

These operating regions are shown in figure 4 to provide one basis against which to assess their optical logic counterparts. Note that, while a substantial range of powers and speeds are covered by both different electronic and optical technologies, any logical processing operation that can be specified in terms of a number of logic gate operations directly implies a minimum energy consumption set by the product of this number and the power–speed product for the chosen device.

A given power consumption thus leads to a minimum processing time and a maximum of gate operations/second. However, one should note that information processing involves not just 'gate operations per second' but also 'numbers of data associations per second'. Thus a two-input OR gate associates two items of data per operation. A neuron may in some sense associate 1000 or 10 000 items per operation. Hence the comparison breaks down if, by using a different technology, a processing algorithm or method becomes possible that is markedly more efficient. In general, the connectivity of the network involved in the processing operation is as important as the number of logic gates and their speed of operation. Optics may be able to make a difference here.

The most widely discussed 'all-optical' logic device is the bistable optical switch. A recent book on optical bistability (Gibbs 1985) lists some 1600 references on the subject, so we will present here only the briefest survey of their principles and properties. Bistable switching behaviour results when some form of positive feedback is present in the control of the transmission or reflection properties of an element. Such feedback can arise from a variety of physical mechanisms.

Purely 'reactive' devices involve intensity-dependent refractive-index effects that follow the light intensity. Such effects arise from the same physical nonlinearities as the optical Kerr effect and are characterised by response times measured in femtoseconds. To produce bistable switching, this material must be formed into an optically phase-sensitive structure such as a Fabry–Perot etalon, which can switch from transmission to reflection, or a resonant waveguide directional coupler which can switch output ports. Changing the incident intensity to the device changes the refractive-index within the resonant structure in such a way as to further enhance the intensity within the device by approaching resonance. Given the correct degree of such positive feedback, switching can occur analogous to that in electronic flip-flops.

Another similar class of devices make use of changes in the refractive index arising from the direct absorption of optical energy. Here, the response time is fundamentally linked to that of the absorption mechanism. Thus, in a semiconductor material, this may be a thermal time constant when the material has been heated, a carrier recombination time when the change in index has been caused by an increase in the numbers of free carriers or a variety of other effects spanning the time frame from picoseconds to milliseconds. However, since the resulting changes in index arise from similar electron–photon interactions, we find that to a first approximation they vary inversely with the characteristic time constant for a given optical power. To produce a power-efficient device, it is desirable to choose a mechanism with an appropriate time constant and then to minimise the volume of material used by decreasing device size.

Such devices are effectively electronic in that they exploit the electrons in a material to change its refractive index although they would not normally be described as electronic since electrical currents and voltages are not involved. By involving the time constants associated with real transitions in the material, frequently electronic effects, they operate at similar or slower speeds to electronic devices.

Quantum-well devices

Another class of bistable optical devices are more overtly electronic in operation. One group is based upon the III–V multiple quantum well (MQW) electro-absorption modulator. Here the intrinsic MQW structure is sandwiched in a PIN structure so that a uniform reverse bias field can be applied across the wells. In the absence of a field, the MQW shows a very pronounced exciton absorption placed on a very sharp band edge. Application of reverse bias fields of a few volts leads to substantial shifts in the position of the absorptions, so that at a wavelength just beyond the zero-voltage band edge, a very large change in absorption can be produced. Once these devices are suitably biased, such absorption leads to photocurrent at high quantum efficiency. Thus they can operate as voltage-controlled absorption modulators and photodetectors simultaneously.

In figure 5 we show a recently obtained set of photocurrent response curves for such a device, showing very clearly the change in absorption with voltage above the band edge and the growth of quantum efficiency for very low bias voltages (0–2 V). The physical mechanisms involved are fast, substantially sub-nanosecond, although they can easily be limited by electronic RC time constants.

Placing such a device in a simple resistive bias circuit offers the possibility of an optically switched bistable element, since changing optical power to the device changes the photocurrent, which in turn varies the bias voltage and can lead to positive feedback through the resulting change in absorption coefficient. Such a device is known as a self electro-optic device (SEED) (Miller et al 1985). It takes full advantage of the strong electron–electron interactions that characterise electronic devices and only uses the electron–photon interaction for absorption. It thus promises good power sensitivity when fabricated into microelectronic component size although because of the details of its operation, it seems unlikely that it will achieve high-contrast switching.

The switching energies of this type of device are fundamentally linked to device capacitance. Assuming a quantum efficiency of near unity, we can expect a detectivity of about 1 A W^{-1}. For a device of $10\,\mu\text{m}$ diameter, we might expect a capacitance of order 0.01 pF. Charging this to say 3 V requires a charge of 0.03 pC and an optical energy of 0.03 pJ or 30 fJ. Such figures are comparable to those for electronic devices, since they are electronic!

Other hybrid logic devices building upon the same basic device offer a variety of different response characteristics (Miller et al 1985) and some seem to be good candidates to overcome the SEED's contrast problem. For example, the monolithically integrated series combination of phototransistor and MQW modulator (Wheatley et al 1987) yields a device response with hard limited 1 and 0 outputs (figure 6) and offers improved signal sensitivity by virtue of the electronic gain in the phototransistor. However, the optical energy taken from the 'power beam' during switching remains essentially the same as for the SEED since, again, the device capacitance must be charged or discharged.

The existence of true gain between signal and output coupled with very uncritical operation is a major advantage for this device. Note that despite the fact that these hybrid devices are electronic in operation, they can be faster than those 'all-optical' devices that involve absorption of energy, since there the time constants involved (thermal, carrier recombination etc) are typically longer than the limiting electron transit times under typical bias fields.

In a class on their own are bistable lasers, being hybrid devices with electrical drive but exhibiting excellent sensitivity by virtue of their inherent electrically driven gain. However, when their total power consumption is taken into account, they again fall into the same power region as the electronic logic elements. Note that in discussing optical wiring, we have highlighted the potential attraction of the third

Figure 5. Plot of the photocurrent response against wavelength for a GaAs/GaAlAs PIN MQW modulator/detector (MV 246 photodiode) for different reverse bias voltages across the structure, showing the clear shift of the exciton absorption and the band edge with voltage, from 827 nm at -6 V to 843 nm at -20 V: ———, -6 V; – – – –, -10 V; — — —, -14 V; – · – · –, -20 V.

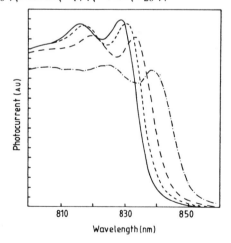

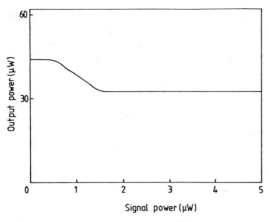

Figure 6. Input/output response curves for the monolithic optoelectronic device from Wheatley (1987), showing the hard limiting ONE and ZERO levels and the high sensitivity. The poor contrast is the result of using a very poor MQW modulator.

dimension in free space. Exploiting that implies fabricating the optical interface elements so that they are addressable normal to the array plane. Whilst surface-emitting lasers do exist (Uchiyama and Iga 1986), they do not at present seem attractive for this application. However, lasers might well be attractive when used in guided wave form in all-optical regenerators.

Device performance

We plotted in figure 4 the typical performance values reported in the literature for a wide range of optically activated logic devices (Midwinter 1985). The picture that emerges is one of a huge spread in performance, not just in terms of speed but also in power–speed product. However, the general conclusion appears to be that, relative to electronic devices, they are normally slow and excessively power-consuming at present. Improvements in performance are anticipated as individual devices are optimised but there seems little likelihood that they will beat discrete electronic devices. It should also be noted that in general they suffer from serious implementation problems stemming from the nature of threshold logic, as discussed below.

A typical input–output response curve for a bistable device is shown in figure 7. Since the device has no intrinsic power gain, restoring the output signal level to the standard logic level implies the use of a separate optical 'bias' beam which, in conjunction with the input signals, produces the switching action and provides the output power. Taking the two output power levels corresponding to the upper and lower branches of the hysteresis loop, we associate with them the logic levels ONE and ZERO respectively. The most practical technique for driving such elements in a logical processing circuit appears to be the 'lock and clock' mode of operation (Smith *et al* 1985). By reference to figure 7, the following sequence of operations is proposed to implement a dual-input OR gate. Using the following notation, it is assumed that three conditions apply:

(a) $P(\text{bias}) > P(\text{SD})$
(b) $P(\text{bias}) + P(1) + P(0) > P(\text{SU})$
(c) $P(\text{bias}) + 2P(0) < P(\text{SU})$

where bias power $P(\text{bias})$
signal power with ONE input $P(1)$
signal power with ZERO input $P(0)$
switch-up threshold power $P(\text{SU})$
switch-down threshold power $P(\text{SD})$.

It follows that application of the bias power alone holds the device in its existing state (ZERO or ONE), application of one or two ONEs to the input switches to the ONE output from the ZERO state, whilst application of two ZEROs leaves the device in its existing state. To carry out the dual-input OR operation, it is thus necessary to go through the following sequence of operations:

(i) remove bias and signal; device resets to ZERO;
(ii) apply bias and signals; device sets to ONE or ZERO in accordance with the OR gate truth table;
(iii) remove signals, retaining bias;
(iv) test output state of device.

Such operation has been demonstrated in very simple three-gate circuits (Smith *et al* 1985). However, it is immediately apparent that threshold logic of this type, in which operation depends upon summing

Figure 7. Typical I/O response for an optical bistable gate. Note that the switch up and down thresholds depend upon the detuning of the device. Hence the contrast, width of hysteresis loop, bias and signal settings are readily varied which is good for experimental purposes but less attractive for serious logic circuit applications.

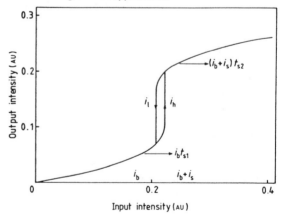

small signals with relatively large bias to exceed (or not) a threshold level, implies tight control both of threshold levels and of power levels. The success of electronic logic rests upon its avoidance of this type of stability problem, relying upon hard clamped levels corresponding to 'ground' and 'supply voltage' which can be common across a complex circuit, with large noise margins in between. It is extremely difficult to see how a large array of bistable gates will achieve sufficient device and optical power uniformity to offer any reasonable prospect of reliable operation.

Furthermore, the comments above concerning the vector nature of optical signals point to further problems in implementing any real optical logic system based upon two-terminal devices. The fact that the switching speed of a bistable device is sensitively dependent upon its exact setting or power level further complicates the design issue, since the requirement for high sensitivity is in conflict with that for fast switching (Wheatley and Midwinter 1987).

Given these problems, there has been a noticeable growth of interest recently in both genuine three- (or more) terminal hybrid devices (see above) as well as optically wired electronic logic. Such approaches focus attention on the design requirement to move from the electronic to the optical domain and vice versa at will and imply the use of a monolithically integrated technology combining electronic and optoelectronic elements. Two material systems seem to be the most likely candidates, III–V semiconductors and III–V grown on silicon.

The III–V GaAs system is by far the most mature and offers a range of electronic and optoelectronic devices. However, since the band edge of GaAs is at a longer wavelength than its GaAlAs matching epitaxial layers, it is difficult to design hybrid devices with a transparent substrate. Using devices based upon InP, with epitaxial layers grown in GaInAs for example avoids this problem but introduces a very immature electronic device technology.

III–V on silicon potentially combines the mature silicon electronic device, probably with silicon detector technology, with III–V sources and/or output modulators. However, it incurs the problems of growing the III–V layers on the silicon substrate, again, a relatively immature technology. Hence we may say that at present a fully suitable technology does not exist in which to implement these proposals.

Detectors are already well developed and a variety of efficient devices exist for use in the wavelength range that falls within the silicon, GaAs and GaInAs band edges, so that entering data from the optical to the electrical circuit is well catered for. The reverse operation presents greater problems. We have already commented that the power efficiency of lasers and LEDs is generally poor. Even the recent development of lasers with very low threshold current, whilst leading to high power conversion efficiency, also requires relatively large powers (many milliwatts). Thus they do not seem well matched to monolithic integration in hybrid processing circuits addressed by space wiring.

The MQW electro-absorption modulator offers an interesting alternative promising acceptably low drive power. Moreover, as a modulator, it can be interrogated using optical power generated elsewhere and can signal, via its photocurrent, information from its interrogation beam. We believe that these or related devices offer the real prospect of an efficient I/O optoelectronic interface technology and thus promise to be the final link in the technology to bring optically interconnected electronic islands to reality.

In part 2 of this article I shall look at architecture and systems considerations.

References

Chown M 1987 Physics in practical fibre optic systems *Phys. Technol.* **18** 101–6

Gibbs H M 1985 *Optical Bistability* (New York: Academic)

Haugen P R, Rychnovsky S and Husein A 1986 Optical interconnects for high speed computing *Opt. Engng* **25** 1076–85

Huang A 1984 Impact of technological advances and architectural insights on the design of optical computers *Phil. Trans. R. Soc.* A **313** 205–11

Huang A and Knauer S 1984 Starlite, a wideband digital switch *Proc. IEEE Global Telecommunications Conf. Atlanta, Georgia, USA* (New York: IEEE) pp 121–5

Keyes R W 1981 Fundamental limits in digital information processing *Proc. IEEE* **69** 267–78

Midwinter J E 1985 Light electronics, myth or reality? *Proc. IEE J* **132** 371–83

Midwinter J E 1986 Optical fibre communications, present and future *Proc. R. Soc.* A **392** 247–77

Midwinter J E 1987 A novel approach to the design of optically activated wideband switching matrices *Proc. IEE J* **134** 261–8

Miller D A B, Chemla D S, Damen T C, Wood T H, Burrus C A, Gossard A C and Weigmann W 1985 The quantum well self-electro-optic effect device: optoelectronic bistability and self oscillation and self linearised modulation *IEEE J. Quantum Electron.* **QE-21** 1462–76

Smith S D, Janossy I, MacKenzie H A, Mathew J G H, Reid J J E, Taghizadeh M R, Tooley F A P, Walker A C 1985 Non-linear optical circuit elements as logic gates for optical computers; the first digital optical circuits *Opt. Engng* **24** 569–74

Tanguay A 1985 Materials requirements for optical processing and computing devices *Opt. Engng* **24** 2–18

Uchiyama S and Iga K 1986 Consideration of threshold current density of GaInAsP/InP surface emitting junction lasers *IEEE J. Quantum Electron.* **QE-22** 302–9

Wheatley P, Whitehead M, Bradley P J, Parry G, Midwinter J E, Mistry P, Pate M A and Roberts J S 1987 A novel non-resonant optoelectronic logic device *Electron. Lett.* **23** 92–3

Wheatley P and Midwinter J 1987 Operating curves for bistable optical devices *Proc. IEE J* **134** 345–50

DIGITAL OPTICS, SMART INTERCONNECT OR OPTICAL LOGIC?

Part 2

J E Midwinter

How can the optical devices described in part 1 of this article (May issue pp 101–8) be applied? Part 2 outlines the new architectures and systems which hold out such promise for optical computing and telecommunications.

Thus far we have examined rather briefly the areas where electronic circuits run into difficulty, the potential offered by optical wiring and the characteristics of optical logic and optical i/o devices. We have seen that optical wiring largely supplements electrical wiring, offering features that electrical wiring would be hard put to achieve but at the same time unable to do simply the local interconnections that electronic circuits take for granted.

Starting from systems needs, we can identify a number of problems in search of solutions. In telecommunications, optical fibre has provided cheap transmission bandwidth, favouring wideband services, but the switching technology is certain to have difficulty with such services. The proliferation of services expected in the coming decade further points to a need for switching systems that can allocate channel bandwidth in a flexible and rapidly varying manner. Coupled with the growing use of packet-type local area networks (LAN), it is tempting to forecast a steady pressure to move towards packetised transmission for many services, providing the ultimate in 'bandwidth on demand'. The 'asynchronous time division' approach favoured under RACE† is a step in this direction. Ultrafast packet or time-slot interchange switches are difficult to construct in electronics.

In general-purpose computing, the pressure for greater processing power has led to intense interest in parallel multiprocessor machines, built around small processors such as transputers. It is suggested

†RACE (Research and development in Advanced Communications technologies for Europe) is the major European Community programme in telecommunications.

that the parallelism inherent in optical space wiring may open up new possibilities here, although how this is to be done does not appear to have been studied fron an engineering point of view in which the practical problems alluded to above are faced. However, some thought has been given to novel architectures that might be able to exploit such a capability if it existed.

The broadband switching matrix is an interesting test case for a hybrid digital optic technology. It places maximum emphasis on huge data throughput, carries out minimal logical processing and is relatively insensitive to the overall time delay taken for data to traverse it. Many switching matrices take the form of simple pipeline processors, with all data flow along the pipeline and the transverse flow restricted to the highly structured reordering of data paths between rows of switches (e.g. figure 1). A further requirement for applications such as fast packet switching is that it should be possible to set very rapidly a new set of pathways through the matrix. In general this involves computing a completely new set of instructions for every cross-point switch in the matrix, although by making inefficient use of the matrix, this task can be simplified.

An alternative approach is to rely upon the use of non-blocking self-routing matrix designs whereby the routing is carried out through the use of logic aligned alongside each cross point and the information to achieve that setting is derived by injecting the desired output port addresses via each input port, (Huang and Knauer 1984). Coupling such concepts to simultaneous optical clocking, wiring and I/O potentially offers the attractive combination of ultrafast routing and massive data throughput (Midwinter 1987). However, many technological problems have to be solved before such systems could be constructed.

Geometrical layouts

If we look at a generalised parallel multiprocessor computer, then a variety of geometrical layouts appear to be of current interest. One is the pipeline processor in which information flow is primarily unidirectional along the pipeline with lateral flow only between processor rows, not across a single

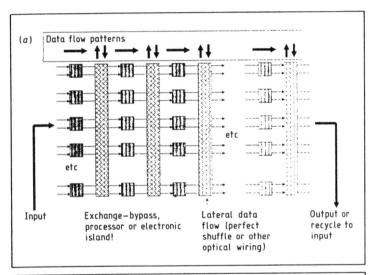
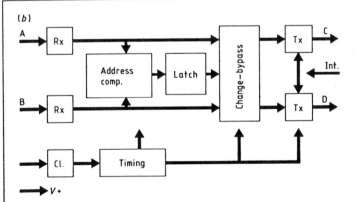

Figure 1. (*a*) Layout for switching matrix based upon perfect shuffle wiring, also showing the data flow patterns inherent in such a pipeline processor. (*b*) The contents of each 'electronic island' to obtain self-routing capability.

row. Such a layout is directly analogous to the switching matrix of figure 1 and is insensitive, to first order, to the time delay in transmission between rows. Thus it can take advantage of the increased communication bandwidth and zero time skew offered by optics along each data pathway and may also benefit from optically delivered clock signals to increase the timing precision.

Another approach is the multiprocessor multimemory machine configured around a fast packet switch. Here, the time delay in obtaining data from memory or writing it to memory is intimately involved in the computational cycle time of the machine. The use of wideband optical paths, perhaps coupled with photonic switching of the network, allows very high-data-rate transfer of information but will not dramatically affect access delay. The time taken for route set-up in an optically switched network is likely to be critically dependent upon the control algorithms used and the nature of the switched network. Moreover, some optical networks, such as wavelength-tuned systems, offer the potential for highly parallel non-blocking communication on a mix of broadcast and point-to-point bases, perhaps allowing different groups of processors to speak with different subsets of memories simultaneously. The switched wideband network at the core of this machine has much in common with a typical telecommunications switch. However, the typical message lengths may be much shorter whilst the sensitivity to path set-up time and path delay is likely to be greater, generating a very different set of constraints. Once again, novel optical solutions have been proposed for this application but fully detailed assessment has yet to be done.

Another multiprocessor geometry of interest is the hypercube in which a large number of processors, each with their own memory, are fully interconnected. Thus, for N processors there are $N-1$ outward and $N-1$ inward connections to and from each of the other processors. Here again, optics can be expected to provide bandwidth but to have little impact on communication time delay and this is seen as a critical limiting factor by some.

Optical computers?

We have discussed these processor concepts as if the 'wiring' were implemented with wires or fibres. However, they could clearly be implemented with optical space wiring of the type discussed in part 1 of this article, accessing processors located in one or more (image) planes. If we now extend this observation to the case when the individual processor is composed of a very small number of gates (one or two) that are optically activated and provided with optical outputs, then we have arrived at what has been much discussed in the press as the 'optical computer'.

We have already noted that optical space wiring systems limited by diffraction could handle significant numbers of parallel interconnections in simple, very regular interconnection patterns. This parallelism has led to the speculation that optical computers might be extremely powerful *vis à vis* their electronic counterparts. However, approaching the concept in this manner highlights the fact that for this gain, a price has been paid, assuming all other things to be equal. The space optical pathways are almost certainly physically longer than most of the electronic pathways in an equivalent LSI chip and will thus involve a greater time delay, even if they have substituted identical delays in place of route-dependent delay (time skew!).

This observation seems to direct one's thoughts directly back to the electronic island concept discussed earlier. One asks whether the penalty from increased communication delay will not greatly outweigh the gain from zero time skew as the unit processor size shrinks below some optimum dimension. Assuming such a processor to be formed in pipeline format, with discrete processors (electronic islands) connected by optical space wiring for across-pipeline data flow (much as in figure 1), then we see that inserting more and more frequent optical space connections will at some point start to increase the total data transit time through the pipeline because of their time delay. In the case of a switching matrix, this may not matter but for computation, it is potentially more serious.

A particular case of interest is any such processor involving loop-back in the logical data path, since the optical space wiring delay then enters directly into the computational cycle time. If it is longer than the 'within island' electronic connection time, then the processor has been slowed down. Moreover, in the case of the all-optical processor in which one space-wired optical loop connection has to be made every clock cycle, a severe upper limit has been set to the clock rate. For example, we note that light travels 10 cm in air in about 300 ps. Assuming logic planes of order 1 cm in lateral (or image) dimension, it is difficult to imagine zero time skew connections in free space optics having a loop length of less than 20–40 cm; this implies a wiring delay of 600–1200 ps and hence clock rates well below 1 Gbit/s (and probably closer to those already achieved in the fastest LSICs).

Logic for optical computers

Leaving aside such considerations, we may reasonably ask how a computer would carry through general-purpose computing operations if it were based upon all-optical wiring between discrete gates with optical I/O. This problem has been examined in

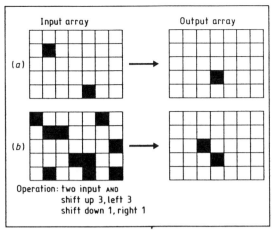

Figure 2. The symbolic substitution operation, showing in (a) the basic pattern substitution, and in (b) the operation applied to an extended array with all pixels occupied by AND gates.

detail by a group of workers at ATT Bell Laboratories who proposed the use of symbolic substitution logic (Huang 1984). In this approach, the data input to the processor is viewed as a two-dimensional image of discrete data points (ON or OFF, 1 or 0). The data position in the array is its address and processing involves performing logical operations on pairs or more of data points with each logic gate also in a meeting array. Because of the highly regular structure of the optical space wiring, it is proposed to perform similar or identical operations with essentially every array logic element on any given clock pulse.

Let us suppose that we have data positions coded by integers i and j where $i=(1-M)$ and $j=(1-N)$. In figure 2 we illustrate a section of this data array and highlight a 4×4 section of it. A typical logical operation might be the following AND operation on points signed as follows:

$$O(i=4m+2, j=4n+2) = I(i=4m, j=4n).\text{AND}.I(i=4m+3, j=4n+3)$$

where O signifies output pixel and I input pixel and the operation is to be carried out for all integer values of m and n. This implies the existence of dual input AND gates at all positions $(i=4m+2, j=4n+2)$. To bring the data array into alignment with the logic gates it is necessary to split it into two equal-intensity arrays, shift one array by three spaces right and three spaces down, and the other array by one space up and one space left. This operation has been named symbolic substitution since it clearly substitutes on a logical basis one pattern for another as shown in figure 2. Note also that if logic gates were placed at all positions in the array, then every occurence of an illuminated pixel four up and four left from another illuminated pixel would result in output one up and one left from the first.

This approach has been used by Murdocca and Streibl (1987) who show that by use of four image operations (copy, shift up, shift down and invert) it is possible to carry out all the basic computational operations required for general-purpose calculation. However, relative to electronic logic, a greater number of clock cycles is sometimes required per operation. Coupled with the comments above on the impact of optical space wiring on the maximum attainable clock cycle and actual parallelism possible in diffraction-limited optics, this seems to suggest that the case of 'single gate per island' is substantially sub-optimum. Such a processor is shown schematically in figure 3. The masks are used to change the connection patterns between cycles of the processor by opening or blocking individual pixel connections. They thus notionally provide a way of feeding a control programme into the processor. Exactly how such programmable masks are to be formed with sufficient complexity and response speed is unclear. Note that in practice it would need at least one further array of latches in the loop to allow for clocked cyclic operation, further slowing down its operation.

All of these processor concepts highlight one key and largely unresolved issue. The efficacy of optics will depend critically upon the spatial nature of data flow required by the computational process. Today's processors and algorithms have been developed to optimise operation with electronic communication constraints. It is far from clear that these still apply in an optical hybrid processor and it is difficult, if not impossible by definition, to obtain fundamental statements on spatial and temporal data flow in computation that are free of specific algorithm considerations.

Neural networks

Finally, the growing interest in all forms of pattern recognition and associative memory is fuelling intense interest in new forms of computing machine that exploit architectures based upon neural networks and their derivatives, although in terms of optical implementation, this work is still at a very

Figure 3. A schematic layout for a general-purpose computer based upon symbolic substitution, as described by Murdocca and Streibl (1987).

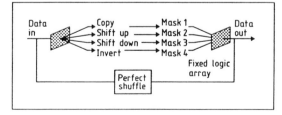

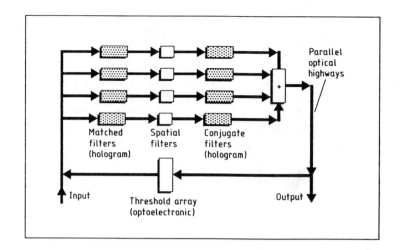

Figure 4. Schematic layout of a 'Hopfield neural network' in the matched filter equivalent model.

early stage of development and many problems remain unsolved. In these machines, the 'program' and 'memory' are indistinguishable and the operation of the machine depends upon its interconnection patterns and their weights associated. Moreover, any given logic element can receive inputs from a very large number of others and contribute to the inputs of a large number. Such machines are thus characterised by extremely complex wiring patterns.

The interest in optical implementations arises primarily from the natural parallelism inherent in imaging optics and already referred to in part 1. Linking this with an element such as a thick hologram, we find another attribute emerging, that of high connectivity. This was illustrated in figure 3 of part 1 (p 104). If we consider light emerging from one pixel in the $N \times N$ input array, then it can impinge on all points of the volume hologram. This can be programmed to direct a predetermined pattern or image from the input plane to an associated image or pattern in the output plane. Thus a large number of points in the input plane can be connected each with varying weights to a large number of points in the output plane. However, a consequence of the discussion earlier on degrees of freedom is that this statement carries some unstated constraints.

An interesting theoretical model of a particular form of neural network has been proposed by Hopfield (1982) and implemented in optics by Psaltis and Farhat (1985). The Hopfield model assumes the existence of a memory matrix of dimension $i \times j$ ($i=j$) elements each having a transmission (or weight) $T(i,j)$. An input vector $I(i)$ is multiplied by the matrix to form the output vector $O(j)$

$$O(j) = \sum_i T(i,j) \cdot I(i).$$

Each output element $O(j)$ is then thresholded and fed back to provide the input for the next cycle ($I(i) = O(i)$). The matrix is programmed to recognise a series of stored vectors according to a simple algorithm. One then observes that, given an input of an incomplete version of the test vector, the system rapidly (two or three cycles) settles down to iterating the pure stored vector most closely matching the supplied test vector. The system has 'recognised' the test vector.

The matrix $T(i,j)$ is very reminiscent of a hologram, except that, as implemented, it operates on intensity only with no phase sensitivity. The resistance of the system to noise and its performance as one scales from small demonstrators to usefully complex scale remain unclear. Recent analysis (Selviah and Midwinter 1988) has shown that the system above is formally identical to a bank of matched filters with common threshold elements, connected as shown in figure 4. The fact that the filters are in parallel highlights the fact that the holographic equivalent element must be formed by multiple exposures, one for each 'filter', although they might not necessarily be fully spatially overlapped and intermingled.

The location of the threshold in this pattern-recognition system seems at first sight odd, since it would seem more natural to place it after the first row of filters, making a logical decision to select the single largest output. However, that would imply a comparative decision-making process between all the filter outputs, in itself a complex process for a large number of filters. The 'neural network' achieves the same result by use of the shared threshold to enhance the largest at the expense of other elements, so that only a single solution dominates after a few iterations. The study also suggests that the threshold form may be critical and points to a need for threshold responses of the form shown in figure 6 of part 1 (p 107).

The engineering of these systems to learn by

experience is critically important and a subject of intense study. Once again, optical approaches to the problem have been attempted (Wagner and Psaltis 1987) but much work remains to be done. The view above clearly implies that the 'holographic element' should somehow self form. This suggests the use of a material such as a photorefractive for its formation (Tanguay 1985). However, despite interesting small-scale demonstrations, major questions concerning the speed, sensitivity and complexity achievable by this route remain to be answered conclusively.

Summary

We have traced many lines of thinking that have as a common base the use of optics to replace electronics, usually in a 'wiring' role. This is a field in which it is easy to be carried away by the beauty of optics and to lose sight of the very tough competition that must be faced from electronics. We have repeatedly highlighted the orthogonality of optics and electronics; each has its strengths and each its weaknesses. For this reason, we believe that in processing it is almost inconceivable that optics will displace electronics. However, optics is penetrating such systems and this will continue. The debate thus centres on where the new equilibrium will be established.

In the course of this article, many statements have been made that involve a large degree of engineering judgement rather than hard physical limitation. Such judgements have tended to be hard on optics, but to displace an established technology, the new one must demonstrate very substantial advantage. However, as with all judgements, we accept that these may prove to be unduly pessimistic. With that warning we leave readers to form their own opinions.

References

Hopfield J J 1982 Neural networks and physical systems with emergent collective computational abilities *Proc. Natl Acad. Sci., USA* **79** 2554–8

Huang A 1984 Impact of technological advances and architectural insights on the design of optical computers *Phil. Trans. R. Soc.* A **313** 205–11

Huang A and Knauer S 1984 Starlite, a wideband digital switch *Proc. IEEE Global Telecommunications Conf, Atlanta, Georgia, USA* (New York: IEEE) pp121–5

Midwinter J E 1987 A novel approach to the design of optionally activated wideband switching matrices *Proc. IEE J* **134** 261–8

Murdocca M and Streibl N 1987 A digital design technique for optical computing *IEEE/OSA Topical Meeting on Optical Computing, Incline Village, Nevada, USA, 16–18 March, 1987*

Psaltis D and Farhat N H 1985 Optical implementation of the Hopfield model *IEEE/OSA Topical Meeting on Optical Computing, Incline Village, Nevada, USA, 18–20 March, 1987*

Selviah D and Midwinter J E 1988 Correlating matched filter model for the analysis and design of neural networks *Proc. IEE* submitted for publication

Tanguay A 1985 Materials requirements for optical processing and computing devices *Opt. Engng* **24** 2–18

Wagner K and Psaltis D 1987 Multilayer optical learning networks *IEEE/OSA Topical Meeting on Optical Computing, Incline Village, Nevada, USA, 16–18 March, 1987*

Architectural Considerations for Photonic Switching Networks

H. SCOTT HINTON, MEMBER, IEEE

(*Invited Paper*)

Abstract—This paper will review some of the photonic technologies that could become important components of future telecommunications systems. It will begin by dividing photonic devices and systems into two classes according to the function they perform. The first class, relational, will be associated with devices, which under external control, maps the input channels to the output channels. The second class, logic, requires that the devices perform some type or combination of Boolean logic functions. After the classes are defined, some of the strengths and weaknesses of the photonic domain will be presented. Relational devices and their applications will then be discussed. Finally, there will be a review of optical logic devices and their potential applications.

I. INTRODUCTION

WITHIN recent years there has been a significant amount of interest in applying the new and developing photonics technology in telecommunications switching systems [1]. As the transmission plant has converted its facilities to fiber, there is an economic interest in completing the optical path through the switching system to the terminal facilities without requiring optical-to-electrical (o/e) conversions. There are several devices that have emerged within the past few years which have the capability of meeting this goal. These devices can be arranged into two major classes according to the function they perform [2]. The first of these classes, called *relational* devices, perform the function of establishing a *relation* or a mapping between the inputs and the outputs. This *relation* is a function of the control signals to the device and is independent of the signal or data inputs. As an example, if the control signal is not enabled, the relation between the inputs and the outputs of a 2 × 2 device might be input port 1 → output port 1 and input port 2 → output port 2. When the control is enabled, the relationship might be input port 1 → output port 2 and input port 2 → output port 1. This change in the *relation* between the inputs and outputs corresponds to a change in the state of the device. Another property of this device is that the information entering and flowing through the devices cannot change or influence the current *relation* between the inputs and outputs. An example of this type of device is the directional coupler as it is used in switching applications. Thus, the strength of relational devices is that they cannot sense the presence of individual bits that are passing through them which allows them to pass high bit rates. The weakness of relational devices is that they cannot sense the presence of individual bits that are passing through them which reduces their flexibility.

The second class of devices will be referred to as *logic* devices. In these devices, the data or information carrying signal that is incident on the device controls the state of the device in such a way that some Boolean function or combination of Boolean functions is performed on the inputs. For this class of device, at least some of the devices within a total system must be able to change states or *switch* as fast or faster than the signal bit rate. This high-speed requirement for logic devices will limit the bit-rates of signals that can eventually flow through their systems to less than those that can pass through relational systems. Thus, the strength of logic devices is the added flexibility that results from their ability to sense the bits that are passing through them while their weakness is that they sense the bits that pass through them which limits the maximum bit-rate that they can handle.

This paper will begin by discussing both the strengths and limitations of the photonic technology. Some of the items to be discussed include power, speed, bandwidth, and parallelism. The next section will discuss optical relational systems with a focus on the directional coupler. The reason for this focus is that directional couplers will most likely be the initial photonic switching component to enter the marketplace. Finally, there will be a discussion of optical logic systems. These systems are still primarily in the research labs and not ready for development.

II. STRENGTHS AND LIMITATIONS OF THE PHOTONIC TECHNOLOGY

Prior to discussing either photonic devices or their applications it is important to understand both their potential and limitations. This section has the purpose of discussing the strengths and weaknesses of the photonics switching technology. It will begin by discussing the power, speed, and bandwidth limitations of photonic devices. Next it will focus on parallelism and how it can be used in photonic systems. Finally, there will be a brief discussion on the size of future devices.

A. Power, Speed, and Bandwidth

There are two speed limitations that must be considered in the design of photonic switching systems. The first of

---

Manuscript received October 16, 1987; revised February 24, 1988. This paper is a combination of papers presented at CLEO'87, Apr. 26–May 1, 1987, Baltimore, MD and OFC'88, Jan. 25–28, 1988, New Orleans, LA.

The author is with AT&T Bell Laboratories, Naperville, IL 60566.

IEEE Log Number 8821378.

these limitations is the time required to *switch* or change the state of a device. *Switching*, in this case, refers to the changing of the present state of a device to an alternate state, as opposed to the "switching" that is analogous to an interconnection network reconfiguration. In the normal operating regions of most devices, a fixed amount of energy, the switching energy, is required to make them change states. This switching energy can be used to establish a relationship between both the switching speed and the power required to change the state of the device. Since the power required to switch the device is equal to the switching energy divided by the switching time, then a shorter switching time will require more power. As an example, for a photonic device with an area of 100 μm^2 and a switching energy of 1 fJ/μm^2 to change states in 1 ps requires 100 mW of power instead of the 100 μW that would be required if the device were to switch at 1 ns. Thus, for high-power signals the device will change states rapidly, while low-power signals yield a slow switching response.

Some approximate limits on the possible *switching* times of a given device, whether optical or electrical, are illustrated in Fig. 1 [3]. In this figure the time required to *switch* the state of a device is on the abscissa while the power/bit required to *switch* the state of a device is on the ordinate. The region of spontaneous switching is the result of a background thermal energy that is present in a device. If the switching energy for the device is too low, the background thermal energy will cause the device to change states spontaneously. To prevent these random transitions in the state of a device, the switching energy required by the device must be much larger than the background thermal energy. To be able to differentiate statistically between two states, this figure assumes that each bit should be composed of at least 1000 photons [4]. Thus, the total energy of 1000 photons sets the approximate boundary for this region of spontaneous switching. For a wavelength of 850 nm, this implies a minimum switching energy on the order of 0.2 fJ.

For the thermal transfer region, Smith assumed that for continuous operation the thermal energy present in the device cannot be removed any faster than 100 W/cm^2 (1 μW/μm^2). There has been some work done to indicate that this value could be as large as 1000 W/cm^2 [5]. This region also assumes that there will be no more than an increase of 20°C in the temperature of the device. Devices can be operated in this region using a pulsed rather than continuous mode of operation. Thus, high-energy pulses can be used if sufficient time is allowed between pulses to allow the absorbed energy to be removed from the devices.

The cloud represents the performance capabilities of current electronic devices. This figure illustrates that optical devices will not be able to *switch* states orders of magnitude faster than electronic devices when the system is in the continuous rather than the pulsed mode of operation. There are, however, other considerations in the use of optical computing or photonic switching devices than

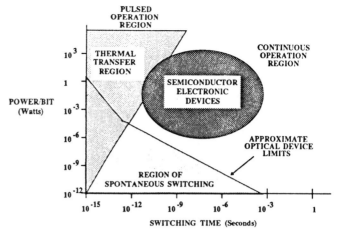

Fig. 1. Fundamental switching limits at 850 nm.

how fast a single device can change states. Assume that several physically small devices need to be interconnected so that the state information of one device can be used to control the state of another device. To communicate this information, there needs to be some type of interconnection with a large bandwidth that will allow short pulses to travel between the separated devices. Fortunately, the optical domain can support the bandwidth necessary to allow bit rates in excess of 100 Gbits/s, which will allow high-speed communication between these individual *switching* devices. In the electrical domain, the communications bandwidth between two or more devices is limited by the resistance, capacitance, and inductance of the path between the different devices. Therefore, even though photonic devices cannot *switch* orders of magnitude faster than their electronic counterparts, the communications capability or transmission bandwidth present in the optical domain should allow higher speed *systems* than are possible in the electrical domain.

The second speed limitation, which applies only to relational devices, will be referred to as the transmission bandwidth. After a relational device has been put into a particular state, it acts like a transmission line to any data entering its inputs. This input data cannot change the state of the device, thus the signal bit-rates passing through a relational device are not limited by the constraints outlined in Fig. 1. For most relational devices, this transmission bandwidth should be able to support bit-rates in excess of 100 Gbits/s.

In summary, networks composed of relational devices will have their signal bit rates limited by the transmission bandwidth and their reconfiguration rates limited by the switching time of the devices, while switching networks based on optical logic will have both their signal bit-rates and reconfiguration rates limited by the switching time of their devices.

B. Parallelism

Another method of increasing the capacity of a system, in addition to operating at higher speeds, is to operate on information in parallel instead of in series. In pursuing this parallelism, attention has recently been placed on free-space optics. These types of systems normally are composed of multiple two-dimensional (2-D) arrays of optical devices that are interconnected through either bulk optics or holography. Fig. 2 shows the optical interconnection between two 2-D arrays of optical elements. The interconnection in this case is a simple lens system. The optical elements, which will be referred to as pixels, could be optical NOR gates, optical light valves, etc. The number of pixels that can be interconnected in this manner is limited by the resolution of the optical interconnection system. Even relatively inexpensive optical imaging systems exhibit resolutions on the order of 10 μm over a 1 mm field. This provides access to 100 × 100 or 10^4 pixels. If each pixel can be equated to a pin-out, then for a 2-D array there can be greater than 10^4 pin-outs. The maximum number of pixels or pin-outs that can be supported by a lens or any optical system is referred to as its space–bandwidth product [6] (SBWP) or the degree of freedom of the system. Satellite imaging systems have been made that have an SBWP of 10^8 pixels.

Fig. 3, on the other hand, illustrates an optical interconnect based on holograms. Holography offers the promise of extremely high SBWP ($>10^{10}$). This will be discussed in more detail in Section IV-B-2.

There is no reason to limit the pixels of these 2-D arrays to all-optical logic gates; they could also be a mixture of electronic and optical devices (smart pixels) as shown in Fig. 4. This mixture of electronic and optical devices is designed to take advantage of the strengths in both the electrical and optical domain. The optical devices include detectors to convert the signals from the previous 2-D array to electronic form and modulators (surface emitting lasers or LED's) to enable the results of the electronically processed information to be transferred to the next stage of 2-D arrays. The electronics does the intelligent processing on the data. Since the electronics is localized with short interconnection lengths, the speed of this electronic island should be fast. The applications of smart pixels will be discussed later in Section IV-A-2.

To take advantage of the parallelism inherent in free-space optics, a device should have the capability of driving other devices (fan-out) in addition to being controlled by more than one device (fan-in). Since fan-out corresponds to a division of the energy emitted from a device output to the inputs of other devices, the output energy must be significantly larger than the energy required by the input to the subsequent device. Another factor that affects the fan-in is the contrast ratio, the contrast ratio being the ratio of the transmitted intensity of both states of a device. For low contrast ratios, unwanted noise is present in the system, thus reducing the ability of a device to sample the input correctly.

C. Device Size

The minimum size of an optical switching device cannot be reached below a volume of $(\lambda/n)^3$ [3] where n is

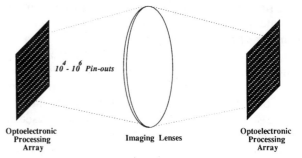
Fig. 2. Optical parallel interconnections.

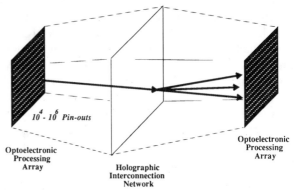
Fig. 3. Holographic interconnects.

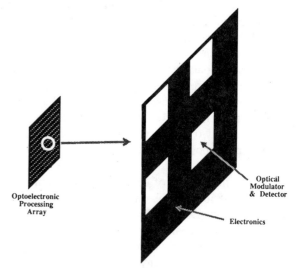
Fig. 4. Smart pixels.

the index of refraction of the device. For the case of a 2-D array of devices, assuming that these devices are fabricated such that there is a separation of λ/n between the devices, then approximately 25 million devices per square centimeter could be possible at $\lambda/n = 1$ μm. As the device switching speed increases, the thermal transfer capacity of the devices will most likely prevent such a large number of devices from ever being realized. As an example, if there are 25 million devices per square centimeter that require 1 pJ to switch, and it is desired to switch these devices in 100 ps, then the total array would require 250 kW/cm² of optical power if all of the devices were to switch at the same time.

III. OPTICAL RELATIONAL SYSTEM

In this section, several photonic switching systems that are based on relational devices will be reviewed. The first type of system to be discussed is based on spatial light modulators (SLM) which are 2-D arrays of devices, each of which has the capability of modulating the light that is incident upon it. The second example of a relational system is based on wavelength-division switching. Finally, the systems based on Ti:LiNbO$_3$ directional couplers will be discussed.

A. Spatial Light Modulators

An SLM is a two-dimensional array of optical modulators [7]. Each of these modulators is independent of the others and has the capability of modulating the incident light. For the applications described in this paper, the modulators will be assumed to be digital, in that they possess two states: transparent to be the incoming light (on) and opaque to the incoming light (off). These arrays are electrically controlled such that an electrically enabled pixel will be transparent while a disabled pixel will block the incident light. An SLM that is currently available in the marketplace is based on the magnetooptic effect [8]. Some other SLM's include the liquid crystal light valves (LCLV) [9], PLZT modulators [10], deformable mirrors [11], and GaAs multiple-quantum well (MQW) modulators [12].

One example of how an SLM can be used as a photonic switch is shown in Fig. 5 [13], [14]. In this figure, the fiber inputs are horizontally aligned as a row of inputs. The inputs are aligned to associate each fiber with a unique column of the SLM. A lens system is used to spread these inputs vertically so that the light emitted from each input is spread over all the elements' of the SLM's associated column. The appropriate pixels of the SLM are enabled before the data pass through the system. The one or more enabled pixels in each column allow the incident light to be transmitted through the device while the remaining pixels block the incident light. The output column of fibers acccepts the light that is passed through the SLM. An important restriction for this type of structure is that only one pixel on each row can be enabled at any time. The relational nature of this structure is evident in that each row of the SLM acts like an $N \times 1$ switch where N is the number of pixels per row. The total structure is topologically equivalent to a nonblocking crossbar interconnection network. The weakness of these systems is the losses that occur in the spreading and collection of the light going to and from the SLM can be greater than $1/N^2$.

As with all relational structures, high signal bit-rates pass through the switch with the speed limitation being the system reconfiguration time.

B. Wavelength-Division Switching Systems

Another type of relational architecture that has received a considerable amount of attention is wavelength-division switching [15]. This is schematically shown in Fig. 6. In this figure, the entering information is used to modulate a light source that has a unique wavelength for each input. All the optical energy is combined and then split so it can be distributed to all the output channels. The tunable filter on each ouput is adjusted such that it only allows the wavelength associated with the desired input channel to pass to the detector. Thus, by varying the tunable filter, an output has access to any or all of the input channels. Obviously, another method of detecting the appropriate wavelength would be through the use of coherent detection techniques [16]. This wavelength-division concept can be generalized to include other orthogonal basis functions as the carrier instead of wavelength [17], [18].

C. Directional Couplers

A directional coupler is a device that has two optical inputs, two optical outputs, and one control input. The control input is electrical and has the capability of putting the device in the *bar* state, input port 1 → output port 1 and input port 2 → output port 2, or the *cross* state, input port 1 → output port 2 and input port 2 → output port 1 [19]. The most advanced implementations of these devices have occurred using the Ti:LiNbO$_3$ technology [20]. The strength of directional couplers is their ability to control extremely high bit rate information. They are limited by several factors: 1) the electronics required to control them limits their maximum reconfiguration rate, 2) the long length of each directional coupler prevents large scale integration, and 3) the losses and crosstalk associated with each device limit the maximum size of a possible network unless some type of signal regeneration is included at critical points within the system [21]. A modest number of these devices have been integrated onto a single substrate to create larger photonic interconnection networks such as an 8×8 crossbar interconnection network [22]. As another example, a 4×4 crossbar interconnection network composed of 16 integrated directional couplers, all having crosstalk less than -35 dB with an average fiber-to-fiber insertion loss of less than 5.2 dB, has been fabricated [23]. The following sections will discuss some of the practical considerations that need to be addressed in the design of a photonic switching system based on directional couplers. Several potential applications will then be discussed.

1) Practical Considerations: There are several practical issues that need to be considered when designing a system based on Ti:LiNbO$_3$ directional couplers. These issues include the required parallel development of polarization maintaining (PM) fiber, optical amplifiers, and packaging required to make the devices reliable and easy to use.

a) Polarization Maintaining Fiber: To minimize the required drive voltages, directional couplers have been optimized to operate with a single linear polarization. This requirement reduces the required switching voltage from approximately 50 V to the 10–15 V range. These lower voltages are desired to allow high-speed switching of the directional couplers. One problem with using single po-

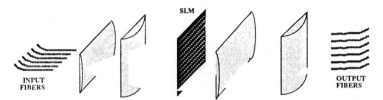

Fig. 5. Optical crossbar interconnection network.

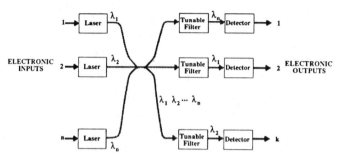

Fig. 6. Wavelength-division switching system.

larization devices is that as light propagates through standard single-mode fiber, the state of polarization can be changed from a linearly polarized wave to a wave having an elliptical polarization. Another complicating factor is that this change in the state of polarization does not remain constant over time. To solve this problem, a PM fiber is required for the interconnection from all the laser sources to the Ti:LiNbO$_3$ substrates and for any substrate-to-substrate interconnection. In addition to PM fiber, there must also be PM fiber connectors. These connectors become important when networks involving the interconnection of multiple LiNbO$_3$ substrates are required.

b) Optical Amplifiers: The eventual size of a switching fabric composed of directional couplers is limited by either the losses through the system or the individual crosstalk terms which degrade the system signal-to-noise ratio (SNR) below an acceptable value [24]. To avoid this problem, thresholding optical amplifiers can be inserted at critical points in the network to both boost the strength of the signal and remove accumulated noise. Unfortunately, most of the optical amplifiers under investigation are linear amplifiers [25], these amplifiers have the disadvantage of amplifying the low-level noise signals as well as the desired signals. It is desired to have an amplifier that will not amplify low-level signals associated with the logic level "0" but will amplify signals above the threshold of the logic level of a "1." The characteristics of a thresholding optical amplifier [26] are shown in Fig. 7. In the ideal case, there would be no amplification for signals below a given intensity level. Once an intensity threshold has been surpassed, a large gain is desired until a saturated or maximum value of output intensity is obtained [27]. Such a device could both amplify a signal and improve the SNR of the signals passing between substrates of a dimensionally large relational photonic switching system.

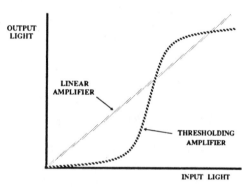

Fig. 7. Optical amplifier characteristic curves.

A component required in conjunction with optical amplifiers is an optical isolator. In addition to isolating lasers from reflections that can occur at connectors, splices, and other interfaces, optical isolators are required to prevent optical amplifiers from lasing. This lasing action can occur since the two requirements for a laser are met: optical gain and positive feedback. The optical amplifier provides the gain and the connectors, splices, or other reflecting interfaces on both sides of the optical amplifier provide the positive feedback required. For high-gain optical amplifiers, this lasing action can occur with small interface reflectivities. By placing an isolator between the unwanted reflections and the optical amplifier, the positive feedback is reduced, thus preventing the undesired lasing.

c) Packaging: To develop both a marketable and reliable system, devices have to be packaged in a useful and reliable manner. This is one of the most overlooked aspects in the development of this technology. At the current time, most of the devices are packaged in their own separate package. For large systems, this could involve an enormous amount of physical space just to house all the individual components. High-speed packaging is another unresolved issue. This subject will require a large amount of attention before this technology can successfully enter the marketplace.

2) Applications of Directional Couplers: Perhaps the key event required to drive the Ti:LiNbO$_3$ directional coupler technology into the marketplace within the next three to five years is a *good* application. Therefore, several potential applications will be outlined in this section. The first application to be discussed utilizes the directional couplers in a space switching environment. Second, the signal formats required to allow these relational devices to operate in a time-division application will be outlined. Finally, there will be brief examples of both time-multiplexed and packet switching.

a) Space Switching: This type of switching will most likely be the first application of the Ti:LiNbO$_3$ directional couplers. It requires long hold times with moderate reconfiguration rates. Once a path has been set up, high-speed data, multiplexed speech, or video can be transferred through the fabric.

The implementation of a large space switch requires the interconnection of many smaller photonic switches that are used as building blocks. These building blocks will most likely have dimensions less than 16 × 16 because of the large size of directional couplers and the large bending radii required in the integrated waveguides. Two examples of topologies for these building blocks are the crossbar interconnection network [21] and the broadcast network proposed by Spanke [28]. For point-to-point networks, the interconnection of these building blocks to construct a larger switching system can be done with Clos, Benes, banyan, omega, or shuffle networks. If video information is to be a main component of the system traffic, then a broadcast environment becomes important. A good topology for a broadcast network is a Richards network [29]. An example of a space-switching experiment is shown in Fig. 8 [30]. In this system, each terminal has the capability of transmitting and receiving one of two wavelengths. Initially a path is set up between two terminals through the folded optical switching network. This network is composed of switching arrays made using Ti:LiNbO$_3$ directional couplers. After the path is set up, an input signal is sent to terminal *B* on a light source of wavelength λ_1. Terminal *B* will then create the return path to terminal *A* using the same physical path but modulating the information onto the light source of wavelength λ_2.

A good application of a directional coupler is a protection switch. In this environment, the only time the switch will need to be reconfigured is when a failure occurs in an existing path. Thus, high bit rates can be passed through the switch with moderate reconfiguration rate requirements. This application matches the capabilities of the directional coupler.

b) Signal Formats: As high bit rate transmission systems are being developed, there is pressure to build switching systems that will switch these complex signals. Transmission systems are designed such that the high bit-rate signal only passes through a few elements as it is

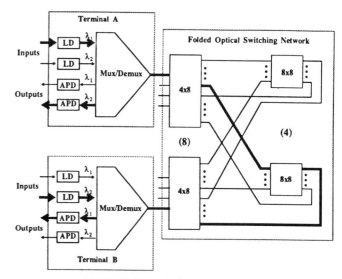

Fig. 8. 32 line optical space switch [30].

generated or processed. These few elements allow all the controlling electronics to be placed close together. In a switching system, the signal must pass through a large number of switching elements which are normally spread over a large physical space; this makes switching difficult at high bit-rates. As bit-rates continue to increase, there will need to be a compromise between the complexity of transmission and the switching systems [31]. One solution to this problem is to change the signal format from a bit-multiplexed to a block-multiplexed format. By changing this format, transmission systems will become more complex because buffers will be required at all the inputs to the multiplexers. Such a new format will simplify switching systems by allowing them to operate directly on the information rather than having to demultiplex the information and then switch the lower bit-rate signals. An example of such a proposed format is SYNTRAN [32]. The concept of this block-multiplexed format is good for future photonic applications but the DS3 bit rate (45 Mbits/s) it proposes is too slow for the photonic domain. These slower bit rates can be handled easily in the electronic domain. This concept does have merit if it is extended to higher bit rates by using individual 125 μs frames of the 45 Mbit/s DS3 information as the basic block of data for the system. As an example, in the FT series G transmission systems, a single 125 μs frame of the 1.7 Gbit/s data stream contains 36 frames of DS3 information. The bits in these frames, plus overhead bits, are interleaved and mixed so that individual DS3 frames cannot be extracted from the stream unless that stream is at least partially demultiplexed down to DS3 channels. By requiring that the DS3 frames be block multiplexed onto the high-speed channel with a small gap between them, individual extraction and insertion of DS3 frames should be possible. The characteristics of a directional coupler (slow reconfiguration rates, but high signal bit rates) make it an ideal device to be used with a block-multiplexed format.

c) Time-Multiplexed Switching: As the bit rates

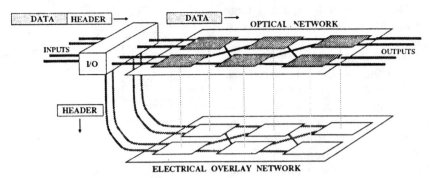

Fig. 9. Directional coupler-based packet switching system [63].

passing through these photonic switches increase, there approaches a point where no single information source can occupy all the bandwidth without some type of time-division multiplexing. As an example, uncompressed digital NTSC television signals require approximately 100 Mbits/s while digital high-definition television will be about 600 Mbits/s. Thus, if the transmission system operates at 1.7 Gbits/s then at least 16 NTSC channels or two high-definition channels can be time-multiplexed together. To build such a system will require a time-slot interchanger, and an elastic store. The elastic store will also be required to remove both frequency and phase jitter from the system inputs. Proposed elastic stores and time-slot interchangers have been demonstrated using fiber delay lines as the memory [33] or bistable laser diodes as bit memories [26].

d) Packet Switching Systems: The problem in implementing a packet switching system based on directional couplers is that a header has to be read to determine the final destination of the packet. This can be done by stripping off a portion of the optical energy and monitoring the header (trailer) electronically [63] as illustrated in Fig. 9. If necessary, the header (trailer) could be at a lower bit rate than the packet data thus allowing the slower electronics to respond to the controlling information. If the header bit rate is at the same bit rate as the data, then the eventual bit-rate upper limit would be governed by the speed of the electronics. It would be possible to have headers encoded at slower bit rates than the data at the cost of reducing the total throughput of the switching system. The cost for such a system is the large amount of electronics that is surrounding the small number of photonic devices. Thus, to be economically justifiable, large packet lengths are required.

IV. Optical Logic Systems

There are certain applications that are not well suited for relational devices. One such application requires the ability to both sense and respond to individual bits of information. A packet switch is a good example of this requirement. A packet entering a network requires a system of devices that can read and understand the header and then reconfigure the network to allow the packet to pass to its desired destination. The ability to interact and sense the individual bits in a stream of information is one of the strengths of optical logic devices [34].

To make a useful system, these devices need to be interconnected, as in current electronic switching systems, to create a large interconnection network. The purpose of the following sections is to discuss several of the optical logic devices that have been fabricated in addition to describing two methods of optically interconnecting these devices. The section on optical logic systems will begin by discussing several optical logic devices. Next, two possible methods of optically interconnecting 2-D arrays of logic devices will be discussed. Finally, there will be a brief discussion of a packet switching system that could be based on these devices.

A. Digital Optical Logic Devices

Digital optical logic devices can be further classified according to the effective number of ports, where a port is either a physical input or output. The discussion on optical logic devices will begin by reviewing several two-port devices and the constraints they impose upon a switching system architecture. Multiport devices and their associated system issues will then be discussed.

1) Two-Port Devices: Two-port devices can be divided into two different types. The first, which will be referred to as combinatorial, are devices whose output is a function of present inputs. These devices include any device that simulates a Boolean logic function. The characteristic curves of these devices are shown in Fig. 10. Part (a) illustrates the characteristic curve of an inverter. Normally, there is a bias beam, separate from the signal beam, which provides the optical energy to be modulated by the device. When the energy from the signal beam is added to the energy of the bias beam, this combined energy is enough to exceed the nonlinear threshold of the device. As the device changes states, the output level goes from a high value or a "one" to the lower value which is equated to a "zero." By reducing the energy in the bias beam and adding more signal beam energy through the use of multiple signal beams, an optical NOR gate can be formed. Part (b) is the characteristic curve of a thresholding optical amplifier. By reducing the bias beam and adding more inputs, this device could be an optical AND gate.

The second type of device is sequential in nature. These

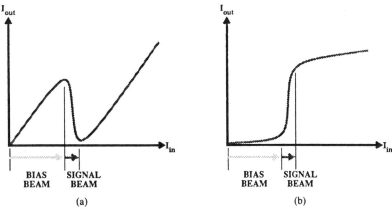

Fig. 10. Characteristics curves of two-port combinational devices.

devices could be smart pixels with one optical input, one optical output, and some high-speed electronics configured as a sequential finite state machine. Perhaps the simplest example of these structures is a bistable device [35]. Since these devices have memory, their outputs are a function of the present inputs and their present state (previous inputs). Another example of this type of two-port device is an optical regenerator. When an optical signal enters the device, the information is both amplified and retimed prior to outputting the reconstructed signal.

a) Proposed Devices: There are two optical logic devices that have received a considerable amount of attention in the past few years. The first of these devices, the self-electrooptic effect device (SEED)[36], is an electrooptical device that requires both optical and electrical energy. The second device, referred to as a nonlinear Fabry–Perot etalon (NLFP) is an all-optical device [34] in that all the energy required to switch the device is supplied optically. Both of these devices can be fabricated into 2-D arrays of devices that are both optically enabled and controlled. The purpose of the 2-D arrays is to provide the opportunity to exploit the parallelism present in the optical domain.

A functional diagram of the SEED is shown in Fig. 11. A p-i-n diode with a multiple quantum well (MQW) material in the intrinsic region is connected in series with a resistor to form the SEED structure. The characteristic curve for the device is shown in part (a) of Fig. 10. A bias beam is required to provide the energy that will be modulated by the signal beam. This modulation of the energy source by the signal beam provides the differential gain that is required in a digital logic system. With low signal beam intensities, the SEED is virtually transparent. This allows nearly all the energy present in the bias beam to pass through the device to become the output beam. When the signal beam is added with the bias beam, the combined energy is enough to force the SEED structure to become an absorber, thus reducing the total amount of energy in the output beam. This characteristic curve is equivalent to the characteristic curve of an optical NOR gate. SEED structures have been integrated into two-dimensional arrays [36]. It is conceivable that within 10

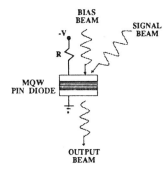

Fig. 11. The self-electrooptic effect device.

years, arrays of 10 000–100 000 individual SEED's per square centimeter will be possible. Currently, 4 fJ/μm^2 of optical energy and 16 fJ/μm^2 of electrical energy are required to change the state of the SEED.

A second possible device for the optical logic domain is a nonlinear Fabry–Perot etalon (NLFP). This device exploits reflection instead of absorption to control or modulate a bias beam. This is shown in Fig. 12. Part (a) of this figure illustrates that without a signal beam, the bias beam passes through the NLFP. With an incident signal beam, part (b) of the figure, the NLFP changes states which forces the bias beam to be reflected instead of transmitted. This device, like the previously described SEED, is operating as a NOR gate. The NLFP can also be designed such that it can operate as an AND, OR, NAND, or XOR gate [34]. These devices, like SEED's, can also be integrated into large arrays [37]. The optical switching energy for NLFP's has been measured to be less than 40 fJ/μm^2 [38] with achieved switching times less than 100 ps [39].

At the current time, there are two types of NLFP etalon structures being studied. The first is based on a thermal nonlinearity which makes it slow, on the order of microseconds [34]. The hope for these devices is that large arrays can be fabricated to exploit the strength of parallelism. The other type of device, referred to as an optical logic etalon (OLE) is a pulsed device that requires two separate wavelengths, one for the bias and one for the signal [40]. The pulsed operation is illustrated in Fig. 13. The two inputs, the data and clock, are separated in both

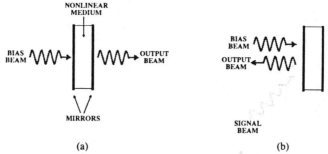

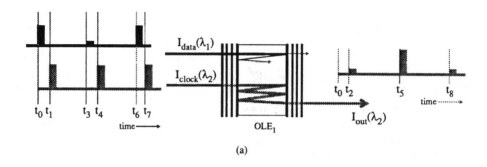

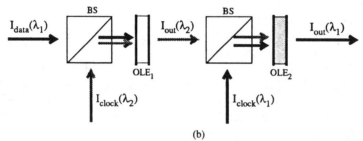

Fig. 12. The nonlinear Fabry–Perot etalon optical gate.

Fig. 13. OLE operation. (a) Input and output signals for OLE operation. (b) Complementary pair of OLE's.

time and wavelength. The input data $I_{data}(\lambda_1)$ is incident on the OLE a short period of time prior to the clock input $I_{clock}(\lambda_2)$. In part (a) of the figure, $I_{data}(\lambda_1)$ is incident upon the device at t_0 while $I_{clock}(\lambda_2)$ arrives at t_1. Ideally, $I_{data}(\lambda_1)$ is chosen such that λ_1 corresponds to an absorption peak of the nonlinear material in the OLE, bulk GaAs. Thus, nearly all of the energy associated with the incoming data is absorbed by the material. This absorbed energy alters the effective index of refraction in the GaAs intercavity material [41]. This change in the index of refraction shifts the resonant peak of the Fabry–Perot etalon by changing the optical path length of the cavity. Through the proper choice of initial detuning, several potential optical logic gates can be implemented. In this figure, the initial detuning was set so that the OLE would behave like an optical NOR gate. If the material relaxation time of the intercavity material is long compared to the pulse duration, the state of the device will not change significantly prior to the arrival of the clock pulse $I_{clock}(\lambda_2)$. When the clock arrives, the OLE will be in either a transmissive or reflective state depending on the value of $I_{data}(\lambda_1)$. In this figure, if there was a logical "one" on the data input,
then the OLE would be reflective forcing the output to be a logical "zero." On the other hand, if the input data was a logical "zero," the device would be transmissive allowing the majority of the clock energy through the OLE creating a logical "one" output. Since λ_2 is not near the absorptive peak of the intercavity material, clock signals much larger than the data signals can provide an effective gain through the device. This separation of the data and clock inputs in either time or wavelength has converted a two-port device into an effective three-port device.

Part (b) of this figures shows what is required to integrate these devices into larger systems. Because the input wavelength of an OLE is different from the output wavelength, the devices are not cascadable. Therefore, to implement any type of system will require a complementary device that will change the output wavelength of the two devices back to the original input wavelength. This part of the figure shows how, through the use of polarization beamsplitters, the two devices can be interconnected. Unfortunately, at the current time a complementary pair has not been demonstrated.

b) System Considerations: When any of the two-port

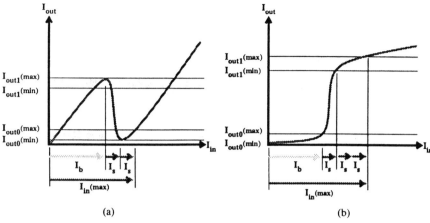

Fig. 14. Input/output levels of critically biased two-port devices.

devices operate with a bias beam as shown in Fig. 14, there are strict requirements on the bias beam's stability [42]-[44]. If the bias beam varies in intensity, it could spontaneously exceed the nonlinear threshold forcing the device to change states. Also, if the intensity of the bias beam decreases significantly, it could prevent the device from changing states when a signal beam is incident upon it. In addition to the bias beam, the point of nonlinearity in the characteristic curve must also remain constant. This required stability puts tight restrictions on thermal control for most devices. As an example, for any structure using a Fabry-Perot cavity, temperature variations will cause changes in the location of the resonant peaks of the cavity. This in turn alters the initial detuning of the device which can alter the characteristic curve of the device. Such variation could cause erroneous device behavior. Also, any device that is based on a material resonance could also suffer detrimental effects with temperature variations.

Another problem associated with the critically biased devices that have been previously described is the size of the fan-in and fan-out of the device. The fan-in is the maximum number of inputs (signal, not clock beams) that can be used as inputs to a given device and can be represented, as shown in Fig. 14, by fan-in = $(I_{in}(\max) - I_b)/I_s$. The fan-out, which is the maximum number of input signals that an output signal can be decomposed into, is represented by fan-out = $(I_{out1}(\min) - I_{out0}(\max))/I_s$. These two equations imply that for a large fan-out and fan-in, which is desirable, a large sharp nonlinearity is required. The problem with a sharp nonlinearity is the strict requirement on the stability of the bias beam. Because of these stability requirements, it is unlikely that for critically biased devices, fan-out greater than five will be practical.

Another area of concern with respect to fan-in occurs when the light in a system has a large coherence length and a single polarization. In this situation, both constructive and destructive interference could occur when two or more beams are incident on the same physical input. As the devices become smaller, which will be required to reduce the switching energy, it will be possible for this interference to prevent a device from operating properly. To avoid this problem, the fan-in could be limited to two inputs with each input being associated with a different polarization. Another alternative is to separate all inputs in time as in the case of the OLE.

2) Multiport Devices: The devices discussed so far can be classified as two-port devices in that they have one input and one output. By adding more inputs and outputs, many of the problems associated with two-port devices can be removed. As an example, by adding more physical input ports, the problem of low device fan-in can be resolved even though the fan-in per port is small (the desired fan-in per port is two, one for each polarization). Also, the effects of critical biasing can be removed by separating the signal input from the bias input thus creating a three-port device.

a) Proposed Devices: There are several devices that have been proposed that could be classified as "multiport devices." The first device to be discussed will be a three-port device in which the signal input is separated from the bias input. The second device is a four-port device referred to as the *symmetric SEED*. This device consists of two electrically interconnected MQW p-i-n diodes. Finally, a five-port device that performs an exchange-by-pass operation will be discussed.

An example of a three-port device is shown in Fig. 15 [45]. In this figure, an MQW modulator is electrically connected to a phototransistor [46]. The objective of this device is twofold: 1) increased sensitivity for the input signal (optical gain) and 2) isolation between the input and output signal. With an input signal present on the phototransistor, a photocurrent is created which is roughly proportional to the input signal power. The optical bias beam also gives rise to a photocurrent through the modulator which is proportional to the power absorbed. These two currents must be equal or a charge will build up on the modulator affecting the absorption such that the two currents will equalize. If the operating wavelength is set just below the band edge, then with no input signal, the majority of the voltage will be dropped across the pho-

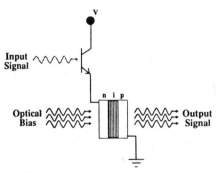

Fig. 15. Photonic three-port device.

totransistor which puts the modulator in the transmissive state. When an input signal is present, the phototransistor will turn on putting the majority of the voltage across the modulator. With this applied voltage, the band edge of the MQW material will shift, putting the modulator in the absorptive state. Thus, a lower power input signal has the capability of modulating the higher power optical source providing gain. The modulation effect provides both gain between the input and output signal and a logical inversion of the input data. A noninverting three-port device has also been demonstrated [47].

The symmetric SEED (S-SEED) is a four-port device with two inputs and two outputs as shown in part (a) of Fig. 16 [48]. This device is the result of electrically interconnecting two MQW p-i-n diodes in series. When the diodes are connected in this fashion, they become complementary, in that, when one of the diodes is "on" the other will be "off." Thus, one of the diodes will be in the absorbing state while the other is in the transmissive state. This is illustrated in the characteristic curves shown in parts (b) and (c). Perhaps the greatest strength of these devices is that changing states is a function of the *ratio* of the two input powers and not the function of the absolute intensity of the input beams. This can be seen in the characteristic curves of part (b) and (c). The optically bistable loop is centered around the point where the two inputs, P_{in1} and P_{in2}, are equal. From these figures, it can be seen that the device will remain in its current state until that ratio exceeds 1.3 or is less than 0.7. The importance of this is that the allowable noise on the signal inputs can be much greater than for the case of a critically biased device such as a SEED.

The S-SEED can be configured such that it can operate as an S-R latch. This is illustrated in Fig. 17. In part (a) of this figure, the inputs are separated into an S (set) input, and R (reset) input, and a clock input, where the clock has approximately the same intensity for both inputs. The S and R inputs are also separated in time from the clock inputs as shown in part (b) of this figure. Note that this clocking scheme is similar to that used by the OLE. The S or R inputs are used to set the state of the device. When the S input is illuminated, the S-SEED will enter a state where the upper MQW p-i-n diode will be transmissive while the lower diode will be absorptive. When the R input occurs, the opposite condition will occur. Since the

energy required to change the state of the devices is a function of the ratio of the S and R inputs, then when only one of those two inputs occurs, low switching intensities should be able to change the device's state. After the device has been put in its proper state, the clock beams are incident on both inputs. Since the two clock beams are roughly equivalent, the ratio between the incident beams should be close to one which will prevent the device from changing states. This higher energy clock pulse will be used to transmit the state of the device to the next stage of the system. Since the S or R inputs are low-intensity pulses and the clock is a high-intensity pulse, a large differential gain may be achieved.

Another example of a multiport device is an exchange-bypass node that has been proposed by Midwinter [49]. This device tries to capitalize on the advanced device fabrication and large scale integration capabilities of the electronic domain and the effective large number of pin-outs possible in the optical domain. Each bypass-exchange node is composed of three optical inputs, one electrical input (it could be optical), and two optical outputs as shown in Fig. 18. The optical inputs and outputs are to be implemented using MQW p-i-n diodes. By sensing the current, these diodes are detectors, while applying a time-varying voltage allows them to modulate light incident upon them. All of the processing on the data is done in the electrical domain. Since all of the interconnections are local, high-speed operation should be possible. The device is a logic device implementation of the directional coupler. If I_c is a logical "one" and the clock is a logical "one," then the inputs I_a and I_b will effectively *bypass* the node being directed to outputs O_a and O_b. On the other hand, if I_c is a logical "zero," then the node will *exchange* the inputs sending I_a to O_b and I_b to O_a with an asserted clock signal. Note that one of the main advantages of this device over a directional coupler is that there is gain and the thresholding nonlinearity of optical logic gates which will reduce the signal-to-noise requirements of the nodes. Eventually, large numbers of these nodes or "smart pixels" can be integrated into two-dimensional arrays to take advantage of the large pin-out capability available in the optical domain. Another example of a multiport device is the 4 × 4 OEIC switch implemented by Iwama et al. [50]. This device, or collection of devices, is composed of an array of optical detectors that receive the optical information. This array is then electrically connected to an electrical GaAs 4 × 4 switch. The output of this switch then drives four laser diodes that have been integrated onto a single substrate. This device has operated at 560 Mbits/s.

When integrating large numbers of these multiport devices into arrays, thermal problems could eventually limit their maximum size. This limitation occurs because both the electrical and optical devices dissipate power, and as the bit rates increase so does the required power.

3) Device Capabilities: In Fig. 19, the two strengths of photonics, bandwidth (data capacity) and parallelism (connectivity), are each assigned to an axis of a graph.

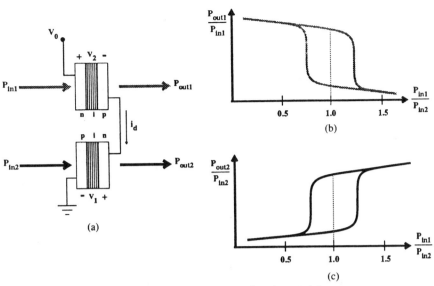

Fig. 16. Symmetric SEED operating characteristics.

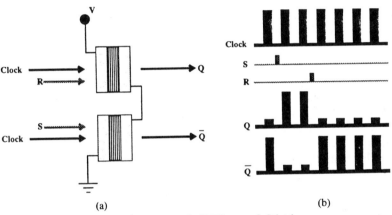

Fig. 17. Symmetric SEED as an S-R latch.

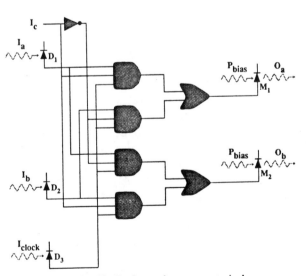

Fig. 18. Exchange-bypass smart pixel.

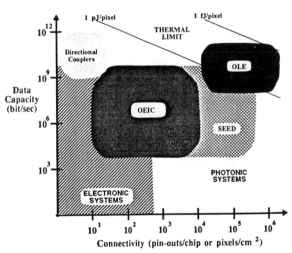

Fig. 19. Photonic device capabilities.

The ordinate, which is the bandwidth component, is labeled as data capacity and has the units of bits per second. The abscissa represents parallelism and is listed as connectivity with the units of either pin-outs per chip or pixels per square centimeter. The capabilities of current electronic systems are located in the lower left hand corner of the graph. The upper limit of electronic systems data capacity is approximately 10 Gbits/s with the maximum number of pin-outs of approximately 500. For the directional coupler, which is a relational device, the upper data capacity limit is in excess of 100 Gbits/s while the number of pin-outs is less than 100. Pin-outs in this case refers to the number of optical fiber inputs and outputs that can be connected to a single $LiNbO_3$ substrate. The lower boundary on the area designated for directional couplers was chosen at a data capacity of approximately 1 Gbit/s where high-speed electronics has been successfully demonstrated. Optoelectronic integrated circuits (OEIC) are devices that combine both the optical and electronic domains; thus, the data capacity limit will be the same as for the electronic domain while the connectivity can approach an effective pin-outs/chip of 10^4. A good example of an OEIC is a smart pixel. The lower boundary was chosen at a data capacity of 64 kbits/s (a single digital voice channel requires 64 kbits/s). The left boundary was chosen to be approximately 10. This boundary includes the possibility of simple linear arrays of optoelectronic regenerators. The self-electrooptic effect device (SEED) is a specific example of a simple OEIC which has the potential for a large number of pin-outs/chip. Finally, the optical logic etalon (OLE) is an all-optical device that has the potential of fabricating pixels (optical NOR gates) with diameters on the order of 1 μm. This small device size could potentially allow more than 10^6 pixels per cm^2 to be fabricated. The left boundary was set in excess of 10^4 pixels since OLE devices need to be small to maintain low switching energies. The lower bound on data capacity was chosen in the Gbits/s region to accommodate the pulsed mode of operation although this boundary could extend to the 64 kbit/s level.

The thermal limit region of this figure illustrates the maximum energy required to change the state of each pixel to maintain thermal stability. This is based on the assumption that 100 W/cm^2 can be removed from an array of devices for continuous operation. As an example, if an array of NLFP's with a pixel density of $10^6/cm^2$ is to have the capability of handling 10^{12} bits per second, it would require the switching energy of the pixels to be less than 1 fJ. For the case of OEIC's, this implies that the combination of electrical and optical energy required to change the state of the pixel be less than the thermal limit.

B. Device-to-Device Interconnection

Once two-dimensional arrays of optical logic gates are available, it will become necessary to interconnect the individual devices on the arrays. This interconnection, which can be thought of as the photonic wires of the network, can be accomplished through the use of either space-variant or space-invariant interconnects. A space-variant network has the property that each input into the network can be redirected (connected) to any or all of the outputs and is dependent on its spatial location. This type of interconnect provides a different interconnection pattern for each spatially separated input. Alternatively, a space-invariant network can interconnect an input to any or all of the outputs, but is independent of its spatial location. These interconnects provide a single pattern for all inputs regardless of their spatial location. This spatial independence implies that the output pattern created by the inputs will be the same only shifted in space. These concepts are illustrated in Fig. 20. In part (a) the space-variant interconnect shows how the optical energy from two spatially separated inputs impinging upon the interconnection network redistribute the energy differently creating different input/output connection patterns. In part (b), the energy from the two inputs is redistributed in the same manner only shifted in spatial dimensions.

The following sections will begin by discussing free-space interconnection networks based on bulk optics and then outline some of the constraints of holographic interconnection networks.

1) Bulk Optics: The first type of free-space interconnection network can be implemented with bulk optics (lenses, prisms, mirrors, etc.). The simplest example of this type of an interconnection network is an optical imaging system composed of conventional lenses (see Fig. 2). Such a system is space-invariant and can be used to transfer the information present on the outputs of one array to the inputs of a second (or the same) array, effectively creating a large pin-out capability. As an example, the minimum resolvable spot size of a lensing system is given by $a = 1.22 \lambda (f/\#)$ where a is the diameter of the minimum resolvable spot size, λ is the wavelength of the light, and $f/\#$ is the f number of the lens (f/D where f = focal length, D = clear aperature of lens). This implies that for a lens system with $f/\# = 8$ and $\lambda = 850$ nm, the minimum resolvable spot size is 8.3 μm. Assuming that the image to be supported by the lens system is square, the SBWP is given by $(F/a)^2$ where F is the size of the unabberated field in one direction. From our previous example, if $F = 1$ cm then the SBWP = 1.45×10^6 pixels or pin-outs. In order to maximize the SBWP of a lens system, the $f/\#$ must be kept small.

Another type of interconnect that is used in several types of multistage networks is the perfect shuffle, [51]-[53]. An example of a perfect shuffle interconnect that can be implemented with bulk optics is shown in Fig. 21 [54]. Part (a) of this figure illustrates the permutation performed by a perfect shuffle network. Part (b) shows how an optical perfect shuffle can be implemented with a beamsplitter, a lens, and two mirrors. Each of the inputs passed through the beamsplitter where their power is divided and directed to mirrors M_1 and M_2. The optical beams incident on M_2 will be shifted upward, pass again through the beamsplitter, and on to the lens where a spatial magnification of the row of inputs takes place. Thus, the lower four rays of information (5-8) will be shifted and magnified imaged onto the pixels associated with the output plane. On the other hand, the light rays reflecting from M_1 will be shifted downward and then magnified. The top four rays (1-4) controlled by M_1 will then be shuffled between the rays imaged by M_2. In this way, the process of perfect shuffling is accomplished by splitting, shifting, and then magnifying the rows of inputs. It should also be pointed out that there are other methods of creat-

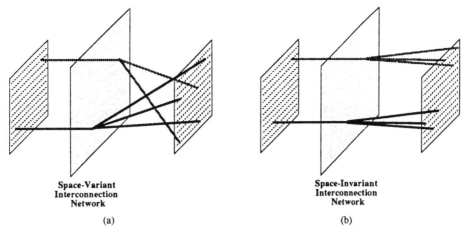

Fig. 20. Space-variant versus space-invariant networks.

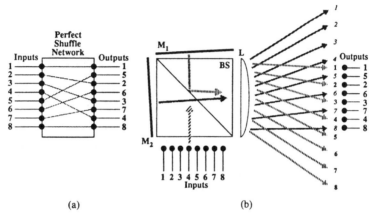

Fig. 21. Optical perfect shuffle [55].

ing a perfect shuffle interconnect using bulk optics [55], [56].

2) Holography: A hologram is a mechanism that can be used to modify and redirect a light wave that is incident upon it [57]. Because of this capability, a hologram or collection of holograms can be thought of as the photonic wires interconnecting optical logic gates [58]. An example of a holographic interconnection network is illustrated in Fig. 22. In this figure, the optical output of a logic device (point a) will be directed by mirror M_1 to the hologram (point b) which will redirect the light via M_2 to one or more other logic devices (points c, d, and e). This redistributed light will be used as the signal beams for the devices they are incident upon. The hologram located at point b could be either space-variant or space-invariant.

There is a price to pay for the flexibility of space-variant connections in terms of the SBWP. Assuming that each of the N^2 subholograms has the capability of addressing any or all of the N^2 pixels of the next stage, this implies a required SBWP $\propto N^4$ [59]. If a computer-generated hologram (CGH) is to be used as the optical interconnect, its SBWP will be determined by dividing the maximum size of the hologram by the minimum feature size and then squaring the result. As an example, for a Fourier hologram, if the minimum linewidth of an elec-

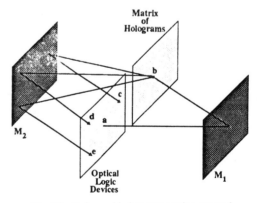

Fig. 22. Holographic interconnection network.

tron-beam system is 0.5 μm and with a maximum hologram size of 10 cm on each side, the maximum SBWP = 4×10^{10}. This implies that the theoretical maximum value of N for a space-variant network must be less than 450. This is the maximum value and for practical systems, N will have to be much smaller than this [59].

For the case of a space-invariant hologram, each point on the hologram should be able to redirect incoming light to all N^2 pixels of the next array and since all the pixels perform the same operation then the SBWP $\propto N^2$. Thus, for space-invariant CGH with the available SBWP of 4×10^{10}, a maximum value of $N \ll 200\,000$ is possible.

A third approach is a hybrid interconnection scheme in which the space-variance requirement is relaxed in an attempt to increase N. If there are N^2 subholograms, each of which can redirect incoming light to $M^2 \ll N^2$ pixels of the next array, then the SBWP $\propto (MN)^2$. Note that M^2 is the effective fan-out of the energy incident on the subhologram. If $M = 3$ (fan-out = 9) the CGH will have $N \ll (4 \times 10^{10}/M^2)^{1/2} = 65\,000$. For networks that require a large N either a space-invariant or a hybrid interconnect will be required.

At the current time, most of the work on holographic interconnects has been theoretical and has yet to be engineered to the point of being practical.

C. Potential Applications of Optical Logic Devices

There are many applications that require the capability of sensing and reacting to each and every bit that passes through the system. As an example, packet switching systems require the ability to sense the information present in the headers and then provide the appropriate network routing. One type of control strategy used for packet systems is to sort the input packets by their destination addresses [51], [60].

This is illustrated by the Batcher bitonic sorting network shown in Fig. 23. In this figure, the entering packets, represented by their destination address, encounter the first rank of nodes. When a node has an up arrow, it means that the largest of the two destination addresses will be directed to the upper output. The lower address will then be directed to the lower output. On the other hand, when the arrow is pointing down, the larger (smaller) address will be directed to the lower (upper) output. After each node determines whether it will provide the exchange or bypass function, the incident packet information will be directed to the appropriate output port. The Batcher network begins with adjacent pairs of input lines, entering the first rank, being ordered in either ascending or descending order depending on the configuration of the node. The upper half (lower half) of rank 1 will be interconnected to the upper half (lower half) of rank 2 through a four-element perfect shuffle network. Rank 2 will then sort the upper half (lower half) of its inputs creating an ascending (descending) sequence which is directed through an eight-element perfect shuffle to rank 3. Finally, rank 3 will sort its interleaved ascending and descending four-element sequences into an ascending eight-element sequence which corresponds to sorting the inputs according to their destination address. As an example, the path corresponding to the output destination address 4 has been highlighted in the figure. Thus, for this type of switching network each node will be required to have a modest amount of intelligence. These nodes must be able to read a packet header, and then reconfigure the node so that the entering packet is directed to the proper output channel. The topology of the Batcher bitonic sorting network can be rearranged such that all the interconnects use the same size of perfect shuffle [51]. An example of this is shown in Fig. 24. This adds more space-invariance to the system, thus reducing the overall required SBWP.

An example of a packet switching system based on sorting networks that could eventually be implemented using optical logic and interconnects is the STARLITE wideband digital switch [61]. The basic architecture of this switch is shown in Fig. 25. The STARLITE switch is a self-routing, nonblocking, constant latency packet switch that has the capability of handling gigabit data rates [62]. The concentrator directs the active inputs, which are much less than the total system inputs, to the sort-to-copy subnetwork. The sort-to-copy and copy subnetworks provide the broadcast capability for this switching system. The output of the copy subnetwork is then sorted according to destination. The expander then redirects the data to its final destination. Each of these basic functions (concentrator, sort-to-copy, copy, etc.) can be decomposed into some type of shuffle network that interconnects 2×2 switching nodes.

It is important to understand that optical logic devices and their associated systems are not at the point of development at this time. The devices that have been discussed are research prototypes that are not ready to be manufactured.

D. Applications of Smart Pixel Devices

Smart pixels are devices that attempt to take advantage of the strengths of both the optical and electrical domain. The strength of the electrical domain is that electrons interact easily which allows an electronic signal to control another electronic signal. Conversely, the strength of the optical domain is that photons do not interact with each other which creates an ideal communications environment. Once a collection of photons are encoded with information, they travel directly to their destination without interacting with other photons. The cost of this communications capability is that photons have a difficult time controlling other photons. Combining these two strengths, in that the electronics will be responsible for the processing of information while the photonics will handle the communications between processing elements, two-dimensional processing based on smart pixels should offer some performance advantage.

An example of such a system has been proposed by Midwinter [49]. This system is illustrated in Fig. 26. For this system, the input data enters as a row of information where each element is a single serial channel. This row of information enters the top row of smart pixels which are multiport devices that functionally behave like exchange–bypass modules. They have the capability of comparing packet addresses and directing the incident packets to the appropriate outputs. After processing, the information from the top row is directed via mirror M_1 to the perfect shuffle optics. Following the perfect shuffle operation, the data are directed by mirrors M_2, M_3, and

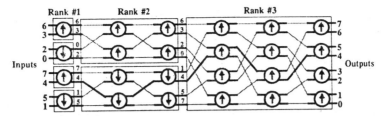

Fig. 23. A Batcher bitonic sorting network [52].

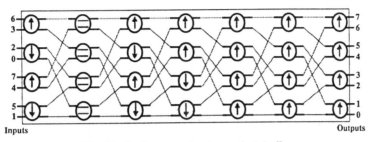

Fig. 24. Sorting network using perfect shuffles.

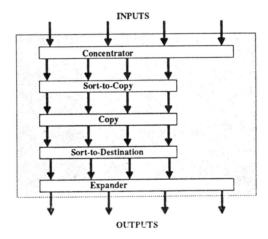

Fig. 25. STARLITE wide-band digital switch.

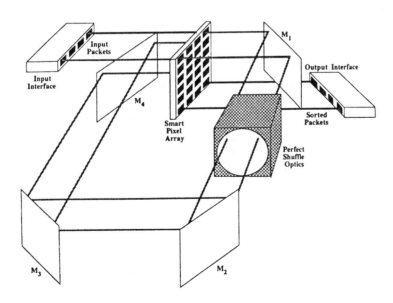

Fig. 26. Photonic sorting network.

M_4 to the second row of the smart pixel array. This looping procedure continues until the information has passed through all rows of the smart pixel array. At this point, the sorted information will be directed to the output interface.

V. Conclusions

This paper has reviewed some of the possible photonic technologies that could become important components of future telecommunications systems. It began by dividing photonic devices and systems into two classes according to the function they perform. The first class, relational, was associated with devices which under external control mapped the input channels to the output channels. The second class, logic, required that the devices perform some type of Boolean logic function. After the classes were defined, the strengths and weaknesses of the photonic domain were shown to be bandwidth and parallelism. Relational devices and their applications were then discussed. It was pointed out that the directional coupler holds the most promise for near-term development. Optical logic devices and systems were then presented. The systems that were outlined were based on SEED, NLFP, or smart pixel devices with either holograms or bulk optics serving as the optical interconnects.

Acknowledgment

I would like to acknowledge the constructive criticism provided by S. J. Hinterlong, A. L. Lentine, and M. E. Prise during the preparation of this paper.

References

[1] P. W. Smith, "On the role of photonic switching in future communications systems," *IEEE Circuits Devices*, pp. 9-14, May 1987.
[2] H. S. Hinton, "Photonic switching technology applications," *AT&T Tech. J.*, vol. 66, pp. 41-53, May/June 1987.
[3] P. W. Smith, "On the physical limits of digital optical switching and logic elements," *Bell Syst. Tech. J.*, vol. 61 pp. 1975-1993, Oct. 1982.
[4] S. L. McCall and H. M. Gibbs, "Conditions and limitations in intrinsic optical bistability," *Optical Bistability*, C. M. Bowden, M. Ciftan, and H. R. Robl, Eds. New York: Plenum, 1981, pp. 1-7.
[5] D. B. Tuckerman and R. F. W. Pease, "High-performance heat sinking for VLSI," *IEEE Electron Device Lett.*, vol. EDL-2, pp. 126-129, May 1981.
[6] J. W. Goodman, *Introduction to Fourier Optics*. New York: McGraw-Hill, 1968.
[7] A. D. Fisher, "A review of spatial light modulators," in *Proc. Top. Meet. Opt. Comput.*, Incline Village, NV, March 18-20, 1985.
[8] W. E. Ross, D. Psaltis, and R. H. Anderson, "2-D magneto optic spatial light modulator for signal processing," in *Proc. SPIE Conf.*, Crystal City-Arlington, VA, May 3-7, 1982.
[9] A. R. Tanguay, "Materials requirements for optical processing and computer devices," *Opt. Eng.*, pp. 2-18, Jan./Feb. 1985.
[10] A. Himeno and M. Kobayashi, "4 × 4 optical-gate matrix switch," *J. Lightwave Technol.*, vol. LT-3, pp. 230-235, Apr. 1985.
[11] D. R. Pape and L. J. Hornbeck, "Characteristics of the deformable mirror device for optical information processing," *Opt. Eng.*, vol. 22, pp. 675-681, 1983.
[12] G. Livescu, D. A. B. Miller, J. E. Henry, A. C. Gossard, and J. H. English, "Spatial light modulator and optical dynamic memory using integrated self electro-optic effect devices," in *Proc. Conf. Lasers Electro-Optics (Postdeadline Paper)*, April 26-May 1, 1987, pp. 283-284.
[13] J. W. Goodman, A. R. Dias, and L. M. Woody, "Fully parallel, high-speed incoherent optical method for performing discrete Fourier transforms," *Opt. Lett.*, vol. 2, pp. 1-3, Jan. 1978.
[14] A. A. Sawchuk, B. K. Jenkins, C. S. Raghavendra, and A. Varma, "Optical crossbar networks," *IEEE Computer*, vol. 20, pp. 50-60, June 1987.
[15] S. Suzuki and K. Nagashima, "Optical broadband communications network architectures utilizing wavelength-division switching technologies," in *Top. Meet. Photonic Switching, Tech. Dig. Series*, vol. 13, Mar. 18-20, 1987, pp. 21-23.
[16] B. S. Glance, K. Pollack, C. A. Burrus, B. L. Kasper, G. Eisenstein, and L. W. Stulz, "WDM coherent optical star network," *J. Lightwave Technol.*, vol. 6, pp. 67-72, Jan. 1988.
[17] T. S. Rzeszewski and A. L. Lentine, "A photonic switch architecture utilizing code-division multiplexing," in *Top. Meet. Photonic Switching, Tech. Dig. Series*, vol. 13, Mar. 18-20, 1987, pp. 144-146.
[18] P. R. Prucnal, D. J. Blumenthal, and P. A. Perrier, "Self-routing switching demonstration with optical control," *Opt. Eng.*, vol. 26, pp. 473-477, May 1987.
[19] H. S. Hinton, "Photonic switching using directional couplers," *IEEE Communications*, vol. 25, pp. 16-26, May 1987.
[20] R. V. Schmidt and R. C. Alferness, "Directional coupler switches, modulators, and filters using alternating $\Delta\beta$ techniques," *IEEE Trans. Circuits Syst.*, vol. CAS-26, pp. 1099-1108, Dec. 1979.
[21] H. S. Hinton, "A non-blocking optical interconnection network using directional couplers," in *Proc. IEEE Global Telecommun. Conf.*, vol. 2, Nov. 1984, pp. 885-889.
[22] P. Granestrand et al., "Strictly nonblocking 8 × 8 integrated optical switch matrix," *Electron. Lett.*, vol. 22, July 17, 1986.
[23] G. A. Bogert, "A low crosstalk 4 × 4 Ti:LiNbO$_3$ optical switch with permanently attached polarization-maintaining fiber arrays," in *Proc. Top. Meet. Integrated Guided-Wave Opt.*, Atlanta, GA, Feb. 1986, pp. PDP 3.1-3.
[24] R. A Spanke, "Architectures for guided-wave optical space switching networks," *IEEE Communications*, vol. 25, pp. 42-48, May 1987.
[25] S. Kobayashi and T. Kimura, "Semiconductor optical amplifiers," *IEEE Spectrum*, pp. 26-33, May 1984.
[26] H. Goto et al., "An experiment on optical time-division digital switching using bistable laser diodes and optical switches," in *Proc. IEEE Global Telecommun. Conf.*, vol. 2, Nov. 1984, pp. 880-884.
[27] Y. Silberberg, "All-optical repeater," *Opt. Lett.*, vol. 11, pp. 392-394, June 1986.
[28] R. A. Spanke, "Architectures for large nonblocking optical space switches," *IEEE J. Quantum Electron.*, vol. QE-22, pp. 964-967, June 1986.
[29] G. W. Richards and F. K. Hwang, "A two-stage rearrangeable broadcast switching network," *IEEE Trans. Commun.*, vol. COM-33, pp. 1025-1035, Oct. 1985.
[30] S. Suzuki et al., "Thirty-two line optical space-division switching experiment," in *Conf. Opt. Fiber Commun./Int. Conf. Integrated Opt. Opt. Fiber Commun. Tech. Digest Series, 1987*, vol. 3, Reno, NV, Jan. 1987, p. 146.
[31] W. A. Payne and H. S. Hinton, "Design of lithium niobate based photonic switching systems," *IEEE Communications*, vol. 25, pp. 37-41, May 1987.
[32] G. R. Ritchie, "SYNTRAN—A new direction for digital transmission terminals," *IEEE Communications*, vol. 23, pp. 20-25, Nov. 1985.
[33] R. A. Thompson and P. P. Giordano, "Experimental photonic time-slot interchanger using optical fibers as reentrant delay-line memories," in *Proc. Conf. Opt. Fiber Commun.*, Atlanta, GA, Feb. 24-26, 1986, pp. 26-27.
[34] S. D. Smith, "Optical bistability, photonic logic, and optical computation," *Appl. Opt.*, vol. 25, pp. 1550-1564, May 15, 1986.
[35] H. M. Gibbs, *Optical Bistability: Controlling Light with Light*. New York: Academic, 1985.
[36] D. A. B. Miller, J. E. Henry, A. C. Gossard, and J. H. English, "Array of optically bistable integrated self-electrooptic effect devices," in *Proc. Conf. Lasers Electro-optics*, San Francisco, CA, June 1986, pp. 32-33.
[37] J. L. Jewell, A. Scherer, S. L. McCall, A. C. Gossard, and J. H. English, "GaAs-AlAs monolithic microresonator arrays," *Appl. Phys. Lett.*, vol. 51, pp. 94-99, July 13, 1987.
[38] J. L. Jewell et al., "3-picojoule 82 MHz optical logic gates in a room temperature GaAs-AlGaAs multiple-quantum-well etalon," *Appl. Phys. Lett.*, vol. 46, p. 918, 1985.

[39] Y. H. Lee et al., "Speed and effectiveness of windowless GaAs etalons as optical logic gates," *Appl. Phys. Lett.*, vol. 49, p. 486, 1986.
[40] J. L. Jewell, M. C. Rushford, and H. M. Gibbs, "Use of a single nonlinear Fabry-Perot etalon as optical logic gates," *App. Phys. Lett.*, vol. 44, pp. 172-174, Jan. 15, 1984.
[41] Y. H. Lee et al., "Room-temperature optical nonlinearities in GaAs," *Phys. Rev. Lett.*, vol. 57, pp. 2446-2449, Nov. 10, 1986.
[42] N. Streibl and M. E. Prise, "Optical considerations in the design of digital optical computers," *Opt. Quantum Electron.*, to be published.
[43] M. E. Prise, N. Streibl, and M. M. Downs, "Computational properties of nonlinear devices," in *Top. Meet. Photonic Switching, Tech. Dig. Series*, vol. 13, 1987, pp. 110-112.
[44] P. Wheatley and J. E. Midwinter, "Operating curves for optical bistable devices," in *Top. Meet. Photonic Switching, Tech. Dig. Series*, vol. 13, 1987, pp. 113-114.
[45] D. A. B. Miller, U.S. Patent 4 546 244.
[46] P. Wheatley et al., "Novel nonresonant optoelectronic logic device," *Electron. Lett.*, vol. 23, pp. 92-93, Jan. 16, 1987.
[47] P. Wheatley et al., "Three-terminal noninverting optoelectronic logic device," *Opt. Lett.*, vol. 12, pp. 784-786, Oct. 1987.
[48] A. L. Lentine et al., "Symmetric self electro-optic effect device," in *Proc. Conf. Lasers Electro-Opt. (Postdeadline Paper)*, April 26-May 1, 1987, pp. 249-250.
[49] J. E. Midwinter, "A novel approach to the design of optically activated wideband switching matrices," *Proc. IEE*, Part J, Opto-electronics, to be published.
[50] T. Iwama et al., "A 4 × 4 GaAs OEIC switch module," in *Conf. Opt. Fiber Commun./Int. Conf. Integrated Opt. Opt. Fiber Commun. Tech. Dig. Series 1987*, vol. 3, Jan. 19-22, 1987, p. 161.
[51] H. S. Stone, "Parallel processing with the perfect shuffle," *IEEE Trans. Comput.*, vol. C-20, pp. 153-161, Feb. 1971.
[52] C.-L. Wu and T.-Y. Feng, "The universality of the shuffle-exchange network," *IEEE Trans. Comput.*, vol. C-30, pp. 324-332, May 1981.
[53] D. S. Parker, Jr., "Notes on shuffle/exchange-type switching networks," *IEEE Trans. Comput.*, vol. C-29, pp. 213-222, Mar. 1980.
[54] K.-H. Brenner, *Appl. Opt.*, submitted for publication.
[55] A. W. Lohmann, W. Stork, and G. Stucke, "Optical perfect shuffle," *Appl. Opt.*, vol. 25, pp. 1530-1531, May 15, 1986.
[56] G. Eichmann and Y. Li, "Compact optical generalized perfect shuffle," *Appl. Opt.*, vol. 26, pp. 1167-1169, Apr. 1, 1987.
[57] R. J. Collier, C. B. Burkhardt, and L. H. Lin, *Optical Holography*. New York: Academic, 1971.
[58] A. A. Sawchuk and T. C. Strand, "Digital optical computing," *Proc. IEEE*, vol. 72, pp. 758-779, July 1984.
[59] B. K. Jenkins, P. Chavel, R. Forchheimer, A. A. Sawchuk, and T. C. Strand, "Architectural implications of a digital optical processor," *Appl. Opt.*, vol. 23, pp. 3465-3474, Oct. 1, 1984.
[60] C. W. Stirk, R. A. Athale, and C. B. Friedlander, "Optical implementation of the compare-and-exchange operation for applications in symbolic computing," *Proc. SPIE*, submitted for publication.
[61] A. Huang and S. Knauer, "Starlite: A wideband digital switch," in *Proc. IEEE Global Telecommun. Conf.*, Atlanta, GA, vol. 1, pp. 121-125, Nov. 1984.
[62] A. Huang, "The relationship between STARLITE, a wideband digital switch and optics," in *Proc. Int. Conf. Commun.*, Toronto, Ont., Canada, June 22, 1986.
[63] A. de Bosio, C. DeBernardi, and F. Melindo, "Deterministic and statistic circuit assignment architectures for optical switching systems," in *Top. Meet. Photonic Switching, Tech. Dig. Series*, vol. 13, Mar. 18-20, 1987, pp. 35-37.

Part 2
Devices and Components

The photonic hardware required to make a photonic switching system consists of devices and the means to interconnect the devices, referred to as optical interconnects. The device papers that have been selected are organized into electro-optic devices, logic devices, optical amplifiers, hybrid switching devices, spatial light modulators, ultrafast devices, optical interconnects, beam-combination, and spot-array-generation.

Electro-optic devices are based on the linear electro-optic effect and provide a transparent channel to the incoming optical signal that can be directed to one of two possible output channels by a separate electronic signal. These devices include the directional coupler, the balanced bridge modulator switch, the BOA-coupler, the X-switch, the total internal reflection switch, and the digital optical switch. Most of these devices are fabricated by using the lithium niobate technology. Such switches, typically several millimeters in length and 10-20 microns in width, can be driven electrically to rates exceeding 10 GHz, and have a huge optical transmission bandwidth when expressed in GHz of optical spectrum! The papers reproduced here (Section 2.1) have been chosen to give the reader a good feel for the strengths and weaknesses inherent in the devices.

The second collection of papers (Section 2.2) reviews devices that are used to implement optical logic functions such as AND, OR, NOR, etc. These devices include the nonlinear Fabry-Perot devices and the self electro-optic devices (SEEDs). They have attracted the most interest from the "optical computing" community, but the groups represented here have telecommunications switching interests. The ways proposed for using these devices in switching are discussed later.

Optical amplifiers (Section 2.3), usually in the form of semiconductor laser-like chips with the end facet reflectors antireflection coated, provide miniature high-gain or high-attenuation optical switches, depending upon whether the drive current is ON or OFF. Hence by assembling arrays of such elements with suitable waveguide interconnections, switching matrices can be constructed.

Hybrid devices (Section 2.4) typically involve a mix of electronics and optics aimed at exploiting the strengths of each medium. While they are limited to the speed of the slowest element, usually assumed to be an electronic device, in practice this may not be a limitation and assemblies of such devices can lead to simple and compact switch designs.

Spatial light modulators (Section 2.5) were first developed for projection TV display and analog coherent optical computing applications. One can envisage them as large two-dimensional arrays of relatively slow (millisecond to microsecond) electrically controlled switches that can be optically addressed via a free-space imaging optical system. A particularly interesting characteristic is the fact that they can offer TV-type resolution, say 500×500 pixels or elements. Some versions also allow for optical control through the use of a photoconductive element per pixel, converting a weak incoming optical image into a 2D control signal for the modulator array.

The ultrafast devices, the final type of device to be reviewed (Section 2.6), may prove to be the most important of all in true optical processing terms, since only they really access a processing regime having no electronic competition. Ultrafast is construed as generally meaning subpicosecond in response time, a time frame almost uniquely accessible to optical systems.

By way of contrast, optical interconnects (Section 2.7) bring to electrical systems the communication capability of optical technology and thereby enhance their processing capability substantially beyond the level we see in VLSICs and WSICs. They can embrace interconnects at board-to-board, chip-to-chip, or within chip levels, although at the latter level, the resultant processor is indistinguishable from the smart-pixel or electronic-island concepts discussed earlier. The section begins with a paper discussing some of the limitations and attributes of optical interconnects. It then reviews the optical perfect shuffle that has been proposed for free-space digital optical switching systems. This review will be followed by a collection of papers discussing star couplers. These interconnects are the backbone of shared-media systems such as wavelength- and code-division switching.

Finally, there are two papers that outline the beam-combination (Section 2.8) and spot-array-generation (Section 2.9) problem that need to be solved for free-space digital optical systems.

Section 2.1: Electro-Optic Devices

Directional Couplers

Electrically switched optical directional coupler: Cobra

M. Papuchon, Y. Combemale, X. Mathieu, D. B. Ostrowsky, L. Reiber,
A. M. Roy, B. Sejourne, and M. Werner

Thomson-CSF, Laboratoire Central de Recherches, B.P. 10, Domaine de Corbeville, 91401 Orsay, France

Since work began on integrated optics, it has been evident that elements capable of switching light from one channel guide to another would play an essential role.[1,2] In this paper we report the realization of such a switch based on an optical directional coupler realized in LiNbO$_3$ by a diffusion technique[3] using titanium.[4] The switch uses an original electrode configuration and we have taken the liberty of calling the element a Cobra (commutateur optique binaire rapide).

If two waveguides are (see Fig. 1) are coupled (by evanescent fields or otherwise), the amplitudes of the electric fields in these guides will evolve according to

$$\frac{dE_1}{dx} = -(\beta_1 + k)E_1 + kE_2, \quad \frac{dE_2}{dx} = kE_1 - (\beta_2 + k)E_2, \quad (1)$$

where β_1 and β_2 are the propagation constants in guides 1 and 2; k is the coupling factor; and E_1 and E_2 are the electric field amplitudes in guides 1 and 2.

If only guide 1 is initially excited, the solution of these coupled equations is

$$|E_1|^2 = \cos^2\left(\frac{\Delta\beta^2}{4c^2} + 1\right)^{1/2} cx + \frac{\Delta\beta^2}{4c^2}\left(\frac{\Delta\beta^2}{4c^2} + 1\right)^{-1} \sin^2\left(\frac{\Delta\beta^2}{4c^2} + 1\right)^{1/2} cx, \quad (2)$$

$$|E_2|^2 = \left(\frac{\Delta\beta^2}{4c^2} + 1\right)^{-1} \sin^2\left(\frac{\Delta\beta^2}{4c^2} + 1\right)^{1/2} cx, \quad \Delta\beta = \beta_1 - \beta_2, \quad ic = k$$

In Fig. 1 we have shown the evolution of the power in guide 1 for two different values of $\Delta\beta$. For the resonant case ($\Delta\beta = 0$) all of the power initially in 1 has been transferred to 2 over a distance L_0 called the coupling length. With $\Delta\beta = c(12)^{1/2}$ less energy has been exchanged and all the power transferred to 2 has returned to 1 over the same distance L_0. Thus by introducing the correct $\Delta\beta$, we can determine the guide by which all the guided energy leaves the coupler. Note that total switching can only occur when the coupler length is an odd multiple of L_0, i.e., all the energy has been transferred out of the guide initially excited. For all other coupler lengths introduction of a nonzero $\Delta\beta$ will lead to incomplete switching.

It has been suggested that switch elements could be realized by fabricating the coupler in an electro-optic material and changing $\Delta\beta$ by changing the guide indices via an applied electric field. The configuration commonly discussed[5,6] is shown in Fig. 2 and is based on the use of initially resonant guides and three electrodes which permits changing the guide indices in opposed directions to maximize the induced $\Delta\beta$. We have used, however, the Cobra configuration, shown in Fig. 2(b). In this configuration the crystal C axis is normal to the guide surface and the opposite direction of the fringing fields in the two guiding regions gives the desired, opposite, index changes (Fig. 3). Elimination of the central electrode is desirable since, due to its small section (2000 Å × 2 μm) and relatively long length (~ several mm) it will be rather fragile and have a high resistance.

With this electrode configuration the principal effect will be to change the $\Delta\beta$ of TM polarized waves due to the r_{zzz} coefficient. Other effects will also occur, due to the complex form of the fringing fields.

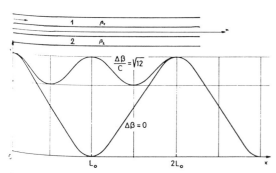

FIG. 1. Power in the initially excited guide vs propagation length when $\Delta\beta = 0$ and $\Delta\beta = c(12)^{1/2}$

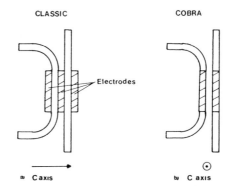

FIG. 2. Two possible electrode configurations for electrically switched directional couplers: (a) Classical configuration (three electrodes); (b) Cobra configuration (two electrodes).

Reprinted with permission from *Appl. Phys. Lett.*, vol. 27, no. 5, pp. 289–291, Sept. 1, 1975.
Copyright © 1975, American Institute of Physics.

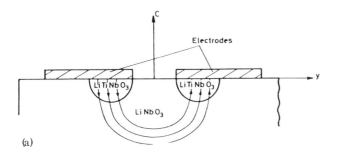

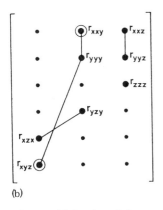

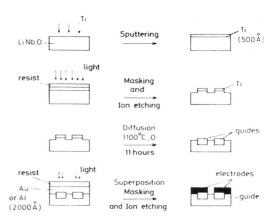

FIG. 4. Cobra fabrication.

FIG. 3. (a) Section of the coupling region showing the crystal orientation, electrode configuration, and electric field lines. (b) The electro-optic tensor of LiNbO$_3$.

FIG. 5. Coupler configuration—launching and observation regions are also shown. $a = 2$ μm; $b = 3$ μm; $c = 3$ mm.

One expects, for example, to observe a different behavior when opposite electrical polarities are applied, due to the r_{yzy} coefficient, which will lead to an asymmetric change of index between the guides.

The fabrication technique is shown in Fig. 4. Approximately 500 Å of Ti is deposited by rf sputtering on the LiNbO$_3$ substrate. This Ti is then masked and etched[7] to obtain the desired guide configuration in the Ti. The Ti is then diffused into the LiNbO$_3$ in an O$_2$ atmosphere at 1100 °C for approximately 11 h to form the desired guide structure.

Directional couplers fabricated in this way and having the configuration shown in Fig. 5 were examined by exciting one of the guides (using a rutile prism coupler) and observing the energy exchange between the guides. The use of the 5145-Å line of an argon laser permitted this observation to be made using either a fluorescence technique[8] or direct observation of diffused light.

It was found that the coupling lengths could be approximated using the model of Marcatili[9] if we consider the guides to have the same widths as the original Ti strip, a 2-μm thickness, and a uniform Δn of about 5×10^{-3}. This is, of course a rather pragmatic description since the guides have a nonuniform index profile and there is certainly some lateral diffusion of Ti leading to a wider structure than that delineated in the Ti. Typical coupling lengths L_0 observed for couplers formed from 2-μm-wide Ti strips were ≈ 500 μm for a 2-μm separation and ≈ 1 mm for a 3-μm separation (which was actually used in the switching experiment).

The electrodes were made by rf sputtering a 2000-Å-thick gold film onto the structure and masking (using a standard optical superposition technique) and ion etching the film to form two 1×3-mm electrodes separated by 3 μm at the coupler center. The electrodes introduce approximately 2 dB/cm loss for the TM modes but since 500 μm–1 mm electrode lengths are possible this is acceptable.

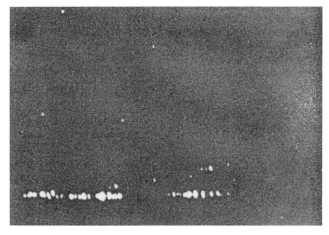

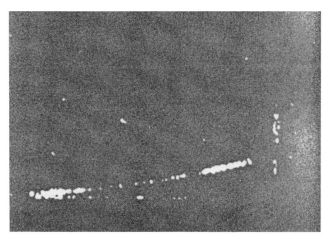

FIG. 6. Photographs of the observation region when (a) 0 V applied, and (b) 6 V applied.

Light from an argon laser was coupled into one of the guides, and the light diffused out of the guides at the output was observed as various voltages were applied to the electrodes. In Fig. 6 we show the light observed at the exit end when an essentially TM guided wave was excited for applied 0 V [Fig. 6(a)] and 6 V [Fig. 6(b)]. We note that with 0 V applied most of the guided light is in the initially unexcited guide (straight guide), as would be expected since $L_0 = 1$ mm and the actual length of the coupling region is 3 mm. With 6 V applied the guided light has been switched back to the initially excited guide (curved guide).

The effect was approximately 3 times more effective for TM guided waves than for TE waves and the expected asymmetrical behavior with respect to the applied polarization was also observed. For these reasons we believe that these experiments validate the Cobra principle and work is continuing to improve these results.

The authors gratefully acknowledge the financial support from the DRME under contract No. 74 433.

[1] S.E. Miller, Bell. Syst. Tech. J. **48**, 2059 (1969).
[2] P.K. Tien, Appl. Opt. **10**, 2395 (1971).
[3] J.M. Hammer and W. Phillips, Appl. Phys. Lett. **24**, 545 (1974).
[4] R.V. Schmidt and I.P. Kaminow, Appl. Phys. Lett. **25**, 458 (1974).
[5] S. Kurazono, K. Iwasaki, and N. Kamagai, Trans. Inst. Electr. Commun. Engr. Jpn. **55-C**, 61 (1972).
[6] H.F. Taylor, J. Appl. Phys. **44**, 3257 (1973).
[7] D.B. Ostrowsky, M. Papuchon, A.M. Roy, and J. Trotel, Appl. Opt. **13**, 636 (1974).
[8] D.B. Ostrowsky, A.M. Roy, and J. Sevin, Appl. Phys. Lett. **24**, 553 (1974).
[9] E.A.J. Marcatili, Bell. Syst. Tech. J. **48**, 2071 (1969).

Switched Directional Couplers with Alternating $\Delta\beta$

HERWIG KOGELNIK, FELLOW, IEEE, AND RONALD V. SCHMIDT, MEMBER, IEEE

Abstract—Coupled waveguide structures with sections of alternating phase mismatch are proposed as switched directional coupler configurations in which complete conversion of light from one guide to the other can be achieved by an electrical adjustment. These structures can be used to make electrooptic switches and amplitude modulators in integrated optics form with improved conversion and on–off ratios. Couplers with 2, 3, 4, and N sections of alternating phase mismatch are analyzed, and diagrams describing their switching characteristics are given.

I. Introduction

OPTICAL directional couplers, such as those formed by two parallel strip waveguides [1], are characterized by 1) the interaction length L, 2) the coupling coefficient κ or the corresponding conversion length $l = \pi/2\kappa$ indicating the minimum length required to obtain complete crossover of light from one guide to the other, and 3) the mismatch $\Delta\beta = \beta_1 - \beta_2$ between the propagation constants β_1 and β_2 of the two guides. Complete crossover is achieved when the guides are phase matched ($\Delta\beta = 0$) and when the interaction length is an exact odd multiple of the coupling length, i.e., when $L = (2\nu + 1)l$. It has been proposed [2]–[5] that an optical switching device can be built by electrically switching the directional coupler from the crossover state to the straight-through state where no net crossover occurs. One way to do this is by fabricating the coupler on electrooptic material and applying a voltage which induces a mismatch $\Delta\beta$ via the electrooptic effect. An early switching experiment of this kind is that of Tada and Hirose [6] who employed planar film guides in GaAs. Recently, Papuchon *et al.* [7] and Campbell *et al.* [8] have demonstrated experimental switches by using Ti-diffused strip guides in LiNbO$_3$ and metal-gap strip guides in GaAs, respectively. When the interaction length L is not made exactly equal to the coupling length l (or an odd multiple thereof), then the crossover is not complete in these devices and crosstalk results. It does not appear to be possible to eliminate the crosstalk in the crossover state of these devices by an electrical adjustment. However, crosstalk requirements are relatively stringent if such optical switches are to be used to switch optical transmission channels (some 25 dB are required for digital systems). This imposes stringent fabrication tolerances which may be hard to meet ([7] does not report measured crossover values and [8] reports crossover values of about 75 percent). In addition, we know that the conversion length l is a function of wavelength [1], which may make it impossible to operate these switches properly beyond the design wavelength.

In this paper we propose switched coupler configurations in which such an electrical adjustment can be made for both the crossover and the straight-through states for just about any interaction length L that is larger than the conversion length l. They are called alternating-$\Delta\beta$ couplers because the technique used to achieve complete crossover in the coupler is to provide along the interaction length two or more sections with a mismatch or asynchronism $\Delta\beta$ of alternating sign.[1] A simple way to induce this alternating $\Delta\beta$ is to provide sectioned electrodes and apply voltages of alternating polarity along the interaction length. There is no requirement for an exact L/l ratio in this configuration, and we find that there is always a voltage that will make the light cross over completely, and another voltage that will make the light go straight through. If the switch has to be operated at another wavelength and l is wavelength dependent, the only adjustment that seems necessary is a change of these voltage values.

In this paper we discuss the principle of operation of alternating-$\Delta\beta$ couplers and present their theory based on coupled-wave theory. The first successful operation of a switched alternating-$\Delta\beta$ coupler with two sections using Ti-diffused strip guides in LiNbO$_3$ is reported elsewhere [9].

To simplify our discussion, we shall call the switching state where the light crosses over completely from one guide to the other the "cross" state, and we associate with this state the symbol $\otimes$; the state in which the light passes straight through, appearing at the output in the same guide that it entered at the device input, we shall call the "bar" state associated with the symbol $\ominus$. We have found it convenient to describe the various switching configurations in a diagram which, for a given L/l value, gives the mismatch values $\Delta\beta$ required to drive the switch into the cross and bar states, respectively. In Section II we work out the transfer matrix and the switching diagram for the conventional switched directional coupler. In Sections III–V we derive the transfer matrices and the switching diagrams for couplers with alternating $\Delta\beta$ with 2, 3, and 4 sections, respectively. In Section VI we discuss alternating-$\Delta\beta$ couplers with N sections.

II. The Switched Directional Coupler

A sketch of a conventional switched directional coupler is shown in Fig. 1. We give here the transfer matrix and the cross-bar diagram of this known coupler in order to provide the basis for the discussion of the following sections. For simplicity, the electrodes are shown in the COBRA configuration [7]. Application of a voltage V to these electrodes induces a mismatch $\Delta\beta(V) = \beta_1 - \beta_2$ between the propagation constants of guides 1 and 2. We describe the coherent light in

Manuscript received December 18, 1975.
The authors are with Bell Laboratories, Holmdel, NJ 07733.

[1] It is interesting to note that in a patent disclosure filed in 1957, Miller [13] proposed reversals in the sign of $\Delta\beta$ or of κ for the purpose of broadening the frequency band of operation of microwave directional couplers.

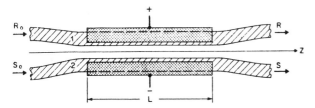

Fig. 1. Switched directional coupler consisting of two optical strip waveguides, with an interaction length L. The electrodes are shown in the COBRA configuration [7].

the two guides by complex amplitudes $R(z)$ and $S(z)$ which vary slowly in the propagation direction chosen coincident with the z axis. We assume that the energy exchange between the two guides can be regarded as a codirectional coupled-wave process which is governed by the coupled-wave equations [10]-[12]

$$R' - j\delta R = -j\kappa S \qquad (1)$$
$$S' + j\delta S = -j\kappa R \qquad (2)$$

where the prime indicates differentiation with respect to z, $\delta = \Delta\beta/2$, and κ is the coupling coefficient. For arbitrary input amplitudes R_0 and S_0 the solution of the coupled-wave equations can be written in matrix form

$$\begin{pmatrix} R \\ S \end{pmatrix} = \begin{pmatrix} A_1 & -jB_1 \\ -jB_1^* & A_1^* \end{pmatrix} \begin{pmatrix} R_0 \\ S_0 \end{pmatrix} \qquad (3)$$

where the asterisk indicates a complex conjugate and the subscript 1 refers to a single section of uniform $\Delta\beta$. The matrix coefficients are [10]-[12]

$$A_1 = \cos z\sqrt{\kappa^2 + \delta^2} + j\delta \sin z\sqrt{\kappa^2 + \delta^2}/\sqrt{\kappa^2 + \delta^2} \qquad (4)$$
$$B_1 = \kappa \sin z\sqrt{\kappa^2 + \delta^2}/\sqrt{\kappa^2 + \delta^2}. \qquad (5)$$

For $z = L$ the matrix

$$M_1^+ = \begin{pmatrix} A_1 & -jB_1 \\ -jB_1^* & A_1^* \end{pmatrix} \qquad (6)$$

may be called the transfer matrix of the coupler. Note that its determinant is unity, B_1 is real for this special case, and that the transfer matrix M_1^- of coupler with a mismatch of $-\Delta\beta$ has the form

$$M_1^- = \begin{pmatrix} A_1^* & -jB_1 \\ -jB_1^* & A_1 \end{pmatrix}. \qquad (7)$$

The switched coupler is in the cross state $\otimes$ when $A_1 = 0$. As is well known, this requires that $\delta L = 0$ and that

$$\kappa L = (2\nu + 1)\pi/2$$

or

$$L/l = 2\nu + 1 \qquad (8)$$

where ν is an integer.

The bar state $\ominus$ is obtained when $B_1 = 0$, which occurs if

$$(\kappa L)^2 + (\delta L)^2 = (\nu\pi)^2$$

or

$$\left(\frac{L}{l}\right)^2 + \left(\frac{\Delta\beta L}{\pi}\right)^2 = (2\nu)^2. \qquad (9)$$

The conditions for the cross and bar states can be graphically represented in a switching diagram where we use the values of L/l and $\Delta\beta L/\pi = 2\delta L/\pi$ as coordinates. This is shown in Fig. 2. According to (8) the cross states are represented by isolated points on the L/l axis, and (9) predicts concentric circles for the bar states as shown in the figure. While in this switch configuration the L/l value is essentially independent of the applied voltage, the $\Delta\beta$ values can be controlled electrically. An increase in the voltage moves the point representing the state of the coupler parallel to the $\Delta\beta L/\pi$ axis to the right. We note that, for any L/l value, there is always a $\Delta\beta L$ (or voltage) value that will drive the switch into the bar state. To obtain the cross state, however, exact odd-integer values of L/l must be produced during device fabrication and no voltage adjustment can compensate for fabrication errors. We shall see in the next sections that an electrical adjustment of the cross state is possible for the alternating-$\Delta\beta$ configurations.

III. SINGLE STEP $\Delta\beta$ REVERSAL

Fig. 3 shows a sketch of a switched coupler configuration in which the interaction length L is divided into two equally long sections. The asynchronism introduced in the two sections is equal in magnitude but reversed in sign, i.e., we have $\Delta\beta$ in the left section and $-\Delta\beta$ in the right section. The figure indicates a simple way to achieve this; one can bisect the electrodes and apply equal voltages of opposite polarity. The transfer matrix M_2 of this alternating-$\Delta\beta$ coupler with two sections can be obtained with the help of the single-section matrices M_1^+ and M_1^- by matrix multiplication

$$M_2 = M_1^- \cdot M_1^+ = \begin{pmatrix} A_2 & -jB_2 \\ -jB_2^* & A_2^* \end{pmatrix} \qquad (10)$$

where we assume that there are no reflections at the interface between the two sections. The matrix elements are calculated to be

$$A_2 = A_1 A_1^* - B_1^2 = 1 - 2B_1^2 \qquad (11)$$
$$B_2 = 2A_1^* B_1 \qquad (12)$$

where A_1 and B_1 are given in (4) and (5) where we put $z = L/2$.

The switch is in the cross state when $A_2 = 0$, i.e., when

$$2B_1^2 = 1. \qquad (13)$$

With (4) this condition can be written in the form

$$\frac{\kappa^2}{\kappa^2 + \delta^2} \sin^2 \frac{L}{2}\sqrt{\kappa^2 + \delta^2} = \frac{1}{2} \equiv \sin^2 \frac{\pi}{4}. \qquad (14)$$

In the switching diagram this condition leads to a family of curves shown in Fig. 4. The curves intersect the vertical axis at the points $L/l = 1, 3, 5, 7$, etc. which are isolated $\otimes$ points in the case of the single section coupler discussed in the preceding section. Near the vertical axis these curves are very nearly circles, and all curves are tangent to a 45° line through the origin (i.e., the line $\delta = \kappa$).

The switch is in the bar state if $B_2 = 0$ which occurs either

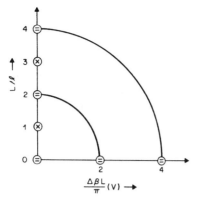

Fig. 2. The switching diagram for a switched directional coupler. The isolated points marking the conditions required for complete energy crossover are marked by an ⊗, and the circles indicating the conditions required for zero net crossover are marked by a ⊖.

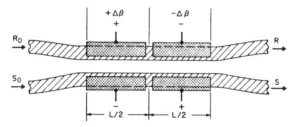

Fig. 3. Split electrode configuration for a switched coupler with stepped $\Delta\beta$ reversal (alternating $\Delta\beta$ with two sections).

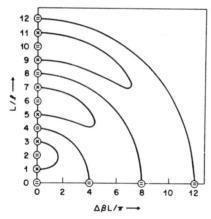

Fig. 4. Switching diagram for a switched coupler with two sections of alternating $\Delta\beta$. The ⊗ sign marks the cross-state conditions and the ⊖ sign marks the bar-state conditions.

for $A_1 = 0$ or for $B_1 = 0$. The first condition implies $\Delta\beta = 0$ and

$$L/l = 2(2\nu + 1) \tag{15}$$

which represents the isolated points at $L/l = 2, 6, 10$, etc., shown for the ⊖ state in Fig. 4. The second condition can be written in the form

$$\left(\frac{L}{l}\right)^2 + \left(\frac{\Delta\beta L}{\pi}\right)^2 = (4\nu)^2. \tag{16}$$

This condition for the ⊖ states corresponds to the concentric circles with radius 4, 8, 12, etc. in the switching diagram. The diagram indicates that the stepped-$\Delta\beta$ configuration makes available ranges of L/l values (e.g., the range from $L/l = 1$ to 3) in which complete crossover can be achieved. The diagram also gives the $\Delta\beta$ or voltage adjustments which are required for this purpose. These adjustment values are of the same order of magnitude as those required for the bar state in the uniform-$\Delta\beta$ configuration. In the next sections we will find that similar L/l ranges for the cross state and the associated possibility for electrical adjustment can be obtained for three or more sections of alternating $\Delta\beta$.

IV. Three Sections of Alternating $\Delta\beta$

In the following we discuss coupler configurations with three and more sections of alternating $\Delta\beta$. As the analysis of these configurations and the associated arguments are very similar in kind to those presented in Section III for couplers with two sections, we can and shall keep the following discussion very brief.

Fig. 5 shows a sketch of a coupler with three sections of alternating $\Delta\beta$. By matrix multiplication we obtain the transfer matrix M_3 of this configuration

$$M_3 = M_1^+ \cdot M_1^- \cdot M_1^+ = \begin{pmatrix} A_3 & -jB_3 \\ -jB_3^* & A_3^* \end{pmatrix}. \tag{17}$$

The matrix elements are

$$A_3 = A_1(1 - 4B_1^2)$$

and

$$B_3 = B_1(3 - 4B_1^2) \tag{18}$$

where we obtain A_1 and B_1 from (4) and (5) by putting $z = L/3$. The condition for the cross state is $A_3 = 0$, which is obtained either when $A_1 = 0$ or when

$$4B_1^2 = 1. \tag{19}$$

The first condition demands that $\Delta\beta = 0$ and that

$$L/l = 3(2\nu + 1) \tag{20}$$

which corresponds to the isolated cross-state points at $L/l = 3, 9, 15$, etc., which are indicated in the diagram of Fig. 6. Condition (19) can be written in the form

$$\frac{\kappa^2}{\kappa^2 + \delta^2} \sin^2 \frac{L}{3}\sqrt{\kappa^2 + \delta^2} = \frac{1}{4} = \sin^2 \frac{\pi}{6}. \tag{21}$$

In the diagram this condition is represented by the curves which intersect the vertical axis at $L/l = 1, 5, 7, 11, 13$, etc., and which are tangent to a 30° line through the origin.

Setting $B_3 = 0$ we obtain the bar-state conditions. The first of these is $B_1 = 0$ or

$$(L/l)^2 + (\Delta\beta L/\pi)^2 = (6\nu)^2 \tag{22}$$

which corresponds to the concentric circles through $L/l = 6, 12$, etc. in the diagram. The second condition is

$$4B_1^2 = 3 \tag{23}$$

which we can write in the form

$$\frac{\kappa^2}{\kappa^2 + \delta^2} \sin^2 \frac{L}{3}\sqrt{\kappa^2 + \delta^2} = \frac{3}{4} \equiv \sin^2 \frac{\pi}{3}. \tag{24}$$

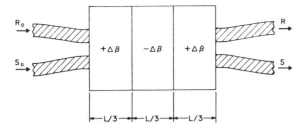

Fig. 5. Sketch of switched coupler with three sections of alternating $\Delta\beta$.

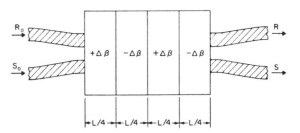

Fig. 7. Sketch of switched coupler with four sections of alternating $\Delta\beta$.

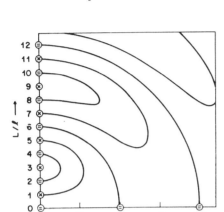

Fig. 6. Switching diagram for a coupler with three sections of alternating $\Delta\beta$. The $\otimes$ and $\ominus$ signs mark the conditions required for the cross and the bar states, respectively.

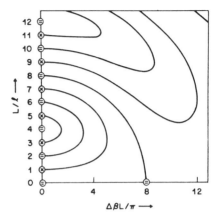

Fig. 8. Switching diagram for a coupler with four sections of alternating $\Delta\beta$.

In the diagram this condition is represented by the curves intersecting the vertical axis at $L/l = 2, 4, 8, 10, 14$, etc., which are tangent to a 60° line through the origin.

V. Four Sections of Alternating $\Delta\beta$

Fig. 7 shows a sketch of a coupler with four sections of alternating $\Delta\beta$. For the transfer matrix M_4 of this configuration we obtain

$$M_4 = M_1^- \cdot M_1^+ \cdot M_1^- \cdot M_1^+ = M_2^2 = \begin{pmatrix} A_4 & -jB_4 \\ -jB_4^* & A_4^* \end{pmatrix} \quad (25)$$

where the matrix elements are

$$A_4 = 2(1 - 2B_1^2)^2 - 1 \quad (26)$$
$$B_4 = 4A_1^* B_1 (1 - 2B_1^2) \quad (27)$$

and A_1 and B_1 are obtained from (4) and (5) with $z = L/4$.

Letting $A_4 = 0$ we get the two conditions for the cross state

$$B_1^2 = \frac{1}{2}\left(1 \pm \frac{1}{\sqrt{2}}\right) \quad (28)$$

which we can write in the form

$$\frac{\kappa^2}{\kappa^2 + \delta^2} \sin^2 \frac{L}{4}\sqrt{\kappa^2 + \delta^2} = \frac{1}{2}\left(1 \pm \frac{1}{\sqrt{2}}\right) \equiv \begin{cases} \sin^2 3\pi/8 \\ \sin^2 \pi/8. \end{cases} \quad (29)$$

These conditions correspond to two sets of curves shown in the diagram of Fig. 8. Taking the $-$ sign we get the curves that intersect the vertical axis at $L/l = 1, 7, 9, 15$, etc., which are tangent to a 22.5° line through the origin; the plus sign yields the curves through $L/l = 3, 5, 11, 13$, etc., which are tangent to the 67.5° line.

Letting $B_4 = 0$ we get three conditions for the bar state. The first is $A_1 = 0$, or $\Delta\beta = 0$ and

$$L/l = 4(2\nu + 1) \quad (30)$$

which yields the isolated points at $L/l = 4, 12$, etc. shown in the diagram. The second condition is $B_1 = 0$ or

$$(L/l)^2 + (\Delta\beta L/\pi)^2 = (8\nu)^2 \quad (31)$$

represented in the diagram by the circles through $L/l = 8, 16,$ etc. The third condition is

$$2B_1^2 = 1 \quad (32)$$

or

$$\frac{\kappa^2}{\kappa^2 + \delta^2} \sin^2 \frac{L}{4}\sqrt{\kappa^2 + \delta^2} = \frac{1}{2} \equiv \sin^2 \frac{\pi}{4}. \quad (33)$$

In the diagram this corresponds to the curves through $L/l = 2, 6, 10, 14$, etc., tangent to a 45° line through the origin.

VI. N Sections of Alternating $\Delta\beta$

From the results of the previous discussion we can already make a fairly good guess at the characteristics of a coupler with an arbitrary number of sections of alternating $\Delta\beta$. Using matrix multiplication and Sylvester's theorem, it is also fairly straightforward to analyze these structures. Assuming N to be an even number, we will distinguish between couplers with an even number (N) of sections and couplers with an odd number ($N + 1$) of sections.

Fig. 9(a) shows a coupler with an even number of sections of

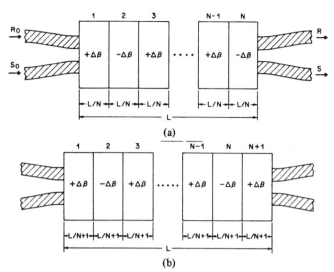

Fig. 9. Sketches of switched couplers with N and $N+1$ sections of alternating $\Delta\beta$. (a) Even number of sections. (b) Odd number of sections.

alternating $\Delta\beta$ and equal length L/N. The mismatch of the first section is $\Delta\beta$ and that of the last section is $-\Delta\beta$. The transfer matrix M_N of this coupler can be written as a power of the two-section matrix M_2 in the form

$$M_N = M_2^{N/2} = \begin{pmatrix} A_N & -jB_N \\ -jB_N^* & A_N^* \end{pmatrix}. \tag{34}$$

The matrix elements are determined with the help of Sylvester's theorem with the result

$$A_N = \cos(N\theta/2) \tag{35}$$

$$B_N = 2A_1^* B_1 \sin(N\theta/2)/\sin\theta \tag{36}$$

where

$$\cos\theta = A_2 = 1 - 2B_1^2 \tag{37}$$

and A_1 and B_1 are determined from (4) and (5) letting $z = L/N$. We rewrite (37) as

$$B_1^2 = \sin^2\theta/2 \tag{38}$$

in order to cast the subsequent results into our standard form.

To determine the cross-state conditions we let $A_N = 0$ and find

$$N\theta/2 = (2\nu + 1)\pi/2 \tag{39}$$

where ν is an integer. Combining this with (5) and (38) we obtain

$$\frac{\kappa^2}{\kappa^2 + \delta^2} \sin^2 \frac{L}{N}\sqrt{\kappa^2 + \delta^2} = \sin^2\left(\frac{2\nu+1}{N} \cdot \frac{\pi}{2}\right). \tag{40}$$

In the switching diagram, these conditions correspond to sets of curves similar to those discussed for couplers with two and four sections. The curves intersect the vertical axis ($\delta = 0$) at the points

$$L/l = 2N\mu \pm (2\nu + 1) \tag{41}$$

where μ is another integer.

For $B_N = 0$, we obtain the bar-state following conditions:

1) $A_1 = 0$ indicating isolated points at $\Delta\beta = 0$ and

$$L/l = N(2\nu + 1) \tag{42}$$

2) $B_1 = 0$ or

$$(L/l)^2 + (\Delta\beta L/\pi)^2 = (2N\nu)^2 \tag{43}$$

corresponding to concentric circles of radius $2N\nu$; and, finally,

$$N\theta/2 = \nu\pi \tag{44}$$

or

$$\frac{\kappa^2}{\kappa^2 + \delta^2} \sin^2 \frac{L}{N}\sqrt{\kappa^2 + \delta^2} = \sin^2 \frac{\nu\pi}{N} \tag{45}$$

corresponding to the now familiar curves which, in this case, intersect the vertical axis of the diagram at the points

$$L/l = 2N\mu \pm 2\nu. \tag{46}$$

For couplers with an odd number of sections, we obtain similar results, with the exception that we get isolated points for the cross states rather than the bar states. Fig. 9(b) shows such a coupler with sections of equal length $L/(N+1)$ starting and terminating with a section of mismatch $+\Delta\beta$. The transfer matrix of this coupler is

$$M_{N+1} = M_1 \cdot M_N = \begin{pmatrix} A_{N+1} & -jB_{N+1} \\ -jB_{N+1}^* & A_{N+1}^* \end{pmatrix} \tag{47}$$

with the matrix elements

$$A_{N+1} = A_1 \cos[\theta(N+1)/2]/\cos(\theta/2) \tag{48}$$

and

$$B_{N+1} = B_1 \sin[\theta(N+1)/2]/\sin(\theta/2). \tag{49}$$

A_1 and B_1 are obtained from (4) and (5) letting $z = L/(N+1)$.

For $A_{N+1} = 0$ we obtain the cross-state conditions, which include now $A_1 = 0$, or $\Delta\beta = 0$ and

$$L/l = (N+1)(2\nu + 1) \tag{50}$$

which are the isolated points mentioned before. The other cross-state condition is

$$\theta(N+1) = (2\nu + 1)\pi \tag{51}$$

or

$$\frac{\kappa^2}{\kappa^2 + \delta^2} \sin^2\left(\frac{L}{N+1}\sqrt{\kappa^2 + \delta^2}\right) = \sin^2\left(\frac{2\nu+1}{N+1} \cdot \frac{\pi}{2}\right). \tag{52}$$

The curves corresponding to this relation intersect the vertical axis in the switching diagram at the points

$$L/l = 2\mu(N+1) \pm (2\nu + 1). \tag{53}$$

Letting $B_{N+1} = 0$ we get the bar-state conditions, which include, first, the condition $B_1 = 0$ or

$$(L/l)^2 + (\Delta\beta L/\pi)^2 = [2\nu(N+1)]^2 \tag{54}$$

which is represented by circles of radius $2\nu(N+1)$ in the diagram. The second condition is

$$\theta(N+1)/2 = \nu\pi \tag{55}$$

or

$$\frac{\kappa^2}{\kappa^2 + \delta^2} \sin^2\left(\frac{L}{(N+1)}\sqrt{\kappa^2 + \delta^2}\right) = \sin^2\left(\frac{\nu}{N+1}\pi\right). \quad (56)$$

The curves representing these conditions in the diagram intersect the vertical axis at the points

$$L/l = 2\mu(N+1) \pm 2\nu. \quad (57)$$

VII. Conclusions

We have examined switched directional couplers in which we have assumed the coupling coefficient κ to be constant and the mismatch $\Delta\beta$ adjustable by an applied voltage, e.g., via the electrooptic effect. We have found that the providing of sections with alternating $\Delta\beta$ (e.g., by providing applied voltages of alternating polarity) makes it possible to induce complete crossover of the light from one guide to the other by an electrical adjustment. This adjustment is possible as long as L/l is larger than unity and can be used to reduce undesired crosstalk. We have used the switching diagram to depict graphically the $\Delta\beta$ adjustments required for each value of L/l to induce either the crossover (cross) or the straight-through (bar) states. In this diagram these states are represented either by isolated points, or by circles, or by curves belonging to a family, all described by equations of the same form. The circles always represent bar states only. The isolated points represent cross states when the number of sections is odd (i.e., 1, 3, $N+1$), and they represent bar states when the number of sections is even (2, 4, N). The isolated point with the lowest L/l value is always the point where L/l equals the number of sections. Beyond the electrical adjustability for both switching states obtained when we divide the coupler into two sections of reversed $\Delta\beta$, nothing essentially new seems to happen when we provide 3 or more sections of alternating $\Delta\beta$. Three or more sections, however, offer a low-voltage (low $\Delta\beta$) adjustment for the cross state when the L/l value is large, with the best choice being a number of sections approximately equal to L/l.

References

[1] E. A. J. Marcatili, "Dielectric rectangular waveguide and directional coupler for integrated optics," *Bell Syst. Tech. J.*, vol. 48, pp. 2071-2102, Sept. 1969.

[2] S. Kurazono, K. Iwasaki, and N. Kumagai, "New optical modulator consisting of coupled optical waveguides," *Electron. Commun. Jap.*, vol. 55, pp. 103-109, Jan. 1972.

[3] H. F. Taylor, "Optical switching and modulation in parallel dielectric waveguides," *J. Appl. Phys.*, vol. 44, pp. 3257-3262, July 1973.

[4] S. Somekh, E. Garmire, H. L. Garvin, and R. G. Hunsperger, "Channel optical waveguides and directional couplers in GaAs—Imbedded and ridged," *Appl. Optics*, vol. 13, pp. 327-330, Feb. 1974.

[5] J. M. Hammer, "Modulation and switching of light in dielectric waveguides," in *Integrated Optics*, vol. 7, T. Tamir, Ed. Berlin, Germany: Springer, pp. 140-198, 1975.

[6] K. Tada and K. Hirose, "A new light modulator using perturbation of synchronism between two coupled guides," *Appl. Phys. Lett.*, vol. 25, pp. 561-562, Nov. 1974.

[7] M. Papuchon et al., "Electrically switched optical directional coupler: COBRA," *Appl. Phys. Lett.*, vol. 27, pp. 289-291, Sept. 1975.

[8] J. C. Campbell, F. A. Blum, D. W. Shaw, and K. L. Lawley, "GaAs electro-optic directional coupler switch," *Appl. Phys. Lett.*, vol. 27, pp. 203-205, Aug. 1975.

[9] R. V. Schmidt and H. Kogelnik, "Electroptically switched coupler with stepped $\Delta\beta$ reversal using Ti diffused LiNbO$_3$ waveguides," *Appl. Phys. Lett.*, vol. 28, pp. 503-506, May 1976.

[10] S. E. Miller, "Coupled-wave throy and waveguide applications," *Bell Syst. Tech. J.*, vol. 33, pp. 661-719, May 1954.

[11] A. Yariv, "Coupled mode theory for guided wave optics," *IEEE J. Quantum Electron.*, vol. QE-9, pp. 919-933, Sept. 1973.

[12] H. Kogelnik, "Theory of dielectric waveguides," in *Integrated Optics*, vol. 7, T. Tamir, Ed. Berlin, Germany: Springer, pp. 15-79, 1975.

[13] S. E. Miller, "Broad-band electromagnetic wave coupler," U.S. patent 2 948 864, Aug. 9, 1960.

Polarization-independent optical directional coupler switch using weighted coupling

R. C. Alferness

Bell Telephone Laboratories, Holmdel, New Jersey 07733

(Received 11 June 1979; accepted for publication 12 September 1979)

We report the first demonstration of a guided-wave polarization-independent electro-optic switch. Using a specially designed weighted Ti-diffused directional coupler with stepped electrodes, we have achieved crosstalk levels below -23 dB for both switch states for arbitrary incident optical polarization.

PACS numbers: 42.80.Lt, 78.20.Dj, 84.40.Ed

We have demonstrated a 2×2 optical switch that, for fixed switching voltages, operates with low crosstalk independent of the polarization of the input optical signal. Using a specially designed Ti-diffused[1] directional coupler for which the coupling strength is carefully weighted along the interaction length[2] and reversed $\Delta\beta$ electrodes,[3,4] we have achieved crosstalk levels below -23 dB for both switch states for arbitrary input polarization.

Guided-wave electro-optical switches demonstrated to date operate effectively for a single linear polarization.[3,5,6] Such switches are not compatible with available low-loss single-mode fibers which do not preserve linear polarization.[7,8] Although specially fabricated single-polarization fibers are currently under investigation,[9] only short lengths have been made and uncertainties about excess loss and splicing ease are yet to be addressed. Several polarization-insensitive switch designs have been proposed,[10,11] but none has been demonstrated. Recently, an interferometric on/off modulator with 13-dB extinction ratio for both TE and TM modes was reported,[12] but it does not perform the switching function.

The difficulty of achieving efficient switching (i.e., low channel crosstalk) for both TE and TM polarizations with the same applied voltage arises because the orthogonal modes see unequal electro-optic coefficients.[13] As a result, for the same applied voltage, the induced phase mismatch $\Delta\beta$ is different for the two polarizations. In addition, because the guide-substrate refractive index difference Δn is generally unequal for the TE and TM modes,[1] the mode confinement and consequently the interguide coupling strength κ depend upon polarization.[13] The values of κ and $\Delta\beta$ together with the interaction length L determine the switching efficiency, and therefore the polarization dependence.

The novel polarization-independent switch, fabricated with Ti-diffused LiNbO$_3$ waveguides, is shown schematically in Fig. 1. It consists of a directional coupler with the coupling coefficient carefully weighted along the length by properly varying the interguide separation distance d.[2] Split electrodes are used (Fig. 1) to allow application of either reversed $\Delta\beta$[3,4] ($V_1 = -V_2$) or uniform $\Delta\beta$ ($V_1 = V_2$) fields. For the Z-cut crystal orientation and electrode placement (Fig. 1)

$$\Delta\beta_{TE} \propto r_{13} V, \quad [1(a)]$$
$$\Delta\beta_{TM} \propto r_{33} V, \quad [1(b)]$$

and $r_{33}/r_{13} \simeq 3$.[14] The two switch states are the crossover or cross state where light incident in one guide emerges in the other and the straight-through or bar state in which light remains in the incident waveguide. An important figure of merit is the switch crosstalk which is the ratio of optical power in the undesired to the desired output waveguides. Polarization-independent switching with the device in Fig. 1 has been achieved in spite of the difficulties listed above through several novel design features. First, essential to the design is the ability to approximately equalize the coupling strength for the TE and TM modes.[14,15] The cross state is achieved without polarization dependence by using reversed $\Delta\beta$ voltages[3,4] and by operating in the region of the switching curve that is least sensitive to differences in $\Delta\beta$. The polarization-independent bar state is obtained using the shaped transfer characteristic of weighted switched couplers.[2]

The coupling strengths for the TE and TM modes can be equalized in spite of the fact that $\Delta n_{TE} \neq \Delta n_{TM}$ by an appropriate choice of waveguide and coupler parameters.[15] No special crystal cut is required.[10] Briefly, we find experimentally that the coupling strength for either polarization can be written approximately as[15]

$$\kappa = \kappa_0 \exp(-d/\gamma), \quad (1)$$

where d is the constant interguide separation and γ is the waveguide lateral evanescent penetration depth. κ_0 and γ, which depend upon waveguide fabrication parameters, are unequal for the two modes primarily because $\Delta n_{TM} > \Delta n_{TE}$.[1] However, since κ_0 increases with increasing Δn, while γ decreases, Eq. (1) indicates that for some value of d (d_c), $\kappa_{TE} = \kappa_{TM}$. Furthermore, κ_{TE}/κ_{TM} is greater (less)

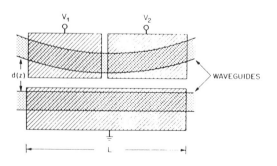

FIG. 1. Schematic drawing of polarization-independent optical switch.

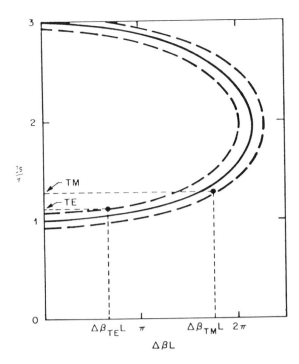

FIG. 2. Theoretical cross-state switching diagram for two sections of reversed $\Delta\beta$ with Hamming function weighting of κ. The large dashed curves indicate the -20-dB crosstalk limits. The measured device values of s_{TE} and s_{TM} are shown.

than 1 for d greater (less) than d_c. The relevant design parameter for weighted couplers is[16,2]

$$s = \int_{-L/2}^{L/2} \kappa(z)\, dz . \qquad (2)$$

Clearly, s_{TE} and s_{TM} can also be approximately equalized for the desired weighting function $\kappa(z)$.

The cross state is achieved with reversed $\Delta\beta$ voltages ($V_1 = -V_2$ in Fig. 1). This method has been shown for single polarization to allow electrical adjustment of the cross state eliminating the untenable fabrication tolerances otherwise necessary.[3,4] However, general application of reversed $\Delta\beta$ voltages will not result in a good dual-polarization cross state even for $\kappa_{TE} = \kappa_{TM}$, because $\Delta\beta_{TE} \neq \Delta\beta_{TM}$.

To understand how, by special design, dual-polarization operation is achieved, the theoretical cross-state switching curve is shown in Fig. 2. This curve specifies, for given s, the required $\pm \Delta\beta L$ over the two sections to achieve a perfect cross state. The results were calculated[17] including a Hamming weighting of κ, but are qualitatively the same as those for uniform coupling[3] if one substitutes κL for s. (Weighted coupling is not required for the cross state, but is included because it is essential for the bar state.) The dashed curves in Fig. 2 indicate the -20-dB crosstalk limits.

The switch is designed with $s_{TM} \gtrsim s_{TE} \simeq \pi/2$ as shown by the actual device values in Fig. 2. For this design the values of $\Delta\beta_{TM}$ and $\Delta\beta_{TE}$ needed to satisfy the switching condition (see Fig. 2) for the respective s values are approximately in the ratio $\Delta\beta_{TE}/\Delta\beta_{TM} \approx r_{13}/r_{33} \approx \frac{1}{3}$ constrained by the relative electro-optic coefficients. Thus for the same applied voltage both modes approximately satisfy the cross-state condition. Furthermore, precise values for s are not required with this design because the large range of allowable $\Delta\beta_{TE}L$ values for $s_{TE} \approx \pi/2$ (Fig. 2) results in a range of s_{TM} for which $\Delta\beta_{TM}L$ is also within the low crosstalk limits for the same applied voltage. Thus tolerances on s are no more severe than $\sim 20\%$ which is within obtainable fabrication limits.[15]

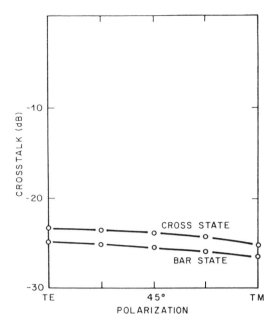

FIG. 4. Measured switch crosstalk versus input optical polarization. For the cross state, $V_1 \simeq -V_2 = 8$ V and for the bar state $V_1 = V_2 = 29$ V.

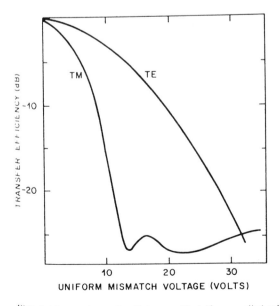

FIG. 3. Measured transfer efficiency with uniform applied voltage ($V_1 = V_2$) for TE and TM modes.

The bar state is obtained by applying a uniform phase-mismatch voltage ($V_1 = V_2$ in Fig. 1). For conventional uniform coupling the transfer response (crossover efficiency versus V) exhibits nodes as well as sidelobes of -9 dB.[14] For small values the crossover efficiency is equivalent to the bar state crosstalk. Again, because for the same applied voltage $\Delta\beta_{TE} \neq \Delta\beta_{TM}$, the voltage required to minimize crossover for the TE mode does not minimize the crossover for the TM mode and may even correspond to a relative maximum.

However, careful weighting of the coupling strength significantly reduces response sidelobes, typically to −25 dB.[2] Thus, with the weighted coupler used here for the voltage required to minimize crossover for the TE mode, the crosstalk for the electro-optically stronger TM mode remains below this acceptable value.

The device was designed and fabricated as follows. A Hamming function weighting of κ was chosen. The corresponding separation function $d(z)$ was designed[2] using measured $\kappa(d)$ data[15] appropriate to the waveguide fabrication parameters used—guide width = 3 μm, metal thickness $\simeq$ 210 Å, and diffusion time and temperature of 4 h at 980 °C. The minimum interguide spacing ($d_0 = 3$ μm) and device length ($L = 3.5$ mm) were then chosen to yield the approximate desired values of s_{TE} and s_{TM}. The measured device values of $s_{TE} \simeq 1.73$ and $s_{TM} \simeq 1.9$ are, according to Fig. 2, suitable for achieving a cross state with better than −20-dB crosstalk. After metal diffusion in a flowing argon atmosphere, a 1200-Å buffer layer of SiO_2 was deposited by chemical vapor deposition to reduce optical loss resulting from the CrAl electrodes[18] which were subsequently evaporated over the waveguides. The crystal ends were polished to allow input/output coupling via lenses. No surface guiding due to lithium out diffusion was observed,[19,20] and no special processing was required.[21]

The device was evaluated at $\lambda = 0.6328$ μm. The measured crossover efficiency versus applied uniform voltage for the TE and TM modes is shown in Fig. 3. Low sidelobes for the TM mode are evident. A good bar state (low crossover) for both modes is obtained for $V \geqslant 29$ V. The measured bar state crosstalk for this voltage versus linear polarization is shown in Fig. 4. The measured cross-state crosstalk versus polarization is also shown in Fig. 4. The reversed $\Delta\beta$ voltage was ~8 V. As shown in Fig. 4, for arbitrary polarization the crosstalk for both switch states remains below −23 dB, a value comparable to the best result obtained for single-polarization switches.[3]

The switching voltages can be reduced considerably by increasing the device length consistent with the design considerations discussed. Precise measurements were not made, but the optical loss appears comparable to reported results, <1.5 dB/cm for both modes.[22] Because of bending loss, the asymmetry between the upper and lower waveguides results in slightly different (~2 dB) crosstalk depending upon which waveguide is excited. This can easily be remedied by symmetrically bending both guides.

In summary, we have demonstrated the first low-crosstalk polarization-independent electro-optic switch. Crosstalk below −23 dB has been obtained for both switch states for arbitrary input polarization. This switch can be used with existing low-loss single-mode fibers that do not preserve linear polarization.

We thank T.G. Vieck for expert device fabrication and R. Bosworth for mask programming. The SiO_2 was deposited by A. Olenginski.

[1] R.V. Schmidt and I.P. Kaminow, Appl. Phys. Lett. **25**, 458 (1974).
[2] R.C. Alferness, Appl. Phys. Lett. **35**, 260 (1979).
[3] R.V. Schmidt and H. Kogelnik, Appl. Phys. Lett. **28**, 503 (1976).
[4] H. Kogelnik and R.V. Schmidt, IEEE J. Quantum Electron. **QE-12**, 396 (1976).
[5] M. Papuchon, Y. Combemale, X. Mathieu, D.B. Ostrowsky, L. Reiber. A.M. Roy, B. Sejourne, and M. Werner, Appl. Phys. Lett. **27**, 289 (1975)
[6] J.C. Campbell, F.A. Blum, D.W. Shaw, and K.L. Lawley, Appl. Phys. Lett. **27**, 202 (1975).
[7] V. Ramaswamy, R.D. Standley, D. Sze, and W.G. French, BSTJ **57**, 635 (1978).
[8] F.P. Kapron, N.F. Borrelli, and D.B. Keck, IEEE J. Quantum Electron **QE-8**, 222 (1972).
[9] R.H. Stolen, V. Ramaswamy, P. Kaiser, and W. Pleibel, Appl. Phys. Lett. **33**, 699 (1978).
[10] B.A. Steinberg and T.G. Giallorenzi, IEEE J. Quantum Electron, **QE-13**, 122 (1977).
[11] R.A. Steinberg, T.G. Giallorenzi, and R.C. Priest, Appl. Opt. **16**, 2166 (1979).
[12] W.K. Burns, T.G. Giallorenzi, R.P. Moeller, and E.J. West, Appl. Phys. Lett. **33**, 944 (1978).
[13] R.A. Steinberg and T.G. Giallorenzi, Appl. Opt. **15**, 2440 (1976).
[14] See, for example, J.M. Hammer, in *Integrated Optics*, edited by T. Tamir (Springer, Heidelberg, 1975).
[15] R.C. Alferness, R.V. Schmidt, and E.H. Turner, Appl. Opt. **18** (1979).
[16] R.C. Alferness and P.S. Cross, IEEE J. Quantum Electron. **QE-14**, 843 (1978).
[17] R.C. Alferness (unpublished). Details of the design analysis will be published in a subsequent paper.
[18] Mosamitu Masuda and Jiro Koyama, Appl. Opt. **16**, 2994 (1977).
[19] T.R. Ranganath and S. Wang, Device Research Conference, Salt Lake City, 1976 (unpublished).
[20] V. Ramaswamy, M.D. Divino, and R.D. Standley, Appl. Phys. Lett. **32**, 644 (1978).
[21] See, for example, W.K. Burns, C.H. Bulmer, and E.J. West, Appl. Phys. Lett. **33**, 70 (1978).
[22] I.P. Kaminow and L.W. Stulz, Appl. Phys. Lett. **33**, 62 (1978).

Directional Coupler Switches, Modulators, and Filters Using Alternating $\Delta\beta$ Techniques

RONALD V. SCHMIDT, MEMBER, IEEE, AND ROD C. ALFERNESS

Invited Paper

Abstract—The principles of operation and current status of single-mode optical directional coupler switches, modulators, and filters using $\Delta\beta$ reversal techniques are reviewed.

INTRODUCTION

INTEGRATED OPTICS has been an active field of research for about 10 years. During this time, considerable progress has been made toward realizing high-performance optical devices using guided-wave techniques [1]. One device which has received considerable attention because of its versatility in performing several important communications functions is the optical switch. As illustrated in Fig. 1, the switch is a four-port device with two inputs, two outputs, and two switch states. It is the analog of the electrical double-pull double-throw reversing switch. In the *straight-through state* a signal entering in the top or bottom input ports (Fig. 1) exits in the top or bottom ports, respectively. In the other state, called the *crossover state*, a signal entering in the top port exits in the bottom port while a signal entering in the bottom port crosses over to the top port. This switch may be used either as an amplitude modulator or as an element for switching the path of optical signals. In addition, specially designed wavelength selective switches can be used for wavelength multiplexing and demultiplexing. In this paper the principles of operation and the current state of the art of optical switches that are formed by waveguide directional couplers and are controlled via the electrooptic effect are reviewed. While no attempt is made to present an exhaustive survey, the major concepts and motivation for these devices are introduced. Many competitive optical modulator and switch devices based on Y junctions [2], [3], diffraction gratings [4], [5], total internal reflection [6], and interferometric bridges [7]–[9] are not discussed because of space limitations. Similarly, we restrict our consideration to single-mode switches; multimode devices [10] are not discussed.

The use of guided-wave techniques has led to switches with increased compactness, lower drive voltages and drive power compared to bulk electrooptical devices. In

Manuscript received June 14, 1979.
R. V. Schmidt was with Bell Telephone Laboratories, Holmdel, NJ. He is now with Hewlett-Packard Laboratories, Palo Alto, CA 94304.
R. C. Alferness is with Bell Telephone Laboratories, Holmdel, NJ 07733.

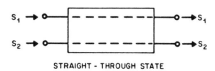
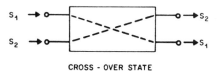

Fig. 1. Diagram showing the straight-through state (top) and crossover state (bottom) of a four-port directional coupler switch.

fact, switches using single-mode optical waveguides have been demonstrated with nanosecond switching times and electrical drive requirements compatible with electrical integrated circuits. Coupling to single-mode waveguide devices can be achieved efficiently only with single-mode fibers or single transverse mode lasers. Such fibers and lasers are now becoming available. The combination of single-mode lasers, fibers, and integrated optical devices which can be controlled by electrical integrated circuits offers new opportunities for the system and circuit designer. It is hoped that this paper will stimulate the identification of some of these opportunities.

GENERAL BACKGROUND

Directional coupler switches utilize single-mode strip optical waveguides. Fig. 2 illustrates the crosssection of a strip waveguide. Typically, three refractive indices define the strip waveguide [11]: the substrate index n_s, the waveguide strip index n_g, and the cladding index n_c. The cladding is normally either air or a thin dielectric layer. To achieve guiding, n_g must be greater than both n_s and n_c. The dimensions of the guide and the value of refractive indices determine the number of guided modes and their propagation constants β. The propagation constant can be written as

$$\beta = 2\pi N/\lambda \qquad (1)$$

where λ is the free-space optical wavelength and N is the so-called effective index of the guided mode. The effective index is determined by waveguide dispersion and in general must be calculated using numerical or approximate

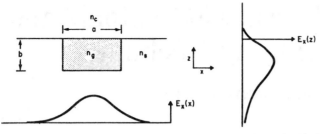

Fig. 2. Illustration of single-mode waveguide cross section and typical modal field distributions.

techniques. The value of N lies between n_s and n_g, being larger (closer to n_g) for more tightly confined modes. Typically, strip waveguides are designed with n_g just slightly larger than n_s so that a single-mode guide has the dimensions of several microns.

One material system which has been of particular interest for integrated optical switches is titanium-diffused LiNbO$_3$ [12]. Lithium niobate is a ferroelectric crystal with excellent electrooptic properties and has been used extensively for the demonstration of integrated optical devices. A strip waveguide is fabricated in LiNbO$_3$ by first photolithographically delineating the waveguide circuit pattern in Ti metal. The metal is then diffused into the LiNbO$_3$ in an oxygen atmosphere at ~1000°C for several hours. For $\lambda = 0.6328$ μm, a 3-μm Ti strip 300-Å thick diffused for 4 h creates a single-mode waveguide with a depth of ~3 μm. The refractive index of the crystal in the volume where the metal is diffused is increased proportionally to the Ti concentration. The guided mode has an intensity distribution similar to that shown in Fig. 2, although the diffused waveguide has graded index boundaries. There is little optical energy at the crystal surface because of the large index difference between LiNbO$_3$ ($n_s = 2.2$) and air ($n_c = 1$). The index difference between the diffused region and the substrate is of the order 10^{-2}. Because the diffusion temperature is below the crystal's Curie temperature there is no need to repole the crystal after diffusion. The loss in Ti-diffused waveguides is typically less than 1 dB/cm at 0.6323 μm [13]. While this is a large loss compared to glass fibers, it is very good for integrated optical waveguide devices which are typically no more than a few centimeters long. Single-mode waveguides have also been made in piezoelectric semiconductors like GaAs using a variety of techniques, but generally have higher losses.

The lack of diffraction effects and the small feature size of single-mode waveguides are the primary reasons for the improved performance of integrated electrooptic devices over their bulk counterparts. The advantages can be seen by considering the guided-wave phase modulator. Phase modulation is a basic electrooptic function which is used in nearly every electrooptic device and is particularly fundamental to switched directional couplers. It is thus worthwhile to review the theoretical operation of a waveguide phase modulator as illustrated in Fig. 3 because the results are applicable to a large variety of waveguide devices.

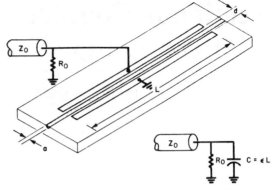

Fig. 3. Illustration of a strip waveguide phase modulator with inset showing the modulator equivalent circuit.

The waveguide phase modulator consists of a strip waveguide which supports a single mode of effective index N. On either side of the guide are placed parallel electrodes of length L separated by a distance d. A voltage V applied to the electrodes creates a field $E \approx V/d$ within the waveguide which electrooptically induces a change in effective index:

$$\Delta N = \tfrac{1}{2} \alpha N^3 r E. \qquad (2)$$

The coefficient α whose value is near unity, accounts for incomplete overlap of the optical and applied electrical fields. The value r is the relevant electrooptic coefficient which depends upon the crystal orientation and the direction of the applied field. The refractive index change produces a phase change in the light propagating in the waveguide by an amount

$$\Delta \phi = \Delta \beta L = 2\pi \Delta N L / \lambda. \qquad (3)$$

Of particular importance to the performance of electrooptic devices is the time and energy required to induce a π-phase change when the electrodes are driven from a transmission line terminated in its characteristic impedance.

The speed of a phase modulator driven in the manner illustrated in Fig. 3 is limited, in the ideal case neglecting parasitics, by either the transit time of the electrical signal down the electrodes $\tau_E \approx \sqrt{\epsilon} \, L/c$ or the capacitor charging time $\tau_{RC} = R_0 C$. The value ϵ is the dielectric permittivity of the crystal, c is the speed of light, and C is the electrode capacitance. In the limit of a coplanar capacitor $C = \epsilon L$. For a 50-Ω system τ_E is slightly less than τ_C for LiNbO$_3$ with an average relative dielectric permittivity of $\epsilon_r \approx 35$. In practice, both electrode capacitance and stray capacitances must be charged and the capacitor charging time is greater than the electrical transit time.

The voltage V_π required to produce a π-phase change, can be found from (2) and (3) and is

$$V_\pi = \frac{\lambda}{\alpha N^3 r} \frac{d}{L}. \qquad (4)$$

Thus the switching energy is

$$\epsilon_\pi = \tfrac{1}{2} C V_\pi^2 = \frac{\epsilon \lambda^2}{2\alpha^2 N^6 r^2} \frac{d^2}{L} \qquad (5)$$

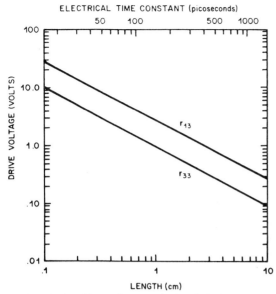

Fig. 4. Plot of the drive voltage and electrical time constant of a LiNbO$_3$ electrooptic phase modulator as a function of length for the r_{13} and r_{33} coefficients. The optical wavelength is taken to be 1 μm and the electrodes are spaced by 3 μm.

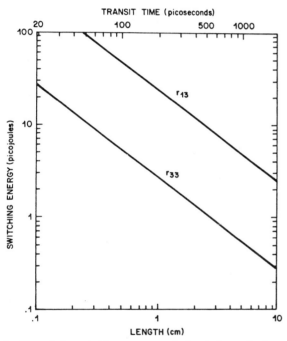

Fig. 5. Plot of the switching energy and electrical transit time of a LiNbO$_3$ electrooptic phase modulator as a function of length for the r_{13} and r_{33} coefficients. The optical wavelength is taken to be 1 μm and the electrodes are spaced by 3 μm.

which is also the electrical drive power per unit bandwidth. These equations are valid only when the electrical transit time is less than the capacitor charging time.

The advantage of strip waveguide phase modulators over bulk devices is apparent from (4) and (5). Low drive voltages and switching energies require a small d/L ratio. In bulk devices where d is the width of the crystal, diffraction limits the interaction length L over which the beam may be focused to approximately d^2/λ. So if d is made small, the interaction length is restricted. For widths comparable to single-mode guides, the maximum interaction length is limited to around 100 μm. Because a waveguide does not allow diffraction or spreading of light, the waveguide modulator does not have this restriction and its length is determined by speed requirements or fabrication convenience.

In Figs. 4 and 5, the theoretical performance of a phase modulator fabricated in LiNbO$_3$ is presented as a function of device length. Results are given for the two largest electrooptic coefficients. The device is assumed to operate at 1-μm wavelength and have an electrode separation of 3 μm. An average relative dielectric permittivity of 35 was used to take into account crystal anisotropy and α was set equal to one. The drive voltage and capacitor charging time in a 50-Ω system are presented in Fig. 4. It can be seen that for nanosecond switching times, drive voltages of less than 1 V can be obtained. The switching energy and electrical transit times for the same devices are presented in Fig. 5. Switching energies below 10 pJ can be obtained with nanosecond switching times. Because the r_{33} electrooptic coefficient is three times larger than r_{13}, it gives superior electrical performance. However, the fabrication of Ti-diffused strip waveguides which allow utilization of the r_{33} coefficient is not well understood because of possible formation of a slab waveguide on the entire surface. This waveguide may result from the out-diffusion of lithium oxide which only affects the extraordinary refractive index. Several methods of suppressing the out-diffusion process are currently under investigation [14], [15].

An experimental phase modulator 4 cm long has been demonstrated at a wavelength of 0.6328 μm with a picojoule switching energy utilizing the r_{33} coefficient [16]. This result is in close agreement with theoretical analysis presented here. The use of guided waves thus makes possible an electrooptical device which can provide modulation with voltages and energies compatible with transistor circuits. Although a phase modulator by itself is of limited practical interest, how electrooptic phase modulation applied to a directional coupler geometry results in amplitude modulation and switching will now be described.

OPTICAL DIRECTIONAL COUPLERS

An integrated optics directional coupler is formed by fabricating two parallel waveguides in close proximity so that light in one waveguide can couple to the other waveguide via the evanescent fields. In Fig. 6, a waveguide directional coupler circuit pattern is illustrated. Two waveguides with propagation constants β_R and β_S are brought within a distance d of one another for a length L. Over this length the waveguides are coupled so that optical energy can transfer between the two guides. If the waveguides have the same propagation constants and energy is incident in only one guide it will transfer completely to the other guide in a distance $l = \pi/2\kappa$ called the transfer length, where κ is coupling coefficient which describes the strength of the interguide coupling. Typically the coupling coefficient is related to separation be-

Fig. 6. Top view of a waveguide directional coupler where L is the coupler length, a is the waveguide width, and d the separation between the waveguides in the interaction region.

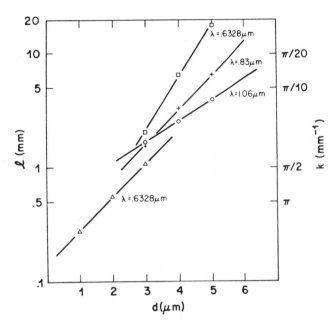

Fig. 7. Coupling coefficient of Ti-diffused LiNbO$_3$ waveguides at 1.06-, 0.83-, and 0.633-μm wavelengths as a function of waveguide separation. For each wavelength the waveguides were fabricated so as to be single mode using the following fabrication parameters: (□) $\lambda = 0.6328\,\mu\text{m}$, τ (Ti metal thickness) $= 300$ Å, $a = 3\,\mu\text{m}$; (△) $\lambda = 0.6328\,\mu\text{m}$, $\tau = 300$ Å, $a = 2\,\mu\text{m}$; (+) $\lambda = 0.83\,\mu\text{m}$, $\tau = 400$ Å, $a = 4\,\mu\text{m}$; and (○) $\lambda = 1.06\,\mu\text{m}$, $\tau = 460$ Å, $a = 5\,\mu\text{m}$.

tween the guides d by an exponential relation of the form

$$\kappa = \kappa_0 \exp(-d/\gamma) \qquad (6)$$

where κ_0 and γ are coefficients that depend upon the various waveguide parameters [17]. In Fig. 7 the coupling coefficient and the transfer length are given for various separation between guides. These measurements were taken on single-mode Ti-diffused waveguides at three different wavelengths. The straight lines are best fits of the exponential relation (6). Transfer lengths as short as 200 μm and as long as 1 cm have been experimentally observed in Ti-diffused waveguides.

Electrically controlled optical switching can be achieved with a directional coupler because the degree-of-light transfer between the waveguides depends upon the difference in propagation constants, $\Delta\beta = \beta_R - \beta_S$, which can be controlled via the electrooptic effect. If the light in the two guides is characterized by complex amplitudes R and S which vary slowly in the propagation direction, the interaction between the two guides is described by the coupled-wave equations [18]:

$$R' - j\delta R = -j\kappa S \qquad (7)$$
$$S' + j\delta S = -j\kappa R \qquad (8)$$

where the primes denote differentiation with respect to the propagation direction, $\delta = \Delta\beta/2$, and κ is the coupling coefficient. For arbitrary input amplitudes R_0 and S_0, as shown in Fig. 6, the solution of (7) and (8) can be expressed in matrix form [19]:

$$\begin{bmatrix} R \\ S \end{bmatrix} = \begin{bmatrix} A & -jB \\ -jB^* & A^* \end{bmatrix} \begin{bmatrix} R_0 \\ S_0 \end{bmatrix} \qquad (9)$$

where the asterisk denotes a complex conjugate. The matrix coefficients are

$$A = \cos\kappa L\sqrt{1+(\delta/\kappa)^2} + j\delta/\kappa \sin\kappa L\sqrt{1+(\delta/\kappa)^2} \qquad (10)$$

$$B = \sin\kappa L\sqrt{1+(\delta/\kappa)^2}\,/\sqrt{1+(\delta/\kappa)^2}. \qquad (11)$$

If light is launched in the R guide at $z = 0$ the power in the two guides at the coupler output $z = L$ is given by

$$SS^* = \frac{\sin^2\kappa L\sqrt{1+(\delta/\kappa)^2}}{1+(\delta/\kappa)^2}\,R_0R_0^* \qquad (12)$$

$$RR^* = 1 - SS^*. \qquad (13)$$

Equation (12) illustrates several directional coupler properties that are important to the design of optical switches. Energy can transfer completely from one guide to the other only if $\Delta\beta = 0$ and

$$\kappa L = (2m+1)\pi/2 \qquad (14)$$

where m is an integer. If the guides do not have the same propagation constants it is impossible for energy to be completely transferred across the coupler and some optical energy always remains in the input waveguide. However, in this phase mismatched case ($\Delta\beta \neq 0$), energy exits entirely in the input waveguide only if

$$(\kappa L)^2 + (\delta L)^2 = (m\pi)^2 \qquad (15)$$

is satisfied.

Directional Coupler Switch/Modulator

We are now ready to explain the operation of directional coupler switches. The first type of directional coupler switch that was experimentally demonstrated is illustrated in Fig. 8. It consists of a directional coupler with electrodes placed over the waveguides in the coupler region. The applied electric fields normal to crystal surface are oppositely directed in each guide so that the linear electrooptic effect produces the phase mismatch between the guides by increasing the refractive index in one guide and decreasing it in the other. The switch is designed so that it is in the crossover state with no voltage applied to the electrodes. This requires fabricating the coupler length to be equal to one transfer length satisfying (14). The straight-through state is obtained by applying an appropriate voltage to the electrodes to satisfy the phase mismatch condition (15). In Fig. 8 the optical power distribution in the two guides is shown as a function of position for both switching states. The results are shown for a coupler length equal to one transfer length $L/l = 1$.

The condition for the crossover and straight-through states can be graphically represented in a switching dia-

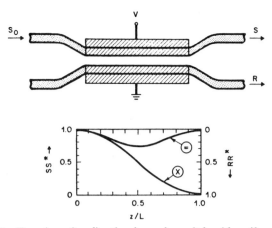

Fig. 8. Top view of a directional coupler switch with uniform electrodes placed over the waveguides in the coupling region. Also shown are the intensity distributions along the coupler in crossover (×) and straight-through (=) switching states.

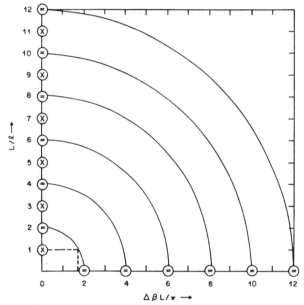

Fig. 9. The crossbar switching diagram for a switch with uniform electrodes. The isolated points marking the conditions required for complete energy crossover are marked by (×), and arcs indicating the conditions required for the straight-through state are marked by (=).

gram or so-called *crossbar diagram* shown in Fig. 9. The coordinates are the device length normalized by the transfer length L/l and the normalized phase mismatch, $\Delta\beta L/\pi = 2\delta L/\pi$. The requirements for a perfect crossover state as specified by (14) are denoted by isolated × points on the L/l axis. The locus of straight-through states, denoted by =, are circles prescribed by (15). For a one transfer length coupler ($L/l = 1$) the required phase mismatch to switch is shown by the dashed line and is

$$\Delta\beta L = \sqrt{3}\,\pi. \qquad (16)$$

Because the electrode configuration is identical to that of the phase modulator previously discussed, the results obtained for phase modulator performance apply also to the directional coupler switch except that the drive voltage and switching energy requirements are increased by a factor of $\sqrt{3}$ and 3, respectively.

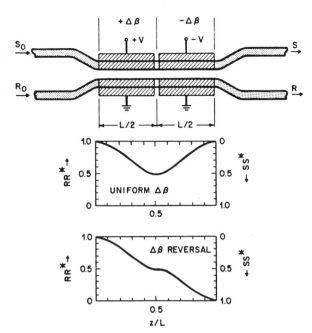

Fig. 10. Top view of stepped $\Delta\beta$ reversal switch requiring two sets of electrodes. Also shown are the intensity distributions along the coupler for a coupler of length $L/l = \sqrt{2}$.

Several switches of this type have been demonstrated [20]–[24]. It has proven difficult, however, to build this directional coupler switch with a good crossover state because of the necessity to fabricate so that $L = l$. It is not possible to significantly alter the value of l by electrooptically changing N. Furthermore, as indicated in (12) and (13) for this directional coupler switch, it is impossible to obtain the cross state via a phase mismatch which can be achieved electrooptically. As a result, the cross-state crosstalk is limited by fabrication tolerances. From (14) we note that to achieve a cross state with better than -20-dB crosstalk requires fabricating such that $L/l = 1$ to within $\sim \pm 6$ percent. As a result, switches of this type typically have reported crosstalk levels greater than -10 dB. Lower crosstalk switches (~ -20 dB) have been demonstrated in GaAs [23], [24] where the crystal may be cleaved to shorten the length and obtain $L \simeq l$. However, such techniques, of course, limit the usefulness of the device because it eliminates the possibility of integration with other devices on the same substrate.

A directional coupler switch that allows electrical adjustment to achieve both switch states with low crosstalk thus eliminating severe fabrication requirements is illustrated in Fig. 10. The switch, which uses a technique called stepped $\Delta\beta$ reversal, requires two sets of electrodes [25]. The electrodes divide the coupler into two equal-length sections in which phase mismatch of opposite sign can be applied via the electrooptic effect. With equal magnitude but opposite sign phase mismatch, electrical adjustment of both switching states is obtained. Electrical adjustment of the straight-through state can also be obtained as before by applying a uniform phase mismatch along the entire length of the coupler.

Electrical adjustability of the crossover state due to stepped $\Delta\beta$ reversal can be understood by considering how the energy is exchanged between the guides. Consider the switch illustrated in Fig. 10 where the R and S waves travel in the top and bottom waveguides, respectively. For simplicity it is convenient to examine the special case where the coupler length is $L/l = \sqrt{2}$. In this case the light is equally divided between the R and S waves in the middle of the coupler when the switch is in the straight-through state with uniform phase mismatch along the entire coupler length. In Fig. 10 this case is illustrated when light is incident in the R waveguide. At this midpoint along the coupler, where the amplitudes are equal, the phases of the R and S waves with respect to the electrooptically induced slow and fast waveguides determine that the light completely returns to the R wave at the end of the coupler. When the switch is used in the stepped $\Delta\beta$ configuration, the fast and the slow waveguides are effectively interchanged at the midpoint of the coupler by virtue of the change in sign of applied voltage. In this case light starting in the R waveguides will exit in the S waveguide, as illustrated in Fig. 10. Similarly, for other L/l ratios ranging from 1 to 3, the crossover states can be obtained by using stepped $\Delta\beta$ reversal. However, in general the voltage required to obtain the crossover state is different from the voltage required to obtain the straight-through state. A more detailed analysis of these switches can be found in [19] and [25].

In Fig. 11 the switching diagram of stepped $\Delta\beta$ reversal switch is shown. The solid lines are the locus of states obtainable with equal and opposite voltages while the dashed lines indicate the state which can be obtained with uniform phase mismatch along the coupler. A coupler whose length is between one and two transfer lengths will provide the lowest switching voltage requirements for both switching states. The fact that both switching states can be obtained by electrical adjustment greatly reduces the precision required to fabricate the switches. Switches with reversed $\Delta\beta$ electrodes have been demonstrated in Ti-diffused $LiNbO_3$ [25] and in GaAs waveguide structures [26], [27] with reported crosstalk levels of -26 dB. Recently, crosstalk as low as -35 dB has been observed [28].

Because the operating wavelength of potential semiconductor laser sources typically can fluctuate with, for example, temperature changes, it is important to determine how the switch crosstalk levels depend upon λ for fixed applied voltage. Such wavelength dependence results because l depends upon λ [17]. Fig. 12 illustrates the measured crosstalk versus wavelength for a 3-mm long $\Delta\beta$ reversed switch fabricated in Ti-diffused $LiNbO_3$. The fixed switch voltage was that required to minimize the crosstalk for $\lambda = 0.6$ μm. As indicated, the crosstalk remains below -20 dB over a spectral bandwidth of 150 Å.

The drive voltage requirements for stepped $\Delta\beta$ reversal switches cannot be as simply expressed as those for the previous cases because there is a range of acceptable coupler lengths. However, both switching states can be obtained for

$$\Delta\beta L < 2\pi. \tag{17}$$

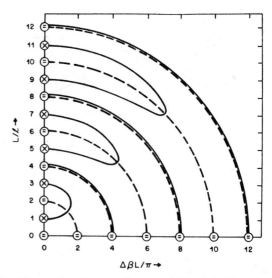

Fig. 11. The crossbar diagram of a stepped $\Delta\beta$ switch. The solid curves are switching states available with equal but opposite polarity voltages applied to the electrodes. The dashed curves are the straight-through states available with the same polarity voltages applied to the electrodes. The solid and dashed curves where slightly separated for clarity are actually identical.

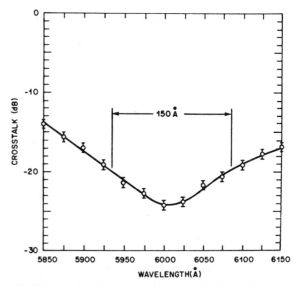

Fig. 12. The wavelength dependence of crosstalk in a 3-mm-long Ti-diffused $LiNbO_3$ stepped $\Delta\beta$ reversal switch.

From this condition it is clear that the drive voltage can be reduced by increasing L. However, (17) is only valid for $l \leq L \leq 2l$; larger L/l ratios required larger phase mismatch (see Fig. 11). Thus, an increase in L requires a similar increase in l which can be achieved most readily by increasing the separation between the waveguides d. This increase in d, however, results in a lower electric field for the same applied voltage (4) which does not allow a voltage reduction proportional to the increase in L. This difficulty can be overcome, however, by using multiple sections of alternating $\Delta\beta$ reversal. For N sections of alternating $\Delta\beta$ reversal, (17) is valid for a coupler length between $N-1$ and $N+1$ transfer lengths, and electrically adjustable crossover and straight-through states can be obtained with full advantage of the increased length [19]. Fig. 13 shows the switching diagram for a switch with six sections of alternating $\Delta\beta$ reversal. The dashed line gives

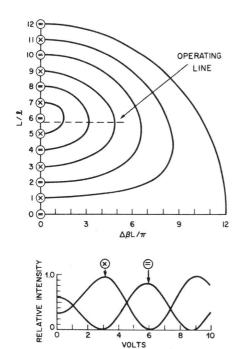

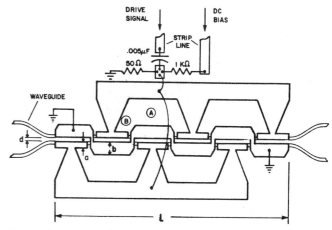

Fig. 14. Diagram of a six-section alternating $\Delta\beta$ switch showing the electrode configuration and the external electrical biasing circuit.

Fig. 13. The crossbar diagram of a six-section alternating $\Delta\beta$ switch. The dashed line is the operating line of a switch 5.75 transfer lengths long. The bottom curves are the output switching characteristics of a six-section switch as a function of applied voltage.

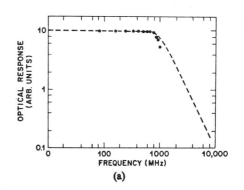

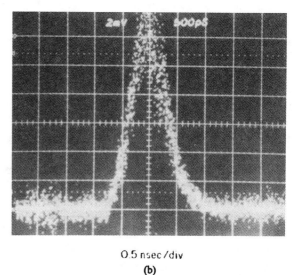

Fig. 15. (a) The measured small signal frequency response of a six-section alternating $\Delta\beta$ switch. (b) Photo of the switch pulse response.

the operating line of a switch 5.75 transfer lengths long. As the voltage is increased, the switch first reaches the crossover state and as the voltage is further increased, the straight-through state is obtained. A detailed analysis of the properties of alternating $\Delta\beta$ switches can be found elsewhere [19].

The actual switching characteristics versus voltage for a six-section switch formed in Ti-diffused $LiNbO_3$ [28] is shown also in Fig. 13. A Z-cut crystal was used with the TE waveguide mode which utilizes the r_{13} electrooptic coefficient. The coupler was 10.5 mm long corresponding to 5.75 transfer length operating line for 0.6328-μm wavelength operation. The guides were 3 μm wide separated by 3 μm. The switch required 3 V for the crossover state and an additional 3 V to switch to the straight-through state. This device can be used either as a switch or as a modulator. As a digital modulator it would be dc biased in the crossover state and switched to the straight-through state with a modulating voltage. Alternatively, for analog modulation, the switch would be biased in the linear region around 4.5 V.

An illustration of the six-section alternating $\Delta\beta$ reversal switch and the electrical biasing circuitry for use in digital modulation applications [29] is presented in Fig. 14. The electrode pattern was made with a 2000-Å thick aluminum layer and consisted of three parts: two outer three-section electrodes to which the modulating signal is applied, and a meandering central ground electrode which produces the required polarity reversal from section to section. The dc bias is applied through a 1-kΩ resistor while the RF drive signal is capacitively coupled to the electrodes from a transmission line and terminated in a 50-Ω resistor in parallel with the electrodes. The bandwidth of the mod-

ulator is nominally the RC time constant of the 50-Ω terminating resistor and the 6.7-pF electrode capacitance, but in practice the electrode resistance and parasitic inductance must be considered [30]. For this modulator the parasitic inductance resonates some of the capacitance and actually extends the bandwidth. The small signal frequency response of the modulator is shown in Fig. 15. The response is flat out to 0.96 GHz after which it falls off sharply, dropping to 1/2 at 1 GHz. Also shown is the pulse response of the modulator when driven by a 400-ps

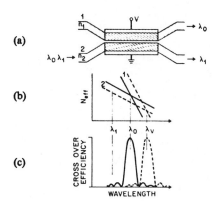

Fig. 16. The directional coupler wavelength filter. (a) A schematic drawing of the device. (b) The waveguide dispersion curves with and without electrooptic tuning. (c) The resulting wavelength filter response.

rise-time pulser and detected by a 200-ps rise-time avalanche photodiode. The rise time of the modulator has been determined to be 590 ps. With 5-V pulses the modulator has about a 10-dB extinction ratio under these pulse operation conditions.

The performance of this switch/modulator is characteristic of the present state of the art in single-mode waveguide switches. Its performance is still somewhat less than ideal as indicated by a comparison with the analysis of the phase modulator. A reduction in drive voltage or increase in switching speed should result from the successful utilization of the large r_{33} electrooptic coefficient. Nevertheless, in terms of drive voltage and speed the performance is greatly improved over that obtained from bulk electrooptic modulators and switches.

Directional Coupler Filter

In addition to switches and modulators, future single-mode optical communication systems will likely utilize wavelength filters to perform multiplexing/demultiplexing. Wavelength multiplexing allows more full utilization of the very wide information bandwidth of single-mode fibers. As noted above, the switching characteristics of the directional coupler devices described thus far are relatively insensitive to wavelength changes and are not suitable for wavelength discrimination. However, a specially designed [31] directional coupler has been demonstrated [32] which allows interguide coupling only for a selected band of wavelengths. The directional coupler filter, as shown in Fig. 16 (a), consists of two strip waveguides of different widths and refractive indices. The wider (narrower) guide has the lower (higher) guide refractive index. Because of the different dimensions and refractive indices, the two guides exhibit distinct modal dispersion characteristics $N(\lambda)$. However, by proper design, the effective indices for the two guides can be made equal at the desired filter center wavelength, λ_0, (Fig. 16 (b)). At this wavelength, the waveguides are phase matched ($\Delta\beta=0$) and complete light transfer between them is possible. However, for λ sufficiently different from λ_0 (e.g., λ_1 in Fig. 16), no transfer occurs. The filter response is shown in Fig. 16 (c) and is given by (12) with $\Delta\beta$ proportional to $\lambda-\lambda_0$. The filter bandwidth is decreased by either increasing the interaction length or by increasing the refractive index difference between the two waveguides [32]. If light of wavelengths λ_0 and λ_1 is input into one guide (Fig. 16 (a)), the light of λ_0 will be in the crossover state while that of λ_1 will pass straight-through providing wavelength demultiplexing.

The filter center wavelength of this device is readily electrically tunable [32]. By applying voltage to electrodes placed over the waveguides (Fig. 16 (a)), the dispersion curves are altered as shown in Fig. 16 (b) resulting in a new filter center wavelength λ_V (Fig. 16 (c)). Furthermore, by the use of the stepped $\Delta\beta$ electrode configuration (not shown in Fig. 16) the transfer efficiency at the filter center wavelength can be electrically adjusted to ~100 percent. This is especially important when tuning over a broad range of wavelengths [32].

The demonstrated filter consisted of Ti-diffused waveguides 1.5 and 3 μm wide separated by 3.5 μm with an interaction length $L=1.5$ cm. The measured filter 3-dB bandwidth was 200 Å with $\lambda_0 \simeq 0.6000$ μm and peak crossover efficiency of ~100 percent. Electrical tunability of the center wavelength was demonstrated over the available source range of 600 Å with a tuning rate of 100 Å/V.

To allow close stacking of wavelength channels without unacceptable crosstalk it is important to reduce filter sidelobes (see Fig. 16 (c)). Such sidelobe reduction can be achieved by carefully weighting the coupling coefficient along the interaction length by, for example, varying the interguide separation d [33]. Sidelobe levels as low as -25 dB have been demonstrated using a Hamming function weighting of κ [34].

Although space does not allow elaboration, we briefly mention some work aimed toward the integration of several directional coupler devices to perform more complex functions. Optical switching networks made by cascading several $\Delta\beta$ reversal switches on the same crystal have been demonstrated in both Ti-diffused LiNbO$_3$ [35] and in GaAs waveguides [27]. Crosstalk levels as low as -18 dB were observed in a 4×4 switching network composed of five switches. Analog-to-digital converters have been demonstrated using interferometric switches [36]. Such A/D converters, which could also be implemented with $\Delta\beta$ reversal switches, offer the potential for very high-speed operation. Optical logic devices have also been proposed [37]. Furthermore, a bistable optical switch which is capable of remote optically controlled switching has been demonstrated [38].

Other Considerations

While much progress has been made in developing high performance single-mode optical switches and other integrated optic devices, there are several question areas relating to system considerations for which answers are only now becoming available and more work is required. These include 1) device polarization dependence, 2) waveguide-fiber coupling, and 3) optical damage.

The devices described thus far operate effectively only for either the TE or TM polarization. Because the orthogonal modes see different electrooptic coefficients and κ depends strongly upon polarization, the required applied voltage for both switch states is polarization dependent [39]. Unfortunately, as light propagates along a typical single-mode fiber, a single linear polarization is not maintained. As a result, the polarization of the received signal that is input to the switch cannot be specified. While fibers with built-in stress induced birefringence that maintain polarization over short lengths have been demonstrated [40], questions concerning loss and cabling and splicing ease are still unanswered. The alternative is to design devices whose performance does not depend upon polarization. Recently, a specially designed directional coupler switch/modulator with weighted coupling was reported that achieved crosstalk levels below -23 dB for both switch states for arbitrary input polarization [41]. An interferometric on/off modulator with 13-dB extinction ratio for both TE and TM modes has also been shown [42]. With more work and careful design it is likely that other guided wave devices, such as filters, can be built that are insensitive to polarization and compatible with existing low-loss single-mode fibers.

A major concern for the implementation of single-mode waveguide devices is the coupling of optical power into the waveguides. Both $LiNbO_3$ and GaAs crystals have cleavage planes which make end-fire coupling an attractive approach. To have efficient end-fire coupling into a single-mode waveguide, the intensity distribution of the incident light must match that of the guided mode. Thus, a buried heterostructure type laser which may have an output mode that matches the switches' mode is one approach which can be used to directly couple a laser to a single-mode waveguide. Little work has been done in this area as of yet. In the case of a single-mode fiber coupling, the modal intensity distribution is axially symmetric while the planar waveguide mode of the switch is not, as shown in Fig. 17. The mode mismatch leads to about a 1-dB coupling loss in the optimum case [44], [45]. Further improvements in the end butting method will require some technique to match the modes better such as a dielectric layer on top of the planar waveguide to make the mode distribution more axial symmetric.

Another area which is of concern in implementing single-mode switches is that of operational stability. A 3 μm wide $\times$ 3 μm deep optical waveguide carrying a 1mW of optical power is subject to an intensity of 10^4 W/cm^2. For visible wavelengths at these intensities Ti-diffused $LiNbO_3$ exhibits an optically induced refractive index change. This change results from the photoexcitation and redistribution of carriers that create a space charge field which electrooptically changes the refractive index. The problem is particularly evident when the crystal has a dc bias field to aid in the redistribution of charge carriers. At 0.6328-μm wavelength and 20-nW power, as much as 15-dB crosstalk drift has been observed in a dc biased switch because of this effect [46]. Fortunately, at the infrared wavelengths, which are of practical interest, the crosstalk stability is much improved probably due to reduced photoexcitation of carriers. In Fig. 18, the cros-

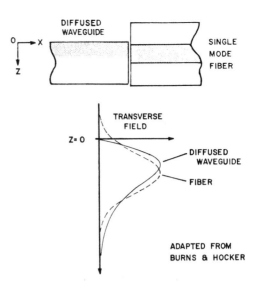

Fig. 17. A diagram illustrating the end-fire coupling of a single-mode fiber and planar waveguide. Also, the field distributions characteristic of fibers and planar waveguides. Adapted from [44].

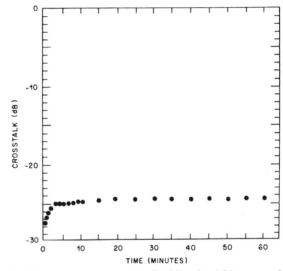

Fig. 18. The crossover state crosstalk drift of a 1.06-μm wavelength switch as a function of time for 1 mW of power output from the waveguides.

talk drift of a switch operating at $\lambda = 1.06$ μm and 1 mW of optical power is presented. In this case, the crosstalk increases from -29 to -24 dB in a few minutes time. The exact mechanism of this crosstalk drift at infrared wavelengths is not yet completely understood [46]. To date, very little attention has been directed towards reducing the optically induced crosstalk drift. It seems likely that attention to device processing which maintains crystal stoichiometry and reduces impurity content will improve the optical stability. Alternatively, one might use the crystal $LiTaO_3$ which is known to be less susceptible to optical damage and yet has electrooptic properties similar to $LiNbO_3$. Optically induced instabilities of this type will probably be a concern in any electrooptic device, dielectric or semiconductor, which requires a bias field.

The implementation of optical switches and the devices based on them which we have discussed rests primarily on further development of single-mode fibers and lasers. Need for improved performance and more information bandwidth will eventually bring about single-mode fiber systems. Alternatively, single-mode devices such as switches offer tremendous opportunities for new and improved systems. The identification of these opportunities may well also stimulate single-mode fiber systems.

REFERENCES

[1] H. Kogelnik, "An introduction to integrated optics," *IEEE Trans. Microwave Theory Tech.*, vol. MTT-23, pp. 2–16, Jan. 1975.
[2] W. K. Burns, A. B. Lee, and A. F. Milton, "Active branching waveguide modulator," *Appl. Phys. Lett.*, vol. 29, pp. 790–792, 1976.
[3] H. Sasaki and R. M. De La Rue, "Electro-optic Y-junction modulator/switch," *Elec. Lett.*, vol. 12, pp. 459–460, Sept. 1976.
[4] J. M. Hammer and W. Phillips, "Low-loss single mode optical waveguides and efficient high-speed modulators of $LiNb_xTa_{1-x}O_3$ on $LiTaO_3$," *Appl. Phys. Lett.*, vol. 24, pp. 545–547, June 1974.
[5] B. Chen and C. M. Meijer, "Bragg switch for optical channel waveguides," *Appl. Phys. Lett.*, vol. 33, pp. 33–35, July 1978.
[6] S. K. Sheem and C. S. Tsai, "Light beam switching and modulation using a built-in dielectric channel in $LiNbO_3$ planar waveguide," *Tech. Digest*, Conf. Laser Eng. and Appl., Washington, DC, June 1977.
[7] W. E. Martin, "A new waveguide switch/modulator for integrated optics," *Appl. Phys. Lett.* vol. 26, p. 562, May 1975.
[8] T. R. Ranganath and S. Wang "Ti-diffused $LiNbO_3$ branched-waveguide modulators performance and design," *IEEE J. Quant. Electron*, vol. QE-13, pp. 290–295, Apr. 1977.
[9] V. Ramaswamy, M. D. Divino, and R. D. Standley "Balanced bridge modulator switch using Ti-diffused $LiNbO_3$ strip waveguides," *Appl. Phys. Lett.*, vol. 32, pp. 644–646, May 1978.
[10] R. A. Soref, D. H. McMahon, and A. R. Nelson, "Multimode achromatic electrooptic waveguide switch for fiber-optic communications," *Appl. Phys. Lett.*, vol. 28, pp. 716–718, June 1976.
[11] H. Kogelnik, "Theory of dielectric waveguides," in *Integrated Optics*, vol. 7, T. Tamir, Ed. Berlin: Springer, 1975, pp. 15–79.
[12] R. V. Schmidt and I. P. Kaminow, "Metal diffused optical waveguides in $LiNbO_3$," vol. 25, pp. 458–460, Oct. 15, 1974.
[13] I. P. Kaminow and L. W. Stulz, "Loss in cleaved Ti-diffused $LiNbO_3$ waveguides," *Appl. Phys. Lett.* vol. 33, pp. 62–65, 1978.
[14] T. R. Ranganath and S. Wang, "Suppression of LiO_2 out-diffusion From Ti-diffused $LiNbO_3$ optical waveguides," *Appl. Phys. Lett.*, vol. 30, pp. 376–379, Apr. 15, 1977.
[15] B. Chen and A. C. Pastor, "Elimination of LiO_2 out-diffusion waveguide in $LiNbO_3$ and $LiTaO_3$," *Appl. Phys. Lett.*, vol. 30, pp. 570–571, 1977.
[16] I. P. Kaminow, L. W. Stulz, and E. H. Turner, "Efficient strip-waveguide modulator," *Appl. Phys. Lett.* vol. 27, pp. 555–557, Nov. 1975.
[17] R. C. Alferness, R. V. Schmidt, and E. M. Turner, "Characteristics of Ti-diffused $LiNbO_3$ optical directional couplers," *Appl. Opt.*, vol. 18, Nov. 1979.
[18] S. E. Miller, "Coupled-wave theory and waveguide applications," *Bell Syst. Tech. J.*, vol. 33, pp. 661–719, May, 1954.
[19] H. Kogelnik and R. V. Schmidt, "Switched directional couplers with alternating $\Delta\beta$," *IEEE J. Quant. Electron.*, vol. QE-12, pp. 396–401, July 1976.
[20] K. Tada and K. Mirose, "A new light modulator using perturbation of synchronism between two coupled guides," *Appl. Phys. Lett.*, vol. 25, pp. 561–562, Nov. 1974.
[21] M. Papuchon *et al.*, "Electrically switched optical directional coupler: COBRA," *Appl. Phys. Lett.*, vol. 27, pp. 289–291, Sept. 1975.
[22] J. C. Campbell *et al.*, "GaAs electrooptic directional coupler switch," *Appl. Phys. Lett.*, vol. 27, pp. 203–205, Aug. 1975.
[23] F. J. Leonberger, J. P. Donnelly, and C. O. Bozler, "GaAs $p^+n^-n^+$ directional coupler switch," *Appl. Phys. Lett.*, vol. 29, pp. 652–654, 1976.
[24] H. Kawaguchi "Directional-coupler switch with schottky barriers," *Electron. Lett.*, vol. 14, pp. 387–388, June 1978.
[25] R. V. Schmidt and H. Kogelnik, "Electro-optically switched coupler with stepped $\Delta\beta$ reversal using Ti-diffused $LiNbO_3$ waveguides," *Appl. Phys. Lett.*, vol. 28, pp. 503–506, May 1, 1976.
[26] F. J. Leonberger and C. O. Bozler, "GaAs directional coupler switch with stepped $\Delta\beta$ reversal," *Appl. Phys. Lett.*, vol. 27, pp. 202–205, 1977.
[27] J. C. Shelton, F. K. Reinhart, and R. A. Logan, "$GaAs-Al_xGa_{1-x}As$ rib waveguide switches with MOS electrooptic control for monolithic integrated optics," *Appl. Opt.* vol. 17, pp. 2548–2555, Aug. 1978.
[28] M. Papuchon, private communication.
[29] R. V. Schmidt and P. S. Cross, "Efficient optical waveguide switch/amplitude modulator," *Opt. Lett.* vol. 2, pp. 45–47, Feb. 1978.
[30] P. S. Cross and R. V. Schmidt, "A 1 G bit/second integrated optical modulator," *IEEE J. Quantum Electron.*, vol. QE-15, pp. 1415–1418, Dec. 1979.
[31] H. F. Taylor, "Frequency-selective coupling in parallel dielectric waveguides," *Opt. Comm.*, vol. 8, pp. 421–425, Aug. 1973.
[32] R. C. Alferness and R. V. Schmidt, "Tunable optical waveguide directional coupler filter," *Appl. Phys. Lett.* vol. 33, pp. 161–163, July 15, 1978.
[33] R. C. Alferness and P. S. Cross, "Filter characteristics of codirectionally coupled waveguides with weighted coupling," *IEEE J. Quantum Electron.*, vol. QE-14, pp. 843–847, Nov. 1978.
[34] R. C. Alferness, "Optical directional couplers with weighted coupling," *Appl. Phys. Lett.*, vol. 35, pp. 260–262, Aug. 1972.
[35] R. V. Schmidt and L. L. Buhl "An experimental 4×4 optical switching network," *Elect. Lett.*, vol. 12, pp. 652–653, Oct. 1976.
[36] H. F. Taylor, M. J. Taylor, and P. W. Bauer, "Electrooptic analog-to-digital conversion using channel waveguide modulators," *Appl. Phys. Lett.*, vol. 32, pp. 559–561, May 1, 1978.
[37] H. F. Taylor, "Guided wave electrooptic devices for logic and computation," *Appl. Opt.*, vol. 17, pp. 1493–1497, May, 1978.
[38] P. S. Cross, R. V. Schmidt, R. L. Thornton, and P. W. Smith, "Optically controlled two channel integrated optical switch," *IEEE J. Quantum Electron.*, vol. QE-14, pp. 517–580, Aug. 1978.
[39] R. A. Steinberg and T. G. Giallorenzi, "Performance limitations imposed on optical waveguide switches by polarization," *Appl. Opt.*, vol. 15, pp. 2440–2453, Oct. 1976.
[40] V. Ramaswamy, I. P. Kaminow, P. Kaiser, and W. G. French, "Single polarization optical fibers: Exposed cladding technique," *Appl. Phys. Lett.*, vol. 33, pp. 814–816, Nov. 1978.
[41] R. C. Alferness, "Polarization independent optical directional coupler switch using weighted coupling," *Appl. Phys. Lett.*, vol. 35, pp. 748–750, Nov. 1979.
[42] W. K. Burns, T. G. Giallorenzi, R. P. Moeller, and E. J. West, "Interferometric waveguide modulator with polarization independent operation," *Appl. Phys. Lett.*, vol. 33, pp. 944–947, Dec. 1978.
[43] H. P. Hsu and A. F. Milton, "Single-mode coupling between fibers and indiffused waveguides," *Elect. Lett.*, vol. 13, pp. 224–233, Apr. 1977.
[44] W. K. Burns and G. B. Hocker, "End-fire coupling between optical fibers and diffused channel guides," *Appl. Opt.*, vol. 16, p. 2048, 1977.
[45] Juichi, Noda *et al.*, "Single-mode optical-waveguide fiber coupler," *Appl. Opt.*, vol. 17, p. 2092–2096, July 1978.
[46] R. V. Schmidt, P. S. Cross, and A. M. Glass, "Optically induced crosstalk in $LiNbO_3$ waveguide switches," to be published in *J. Appl. Phys.*, Jan. 1980.

Low-Loss Polarization-Independent Electrooptical Switches at λ = 1.3 μm

LEON MCCAUGHAN

Abstract—Polarization-independent single-mode optical cross points with low loss have been fabricated for λ = 1.3-μm wavelength. Waveguide design, coupling strength equalization for the two polarizations, fabrication tolerances, and the performance of several Ti:LiNbO$_3$ directional-coupler switches are presented.

Two 2 × 2 switches have been made with fiber pigtails. The switches have an insertion loss of ~3- and ~ -14-dB crosstalk isolation with a 70-V operating voltage. A 4 × 3 switch had similar performance but higher insertion loss.

I. Introduction

POLARIZATION-INDEPENDENT Ti:LiNbO$_3$ directional-coupler switches have been demonstrated [1], [2] at short (0.63- and 0.83-μm) wavelengths. More recently, Ti:LiNbO$_3$ waveguides with fiber-waveguide-fiber insertion losses of 1-2 dB for both polarizations have been fabricated for λ = 1.3-μm wavelength light [3], [4]. In principle, therefore, it should be possible to design and build an efficient optical cross point which is compatible with λ = 1.3-μm single-mode fiber systems.

Waveguide fabrication conditions were first chosen to produce low-loss waveguides compatible with current single-mode optical fibers (approximately 4-μm waist). The coupler geometry was then designed around these waveguides to produce equal coupling coefficients for the TE and TM modes (polarizations).

II. Cross-Point Design

Previous polarization-independent devices [1], [2] have been fabricated with an unequal and nonintegral number of coupling lengths. Electrode segments with alternating polarity (reverse $\Delta\beta$) were used to improve the imperfect crossed state. The devices described in this paper were designed to have an interaction region exactly three coupling lengths long. This eliminates the need to electrically reconfigure the electrodes for the crossed state. Fabrication tolerances should be equally stringent for the two cases since the quality of the crossed state (whether electrically adjusted or not) depends on the control of both coupling coefficients.

The uncrossed (bar) state of the switch was achieved by mismatching the coupler with a uniform electric field. Maximum mismatch of the two polarizations does not occur at a common voltage, however, since their electrooptic coefficients are unequal. Tapered coupling (Fig. 1) can be used to suppress the side lobes in the response function by about 25 dB [1]. A Hamming function [5] was, therefore, used to spatially weight the coupling coefficients.

III. Fabrication

Approximately 6-μm-wide Ti-diffused channel waveguides were photolithographically fabricated (Fig. 1) in C-cut LiNbO$_3$. The Ti was deposited by electron-beam evaporation (~95-percent bulk metal density) to a depth of ~650 Å. The ridge height was measured with a profileometer before thermal diffusion at 1050°C for 6 h. These Ti strip dimensions and diffusion conditions have been shown [4] to produce waveguides with low loss for both polarizations (TE and TM).

A 2000-Å-thick SiO$_2$ buffer layer was deposited (by chemical vapor deposition) over the LiNbO$_3$ surface after the waveguide diffusion. Cr/Al electrodes were patterned directly over the waveguides (Cobra [6] configuration). The electrodes were extended over the portion of the waveguide bends adjacent to the coupler which were calculated to contribute significantly to the coupling strength.

Fibers were attached to the 2 × 2 switches with a small

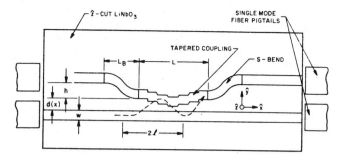

Fig. 1. Schematic diagram of Ti strip pattern for tapered coupling. Dashed line shows the effect of three coupling lengths on light intensity position.

amount of UV-curable cement applied to the fiber-LiNbO$_3$ joint. After the cement cured, the fiber was secured to a glass base with epoxy.

IV. Results

A. Waveguide Loss Minimization

Fiber-waveguide-fiber insertion loss has been measured as a function of the two Ti strip dimensions [4]. For ~6-μm-wide Ti strips diffused at 1050°C for 6 h, a range of Ti dopant thicknesses has been found which produces low insertion loss (~2 dB/2.5 cm) with respect to a single-mode fiber with a 4-μm waist (Fig. 2). For these diffusion conditions the loss minimum coincides with nearly maximum confinement of the fundamental mode (just beyond cutoff of the next order mode).

S-curve segments (Fig. 1) in the waveguide are necessary to separate the adjacent ports by about 300 μm in order to attach fiber pigtails (125-μm-diam cladding). Bend shape was described by a modified sine function

$$y(x) = \frac{dx}{L_B} + \frac{h}{2\pi} \sin \frac{2\pi x}{L_B} \quad (1)$$

where h and L_B are the height and length of the S-curve. To keep both the TE and TM bend losses below 0.1 dB/bend, S-curves with an offset of $h = 300$ μm were calculated [7] to require a length of $L_B = 8$ mm. Observed bend losses were 0.25 dB (TE) and 0.1 dB (TM).

The distribution of loss in the waveguides is listed in Table I. The coupling loss (due to fiber-waveguide mode mismatch) and Fresnel loss components are calculated [4] values. The data can be used to predict an average insertion loss of 2.6 dB for a 2 × 2 switch with an ideal directional coupler.

B. Coupling Coefficient Equalization

The most straightforward polarization-independent switch design incorporates an equal and odd number of coupling lengths for both the TE and the TM polarizations. Electrically induced uniform $\Delta\beta$ mismatch (with tapered coupling) is used to produce the uncrossed state [1].

Coupling length l and interwaveguide separation d (see Fig. 1) obey the exponential relationship [8]

$$l = l_0 \exp(d/\gamma). \quad (2)$$

There is no direct method, however, for predicting the constants l_0 and γ from the fabrication (geometric and diffusion) parameters. The coupling lengths of each polarization were,

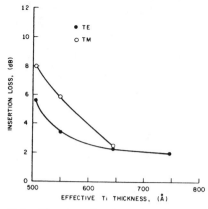

Fig. 2. Fiber-device-fiber insertion loss plotted as a function of Ti strip thickness (6-μm strip width). Thickness is corrected to bulk metal density. Waveguide length is 2.5 cm. Diffusion conditions: 1050°C for 6 h. Above 750 Å the waveguide supports higher order modes.

therefore, experimentally determined for a range of Ti thicknesses consistent with low-loss waveguides.

Relative optical power transfer (η) was measured as a function of the edge-edge Ti strip gap (d) and the interaction length (L). The characteristic coupling length $l(d)$ of low-loss waveguide pairs was calculated [8] from

$$\eta = \sin^2 [\kappa(d) L + \text{constant}] \quad (3a)$$

where

$$\kappa(d) = \pi/2l(d) \quad (3b)$$

is the coupling coefficient.

Coupling length is plotted versus interwaveguide gap in Fig. 3 for two Ti strip thicknesses and at two strip widths. For each fabrication condition the polarization pairs (TE and TM) of the curves intersect. A low-loss polarization-independent directional coupler can, therefore, be made by choosing the appropriate strip gap (and, therefore, the coupling length). The crossed state of the switch is produced by making the interaction length (L) and odd number of coupling lengths (l).

Trends in the coupling length data (Fig. 3) are consistent with a mode confinement model: increasing dopant levels (i.e., Ti strip dimensions) increases the refractive index difference and, therefore, the confinement of the light. The optical field overlap between the two waveguides is reduced, increasing the coupling length. For these tightly confined waveguides the change in coupling length with Ti concentration will be larger for the extraordinary (TM) polarization [9]. The marked increase in TM coupling length with Ti strip width is also consistent with the model.

The S-bends on either side of the interaction region of a directional coupler will contribute

$$S_B = \int \kappa[d(x)] \, dx \quad (5)$$

to the total coupling strength. The coupling coefficient $k(d)$ is derived from (2) and (3b). Using nominal values of l_0 and γ from Fig. 3, each pair of waveguide bends will contribute $S_{TE} \cong$

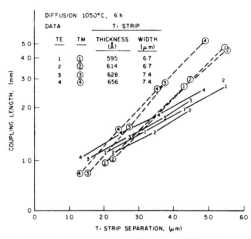

Fig. 3. Coupling length plotted as a function of Ti strip gap for several strip dimensions. Strip thickness is corrected to bulk metal density. Diffusion conditions were 1050°C for 6 h.

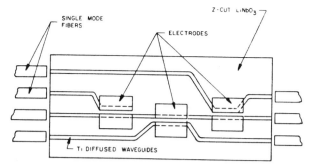

Fig. 4. Schematic representation of the polatization-independent 4 × 3 switch.

TABLE I
WAVEGUIDE LOSS DISTRIBUTION

	TE	TM
1. INSERTION LOSS		
— FRESNEL LOSS (2 FACES)	0.4	0.4 dB
— COUPLING MISMATCH (2 FACES)	1.0	1.4 dB
— PROPAGATION LOSS (2.5 cm.)	0.8	0.5 dB
2. BEND LOSS (2 BENDS)	0.5	0.2 dB
AVE. TOTAL		2.6 dB

0.2π and $S_{TM} \cong 0.03\pi$ to the coupling. This unequal contribution is compensated for by adjusting the strip gap (Fig. 3) to make the coupling length of the TE mode slightly larger than that of the TM mode.

V. DEVICE RESULTS

The 2 × 2 and 4 × 3 (Fig. 4) polarization-independent switches were fabricated with tapered directional couplers designed to be three coupling lengths long. Light crossover efficiency is plotted as a function of voltage in Fig. 5 (uniform $\Delta\beta$ mismatch) for one of the 2 × 2 switches. The switching curves (TE and TM) are symmetric around −3.5 V; the offset was caused by a small difference in the propagation constants of the two guides. The TM mode reaches its first crossover minimum at 40 V, but has a better extinction (−21.6 dB) at the second node (70 V). The crossover minimum of the TE mode is above 80 V.

No hysterisis or drift was observed in the response of the switch, either at short or long cycle times. When observed by others [2], drift in the light output at constant voltage has been attributed to an imperfect buffer layer.

Insertion loss and crosstalk isolation of the devices were measured relative to an input and an output fiber. The performance of the crossed and bar states are tabulated for each polarization of the three devices in Table II. The contribution of the waveguides to the average insertion losses of the 2 × 2 switches are about what was calculated in Table I. The insertion loss of the longer 4 × 3 switch is about 1 dB higher than is calculated from the estimated propagation loss (Table I).

Because of slightly oversized features on the photolithographic mask, the number of coupling periods was measured to be about 10 percent above their designed values ($L/l = 3$). As a consequence the incomplete coupling contributed to the insertion losses and reduced the crosstalk isolation in the crossed states. Incomplete (uniform $\Delta\beta$) waveguide mismatch (Fig. 5) lowered the crosstalk isolation and slightly raised the insertion loss of the uncrossed state of the TE mode (Table II). The observed crosstalk isolation of the TM mode is less than predicted (~−25 dB). Nonuniform mismatch would account for the reduction.

Cementing single-mode fibers to the four ports of the 2 × 2 switches added less than 0.2 dB to the measured fiber-device-fiber insertion losses (Table II). Experience showed that pairs of fibers (125-μm-diam cladding) could not be easily cemented to the same side of the LiNbO₃ substrate if separated by less then 300 μm. A 2 × 2 switch with fiber pigtails mounted in an aluminum housing is shown in Fig. 6.

VI. DISCUSSION

The thickness of the Ti strip is the most difficult to control of the fabrication parameters. Fig. 3 can be used to estimate the accuracy necessary to produce a directional coupler with, say, a 99-percent efficient (−20-dB) crossover. Consider a device made up of three 1-mm coupling lengths. From (2a) and the expansion

$$\Delta(L/l) \cong -(L/l^2)\Delta l \qquad (6)$$

a −20-dB crosstalk isolation translates to about 42-μm permissible error in the coupling length l. This equates to $\Delta\tau \simeq 15$-Å allowed error in the Ti thickness for $l \cong 1$ mm. This restriction can be relaxed while maintaining the same interaction (and therefore electrode) length if the device is made with one coupling period. For example if $L = l = 3$ mm, to retain a −20-dB crosstalk isolation, $\Delta\tau \lesssim 50$ Å.

Fabricating a polarization-independent switch with one coupling period (same interaction length) would also lower the switching voltage (for uniform mismatch of a tapered coupler) by about 30 percent [5]. Coupling lengths can be equalized ($l_{TE} = l_{TM}$) at larger values by using waveguides

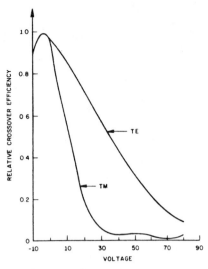

Fig. 5. Measured crossover efficiency versus applied (uniform $\Delta\beta$) voltage for a 2 × 2 directional-coupler switch with tapered coupling.

TABLE II
POLARIZATION-INDEPENDENT SWITCH RESULTS

SWITCH TYPE	T(Å)[1]	L(cm)[2]	V_0[3] (VOLTS)		CROSSED STATE			UNCROSSED (BAR) STATE		
					TE	TM	AVE	TE	TM	AVE
2×2	632	2.3	70	INSERTION LOSS[4] (dB)	3.3	3.7	3.5	3.1	3.3	3.2
				-WAVEGUIDE	2.4	3.3		2.4	3.3	
				-COUPLER	0.9	0.4		0.7	0.0	
				CROSSTALK ISOLATION[5] (dB)	-9.7	-13.9	-11.3	-10.9	-24.9	-13.7
2×2	618	2.3	80	INSERTION LOSS (dB)	3.0	3.0	3.0	2.4	3.1	2.7
				-WAVEGUIDE	2.1	2.8		2.1	2.8	
				-COUPLER	0.9	0.2		0.3	0.3	
				CROSSTALK ISOLATION (dB)	-9.4	-16.3	-11.6	-13.6	-15.1	-14.3
4×3	614	5.5	60	INSERTION LOSS (dB)	5.5	6.2	5.8	4.7	5.5	5.1
				-WAVEGUIDE	4.0	5.4		4.0	5.4	
				-COUPLER	1.5	0.8		0.7	0.1	
				CROSSTALK ISOLATION (dB)	-9.3	-13.1	-10.8	-12.0	-20.7	-14.5

NOTES
(1) Ti STRIP THICKNESS (NORMALIZED TO BULK METAL DENSITY)
(2) OVERALL DEVICE LENGTH
(3) OPERATING VOLTAGE (BAR STATE)
(4) FIBER-DEVICE-FIBER
(5) WAVEGUIDE PLUS COUPLER

Fig. 6. Photograph of 4 × 3 optical switch with attached fibers. Refer to Table II for performance details.

with less well-confined modes [10]-[12]. Unfortunately weaker mode confinement will increase the waveguide insertion loss both in straight guides (Fig. 2) and in bends [7]. The improvement in the switching voltage may also be reduced because a larger mode will have poorer optical/electrical field overlap.

The measured switch response to a uniform $\Delta\beta$ field (Fig. 5) was poorer than would have been first estimated from earlier results at shorter wavelengths [1], [2]. The electrical/optical field overlap calculations of Marcuse [13] and the measured mode dimensions[1] were used to compute the electrooptic phase mismatch. The calculations predict the first crossover minimum (Fig. 5) at 31 V for the TM mode (compared to the observed 40 V) and 89 V for the TE mode. The calculated

[1]Mode profiles were measured as per [4]. Amplitude profiles were fit to a Gaussian (half-width σ_3) parallel to the LiNbO$_3$ surface and 2 half-Gaussians (half-widths σ_1 and σ_2) perpendicular to the surface. Measured mode dimensions (half-width at e^{-1}) were $\sigma_1 = 1.5$ μm, $\sigma_2 = 3.3$ μm, and $\sigma_3 = 3.7$ μm.

electric field is equivalent to that of a pair of parallel plates with a 6-μm separation and an 8-V potential difference.

VII. CONCLUSIONS

Several of the first polarization-independent optical switches for the $\lambda = 1.3$-μm wavelength have been fabricated with fiber pigtails. The better of the two 2 × 2 switches have a fiber-device-fiber insertion loss of ~3 dB, which is substantially smaller than previously reported [14], [15].

The crosstalk isolation all of the devices was about -12--14 dB and was not symmetric for the crossed and uncrossed states. With minor improvements in the design and in the fabrication accuracy of the devices, the insertion loss should be reduced to 2-2.5 dB and the crosstalk isolation improved to -25 dB. It should be remembered that because optical detectors are square-law devices, a -25-dB optical isolation translates to -50-dB electrical power isolation.

The operating voltages for the switches were larger than have been observed at shorter wavelengths [1], [2], where the waveguides are smaller and closer to the surface electrodes. It may be possible to locally constrict the waveguide modes (TE and TM) in the coupler region without sacrificing too much fiber-device insertion loss.

ACKNOWLEDGMENT

The author wishes to thank E. J. Murphy for fabricating the devices and R. Meacham who assisted with the measurements.

REFERENCES

[1] R. C. Alferness, "Polarization-independent optical directional coupler switch using weighted coupling," *Appl. Phys. Lett.*, vol. 35, pp. 748–750, 1979.
[2] O. G. Ramer, C. Mohr, and J. Pikulski, "Polarization-independent optical switch with multiple sections of $\Delta\beta$ reversal and Gaussian taper functions," *IEEE J. Quantum Electron.*, vol. QE-18, pp. 1760–1767, 1982.
[3] R. C. Alferness, V. K. Ramaswamy, S. K. Korotky, M. D. Divino, and L. L. Buhl, "Efficient single-mode fiber to titanium diffused lithium niobate waveguide coupling for $\lambda = 1.32$ μm," *IEEE J. Quantum Electron.*, vol. QE-18, pp. 1807–1813, Oct. 1982.
[4] L. McCaughan and E. J. Murphy, "Influence of temperature and Ti dimensions on fiber-Ti:LiNbO$_3$ waveguide insertion loss at $\lambda = 1.3$ μm," *IEEE J. Quantum Electron.*, vol. 19, pp. 131–136, 1983.
[5] R. C. Alferness and P. S. Cross, "Filter characteristics of codirectionally-coupled waveguides with weighted coupling," *IEEE J. Quantum Electron.*, vol. QE-14, pp. 843–847, 1978.
[6] M. Papuchon, Y. Combemale, X. Mathieu, D. B. Ostrowsky, L. Rieber, A. M. Roy, B. Sejourne, and M. Werner, "Electrically switched optical directional coupler: Cobra," *Appl. Phys. Lett.*, vol. 17, pp. 289–291, 1975.
[7] W. J. Minford, S. K. Korotky, and R. C. Alferness, "Low-loss Ti:LiNbO$_3$-waveguide bends at $\lambda = 1.3$ μm," *J. Quantum Electron.*, vol. 18, pp. 1790–1794, 1982.
[8] R. C. Alferness, R. V. Schmidt and E. H. Turner, "Characteristics of Ti-diffused LiNbO$_3$ optical directional couplers," *Appl. Opt.*, vol. 18, pp. 4012–4016, 1979.
[9] M. Minakata, S. Saito, M. Shibata and S. Miyazawa, "Precise determination of refractive index changes in Ti:LiNbO$_3$ waveguides," *J. Appl. Phys.*, vol. 49, pp. 4677–4682, 1978.
[10] C. H. Bulmer and W. K. Burns, "Polarization characteristics of LiNbO$_3$ channel waveguide directional couplers," *IEEE J. Lightwave Tech.*, vol. LT-1, pp. 227–235, 1983.
[11] R. C. Alferness, private communication.
[12] L. McCaughan, unpublished.
[13] D. Marcuse, "Optimal electrode design for integrated optics modulators," *IEEE J. Quantum Electron.*, vol. QE-18, pp. 398–401, 1982.
[14] M. Kondo, Y. Ohta, M. Fujiwara, and M. Sakaguchi, "Integrated optical switch matrix for single mode fiber networks," *IEEE J. Quantum Electron.*, vol QE-18, pp. 1747–1752, 1982.
[15] O. G. Ramer, C. C. Nelson, C. M. Mohr, "Experimental integrated optic circuit losses and fiber pigtailing of chips," presented at the Third Int. Conf. Integrated Opt. Fiber Commun. (San Francisco, CA), Apr. 27–29, 1981.

STRICTLY NONBLOCKING 8 × 8 INTEGRATED OPTICAL SWITCH MATRIX

P. GRANESTRAND
B. STOLTZ*
L. THYLEN
K. BERGVALL*

Ericsson Telecom
S-126 25 Stockholm, Sweden

*RIFA AB
S-163 81 Stockholm, Sweden

3rd June 1986

W. DÖLDISSEN
H. HEINRICH
D. HOFFMANN

Heinrich Hertz Institut für Nachrichtentechnik GmbH
D-1000 Berlin, W. Germany

Indexing terms: Integrated optics, Optical switching

We report on a strictly nonblocking 8 × 8 integrated optics switch matrix in Ti : LiNbO$_3$. The matrix comprises 64 directional couplers on one chip. Insertion losses less than 7 dB have been measured. The extinction ratio has an average value of 30·5 dB for the directional couplers.

Introduction: The integrated optical switch matrix is expected to play an important role in future telecommunication systems. The fact that the signal is switched in optical form makes it possible to route information with bandwidths exceeding 1 THz such as frequency-multiplexed channels in coherent communication systems. Switch matrices up to 4 × 4 in size have been implemented in Ti : LiNbO$_3$.[1–3] We report on a 8 × 8 switch matrix; it is strictly nonblocking and comprises 64 2 mm-long directional couplers.

Fig. 1 shows a schematic view of the totally symmetric matrix. This is the so-called busbar structure. It is a straightforward translation of the electrical square array to the optical domain.

The crosspoints are, as previously mentioned, directional couplers. They are of stepped $\Delta\beta$ reversal type, each directional coupler having three electrodes connected (Fig. 1) to make both $\Delta\beta$ reversal operation and operation with equal voltages applied to the electrodes (unioperation) possible. By driving the directional couplers in $\Delta\beta$ reversal operation for the cross-state and unioperation for the barstate it is possible to keep the absolute values of the drive voltages down. The short directional coupler length and the ensuing relatively high drive voltages might otherwise cause electrical breakdown. To use this type of operation advantageously it is required that the physical length L of the directional coupler is between one and two coupling lengths l_c.[4]

The matrix sample described in Reference 5 did not meet this requirement. In this letter complete measurement results are presented on a switch matrix chip which fulfils $l_c < L < 2l_c$ and has considerably lower insertion loss than the device of Reference 5.

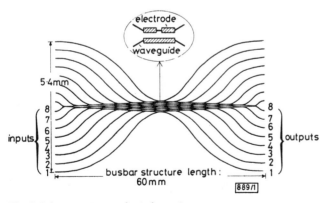

Fig. 1 *Schematic picture of switch matrix structure*

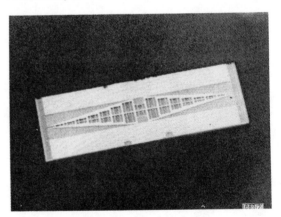

Fig. 2 *Photograph of device*

Reprinted with permission from *Electron. Lett.,* vol. 22, no. 15, pp. 816–818, July 17, 1986.
Copyright © 1986, The Institute of Electrical Engineers, U.K.

Design data: The matrix is designed for operation at a wavelength of 1·3 μm. The LiNbO$_3$ chip uses the Z-cut, Y-propagating configuration. The lateral offsets of the waveguides are implemented by circular arcbends and straight lines not parallel with the Y-axis. The radius of curvature is 30 mm. The 60 mm-long matrix has 360 μm between incoming fibres and 50 μm between directional couplers.

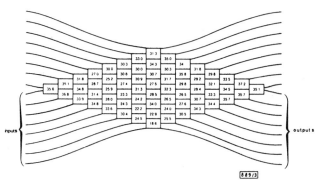

Fig. 3 *Experimental results: extinction ratio for each directional coupler*

Maximum 18·6, minimum 37·2, average 30·5, standard deviation 4·1 dB

Between adjacent directional couplers, guard electrodes are inserted to reduce electrical crosstalk, as a computer simulated by solving the pertinent electrostatic problem.[6] The electrodes are extended longitudinally outside the 2 mm interaction length to a total length of 3 mm.

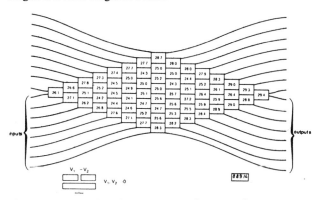

Fig. 4 *Experimental results: cross-state voltages in volts*

$(V_1 + V_2)/2$ is shown for each directional coupler; average 26·4 V, standard deviation 1·6 V

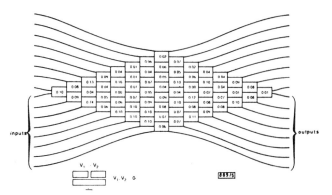

Fig. 5 *Experimental results: cross-state voltages: degree of imbalance*

$(V_1 - V_2)/(V_1 + V_2)$ is shown for each directional coupler; average −0·07, standard deviation 0·09

The device is fabricated by standard indiffusion of Ti strips of 5 μm width and 850 Å thickness into the LiNbO$_3$ substrate. The diffusion depth is 4 μm. An SiO$_2$ buffer layer is deposited between the waveguides and the electrodes. Fig. 2 shows a photograph of the chip.

Experimental results: In Fig. 3 the extinction ratio is shown for each switch. The worst-case value, 18·6 dB, is due to a slight damage in the waveguides for the corresponding directional coupler.

Figs. 4 and 5 show the cross-state voltages for each directional coupler. The matrix is relatively uniform in this respect. The higher voltages at the edges of the matrix are due to the fact that the guard electrodes between the directional couplers inside the matrix make these directional couplers (inside the matrix) more efficient.[5]

In Fig. 5 it can be seen that the voltages on the two sections in Δβ operation are usually not symmetrical. The reason for the imbalance is probably a longitudinal misalignment of the electrode mask.

The drive voltages in barstate have an average value of 18·4 V and a standard deviation of 2·4 V. For cross-states the average value is 26·4 V and the standard deviation is 1·6 V.

The insertion loss for different signal paths is shown in Fig. 6. The results imply an excess loss of 0·1 dB per directional coupler. The drastic improvement in insertion loss compared to Reference 5 is mainly due to an increase in buffer layer thickness and also to some extent lower bending losses.

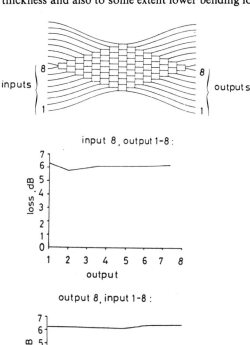

Fig. 6 *Insertion losses*

Losses: longest way (input 8–output 8) 6·8 dB, shortest way (input 1–output 1) 5·3 dB

The light source is a semiconductor laser and polarisation-maintaining fibres are used at the input and output ends of the chip. The insertion loss values include Fresnel loss.

Summary: An 8 × 8 strictly nonblocking switch matrix has been fabricated in Ti:LiNbO$_3$. The 60 mm-long device showed relatively low loss and uniform switch voltages. By optimising the design and processing, including, for example, an AR coating, it should be possible to reduce the loss and also to increase uniformity across the matrix.

References

1 MCCAUGHAN, L., and BOGERT, G. A.: '4 × 4 strictly nonblocking integrated Ti:LiNbO$_3$ switch array'. Technical digest, Conference on optical fiber communication (Optical Society of America, Washington, DC, 1985), Paper TuQ20
2 KONDO, M., TAKADO, N., KOMATSU, K., and OTHA, Y.: '32 switch-elements integrated low-crosstalk LiNbO$_3$ 4 × 4 optical matrix switch'. IOOC-ECOC'85, p. 361
3 NEYER, A., MEVENKAMP, W., and KRETCHMANN, B.: 'Nonblocking 4 × 4 switch array with sixteen X-switches in Ti:LiNbO$_3$'. IGWO'86, Paper WAA2
4 KOGELNIK, H., and SCHMIDT, R.: 'Switched directional couplers with alternating $\Delta\beta$', *IEEE J. Quantum Electron.*, July 1976, **QE-12**
5 GRANESTRAND, P., THYLEN, L., STOLTZ, B., BERGWALL, K., DÖLDISSEN, W., HEINRICH, H., and HOFFMANN, D.: 'Strictly non-blocking 8 × 8 integrated optics switch matrix in Ti:LiNbO$_3$'. IGWO'86, Paper WAA3
6 THYLÉN, L., and GRANESTRAND, P.: 'Integrated optic electro-optic device electrode analysis: the influence of buffer layers', *J. Opt. Commun.*, 1986, **1**, pp. 11–14
7 DUCHET, C., and MARTIN, R.: 'Electro-optic modulator in Ti:LiNbO$_3$ with very low drive voltage', *Electron. Lett.*, 1984, **20**, pp. 567–568

Performance of Two 4 × 4 Guided-Wave Photonic Switching Systems

JOHN R. ERICKSON, GAIL A. BOGERT, ROGER F. HUISMAN, AND RON A. SPANKE

Abstract—Data from two photonic switching systems is presented, the first is a 4 × 4 crossbar, and the second is a 4 × 4 passive splitter/active combiner broadcast switch. The 4 × 4 crossbar has operated continuously and reliably for 18 months with no noticeable performance degradation. The 4 × 4 passive splitter/active combiner has operated for six months and yields excellent bit-error-rate performance at 1.7 Gbits/s, showing that the influence of optical crosstalk in a small guided-wave optical switching system does not significantly affect the BER. As guided-wave optical switches advance toward products, more extensive testing must be performed.

I. Introduction

INTEGRATION of directional coupler switches on substrates of Ti:LiNbO$_3$ has progressed to the point of considering photonic switching systems based on this technology. A critical step in transferring this technology from a laboratory interest to a useful product is the testing of Ti:LiNbO$_3$-based switching subsystems. This paper presents data from testing of two guided-wave photonic switching systems.

The first system is a 4 × 4 crossbar switch that is in use in an advanced technology display [1]. This switch has consistently permuted inputs to outputs for 18 months with no noticeable degradation in the switching performance.

A 4 × 4 passive splitter/active combiner (PSAC) broadcast switch is the second switch tested. The data taken include: initial contrast ratios, drift exhibited in individual couplers, and bit-error-rate (BER) performance at 1.7 Gbits/s [2]. This switch architecture will be used in a 1.7 Gbit/s time-multiplexed photonic switching experiment [3].

We briefly conclude with suggestions for future testing.

II. Ti:LiNbO$_3$ Photonic Switching System Overview

Directional couplers in Ti:LiNbO$_3$ are 2 × 2 optical switches that operate by using the linear electrooptical effect to affect the coupling between two waveguides [4]. Given a 2 × 2 photonic switch as a building block, networks of various sizes and topologies have been proposed and fabricated [5]–[8], including some clever approaches

Manuscript received November 17, 1987; revised May 5, 1988.
J. R. Erickson, R. F. Huisman, and R. A. Spanke are with AT&T Bell Laboratories, Naperville, IL 60566.
G. A. Bogert is with AT&T Bell Laboratories, Allentown, PA 18103.
IEEE Log Number 8822446.

to architect for improved signal-to-noise ratio (SNR) [9], [10]. Several switching systems based on these and other architectures have been proposed and a few have reached realization [1], [11], [12].

As system applications increase, it becomes important to supplement the knowledge of materials and devices with real system performance tests of real Ti:LiNbO$_3$ switching systems.

III. A Photonic Switching Demonstration Display

To demonstrate the feasibility of photonic switching for video, we constructed a photonic switching demonstration display that routes four camera images to four monitors [1]. All switching in the display is done through a 4 × 4 Ti:LiNbO$_3$ crossbar switch shown schematically in Fig. 1.

A. 4 × 4 Crossbar Design and Fabrication

The crossbar switch matrix consists of 16 individually addressable directional coupler unit cells integrated on a single substrate. The directional couplers were interconnected with low-loss s bends [13]. The devices were fabricated in z-cut y-propagating LiNbO$_3$ and were designed for the TM polarization at $\lambda = 1.3$ μm. The waveguides were formed by diffusing 6.5 μm wide titanium stripes at 1050°C for 6 h. A LPCVD SiO$_2$ buffer layer was deposited and 5000 Å thick aluminum electrodes photolithographically defined. We used the split electrode in the uniform $\Delta\beta$ configuration for the bar state and the reverse $\Delta\beta$ configuration for the cross state. The directional couplers were ~5.5 mm long which corresponded to 1.3 coupling lengths.

The device was packaged for convenient electrical and optical interconnection. The electrical connections were made by mounting and wirebonding the device on a printed circuit board which was compatible with standard card edge connectors. The output fiber array was a silicon V-groove array of standard single-mode fibers [14]. For the input array, flattened polarization maintaining fiber was used for ease of alignment [15], [16]. The fibers were lensed and aligned in standard laser packages.

B. Long-Term System Performance of a 4 × 4 Crossbar

Over its 18 months of operation, we have recorded the contrast ratio nearly every month for each of the 16 couplers in the crossbar array. For the crossbar, the contrast

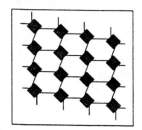

Fig. 1. 4 × 4 crossbar architecture using directional couplers.

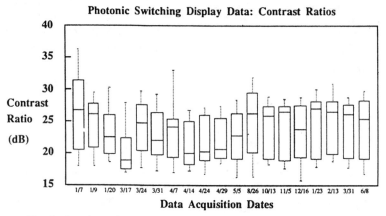

Fig. 2. 4 × 4 crossbar contrast ratio distributions (16 couplers per boxplot).

ratio of the coupler at the intersection of the ith row and jth column in the crossbar is found by launching power into input i, putting all couplers in the cross state and coupler i, j in the bar state, measuring power at output j, then putting coupler i, j in the cross state and again measuring power at output j. The ratio (in decibels) of the output power measurements is the bar-to-cross state contrast ratio for the lower output of each coupler. Fig. 2 summarizes the data. Each boxplot represents one reading of the contrast ratios of 16 directional couplers. The center line of each box is the median contrast ratio, the box upper half is the upper quartile of data and the box lower half is the lower quartile, with whiskers extending to the best and worst contrast ratios. The boxplots in Fig. 2 show that although there is variation between the best and worst contrast ratios in any individual reading and variation between readings, there is no monotonic decrease in the switch performance over a year and a half of continual operation. Optical power of -7 dBm per channel is continually traversing the switch and the switch is continually being reconfigured every few seconds.

IV. Performance of a 4 × 4 Passive Splitter/Active Combiner Switch

The crossbar architecture, while providing nonblocking point-to-point operation with the use of a trivial algorithm, does not have convenient broadcast capability. We next report on a 4 × 4 Ti:LiNbO$_3$ device which has full broadcast capabilities because it uses a passive splitter/active combiner (PSAC) architecture (see Fig. 3) [2], [17].

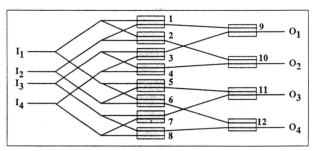

Fig. 3. A 4 × 4 passive splitter/active combiner architecture.

A. Design and Fabrication of the PSAC

Several alternatives are available for the splitters: Y's, passive 3 dB directional couplers, and active couplers tuned to the 3 dB point. Significant electrical control will be required for the combiner stage, so avoiding it in the splitter stage will reduce the number of electrical contacts necessary. We chose to utilize Y's, since they provide a passive, fabrication-tolerant, high-quality splitting ratio.

The coupler states required to complete a given permutation of inputs to outputs are not as readily recognizable as those of the crossbar. Fig. 4 shows coupler states required for routing different paths. For example, routing input 1 to output 1 requires that couplers 1 and 9 both be in the cross state. Each path involves two directional couplers.

The device was fabricated using the process detailed in Section III-A. The Y's within the array (used to supply the splitting function) were made with reflected s bends. The intersecting waveguides contained single-Δn inter-

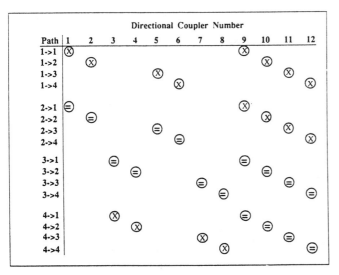

Fig. 4. Coupler states for designated connections.

TABLE I
INSERTION LOSS (dB): FIBER-WAVEGUIDE-FIBER

		OUTPUTS			
		1	2	3	4
INPUTS	1	11.9	13.9	13.1	13.7
	2	12.8	13.0	13.1	11.8
	3	12.3	12.5	13.3	12.9
	4	13.1	12.1	13.7	11.6

section regions. The required angle of intersection for negligible crosstalk and insertion loss was taken from previous work [18].

B. PSAC Loss

Fiber-to-fiber insertion loss measurements of the device were made. The fibers were mounted in capillary tubes and placed against the edge of the crystal. Index matching fluid was used. The values for the insertion loss are shown in Table I. The average value is 12.8 dB. The broadcast capability of this design requires two 3 dB splitters in each path. This contributes 6 dB of the 12.8 dB total insertion loss.

Features to the side of the array provided a means of determining the loss of the individual components within the array. Straights averaged 2 dB, Y's were 1.35 dB, and the smallest angle crossovers (7°) exhibited 0.3 dB. Additional loss was suffered by couplers with electrodes patterned above them. Whether this is due to loading or a strain effect in the gap region is being investigated.

Two additional arrays were fabricated using a modified waveguide design. The design contained lower loss Y's (0.23 dB excess loss). The average array loss for the redesigned arrays was 11.0 dB (6 dB inherently due to broadcasting). Elimination of metal loading should result in arrays with improved loss values of ~9.5 dB.

C. PSAC Initial Contrast Ratios and Voltages

The device contains 12 directional couplers each with reverse $\Delta\beta$ electrodes and a separate ground. The cross state was achieved by applying reverse polarities to the pads. Uniform $\Delta\beta$ voltage was utilized for the bar state. To determine the efficiency of each coupler, both the bar-to-cross and cross-to-bar state contrast ratios were measured. The contrast ratios for all but one of the couplers, both cross and bar, were measurement limited at >35 dB. For one coupler in one of the arrays, full flexibility of the voltage polarities was required, providing measurement-limited cross-to-bar state contrast ratios and 25-30 dB bar-to-cross state contrast ratios.

The voltages applied to the device to achieve these values for contrast ratio are shown in Table II. The various directional couplers have been assigned the numbers shown in Fig. 3.

D. Applications

Once designed and fabricated, the devices were mounted on printed circuit boards compatible with standard edge connectors. Single-mode fiber arrays were attached using silicon V-groove technology. One of the devices was then integrated into a video switching display, where four 90 Mbit/s digital video signals from cameras are switched to four monitors. The electrodes are driven by digital-to-analog converters and connections are determined by microcomputer software. In all connections of inputs to outputs, including broadcast connections, there is no noticeable picture degradation on the monitors. This device has run reliably in the display for six months.

In addition, this device is the forerunner to a device that will be used in a 1.7 Gbit/s time-multiplexed photonic switching experiment [3].

E. Contrast Ratio Drift under Constant Voltage

Since some potential applications of Ti:LiNbO$_3$ include protection switching and infrequent reconfiguration in a cross connect, we tested one 4 × 4 PSAC for contrast ratio over time with voltage held constant. Normal operation of the PSAC allows cross-to-bar and bar-to-cross state contrast ratios for one output. First the input polarizations and coupler voltages were adjusted to peak the contrast ratios of each coupler. These peaked values are represented in the boxplots of Figs. 5 and 6 above the "1/8" reading. Note the better operation of the cross-to-bar state contrast ratios. The next day, "1/9," the plots show a degradation in the contrast ratios but subsequent readings throughout the month indicate there is no monotonic decrease in performance. This supports the data taken from the crossbar switch cited above, although the PSAC switch was left static between measurements whereas the crossbar was constantly switching. Since we did not adjust input polarization during the month, there may be degradation due to polarization drift in the single-mode fiber connecting the lasers to the switch. A scrutiny

TABLE II
ELECTRODE VOLTAGES

Directional Coupler Number	Uniform Δβ		Reverse Δβ	
	$+V_1$	$+V_2$	$+V_3$	$-V_4$
1	14.6	14.5	9.9	-6.2
2	13.6	13.4	8.9	-6.4
3	12.2	11.2	6.8	-8.9
4	13.3	12.2	8.9	-7.4
5	14.7	13.7	9.5	-7.1
6	14.2	14.0	11.4	-11.0
7	13.1	13.3	8.0	-6.9
8	14.2	13.9	9.3	-6.7
9	3.0	-11.0	17.0	7.0
10	13.6	13.8	10.6	-10.0
11	12.9	12.2	10.0	-10.8
12	14.4	11.5	8.5	-9.8
Average (Excluding Coupler No. 9)	13.7	13.1	9.3	-8.3

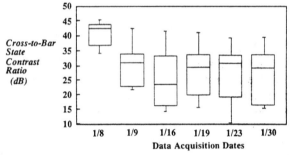

Fig. 5. Cross-to-bar state contrast ratios for 4 × 4 PSAC with no voltage adjustment.

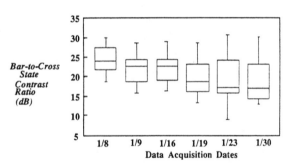

Fig. 6. Bar-to-cross state contrast ratio drift for 4 × 4 PSAC with no voltage adjustment.

of the individual coupler data shows, however, that some couplers drifted significantly more than others that received power from the same laser. Even from the box-plots, we see that the best coupler contrast ratio (top whiskers) varied little throughout the month of testing. This unequal distribution of drift suggests that further refinement of the processing may decrease the tendency to drift in the more unstable of the elements in an array. Preliminary data on the most recent fabrications of this topology seem to confirm that stable couplers can be more uniformly fabricated.

F. Output Power Drift over Time

Shifting to individual drift-over-time measurements for each coupler, we measured received power change over time after having changed the voltage on a single coupler. Data for one coupler in the array are shown here. The whole switch was initially left in a static state for a time equal to the measurement time. The voltage on a single coupler was changed, thereby changing the state of the coupler from cross to bar or bar to cross, depending on the reading taken. Four such readings were taken: cross and bar state drift with optical power of -7 dBm directed to one coupler input, then the other coupler input. The output monitored depends on which coupler is tested; for example, Fig. 3 shows that only the one output of any coupler can be measured, but that light can be directed to either input. The following data are meant to show only power drift and were taken with nonoptimized voltages at the outset, thus accounting for contrast ratios lower than those given in the previous data.

Fig. 7 shows the 5 min drift of coupler 4. As a typical case, coupler 4 exhibits some drift in the first few seconds, but little drift thereafter.

Examining the long-term drift, we performed the same test but over 24 h. As shown in Fig. 8, coupler 4 shows some meandering but eventual leveling in its power output curve.

In the array of 12 couplers, coupler 4 represents a typical case of drift severity with some couplers more stable and some with worse drift characteristics. Coupler 8, singular in its drift severity, appears to drift without bound over a full 24 h period. Severe drift may be attributed to local contamination.

G. Bit-Error-Rate Performance

The final tests were taken to simulate the performance of the switch in a system carrying digital data. We give here the results of four tests of the BER performance of the switch with a pseudorandom bit stream of $2^{15} - 1$ bits at 1.667 Gbits/s (~1.7 Gbits/s). The four tests are described in the next section, with the data explained in following sections.

1) Experimental Setup: Four sets of BER measurements were taken. We label each in capital letters for future use in presenting the data.

BASELINE: Fig. 9 shows the configuration used in performing the 1.7 Gbit/s BER tests. Baseline (reference) tests were first run to determine the performance of all components without the optical switch. Laser transmitter 4 (Xtmr4 in Fig. 9) was connected directly to the input of the variable optical attenuator. Optical power was reduced toward the receiver noise as the BER was recorded at various levels.

NO-INTERFERENCE: The optical switch was inserted for the second set of BER tests. Input 4 of the switch was connected to transmitter 4 and output 1 to the variable optical attenuator. The switch was configured to connect input 4 to output 1 and the BER was recorded at various optical power levels with no other signals within the switch.

NONSWITCHED INTERFERENCE: Three additional lasers driven from the 1.7 Gbit/s pulse pattern gen-

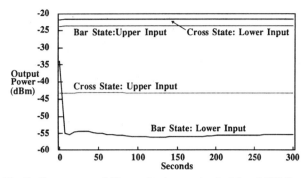

Fig. 7. Output power drift over 5 min, coupler 4 of 4 × 4 PSAC.

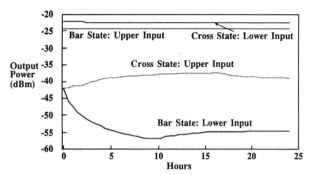

Fig. 8. Output power drift over 24 h, coupler 4 of 4 × 4 PSAC.

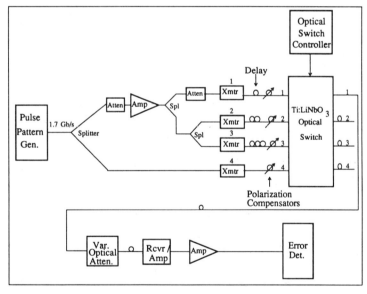

Fig. 9. Bit-error-rate test configuration.

erator through splitters were connected via various delays (loops in Fig. 9) to the three remaining inputs of the switch. The switch was configured to connect input 4 to output 1 (a worst case crosstalk path) and interference inputs 3, 2, and 1 to outputs 4, 3, and 2, respectively. The BER was recorded at various optical power levels with these interference signals passing through the switch.

SWITCHED INTERFERENCE: Finally the effect of switching interference was measured. The three interference channels were continuously switched to the outputs (through all possible combinations) while the BER from path 4 → 1 was recorded at various optical power levels.

2) Bit-Error-Rate Power Penalty for Crosstalk: Fig. 10 shows the BER data for the 4 × 4 PSAC at 1.7 Gbits/s. The horizontal axis is the received power in dBm, the vertical axis is the log of the BER. The "*" character plots the BASELINE, the "X" marks the NO-INTERFERENCE measurement, the "+" marks the NONSWITCHED INTERFERENCE measurement, and the "O" marks the SWITCHED INTERFERENCE measurement.

In the error rates of interest for communication systems

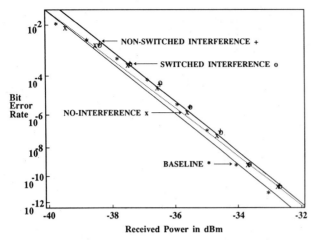

Fig. 10. Bit-error-rate performance of 4 × 4 PSAC.

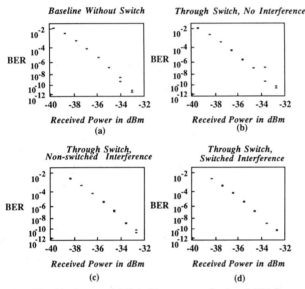

Fig. 11. Range of drift in bit-error-rate for 4 × 4 PSAC.

These BER data were taken on a switch where the measured path had been established for over an hour. We do not see here the BER drift immediately after a switching transition in the signal path, but rather settled-state BER drift characteristics.

The baseline measurement shows some settling occurring in the system in the -34 dBm measurement, likely due to migration towards thermal equilibrium in the measurement instruments, laser diode, and avalanche photodiode. The high-power, low-BER measurements were taken first, progressing toward the low-power, high-BER measurements. Introduction of the switch in the graph (b) of Fig. 11 again exhibits some settling, and some noise interference outside the system causing the orders-of-magnitude jump near -34 dBm. Once the switch is in place and the interference introduced, there is no order-of-magnitude shift in the BER performance over the hour-long measurement period.

V. FUTURE TESTING

Although the testing reported here shows some encouraging trends pointing toward the use of Ti:LiNbO$_3$ in systems, this testing is far from comprehensive. To use Ti:LiNbO$_3$ with confidence we must test Ti:LiNbO$_3$ arrays and systems under varying temperature and humidity and under increasingly rigorous switching frequencies. We must also increase the number of testing samples to a number large enough for statistical analysis. This requires continued efforts in design, fabrication, and system testing.

ACKNOWLEDGMENT

We would like to acknowledge the contributions of M. Dautartas, R. J. Holmes, C. T. Kemmerer, Y. S. Kim, and S. H. Kravitz for process development; F. Sandy, L. Cheese, and R. Higgins for fabrication of the devices; K. Bahadori and Y. Chen for fiber array attachment, and F. T. Stone and W. A. Payne for project support.

R. O. Miller and A. F. Ambrose provided 1.7 Gbit/s lasers with polarization-maintaining fiber pigtails and F. A. Serino packaged these lasers into transmitter modules. Without their timely assistance, no high-speed testing could have been done.

REFERENCES

[1] J. R. Erickson et al., "A photonic switching demonstration display," in *Top. Meet. Photon. Switching, Tech. Dig. Series*, OSA, vol. 13, 1987, pp. 30–32.
[2] R. A. Spanke, "Architectures for large nonblocking optical space switches," *IEEE J. Quantum Electron.*, vol. QE-22, p. 964, June 1986.
[3] J. R. Erickson et al., "A 1.7 gigabit-per-second, time-multiplexed photonic switching experiment," *IEEE Communications*, vol. 25, pp. 56–58, May 1987.
[4] R. C. Alferness, R. V. Schmidt, and E. H. Turner, "Characteristics of Ti-diffused lithium niobate directional couplers," *Appl. Opt.*, vol. 18, p. 4012, 1979.
[5] G. A. Bogert, E. J. Murphy, and R. T. Ku, "A low crosstalk 4 × 4 Ti:LiNbO$_3$ optical switch with permanently attached polarization-maintaining fiber arrays," in *Proc. Top. Meet. Integrated Guided-Wave Opt.*, Atlanta, GA, Feb. 1986, pp. PDP 3.1–3.3.

($<10^{-9}$), we see that traversing the switch required an increase in received power of up to 0.5 dB for the same BER, but the interfering channels in measurements curves "+" and "O" of Fig. 10 required only a few tenths of a decibel of additional power increase; the crosstalk in this switch did not seriously degrade the BER performance. As expected with the PSAC architecture, switching the other channels had insignificant additional interference effects over nonswitched interference.

With -32 dBm of received power, even a fully loaded, reconfiguring switch allows a BER better than 10^{-11}, and could meet the requirements of most communications systems.

3) Bit-Error-Rate Drift over Time: The curves in Fig. 10 show juxtaposed BER data of the four tests with a single power reading per plotted point. Fig. 11 shows the range of all BER measurements taken. As before, the horizontal axis is the received power and the vertical axis is the BER. The two dashes in a column represent the maximum and minimum BER readings taken during an hour at the same nominal received power.

[6] P. Granestrand et al., "Strictly nonblocking 8 × 8 integrated-optic switch matrix in Ti:LiNbO$_3$," in *Proc. Top. Meet. Integrated Guided-Wave Opt.*, Atlanta, GA, Feb. 1986, paper WAA3, p. 4.

[7] P. J. Duthie, M. J. Wale, and I. Bennion, "A new architecture for large integrated optical switch-arrays," in *Proc. Top. Meet. Photon. Switching, Tech. Dig. Series*, OSA, 1987, vol. 13, 1987, pp. 71-73.

[8] A. Neyer, W. Mevenkamp, and B. Kretzschmann, in *Proc. Top. Meet. Integrated Guided-Wave Opt.*, Atlanta, GA, Feb. 1986, paper WAA2.

[9] M. Kondo et al., "32 switch-elements integrated low-crosstalk LiNbO$_3$ 4 × 4 optical matrix switch," in *Proc. IOOC-ECOC'85*, Venice, Italy, 1985, p. 361.

[10] K. Padmanabhan and A. Netravali, "Dilated networks for photonic switching," in *Top. Meet. Photon. Switching, Tech. Dig. Series*, OSA, vol. 13, 1987, pp. 38-41.

[11] J. Saniter, F. Schmidt, and W. Werner, "LOCNET: An all-optical wideband local area network," in *Proc. Opt. Fiber Conf. '85*, Feb. 1985, pp. 44-45.

[12] S. Suzuki et al., "Thirty-two-line optical space-division switching system," in *Proc. Opt. Fiber Conf./Integrated Opt. Fiber Commun. '87*, Jan. 1987, p. 146.

[13] W. J. Minford, S. K. Korotky, and R. C. Alferness, *IEEE J. Quantum Electron.*, vol. QE-18, p. 1802, 1982.

[14] E. J. Murphy et al., *J. Lightwave Technol.*, vol. LT-3, p. 795, 1985.

[15] R. H. Stolen, W. Pleibel, and J. R. Simpson, *J. Lightwave Technol.*, vol. LT-2, p. 639, 1984.

[16] E. J. Murphy et al., *Proc. Opt. Fiber Commun. Conf.*, Atlanta, GA, 1986, paper PDP-2.

[17] K. Habara and K. Kikuchi, *Electron. Lett.*, vol. 21, p. 631, July 1985.

[18] G. A. Bogert, *Electron. Lett.*, vol. 23, p. 72, Jan. 1987.

Balanced bridge modulator switch using Ti-diffused LiNbO$_3$ strip waveguides

V. Ramaswamy, M. D. Divino, and R. D. Standley

Bell Laboratories, Crawford Hill Laboratory, Holmdel, New Jersey 07733

An experimental integrated optical version of the Mach-Zehnder interferometer switch, analogous to the microwave balanced bridge, is reported. The bridge is formed by 3-μm-wide Ti-diffused strip waveguides in LiNbO$_3$ and the switch utilizes electro-optical tuning to achieve 3-dB operation of the directional couplers. The switching voltage required to switch between the states, corresponding to a π phase shift in one arm, equals 14.8 V. The cross talk between channels that corresponds to the extinction ratio when operated as an on-off modulator is −21.6 dB.

Several switches using optical waveguide directional couplers have been proposed[1-3] and many have already been realized.[4-6] Recently, Schmidt and Kogelnik[6] reported a switched directional coupler with two sections of alternating $\Delta\beta$. This switch requires two equal voltages for switching to the straight-through state and slightly different equal and opposite voltages for switching to the crossover state. The voltage difference is necessary to compensate for the slight asymmetry introduced during fabrication. We present the first experimental results on a balanced bridge modulator switch, which requires only one voltage to switch between the two output states even if there are asymmetries introduced in the directional couplers during fabrication. This is accomplished by interconnecting the two directional couplers through an electro-optically controllable phase shifter. More importantly, this switch demonstrates the balancing of both amplitude and phase in an integrated optical device, 38 mm long, incorporating several bends that interconnect waveguides of slightly different widths.

In optics, the Mach-Zehnder interferometer provides a means of converting phase changes into intensity changes and is analogous to the conventional balanced bridge scheme at microwave frequencies. Several authors have suggested the integrated optical circuit adaptation of the same using diffused strip waveguides.[7,8] The balanced bridge modulator switch consists of two directional coupler pairs connected together via an intermediate electro-optic phase shifter. In a previous paper,[8] it was established that the coupler pairs consisting of asynchronous guides yielding a coupling strength not equal to 3 dB can be tuned electro-optically for 3-dB operation by adjusting the phase mismatch between the guides in each of the directional couplers. Then, after the initial adjustment of the voltage of the phase shifter to turn on one of the states, the application and removal of a voltage corresponding to π phase shift switches the output states.

The balanced bridge illustrated in Fig. 1 was fabricated in y-cut LiNbO$_3$ with Ti-diffused strip waveguides. The master mask of the pattern used in the fabrication was generated by an electron-beam exposure system (EBES). The electron beam is controlled by a digital computer in order to ensure flexibility and repeatability in the generation of complex patterns.

The input guides, waveguide sections of the directional couplers, and the intermediate phase shifter sections are all 3 μm wide. The separation between the guides in the coupling region is also 3 μm. However, the phase shifter guide was displaced by 12 μm from the coupler guide and these two offset straight waveguides were connected by diagonal sections of slightly smaller widths, resulting in tilts of about 0.1°. The digital computer controlled electron beam moves in discrete steps in both x and y directions and as a result cannot draw smooth straight lines at arbitrary angles. Due to this inherent property of the EBES, the interconnecting diagonal section comprises of a staircase structure (not shown) of $\frac{1}{2}$-μm steps. Besides the tilt losses, this staircase structure creates additional scattering loss in the arm containing the diagonal sections. Therefore, the two arms must be symmetrical to facilitate balancing of the bridge. In the previously suggested scheme,[8] where one arm of the bridge is a straight-through section, we obtain a much smaller extinction ratio.

The pattern shown in Fig. 1 was regenerated on a 40-mil-thick LiNbO$_3$ crystal with 300 Å of evaporated Ti using conformable mask[9] and conventional photoresist lift-off techniques. Ti was diffused into the substrate at 960 °C for 6 h in an argon atmosphere and

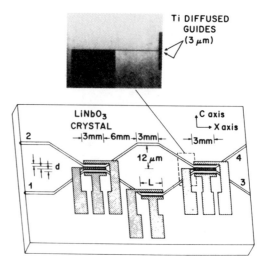

FIG. 1. Balanced bridge scheme with a photograph of an actual device showing the alignment of the electrodes and the guides in the coupling region.

Reprinted with permission from *Appl. Phys. Lett.*, vol. 32, no. 10, pp. 644-646, May 15, 1978.
Copyright © 1978, American Institute of Physics.

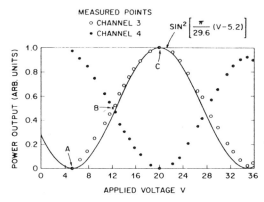

FIG. 2. Normalized power output in channels 3 and 4 versus the electro-optic phase shifter voltage V.

flushed with oxygen for about 1 h during the cooling process. The substrate ends were polished after epoxying two small slices of LiNbO$_3$ of comparable thickness at each end. Polarized light at 6328 Å was coupled into the end of light guide 1 using a lens and the output of the device was detected through an appropriate lens. Provision was also made for viewing by a TV camera.[5] This technique makes studying the modal behavior of the diffused waveguide easier; it also simplifies the measurement of the relative power levels in both channels of a switch or a directional coupler.

The electrode patterns, also illustrated in Fig. 1, were generated by EBES, and using registration marks in both the pattern and electrode masks, titanium-gold electrodes were fabricated onto the diffused LiNbO$_3$ samples as shown. The alignment of the 3-μm center electrode between the guides in the coupling region is rather critical in that any slight displacement away from the center increases enormously the TM mode losses in one arm, thereby reducing the on-off ratio and the efficiency of the switch. The electrode alignment is shown in a rectangular cut-out in Fig. 1. The electrodes were wired so that either a positive or a negative voltage could be applied to each of the guides in the couplers as well as to the phase shifter.

To ensure 3-dB operation of the couplers, sufficiently large dc bias was applied at first to the coupling region of one of the couplers, say the first coupler, to increase the phase mismatch, so that practically no coupling occurred through this section of the device. Then the second coupler was tuned for 3-dB operation and the voltage V_2 was noted. The procedure was repeated by reversing the role of the couplers to obtain the voltage V_1 applied to the first coupler for 3-dB operation. Then the two voltages V_1 and V_2 were applied simultaneously and the phase shifter voltage V_ϕ was adjusted to minimize the output in one channel. Then all the voltages were tuned to obtain a deep null in this channel, simultaneously maximizing the output on the other. Alternatively, it was found possible to adjust all the control voltages in sequence to obtain a null in one channel and a maximum at the other by visually observing the output in the TV monitor. Voltages V_1 and V_2 depend on the asymmetries between the directional coupler guides of the device introduced during fabrication and have varied between $\frac{1}{2}$ and 7 V for our experimental devices. The voltage in the connecting arm to turn on channel 4 is 5.2 V for the device reported here and the switching voltage, introducing $\phi = \pi$, is 14.8 V. Figure 2 shows the normalized power output in both channels 3 and 4 as a function of the phase shifter voltage. Points A, B, and C correspond to states where all the power is in channel 4, equal amounts in 3 and 4, and all the power is in channel 3, respectively. The near-field output of the device viewed by the TV monitor is shown in Figs. 3(a)–3(c) and correspond to points A, B, and C, respectively. The cross talk between channels that correspond to the extinction ratio when operated as an on-off modulator is −21.6 dB.

With the initial voltage (5.2 V) required for switching on one state, the power on arms 3 and 4 are[8]

$$P_3 = 0$$
and
$$P_4 = 1. \tag{1}$$

If we now vary the phase shift ϕ, by changing the voltage, then[8]

$$P_3 = \sin^2(\tfrac{1}{2}\phi)$$
and
$$P_4 = \cos^2(\tfrac{1}{2}\phi), \tag{2}$$

where

$$\phi = \frac{2\pi}{\lambda}\left[\frac{1}{2}n_0^3 r_{13}\left(\frac{2}{\pi}\right)\frac{V}{d}L\right]. \tag{3}$$

The factor $(2/\pi)$ is the effective field amplitude coefficient[10] for semi-infinite electrodes separated by a distance d. This field E_z increases away from the center,

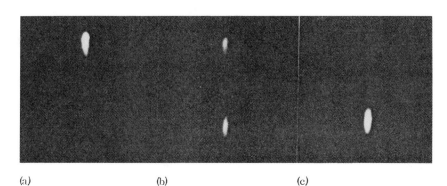

(a)   (b)   (c)

FIG. 3. Photographs of the near-field pattern of the output channels when the energy is (a) entirely in channel 4, (b) equally split between 3 and 4, and (c) entirely in channel 3.

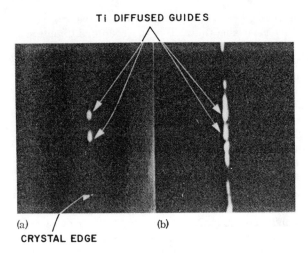

FIG. 4. Photographs of the output (near field) of a diffused LiNbO$_3$ bridge with (a) TM mode and (b) TE mode excitation at the input.

reaching a maximum at the electrodes, and decreases with depth inside the crystal. For the purpose of calculation, the field E_z is assumed to be uniform and equal to $(2V/\pi d)$. The calculated voltage required for switching the states corresponding to $\phi = \pi$ is 10.9 V and compares favorably with the measured values of 14.8 V. The reported results were obtained with TM excitation for which the guides are single moded, as illustrated by the output of a typical sample in Fig. 4(a); however, for TE mode excitation the energy is no longer confined to the Ti-diffused guiding region [Fig. 4(b)] because of the Li out diffusion. Our independent attempts to duplicate other approaches[11,12] to provide single mode strip waveguides for the TE modes due to Ti in diffusion while blocking Li out diffusion were not successful. Thus we operated the switch with single TM mode utilizing the r_{13} electro-optic coefficient; a successful scheme to fabricate and operate this device with a single TE mode strip guide utilizing the r_{33} electro-optic coefficient should result in a switching voltage of 4.2 V requiring nine times less switching power than for the TM mode case. By incorporating the phase shifter in both arms, the switching voltage should be reduced further by a factor of 2.

We also observed that the cross-talk level between the channels increased with time and was dependent on the original light intensity in the guides. This temporal drift phenomenon[13] due to optical damage can always be compensated by an increase in the bias voltages and the cross-talk level can be returned to the original value. One could eliminate this effect by alternating the polarity of the bias voltages V_1 and V_2.

In conclusion, we have demonstrated an integrated optical bridge modulator switch utilizing the concepts of Mach-Zehnder interferometer analogous to the balanced bridge scheme at microwave frequencies. The interferometer might prove useful in protection switching, PCM detection, frequency multiplexing, and other obvious extensions of prior microwave art. This switch clearly illustrates that both the amplitude and the phase, in a 38-mm-long integrated optical device, incorporating several bends that interconnect waveguides of slightly different widths, can indeed be balanced so as to achieve an on-off ratio of almost 22 dB.

The authors are grateful to R.H. Bosworth for programming the EBES masks and to L.L. Buhl for the assistance provided in making the conformable mask.

[1] S. Kurazano, K. Iwasaki, and N. Kumagai, Electron. Commun. Jpn. **55**, 103–109 (1972).
[2] H.F. Taylor, J. Appl. Phys. **44**, 3257 (1973).
[3] J.M. Hammer, in *Integrated Optics*, edited by T. Tamir, (Springer, Berlin, 1975), pp. 140–198.
[4] M. Papuchon, Y. Combemaie, X. Mathieu, D.B. Ostrowsky, L. Reiber, A.M. Roy, B. Sejourne, and M. Werner, Appl. Phys. Lett. **27**, 289–291 (1975).
[5] J.C. Campbell, F.A. Blum, D.W. Shaw, and K.L. Lawley, Appl. Phys. Lett. **27**, 203–205 (1975).
[6] R.V. Schmidt and H. Kogelnik, Appl. Phys. Lett. **28**, 503–506 (1976).
[7] I.P. Kaminow, IEEE Trans. Microwave Theory Tech. MTT-**23**, 57–70 (1975).
[8] V. Ramaswamy and R.D. Standley, Bell Syst. Tech. J. **55**, 767–775 (1976).
[9] H.I. Smith, Proc. IEEE **62**, 1361–1387 (1974).
[10] D. Marcuse, IEEE J. Quantum Electron. QE-**11**, 759–767 (1975).
[11] T.S. Ranaganath and S. Wang, Appl. Phys. Lett. **30**, 376–379 (1977).
[12] B. Chen and A.C. Pastor, Appl. Phys. Lett. **30**, 570–571 (1977).
[13] R.V. Schmidt and L.L. Buhl, Electron. Lett. **12**, 575–577 (1976).

BOA-Coupler

Electrically active optical bifurcation: BOA

M. Papuchon and Am. Roy

Thomson CSF—L. C. R., B. P. 10, Domaine de Corbeville, 91401 Orsay, France

D. B. Ostrowsky

Université de Nice, Parc Valrose, 06340 Nice, France

We report the realization of an electro-optical switch based on the interference between modes in a two-mode waveguide (equivalent to a directional coupler with no gap between the guides). Switching is achieved by uniformly changing the refractive index of the guide. This permits attaining the two switching states even in the presence of fabrication errors, with a very simple electrode configuration. Crosstalks of −16 and −18 dB for command voltages of 8 and −18 V have been obtained for such a structure realized in Ti:diffused lithium niobate.

Electro-optical switching has already been demonstrated using integrated optical directional couplers[1-3] or other configurations.[4-6] The integrated optical directional couplers realized to date have all been based on the use of separate disjoint waveguides. The efficient operation of such switches requires the use of *asymmetric* electrically induced index changes in the two guides[7] to destroy their resonance. This imposes the use of a three-electrode configuration[8] to allow compensation of fabrication errors. In this paper we report a simple single waveguide directional coupler configuration which can use a *symmetric uniform* change in guide indices for switching, which permits compensation of fabrication errors with two electrodes.

The structure of the BOA (bifurcation optique active) is shown in Fig. 1. It consists of a central waveguide which has two transverse modes coupled with single-mode entrance and exit waveguides.

The structure can be considered to be a directional coupler in which the gap between the guides in the coupling region has been reduced to zero. As for an ordinary directional coupler, the eigenmodes of the "coupling section" consist of one symmetric and one antisymmetric mode. If one excites these two eigenmodes properly the energy distribution evolves as shown in Fig. 2. The energy which was originally concentrated in the "top" half of the guide is found in the "bottom" half at distances given by

$$L_0 = \frac{m\pi}{(\beta_s - \beta_{as})} = \frac{m\pi}{\Delta\beta}, \quad (1)$$

where m is a positive odd integer, β_s is the propagation constant of the symmetric mode and β_{as} that of the antisymmetric mode. It is then clear that the energy at the output of the two-mode section can be switched from the "top" half to the "bottom" half of the guide if we are able to change $\Delta\beta$.

This can be done by introducing a change δn in the waveguide index. Thus if L is the physical length of the "coupling section", the required value $\delta(\Delta\beta)$ is given by

$$\delta(\Delta\beta) L = \pi$$

or

$$\delta n \frac{\partial \Delta\beta}{\partial n} L = \pi$$

leading to

$$\delta n = \frac{\pi}{L} \left(\frac{\partial \Delta\beta}{\partial n} \right)^{-1}. \quad (2)$$

As δn is proportional to the applied voltage in linear electro-optical materials, we see that the voltage required to switch the light from the top half to the bottom is inversely proportional to $(\partial \Delta\beta/\partial n)$. This ratio depends strongly on the separation of the guides of a directional coupler and increases as the spacing is decreased. Thus its maximum value (and the minimum value of the command voltage) is obtained by taking a zero spacing, i.e., a two-mode waveguide. This is shown in Fig. 3 where $\partial(\Delta\beta/K)/\partial n$, calculated using the theory of Marcatili,[9] has been plotted for different values of the gap between the guides of a coupler whose parameters are also shown in Fig. 3. In this case a change Δn has been assumed in the two guides and in the intermediate region that forms the gap to simulate an electro-optical change created by two electrodes, realized on each side of the coupling section.

The value for a zero gap has been calculated[9] by considering a waveguide with a width equal to twice that of the separate guides of the coupler. We see that the value of $\partial(\Delta\beta/K)/\partial n$ can vary by an order of magnitude between the ordinary directional coupler and the zero gap coupler.

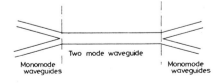

FIG. 1. Configuration of the switch.

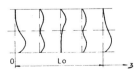

FIG. 2. Energy distribution along the propagation direction in a two-mode waveguide due to the interference between modes.

Reprinted with permission from *Appl. Phys. Lett.*, vol. 31, no. 4, pp. 266-267, Aug. 15, 1977.
Copyright © 1977, American Institute of Physics.

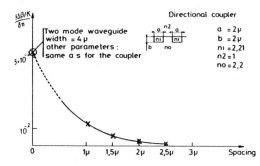

FIG. 3. $\partial(\Delta\beta/K)/\partial n$ versus the gap between the guides in a directional coupler. For zero gap the calculation has been made for a two mode waveguide.

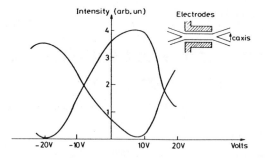

FIG. 4. Experimental results showing the energy in the two output monomode guides versus applied voltage. The switch configuration with its electrodes and orientation of the c axis of LiNbO$_3$ is also shown.

This explains why a symmetrical change of Δn permits switching with reasonable voltages in the zero gap case, whereas in the classical directional coupler this will lead to unacceptably high switching voltages.

To sum up, the use of a two-mode waveguide permits maximization of $\partial(\Delta\beta/K)/\partial n$ and hence allows us to use a symmetrical index change to switch light from one output guide to the other. Note that this method allows the compensation of fabrication errors because practically for any value of L we can always induce the Δn necessary to place the switch in one state or the other.

The switch used in the experiments consists of monomode branches for input and output made with 2-μm wide waveguides having a branching angle of 1°. The two-mode section has a width of 4 μm and a length of 5 mm and the electrodes are 5 μm apart and 5 mm long.

The guides have been realized by Ti diffusion in the usual fashion[10-11]:

500 Å of titanium is sputtered onto a LiNbO$_3$ Y plate (so that the r_{33} electro-optical coefficient will be used). After masking and etching, the sample is heated for 4 h at 1000 °C in an oxygen atmosphere to perform the diffusion process; gold electrodes are then realized by sputtering and standard superposition masking techniques.

A laser ($\lambda = 5145$ Å) is then coupled into one of the input monomode waveguides via a rutile prism and the light emerging from the output monomode waveguides is detected and sent to an oscilloscope.

The experimental results of the measurements on the two guides of the output branch are shown in Fig. 4 as well as the switch configuration. For $V = 0$ the switch is in an intermediate state but, by applying +8 and -18 V we can switch the light from one output guide to the other. The measured crosstalks are -16 and -18.3 dB for the two states of the switch. Note the quasi-sinusoidal form of the variation of the output intensity resulting from the interference character of the switch.

In conclusion a zero gap directional coupler (two-mode waveguide) switch has been realized. This permits the switching by inducing a symmetrical change in the refractive index with two electrodes. The different optogeometrical parameters are not critical. The electrical selection of the two output states with command voltages of 8 and -18 V have been demonstrated and crosstalks of -16 and -18.3 dB have been obtained for the two states.

It is a pleasure to thank M. Jordan for cutting and polishing the substrates, X. Mathieu for preparing the films, M. Werner for the realization of the masks with the "masqueur electronique", and B. Puech for technical assistance.

[1] M. Papuchon, Y. Combemale, X. Mathieu, D.B. Ostrowsky, L. Reiber, A.M. Roy, B. Sejourne, and M. Werner, Appl. Phys. Lett. **27**, 289 (1975).
[2] J.C. Campbell, F.A. Blum, D.W. Shaw, and K.L. Lawley, Appl. Phys. Lett. **27**, 203 (1975).
[3] O. Mikami and J. Noda, Appl. Phys. Lett. **29**, 555 (1976).
[4] H. Sasaki and R.M. DeLarue, Electron. Lett. **12**, 460 (1976).
[5] F. Zernike, Dig. of Tech. Papers, Topical meeting on Integrated Optics, New Orleans, 1974 (unpublished).
[6] W.E. Martin, Appl. Phys. Lett. **26**, 562 (1975).
[7] S. Kurazono, K. Iwasaki, and N. Kamagai, Trans. Inst. Electr. Commun. Engr. Jpn. C **55**, 61 (1972).
[8] R.V. Schmidt and H. Kogelnik, Appl. Phys. Lett. **28**, 503 (1976).
[9] E.A.J. Marcatili, Bell Syst. Tech. J. **48**, 2071 (1969).
[10] J.M. Hammer and W. Phillips, Appl. Phys. Lett. **24**, 545 (1974).
[11] R.V. Schmidt and I.P. Kaminow, Appl. Phys. Lett. **25**, 458 (1974).

X-Switches

ELECTRO-OPTIC X-SWITCH USING SINGLE-MODE Ti:LiNbO₃ CHANNEL WAVEGUIDES

A. NEYER   1st June 1983

Universität Dortmund
Lehrstuhl für Hochfrequenztechnik
Postfach 500500, D-4600 Dortmund 50, W. Germany

Indexing terms: Optoelectronics, Optical waveguides

A new type of single-mode four-port electro-optic switch is investigated. The operation principle is based on the electrically controlled two-mode interference in the intersection region of two crossing single-mode channel waveguides. The switch is analysed numerically by the beam-propagation method, and is realised experimentally by using Ti:LiNbO₃ channel waveguides.

Introduction: Wideband multichannel fibre-optic systems require compact and high-speed optical switching networks for the processing of optical information, e.g. for distributing, time-multiplexing and signal-routing. Several integrated optical switching networks have been realised by using either directional couplers[1] or multimode total internal reflection switches.[2]

In this letter we report a new single-mode four-port electro-optic switch (X-switch), which is well suited to be integrated in compact and highly efficient optical switching networks.

Principle of operation: A schematic drawing of the X-switch is shown in Fig. 1a. The structure consists of two single-mode channel waveguides of width w, which are crossing each other at an angle $\alpha \lesssim 1°$. The maximum change of the refractive index in the intersection area ($2\Delta n$) is twice that in each channel waveguide (Δn).

The operation principle of the X-switch is explained by the schematic illustration of the mode propagation in Fig. 1b. The input as well as the output waveguides of the switch (regions I and III) represent linearly tapered directional couplers. The light power initially launched into the fundamental mode of one of the input waveguides excites the symmetric and antisymmetric modes of the two coupled waveguides. It is of crucial importance for the nearly lossless operation and the low crosstalk of the switch, that the tapered coupling region adiabatically converts these modes into the two lowest-order lateral modes of the intersecting waveguides.

The intersection region (II) is characterised by the fact that the width of the $2\Delta n$-area linearly increases, whereas the total waveguide width decreases to the minimum value w in the middle of the intersection. The waveguide parameters w and Δn are chosen in such a way that the two lowest-order lateral modes are guided throughout the whole intersection. The difference $\Delta \beta$ between the propagation constants of these two modes leads to an oscillation of their relative phase along the propagation direction. The phase difference at the end of the intersection region determines the amplitudes of the funda-

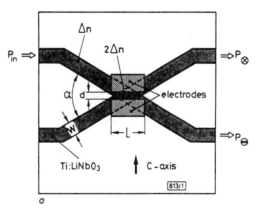

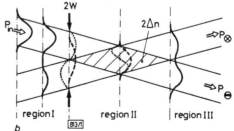

Fig. 1

a Schematic configuration of X-switch
b Illustration of mode propagation
———— optical field distribution
– – – – symmetric
·········· antisymmetric mode

mental modes coupled adiabatically into the two output waveguides. This mode conversion is again almost lossless.

Thus, by the principle of mode superposition, the relative light output power coupled into the output states is given approximately by

$$P_{\otimes} \simeq P_{in} \cos^2(\Delta\beta l/2), \quad P_{\ominus} \simeq P_{in} - P_{\otimes} \quad (1)$$

where $l = w/\sin(\alpha/2)$ is the length of the intersection. The straight-through state of the X-switch is denoted by the symbol $\ominus$, and the crossover state by $\otimes$.

Electro-optic switching is based on the fact that the parameter $\Delta\beta$ can be controlled electrically. For y-cut LiNbO$_3$, the electrode separation d (Fig. 1a) should be much smaller than the waveguide width w. Then, only the propagation constant of the fundamental mode is modified strongly. This yields low switching voltages even for short electrode lengths ($L < 1$ mm).

In order to get a detailed insight into the switching characteristic of the X-switch, the light propagation through the structure is calculated numerically by using the beam propagation method (BPM).[3] In the calculations a lateral effective index profile given by $\Delta n/\cosh^2(2x/w)$ is assumed.[4]

Fig. 2 shows the result of the calculated switching characteristic, where the following data are used: $\alpha = 0.6°$, $w = 3$ μm,

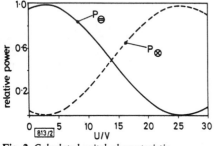

Fig. 2 *Calculated switch characteristic*

$\Delta n = 6 \times 10^{-3}$, $L = 1$ mm, $d = 1$ μm and 1.3 μm wavelength. With a voltage of 2.5 V the switch is tuned to the maximum transmission into the $\ominus$ state. At this point, the light power in the $\otimes$ state is down to $10^{-3} P_{in}$, which corresponds to a crosstalk between the two channels of -30 dB. For a switching voltage of 25 V, these states are interchanged. As it is seen from eqn. 1, and from the numerical results in Fig. 2, both switch states can be achieved electrically, independent of the intersection length l.

The low radiation loss of about 2% and the low crosstalk levels of -30 dB of both switch states are evident consequences of the adiabatic mode conversion within the X-switch.

Experimental results: The optical channel waveguides of the X-switch are fabricated by the conventional Ti-indiffusion technique in a water rich atmosphere.[5] The Ti structure of the X with a crossing angle $\alpha = 1°$ is realised by a two-step process in order to obtain a double Ti thickness in the intersection region. The 40 nm-thick and 4 μm-wide Ti stripes are indiffused for 6 h at 980°C resulting in single-mode waveguides at $\lambda = 0.63$ μm wavelength. A coplanar electrode with a gap of $d = 3$ μm, and a length $L = 1$ mm is adjusted along the X as indicated in Fig. 1.

TE-polarised light of an HeNe laser is endfire-coupled into one of the two input ports of the switch. The measured light power switched into the $\otimes$ state is shown in Fig. 3. The experiment verifies the theoretically predicted sinusoidal behaviour of the light-output power of the X-switch as a function of the applied voltage. With a switching voltage of 12.5 V an on/off ratio of 1:200 (-23 dB) is measured. At maximum light power switched into the $\otimes$ state, the ratio between $P_{\otimes}$ and $P_{\ominus}$ is about 5:1, which corresponds to a switching efficiency of 90%. This relatively high crosstalk of -10 dB between the two channels is explained by the nonoptimised overlap between the optical and electrical fields. An improvement of the interchannel crosstalk down to values of -30 dB is expected.

Conclusion: A novel single-mode electro-optical switch is reported. It combines several features, which are important for the integration in complex optical switching networks: simple waveguide and electrode structures, compact size, short electrode lengths (0.5–1 mm), low capacitance with a resulting

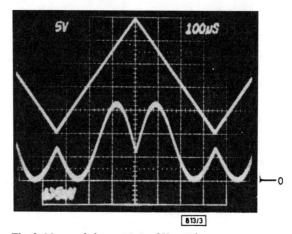

Fig. 3 *Measured characteristic of X-switch*

Upper trace: sawtooth drive voltage (5 V/div.)
Lower trace: light-output power of the $\otimes$ state

high bandwidth (> 1 GHz), moderate switching voltages (10–20 V), low crosstalk levels (< -20 dB), and a simple tuning of the switch states.

References

1 KONDO, M., OHTA, Y., MASAHIKO, F., and SAKAGUCHI, M.: 'Integrated optical switch matrix for single-mode fiber networks', *IEEE J. Quantum Electron.*, 1982, **QE-18**, pp. 1759–1765
2 CHANG, C. L., and TSAI, C. S.: 'GHz bandwidth optical channel waveguide TIR switches and 4 × 4 switching networks'. 1982 Topical Meeting on integrated and guided-wave optics, ThD2
3 FEIT, M. D., and FLECK, J. A.: 'Light propagation in graded-index optical fibers', *Appl. Opt.*, 1978, **17**, pp. 3990–3998
4 KOGELNIK, H.: 'Theory of dielectric waveguides' in TAMIR, T. (Ed.): 'Integrated optics' (Springer Verlag, Berlin, 1975)
5 JACKEL, J. L., RAMASWAMY, V., and LYMAN, S. P.: 'Elimination of outdiffused surface guiding in titanium-diffused LiNbO$_3$', *Appl. Phys. Lett.*, 1981, **38**, pp. 509–511

Optical Channel Waveguide Switch and Coupler Using Total Internal Reflection

CHEN S. TSAI, SENIOR MEMBER, IEEE, BUMMAN KIM, STUDENT MEMBER, IEEE, AND FATHI R. EL-AKKARI

Abstract—Light beam switching and coupling in a four-port channel waveguide-horn structure has been accomplished using electrooptic modulation of the critical angle of a refractive index interface in a Y-cut LiNbO$_3$ substrate. The resulting double-pole double-throw switch/coupler is potentially capable of simultaneously providing a combination of desirable characteristics.

A NUMBER of optical channel waveguide double-pole double-throw (DPDT) switches and couplers have been reported recently [1]-[12]. Devices potentially capable of forming such type of switches and couplers have also been reported [13], [14]. In this paper we report the theoretical and experimental results on a simple DPDT switch/coupler that utilizes a channel waveguide-horn structure and the electrooptic modulation of the refractive index of a narrow layer in a Y-cut LiNbO$_3$ substrate. Specifically, a channel-guided light beam is incident upon this refractive index layer at an angle close to the *critical angle* which, in turn, is field-dependent because of electrooptic effects. The ratio of the reflected light power to the transmitted light power is then modulated by varying the applied voltage. A class of acoustic components based on a somewhat similar working principle has been proposed [15]. Two versions of the device, namely, the one with a fixed bias in the refractive index and the other without such a bias are discussed in this paper. We have shown that such a four-port device is potentially capable of *simultaneously* providing a combination of desirable characteristics: subnanosecond switching speed, simple electrode configuration, moderate device dimensions, moderate to low drive voltage requirement, electronically variable coupling, relaxed requirements in design and fabrication tolerances, moderate to small crosstalks, small optical insertion loss, and preservation of the light beam quality.

Let us imagine that two pairs of channel waveguides with identical horn structures have been formed in a Y-cut LiNbO$_3$ substrate as shown in Fig. 1. These two pairs of channel waveguides are tilted symmetrically with respect to the $x(a)$ axis of the substrate, and are to function as the input and output ports of the switch/coupler. The intersection region between the horn apertures contains a planar waveguide. A pair of parallel electrodes, aligned with the x-axis, are located in the middle of the intersection region. Note that this channel waveguide-horn structure is similar to that of Bragg switch [9], [10]. However, for the device to be discussed in this paper, the tilt angle θ_i is determined by the critical angle

Manuscript received February 9, 1978; revised March 13, 1978. This work was supported in part by the AROD and in part by the NSF. Part of this work was presented as a post-deadline paper at the 1978 Topical Meeting on Integrated and Guided-Wave Optics.
The authors are with the Department of Electrical Engineering, Carnegie-Mellon University, Pittsburgh, PA 15213.

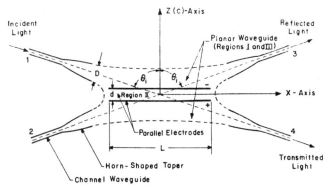

Fig. 1. Configuration of optical channel waveguide double-pole double-throw switch and coupler in Y-cut LiNbO₃ substrate.

rather than the Bragg angle [10]. Consequently, the drive voltage for this device can be much lower than that of the Bragg device which employs a single pair of parallel electrodes [10]. Now, in the first version, a fixed variation in the refractive index is imbedded in the region defined by the parallel electrodes. This bias in refractive index can be introduced by means of titanium-diffusion [16], [17]. In the second version, no such a bias in refractive index is introduced.

For simplicity, the second version is described first. In the absence of an applied voltage, an incident guided-light beam, from port 1, for example, will encounter no refractive index interface, will transmit freely (assuming negligible effect by the electrodes) through the region defined by the parallel electrodes (Region II), and will enter port 4. Since it is possible to design and fabricate horn-shaped tapers of minimal insertion loss and mode conversion [18], [19], only a small leakage of the incident light and thus only a small crosstalk are expected to appear at port 3. However, when a voltage v of the indicated polarity is applied, the refractive index of the layer is reduced due to the linear electrooptic effect. Clearly, two refractive index interfaces have been created electrically. These two interfaces in turn cause reflection of the incident light. Note that total internal reflection of the light may occur at the first interface [16], [17], [20] if the incident angle is equal to or larger than the critical angle. As a result, a portion or all of the incident light is switched to port 3. Due to the symmetry of the device, an identical result can be expected when the light beam is incident from port 2.

For simplicity, we assume that the electric field distribution in the layer is uniform and that the light beam propagates in a TE mode. It can readily be shown that the critical angle for the interface separating the layer and region I is given by

$$\theta_c = \sin^{-1}\left\{1 - \frac{1}{2} n_1^2 r_{33}\left(\frac{V}{d}\right)\right\} \quad (1)$$

where n_1 is the effective refractive index in the regions (Region I and III) outside of the layer, r_{33} is the relevant electrooptic coefficient, and d is the separation of the electrode pair. For the case with a Y-cut LiNbO₃ substrate we have $n_1 = 2.2$,

$r_{33} = 30.8 \times 10^{-10}$ cm/V, and $\theta_c \simeq 87.8$ degrees at an electric field intensity of 1×10^5 V/cm, or 10 V/μm. Clearly, the angle decreases as the applied voltage is increased. Therefore, if the tilt angle of the channel waveguides and, consequently, the incident angle of the light beam θ_i is chosen to be in the neighborhood of the critical angle, the ratio of the reflected (switched) light power to the transmitted (unswitched) light power can become a sensitive function of the applied electric field intensity. The light beam will encounter total internal reflection (TIR) at the first interface as the electric field intensity reaches the following value:

$$\left.\frac{V}{d}\right|_{TIR} = \frac{2(1 - \sin\theta_i)}{n_1^2 r_{33}} \simeq \frac{1}{n_1^2 r_{33}}\left(\frac{\pi}{2} - \theta_i\right)^2 \quad (2)$$

The above approximation is valid because for all practical cases θ_c and, thus, θ_i are very close to 90 degrees as indicated in the above example. We see that the closer the tilt angle is to 90 degrees, the smaller is the drive voltage needed for total internal reflection at the first interface. At total internal reflection, a portion of the incident light power will be transmitted through the layer by means of evanescent wave coupling. Using the above uniform refractive index model, it is convenient to calculate the ratio of the reflected light power to the transmitted light power, and thus the crosstalks, as a function of the wavelength and the incident angle of the light beam, the electrode separation, and the applied voltage. For example, a set of plots showing the drive voltage requirement versus the electrode separation, with the crosstalk as a parameter, has been obtained for the case of $\theta_i = 89$ degrees and a TE₀-mode light wave at 0.6328 μm (see Fig. 2). These plots clearly identify the optimum drive voltage and electrode separation for a given crosstalk figure. Since the plots have also indicated that a relatively wide range of electrode separation can be used at this optimum drive voltage, relaxed requirements in design and fabrication tolerances for the proposed device are expected. This important advantage has been demonstrated experimentally.

We now turn to the device of the first version. Since the working principle is essentially the same as in the second version, it is not repeated here. It suffices to observe, however, that by means of the Ti-diffusion technique [21], the lateral variation of the imbedded refractive index can be made gradual and the peak index change can be as much as a few percents. As a result, the light beam continuously vary its angle of incidence (θ_i) as it propagates through the index layer. At the region between the parallel-electrode pair, the incident angle and thus the critical angle (in the absence of an applied voltage) may approach 90 degrees. This will in turn result in a reduction in the drive voltage for a given crosstalk figure [see (2)].

In the experimental study, both versions of the proposed optical DPDT were fabricated in Y-cut LiNbO₃ plates. The devices of the second version were fabricated by first using a photomask of the pattern shown in Fig. 1 and the titanium-

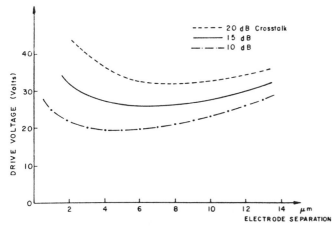

Fig. 2. Drive voltage versus electrode separation with crosstalk as a parameter for devices of the second version (the one without fixed bias).

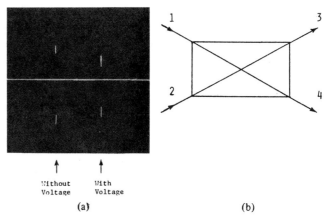

Fig. 3. Photographs of unswitched (transmitted) and switched (reflected) light beams for a device of the second version: (a) light incident from port 2; (b) light incident from port 1.

diffusion technique [21] to form the channel waveguides, the horns, and the planar waveguide in a Y-cut LiNbO$_3$ plate. A pair of parallel electrodes was subsequently deposited in the intersection region. Identical designs were chosen for the input and output channel waveguides and the horns using the established design methodology [19]. Each channel waveguide has a horn of 4.7 mm long, tapering from 4 to 40 μm. The tilt angle of each channel waveguide from the x-axis is 1 degree. The length of the electrode pair is 3.4 mm and the separation is 4 μm. The total length of the device is 1.3 cm. In an optimum design, the device length can be shortened considerably because the length of each horn may be much smaller than 4.7 mm.

Switching and coupling experiments were carried out using a He-Ne laser light at 6328 Å. A pair of rutile prisms were used to couple in and couple out the incident and the reflected/transmitted light beams. Fig. 3 shows the composite photographs of the transmitted (unswitched) and reflected (switched) light beams after propagating through the output prism obtained with one device of the second version, with and without 40 V applied. Note that Figs. 3(a) and (b) are, respectively, for the cases in which the light beam is incident at port 1 and port 2. Clearly, the switched light beam is completely separated from the unswitched beam. It is seen that a high and nearly identical degree of transmission and reflection could be obtained at both ports. The maximum reflected light power is smaller than the maximum transmitted light power only by 5 percent. It is also seen that transmitted and reflected light beams of good quality are achievable. Good optical beam quality is attributable to the utilization of only one electrode pair in this device. It is a common experience that the quality and through-put of a light beam after propagating through a device with a large number of electrode pairs, such as in a Bragg device, tend to degrade considerably.

Fig. 4 depicts the relative power of the reflected light (P_r) and the transmitted light (P_t) that was measured with the same device using a dc voltage when the light was incident

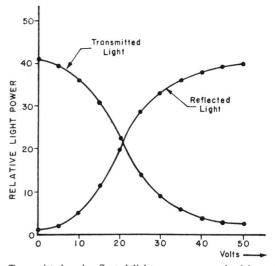

Fig. 4. Transmitted and reflected light power versus dc drive voltage measured with a device of the second version (the one without fixed bias).

at port 1. Clearly, power coupling of good linearity has been demonstrated in both the transmitted light and the reflected light. The measured crosstalk at the switched port 3, defined as 10 log (P_r/P_t) in dB in the absence of applied voltage, is -15 dB. This crosstalk was largely caused by the leakage of light from port 2. The aperture of the light beam that was incident upon the input prism coupler partially spread over to channel waveguide 2 due to insufficient separation between channel waveguide 1 and channel waveguide 2. As a result, during the process of exciting channel waveguide 1, channel waveguide 2 was also weakly excited. The measured crosstalk at the unswitched port 4, defined as 10 log (P_t/P_r) in dB at the applied voltage of 50 V minus the leakage of the light from port 2, is -14.7 dB. Note that a -14.7 dB crosstalk is equivalent to a 97 percent intensity modulation. Most recently, the drive voltage has been reduced from 50 to 30 V in a separate device. These measured values of crosstalks and drive voltage are in reasonable agreement with the calculated values shown in Fig. 2. From (2) it is clear that this drive voltage may be reduced by a factor of four if the tilt angle of each channel waveguide from the x-axis is reduced by a factor of two. Therefore, it is concluded that this type of DPDT is capable of providing small crosstalks at a moderate to small drive voltage. Finally, the calculated capacitance of the parallel electrode pair of the device is 1.1 pF, and the corresponding base bandwidth with a 50 Ω termination would be 5.9 GHz. Again, this large bandwidth is attributable to the utilization of only one electrode pair in the device. Switching and coupling experiments using the device of the first version were also carried out in the manner just described. The measured drive voltage ($\simeq$25 V) is considerably higher than the calculated value. This discrepancy was found to be caused mainly by the misalignment between the electrode pair and the edges of the layer with the fixed refractive index variation. Further experiments are in progress and the results will be reported in a future paper.

In summary, electrooptic modulation of the critical angle of a refractive index layer in a Y-cut $LiNbO_3$ substrate has been utilized to perform electronically variable light beam switching and coupling in a four-port channel waveguide-horn structure. The resulting channel waveguide double-pole double-throw switch/coupler has been shown to be potentially capable of *simultaneously* providing a combination of desirable characteristics. Those that have been demonstrated experimentally include simple electrode configuration, moderate device dimensions, moderate drive voltage requirement, electronically variable coupling, relaxed requirements in design and fabrication tolerances, moderate to small crosstalks, and preservation of the light beam quality. Those that have been theoretically predicted include very fast switching speed, small device dimensions, low drive voltage requirement, and small optical insertion loss. This optical DPDT, when fully developed, should be useful in future wideband multichannel integrated and fiber optic systems. For examples, this device would constitute a basic building block for optical switching networks [22], [23] optical time-division demultiplexers [24], [25] and optical computers [26]. Finally, it is clear that the electrooptically induced refractive index layer may also be employed to switch/couple planar guided-light beams. One disadvantage of the resulting planar waveguide devices results, however, from the fact that the light beam spreads (due to diffraction) in the plane of the waveguide as it propagates. Consequently, it is difficult to realize optical switching networks involving a large number of stages.

REFERENCES

[1] E. A. J. Marcatili, "Dielectric rectangular waveguide and directional coupler for integrated optics," *Bell Syst. Tech. J.*, vol. 48, pp. 2071-2102, 1969.

[2] H. F. Taylor, "Optical switching and modulation in parallel dielectric waveguides," *J. Appl. Phys.*, vol. 44, pp. 3257-3262, 1973.

[3] K. Tada and K. Hirose, "A new light modulator using perturbation of synchronism between two coupled guides," *Appl. Phys. Lett.*, vol. 25, pp. 561-562, 1974.

[4] M. Papuchon, Y. Combemale, X. Mathieu, D. B. Ostrowsky, L. Reiber, A. M. Roy, B. Sejourne, and M. Werner, "Electrically switched optical directional coupler: COBRA," *Appl. Phys. Lett.*, vol. 27, pp. 289-291, 1975.

[5] J. C. Campbell, F. A. Blum, D. W. Shaw, and K. L. Lawley, "GaAs electrooptic directional coupler switch," *Appl. Phys. Lett.*, vol. 27, pp. 203-205, 1975.

[6] R. V. Schmidt and H. Kogelnik, "Electrooptically switched coupler with stepped $\Delta\beta$ reversal using Ti-diffused $LiNbO_3$ waveguides," *Appl. Phys. Lett.*, vol. 28, pp. 503-506, 1975.

[7] H. Kogelnik and R. V. Schmidt, "Switched directional couplers with alternating $\Delta\beta$," *IEEE J. Quantum Electron.*, vol. QE-12, pp. 396-401, 1976.

[8] F. Auracher, H. Boroffka, and R.Th. Kersten, "Planar branching networks for multimode and monomode glass fiber systems," *1976 Topical Meeting Integrated Optics, Tech. Dig.*, MD7-1, Salt Lake City, UT, Jan. 12-14.

[9] See, for example, J. F. St. Ledger and E. A. Ash, "Laser beam modulation using grating diffraction effects," *Electron. Lett.*, vol. 4, pp. 99-100, 1968; M. A. R. P. De Barros, "High-speed electro-optic diffraction modulator for baseband operation," *Proc. Inst. Elec. Eng.*, vol. 119, pp. 807-814, 1972; J. N. Polky and J. H. Harris, "Interdigital electro-optic thin-film modulator," *Appl. Phys. Lett.*, vol. 21, pp. 307-309, Oct. 1972; J. M. Hammer and W. Phillips, "Low-loss single-mode optical waveguides and efficient high-speed modulators of $LiNb_xTa_{1-x}O_3$ on $LiTaO_3$," *Appl. Phys. Lett.*, vol. 24, pp. 545-547, 1974.

[10] B. Chen, G. L. Tangonan, and A. B. Lee, "Horn structures and thin film optical switch," *1977 IEEE/OSA Conf. Laser Engineering and Applications, Dig. Tech. Papers*, Washington, DC, June 1-3; pp. 8-9; B. Chen M. K. Barnoski, and C. M. Meijer, "Thin-film Bragg switch," *1978 Topical Meeting Integrated and Guided-Wave Optics, Tech. Dig.*, Salt Lake City, UT, Jan. 16-18, pp. TuB3-1 to -4.

[11] R. A. Steinberg, and T. G. Giallorenzi, "Performance limitations imposed on optical waveguide switches and modulators by polarization," *Appl. Opt.*, vol. 15, pp. 2440-2453, 1976.

[12] J. C. Shelton, F. K. Reinhart, and R. A. Logan, "Single mode GaAs-$Al_xGa_{1-x}As$ rib waveguide switches," presented at 1977 IEEE/OSA Conf. Laser Engineering and Applications, post-deadline paper 2.10 PD-A, Washington, DC, June 1-3; also, "Electrooptic $Al_yGa_{1-y}As$-Al $Ga_{1-x}As$ rib waveguide switching modulators for monolithic integrated optics," *1977 IEEE IOOC, Tech. Dig. post-deadline papers*, pp. 41-44.

[13] F. Zernike, "Integrated optics switch," *1974 Topical Meeting Integrated Optics, Tech. Dig.*, pp. WA5-1 to -4, New Orleans, LA, Jan. 21-24, pp. WA5-1 to -4; W. E. Martin, "A new waveguide switch/modulator for integrated optics," *Appl. Phys. Lett.*, vol. 26, pp. 562-546, 1975; V. Ramaswamy and M. D. Divino, "A balanced bridge modulator switch," paper TuA 4; and T. G. Giallorenzi, W. K. Burns, and R. A. Steinberg, "New microoptical switches for use with fiber optical transmission lines," paper TuA 5, presented at the 1978 Topical Meeting Integrated and Guided-Wave Optics, Salt Lake City, UT, Jan. 16-18.

[14] See, for example, Y. Ohmachi and J. Noda, "Electro-optic light modulator with branched ridge waveguide," *Appl. Phys. Lett.*, vol. 27, pp. 544-546, Nov. 1975; W. K. Burns, A. B. Lee, and A. F. Milton, "Active branching waveguide modulator," *Appl. Phys. Lett.*, vol. 29, pp. 790-792, Dec. 1976; and T. R. Ranganath and S. Wang, "Ti-diffused $LiNbO_3$ branched waveguide modulators: Performance and design," *IEEE J. Quantum Electron.*, vol. QE-13, pp. 290-295, Apr. 1977.

[15] A. A. Oliner and K. H. Yen, "A new class of components which do not require piezoelectric substrates," *1974 Ultrasonics Symp. Proc.*, IEEE Cat. 74 CHO 896-1SU, pp. 108-113.

[16] S. K. Sheem and C. S. Tsai, "Light beam switching and modulation using a built-in dielectric channel in $LiNbO_3$ planar waveguide," *1977 IEEE/OSA Conf. Laser Engineering and Applications, Dig. Tech. Papers*, Washington, DC, June 1-3, pp. 7-8.

[17] T. Nakayama, Y. Nomura, H. Naito, K. Muto, and H. Kashiwagi,

"Monolithic LiNbO$_3$ integrated optical circuit," *1977 Int. Conf. Integrated Optics and Optical Fiber Communication, Tech. Dig.*, Tokyo, Japan, July 18-20, pp. 225-258.
[18] R. K. Winn, and J. H. Harris, "Coupling of multimode to simple-mode linear waveguides using horn-shaped structures," *IEEE Trans. Microwave Theory Tech.*, vol. MTT-23, pp. 92-97, 1975; A. R. Nelson, "Coupling optical waveguides by tapers," *Appl. Opt.*, vol. 14, pp. 3012-3015, 1975.
[19] W. K. Burns, A. F. Milton, and A. B. Lee, "Optical waveguide parabolic coupling horns," *Appl. Phys. Lett.*, vol. 30, pp. 28-30, 1977.
[20] P. K. Tien, S. Riva-Sanseverino, and A. A. Ballman, "Light beam scanning and deflection in epitaxial LiNbO$_3$ electro-optic waveguides," *Appl. Phys. Lett.*, vol. 25, pp. 563-565, 1974.
[21] R. V. Schmidt and I. P. Kaminow, "Metal-diffused optical waveguides in LiNbO$_3$," *Appl. Phys. Lett.*, vol. 25, pp. 458-460, 1974.
[22] H. F. Taylor, "Optical waveguide connecting networks," *Electron. Lett.*, vol. 10, pp. 41-43, 1974.
[23] R. V. Schmidt and L. L. Buhl, "Experiment 4 × 4 optical switching network," *Electron Lett.*, vol. 12, pp. 575-577, 1976.
[24] C. S. Tsai, S. K. Yao, A. M. Mohammed, and P. Saunier, "High-speed guided-wave electrooptic and acoustooptic switches," *1974 Int. Electron Devices Meeting, Tech. Dig.*, Washington, DC, Dec. 9-12, IEEE Cat. 74CHO906-8ED, pp. 85-87.
[25] H. Wichansky, L. U. Dworkin, and D. H. McMahon, "Multi-mode optic devices for signal processing," *Proc. 1976 Electro-Optical Systems Design Conf.*, New York, NY, Sept. 14-16, pp. 561-568.
[26] Y. Tsunoda, A. Huang, and J. W. Goodman, "Proposed optoelectronic residue matrix-vector multiplier," *Appl. Opt.*, to be published.

Digital optical switch

Y. Silberberg, P. Perlmutter, and J. E. Baran

Bell Communications Research, Navesink Research and Engineering Center, Red Bank, New Jersey 07701

> We propose and demonstrate a novel polarization- and wavelength-independent digital electro-optic switch in Ti:LiNbO$_3$. This four-port integrated optics switch is characterized by a steplike response to the applied voltage. Switching is achieved through adiabatic eigenmode transformation in an asymmetric waveguide junction. We demonstrate switching of both polarization components at two wavelengths (1.32 and 1.52 μm) with a crosstalk of -20 dB.

We propose and demonstrate a novel four-port (2×2) digital electro-optic switch where the light output exhibits a steplike response to the switching voltage. This response characteristic eliminates the need for precise voltage control for switching and it therefore enables the operation of many such elements by a single voltage source, as required for switching arrays. Moreover, the new device can switch both polarization components simultaneously and is insensitive to the precise wavelength of light. The performance of this switch does not depend critically on any of the design parameters, and therefore it is quite tolerant to variation in the fabrication process.

Most electro-optic switches are interferometric in nature, i.e., they require a precise phase shift to achieve a switched state with low crosstalk. The directional coupler switch, for example, requires a phase shift of $\sqrt{3}\,\pi$ between its two waveguides to switch.[1] Because of small fabrication variations, this phase shift requires slightly different voltages for each switching element in a switch array. It is also very difficult to switch the two orthogonal polarization components simultaneously. Another class of 2×2 switches is based on modal interference. This class includes the bifurcation optique active (BOA) switch,[2] the X switch,[3] and the symmetric directional coupler switch.[4] All exhibit a sinusoidal response to the applied voltage and therefore require precise voltage control as well. A third class of switches based on an asymmetric Y junction was investigated by several groups in the mid 1970's.[5,6] These were 1×2 switches which operated on principles similar to the switch described here. Possibly due to the early state of fabrication technology these studies yielded devices with less then perfect characteristics, and subsequently the interest in them declined. We believe that the 2×2 switch proposed here and its demonstrated advantages will revive the interest in this class of structures.

The digital switch is shown schematically in Fig. 1 for two possible realizations in lithium niobate. The structure is based on an asymmetric waveguide junction, composed of two unequal input guides, a double-moded central region, and a symmetric output branching. The symmetry of the output branching can be broken by applying an external electric field. An asymmetric waveguide branching is known to perform mode sorting.[7] The fundamental or first-order mode of the central region can be excited by launching light through the wider or the narrower input guides, respectively. These structures were studied extensively by Burns and Milton.[8,9] They have shown that mode sorting is obtained if the angle θ satisfies

$$\theta \ll \delta\beta/\gamma, \qquad (1)$$

where $\delta\beta$ is the average difference between the propagation constants of the two normal modes and γ is their transverse propagation constant in the cladding region. The salient feature of the asymmetric Y junction is the fact that mode sorting is obtained not at a particular point of operation, but for a range of parameters satisfying Eq. (1). By making the angle small enough one can assure mode sorting. The two normal modes in the junction area can similarly be routed to the required output guide by properly biasing the output branching. The fundamental mode will be directed always to the arm with higher index of refraction if the bias is high enough.

The device switches by symmetry breaking: switching is not periodic or quasiperiodic, but depends only on the direction of the bias. Each input arm excites an orthogonal normal mode at the input plane. This mode evolves adiabatically so that it remains a local normal mode at each location

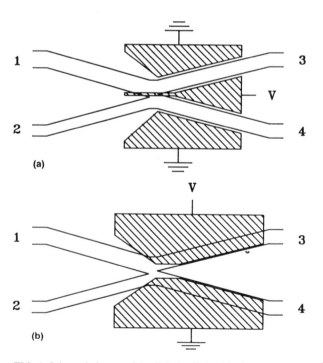

FIG. 1. Schematic layout of the digital switch with electrode pattern suitable for (a) x-cut and (b) z-cut LiNbO$_3$.

Reprinted with permission from *Appl. Phys. Lett.*, vol. 51, no. 16, pp. 1230–1232, July 1978.
Copyright © 1978, American Institute of Physics.

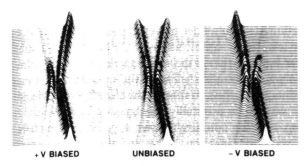

FIG. 2. Simulations of light propagation along the switch. The input guides are 2.75 and 3.25 μm wide, the output guides are 3 μm wide each, the index of refraction of the guides without bias is 2.205, and the background index is 2.200. The angle between the guides is 1 mrad. Light is coupled through the narrow input guide at the bottom.

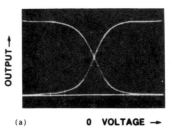

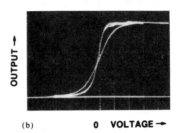

FIG. 4. Experimental results. (a) Superimposed light outputs from the two output waveguides vs bias voltage for TE polarized light at 1.32 μm. (b) Superimposed traces of TE and TM polarized light through one of the output guides vs bias voltage. The horizontal scale is 20 V/div.

along the structure. Switching is therefore achieved without resorting to modal interference. Care must be taken to ensure that the applied field does not disrupt the adiabatic propagation. For this purpose the electrodes in Fig. 1 are shaped so that the electric field increases gradually towards the junction area.

Figure 2 shows a simulation of light propagation in such a switch. The simulation assumes a step index slab waveguide configuration, and it uses the local normal mode approach for the propagation analysis.[8] Light is assumed to be coupled to the narrow input guide of the structure. As can be seen, the first order mode (double-humped intensity distribution) is excited at the center of the device, which is then routed to the arm with decreased index of refraction. Using a similar analysis we calculated the switch response as a function of the induced index imbalance between the output arms. Figure 3 shows the power at each of the output guides when the narrow guide is excited. As expected, the output is evenly split between the output guides without any bias. As the bias is increased, light is coupled preferentially to the guide with lower index of refraction. Switching with crosstalk better than − 20 dB is obtained, for this specific param-

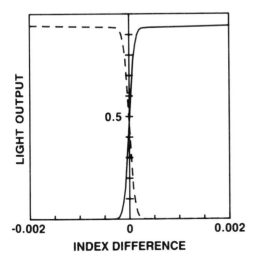

FIG. 3. Calculated output intensity through the two output guides in a switch with parameters of Fig. 2.

eter set, when an index imbalance of 0.0002 is induced between the output guide. This crosstalk drops below − 30 dB and does not increase above this value as the switching voltage is increased by an order of magnitude above the minimal switching voltage. Note that the required index imbalance, and hence the switching voltage, depends on the design parameters. Unlike most electro-optic switches, there is no precise voltage-length product which is required for switching.

The steplike response eliminates the need for precise voltage control, as any voltage in this wide range will be effective. Moreover, this response can be used to generate polarization-independent switching in lithium niobate devices. In common z- or x-cut lithium niobate one polarization component is affected by an electro-optic interaction which is three times weaker than the other component. It is obvious that in a switch with the above response both polarizations can be switched by a strong enough bias. Another advantage of this switch is its insensitivity to the precise wavelength. Indeed, the switching voltage may change with wavelength due to changes in optical mode size. However, switching with a good crosstalk figure should still be obtained, provided that the waveguides involved are still single mode. It is conceivable then that a single element will switch a broad wavelength range, for example, 1.3–1.55 μm, simultaneously. Another scheme for polarization-independent switching using directional couplers has been demonstrated[10]; however, the weighted coupling required is sensitive to the fabrication process and is accompanied by a substantial penalty in switching voltage.

We have fabricated this switch using Ti:LiNbO$_3$ technology. The waveguides are formed by diffusing 675 Å of titanium for 6 h at 1050 °C in a wet Ar atmosphere. The input waveguides were 6 and 8 μm wide, the output waveguides were 7 μm wide, and the junction angle was 1 mrad. We have used a white-light measurement technique[11] to verify that all these waveguides carried a single mode in the wavelength range of 1.3–1.6 μm. We have used a z-cut substrate with the electrode structure shown in Fig. 1(b). Note that because the electrodes separate along the output branching, their effectiveness is reduced compared with the x-cut version of Fig. 1(a). The gold electrodes were defined using a lift-off technique, and they extended 1.5 cm beyond the junction region.

Figure 4(a) shows the measured output of TE polarized light at 1.32 μm as a function of voltage from both output ports. Note that this is the least sensitive polarization component in a z-cut crystal. The steplike response is clearly demonstrated. Figure 4(b) shows the superimposed responses of both TM and TE polarizations at one output port. It is obvious that by applying a strong enough field both polarizations can be switched simultaneously. Identical results were obtained for light at a 1.52-μm wavelength. The measured crosstalk was better than -20 dB for the TE polarization and was slightly worse for TM because of residual surface guiding. The switching voltages to obtain at least -15 dB crosstalk in this first device were ± 15 V for TM, ± 45 V for TE for both wavelengths. Note that the switching voltage is not increased for the longer wavelength, probably because of longer effective interaction length. This switch can therefore switch simultaneously randomly polarized light at both wavelengths.

Although the switching voltages obtained were relatively high, we believe that they can be lowered considerably by improving the design. A simple and straightforward modification would be to fabricate the switch on x-cut $LiNbO_3$ as explained before. This would enable bigger and uniform refractive index changes along the output guides for the same applied voltage. We believe, however, that this structure may be optimized in a wider sense. Unlike electro-optic switches that operate by interference, the digital switch cannot be characterized by a single phase shift required for switching. Our simulations show that approximately 2π rad were required in our model; however, this number can probably be lowered by optimizations of the design parameters. Note for example that as the voltage is increased the output waveguides are decoupled closer to the junction area. It may be advantageous, then, to shorten the electrodes and decouple the waveguide externally by changing their width. In addition, one may optimize the switch parameters by using curved waveguide junctions. We would also like to comment that this switch can be easily adapted to high-speed operation by a proper design of traveling wave electrode.

In conclusion, we have proposed and demonstrated a polarization- and wavelength-independent electro-optic switch that exhibits a steplike response to the switching voltage. It should be useful in switching arrays and as a polarization- and wavelength-independent switch.

The authors wish to acknowledge helpful discussions on these switching structures with E. Kapon.

[1] R. V. Schmidt and R. C. Alferness, IEEE Trans. Circuits Syst. **CAS-26**, 1099 (1979).
[2] M. Papuchon, A. Roy, and D. B. Ostrowsky, Appl. Phys. Lett. **31**, 266 (1977).
[3] C. S. Tsai, B. Kim, and F. R. El-Akkari, IEEE J. Quantum Electron. **QE-14**, 513 (1978).
[4] R. A. Forber and E. Marom, IEEE J. Quantum Electron. **QE-22**, 911 (1986).
[5] W. K. Burns, A. B. Lee, and A. F. Milton, Appl. Phys. Lett. **29**, 790 (1976).
[6] H. Sasaki and R. M. De La Rue, Electron. Lett. **12**, 459 (1976); 636(E) (1976); H. Sasaki and I. Anderson, IEEE J. Quantum Electron. **QE-14**, 883 (1978).
[7] H. Yajima, Appl. Phys. Lett. **22**, 647 (1973).
[8] W. K. Burns and A. F. Milton, IEEE J. Quantum Electron. **QE-11**, 32 (1975).
[9] W. K. Burns and A. F. Milton, IEEE J. Quantum Electron. **QE-16**, 446 (1980).
[10] R. C. Alferness, Appl. Phys. Lett. **35**, 748 (1979).
[11] J. E. Baran, Y. Silberberg, and P. Perlmutter, Conference on Optical Fiber Communication, Reno, Nevada, January 1987, paper TUO2.

Section 2.2: Logical Switching Elements

Nonlinear Fabri-Perot Structures

Optical bistability, photonic logic, and optical computation

S. D. Smith

> This paper presents new results and reviews the latest state of research in all-optical nonlinear logic switches, amplifiers, and memories. Optical circuit elements that perform the logic functions of the electronic computer are described. Switching speed on a picosecond time scale, the availability of fast high bandwidth consistent communication, and the application of optical parallelism in free space optical wiring are some advantages of the optical computing elements.

I. Introduction

The main purpose of this paper is to present new results and review the latest state of research in all-optical nonlinear logic switches, amplifiers, and memories.

The response of semiconductor microelectronics to the high data rate requirements of digital signal processing and computing has been to increase switching speeds and further miniaturize components in the form of very large scale integration (VLSI). The precepts of information storage in the form of energy are maintained[1] along with the need for power gain in switching and logic restoration, while switching speed has been increased. Transistors made from GaAs have been reported with effective switching times as fast as 12 ps.[2] However, this will not necessarily solve the problem of coping with a high data rate since the processing time in conventional computers is many times the logic switching time due to the necessity to transfer information to the next part of the circuit. This involves capacitance time constant limits. As noted, for example, by Huang[3] and Mead and Conway,[1] VLSI does not solve the RC time constant problem. As the length of a wire shrinks by a factor of α and the cross-sectional area of the wire is reduced by a factor of α^2, the capacitance of the wire decreases by this factor of α while the resistance increases by the same amount. Thus the time constant remains the same, and the input charging time remains unaltered independent of scaling.[1]

The standard method of computer communication in use today connects the logic unit with the memory through an address device. This reduces the number of interconnections but can only address one storage element at one time. This widely used scheme was first suggested by John von Neumann, but, rather than receiving credit for this most practical innovation, he is now rather undeservedly blamed for this so-called von Neumann bottleneck (Fig. 1). The timing problems associated with circulating logic signals around a 1-D processor of this type (clock skew) combine to indicate that future problems in digital computers are likely to be those of communication. This may apply at the architectural, bus, and chip level and stems from use of time multiplex to compensate for the inability of electrical methods to communicate many channels of information in parallel.

If optics can help to further high speed computing, generic terms are required to describe the ideas of optical logic, nonlinear optical switching, optically bistable memories, optical circuit elements, the design of digital optical circuits, and a new optical computer architecture. The starting point may be in the electronically based communication by optical methods through the use of optical fibers in long-range telephone lines. The higher carrier frequency used gives potentially higher bandwidth, albeit electronic limitations on modulation techniques have restricted our ability to exploit fully this greater information carrying capacity. Currently it is the long range made possible by low signal attenuation that has been utilized. Optics is responsible for the man–machine interface at the input and output of a computer. Further use of optics for processing information depends on optical circuit elements that will provide the advantages of noninterfering propagation at the speed of light, high available bandwidth, and the ease of use of parallelism. A simple lens costing very little can readily transmit millions of resolvable spots, equivalent when combined with optical circuit elements to

The author is with Heriot-Watt University, Physics Department, Riccarton, Edinburgh, EH14 4AS, U.K.
Received 2 October 1985.

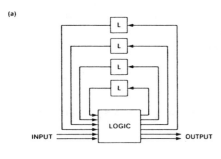

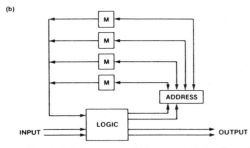

Fig. 1. (a) Classical finite state machine not suffering from the Von Neumann bottleneck, since it can update all its memory in parallel without the need for addresses. (b) Modified finite state machine suffering from the Von Neumann bottleneck since it can only update one memory element at a time and consequently needs an address to do so.

millions of electrical wires. Optical methods promise advantages in image processing and recognition, sorting, radar array signal processing, machine vision, and artificial intelligence.

Optical computing, practiced for the last 20 years by optical researchers using linear processing devices, e.g., spatial filters and Fourier transform processes in image processing, has recently been extended to numerical optical computing techniques. A collection of relevant papers[4-8] contains ideas on the use of symbolic substitution, residue arithmetic, vector–matrix multiplication, and primitive processing experiments using a liquid crystal light valve as a hybrid (i.e., optics with electronics) spatial light modulator but not itself capable of all-optical logic action. Most of the reports are theoretical proposals of what might be done.[9]

The missing link has been the absence of nonlinear all-optical circuit elements preferably capable of fabrication in the form of 2-D arrays and of small enough size to have small switching energies and high speeds.

In addition to the purpose noted early in Sec. I, it will further be shown that, with the experimental realization of the early implementation of all-optical circuits,[8] it is time to combine these two previously separate research efforts.

A comparison of optical circuit elements with electronic components shows that electrons interact with each other very strongly at short range, needing insulation to separate their motion, whereas photons only interact in the presence of nonlinear optical material. Until fairly recently it was thought that nonlinear optical effects implied the use of powerful laser beams with intensities of the order of MW/cm^2 implying oscillating electric fields of the order of 10^7 V/cm—comparable to interatomic fields. It has now been shown that nonlinear devices can be operated with power densities less than W/cm^2 and input powers for an individual device of the order of microwatts or less.[8] This has led to recent progress in all optical circuit elements. This progress is mostly based on the combination of optical nonlinearity and feedback and the concept of optical bistability with a whole family of devices based on a common set of physical and mathematical principles. The series of devices, recently described by Smith,[10] includes optical logic gates, bistable memories, amplifiers, sometimes known as optical transistors or transphasors, and power limiters.

Historically, Szoke et al.[11] in 1969 proposed that a Fabry-Perot optical resonator containing a saturable absorber as its spacer layer could exhibit two states of transmission for the same input intensity. This simple bistable action was predicted to arise from the existence of a high internal optical field at constructive interference given that sufficient intensity has been incident on the resonator to bleach the absorber. To reach this condition required a greater input intensity than that required to maintain it. By contrast, at low input intensity the nonbleached absorption held the transmission of the device at a low level. In practice this condition is quite hard to achieve experimentally, and the experiments described do not in fact show optical bistability. Observation of optical bistability was not made until 1976 when Gibbs et al.,[12] using an interferometer containing sodium vapor, observed bistable transmission but deduced that the dominant mechanism was refractive—involving a shift in resonator frequency—rather than absorptive. Effective refractive nonlinearity nevertheless resulted from saturation of the atomic absorption. Such a device, although using only milliwatts of power, was relatively large (centimeters in length) and relatively slow (milliseconds) compared to electronic circuit components.

In the same year, 1976, the laboratory at Heriot-Watt University discovered a giant nonlinear refraction defined from

$$n = n_0 + n_2 I \qquad (1)$$

(where the nonlinear refractive index n_2 can be measured in units of cm^2/kW) present in the narrow band gap semiconductor InSb. This came about as a result of studying troublesome beam propagation in connection with a tunable laser device, the spin flip Raman laser, which used the same material and the same pump laser—a CO laser capable of emitting 70 closely spaced (and selectable) lines in the 5–6-μm infrared spectral region. A simple investigation of beam propagation by Weaire et al.[13] indicated that a Gaussian beam developed a twin peak output and doubled its output width with 30-mW input power. This surprising result led us to deduce the existence of a very large negative nonlinear refractive index n_2, of the order of 0.1–1 cm^2/kW. The immediate implication was that a bistable resonator could be made of micron dimension and, since the effect was shown to be electronic, would

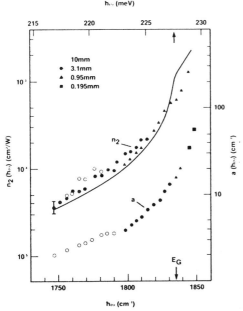

Fig. 2. Nonlinear refractive-index coefficient n_2 and absorption coefficient α plotted against wave number. The solid line is a semiempirical fit similar to Eq. (4), Sec. II (after Ref. 20).

be fast, probably on a nanosecond time scale. A second implication was that one beam could modulate the optical properties of a small slice of semiconductor and affect a second beam thus making an optical modulator or optical transistor. The details of both the refractive and associated absorptive nonlinearities at milliwatt powers are described in Miller et al.[14] but retrospectively can be recognized in the earlier work of Scragg and Smith[15] and Ironside.[16] Both the device possibilities described above had been practically realized in InSb by 1979[17] in which continuous wave laser beams were used leading to steady state operation and true optical bistability as well as the observation of gain in an optical transistor.[18] Simultaneously and independently optical bistability was reported by Gibbs et al.[19] using pulsed dye laser radiation in the larger gap semiconductor GaAs. Larger absorption and smaller nonlinearity in this material, however, prevented steady state operation, and this observation was, therefore, quasi-dynamic and did not permit the demonstration of logic levels or differential amplification.

II. Origin of Giant Nonlinearities

The physical explanation of the large nonlinearities in both these semiconductors involves the excitation of electrons to give some degree of saturation or blocking. In the case of InSb, exciton effects are negligible in the conditions of the experiment, and a plausible explanation has been given by Miller et al.[20] in terms of a dynamic Burstein-Moss shift of the band edge. Physically a number of electrons ($\sim 10^{15}-10^{16}$ cm^{-3}) are excited into lower conduction states by the laser photons. The incident intensity I generates an equilibrium number of electrons,

$$N = \frac{I}{\hbar\omega}\alpha_{\text{eff}}\tau_R, \quad (2)$$

where α_{eff} is the absorption coefficient generating carriers, and τ_R is the carrier lifetime. Subsequent band filling following thermalization modifies the absorption edge, and by application of the Kramers-Kronig relationship causes a change in refractive index Δn:

$$\Delta n = \frac{\hbar c}{\pi}\int_0^\infty \frac{\Delta\alpha(\hbar\omega')}{(\hbar\omega')^2-(\hbar\omega)^2}d(\hbar\omega'), \quad (3)$$

where $\alpha(\hbar\omega)$ is the interband absorption coefficient. Hence using Eqs. (1) and (3) and standard semiconductor band theory with the approximation $n_2 = \Delta n/I$ we obtain

$$n_2 = \frac{-4}{3}\sqrt{\pi}\,\frac{e^2 P^2}{n_0 kT}\frac{\alpha_{\text{eff}}\tau_R}{(\hbar\omega)^3}F(\hbar\omega/E_G), \quad (4)$$

where P^2 is a constant for most semiconductors and F is a function resonating at the band gap E_G.

Recent experimental results are shown in Fig. 2 and indicate a good fit to this semiempirical theory. The magnitude of n_2 and the coincident bandtail absorption α are important. n_2 is ~ 0.1 cm^2/kW, and α is ~ 10 cm^{-1} at a typical operating point. This means that (if the nonlinearity is constant) the intensity of 1 W/cm^2 can change n by 0.001 and can be obtained from a milliwatt power laser focused on a spot of diameter of about a third of a millimeter: a very modest power density. This change in n of only $\sim 10^{-3}$ is sufficient for device action. The absorption level means that in a sample of 100 μm thick, negligible temperature rises occur. In GaAs the mechanism seems to involve the saturation of a discrete exciton peak and leads to an $n_2 \sim (10^{-4}-10^{-5})$ cm^2/kW, also usefully large. This is, however, accompanied by a high absorption level ($\alpha \sim 10^3-10^4$ cm^{-1}), which means in practice that devices are restricted to a thickness of a few microns and that thermal runaway is a problem. Since this early work a variety of materials have been shown to show giant nonlinearity (see Table I); these large values should be compared with the more conventionally accepted quantities known before 1976. Comparison can best be made by noting that n_2 is related to the third-order susceptibility $\chi^{(3)}$ through the relationship

$$\text{Re}\chi^{(3)} = \frac{n^2 c}{4\pi^2}n_2 \text{ (esu)},$$

$$\left\{\chi^{(3)}(\text{esu}) = \frac{9\times 10^8}{4\pi}\chi^{(3)}[\text{SI}(\text{V/m})]\right.$$
$$\left. = \frac{9\times 10^8}{4\pi}4\epsilon_0 n_0^2 c n_2[\text{SI}(\text{cm}^2/\text{W})]^2\right\}. \quad (5)$$

This quantity defines the third-order polarization in the expansion $\mathbf{P} = \chi^{(1)}\mathbf{E} - \chi^{(2)}\mathbf{E}\cdot\mathbf{E} - \chi^{(3)}\mathbf{E}\cdot\mathbf{E}\cdot\mathbf{E}$. The usual values of $\chi^{(3)}$ in solids are in the $10^{-9}-10^{-12}$-esu range. A simple explanation of this very large range of nonlinearity has been given by Wherrett.[47,48]

A quasi-dimensional analysis provides a first assessment of $\chi^{(3)}$. Consider N_0 discrete atoms/unit volume with just two energy states with energy difference E

Table I. Materials in which Intrinsic Optical Bistability has been Observed

Material	Operating Temperature (K)	Bandgap λ_g (μm) (or Exciton energy)	Operating Wavelength (μm)	Irradiance (w/cm^2)	Recombination Time (ns)	Mechanism	Reference
InSb	5	5.3	5.4	300	500	a	17
InSb	77-120	5.4	5.5	20	500	a	21
InSb	77	5.4	5.5	60	--	b	22
InSb	300	7	10.6	200 K	47	a	23
GaAs	5-120	0.8	0.82	100 K	40	a	19
GaAs	300	0.85	0.857	100 K	40	a	24
GaAs/GaAlAs	300	0.806	0.83	100 K	40	a	24
GaAs/GaAlAs	300	0.845	0.857	900	--	b	25
CdS	2	0.481	0.487	300	.5	a	26
CdS	2	0.481	0.489	600 K	.5	c	27
CdS	2-50	0.481	0.487	1 K	--	b	28
Cd$_{0.23}$Hg$_{0.77}$Te	77	10	10.6	4 K	2500	a	29
Cd$_{0.23}$Hg$_{0.77}$Te	300	6.5	10.6	100 K	20	a	30
InAs	77	3	3.1	75	200	c	31
CaCl	7	0.3849	0.3900	10 M	.5	a	32
Ta	300	2.5	10.6/5.3	10 M	100	a	33
Si	300	1.11	1.06	1 K	--	d	34
GaSe$_z$	300	0.5	0.6328	100	--	b	35
ZnSe	300	0.48	0.5145	250	--	d	36
Znse	300	0.48	0.5145	1 K	--	b	37
ZnS	300	0.3	0.5145	10 K	--	d	38
Ma	400	--	0.5890	0.5	--	a	12
Ruby	77	--	0.6328	2 K	--	a	39
Liquid Crystal	300	--	0.5145	10	--	d	40
Rb	450	--	0.7779	--	--	a, c	41
Glass	300	--	0.6471	1 K	--	d	42
MBBA	300	--	0.694	4 M	--	a	43
CS$_2$	300	--	0.694	12 M	--	a	43
Nitrobenzene	300	--	0.694	10 M	--	a	43
Dye	300	--	0.5	100 πW	--	d	44
SF$_6$	300	--	10.6	1 M	--	e	45
NH$_3$	300	--	10.6	1 M	--	e	46

(a) Electronic Refractive
(b) Thermal Absorptive
(c) Electronic Absorptive
(d) Thermal Refractive
(e) Vibronic Refractive

and transition dipole moment er. Introducing the electromagnetic field interaction 3 times to obtain the third-order polarization $N_0 er$, one obtains

$$\chi^{(3)} \propto N_0 e^4 r^4 E^{-3} F\left(\frac{\hbar\omega}{E}\right), \quad (6)$$

the factor F, determined by the closeness to resonance, can be set to unity for purposes of comparison, and with N_0 representing the density of dipoles at NTP (10^{18} cm^{-3}) we obtain a value of $\chi^{(3)} \sim 10^{-18}$ esu—a very small nonlinearity.

Transferring the argument to solids, the density of dipoles rises to $\sim 10^{23}$ cm^{-3} with N_0 replaced by the k-dependent energy gap $E(k) = E_G + \hbar^2 k^2/2m_r$. In semiconductor notation it is usual to express the dipole moment through a momentum operator described by the Kane P-parameter proportional to the momentum matrix element for interband transitions; thus expressing the reduced effective mass m_r in terms of P through $k \cdot p$ theory we have

$$\chi^{(3)}_{\text{semiconductor}} = e^4 P E_G^{-4} F(\hbar\omega/E_G). \quad (7)$$

If we again assume $F = 1$, $\chi^{(3)} \sim 10^{-9}$ esu. The advantage of this notation is that the quantity P is essentially constant over a large range of semiconductor materials. The dependence of $\chi^{(3)}$ as E_G^{-4} shows that small gap semiconductors are strongly favored: between visible wavelengths and (say) 5 μm in the IR, factors of approximately thousands are involved. The quantity F will contain resonance enhancement factors. A four-stage transition scheme describes the three electromagnetic field interactions and the emission of a photon via the generated polarization.[47] For the case of refractive nonlinearity [real part of $\chi^{(3)}$], all frequencies are the same leading to multiple resonance. The magnitude of the resonance enhancement will depend on the energy mismatch (ΔE) between the intermediate (virtual) state of the system and the initial state. With degenerate frequencies $\Delta E = 0$, and the contributions diverge to infinity. In practice each intermediate state exists for a time Δt leading to a state broadening following the uncertainty principle ($\Delta t \sim \hbar/\Delta E$). Up to a nine-order-of-magnitude increase in $\chi^{(3)}$ can be derived from F.

Near to resonance, the nonlinearity becomes active with real excitations which modify the optical properties of the material over a range of time scales from picosecond to microseconds. The virtual nonlinearities are enhanced in the ratio of the carrier lifetime τ_R to the state-broadening interval Δt, the latter being of the order of picoseconds.

An analytical expression for n_2 has been given by Miller et al.[20] [Eq. (4)]. This expression gives a good description of the resonant refractive nonlinearity in InSb, InAs, and CdHgTe; α is determined empirically from experiment.

For experiments which integrate over thermal relaxation times (tens of picoseconds) the mechanism then becomes the same as discussed above as band filling. More recently Walker et al.[21] and Miller et al.[49] have improved the description of intensity dependence of the refractive nonlinearity by including the dependence of carrier lifetime on excess carrier density.

III. Device Physics

The simplest configuration which provides optical feedback is a simple Fabry-Perot etalon containing a nonlinear refractive material (Fig. 3). Its optical thickness is given by

$$nL = (n_o + n_2 I_c)L, \quad (8)$$

and this changes with the internal intensity I_c. Consider now (Fig. 3) the transmission of such an interferometer as a function of intensity: if we start in an initial condition where the illuminating laser wavelength is detuned from maximum transmission by a wavelength increment $\delta\lambda$ [Fig. 3(a)], we see from Fig. 3(b) that the relation between output and input would give rise to, in the case of a linear device, a straight line of low slope: if the device were tuned to resonance and there were no absorption loss, output would be related to input by a line at 45°. If we now increase the intensity from the initial condition, the nonlinear resonator tends toward resonance as its optical thickness changes with intensity. This would give rise to a nonlinear relation between output and input. However, as we approach resonance the internal field circulating within the resonator itself builds up according to

$$I_c = I_i T(1 + R)/(1 - R), \quad (9)$$

where I_i is the incident intensity, T is a function of frequency [as in Fig. 3(a)], and R is the (constant) reflectivity of the resonator mirrors; thus at resonance the internal intensity is at its maximum where $T = 1$ and is amplified by the term $(1 + R)/(1 - R)$. This gives rise to positive feedback. As resonance is approached the internal field builds up, and the rate of approach to resonance depends on the change in optical thickness, which itself depends on the magnitude of the internal field. The rate of approach to resonance thus speeds up. This can be readily expressed through the expression

$$d\frac{(\delta\lambda)}{dI_i} = T(\lambda) \bigg/ \left(\frac{T_{\max}}{2n_2 L} - I_i \frac{dT}{d\lambda}\right). \quad (10)$$

Figure 3(c) shows a plot of the rate of approach to resonance as a function of either incident intensity I_i {or as a function of time if we assume a linear ramp of I_i against time [Fig. 3(d)]}. As I_i increases $(dT)/(d\lambda)$ also

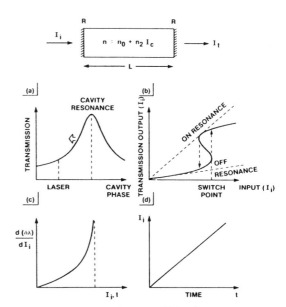

Fig. 3. Dynamics of switching in a Fabry-Perot etalon—refer to text for individual diagrams.

varies, the denominator of Eq. (10) tends to zero, and the rate of approach to resonance becomes infinite leading to rapid switching. This gives a physical feel for the dynamics of the switching process.

A simple description of bistable switching can be made by considering a plane wave model of the linear Fabry-Perot resonator. In this model the intensity is assumed constant throughout the resonator both transversely and longitudinally. The Airy sum of multiple beams for such a resonator, modified for intensity-dependent optical thickness, is given:

$$T(\lambda, I_i) = \frac{I_t}{I_i} = \frac{1}{1 + F \sin^2 \delta/2}, \quad (11)$$

where I_t is the transmitted intensity and the intensity-dependent phase change $\delta = 2\pi n L/(\lambda/2)$ so that $\delta(I_c) = 4\pi n_0 L/\lambda + 4\pi n_2 L I_c/\lambda$.

A second representation for T can also be obtained by relating the internal and external intensities I_c and I_i:

$$T(\lambda, I_i) = \frac{I_c}{I_i} \frac{(1 - R)}{(1 + R)}. \quad (12)$$

Only certain values of I_c for certain I_i have satisfy self-consistency for Eqs. (11) and (12). The solution is best found graphically (Fig. 4); Eq. (11) gives a straight line relationship of vertical slope at zero intensity and lessening slope as the intensity is increased. Intersections with Airy's sum, Eq. (12), give the solutions. As the slope is reduced intersections eventually show two simultaneous solutions. This represents optical bistability, i.e., two values of the transmission for one of the intensity. Two features are noteworthy:

(1) a large value of finesse (ratio of spacing to half-width of the transmission peaks) makes it easier to obtain bistability in first order—otherwise sufficient

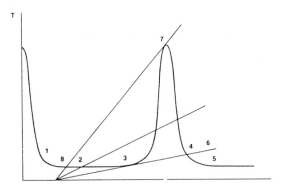

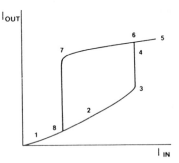

Fig. 4. Graphic solution of Eqs. (11) and (12) in Sec. III, the sequence 1–8 is equivalent to increasing and decreasing the incident power.

intensity and nonlinearity to reach higher orders are necessary; and

(2) the positions of intersection and the form of the curve implied by intersections change with the initial condition, i.e., depend on $\delta\lambda$.

We illustrate such a set of characteristics in Fig. 5, which is in fact an early experimental result using the CO laser and a plane-parallel interferometer constructed from InSb. The family of curves shows the characteristics obtained for different values of the initial detuning parameter $\delta\lambda$ if the initial condition is close to resonance; power limiting action is obtained (curve a) since the transmission can only fall as intensity rises. As $\delta\lambda$ is increased, curve b begins to kink, and c shows a greater change in output than for an incremental change in input, thus exhibiting differential gain. This is responsible for optical transistor or transphasor action. A further increase of $\delta\lambda$ leads to bistable loops of varying width.

Thus by simple change of initial conditions a whole family of optical devices can be produced.

IV. Figure of Merit for Optical Circuit Elements

Analysis of the factors concerned in the design of such optical circuit elements has been given by Miller.[50] He shows that the quantity which gives a figure of merit (in terms of nonlinear Fabry-Perot cavity optical switching in the presence of linear absorption α) is $n_2/\lambda\alpha$, where

$$I_s = \frac{\lambda\alpha}{n_2} f(R,\alpha L). \quad (13)$$

This equation determines the lowest critical value of

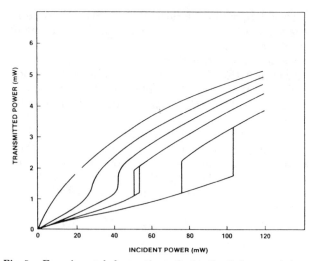

Fig. 5. Experimental observations of a family of characteristics of an InSb bistable device obtained by changing the initial detuning from resonance of the etalon from 0 to π.

input irradiance I_s for a device of given size to obtain bistable switching or nonlinear characteristic, where $f(R,\alpha L)$ is a function describing the cavity properties.

The result is physically sensible since the switching power will be lower for a larger nonlinearity n_2; the shorter the wavelength the smaller the refractive change $\Delta(nL)$ that will be required to effect a change from constructive to destructive interference (i.e., $\Delta nL = \lambda/2$), and for smaller absorption (assumed linear in this analysis) the longer the device may be for a given loss.

We have found in our experimental work that a value of $n_2 \sim 0.1 - 1$ cm^2/kW and $\alpha \sim 1$–100 cm^{-1} gives useful devices of thickness L 50–200 μm, bearing in mind that a fractional change $\Delta n/n \sim 10^{-3}$ is usually required.

The nonlinearity described is for the case of electronic origin and not dependent on material extent.

For nonlinearities of thermal origin I_s may take the form[51]

$$I_s = \frac{\lambda\,\alpha\,\kappa_s}{\left(\frac{\delta n}{\delta T}\right)L'} \frac{f(R,\alpha L)}{\alpha L},$$

where κ_s is the thermal conductivity of the substrate, L is the cavity (or active layer) thickness, and L' is the substrate thickness. Hence in this case

$$n_2 = \frac{\delta n}{\delta T} \frac{aLL'}{\kappa_s},$$

and device dimensions are included.

In our work, to obtain favorable values for $n_2/\lambda\alpha$, we make particular use of two effects resonant with the energy gap E_G in semiconducting materials.

In designing such devices with electronic origin n_2, we must combine these trends of the figure of merit with a development of Sec. II: that $n_2 \sim 1/E_G{}^3$ so that $I_s \sim \lambda^2$. In addition the considerations of Sec. II suggest that the smallest possible devices should be constructed. Diffraction limits suggest that the area limit will be $\sim(\lambda/n)^2$. Thus, although the nonlinearity is clearly large at longer wavelengths for small gap materials, the interference conditions and device size favor shorter wavelengths. Analysis of the detail of the frequency dependence of the nonlinearity, together with the attending unwanted losses, does not at this time indicate an optimum wavelength.

Two materials have been extensively investigated: InSb (Fig. 2) with a typical working wavelength (5.5 μm) corresponding to 1820 cm^{-1} yields excellent devices with $n_2 = 0.1$ cm^2/kW and $\alpha = 10$ cm^{-1} at 77 K. This material is one of the few in which there are sufficient available laser frequencies to undertake detailed examination of the frequency and hence resonant behavior of n_2 and α near the band gap.

GaAs has also been much investigated[19,24,25] but differs by having a strong exciton feature near the absorption edge. The nonlinearity $n_2 \sim 10^{-3}$ cm^2/kW is quite practicable for devices, but the absorption coefficient in epitaxially grown material is so far giving values of $\alpha \sim 10^4$ cm^{-1}. Thus thicknesses are restricted to a few microns, and thermal stability poses a problem.

Both the above materials show negative values of n_2 caused by electronic effects. Carrier lifetimes are in the range of tens to hundreds nanoseconds. The nonlinearities can be switched on more quickly than this time interval by rapidly introducing carriers with relatively intense pulses. The relative figures of merit between GaAs and InSb favor InSb by a factor of 10^5, which has given InSb the advantage of steady state operation but the disadvantage of 77 K temperature, whereas GaAs has operated at room temperature and at shorter wavelengths. So far there are no reports of true steady state operation for GaAs material. Other materials, which have been used in purely optical switching devices, are listed in Table I.

There is a second useful form of nonlinearity involving thermal shift of the band edge due to temperature rise of the bulk of the nonlinear material. This effect also resonates with the band edge and is associated with moderate values of absorption coefficient. Examples in bulk material include silicon, GaAs, CdHgTe, InSb, CdS, and ZnSe.

Particularly useful, however, is when such thermal effects are associated with thin film devices: these take the form of interference filters, and film thickness (~ 2 μm) implies a very small heat capacity so that in small devices nanosecond time constants are not impossible.

The important physics lies in the control of heat sinking by means of the conductivity of a relatively massive substrate. Figure 6 shows the very promising results obtained using ZnSe as the active film which has a band gap conveniently resonant with argon-ion laser wavelengths at 514 and 528 nm.

A fairly complete list of practicable materials for optically bistable and optical logic devices is shown in Table I. The table includes the earliest results in sodium vapor for historical reasons but otherwise presumes that atomic gaseous devices will be too bulky

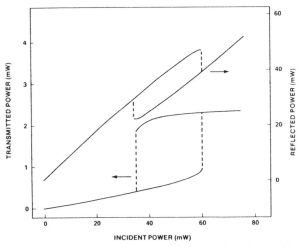

Fig. 6. Power transmitted and reflected from an interference filter containing ZnSe plotted against incident power showing optical bistability.

and too slow for practical applications. In general the semiconducting materials using both electronic and thermal nonlinearity and organic materials such as liquid crystals seem at the moment to be promising for a new technology.

V. Requirements for Optical Computing

Since our proposition is to undertake logic operation by means of these optical circuit elements, we can define some of the requirements if they are to be put together in the form of optical circuits to construct an optical computer. Requirements are as follows:

A. High Contrast

A logic device needs to show a large change between logic 0 and logic 1 levels.

B. Steady State Bias

To make various different logic gates it is necessary to be able to alter controllably optical bias levels. In terms of optical bistability this means that the device can be held indefinitely at any point on the characteristic with a cw laser beam—the holding beam—and implies a degree of thermal stability. Devices based on InSb at 77 K and on ZnSe at 300 K have been the first to show this behavior.

C. External Address

For logic functions it is clearly necessary that separate external signal beams can be combined with the holding beam to switch the device. The switching energy in fact is derived from the holding beam, and this switched beam propagates in transmission or reflection as the output beam to the next devices in an optical circuit.

D. Elements Must be Cascadable

This means that the output of one device must be sufficient to switch at least one succeeding device. The ability to set a cw holding beam near the switch point in fact fulfills this condition, since the extra increment is then small compared with the change in output even in the presence of loss. Since each device has its own power supply, holding beam, logic levels are restored.

E. Fan-Out and Fan-In

The probable advantages of parallel processing in optical devices place emphasis on the ability of one device to drive a large number of succeeding devices probably using free space propagation for address. The summed effect of several elements can readily be focused on one device to achieve fan-in.

F. High Gain

Items in Sec. V.D and E demand that the elements show a value of differential gain >1.

G. Arrays

The technology shall ideally be such that 2-D arrays are easily constructed.

H. Power and Speed

Low power per device is a necessity, and this should preferably be approximately milliwatts or less. Speed and power will be interchangeable, but speed itself will vary with use: for parallel arrays microseconds will suffice, and for 1-D circuits subnanosecond or picosecond switching times are desirable.

VI. Experimental Results

Figure 7 shows optically bistable switching in a 210-μm thick InSb resonator operating in reflection and utilizing the natural reflection ($R = 0.36$) of the material. The power is provided by an Edinburgh Instruments PL3 CO laser grating tuned to operate on a line at 1820 cm^{-1}. Wherrett[52] has shown theoretically that for the parameters of such a low finesse device, the reflection mode is favorable: as the figure shows this bistable switch has a contrast ratio of nearly 5:1 and a change in output power of 4 mW on switching. This indicates the required high contrast between logic states of Sec. V.A.

Figure 8 shows an earlier result[53] for an InSb resonator operating in transmission. Although the critical switching power is 20–25 mW incident on a diameter of ~200 μm, the illustrated loop is extremely stable, and the holding beam can be adjusted to hold stably to within 3 mW of the switch point. This remarkable result indicates that the complete system is stable to within nearly one part in 10^4. If the device is set up in this way with the holding beam adjusted in intensity to be just short of the switch point, the device may be externally addressed. In Fig. 8 the result is shown of addressing with a single 35-ps long pulse of 5-nJ energy from a Nd:YAG laser. The arrival of the single pulse is sufficient to trigger the switching, and the device remains in the on state. Interrupting the holding beam returns the switch to the off state, and as can be seen the logic levels are extremely stable. The device was further developed[53] to show explicit AND gate operation by dividing the switching pulse with a beam split-

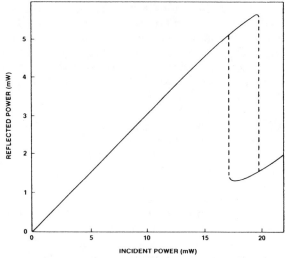

Fig. 7. Bistable action observed in an uncoated etalon of InSb when illuminated by a cw CO laser operating at a wavelength of 5.5 μm.

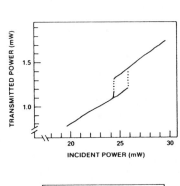

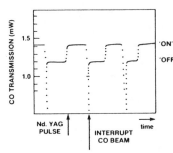

Fig. 8. (a) Transmission characteristic of an InSb etalon; (b) change in transmission induced by the absorption of one 35-ps long pulse from a Nd:YAG laser (1.06 μm).

ter and observing that the switching energy was quite definite so that the device could be set to switch only when both pulses were incident (in addition to the steady state holding beam) and would not switch with a single pulse. The device is, therefore, acting as an AND gate and an optical memory.

Using the theory and experiment described in Sec. III, we may deduce that sufficient free carriers were induced near the surface of the InSb resonator to cause sufficient change in optical thickness to initiate switching within ~3 ps. Elsaesser et al.[54] have recently shown that this nonlinearity can be switched on in <4 ps. We may deduce, therefore, that the probable switching time is limited only by the macroscopic effect of internal field buildup within the resonator. This depends only on the round-trip time for the 210-μm thick etalon and amounts to ~8 ps. We may infer that picosecond time scale optical logic has been demonstrated: this proposition is supported by time resolved experiments by Migus et al.[55] on GaAs etalons.

The AND gate[53] operation was also carried out with a delay introduced on the second pulse. This suggested an effective carrier lifetime of ~90 ns. Such a time scale would characterize the switching from on to off following a transient reduction in the holding beam. Thus at present such logic gates can only be returned to their initial states substantially more slowly than they can be switched up. There exist, however, several possible methods for increasing this recovery switching.

A. First Digital Optical Circuits: Cascadable Optical Logic

Since the result illustrated in Fig. 8 shows that with a cw steady state hold an InSb logic device can be switched with incremental power of as little as 3 μW, the output change of the device operating in reflection (Fig. 7), ~4 mW, should be quite sufficient to switch a succeeding device. A first experiment is illustrated in Fig. 9. Optical switch A is addressed by a holding beam which is reflected from the element and focused on to a second element, 3500 μm away on the same crystal slice. This second element is held near the switch point by a second holding beam illuminating it from the opposite side. The output of A then acts as the address beam for B, and the optical circuit output is observed as the reflection from B. Figure 9 also depicts the circuit inputs and outputs as the input to A is steadily increased until a point is reached where its reflected output acting as an address on B causes this second gate to switch to a lower state in reflection. B remains in this lower state until A itself switches down in reflection when B simultaneously switches up in reflection. B remains in the up state as the input to A is further increased until it again switches down when the reflected input from A is high enough. The first example of an optical logic circuit is equivalent to an XNOR gate. If A were to be held close to the switch point and addressed with a pulse, information would be stored in A. If the holding beam on B is at the same time too far from the switch point, the information will not be transferred from B. If this holding beam is then programmed to come within the range where B can be switched, the information can be transferred to B, and by appropriate reduction of the holding beam on B the information can be moved from A. The device can, therefore, be made to act as a shift register. With further holding beams defining optical circuit elements across the crystal these forms of optical logic become indefinitely extensible.

B. Optical Transistors or Transphasors

The steepness of the output/input characteristic shows that the output can be made to change by more

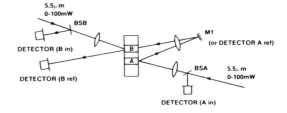
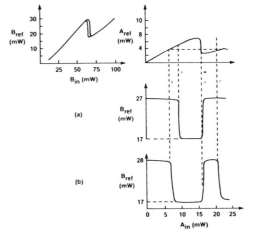

Fig. 9. First optical circuit: the experimental arrangement used in fabrication of an XNOR gate by coupling two InSb logic gates and the characteristics of the individual and coupled gates.

than the input if the initial detuning is adjusted to give a single valued form of the characteristic. Adjusting the holding beam to be within this differential gain region an external beam can be introduced, which in effect changes the phase thickness of the resonator. This transferred phase effect is the analog of the transferred resistance effect, well known in the transistor. Transphasor action is defined in the sense that amplification takes place by this mechanism. Using CO laser beams modulated at low frequency with the aid of a mechanical chopper, light amplification as high as 13,000 has been demonstrated,[56] consistent with the ability to hold within 3 μW of a 30-mW critical switch power.

To extend the gain measurements to higher frequencies, a semiconductor diode laser address can be used. Diodes at 1.3 μm (STL) and 5.5 μm (IAF Freiburg) have been successfully used for modulation. Gain and faithful amplification at 200 kHz have been obtained. The high frequency response is limited by carrier lifetime.[57]

C. Incoherent Address

The ability of this system to be held close to the switching power has been used in the first demonstration of incoherent to coherent conversion using an optically bistable device.[8,10,21] It was observed that the partial absorption of a pulse of white light was sufficient to switch a bistable device between its two states.

The source was a small photographic flashlamp which was unfocused and placed 15 cm from the sample. The energy absorbed by the sample when the flashlamp was triggered (1 nJ) was sufficient to cause the bistable system to switch from its off state to its on state and remain there. The intensity dependence of the switching was investigated, and the results were found consistent with an electronic mechanism.

The flashgun pulses were also used to demonstrate that the same bistable system could be switched from its on state to its off state. The absorption of around 1 μJ was sufficient to heat the sample by ~2 K. The change in the characteristic induced by this temperature change was sufficient to change the state of the bistable system. The energy dependence of the results is consistent with a thermal effect.

These observations are particularly effective in illustrating the full potential of bistable systems of this type since they imply that an array of bistable switches could be addressed by an image in incoherent light.

D. Optical Circuits from Nonlinear Interference Filter Structures

Thin-film interference structures comprising alternate layers of high index and low index material deposited on a flat substrate such as float glass are a convenient way of constructing a resonator formerly equivalent to the devices described in Sec. III. If sufficient nonlinearity can be induced in the very thin interference layers the technology has many advantages, particularly that very uniform devices covering many square centimeters can be easy to fabricate. Thus large area arrays are readily possible. The typical total thickness of films for a device with 15 layers is only ~2–3 μm. The total heat capacity of pixels of micron dimension is low and implies that the incident power of ~10 mW can cause a temperature rise of 50°C in ~1 ns.

The heat conductivity through the thin-film structure is relatively fast compared to the rate of heat sinking into the substrate. Thus for a given input power the temperature rise is inversely proportional to the thermal conductivity of the substrate. This gives a readily controllable parameter which controls both sensitivity and speed. In Sec. II the band gap resonance of nonlinearities with electronic origin was discussed; absorption as well as dispersion shows resonant behavior. The absorption of laser power is always associated with thermal effects in which the crystal lattice heats up and, therefore, changes the interatomic distance and hence the position of the absorption edge. This feature usually moves to longer wavelengths. Thus the phenomenon of increased absorption occurs with increase in laser power. This property itself can be used for optical bistability (see Table I). Associated with this absorption edge shift is a change in refractive index which also resonates at the band edge. It is this latter effect that we make use of with the interference structures noting that with the small heat capacities and the ability to heat sink to the substrate such devices can be quite fast.

The first report of effects in interference filters was given by Karpushko and Sinitsyn.[38] These authors

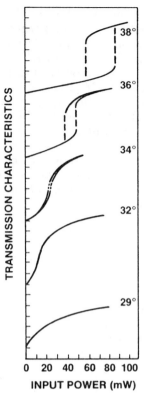

Fig. 10. Series of characteristics obtained by varying the initial detuning of an interference filter containing ZnSe.

claimed very low switching power and quite high speeds and invoked an electronic mechanism concerned with film impurities or defects. This interpretation seemed unlikely and indeed was not confirmed in experiments reported by Olbright et al.[58] The thermal mechanism described above seemed a much more likely origin and in both ZnSe and ZnS. We designed to optimise both the absorption and heat sinking properties. The filters immediately gave optical switching results with stable operation and the capability of steady state cw holding with powers of ~7 mW for 35-μm diam spot sizes. A series of characteristics with varying initial detuning is easily obtained by simply varying the angle of incidence (Fig. 10). Steady state cw hold and external address were readily possible (Fig. 11); it will be observed that the device can be switched with an external beam and remain switched on removal of the beam. The time constant of the devices is illustrated in Fig. 11, and with 35-μm spot size and 7-mW critical holding power a slow change in input signal causes switching in ~100 μs. Reduction of spot diameter to 8 μm reduces the switching time to ~4 μs, and changing the substrate from float glass to sapphire (which has a thermal conductivity an order of magnitude greater than float glass) may reduce the response time to submicroseconds, consistent with recent results of Apanasevich et al.[59]

The relationship between critical holding power and spot diameter has been investigated.[60] It is theoretically predicted that this will be proportional to the

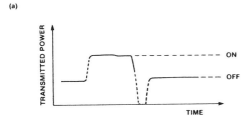

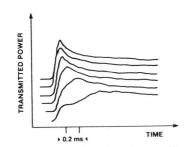

Fig. 11. External switching of an interference filter containing ZnSe by an argon laser. The lower figure shows the response time observed with a large spot size in samples coated onto float glass.

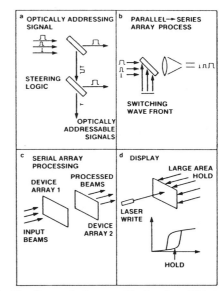

Fig. 12. Proposals for uses of logic gates of this type.

diameter, and this is confirmed by recent experimental results.

The device has been shown to give a differential gain of 4, and the contrast and output change are sufficient to switch a suitably biased succeeding device. Visible wavelength room temperature optical logic circuits have thus been demonstrated with switching on microsecond time scales.

All the parameters referred to above give considerable flexibility: the technology is sufficiently close to existing interference filter methods to allow reliable construction of the optical logic devices. With appropriate materials the method can also be extended to other wavelengths; it should be noted, however, that the argon-ion laser lines at 514 and 528 nm are particularly convenient for the absorption edge of thin-film ZnSe. ZnS can also be used but with higher power levels.

The ease of use and proved stable performance of these devices give perhaps the best present possibility of constructing digital optical circuits in the form of a simple computer. With the possibility of devices as small as in the diffraction limit $(\lambda/n)^2$, holding powers as low as 0.1 mW on practically realizable elements seem quite possible by extrapolation of present results. An efficient device to make multiple spots from the 10 W available in the argon-ion 154-nm line would suggest that more than 10^4 elements could be simultaneously driven by one laser source. Interaction studies between neighboring spots suggest that a single device of 2×2 cm might well contain this number of elements without crosstalk.

The simplicity of operation of these ZnSe interference devices allows for an early demonstration of visible wavelength optically bistable switching. By defin-

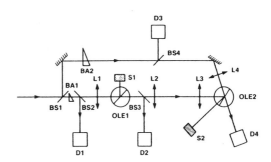

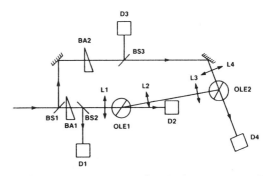

Fig. 13. Optical arrangements used in the demonstrations that the output changes in both reflection and transmission are sufficient to induce another similar gate to also switch.

ing separate elements and activating them by external address, the projection and display possibilities of this technology are apparent. It might be called laser pen writing on the optical blackboard (Fig. 12).

The first optical circuits to be demonstrated utilized IR wavelengths (Fig. 9). The visible devices have now shown similar capability: a two-element circuit is illustrated in Fig. 13.

VII. Optical Computers

The contemporary performance of a number of optical circuit elements has been described showing that in principle all logic functions familiar in electronics can be reproduced by these nonlinear optical devices. It should, therefore, be possible to construct an all-optical computer.

There exists a considerable body of literature reporting the research of a number of groups on the subject of optical computing. To date this has not included use of such nonlinear elements as discussed in this paper. Nevertheless, the advantages of using optical methods for a variety of computations have been apparent for a number of years. Historically, most optical computing systems have been of the analog type in which information is stored and processed as a continuum of signal levels. There are major drawbacks to such analog systems including limited flexibility, noise accumulation, and input/output device limitations. The advantages of optical binary logic, now in principle possible with the devices described here, may prove crucial in progressing optical computing into a new practical technology. Photon–electron or electron–photon conversions should be avoided. The considerations can be made at several levels. These include individual optical logic devices (gates and arrays of devices) as the first level. The second level considers optical communication interconnections and input/output among the logic gates, among arrays, and among circuit boards or processors. The third level considers the possibilities of new computer architecture to take advantage of the inherent parallelism of optics.

Advantages of optical methods can be examined in two separate ways:

A. Speed of Switching and Communication

It has been shown that individual optical logic operations can almost certainly be performed on a picosecond time scale. Combining this with the output characteristics of cw mode-locked lasers, it is possible to conceive bit rates approaching terahertz for an exact analog of electronic digital logic. It is not yet clear that recovery times can be imposed on the optical logic switches at acceptable power levels to ensure continuous operation. The possibility exists of series-to-parallel and parallel-to-series conversion on picosecond time scales (Fig. 12). Fast high bandwidth communication will also be consistently available.

B. Parallelism and Communication

The logic devices described in this paper are capable of providing steady state information holding and cascadable operation. They have time constants in the range of 100 ns to 100 μs. This suggests that the first approach to the use of optics should emphasize the use of parallelism and accept cycle rates similar to existing electronic speeds. The intriguing intellectual challenge is to use the flexibility of optics to provide appropriate interconnections: requirements could be to shift a logic array pattern by one element on each cycle or to achieve a perfect shuffle. Other schemes include a vector–matrix multiplication involving a fan-in from a matrix array to a column followed by rotation to this

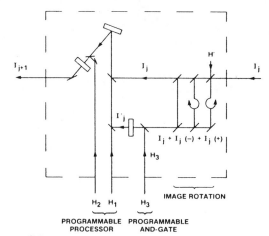

Fig. 14. Schematic of an optical parallel processor proposed to solve a specific physical problem, the 1-D Ising model, after Wherrett.[61]

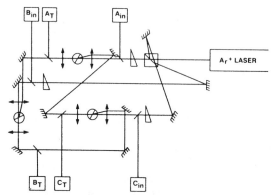

Fig. 15. Optical arrangement used in the first demonstration of a loop delay consisting of three interacting gates.

column between output and input for successive cycles.[4]

Wherrett[61] has designed a number of optical-computer architectural components including memory units, clocks, a programmable processing stage, and a simple full adder utilizing simultaneously the transmission and reflection from an optical gate (Fig. 14). The necessity to store the calculation before communicating to the next element is important in the optical case: propagation at the speed of light otherwise would mean that all elements would be simultaneously addressed (admittedly avoiding clockscrew[1]). Delays can be readily implemented by suitably programming the holding beams so that the optical bias is only set to receive a signal when the device is clear of other signals. As a first step, consistent with the elements described, this could be achieved electronically making an interesting interface to existing computer technology. At the time of writing, a three-element loop processor with optical bias delay loop clocking (Fig. 15) has been implemented. Acoustooptic modulators, controlled by a microcomputer, provide both optical bias levels and input data pulses (Fig. 16). In principle, indefinitely extensible optical logic has been demonstrated. A generalized optical computer employing these ideas may take the form shown in Fig. 17.

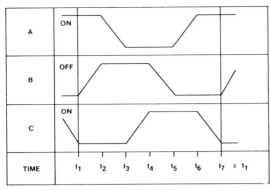

Fig. 16. Delay loop is achieved by clocking the individual bias levels to the three gates: A, B, and C in Fig. 15 in this sequence.

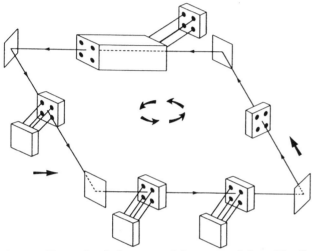

Fig. 17. Types of optical computer being proposed that utilize the parallelism of optics by using arrays of elements. The iteration present in the type of problem which could be solved on such a computer is achieved through use of a loop processor architecture.

VIII. Conclusion

Practical optical circuit elements have been demonstrated, all binary logical operations can be performed, and digital optical circuits have been shown practical.

At this time the most favorable method of applying optical methods to digital optical computing would seem to be by the use of digital arrays of gates speaking to (in an optical sense) further arrays of gates with the possibility of fixed or variable intercommunications, fan-out, and fan-in using free space propagation as efficient ways of optical wiring. It remains to be shown that holding powers and switching energies are sufficiently small to allow operation with existing lasers and devices. Scaling from experimental results, one may deduce that, for a visible wavelength device, a single logic element area of ~ 1 μm^2 should be practical, which at 1-μs switching time should require ~ 1 mW of holding power. Perhaps 10% of this would be absorbed. The 10 W available in the 514-nm argon-ion line from a typical commercial laser would allow operation of 10^4 gates assuming perfect optics. This would give the array a rate of information processing equivalent to 10^{10} logic operations per second. A rather similar number of logic operations could be achieved by using small gap semiconductors in the IR where the electronic processes would allow faster switching, but the packing density and device size would be greater. Even allowing for imperfections, which would reduce these numbers by a factor of 100–1000 in the present state of the art, the data rates are sufficiently high to encourage exploration of their use for tasks where conventional 1-D sequential digital electronics finds difficulties—such as pattern recognition, artificial intelligence, sorting, and specialized computational problems. Equally interesting will be applications such as power limiters, optical noise reduction, and laser projectors and displays.

This work was partially supported by the U.K. Joint Optoelectronics Research Scheme, the Procurement Executive of the U.K. Ministry of Defense sponsored by DCVD, Rome Air Development Center/ESO U.S. Air Force grant AFOSR 82-0149, and the Science and Engineering Research Council. It was carried out within the framework of the European Joint Optical Bistability Project (EJOB) of the Commission of the European Communities.

In addition to the work cited in the references, the input and assistance of E. A. Abraham, W. J. Firthy, I. Janossy, A. K. Kar, H. A. MacKenzie, J. G. H. Mathew, M. R. Taghizadeh, F. A. P. Tooley, A. C. Walker, and B. S. Wherrett are acknowledged.

References

1. C. Mead and L. Conway, *An Introduction to VLSI Systems* (Addison-Wesley, Reading, MA, 1980).
2. Z. C. P. Lee et al., *Technical Digest, IEEE GaAs IC Symposium* (IEEE, New York, 1983), p. 162.
3. A. Huang, "Architectural Considerations Involved in the Design of an Optical Digital Computer," Proc. IEEE **72**, 780 (1984).
4. J. N. Lee and R. A. Athale, "Optical Implementation of the Triple-Matrix Product," in *Conference Digest, ICO-13*, Sapporo (1984), paper A1-6.
5. A. A. Sawchuk, "Numerical Optical Computing Techniques," in *Conference Digest, ICO-13*, Sapporo (1984), paper A1-1.
6. D. Psaltis and N. Farhat, "A New Approach to Optical Information Processing based on the Hopfield Model," in *Conference Digest, ICO-13*, Sapporo (1984), paper A1-9.
7. J. L. Jewell, A. C. Gossard, and W. Wiegmann, "Optical Logic in GaAs Fabry-Perot Etalons," in *Conference Digest, ICO-13*, Sapporo (1984), paper A2-3.
8. S. D. Smith and A. C. Walker, "Optical Bistability and its Application to Computing," in *Conference Digest, ICO-13*, Sapporo (1984), paper B5-10.
9. S. D. Smith, I. Janossy, H. A. MacKenzie, J. G. H. Mathew, J. J. E. Reid, M. R. Taghizadeh, F. A. P. Tooley, and A. C. Walker, "Nonlinear Optical Circuit Elements, Logic Gates for Optical Computers: The First Digital Optical Circuits," Opt. Eng. **24**, 4 (1985).
10. B. S. Wherrett and S. D. Smith, Eds., "Optical Bistability, Dynamical Nonlinearity and Photonic Logic," Royal Society, London, 1985; also published in Philos. Trans. R. Soc. A **313**, 191 (1984).
11. S. Szöke, V. Danue, J. Goldhar, and N. A. Kurnit, "Bistable Optical Element and its Applications," Appl. Phys. Lett. **15**, 376 (1969).
12. H. M. Gibbs, S. L. McCall, and T. N. C. Venkatesan, "Differen-

tial Gain and Bistability using a Sodium-filled Fabry-Perot Interferometer," Phys. Rev. Lett. **36**, 1135 (1976).

13. D. Weaire, B. S. Wherrett, D. A. B. Miller, and S. D. Smith, "Effect of Low-Power Nonlinear Refraction on Laser-Beam Propagation in InSb," Opt. Lett. **4**, 331 (1979).

14. D. A. B. Miller, M. H. Mozolowski, A. Miller, and S. D. Smith, "Nonlinear Optical Effects in InSb with a c.w. CO Laser," Opt. Commun. **27**, 133 (1978).

15. T. Scragg and S. D. Smith, "External Cavity Operation of the Spin-Flip Raman Laser," Opt. Commun. **15**, 188 (1975).

16. C. N. Ironside, "The Spin Flip Laser Spectrometer and Aspects of the Nonlinear Optics in InSb," Ph.D. Thesis, Heriot-Watt U., Edinburgh (unpublished) (1979).

17. D. A. B. Miller, S. D. Smith, and A. Johnston, "Optical Bistability and Signal Amplification in a Semiconductor Crystal: Applications of New Low-power Nonlinear Effects in InSb," Appl. Phys. Lett. **35**, 658 (1979).

18. D. A. B. Miller and S. D. Smith, "Two Beam Optical Amplification and Bistability in InSb," Opt. Commun. **31**, 101 (1979).

19. H. M. Gibbs, S. L. McCall, T. N. C. Venkatesan, A. C. Gossard, A. Passner, and W. Wiegmann, "Optical Bistability in Semiconductors," Appl. Phys. Lett. **36**, 6 (1979).

20. D. A. B. Miller, S. D. Smith, and B. S. Wherrett, "The Microscopic Mechanism of Third-Order Optical Nonlinearity in InSb," Opt. Commun. **35**, 2 (1980); D. A. B. Miller, C. T. Seaton, M. E. Prise, and S. D. Smith, "Band-Gap-Resonant Nonlinear Refraction in III–V Semiconductors," Phys. Rev. Lett. **47**, 197 (1981).

21. A. C. Walker, F. A. P. Tooley, M. E. Prise, J. G. H. Mathew, A. K. Kar, M. R. Taghizadeh, and S. D. Smith, "InSb Devices: Transphasors with High Gain, Bistable Switches and Sequential Logica Gates," in Proceedings, Optical Bistability, Dynamical Nonlinearity and Photonic Logic; also published as Philos. Trans. R. Soc. London A **313**, 249 (1984).

22. B. S. Wherrett, F. A. P. Tooley, and S. D. Smith, "Absorption Switching and Bistability in InSb," Opt. Commun. **52**, 4 (1984).

23. A. K. Kar, J. G. H. Mathew, S. D. Smith, B. Davis, and W. Prettl, "Optical Bistability in InSb at Room Temperature with Two-Photon Excitation," Appl. Phys. Lett. **42**, 4 (1983).

24. H. M. Gibbs et al., "Semiconductor Nonlinear Etalons," in Proceedings, Optical Bistability, Dynamical Nonlinearity and Photonic Logic; Philos. Trans. R. Soc. London A **313**, 245 (1984).

25. D. A. B. Miller, A. C. Gossard, and W. Wiegmann, "Optical Bistability due to Increasing Absorption," Opt. Lett. **9**, 162 (1984).

26. M. Dagenais and W. F. Sharfin, "Picojoule, Subnanosecond, All-optical Switching using Bound Excitons in CdS," Appl. Phys. Lett. **46**, 3 (1985).

27. K. Bohnert, H. Kalt, and C. Klingshirn, "Intrinsic Absorptive Optical Bistability in CdS," Appl. Phys. Lett. **43**, 12 (1983).

28. M. Dagenais and W. F. Sharfin, "Cavityless Optical Bistability due to Light-Induced Absorption in Cadmium Sulphide," Appl. Phys. Lett. **45**, 3 (1984).

29. A. Miller, G. Parry, and R. Daley, "Low-Power Nonlinear Fabry-Perot Reflection in CdHgTe at 10 μm," IEEE J. Quantum Electron. **QE-20**, 7 (1984).

30. J. G. H. Mathew, D. Craig, and A. Miller, "Optical Switching in CdHgTe Etalon at Room Temperature," Appl. Phys. Lett. **46**, 2 (1985).

31. C. D. Poole and E. Garmire, "Optical Bistability at the Bandgap in InSb," Appl. Phys. Lett. **44**, 4 (1984).

32. N. Peyghambarian, H. M. Gibbs, M. C. Rushford, and D. A. Weinberger, "Observation of Biexcitonic Optical Bistability and Optical Limiting in CuCl," Phys. Rev. Lett. **51**, 16 (1983); R. Levy, J. Y. Bigot, B. Hönerlage, F. Tomasini, and J. B. Grun, "Optical Bistability due to Biexcitons in CuCl," Solid State Commun. **48**, 8 (1983).

33. G. Staupendahl and K. Schindler, "A New Optical-Optical Modulator," Opt. Quantum Electron. **14**, 157 (1982).

34. H. J. Eichler, "Optical Multistability in Silicon observed with a cw Laser at 1.06 μm," Opt. Commun. **45**, 1 (1983).

35. J. Hajto and I. Janossy, "Optical Bistability observed in Amorphous Semiconductor Films," Philos. Mag. B **47**, 4 (1983).

36. S. D. Smith, J. G. H. Mathew, M. R. Taghizadeh, A. C. Walker, B. S. Wherrett, and A. Hendry, "Room Temperature, Visible Wavelength Optical Bistability in ZnSe Interference Filters," Opt. Commun. **51**, 357 (1984).

37. M. R. Taghizadeh, I. Janossy, and S. D. Smith, "Optical Bistability in Bulk ZnSe due to Increasing Absorption and Self-Focussing," Appl. Phys. Lett. **45**, 4 (1985).

38. F. V. Karpuskho and G. V. Sinitsyn, "An Optical Logic Element for Integrated Optics in a Nonlinear Semiconductor Interferometer," J. Appl. Spectrosc. USSR **29**, 1323 (1978).

39. T. N. C. Venkatesan and S. L. McCall, "Optical Bistability and Differential Gain between 85 and 296°K in a Fabry-Perot Containing Ruby," Appl. Phys. Lett. **30**, 282 (1977).

40. J-W Song, S.-Y. Shin, and Y-S. Kwon, "Optical Bistability, Regenerative Oscillation, and Monostable Pulse Generation in a Liquid Crystal Bistable Optical Device," Appl. Opt. **23**, 1521 (1984).

41. D. Grischowsky, "Nonlinear Fabry-Perot Interferometer with Subnanosecond Response Times," J. Opt. Soc. Am. **68**, 641 (1978).

42. S. L. McCall and H. M. Gibbs, "Optical Bistability via Thermal Effects in a Glass Filter," J. Opt. Soc. Am. **68**, 378 (1978).

43. T. Bischofberger and Y. R. Shen, "Transient Behaviour of a Nonlinear Fabry-Perot," Appl. Phys. Lett. **32**, 156 (1978); "Nonlinear Fabry-Perot Filled with CS_2 and Nitrobenzene," Opt. Lett. **4**, 40 (1979); "Nonlinear Fabry-Perot Filled with CS_2 and Nitrobenzene (Errata)," Opt. Lett. **4**, 175 (1979).

44. J. L. Jewell, M. C. Rushford, and H. M. Gibbs, "Use of a Single Nonlinear Fabry-Perot Etalon as Optical Logic Gates," Appl. Phys. Lett. **43**, 2, 1975 (1984).

45. R. G. Harrison, I. A. Al-Saidi, E. J. D. Cummins, and W. J. Firth, "Evidence for Optical Bistability in Millimeter Gas Cells," Appl. Phys. Lett. **46**, 532 (1985).

46. R. G. Harrison, W. J. Firth, C. A. Emshary, and I. A. Al-Saidi, "Observation of Optical Hysteresis in an All-Optical Passive Ring Cavity Containing Molecular Gas," Appl. Phys. Lett. **44**, 716 (1984).

47. B. S. Wherrett, "A Comparison of Theories of Resonant Nonlinear Refraction in Semiconductors," Proc. R. Soc. London A **390**, 373 (1983).

48. B. S. Wherrett, "One Electron Theory of Nonlinear Refraction," in Proceedings, Optical Bistability, Dynamical Nonlinearity and Photonic Logic, Philos. Trans. R. Soc. London A **313**, 213 (1984).

49. A. Miller, G. Parry, and R. Daley, "Optical Bistability in Semiconductors with Density dependent Carrier Lifetimes," Opt. Quantum Electron. **16**, 339 (1984).

50. D. A. B. Miller, "Refractive Fabry-Perot Bistability with Linear Absorption," IEEE J. Quantum Electron. **QE-17**, 3 (1981).

51. B. S. Wherrett, D. Hutchings, and D. Russels, Heriot-Watt U.; private communication (1985).

52. B. S. Wherrett, "Fabry-Perot Bistable Cavity Optimisation," IEEE J. Quantum Electron. **QE-20**, 646 (1984).

53. C. T. Seaton, S. D. Smith, F. A. P. Tooley, M. E. Prise, and M. R. Taghizadeh, "The Realization of an InSb Bistable Device as an Optical AND Gate and its use to measure Carrier Recombination Times," Appl. Phys. Lett. **42**, 131 (1983).

54. T. Elsaesser, H. Lobentanzer, and W. Kaiser, "Self-Defocusing and Self-Phase Modulation in InSb measured with Picosecond Infrared Pulses," Appl. Phys. Lett. **47**, 11 (1985).

55. A. Migus, A. Antoneti, D. Hulin, A. Mysyrowscly, H. M. Gibbs, N. Peyghambarian, and J. L. Jewell, "One-picosecond Optical

NOR Gate at Room Temperature with a GaAs-AlGaAs Multiple-quantum-well Nonlinear Fabry-Perot Etalon," Appl. Phys. Lett. **46,** 1 (1985).

56. F. A. P. Tooley, S. D. Smith, and C. T. Seaton, "High Gain Signal Amplification in an InSb Transphasor at 77 K," Appl. Phys. **43,** 9 (1983).
57. F. A. P. Tooley, W. J. Firth, A. C. Walker, H. A. MacKenzie, J. J. E. Reid, and S. D. Smith, "Measurement of the Bandwidth of an Optical Transphasor," IEEE **QE-21,** 1356 (1985)
58. G. R. Olbright, N. Peyghambarian, H. M. Gibbs, H. A. Macleod, and F. van Milligen, "Micro-second Room-temperature Optical Bistability and Crosstalk Studies in ZnS and ZnSe Interference Filters with Visible Light and Milliwatt Powers," Appl. Phys. Lett. **45,** 10 (1984).
59. S. P. Apanasevich, F. V. Karpushko, and G. V. Sinitsyn, "Response Time of Bistable Devices Based on Evaporated Thin-Film Interferometers," Sov. J. Quantum Electron. **14,** 873 (1984).
60. I. Janossy, M. R. Taghizadeh, J. G. H. Mathew, and S. D. Smith, "Thermally Induced Optical Bistability in Thin Film Devices," IEEE **QE-21,** 9 (1985).
61. B. S. Wherrett, "All-Optical Computation: a Design for Tackling a Specific Physical Problem," Appl. Opt. **24,** (1985).

Use of a single nonlinear Fabry–Perot etalon as optical logic gates

J. L. Jewell, M. C. Rushford, and H. M. Gibbs

Optical Sciences Center, University of Arizona, Tucson, Arizona 85721

The all-optical operation of many logic elements (NOR, etc.) on a single passive nonlinear Fabry–Perot etalon is discussed. An experiment using dye-filled etalons verifies the prediction and agrees with a computer simulation. This kind of Fabry–Perot gate should be capable of very high speed operation with minimum energy per cycle.

Digital processing of many (on the order of 10^6) information bits in parallel may be possible with arrays of optical logic gates.[1] We report a technique for operating a single nonlinear Fabry–Perot etalon which yields the decisions NOR, NAND, XOR, OR, AND, and $\overline{\text{XOR}}$, simultaneously if desired. The predicted transmission versus time characteristics are verified by an experiment with a dye-filled etalon, to some extent by previous work in a GaAs device, and by computer simulation. The information can be pulsed rather than cw, allowing the logical decision to be made in <1 ps for some materials (e.g., GaAs[2]), and the device will self-reset in a few τ_R, where τ_R is the medium relaxation time. In this pulsed mode both the time and energy per cycle are minimized.

Consider both inputs and the probe to be pulses of duration short compred to τ_R. The nonliner medium must be such that absorption of one input pulse changes the refractive index at the probe wvelength enough to shift the Fabry–Perot transmission peak by about one full width at half-maximum (FWHM). The peak will of course return to its original wavelength in a few τ_R, but if the probe pulse is incident immediatley after the input(s) only this instantaneous transmission determines the output. With certain initial detunings of the probe, the various gates are obtained. This is shown graphically in Fig. 1 from which the approximate transmissions are found for five "standard" probe detunings with 0, 1, or 2 inputs. Desirable characteristics (e.g., contrast, reliability, low power) can be improved by fine adjusting the detuning and input strength. It is also desirable in these gates that the probe not affect the nonlinear medium thereby complicating its operation. Preferably the medium would have only very low absorption at the probe wavelength thus allowing high transmission, and the reflected signal would be the complement. Thus, if the device is probed at five selected wavelengths the result would be all the gates of Fig. 1 simultaneously, although all this information would be difficult to utilize in a working system.

The transmission versus time was calculated using the standard Fabry–Perot formula and a refractive index that suddenly changes (as from an input pulse), then exponentially relaxes to its original value. It is assumed that two inputs produce twice the phase shift as one. These characteristics (Fig. 2) not only give the transmission at the time of the probe, but also show that they are *self-resetting* in only a few τ_R.

Figure 3 shows the experimental layout used for focusing two separate inputs and a probe onto an etalon. The input pulses heat the dye changing its refractive index at the probe wavelength. The weak probe beam was cw in order to observe the relaxation characteristics. The response for five initial detunings is seen (Fig. 4) to resemble the computer

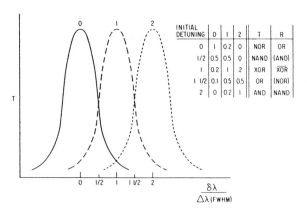

FIG. 1. Position of the transmission peak after 0, 1, or 2 input pulses are incident. With the probe wavelength at one of the five labeled values (expressed by the initial detuning in FWHM's of the transmission peak) the gates in the table are obtained. The fractional values in the columns below 0, 1, and 2 (of inputs) are the approximate transmissions when each input shifts the peak by 1 FWHM. In reflection the AND and NOR have poor contrast.

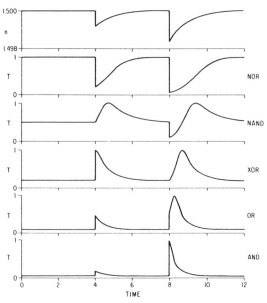

FIG. 2. Computer calculated transmission vs time for five initial detunings with 0, 1, and 2 inputs incident at times 0, 4, and 8. Time units are ($1/e$ recovery times for the refractive index. Top trace is refractive index vs time. Mirror reflectivities are 90% with 10-μm spacing, wavelength 500 nm. and losses are zero.

Reprinted with permission from *Appl. Phys. Lett.*, vol. 44, no. 2, pp. 172–174, Jan. 15, 1984.
Copyright © 1984, American Institute of Physics.

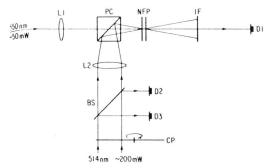

FIG. 3. Experimental layout for monitoring gate transmission. Components: $L1$, $L2$—lenses; $D1$, $D2$, $D3$—detectors; PC—polarizing cube; NFP—nonlinear Fabry–Perot; IF—interference filter; CP—chopper; BS—beamsplitter.

simulated responses except in the OR gate which showed unexpected behavior. For Figs. 4(a)–4(c) the two input beams are replaced by a single beam in which the pulse for the two-input case is twice as long as that for one input. Although the dye etalons are slow, making them smaller can yield microsecond response times, and they demonstrate a principle that should extend to many other nonlinearities. Modulation experiments[3] with GaAs showed recovery times less than 10 ns with saturation energy densities of 10 $\mu J/cm^2$ (0.1 $pJ/\mu m^2$). Waveforms resembling the NOR and the XOR with zero and one input are shown in Fig. 1 of that article. The recovery times can be shortened to the picosecond regime by doping, proton bombardment,[4] or surface recombination.

Experimentally, the NOR gate is most easily and reliably achieved, and it is least sensitive to input pulse energy. Increasing the input strength to cause a shift of more than one FWHM merely improves contrast (and slightly lengthens recovery time). Since NOR gates can form a complete logical set,[1] this gate appears most promising. Although the other gates might also be easily achieved, the NOR will be least sensitive to input pulse strength.

The pulsed operation allows both the time and energy per cycle to be minimized because it *does not require keeping the medium excited*. Thus no extra energy is spent holding the device on and it *relaxes in the dark* providing minimum relaxation time. Since optical information transmission gen-

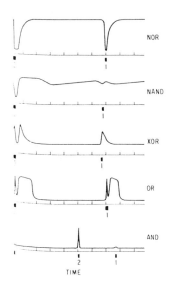

FIG. 4. Transmission vs time for the dye-filled etalon. The marks below time axis indicate the time and duration of the input(s) with the number directly below telling how many inputs are present. Time/division: a,b,c 2 ms; d,e 5 ms. The OR does not agree with predictions.

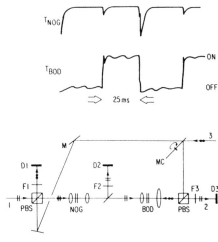

FIG. 5. (a) Optical flip-flop. (1) 100-mW cw horizontally polarized yellow input beam from dye laser. (2) 4-mW transmission of yellow beam through bistable optical device (BOD) switched on or off by (3) 1 W of vertically polarized cw Ar laser at 514.5 nm, directed by rotating mirror (90% reflective)/chopper MC. Filter $F1$ allows 514.5-nm beam to hit detector $D1$ for scope triggering. Filter $F2$ allows yellow beam to hit detector $D2$ for negative optical gate (NOG) monitoring. Filter $F3$ allows yellow beam output to hit detector $D3$ for on or off state monitoring. Yellow and green beams are combined by polarization beam splitters (PBS). (b) Response of optical flip-flop. Top trace is transmission of NOG monitored by $D2$. The large pulse switches BOD off; the small pulse is timing reference from mirror/chopper transmission and BOD transmission affecting the NOG. Bottom trace is on or off intensity monitored by $D3$. The on-to-off intensity ratio is ≈ 10. The ripple is due to room lights.

erally takes place in the form of pulses, this mode of operation seems well suited for optics. Although timing becomes important, it is easily within the capability of present day optical technology.

These gates can also work in a cw or mixed mode (cw inputs, pulse train probe) although accumulative heating would increase. The etalon could then be considered a kind of flip-flop. Another flip-flop that was conceived and tested here[5] employs two etalons in series in a cw beam [Fig. 5(a)]. The first etalon is tuned for maximum transmission and the second uses the throughput to operate as a bistable optical device[6] in the usual sense. The second etalon alone can be switched on by an optical pulse,[7] but switch-off requires a second nonlinearity or a drop in the incident intensity. The first etalon accomplishes such a drop if it is hit by a switch-off pulse which detunes its transmission peak away from the incident wavelength. This process is identical to that of the NOR gate except there is only one input pulse and the probe is cw. This flip-flop was also tested with dye-filled eatlons yielding light-by-light control [Fig. 5(b)]. An optical memory or pulse-to-cw conversion are two possible applications for the flip-flop. Another application is to generate short rectangular-shaped optical pulses of precisely controllable length. Since switching on and off are both accomplished by *excitation* of the nonlienar media the rise and fall times can be extremely short.[2,7] The relative time separation between the "on" and "off" pulses gives the length of the transmitted pulse.

A fundamental requirement for any computer is that the output of a gate be of sufficient strength to drive other gates. For an optical gate held on by only one wavelength this poses serious problems. The transmission is limited because there must be absorption to drive the nonlinearity. Thus it appears that a single wavelength gate would require gain or suffer unwanted absorption to have any hope of driving other gates. Switches with gain have been analyzed theoretically,[8] and lasing in ultrashort etalons is not hard to achieve,[8] however gates with gain would be more complex and require more power than those described here. Furthermore, negative logic is more straightforwardly obtained from a device in which the probe does not affect its own transmission.

These fundamental problems can be overcome in a passive device if the probe wavelength differs from that of the inputs. Before tackling the problem of signal propagation through many gates let us first consider the advantages of a two-wavelength gate. Efficiency of the input pulses can be maximized by tuning them for maximum absorption by the nonlinear medium. The probe can in turn be tuned for low absorption allowing high transmission and finesse. The increased finesse means that only a very small change of refractive index Δn, produces the desire change in the output. In the low absorption region of some nonlinearities (e.g., saturation of an absorption peak) the linear absorption α falls off as $\approx 1/\Delta^2$ (Δ = detuning from resonance) while Δn for a given degree of saturation decreases as $\approx 1/\Delta$. For nearly 100% reflecting mirrors the finesse is $\approx \pi/(\alpha L + \alpha_B L + S)$, where L is the cavity length, α_B is the background absorption coefficient, and S is the scattering from each mirror. Thus for $\alpha L \gg \alpha_B L, S$ the required input energy is approximately inversely proportional to detuning, and best performance should be obtained when $\alpha L \lesssim \alpha_B L + S$. The terms $\alpha_B L$ and S represent technological limits to the performance of these gates. Note that lowering power requirements by increasing finesse should not help thermal stability since the device is correspondingly more sensitive to temperature change. However, when you consider using 10^6 devices at once, it is very advantageous to minimize total power. More importantly, the probe can have many times the energy of an input while still not affecting the gate's operation and thus be able to drive many other similar devices. The output would be less affected by any spontaneous or stimulated emission from energy injected by the input(s) since this energy is small compared to that of the probe.

By operating on the approximately linear portion of the transmission peak in the cw mode, a two-wavelength device could achieve amplification of an input or sum of inputs (with some zero-input bias present) as long as the modulation is slow compared to τ_R. This action achieves total gain rather than the differential gain observed in bistable devices.[6]

In a system of two-wavelength gates, the gates cannot be identical of course, because the output from the first one could not affect any others. What is possible is a system using two similar nonlinear materials where the probe wavelength for one device is on the absorption peak for the succeeding gate. This gate should also have low absorption at the first gate's input wavelength. Thus the wavelengths of input and probe are reversed and propagation through an even number of gates returns the original wavelength. An isolated saturable line absorption feature which is shifted in one material relative to the other should provide the desired characteristics. In semiconductors such as GaAs or CdS the absorption might be an exciton feature well resolved from the band edge. In GaAs, the peak can be shifted by a small amount of aluminum in bulk material or by varying the well thickness in a multiple quantum well structure.[10]

A technique for performing logic operations on a single etalon has been presented. Some other optical logic gates have been proposed[8] or tested[11] but those presented here are simple, utilize the high-speed capability of optics with low power requirements, and represent a new mode of operation for existing devices. A more in-depth study of these gates will be presented later.

The authors gratefully acknoweldge helpful discussions with A. Huang, N. Peyghambarian, and R. Shoemaker. This work was supported by the Air Force Office of Scientific Research, the Army Research Office, and the National Science Foundation.

[1] A. Huang, 10th International Optical Computing Conference, Cambridge, Mass 1983, p. 13; A. W. Lohman, ibid. p. 1; P. Chavel, R. Forchheimer, B. K. Jenkins, A. A. Sawchuk, and T. C. Strand, ibid. p. 6; A. Huang, SPIE **232**, 119 (1980); R. A. Athale and S. H. Lee, Opt. Eng. **18**, 513 (1979); D. H. Schaefer and J. P. Strong III, Proc. IEEE **65**, 129 (1977).
[2] C. V. Shank, R. L. Fork, B. I. Greene, C. Weisbuch, and A. C. Gossard, Surf. Sci. **113**, 108 (1982); C. V. Shank, R. L. Fork, R. F. Leheny, and J Shah, Phys. Rev. Lett. **42**, 112 (1979).
[3] H. M. Gibbs, T. N. C. Venkatesan, S. L. McCall, A. Passner, A. C. Gossard, and W. Wiegmann, Appl. Phys. Lett. **34**, 511 (1979).
[4] D. H. Auston and P. R. Smith, Appl. Phys. Lett. **41**, 599 (1982).
[5] J. L. Jewell, S. S. Tarng, H. M. Gibbs, K. Tai, D. A. Weinberger, S. Ovadia, A. C. Gossard, S. L. McCall, A. Passner, T. Venkatesan, and W. Wiegmann. Topical Meeting on Optical Bistability, Rochester, New York 1983; M. C. Rushford, H. M. Gibbs, J. L. Jewell, N. Peyghambarian, D. A. Weinberger, and C. F. Li, ibid.
[6] H. M. Gibbs, S. L. McCall, and T. N. C. Venkatesan, Opt. Eng. **19**, 463 (1980); D. A. B. Miller, Laser Focus **18**, 79 (1982).
[7] S. S. Tarng, K. Tai, J. L. Jewell, H. M. Gibbs, A. C. Gossard, S. L. McCall, A. Passner, T. N. C. Venkatesan, and W. Wiegman, Appl. Phys. Lett. **40**, 205 (1982).
[8] R. L. Fork, Phys. Rev. A **26**, 2049 (1982).
[9] J. M. Weisenfeld and J. Stone, Opt. Lett. **8**, 262 (1983); J. Stone, C. A burns, and J. C. Campbell, J. Appl. Phys. **51**, 3038 (1980); J. P. van der Ziel, R. Dingle, R. C. Miller, W. Wiegmann, and W. A. Nordland, Jr. Appl. Phys. Lett. **26**, 463 (1975).
[10] R. Dingle, Festkorperprobleme **15**, 21 (1975); A. C. Gossard, Thin Solid Films **57**, 3 (1979).
[11] C. T. Seaton, S. D. Smith, F. A. P. Tooley, M. E. Prise, and M. R. Taghizadeh, Appl. Phys. LSett. **42**, 131 (1983); S. A. Collins, Jr., M. T. Fatehi, and K. C. Wasmundt, SPIE **232**, 168 (1980); B. H. Soffer, D. Boswell, A. M Lackner, P. Chavel, A. A. Sawchuk, T. C. Strand, and A. R. Tanguay, Jr. ibid. **232**, 128 (1980).

GaAs-AlAs monolithic microresonator arrays

J. L. Jewell
AT&T Bell Laboratories, Holmdel, New Jersey 07733

A. Scherer
Bell Communications Research, Red Bank, New Jersey 07701

S. L. McCall, A. C. Gossard, and J. H. English
AT&T Bell Laboratories, Murray Hill, New Jersey 07974

Monolithic optical logic devices 1.5–5 μm across are defined by ion-beam assisted etching through a GaAs/AlAs Fabry–Perot structure grown by molecular beam epitaxy. They show reduced energy requirements (more than an order of magnitude smaller than the unetched heterostructure), uniform response over small arrays, negligible crosstalk at 3 μm center-center spacing, ~150 ps recovery time, and thermal stability at 82 MHz operating frequency. All experiments were performed at room temperature.

Miniaturization of optical logic devices has long been considered a key to minimizing their energy requirements, and has also been considered difficult (at best) to achieve.[1-4] We have demonstrated a straightforward technique for fabricating arrays of integrated GaAs/AlAs Fabry–Perot étalon devices (microresonators) as small as 1.5 μm in diameter with densities around 10^7 devices/cm^2. Furthermore, submicron devices with densities $\sim 10^8$/cm^2 should be possible. The growth of integrated GaAs-AlAs nonlinear étalons by molecular beam epitaxy[5] (MBE) offers much improved manufacturability and uniformity over previous semiconductor étalon fabrication techniques[6] which involved sandwiching semiconductor films between dielectric mirrors. From the sample of Ref. 5 we have formed close-packed arrays of monolithic "posts" or microresonators 1.5–5 μm across by ion-beam-assisted etching[7] and have performed optical NOR/OR gating experiments on them using picosecond pump and probe pulses. These microresonators represent a qualitative advance over the GaAs devices reported by Lee et al.[8] In that work pixels 9×9 μm square were formed by etching the active material only, and then sandwiching between dielectric mirrors in the usual way.[6] Growth of integrated devices by epitaxial techniques such as MBE allows us to etch right through both mirrors and the active material. This is critical since in an optimized nonlinear étalon the dielectric mirrors comprise most of the total thickness. Diffractive beam spreading, normally to 5–10 μm diameter in GaAs étalons, is therefore defeated throughout the entire device. The lateral optical confinement in these waveguiding structures allows efficient operation with diameters as small as one can focus the light. Reduction of energy requirements is expected due to the decreased volume of interaction. Further energy reduction may occur since restriction on the allowed transverse modes of the device should result in a narrower transmission peak. Elimination of carrier diffusion[9,10] out of the devices should allow them to be spaced very closely. Finally, surface recombination[8,11] on the *sidewalls* of the microresonators should produce fast relaxation times. Our experimental results show more than an order of magnitude reduction in energy requirements, essentially uniform response over arrays at least several pixels across in each dimension, practically no crosstalk with 3 μm center-center spacing, ~150 ps full recovery time, and thermal stability at 82 MHz operating frequency.

The MBE growth upon a GaAs substrate comprised 9 1/2 pairs of AlAs/GaAs layers 813 Å/594 Å thick (quarter-wave-stack mirror) followed by a 1.6-μm GaAs spacer and 7 more AlAs/GaAs pairs for a total thickness of 4 μm. This design yields approximately equal mirror reflectivities (~90%) when the structure is intact on the substrate. Since the GaAs substrate has significant absorption at the wavelengths used, transmission was not monitored and the reflection was taken as the output. No attempt was made to achieve uniformity in the growth and the thicknesses varied over the sample area. The etching was accomplished in a 5:2 Ar:Cl$_2$ gas mixture at 8×10^{-4} Torr. With 1500 V between the electrodes the etch rate was ~1 μm/min. As seen in Fig. 1 the devices have vertical walls despite the deep etch and extreme variation (0–100%) in Al concentration. The mask contained circular and square features 1.5–5 μm across which were transferred to the wafer by contact optical lithography.

For the optical measurements we used an 82-MHz mode-locked Nd:YAG laser and two frequency doublers

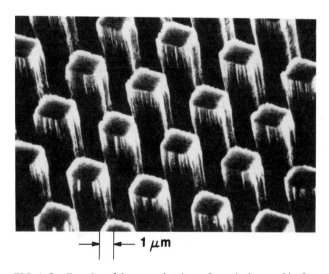

FIG. 1. Small section of the array showing ~2 μm devices used in the experiments.

Reprinted with permission from *Appl. Phys. Lett.*, vol. 51, no. 2, pp. 94–96, July 13, 1987.
Copyright © 1987, American Institute of Physics.

(the second acting on the leftover fundamental beam from the first doubler, thus yielding two second-harmonic beams) to achieve synchronously pumped mode-locked operation of two LDS 821 dye lasers. The pulses were 5–10 ps long and the wavelengths were tunable from ~780 to 950 nm (with mirror exchanges). A Burleigh piezoelectrically driven "inchworm" XYZ translator positioned the array of microresonators with high precision. The pump beam could be temporally varied over 330 ps to measure relaxation phenomena. A fast avalanche photodiode (APD) monitored the reflected output; however, the temporal resolution was ~15 ps determined by the pulse durations. Our $f/1.1$ (numerical aperture NA = 0.41) Fujinon LSR-F35B lens has a theoretical full width from zero to zero (FWZZ, i.e., the *entire* central lobe of the Airy disk) of 2.64 μm for the focused beam at 890 nm wavelength.

This MBE-grown étalon was not designed for high performance or even for this kind of gating. Prior to etching the lowest pump energy required for 5:1 contrast was ~20 pJ.[5] So far we have achieved similar or better response with a 1.5-pJ pump, more than an order of magnitude lower. The probe energy was 10 pJ or ~7 times *larger* than the pump. Figure 2 (upper trace) shows a NOR gate response to the 1.5 pJ input in a ~1.75 μm pixel. In this and the following traces the left side of the pulse train shows the gate output with no input while the right side has the input present. The 35 ns "slope" in the response is due entirely to the acousto-optic modulator and has nothing to do with the actual device speeds. Wavelengths are typically 850 nm for the input (pump) beam and 890 nm for the probe due to this étalon design. Pulsed operation such as this generally has the input and output at different wavelengths and a pair of complementary gates must be achieved for use in a practical system. Complementarity is most conveniently achievable using an isolated resonance but is also possible by working on the bandtail of conventional semiconductors. With a "sandwiched" GaAs étalon we have achieved 5:1 contrast gating with the probe beam 30–50 Å *shorter* and ~8 times *more* intense than the input beam. The wide range of pixel sizes was chosen for this first experiment since we did not know how well the light would interact with or couple into the microresonators. Results indicate that a ~2.5 μm diameter is optimum for this setup, consistent with the expected focusing ability. Larger pixels tend to require more energy, and smaller ones require essentially the same *incident* energies. Since the smallest ones are smaller than the FWZZ of the focused beams, much light is scattered and the amount of light actually coupled into the device is reduced. Preliminary measurements on these devices show maximum reflectivity from ~15 to >50% and gain factors up to 3 or 4. Tighter focusing into smaller posts of better design should yield much better performance.

The post sizes were designed to be constant in one direction which was parallel to the gradient of the étalon thickness. Despite the fact that this direction gives the fastest variation in response, for medium to large pixels (i.e., >2.5 μm) we could move at least several pixels in either direction (totals are ≥8 pixels or ≥40 μm) without significant change in response. In the orthogonal direction (smallest thickness variation) the pixel size changed every four pixels, but uniform responses were often obtained for same-size pixels. To look for crosstalk, the pump and probe were first aligned on one pixel yielding a 5:1 contrast NOR gate. Then with the pump centered on an adjacent pixel *only 3 μm away* on either side practically no response could be seen (Fig. 3). This was expected[9] since the vertical boundaries should eliminate long range carrier diffusion and also reduce diffraction. Previously, spacings of ≥10 μm showed comparable signs of crosstalk.[9]

Prior to etching it was not possible to employ surface recombination[8,11] to speed up the recovery of the MBE-grown étalon and with >300 ps delay (between pump and probe) *no sign of recovery* was seen. It presumably took several nanoseconds as in earlier GaAs devices.[12] Recombination on the sidewalls of the microresonators, however, yielded *nearly full recovery* in as little as 150 ps (Fig. 2, lower trace has the probe delayed 150 ps relative to the upper

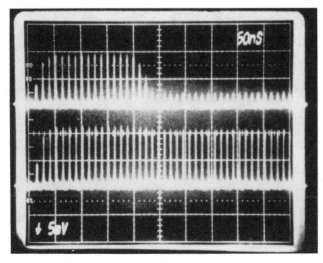

FIG. 2. (Upper) Response of a ~1.75 μm device to 1.5 pJ input pulses. The left side is the output with no input (data) present and the right side is the output *with* input present. (Lower) Same as upper but with probe pulses delayed 150 ps.

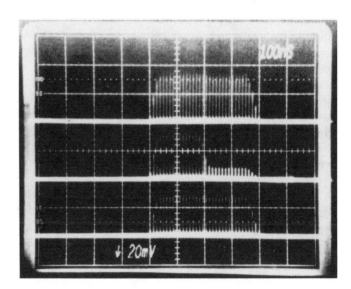

FIG. 3. Response of a 1.5-μm pixel with the probe beam constantly on one pixel and the input centered on the adjacent pixel to the left (top), on the same pixel (center), one pixel to the right (bottom). Center-center separation is 3 μm.

trace). As expected the recovery was fastest for the smallest pixels, but it also varied with the etching parameters. Probably sidewall surface recombination is quite sensitive to the etching process and consequent material damage on the sides. It is worthwhile to note that sidewall recombination should also work for multiple quantum well semiconductors, not just in bulk material.

Most data were taken with offset modulated pulse trains ~300 ns long (to observe both high and low outputs at a glance) and widely spread to avoid thermal effects. However, heating effects appear to be minimal since the response of a 5:1 contrast NOR was very similar with the pulses on *continually* at the 82 MHz mode-locked repetition rate as it was in the "thermal-free" (~300 ns pulse train envelope) case (Fig. 4). The device still functions as a NOR gate; however, the detector output shows a noticeable drop for the "high" signal when going from low to high duty cycle as seen in the figure. Saturation of the APD is believed to account for at least most of the difference since a similar drop was seen when the device was replaced by a highly reflecting mirror.

In conclusion, we have demonstrated a qualitative advance in étalon optical logic device technology which improves most of the key device characteristics such as energy, speed, uniformity, and crosstalk. In no way does the performance appear to be degraded; only good focusing is required. Since the diameter of a diffraction-limited spot scales with wavelength, submicron-diameter devices *still using* GaAs should be achievable by focusing through a high refractive index material. For example, light with a vacuum wavelength, λ, of 0.88 μm, has in GaP (with an index $n > 3.1$) a wavelength in the medium of only $< 0.3 \mu$m. This offers another *order-of-magnitude reduction* in device areas, allowing them to approach $(\lambda /n)^2$. These first experiments with microresonators have confirmed our expectations of reductions in energy requirements, crosstalk, and recovery times. The thermal stability at 82 MHz gives additional encouragement to their eventual practicality. Of course, we will eventually want to run them much faster and will undoubtedly have to deal with thermal dissipation. Exactly what percentage of the light actually coupled into the devices is not known. Microresonators designed for transmitted outputs (with the substrate removed) should allow much more accurate quantitative measurements. It would be quite surprising, however, if these first attempts resulted in optimized coupling. High performance étalon designs, optimized coupling, good fabrication of still smaller devices, and possibly some reduction of operating temperature should allow us to push these devices close to the fundamental and statistical limits of performance.[1,3]

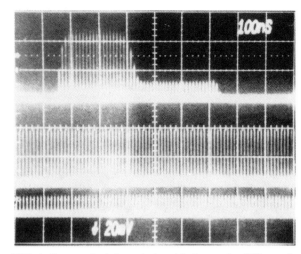

FIG. 4. Output of a 1.5-μm device with the usual ~300 ns pulse trains (top), probe on continually with the input blocked (center), both input and probe on continually (bottom). The decrease in the "high" signal level when going from low to high duty signal is mostly due to detector saturation rather than device heating.

[1]S. L. McCall and H. M. Gibbs, in *Optical Bistability*, edited by C. M. Bowden, M. Ciftan, and H. R. Robl (Plenum, New York, 1981), p. 1. S. L. McCall and H. M. Gibbs, in *Dissipative Systems in Quantum Optics*, edited by R. Bonifacio (Springer, Berlin, 1982), p. 93.
[2]R. L. Fork, Phys. Rev. A **26**, 2049 (1982).
[3]P. W. Smith, Bell Syst. Tech. J. **61**, 1975 (1982).
[4]Robert W. Keyes, Opt. Acta **32**, 525 (1985).
[5]S. L. McCall, A. C. Gossard, J. H. English, J. L. Jewell, and J. F. Duffy. *CLEO '86* Tech. Dig. (Optical Society of America, San Francisco, CA). paper FK3.
[6]J. L. Jewell, H. M. Gibbs, A. C. Gossard, A. Passner, and W. Wiegmann. Mater. Lett. **1**, 148 (1983).
[7]M. W. Geis, G. A. Lincoln, N. N. Alfremow, and W. J. Piazentini, J. Vac. Sci. Technol. **90**, 1390 (1981).
[8]Y. H. Lee, M. Warren, G. R. Olbright, H. M. Gibbs, N. Peyghambarian. T. Venkatesan, J. S. Smith, and A. Yariv, Appl. Phys. Lett. **48**, 754 (1986).
[9]J. L. Jewell, Y. H. Lee, J. F. Duffy, A. C. Gossard, and W. Wiegmann. Appl. Phys. Lett. **48**, 1342 (1986).
[10]D. J. Hagan, I. Galbraith, H. A. MacKenzie, W. J. Firth, A. C. Walker, J. Young, and S. D. Smith, in *Optical Bistability III*, edited by H. M. Gibbs, P. Mandel, N. Peyghambarian, and S. D. Smith (Springer, Berlin, 1986). p. 189. W. J. Firth, I. Galbraith, and E. M. Wright, *ibid.*, p. 193.
[11]Y. H. Lee, H. M. Gibbs, J. L. Jewell, J. F. Duffy, T. Venkatesan, A. C. Gossard, W. Wiegmann, and J. H. English, Appl. Phys. Lett. **49**, 486 (1986).
[12]J. L. Jewell, Y. H. Lee, M. Warren, H. M. Gibbs, N. Peyghambarian, A. C. Gossard, and W. Wiegmann, Appl. Phys. Lett. **46**, 918 (1985).

Novel hybrid optically bistable switch: The quantum well self-electro-optic effect device

D. A. B. Miller, D. S. Chemla, and T. C. Damen
AT&T Bell Laboratories, Holmdel, New Jersey 07733

A. C. Gossard and W. Wiegmann
AT&T Bell Laboratories, Murray Hill, New Jersey 07974

T. H. Wood and C. A. Burrus
AT&T Bell Laboratories, Crawford Hill, New Jersey 07733

Switching devices using optical inputs and outputs have been the subject of considerable research in recent years. Optical bistability (OB) in particular has received much attention,[1] both as an interesting physical phenomenon and as an attractive method for making optical logic elements. Many methods have been proposed for attaining OB; usually they require a combination of a microscopic nonlinearity with some macroscopic feedback. One problem with most forms of OB has been the difficulty of achieving adequately fast switching times at sufficiently low powers (i.e., low switching energy) to make useful devices. Several approaches have been taken to minimize switching energy. The use of a resonant cavity (such as the Fabry–Perot) helps, but at the expense of increased difficulty of fabrication. Research has also been undertaken to find novel nonlinear materials, especially semiconductors with large nonlinear refractive indices.[2-6] "Hybrid" methods of synthesizing nonlinear optical behavior by combining electro-optic materials with electrical detection and electrical feedback have also been investigated,[7] but these methods usually work at the expense of considerable electrical complexity; even with the best conventional electro-optic materials, devices with interestingly low switching energies[8] have been relatively long (consequently requiring waveguide confinement). Some attempts have also been made to reduce electrical complexity by combining the electrical detector and electro-optic material.[9]

Two recent developments have prepared the way for the novel optically bistable device described in this paper. First, a new type of OB has recently been discovered which relies only on a material whose absorption increases as the material becomes more excited.[10,11] No mirrors or other external feedback are required; the feedback is internal and positive (increasing incident power gives more absorbed power, which excites the material, resulting in increased absorption and hence more absorbed power, and so on). Second, new electroabsorptive processes, much larger than the usual Franz–Keldysh effect seen in conventional semiconductors, are seen in room-temperature GaAs/GaAlAs multiple quantum well (MQW) material.[12-14] In particular, when an electric field is applied perpendicular to the quantum well layers, the band edge absorption, including any exciton resonance peaks, can be shifted to lower photon energies. In a few microns of material, changes in transmission of $\sim 50\%$ can be achieved at modest drive voltages (~ 8 V).[13] When this field is applied using a reverse-biased p-i-n diode with the MQW inside the intrinsic (i) region, the structure also is an efficient photodetector.[14] These recent discoveries are combined in this work to make a hybrid version of OB due to increasing absorption in which the same micron-thick piece of MQW is used as both detector and electroabsorptive modulator. The high sensitivity of this new electroabsorptive effect enables attractive switching energies and speeds without any resonant cavities.

Only a series resistor (R) and a constant voltage bias supply (V_0) must be added to the p-i-n diode to make the optically bistable device. The configuration is shown schematically in Fig. 1. We call this combination of MQW modulator and detector a self-electro-optic effect device (SEED). To make the device switch, the incident light wavelength is chosen to be near the exciton resonance position for zero

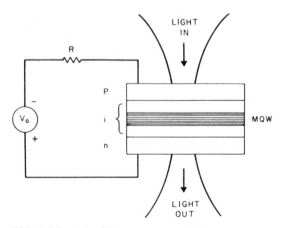

FIG. 1. Schematic of the quantum well SEED.

Reprinted with permission from *Appl. Phys. Lett.*, vol. 45, no. 1, pp. 13-15, July 1, 1984.
Copyright © 1984, American Institute of Physics.

voltage across the diode. With low optical power, nearly all the supply voltage is dropped across the diode because there is little photocurrent. This voltage shifts the exciton absorption to longer wavelengths (lower energies) and the optical absorption is relatively low. Increasing the optical power increases the photocurrent, reducing the voltage across the diode. However, this reduced voltage gives increased absorption as the exciton resonances move back, resulting in further increased photocurrent and consequently leading, under the right conditions,[10,11] to regenerative feedback and switching.

In our experiments we have used a krypton-pumped LDS 821 cw dye laser, although the powers and wavelengths are clearly compatible with a diode laser source. The sample used has fifty 95-Å GaAs quantum well layers separated by 98-Å GaAlAs barrier layers, all with no intentional doping. This MQW material is within the intrinsic region of a p-i-n diode structure grown by molecular beam epitaxy on a GaAs substrate; contact and buffer regions are made of GaAlAs and GaAs/GaAlAs superlattice materials which are substantially transparent at the wavelengths of interest here. The material is etched to give a 600-μm-diam mesa, a gold ring contact with a 100-μm-diam hole is applied to the top surface, and the GaAs substrate is removed underneath this hole to give a clear optical path perpendicularly through the layers. This sample is similar to those previously used for modulators[13] and is described in greater detail elsewhere.[13] When a reverse bias is applied to this structure, the absorption spectrum near the band-gap energy shifts to lower energies with some broadening as has been described.[13,14] However, our measurements also show that, with reverse bias, the internal quantum efficiency of the structure as a detector is high with one carrier pair collected for each absorbed photon within experimental error.[14] Only at reverse bias $\lesssim 2$ V does the quantum efficiency drop off, probably because the depletion region then does not extend completely through the MQW.

Figure 2(a) (solid line) shows the measured (external) responsivity S of the sample as a function of reverse bias; this measurement is made at 1.456 eV (851.7 nm) which is approximately the zero bias energy of the heavy hole exciton resonance. As the reverse bias is increased, the responsivity first increases as depletion becomes complete and then decreases as the exciton absorption peak moves to lower energy. The subsequent feature between 8 and 16 V is due to the light hole exciton resonance similarly moving past the measuring wavelength.

To calculate the input/output characteristic, we solve two simultaneous equations. The first is

$$S = S(V), \quad (1)$$

where $S(V)$ is the measured responsivity [the solid line in Fig. 2(a)]. The second is $V = V_0 - RSP$, where P is the optical input power and V is the voltage across the diode, i.e.,

$$S = (V_0 - V)/RP. \quad (2)$$

The graphical solution is straightforward [Fig. 2(a)] with Eq. (2) giving straight lines (shown dashed) of decreasing (negative) slope for increasing P. Bistability results from the multiple intersections of straight line and curve. We can also cal-

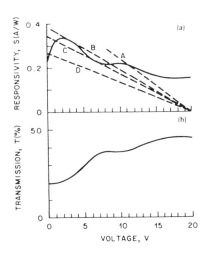

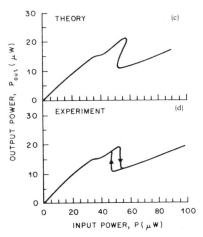

FIG. 2. (a) The solid line is the measured responsivity at 851.7 nm with reverse bias, V [Eq. (1)]. The dashed lines A–D correspond to the constraint imposed by the external circuit [Eq. (2)] for increasing power P ($V_0 = 20$ V). Lines A and D show only one intersection (i.e., only one solution); all lines between B and C show three intersection points, corresponding to bistability (the middle solution is unstable). (b) shows the measured optical transmission of the device with voltage at 851.7 nm (c) shows the theoretical optical input/output characteristic calculated using the measured responsivity and transmission with $R = 1$ MΩ and $V_0 = 20$ V as described in the text; there are no fitted parameters. (d) shows the measured optical input/output characteristic at 851.7 nm with $R = 1$ MΩ and $V_0 = 20$ V.

culate S and V as a function of P by choosing V, deducing S from Eq. (1) and P from Eq. (2). For reverse bias $\gtrsim 2$ V, the optical absorption closely follows the responsivity; to make a more accurate calculation, however, we have measured the transmission $T(V)$ [Fig. 2(b)] from which we can deduce the output power $P_{out} (\equiv PT)$ for each V and P and, hence, the whole theoretical input/output characteristic [Fig. 2(c)].

We have measured the input/output characteristics by amplitude modulating the incident light beam with an acousto-optic modulator and displaying the input and output powers, as monitored by silicon photodiodes, on the X and Y axes of an oscilloscope. Figure 2(d) shows the experimental result with $R = 1$ MΩ and $V_0 = 20$ V at 851.7 nm [the same parameters were used for the theory, Fig. 2(c)]. Bistability is clearly seen, and the result is in very good agreement with the theory with no fitted parameters.

The capacitance of this device[13] is $C \sim 20$ pF and the

TABLE I. Measured switching times and powers with $V_0 = 20$ V at 853.0 nm using various resistors. RC time constants are calculated using $C = 20$ pF. τ_a is the 10%–90% switching time for high-to-low transmission, measured with slowly varying input powers. P_s is the incident power at the high-to-low transition. Switching times are accurate only within a factor of 2 due to laser noise.

Resistance (R)	Switching power (P_s)	Switching time (τ_a)	RC time constant	Switching energy ($P_s \times \tau_a$)
100 MΩ	670 nW	1.5 ms	2 ms	1.0 nJ
10 MΩ	6.5 μW	180 μs	200 μs	1.2 nJ
1 MΩ	66 μW	20 μs	20 μs	1.3 nJ
100 kΩ	660 μW	2.5 μs	2 μs	1.7 nJ
22 kΩ	3.7 mW	400 ns	440 ns	1.5 nJ

measured switching time from high to low transmission states, τ_a, is dominated by the RC time constant as is shown in the results in Table I. Switching times were measured by ramping the input optical power up and down and measuring the optical rise and fall times of the transmitted light at the switching transitions. The switching power is proportional to $1/R$ (from the above theory), which leads to the measured speed/power trade off seen in Table I. The switching transition from low-to-high transmission also shows "critical slowing down,"[15] with switching being delayed by up to five times the switch-on time τ_a, when the input power is ramped rapidly. Critical slowing down was also observed in the thermal OB due to increasing absorption.[10] At $V_0 > 25$ V we can also observe OB due to the light hole exciton resonance and can resolve two overlapping bistable regions.

One of the remarkable features of this device is the broad range of parameters with which it will operate. The switching power and speed can be chosen over a range of nearly 10^4 as shown in Table I. We also find that OB can be seen for all V_0 from 15 V to the highest voltage used (40 V). The operation is insensitive to incident light spot size from $<10\,\mu$m diameter up to the maximum mechanically allowed ($\sim 100\,\mu$m); this is to be expected since the device operation depends on power, not intensity. The switching is remarkably insensitive to wavelength; we observe OB from 850.1 to 861.8 nm with $V_0 = 20$ V with less than 40% shift in the switching powers. The device also works equally well with multiline output from the laser (e.g., 850–853 nm); this is to be expected as theoretically this device should not require coherent light. Furthermore, it is possible to set and reset the device electrically; increases and decreases ~ 0.8 V can switch the device to high and low transmission states respectively near 20-V bias and the device is also electrically bistable at constant optical power.

Switching power and speed limits are clearly important for any practical switch. The present device is large (600 μm) by optical standards and consequently has a large capacitance and only a moderately low switching energy, i.e., power $\times$ time (~ 1 nJ incident optical energy and $\gtrsim 4$ nJ calculated dissipated electrical energy at 20-V bias during the switching. However, the switching energies per unit area are remarkably low (~ 4 fJ/μm^2 incident optical and $\lesssim 14$ fJ/μm^2 electrical); optical and total switching energies per unit area (comparing our total device area to optical spot sizes in other devices when appropriate) are lower by factors of ~ 30 and ~ 6, respectively, than the lowest measured optical energy per unit area for any OB device at a comparable wavelength.[8] The scaled total energy requirement of our device in the physical limit (i.e., ~ 1 fJ in a λ^2/n^2 device) is comparable to or better than the best scaled high-finesse resonant cavity OB devices[16] despite the absence of a resonant cavity to reduce switching energy. The low energies and absence of cavities make this device particularly attractive for switching arrays. Our maximum switching speed was dictated by a conservative maximum operating power rather than any fundamental limit. With smaller devices, therefore, faster, lower energy operation should be possible. The device has already been operated as a modulator with response times in the range of 2 ns,[13] and as a detector with $\lesssim 10$-ns response times.

In conclusion, we have demonstrated a novel hybrid optically bistable device utilizing multiple quantum well material simultaneously as both a modulator and detector. The circuit is electrically very simple, the device requires no mirrors or other external optical feedback, it operates over a broad range of conditions readily compatible with semiconductor light sources, it shows inverting logic operation, and smaller, scaled devices offer the possibility of attractively low switching energies.

[1]For a recent review, see for example D. A. B. Miller, Laser Focus **18**, No. 4, 79 (1982).
[2]H. M. Gibbs, A. C. Gossard, S. L. McCall, A. Passner, W. Wiegmann, and T. N. C. Venkatesan, Solid State Commun. **30**, 271 (1979).
[3]D. A. B. Miller, C. T. Seaton, M. E. Prise, and S. D. Smith, Phys. Rev. Lett. **47**, 197 (1981).
[4]D. A. B. Miller, D. S. Chemla, D. J. Eilenberger, P. W. Smith, A. C. Gossard, and W. T. Tsang, Appl. Phys. Lett. **41**, 679 (1982); D. A. B. Miller, D. S. Chemla, D. J. Eilenberger, P. W. Smith, A. C. Gossard, and W. Wiegmann, Appl. Phys. Lett. **42**, 925 (1983); D. S. Chemla, D. A. B. Miller, P. W. Smith, A. C. Gossard, and W. Wiegmann, IEEE J. Quantum Electron. QE-20, 265 (1984).
[5]M. Dagenais, Appl. Phys. Lett. **43**, 742 (1983).
[6]S. W. Koch and H. Haug, Phys. Rev. Lett. **45**, 450 (1981); E. Hanamura, Solid State Commun. **38**, 939 (1981).
[7]See for example, P. W. Smith and E. H. Turner, Appl. Phys. Lett. **30**, 280 (1977); E. Garmire, J. H. Marburger, and S. D. Allen, Appl. Phys. Lett. **32**, 320 (1978).
[8]P. W. Smith, I. P. Kaminow, P. J. Maloney, and L. W. Stulz, Appl. Phys. Lett. **34**, 62 (1979).
[9]B. S. Ryvkin, Sov. Phys. Semicond. **15**, 796 (1981) [Fiz. Tekh. Poluprovodn. **15**, 1380 (1981)]; B. S. Ryvkin and M. N. Stepanova, Sov. Tech. Phys. Lett. **8**, 413 (1982) [Pis'ma Zh. Tekh. Fiz. **8**, 951 (1982)].
[10]D. A. B. Miller, A. C. Gossard, and W. Wiegmann, Opt. Lett. **9**, 162 (1984).
[11]D. A. B. Miller (unpublished).
[12]D. S. Chemla, T. C. Damen, D. A. B. Miller, A. C. Gossard, and W. Wiegmann, Appl. Phys. Lett. **42**, 864 (1983).
[13]T. H. Wood, C. A. Burrus, D. A. B. Miller, D. S. Chemla, T. C. Damen, A. C. Gossard, and W. Wiegmann, Appl. Phys. Lett. **44**, 16 (1984).
[14]D. A. B. Miller, D. S. Chemla, T. C. Damen, A. C. Gossard, W. Wiegmann, T. H. Wood, and C. A. Burrus (unpublished).
[15]See, for example, E. Garmire, J. H. Marburger, S. D. Allen, and H. G. Winful, Appl. Phys. Lett. **34**, 374 (1979).
[16]P. W. Smith, Bell Syst. Technol. J. **61**, 1975 (1982).

The Quantum Well Self-Electrooptic Effect Device: Optoelectronic Bistability and Oscillation, and Self-Linearized Modulation

DAVID A. B. MILLER, MEMBER, IEEE, DANIEL S. CHEMLA, THEODORE C. DAMEN, THOMAS H. WOOD, MEMBER, IEEE, CHARLES A. BURRUS, JR., FELLOW, IEEE, ARTHUR C. GOSSARD, AND WILLIAM WIEGMANN

Abstract—We report extended experimental and theoretical results for the quantum well self-electrooptic effect devices. Four modes of operation are demonstrated: 1) optical bistability, 2) electrical bistability, 3) simultaneous optical and electronic self-oscillation, and 4) self-linearized modulation and optical level shifting. All of these can be observed at room-temperature with a CW laser diode as the light source. Bistability can be observed with 18 nW of incident power, or with 30 ns switching time at 1.6 mW with a reciprocal relation between switching power and speed. We also now report bistability with low electrical bias voltages (e.g., 2 V) using a constant current load. Negative resistance self-oscillation is observed with an inductive load; this imposes a self-modulation on the transmitted optical beam. With current bias, self-linearized modulation is obtained, with absorbed optical power linearly proportional to current. This is extended to demonstrate light-by-light modulation and incoherent-to-incoherent conversion using a separate photodiode. The nature of the optoelectronic feedback underlying the operation of the devices is discussed, and the physical mechanisms which give rise to the very low optical switching energy (~ 4 fJ/μm^2) are discussed.

I. INTRODUCTION

MULTIPLE quantum well structures (MQWS) consisting of alternate thin layers of GaAs and AlGaAs have recently become interesting for their optical properties. One exceptional property of the MQWS is the existence of clearly-resolved exciton absorption peaks near the optical absorption edge at room temperature [1]–[4]. Most normal semiconductors show well-resolved peaks only at low temperature; with increasing temperature the collisions with the increasing density of optical photons result in severe broadening. However, the quantum confinement of the electrons and holes in the GaAs layers within a thickness (e.g., 100 Å) much less than the normal exciton diameter in GaAs (about 300 Å) makes the exciton binding energy (the separation of the resonances from the bandgap) larger without further increasing the phonon broadening; this and other consequences of the quantum confinement can explain the persistence of the resonances to room temperature [2], [5], [6]. (One incidental consequence of the confinement is that it also removes the degeneracy in the valence bands of the semiconductor resulting in two exciton resonances, the "light hole" and "heavy hole" excitons.) The existence of these resonances has led among other things to the study of their nonlinear optical properties [1], [2], [5], [6], such as absorption saturation and the associated nonlinear refraction.

However, the present paper is concerned with another consequence of the excitons which appears to be even more dependent on the quantum confinement of the carriers. When an electric field is applied perpendicular to the quantum well layers, the whole absorption edge (including the exciton resonances) moves to lower photon energies [7]–[9]. This is remarkable because normal semiconductors do not show this effect at any temperature. The consequence of applying electric fields to conventional semiconductors is the Franz–Keldysh effect [10], which is primarily a broadening of the bandedge with comparatively little shift; the exciton peaks broaden and disappear at low fields. However for the perpendicular fields in the MQWS, the excitons remain resolved to high fields. This effect in the MWQS has recently been explained through a novel mechanism called the quantum-confined Stark effect (QCSE) [8], [9].

The cause of the broadening of the exciton resonances with increasing field is basically the reduction of exciton life time due to field ionization; the electron and hole are ripped apart by the applied field. In the MQWS, there are two consequences of confinement which inhibit this ionization: 1) the potential barriers presented by the walls of the wells (i.e., the larger bandgap AlGaAs layers) inhibit the electron and hole from tunneling completely away from one another; 2) even if the electron and hole are pulled almost completely to opposite sides of the GaAs wells, the wells are so thin that there is still strong Coulomb interaction and strong bound electron-hole states (excitons) still exist. As a consequence, very large fields can be applied without broadening the exciton. The formalism for solving the problem of the shift of the exciton resonances reduces to that of the Stark shift of a confined hydrogen atom [8], [9], hence the title QCSE. Shifts of 2.5 times the zero-field binding energy are possible at 50 times the classical ionization field.

Manuscript received January 24, 1985.
D. A. B. Miller, D. S. Chemla, and T. C. Damen are with AT&T Bell Laboratories, Holmdel, NJ 07733.
T. H. Wood and C. A. Burrus, Jr., are with AT&T Bell Laboratories, Crawford Hill, NJ 07733.
A. C. Gossard and W. Wiegmann are with AT&T Bell Laboratories, Murray Hill, NJ 07947.

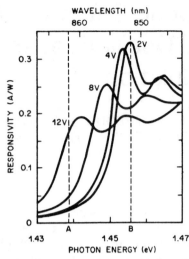

Fig. 1. Responsivity as a function of wavelength for various reverse bias voltages as shown for the p-i-n structure. Point A is 861.8 nm; point B is 851.7 nm; 600 µm diameter sample.

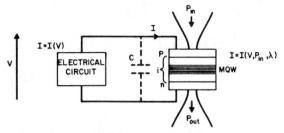

Fig. 2. Generalized schematic diagram for the SEED configuration. When it is necessary to consider the time dependence of the circuit, we will separate the capacitance of the p-i-n diode as if it were an external component, as shown by the dashed capacitor C.

Because the QCSE gives a shift of a large absorption, modulators can be made with only microns of material [7], [11]. These devices are also capable of high-speed operation, so far tested down to 131 ps [11]. The same structures used to demonstrate modulation can also, however, function as optical detectors. In fact, we find that we can obtain approximately unit internal quantum efficiency in the p-i-n structure with the quantum wells in the i region as discussed in detail below, i.e., one carrier (pair) is collected for every photon absorbed [9]. One incidental consequence of this is that the QCSE can equally well be monitored through the photocurrent spectra; a set of spectra are shown in Fig. 1 for several different voltages across the structure. The shift of the exciton peaks can be clearly resolved. The peak at lower photon energy is the heavy hole exciton peak, and the smaller peak at higher photon energy is the light hole exciton peak. This simultaneous operation as a modulator and photo detector is also the basis of the self electrooptic device (SEED), which we discuss in this paper.

The general principle of the SEED is that the photocurrent passing through an electronic circuit influences the voltage across the modulator; the voltage across the modulator influences the absorption of light by the modulator and hence, influences the photocurrent. Thus a feedback is established. This feedback is truly optoelectronic: without the light shining on the device, there is no photocurrent and no change in photocurrent with voltage due to the change in optical absorption; in turn, the electrical circuit determines the way in which the change in photocurrent changes the voltage. The general schematic of the SEED is shown in Fig. 2. While the SEED configurations discussed here are certainly hybrid in that they involve both optics and electronics, the SEED is intrinsically an integrated form of device, and furthermore these configurations do not require any active electronic components or external gain for their operation.

The behavior of the SEED depends greatly on the nature of the electronic circuit and on the sign of the feedback. Under positive feedback we obtain bistability or oscillation. The bistability has previously been reported by us briefly [12]. The oscillation is a new mode of operation which, from constant electrical and optical bias, gives both electrical and optical oscillation. Under negative feedback, totally different behavior is obtained; so far, self-linearized modulation, linear light-by-light modulation, and optical level shifting have been demonstrated [13].

The SEED's are interesting for a number of reasons. First, they represent a novel class of optoelectronic devices without a close precedent [14]–[17]; it is therefore interesting to understand how they operate and the physical principles underlying their behavior, and that understanding is the main purpose of this paper. Second, they represent a new opportunity for devices, and they may offer novel solutions to existing or future problems. Third, they operate under very practical conditions: they run at room-temperature; they operate at wavelength and powers compatible with light-emitting or laser diodes and at voltages compatible with semiconductor electronics; they are compatible in materials and growth technology with both electronics and semiconductor light sources, suggesting the possibility of integration, and they offer very low energy operation.

The layout of this paper is as follows. In Section II the experimental samples, apparatus, and methods will be briefly discussed. In Section III we will present the general theoretical framework for the SEED modeling. The nature of the optical bistability will be investigated in Section IV, and the associated electrical bistability will be discussed in Section V. In Section VI, the oscillator will be described. The operation as a self-linearized modulator and optical level shifter will be examined in Section VII. The scaling arguments on operating energies will be presented in Section VIII, and the overall conclusions of the work will be given in Section IX.

II. Experimental Details

To apply an electric field perpendicular to the MQWS layers we grew a p-i-n diode structure with the quantum wells in the intrinsic region of the diode. There are two reasons for choosing this structure to implement the SEED's. First, reverse biasing the diode can give large fields across the wells without large currents flowing,

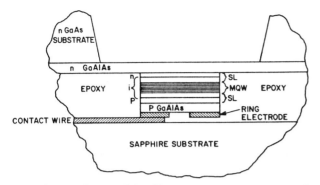

Fig. 3. Schematic diagram of the 100 μm mesa sample (not to scale) showing the layer structure of the material. The n-contact is taken off the substrate. The p-contact is through the gold ring electrode and the contact wire as shown. The total epitaxial layer thickness is 3.9 μm and the thicknesses of these layers are shown greatly exaggerated for clarity. The aluminum mole fraction is 0.32 in all GaAlAs layers. p- and n-doping levels were both 5×10^{17} cm^{-3}. The doped GaAlAs layers are both 0.98 μm thick. The multiple quantum well (MQW) region is 0.97 μm thick, and consists of 50 periods of 95 Å GaAs and 98 Å GaAlAs. The superlattice (SL) consists of 28.5 Å GaAs layers alternating with 68.5 Å GaAlAs layers. Doped superlattice regions are 0.19 μm (20 periods) thick. Undoped (i) superlattice regions are 0.29 μm (30 periods) thick. The mesa diameter is 100 μm, and the hole in the ring electrode is 25 μm diameter. The sample is epoxied to the sapphire substrate after the mesa etch.

which allows efficient operation and eliminates possible thermal problems. Second, this structure is ideally suited for simultaneous operation as a detector. We have found empirically [9] that the responsivity of this structure corresponds to unit internal quantum efficiency, within experimental error, for all reverse bias voltages greater than 2 V. This agrees well with our estimate that above about 2 V the depletion region extends throughout the MQWS region [7], [9].

We will not discuss the material structure and growth in great detail here as it is more extensively discussed elsewhere [7], [9]. The detailed layer thicknesses and compositions are given in the caption to Fig. 3. The material used in these SEED experiments is grown by molecular beam epitaxy and is the same as that used by us for modulator experiments [7] and investigations of the QCSE [8], [9]. The layer sequence is as shown in the sample schematic diagram of Fig. 3. Starting with a silicon—(i.e., n-doped) GaAs substrate, a transparent n-doped GaAlAs layer is grown. Above this a transparent superlattice buffer region is grown, first n-doped and then undoped to form the start of the intrinsic region of the diode. Next, the 50 95 Å GaAs quantum well layers alternate with 98 Å GaAlAs barriers to form the MQWS. Then, the sequence of undoped superlattice, doped superlattice and doped GaAlAs is grown, this time with p-doping. The thinness of the GaAs layers in the superlattice (28.5 Å) ensures that the optical absorption of this material is quantum-shifted to wavelengths shorter than those of interest here.

To make the samples for the experiments, we have formed a mesa in the epitaxial layers as shown in Fig. 3 to define a small area diode. The electrical contact on the p-doped top contact is formed by zinc diffusion, gold alloying, and plating to leave a gold-plated ring electrode with a hole in the center. To allow optical access, the GaAs substrate is etched away beneath the mesa using a selective etch. Two samples are used in the present study. One has a 600 μm diameter mesa with a 100 μm diameter hole in the electrode; this sample is used for all experiments unless otherwise stated. The second sample (shown schematically in Fig. 3) has a 100 μm diameter mesa with a 25 μm diameter electrode hole; this sample is epoxied to a sapphire substrate as shown.

An LDS 821 continuous-wave dye laser pumped by a krypton-ion laser was used for all experiments unless otherwise stated. We also used a Hitachi HLP 1400 diode laser for all the experiments with the 100 μm sample. The diode laser was temperature-tuned over a range of 5 nm. For bistability experiments with the dye laser, the power was modulated using an acoustooptic modulator. With the diode laser, the power was modulated by varying the current of the diode. In both cases, neutral density filters were used as necessary to reduce the overall power in the optical beam. Detection was with silicon photodiodes, with an avalanche diode used for high-speed measurements. Powers were measured with a UDT 161 power meter.

III. Formal Description of SEED Behavior

In general, the feedback behavior of the SEED devices can be described in terms of two simultaneous relations. One relation describes the detection properties of the quantum well structure (as shown in Fig. 1), giving the current I through the modulator as a function of the voltage V on the modulator, the optical input power P_{in}, and the wavelength λ of the incident light,

$$I = I(V, P_{in}, \lambda). \quad (1)$$

The other relation describes the behavior of the circuit to which the modulator is connected, i.e., the relation between I and V imposed by the circuit

$$I = I(V). \quad (2)$$

In this description, (2) is the "load line" of the circuit, and the current I will depend on other circuit parameters not explicitly stated here, such as supply voltage or component values. Solving these two equations simultaneously to eliminate I will give the relation between P_{in} and V at a given wavelength, and the internal behavior of the device will be understood in terms of both optical and electrical parameters.

One further simple relation is required to deduce the output power from the input power, namely the modulator transmission function for the output optical power P_{out},

$$P_{out} = P_{out}(P_{in}, V, \lambda). \quad (3)$$

For all the situations in this paper the detector equation (1) reduces to the simple relation

$$I = S(V, \lambda) P_{in} \quad (4)$$

where S is the measured responsivity of the detector (in A/W) as a function of voltage and wavelength. We can make this simplification because the leakage current of the diode is negligible, and the responsivity of the diode (at fixed wavelength and voltage) empirically does not depend on power over the range of powers we use. For (3), the transmission function also simplifies to give

$$P_{out} = T(V, \lambda) P_{in} \quad (5)$$

where T is the measured transmission as a function of voltage and wavelength, because the transmission also empirically does not depend on power in our conditions. In fact, because the internal quantum efficiency is constant (approximately unity) for all reverse bias voltages above 2 V, the absorption closely follows the absorbed power and the transmission looks like a mirror image of the responsivity (see Fig. 4).

The behavior of all of the SEED configurations so far demonstrated can be described by these relations. The feedback in the system is contained in the interplay between relations (1) [or (4)] and (2). When $I(V)$ in (2) is not explicitly time-dependent (e.g., it contains no capacitors, or inductors, or time varying drive voltages or currents) we can formally deduce whether the feedback is positive or negative at a given (equilibrium) solution of (4) and (2) by performing a linearized stability analysis. This is a valid approach for the bistability and self-linearized modulation (we will consider the oscillator separately below in Section VI). Consider for example the change in voltage v about the equilibrium point. We presume there is a finite capacitance C associated with the device for mathematical convenience as shown dashed in Fig. 2; this is also physically realistic as the device does have capacitance due to the depletion region. We obtain, by expanding to first order about the equilibrium solutions,

$$C \frac{dv}{dt} = \left[\frac{dI}{dV} - P_{in} \frac{dS}{dV} \right] v. \quad (6)$$

If the term in square brackets is positive, v will diverge exponentially, the solution is unstable, and the feedback is positive. Conversely, if this term is negative, so also is the feedback, v settles exponentially and the solution is stable.

In all the configurations considered here, the electrical circuit to which the quantum wells are connected will have positive slope resistance corresponding to negative dI/dV with the current convention shown in Fig. 2. The sign of the feedback can only be positive if $P_{in} dS/dV$ is negative, and then only if

$$P_{in} \frac{dS}{dV} < \frac{dI}{dV} \quad (7)$$

which becomes the condition for instability. In other words, as long as S increases with V, the solutions are stable, but if S decreases with V, the solutions can be unstable depending on the rest of the cirucit.

Illustrative operating wavelengths (photon energies) for the two feedback modes are shown on Fig. 1. At a photon energy below the zero-field heavy-hole exciton peak position such as at point A in Fig. 1, increasing voltage V will

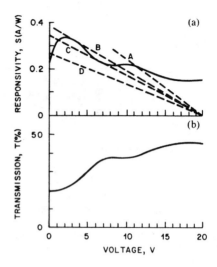

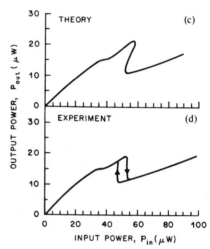

Fig. 4. Graphical solution method for bistable SEED behavior with a simple resistive load. (a) The solid line is the measured responsivity at 851.7 nm with reverse bias, V [see (1)]. This is the same wavelength as point B on Fig. 1. The dashed lines A–D are the load lines imposed by the circuit (equation (2)) for various increasing values of the input optical power, P_{in}. The intercept of these lines with the horizontal axis is the supply voltage, V_0, which is 20 V in this example. Lines A and D show only one intersection (i.e., only one solution); all lines between B and C show three intersection points, corresponding to bistability. As discussed in the text, the middle of these three points is unstable. (b) The measured optical transmission of the p-i-n diode, also at 851.7 nm. (c) The theoretical optical input/output characteristic calculated using the measured responsivity and transmission with $R = 1$ MΩ and $V_0 = 20$ V as described in the text; there are no fitted parameters. (d) The measured optical input/output characteristic at 851.7 nm with $R = 1$ MΩ and $V = 20$ V. All measurements are for the 600 μm diameter sample.

give increasing responsivity S over the entire voltage range shown, so all solutions at this wavelength are stable over this range. However at point B in Fig. 1, as the exciton peaks move past the operating wavelength, the responsivity S can decrease with increasing voltage V and instability is possible.

Another way of viewing the positive feedback in the device is to note its close connection with negative differential conductance. At constant power the differential conductance of the p-i-n diode is $P_{in} dS/dV$, so that the diode shows negative differential conductance whenever S decreases with V. This is true over several ranges of voltages and wavelengths as can be seen in Fig 4(a). The operation of electrical bistability (Section V) and the oscillator (Section VI) can both be understood as consequences of this negative conductance in a manner analogous to purely electrical negative conductance circuits such as those utilizing tunnel diodes [18]. It must be remembered, however, that this negative conductance results from an optoelectronic feedback. Unlike the tunnel diode, the whole scale of the electrical characteristic (including the negative conductance) can be altered by changing the optical input power, and the optical bistability is not, therefore, analogous to any simple electrical circuit; it is, however, true that the optical bistability cannot exist without the existence of negative differential conductance at some power and voltage.

That optical and electrical bistability should be so interrelated is a consequence of a more general principle. Bistability is basically a simple cusp catastrophe [19]. One consequence of this is that, if the system is bistable with respect to variation of one parameter, we can expect that it will be bistable with respect to others. Thus, for example in Fabry–Perot bistability, the system is bistable under variation of the input power, and also under variation of wavelength and cavity tuning or some combination of all three. In the present case, since supply voltage is a parameter of the system, we should expect to see bistability with respect to supply voltage when the optical input conditions are held fixed. This will be discussed in Section V. In another case where the circuit parameters are altered with a second light beam, we also see bistability with respect to the variation of power in the second light beam. This is discussed in Section IV-B.

IV. SEED Optical Bistability

A. Resistive Load

The simple circuit that we use for optical bistability with a resistive load is shown schematically in Fig. 5. A constant reverse bias voltage V_0 is connected through a series resistor to the p-i-n diode. The current equation (2) becomes

$$I = (V_0 - V)/R \quad (8)$$

where R is the value of the series resistance. Using (4) we may rewrite this as

$$S = \frac{(V_0 - V)}{RP_{in}}. \quad (9)$$

For various P_{in}, this gives us the straight lines in Fig. 4(a). The detector equation (1) at constant wavelength can be formally rewritten using (4) as

$$S = S(V) \quad (10)$$

(i.e., the measured responsivity function for a given wavelength). A specific measured function appears as the curve in Fig. 4(a). The particular curve shown is the measured responsivity at the wavelength of the heavy hole exciton peak at zero field. At first, the responsivity rises with increasing voltage because the quantum efficiency is rising as the depletion region extends throughout the device. Then the responsivity falls because the exciton peak moves to lower photon energies. The subsequent peak near 10 V is due to the light hole exciton peak moving through the operating wavelength. This behavior is consistent with the transmission curve of Fig. 4(b).

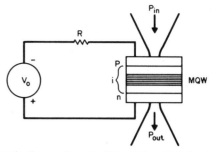

Fig. 5. Schematic diagram for optical bistability with a simple resistive load.

The simultaneous solution of (1) and (2) then reduces to the graphical solution on Fig. 4(a); the solutions are the intersections of line and curve. The lines with triple intersections (between lines B and C) correspond to the powers for which bistability is possible. Applying the stability criterion equation (7) it can be seen that the middle intersection is an unstable solution and the others are stable. The criterion for the existence of bistability can be deduced directly from this graphical method, from the criterion in [20] (the relation to [20] is discussed below) or from the instability criterion equation (7) (above) since the bistability cannot exist without one unstable point. Hence, using (7) and (4)

$$\frac{dS}{dV} < \frac{S}{V - V_0}. \quad (11)$$

For bistability to exist at a given V_0 and λ, this criterion must be satisfied for some range of V.

A nongraphical modeling method is the simple "backwards" technique of choosing a voltage V, deducing S from the measured responsivity (8), and hence deducing P_{in} from (7). By either method, the relation between P_{in} and V is obtained. Finally, the relation between P_{out} and P_{in} as shown in Fig. 4(c) is obtained for the particular wavelength from the measured transmission function (5) [Fig. 4(b)]. The curve shows the classic "S" shape for optical bistability.

In Fig. 4(d), we show an experimental result taken using the same parameters as used in the theoretical curve [Fig. 4(c)]. There are no adjustable parameters in the theory, and the agreement with experiment is good. Clear bistability is seen, with switching transitions at the edges of the hysteresis region.

As has been pointed out [12], [20], this bistability is an example of optical bistability from increasing absorption. This class of bistability requires no mirrors, cavities, or other external optical feedback. Various examples of this bistability class have been independently demonstrated or proposed; many of these have recently been summarized

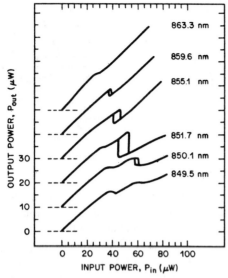

Fig. 6. Input/output characteristic as a function of operating wavelength with the circuit as shown in Fig. 5. The zeros are displaced on successive spectra for clarity as shown by the dashed lines. $V_0 = 20$ V; $R = 1$ MΩ, 600 μm diameter sample.

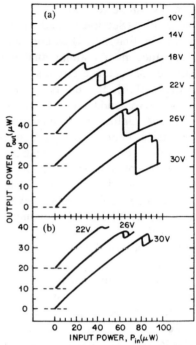

Fig. 7. Input/output characteristics as a function of supply voltage V_0 with the circuit as shown in Fig. 5. The zeros are displaced on successive spectra for clarity as shown by the dashed lines. (a) Curves without restrictions on input power, showing overlapping bistable regions above 26 V; (b) Curves with maximum input power restricted to show light hole bistability while avoiding second (heavy-hole) bistable transition. $\lambda = 851.7$ nm; $R = 1$ MΩ; 600 μm diameter sample.

[20]. The general feedback mechanism of this class relies on having a material (or system) in which the fraction of the incident light absorbed (the absorption) increases as the material becomes more excited. Then as the incident power increases the material becomes more excited, thus further increasing the absorption, which in turn makes the material even more excited, and so on. Thus a positive feedback is established which can lead to switching under the right conditions.

The relations (9) and (10) are of exactly the same functional form as the general relations (2) and (1) of [20] if we choose for the measure of the degree of excitation of the system the voltage change $V_0 - V$, and if instead of the actual fractional absorption A we use the responsivity S. This latter slight formal departure is of little importance; the responsivity is proportional to the absorption over most of the voltage region of interest.

The behavior of the input/output characteristic as a function of operating wavelength is shown in Fig. 6. Bistability is seen over a comparatively large range of wavelengths (861.8–850.1 nm at 20 V bias). The width of the bistable region also exemplifies one of the predictions of the theory of this class of bistability [20], that the width of the bistable region is largely determined by the ratio between the absorption in the high and low transmission states; larger contrast between the two states should give larger width.

The behavior of the optical input/output characteristic as a funciton of bias voltage V_0 is shown in Fig. 7. Bistability is seen for all voltages above 14 V for this wavelength (851.7 nm). With increasing voltage, the "kink" in the characteristic seen just below the bistable region at, for example, 22 V on Fig. 7(a) develops into a bistable region itself. This second bistability results from the light-hole exciton peak; as can be visualized from Fig. 4(a), if the voltage is increased enough eventually multiple intersections of straight lines and curve will be possible through the light-hole peak. This bistability can overlap the other bistable region. At 26 and 30 V, two "switch-down" transitions are seen although only one "switch-up" is apparent; this is a consequence of the precise functional shape of the responsivity curve. The light-hole bistability can be viewed separately by restricting the maximum input power so that the heavy-hole bistability threshold is not reached as shown in Fig. 7(b).

The curves in Figs. 4, 6, and 7 are deliberately taken at slow scan rates of the incident power. With increasing scan rates, the switching transitions are no longer sharp due to the finite switching times. The switching times depend very much on how rapidly the input power is ramped, with rapid ramping of the input giving sharper switching transitions. The hysteretic regions also broaden as the input power is ramped more rapidly due to delays in the switching. Both of these phenomena are characteristic of the effect known as "critical slowing down" [21] seen in bistable switching systems. (One simple mathematical manifestation of critical slowing down can be seen in the linearized limit in (6); at the critical points for switching the term in square brackets vanishes; thus although the point is not truly stable, the system will take an infinite time to switch.) There is thus no simple definition for the switching time of the system.

The switching down (i.e., from high to low transmission) is also generally less affected by the critical slowing-down and all the times we discuss will be for this transition. The switch up is typically about a factor of two

TABLE I

Resistance (R)	Switching Power	Switching Time	RC Time Constant	Switching Energy Per Unit Area (fJ/μm^2)
600 μm sample				
100 MΩ	670 nW	1.5 ms	2 ms	3.5
10 MΩ	6.5 μW	180 μs	200 μs	4.2
1 MΩ	66 μW	20 μs	20 μs	4.6
100 kΩ	660 μW	2.5 μs	2 μs	6.0
22 kΩ	3.7 mW	400 ns	440 ns	5.3
100 μm sample				
1.1 GΩ	107 nW	1 ms	660 μs	14
22 MΩ	4.9 μW	20 μs	13 μs	12
47 kΩ	1.6 mW	30 ns	28 ns	6.1

slower. We find empirically that there is a limit to how much the switching can be speeded up by overdriving the input, thus enabling us to define a limiting switching time. With slowly varying input powers, the switching time typically lengthened by a factor of 4–5. For example, the 30 ns switching time measured for the 100 μm diameter sample listed in Table I was taken with the input power ramped up and down in 600 ns. Ramping up and down in 8 μs gave a switching time of 140 ns. This behavior is characteristic of critical slowing down, with the switching time tending to limit towards the underlying time constant of the system as the switching is strongly over-driven [21].

Table I lists (limiting) switching times for both of the samples with a variety of different resistors. These times were measured with 20 V bias at 853 nm. All of the measurements on the 100 μm sample were taken using the diode laser as the light source. To emphasize the scaling, we have listed the switching energy per unit area, this being the product of the limiting switching time and the switching power divided by the area of the mesa. We have also tabulated the RC time constants in each case. We use a measured capacitance of 20 pF for the 600 μm sample and an estimated capacitance of 600 fF for the 100 μm sample (stray capacitances in the circuit make any more accurate measurement of the capacitance difficult for this sample). The overall accuracy in switching time measurements is only about a factor of two due to laser noise.

For both samples, the switching time clearly scales with the RC time constant of the system, and the switching power scales inversely with the resistance, as would be expected theoretically. The switching energy per unit area remains remarkably constant over the wide range of switching times reported; the variation in switching energy per unit area is within experimental error when stray capacitance is taken into account for the smaller sample. Slower switching was not possible for the 600 μm sample because the leakage current of the diode then became dominant. The proportionality between switching time and total RC time constant was, however, retained when external capacitors were added in parallel with the sample to slow down the switching.

With the 100 μm sample we were able to obtain optical bistability with 18 nW of incident power by using three reverse-biased silicon diodes in series instead of the load resistor. The leakage current of the diodes in series was about 3 nA with 20 V bias, giving an effective resistance of about 6 GΩ. The switching time was 40 ms; this is somewhat longer than expected from scaling, but we attribute this to the capacitance of the diodes.

B. Constant Current Load

As can be seen from Fig. 7, with increasing bias voltage V_0 the bistability becomes easier to achieve and the loop becomes broader. The bistability occurs at higher input powers, but if we also proportionately increased the load resistance R as we increased V_0, the switching power would remain approximately constant. This behavior is easily understood from the graphical solution of Fig. 4. With V_0 and R increasing in proportion, the intercept of the straight load lines with the horizontal axis moves further to the right and the load lines become more horizontal while still retaining exactly the same intercept with the vertical axis. The more horizontal the load lines become, the easier it is to satisfy the condition for bistability, (7), because dI/dV becomes less negative. This suggests that, for the best bistability, we should use very large voltages and resistance. In the limit of infinite voltage and resistance, $dI/dV = 0$, bistability is possible in principle with even the slightest negative slope dS/dV. High voltages are not, however, desirable for various reasons. One disadvantage of high voltages is high electrical switching energy because of the requirement to charge the capacitance of the p-i-n diode to a large fraction of this voltage for switching.

An alternative method which embodies many of the advantages of the above method without some of the disadvantages is to use a constant current bias circuit. An idealized circuit is shown in Fig. 8(a) and two of the many possible practical circuits for synthesizing a constant current source are shown in Fig 8(b) and (c). For the moment we will only consider the circuit in Fig. 8(b); we will return to use both circuits in the discussion of negative feedback operation with current bias (Section VII). Here we have used a tungsten bulb to illuminate the silicon photodiode in Fig. 8(b) with incoherent light. With this control light shining on the photodiode, this diode passes an ap-

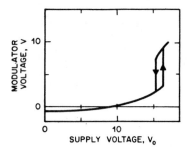

Fig. 8. Constant current circuits. (a) Idealized schematic for the SEED under current bias. (b) Constant current source with current value controlled by the amount of light shining on the silicon photodiode. (c) Constant current source with current value controlled by the control voltage V_c.

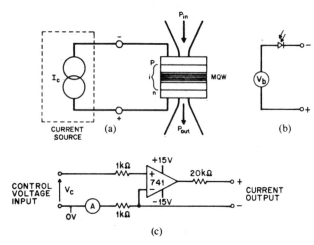

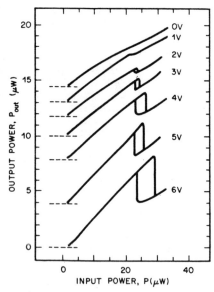

Fig. 10. Voltage V across the p-i-n modulator as supply voltage V_0 is varied. The electrical circuit is as shown in Fig. 5. Optical input power and wavelength are held constant at 46 μW and 852.4 nm respectively. $R = 1$ MΩ; measuring probe impedance (in parallel with p-i-n) 10 MΩ; 600 μm diameter sample.

Fig. 9. Optical input/output characteristics at 853 nm for the 100 μm sample using the constant current circuit of Fig. 8(b) with reverse bias supply voltages V_b between 0 and 6 V as shown, 9.3 μA constant current bias. The zeros are displaced on successive spectra for clarity as shown by the dashed lines.

proximately constant current as long as it is reverse biased between 1 and 6 V. Within these limits therefore, dI/dV is approximately zero, and bistability should be easily achieved. (We restricted the supply voltage to 6 V or less in our experiments.) This constant current is the photocurrent, and its value can be controlled by adjusting the intensity of the tungsten bulb light shining on it. This circuit, in common with the many other methods of generating constant currents, can only operate up to a finite voltage output; when the output voltage from the circuit approaches the supply voltage, the current starts to drop off. (In the present case the photocurrent decreases as the voltage across the diode drops below 1 V.) For our present purposes, we can treat the load line of the constant current source as being horizontal for output voltages up to about 1 V less than the supply voltage, and as falling off nearly vertically towards the horizontal axis thereafter. Thus, even for low supply voltages, we can achieve the three intersections of load line and curve necessary for bistability.

Typical results are shown in Fig. 9 for bistability with the photodiode load using the laser diode as the light source for the 100 μm diameter p-i-n modulator with 6 V supply voltage. Note the total absence of any kink in the curve from the light hole resonance; this is because the voltage is never high enough to move the light hole resonance to the operating wavelength. With this circuit, we were able to observe bistability with as little as 2 V supply voltage. By contrast, with the resistive load discussed above, 14 V were necessary. This type of constant current loading therefore can greatly reduce the operating voltage for bistability, which is ultimately important for limiting electrical energies in switching applications. The present circuit does not show particularly fast switching compared to that seen at similar switching powers with the resistive load, presumably because of the depletion capacitance of the photodiode.

One incidental feature of this particular configuration is that the switching power can be controlled optically by adjusting the amount of control light (in this case from the tungsten bulb) landing on the silicon photodiode. The bistability can also be set and reset by slight variations of this control power, and the system is also therefore bistable with respect to variation of the control power at constant laser power. Note that the control power can be from a broad-band incoherent light source as in this case, or from any other convenient source.

V. Electrical Bistability

As discussed in Section III, we should expect bistability with respect to variation of supply voltage because of the existence of negative differential conductance and for fundamental reasons. In Fig. 10 we show electrical bistability observed with this system. We use the circuit of Fig. 5 except that now we vary the supply voltage V_0 and hold the optical input power constant. This bistability can be observed either in the voltage across the device (as in Fig. 10) or in the optical transmission as the supply voltage is varied. (The optical bistability as a function of input power can, of course, be monitored equally well through the voltage across the device.)

The behavior of the electrical bistability can be understood through a graphical analysis similar to that of Fig. 4; now instead of varying the slope of the straight lines as a constant intercept (the fixed V_0), the slope of the lines is held fixed and the intercept is varied (the varying V_0). As can be seen in Fig. 10, the voltage V starts out below the axis; this corresponds to the p-i-n being in forward bias from its own photocurrent. (The responsivity S falls off in forward bias (not shown on Fig. 4) to reach zero near the built-in voltage of the diode.) With increasing V_0, the system eventually switches to the high-voltage, low-optical absorption state.

The existence of electrical bistability also implies that optical bistability can be set and reset electrically. We have tested this directly. With the various parameters set so that the system is in the middle of a bistability loop, a momentary decrease of V_0 switched the system from the high (optical) transmission state to the low transmission state, and a momentary increase would switch it back. Voltage changes of ~1 V were usually sufficient to achieve this, corresponding to the width of the electrical bistable region as in Fig. 10.

VI. Negative Resistance Optoelectronic Oscillation

As discussed in Section III, because of the existence of negative differential conductivity, we should expect to be able to demonstrate a negative resistance (or conductance) oscillator by an analogy with similar circuits using tunnel diodes. A simple formal schematic for such a circuit is shown in Fig. 11. For formal convenience we have shown the capacitance C of the p-i-n diode as if it were separate from the diode. This fiction enables us to retain relation (4) to describe the diode, even in this time-dependent case, as there are no other time-dependent terms expected in the diode behavior on the timescales of interest here. It also means that we can treat any additional capacitance added in parallel with the diode with the same theory. Using (4) and analyzing the circuit of Fig. 11, we obtain

$$\frac{d^2V}{dt^2} + \left(\frac{R_l}{L} + \frac{P_{in}}{C}\frac{dS}{dV}\right)\frac{dV}{dt} + \frac{1}{LC}$$
$$\cdot (V - R_l P_{in} S(V) - V_b) = 0 \quad (12)$$

at constant optical power and wavelength. We will not attempt to solve this large signal relation. However, for small signals (12) reduces to

$$\frac{d^2v}{dt^2} + \left(\frac{R_l}{L} + \frac{P_{in}}{C}\frac{dS}{dV}\right)\frac{dv}{dt} + \frac{1}{LC}\left(1 + P_{in}R_l\frac{dS}{dV}\right)v = 0 \quad (13)$$

where we have expanded S to first order about V_b and v is the departure of the voltage V from equilibrium. For this standard equation for a damped oscillator the condition for self-oscillation is

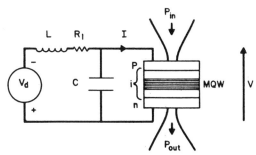

Fig. 11. Idealized circuit for optoelectronic oscillator, with C as the capacitance of the diode (and other parallel capacitance), L as the inductance, R_l as the series loss resistance of the circuit, and V_d as the supply voltage. V is the voltage across the diode.

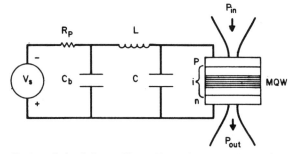

Fig. 12. Actual circuit for oscillator. The series loss resistance is not explicitly shown. The supply voltage V_s is somewhat larger than the idealized mean reverse bias voltage V_d of Fig. 11 because of the voltage drop across series protection resistor R_p from the mean dc bias current of the device.

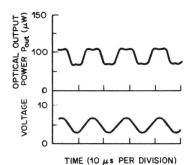

Fig. 13. Simultaneous optical and electronic self-oscillation of the SEED circuit of Fig. 12. The voltage is the voltage V as shown in Fig. 12 as measured using a ×100 10 MΩ 2.2 pF probe. The 100 μm diameter sample is used with 430 μW incident power P_{in} from the laser diode at 852 nm. Inductance L: 97 mH. Bypass capacitor C_b: 100 nF. Oscillator capacitance C: stray capacitance of inductor in parallel with other stray capacitance and probe and modulator capacitances. 100 μm diameter sample. Supply voltage V_s: 7 V. Mean dc reverse bias voltage V_d (i.e., dc voltage measured across bypass capacitor C_b): 4.9 V.

$$P_{in}\frac{dS}{dV} < -R_l\frac{C}{L}. \quad (14)$$

We were able to observe oscillation with the practical circuit shown in Fig. 12. The supply voltage now has a series protection resistor and an additional large bypass capacitor has been added as an ac short to give the dc power supply negligible ac impedance despite the presence of the series resistor; for ac purposes the two circuits are essentially identical. A typical pair of traces of the optical and electronic oscillations is shown in Fig. 13, in this case taken with the 100 μm sample and the CW laser diode.

We performed an extended series of experiments with the 600 μm sample and the dye laser source. It was not possible to resolve the power for the onset of true oscillation because of laser noise; fluctuations in the laser power can cause the circuit to "ring" even if it is below the true threshold for self-oscillation. At 851.7 nm, oscillation was clear above 40 μW between 5.3 and 5.9 V supply voltage V_s; this observation is consistent with the form of the responsivity curve in Fig. 4(a), which shows its largest negative slope, and hence its largest negative conductance, near 5–6 V. With increasing power the range of the supply voltage over which oscillation was observed increased; this is also consistent with the form of the responsivity since, with increasing power, the whole scale of the negative conductances is correspondingly increased, hence permitting oscillation over a larger voltage range. With 300 μW we were able to observe oscillation between 11.3 and 25.2 V supply voltage. (At this power the voltage drop across the series protection resistor is not negligible, and we estimate it at 4–8 V. The actual bias voltage at the p-i-n is correspondingly smaller.)

In all cases in which the oscillation was clear it was a large signal oscillation; as can be seen on Fig. 13, the optical modulation is consequently deep and not sinusoidal, although the electrical oscillation is more nearly sinusoidal. With the 600 μm sample, the electrical oscillation was generally so large that it extended into forward bias (the trace goes below the axis).

With incident powers of 150 to 300 μW and optimization of the supply voltage, oscillation could be observed with input wavelengths from 840.4 to 861.1 nm. At higher voltages and powers (e.g., at 150 μW between 14.5 and 17.0 V) it was also possible to resolve a second region of oscillation of smaller amplitude which we ascribe to the negative resistance associated with the light hole exciton.

Using a variety of inductors from 97 down to 0.12 mH we observed oscillation from 53 kHz to 1.88 MHz using the 600 μm sample without any added capacitance. These frequencies were somewhat lower than the oscillation frequency expected using 20 pF for the internal capacitance of the diode; the effective capacitance required to explain the observed frequency varied from inductor to inductor, and the explanation may be stray capacitance in the inductors themselves. With large amounts of added capacitance, the frequency approaches the expected value. We expect that higher frequecies would be attainable with improved inductors and packaging.

VII. Self-Linearized Modulation and Optical Level Shifting

One simple characteristic of any equilibrium state (stable or unstable) of the general SEED circuit in Fig. 2 is that the photocurrent as given by (1) or (4) equals the current through the electrical circuit, as can be deduced directly from conservation of current. As mentioned above, the p-i-n diode used shows, empirically, unit internal quantum efficiency above about 2 V reverse bias. In this range, therefore, the absorption A can be written

$$A = \frac{\hbar\omega}{e} S. \quad (14)$$

The optical power absorbed in the device is $P_a = P_{in}A$, and the (photo)current is $I = SP_{in}$. Hence, using (14),

$$P_a = \frac{\hbar\omega}{e} I. \quad (15)$$

Consequently, in any equilibrium state the absorbed power is proportional to the current flowing round the circuit, regardless of the nature of the circuit.

As was discussed in Section III, if the device is operated under conditions in which dS/dV is positive and dI/dV for the circuit is negative (corresponding to positive resistance in the circuit), the optoelectronic feedback is negative and the equilibrium state is stable. Consequently, the system will stay in the equilibrium state and the absorbed power will be proportional to the circuit current. If we can control the current linearly with some external parameter, we can utilize this property to make linear optical modulators, with absorbed power proportional to that parameter.

To make the current dependent only on some external control parameter requires that it be independent of the internal voltage V; we must therefore design "constant current" circuits in which the current is independent of voltage and whose current setting I_c [see Fig. 8(a)] is controlled by an appropriate external parameter. There are many ways of achieving this; two such circuits are shown in Fig. 8. Fig. 8(b) shows a constant current circuit controlled by light and Fig. 8(c) shows one controlled by voltage V_c.

With a constant current source, it is particularly easy to understand the operation of the circuit directly. In the generalized circuit shown in Fig. 8(a), the action of the current from the constant current source is to charge up the capacitance of the diode, whereas the action of the photocurrent generated in the diode is to discharge it. If the photocurrent is less than the source current I_c, the voltage across the diode starts to rise. However, because dS/dV is positive, the photocurrent rises. Conversely, if the photocurrent exceeds I_c, the voltage starts to fall, which in turn makes the photocurrent decrease. Thus, the point with equal photocurrent and source current is stable, and because the absorbed power and photocurrent are proportional, the absorbed power is set proportional to the source current. This proportionality exists because of the feedback and not because of any intrinsic linearity in the modulation; the modulation with voltage is far from linear. Because the linearity results from feedback, we call this mode of operation self-linearized modulation.

Fig. 14 shows the transmitted optical power as a function of current for various input powers. Beyond a "knee" in the characteristic the output power decreases linearly with increasing current with coefficient $\hbar\omega/e$ as expected from (15). This set of data was actually taken using the voltage-controlled constant current circuit shown in Fig. 8(c), and the form of the curves is identical within exper-

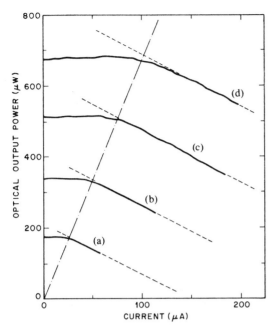

Fig. 14. Optical output power as a function of the current at a wavelength of 855.8 nm for four different optical input powers with the 600 μm diameter sample. (a) 330 μW, (b) 650 μW, (c) 980 μW, and (d) 1.3 mW. The dashed lines have slope $-\hbar\omega/e$ as expected from (15) for unit internal quantum efficiency. The dotted-dashed line has a slope of 150 mA/W, corresponding to the responsivity of the device near 2 V reverse bias at 855.8 nm.

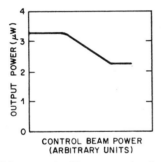

Fig. 15. Control of the transmitted laser power (vertical axis) by the incoherent light power (horizontal axis) incident on a silicon photodiode using the circuit of Fig. 8(b). Incident diode laser power: 9.4 μW. Laser wavelength: 855 nm. 100 μm diameter sample. Reverse bias supply voltage, V_b: 6 V.

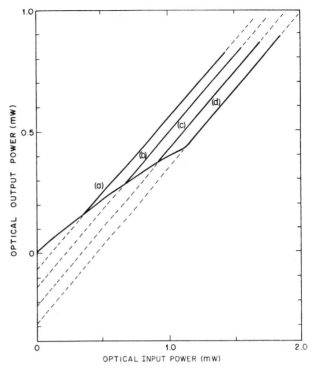

Fig. 16. Optical level shifter action (subtraction of constant optical base lines) for the SEED under constant current bias with the circuit of Fig. 8(c). Bias currents: (a) 50 μA, (b) 100 μA, (c) 150 μA and (d) 200 μA. Dye laser wavelength 858.0 nm. The dashed lines are parallel, with intercepts on the output power axis of $-\hbar\omega I_c/e$, where I_c is the appropriate constant bias current, as predicted by (15).

imental error if the transmitted power is plotted against control voltage. The reason for the "knee" and the constant transmitted power below this point on each curve is that the modulator at any given operating wavelength has a minimum absorption; once this minimum is reached the transmitted power cannot be increased further. Below the "knee," the voltage across the modulator reduces and eventually goes into forward bias. This reduction results in the breakdown of the proportionality between absorbed power and photocurrent [see (15)] because the quantum efficiency drops off. The "knee" actually corresponds to the point at which this proportionality starts to break down (i.e., below about 2 V reverse bias), and the slope of the dashed line that indicates the "knee" in Fig. 14 is, correspondingly, approximately the reciprocal of the responsivity at 2 V reverse bias.

Fig. 15 shows a result similar to those of Fig. 14, this time taken with the light-controlled constant current source of Fig. 8(b). In this case, the silicon photodiode produces a photocurrent, almost independent of reverse bias supply voltage V_b between 1 and 6 V, that is linearly proportional to the amount of light shining on it. The horizontal axis in Fig. 15 is the light power from a tungsten filament bulb as measured with a separate detector. Light from the same bulb is focused onto the silicon photodiode and the brightness of the bulb is altered to give the characteristic as shown in Fig. 15. The vertical axis is the laser beam power transmitted through the modulator. This system also behaves as a self-linearized modulator, this time controlled by an incoherent light source. Increasing tungsten bulb power gives decreasing transmitted laser power; hence this system behaves as a linearized, inverting, light-by-light modulator and incoherent-to-coherent converter. With increasing tungsten bulb power, this characteristic shows a second "knee" after the self-linearized region in which the transmitted laser power is again independent of the incident tungsten bulb power. In this region, increasing bulb power gives no further increase in bias current as the voltage across the silicon photodiode is low (e.g., less than 1 V reverse bias) and its responsivity decreases.

The final mode of operation of the SEED device that will be considered here is the optical level shifter configuration. This is the same as the self-linearized modulator except that the bias current is held constant and the optical input power is varied. A typical set of optical input/output characteristics taken with several different bias currents is shown in Fig. 16. To the right of the "knees" in these characteristics, the operation is exactly the same as in the self-linearized modulator, in that the optical power absorbed is proportional only to the bias current; this proportionality remains when the optical power is varied and the bias current is held constant (the opposite situation to that described for the self-linearized modulator above) as can be deduced from (15). Thus, for constant bias current [derived in this case from the circuit of Fig. 8(c)] a constant optical power is absorbed. When the characteristics above the "knee" are projected back to the vertical axis, they intercept at the (negative) powers predicted by (15) for unit internal quantum efficiency. The "knee" in the characteristics occurs in this case because there is a limit to the maximum absorption of which the modulator is capable at a given wavelength; below the "knee" either the constant current source can give no further increase in voltage or (as in this case) the modulator can give no further increase in absorption regardless of any further increase in voltage. The effect of this configuration is to subtract a constant optical "base line" from the transmitted power above the "knee," hence the title of optical level shifter. Better contrast between the two regions of the characteristic would be possible for devices with higher peak absorption (e.g., thicker modulators).

The behavior of the optical level shifter is related to the bistability under current bias discussed above in Section IV-B. In fact, the region to the left of the bistable loops in the characteristics shown in Fig. 9 actually starts out as an optical level shifter characteristic at powers too low to be clearly resolved on Fig. 9. The optical level shifter behavior breaks down at powers somewhat higher than those in Fig. 15 because there is a maximum transmission of the modulator at any given wavelength, and all the curves in Fig. 15 eventually limit to the maximum transmission line. This causes the negative feedback to break down; positive feedback then becomes possible, and bistability can take place.

VIII. Performance Scaling

One of the features that makes the SEED devices of practical interest is their low operating energy. Operating energies are important for two reasons: 1) some power source is required to run the devices (e.g., a laser), and low operating energies help reduce these power requirements; 2) thermal conduction of the dissipated energy sets a limit to how often, and how many, devices can be operated. High operating energies have long been a problem for optical switching devices. In this section, we will briefly discuss the physics behind the operating energy and the scaling of the energy with various parameters. We also consider the scaling of the switching time.

There are several ways of looking at operating energy of an optical switch. First, we could consider the incident energy in a light beam required to make enough change in the optical properties of the material to swing the output of the system from one logic level to the other; this is essentially the energy considered by Smith [22] in his scaling arguments on optical switching energies. The absorbed or dissipated optical energy may be less than this estimate. The extent to which critical slowing down increases these energies depends on how the system is driven, and we will not consider this in any detail. Second, we should also consider energy requirements and dissipation from any other energy source; in the present case, we must consider the electrical energy and dissipation.

In the case of the SEED, we can estimate, from first principles and from the known properties of the QCSE, the requirements on absorbed optical energy, stored electrostatic energy, and dissipated electrical energy during a switching transition.

To make the devices operate in any logic mode, we must substantially change the optical properties of the material. All the present device configurations rely on the change of the optical absorption. These configurations also rely on only a single pass of the light beam through the material. While multiple passes and/or resonant cavities are possible, we will not consider these here. Hence we must make the material sufficiently thick so that the change in absorption is strong as the device goes from one state to the other (e.g., one absorption length). Because the absorption changes from the QCSE can be of the order of 5000–10 000 cm^{-1}, we therefore require at least 1 μm material thickness. To make these changes in absorption, we require fields of ~ 50 kV/cm in the quantum well material. Thus electrostatic energy must be stored in the material to operate the device. An equivalent way to view this energy is as the energy required to charge up the capacitance of the device. To charge a capacitor C to a steady-state voltage V, in addition to storing $(\frac{1}{2}) CV^2$ energy in the capacitor, dissipates at least as much energy in the (resistive) circuit used to charge it up, regardless' of the resistance of the circuit. (With an inductor, it is possible to charge up the capacitor without dissipating this energy, but the voltage oscillates and the state is not steady.) Hence for any logical operation, at least this amount of stored electrostatic energy must be dissipated. While we could use more material and charge it with a lower field, this would be less efficient in stored electrostatic energy because the change in absorption with much lower fields is more than proportionately lower. Since the capacitance of a 1 μm sheet of this material is ~ 100 aF for every square square micron of area, we expect a minimum dissipated energy of ~ 1.3 fJ/μm^2 to operate the device, and the total dissipated electrical energy should scale linearly with the area of the device because the capacitance does.

In the device configurations considered here, the only means of discharging the capacitance of the device is the photocurrent; the current from the power supply always charges the capacitance of the device (to increasing reverse bias). Hence we can deduce a minimum optical energy to operate the device. The charge density required to discharge the capacitance is $Q = CV$, which for the numbers used here is 500 aC/μm^2. One photon must be absorbed for every electronic charge on (either) plate. The photon energy is about 1.5 eV; consequently about 750 aJ/μm² optical energy must be absorbed in every switching cycle. The incident optical energy must be somewhat larger (e.g., by a factor of 2) as a useful amount of light must be transmitted. Hence we might expect an incident optical energy limit of about 1.5 fJ/μm². Note that this energy density refers to the area of the device, not the area of the optical beam. Again we expect this energy to scale with device area.

These physical arguments therefore predict limiting total operating energies of ~3 fJ/μm². Scaling to a hypothetical limiting device of area $(\lambda)^2/n^2$ (where n is the refractive index) to compare with the limits of [22] gives an incident optical switching energy of 120 aJ (about 500 photons). This very low energy is achieved despite the absence of a resonant cavity to reduce switching energy; other scalings in [22] assume a resonator finesse of 30 with a proportionate reduction in scaled switching energy. We will now compare the behavior of the actual devices with these limits. First, as can be seen from Table I, the switching energy is independent of the switching time over a large range. This is expected from the switching energy argument above and also from the general theory of the bistable switching. Switching power should be inversely proportional to the resistance R, switching time should be proportional to the RC time constant, and hence their product should be constant. Second, the optical and electrical switching energies in Table I compare well with the energy densities predicted above when the increased operating voltage of the bistability with a resistive load is taken into account. Instead of a field of 50 kV/cm, the modulator is run at up to 125 kV/cm. Consequently, 2.5 times as much optical energy should be required because 2.5 times as much charge must be created to neutralize this field. Hence we should expect incident optical energies of about 4 fJ/μm², in agreement with the measurements. The electrical energies are larger because of the increased field (by a factor of $(2.5)^2$) and also because of the superlattice buffer regions which increase the field volume by a factor of 1.6. Hence we should expect electrical energies of about 13 fJ/μm². Again this is in reasonable agreement with the results in Table I. Furthermore, we have demonstrated that switching can be obtained with lower bias voltages (e.g., 6 V) with the constant current load, so that high voltages are not a fundamental requirement for switching.

These switching energies per unit area compare very favorably with other demonstrated optically bistable switches at comparable wavelengths. In contrast to other optically bistable switches, because of the nature of our device we find no variation in switching power with optical spot size up to the maximum optically allowed, and consequently we compare the size of our device with the optical spot size in other systems in evaluating switching energy per unit area. GaAs–AlGaAs etalon devices [23] may show switching energies of the order of hundreds of fJ/μm². Only the very high finesse (~420) Na atomic beam system [24] shows energies per unit area comparable to those of the SEED, to our knowledge.

The lower limit on switching time with the SEED devices is not currently clear. The results with the two different samples do show the scaling expected with decreasing capacitance. We have tested the operation of the device as a modulator down to 131 ps, and tests of the samples as detectors showed circuit limited response times of <2 ns. However, 20 ns was the fastest switching time we could observe with the 100 μm sample. With more power the switching transition disappeared. One possible cause of the disappearance is local heating of the sample, which may move the operating point of the device and defeat the bistability. Smaller samples and better heat sinking may improve this performance.

IX. Conclusions

Using the SEED concept of internal optoelectronic feedback combined with the novel quantum-confined Stark effect modulator, we have been able to demonstrate a variety of devices under practical operating conditions, and have been able to understand their operation from simple arguments.

The SEED's have a number of attractive practical features. Their operating power can be scaled over a large range with reciprocal variation in switching time. They run at room temperature. No optical resonators are used, simplifying fabrication and tuning of the devices. The devices are insensitive to the form or area of the light beam as long as it all lands on the device. Although we have not discussed it here, it should also be possible to operate these devices with multiple input beams for logic functions. It is not necessary that the beams spatially overlap on the device; they can be incident on different areas, which incidentally can prevent problems due to interference of mutually coherent beams. Furthermore, because no optical resonators are required, it is not even necessary that the beams be coherent; we have been able to operate all of the configurations here with the dye laser running broad-band (e.g., 3 nm bandwidth) or the diode laser running below threshold. Operation is, however, certainly better if the incident light is narrow-band (e.g., the bistable loops narrow with increasing spectral bandwidth).

All of the configurations presented here can be operated with a commercial cw diode laser. The devices interface readily with semiconductor electronics, and they are truly optoelectronic in that they can operate with optical and/or electronic control inputs and optical and/or electronic signal outputs. The GaAs-based material system is compatible with both electronics and diode lasers, suggesting the possibility of integration.

Finally, the switching energy of the devices is extremely low; this arises from the magnitude of the recently-discovered QCSE and the high quantum efficiency of the internal optical detection. While there are many features to be considered in interpreting switching energies, the basic operating energy of the SEED is arguably so low that high operating energy is no longer a fundamental argument against optical switching.

Acknowledgment

We acknowledge the expert assistance of J. E. Henry in sample preparation.

References

[1] D. A. B. Miller, D. S. Chemla, P. W. Smith, A. C. Gossard, and W. T. Tsang, "Room-temperature saturation characteristics of GaAs/GaAlAs multiple quantum well structures and of the bulk GaAs," *Appl. Phys.*, vol. B28, pp. 96-97, 1982.

[2] D. A. B. Miller, D. S. Chemla, D. J. Eilenberger, P. W. Smith, A. C. Gossard, and W. T. Tsang, "Large room-temperature optical nonlinearity in $GaAs/Ga_{1-x}Al_xAs$ multiple quantum well structures," *Appl. Phys. Lett.*, vol. 41, pp. 679-681, 1982.

[3] T. Ishibashi, S. Tarucha, and H. Okamoto, "Exciton associated optical absorption spectra of AlAs/GaAs superlattices at 300 K," in *Proc. Int. Symp. GaAs Related Compounds, Inst. Phys. Conf.*, Japan, 1981, pp. 587-588.

[4] S. W. Kirchoefer, N. Holonyak, K. Hess, D. A. Gulino, H. G. Drickamer, J. J. Coleman and P. D. Dapkus, "Absorption measurements at high pressure on $AlAs-Al_xGa_{1-x}As$-GaAs superlattices," *Appl. Phys. Lett.*, vol. 40, pp. 821-824, 1982.

[5] D. S. Chemla, D. A. B. Miller, P. W. Smith, A. C. Gossard, and W. Wiegmann, "Room temperature excitonic nonlinear absorption and refraction in GaAs/AlGaAs multiple quantum well structures," *IEEE J. Quantum Electron.*, vol. QE-20, pp. 265-275, 1984.

[6] D. S. Chemla and D. A. B. Miller, "Room-temperature excitonic nonlinear-optical effects in semiconductor quantum well structures," *J. Opt. Soc. Amer.*, vol. B2, pp. 1155-1173, 1985.

[7] T. H. Wood, C. A. Burrus, D. A. B. Miller, D. S. Chemla, T. C. Damen, A. C. Gossard, and W. Wiegmann, "High-speed optical modulation with GaAs/GaAlAs quantum wells in a p-i-n diode structure," *Appl. Phys. Lett.*, vol. 44, pp. 16-18, 1984.

[8] D. A. B. Miller, D. S. Chemla, T. C. Damen, A. C. Gossard, W. Wiegmann, T. H. Wood, and C. A. Burrus, "Bandedge electroabsorption in quantum well structures: The quantum confined Stark effect," *Phys. Rev. Lett.*, vol. 53, pp. 2173-2177, 1984.

[9] —, "Electric field dependence of optical absorption near the bandgap of quantum well structures," *Phys. Rev.*, vol. B32, p. 1043, 1985.

[10] For a discussion of the Franz-Keldysh effect and the broadening of exciton resonances with field see J. D. Dow and D. Redfield, "Electroabsorption in semiconductors: The excitonic absorption edge," *Phys. Rev.*, vol. B1, pp. 3358-3371, 1970.

[11] T. H. Wood, C. A. Burrus, D. A. B. Miller, D. S. Chemla, T. C. Damen, A. C. Gossard, and W. Wiegmann, "131 ps optical modulation in semiconductor multiple quantum wells," *IEEE J. Quantum Electron.*, vol. QE-21, pp. 117-118, 1985.

[12] D. A. B. Miller, D. S. Chemla, T. C. Damen, A. C. Gossard, W. Wiegmann, T. H. Wood, and C. A. Burrus, "Novel hybrid optically bistable switch: The quantum well self electro-optic effect device," *Appl. Phys. Lett.*, vol. 45, pp. 13-15, 1984.

[13] D. A. B. Miller, D. S. Chemla, T. C. Damen, T. H. Wood, C. A. Burrus, A. C. Gossard, and W. Wiegmann, "Optical level shifter and self-linearized optical modulator using a quantum-well self-electrooptic effect device," *Optics Lett.*, vol. 9, pp. 567-569, 1984.

[14] Perhaps the closest precedent to the SEED devices is in the interesting work that has been performed on devices using the Franz-Keldysh effect in diodes [15]-[17]. One configuration utilizes the diode in a resonator; this has been considered both theoretically [15] and experimentally [16]. As in the SEED devices reported here, the modulator and photodiode are the same p-n junction. The Franz-Keldysh effect can however only usefully give an increase in absorption with increasing voltage; consequently, the mechanism of bistability from increasing absorption (with decreasing voltage) used in the SEED device (see Section III) is not available and instead a resonator is used as in conventional absorptive bistability to achieve bistability with decreasing absorption. The Franz-Keldysh bistability is therefore a hybrid implementation of conventional absorptive bistability. Another configuration using the Franz-Keldysh effect, with an additional separate photodiode for detection and an external transistor to give the gain for bistability, has also recently been reported [17]. Because of the transistor, the cavity is not necessary, and, as in the SEED devices reported here, coherent light is no longer required.

[15] B. S. Ryvkin, "Falling current-voltage characteristic and optical bistability of a resonator photocell in the Franz-Keldysh effect," *Sov. Phys. Semicond.*, vol. 15, pp. 796-798, 1981 (translation of *Fiz. Tekh. Poluprovodn.*, vol. 15, pp. 1380-1384, 1981).

[16] B. S. Ryvkin and M. N. Stepanova, "Bistable optical characteristics of a resonator photocell with two-step optical transitions," *Sov. Tech. Phys. Lett.*, vol. 8, pp. 413-414, 1982 (translation of *Pis'ma Zh. Tekh. Fiz.*, vol. 8, pp. 951-954, 1982).

[17] M. I. Nemenov, B. S. Ryvkin, and M. N. Stepanova, "Optical bistability due to the Franz-Keldysh effect with incoherent unpolarized light," *Sov. Tech. Phys. Lett.*, vol. 9, pp. 260-262, 1983 (translation of *Pis'ma Zh. Tekh. Fiz.*, vol. 9, pp. 604-609, 1983).

[18] M. A. Lee, B. Easter, and H. A. Bell, *Tunnel Diodes*. London, England: Chapman and Hall, 1967.

[19] G. P. Agrawal and H. J. Carmichael, "Optical bistability through nonlinear dispersion and absorption," *Phys. Rev.*, vol. A19, pp. 2074-2086, 1979.

[20] D. A. B. Miller, "Optical bistability and differential gain resulting from absorption increasing with excitation," *J. Opt. Soc. Amer.*, vol. B1, pp. 857-864, 1984.

[21] E. Garmire, J. H. Marburger, S. D. Allen, and H. G. Winful, "Transient response of hybrid bistable optical devices," *Appl. Phys. Lett.*, vol. 34, pp. 374-376, 1979.

[22] P. W. Smith, "On the physical limits of digital optical switching and logic elements," *Bell Syst. Tech. J.*, vol. 61, pp. 1975-1993, 1982.

[23] S. S. Tarng, H. M. Gibbs, J. L. Jewell, N. Peyghambarian, A. C. Gossard, T. Venkatesan, and W. Wiegmann, "Use of a diode laser to observe room-temperature, low-power optical bistability in a GaAs-AlGaAs etalon," *Appl. Phys. Lett.*, vol. 44, pp. 360-361, 1984.

[24] D. E. Grant and H. J. Kimble, "Transient response in absorption bistability," *Optics Commun.*, vol. 44, pp. 415-420, 1983

Integrated quantum well self-electro-optic effect device: 2×2 array of optically bistable switches

D. A. B. Miller and J. E. Henry
AT&T Bell Laboratories, Holmdel, New Jersey 07733

A. C. Gossard and J. H. English
AT&T Bell Laboratories, Murray Hill, New Jersey 07974

We demonstrate 2×2 arrays of optically bistable devices with very uniform optical characteristics. They are fabricated from an integrated self-electro-optic effect device structure consisting of a quantum well p-i-n diode grown in series with a load photodiode. Operating power can be optically controlled with a separate beam between ~40 pW and >470 μW with associated switching times of ~10 s and <2 μs.

One attractive feature of optics for switching and processing applications is its ability to communicate large amounts of information in parallel. To exploit this fully requires uniform, two-dimensional arrays of fast optical devices with low enough energy dissipation that their switching speed can be utilized in dense arrays without thermal problems. Some recent approaches use Fabry–Perot étalons containing semiconductor nonlinear refractive materials (see Ref. 1 for a general discussion) in which several different parts of the area of the étalon can be used simultaneously and independently.[2–4] Uniform operation over usable areas has been obtained.[3] Etching of mesas to overcome diffusion limits and shorten switching times has also been demonstrated.[4]

Another type of device, the quantum well self-electro-optic effect device (SEED),[5–9] relies first on the changes in optical absorption that can be induced by electric fields perpendicular to the thin semiconductor layers in quantum well materials, a mechanism called the quantum-confined Stark effect (QCSE).[10,11] Strong so-called "exciton" absorption resonances (e.g., $\sim 10^4$ cm^{-1} absorption coefficient) can be seen near the band-gap energy (e.g., at ~850 nm wavelength) at room temperature in quantum wells. The QCSE is an effect in which, among other things, these resonances can be moved spectrally with field. Combining the QCSE with optical detection within the same structure causes optoelectronic feedback that, when positive, gives optical bistability.[5–9] An optically bistable SEED generally consists of a diode containing quantum wells, an electrical load (e.g., a resistor), and, usually, a reverse bias voltage supply. The quantum well diode functions simultaneously as an absorption modulator and a detector. The wavelength is chosen so that the absorption (and hence the photocurrent) increases with decreasing reverse bias voltage on the diode; this behavior is seen near to the zero-field wavelength of the so-called "heavy-hole" exciton resonance. Hence increasing incident light gives increasing photocurrent that in turn results in reduced voltage on the diode because of the voltage drop across the resistor. Consequently the absorption and photocurrent increase, giving a positive feedback that can lead to switching into a high absorption state.

Other device functions are also possible.[6–9] The SEED does not require Fabry–Perot resonators. Consequently neither optical thickness nor operating wavelength is very critical. Because of the strength of the QCSE, SEED's also have the potential for very low total switching energy densities,[5,7] despite the absence of resonators to reduce switching energy.

The capacitances of the diode and the load, and any stray capacitances (except those across the power supply) must be charged or discharged for switching. Although internal capacitance could be made small in a small integrated structure, devices demonstrated to date have used external electrical components for the load, and there is little prospect of constructing such a circuit without ~1 pF capacitance. Arrays also require separate loads for each device. In this letter, we describe a method for making arrays of SEED's with an integrated load that has the potential for overcoming these problems, and we present results for 2×2 SEED arrays. The structure used also has an optically controllable load that adds flexibility to the device.

The integrated SEED structure used here consists of a p-i-n diode with quantum wells in the intrinsic (i) region, vertically integrated in series with *another* photodiode (also a p-i-n structure) that is used instead of a resistor as the load.[7] This load photodiode is transparent to the infrared wavelengths used for the quantum wells, but is essentially opaque for all wavelengths <750 nm. In operation, a second short-wavelength control light beam (e.g., from a HeNe laser) is shone on this photodiode. The single device is shown schematically in Fig. 1, together with its equivalent circuit. Crudely speaking, this load photodiode can be regarded as a transparent resistor whose value can be externally set with another light beam. The operation of this device is therefore qualitatively similar to that of the resistor-loaded device. In actual fact, the load photodiode behaves more like a constant current load, and this kind of load can give better bistability compared to the resistor load.[7] The speed of switching is then set by the time taken for the difference in photocurrents in the two diodes to charge their capacitance rather than by a resistor-capacitor time constant. Switching occurs when the photocurrents in the load diode and in the quantum well diode are comparable. The control light power therefore sets the switching power and speed of the devices. The detailed behavior can be quantitatively understood by the use of load lines.[7]

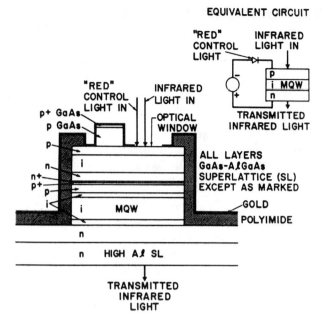

FIG. 1. Schematic diagram of the integrated SEED structure for a single device in an array, showing also the equivalent electrical circuit.

TABLE II. Quantum well and superlattice material parameters.

	Sample 1	Sample 2
Al fraction in AlGaAs	0.297	0.345
Superlattice (SL)	18.4 Å GaAs	20.5 Å GaAs
	11.1 Å AlGaAs	13.4 Å AlGaAs
High Al SL	9.2 Å GaAs	10.3 Å GaAs
	84.0 Å AlGaAs	100 Å AlGaAs
Quantum wells	86.5 Å GaAs	92.4 Å GaAs
	51.0 Å AlGaAs	61.3 Å AlGaAs
Doping (10^{18} cm^{-3})		
p^+	10	10
p	1	1
n^+	3	3
n	1	1

The structures are grown by molecular beam epitaxy starting with silicon-doped n-GaAs substrates. The sequence of regions is detailed in Fig. 1, and Tables I and II. We chose superlattice material of very fine period alternate layers of GaAs and AlGaAs in the structure rather than thick layers of AlGaAs because the use of superlattice may improve the quality of the material,[12] and because the average aluminum content of the layers can be adjusted without adjusting the temperature of the aluminum oven. Such fine superlattices behave approximately like the average alloy.

The first, high Al region is a stop layer for the substrate selective etch. In the quantum well p-i-n diode, in sample 1 only we included intrinsic buffer layers around the quantum well region as used previously (see, e.g., Ref. 11). The highly

TABLE I. Layer structure of samples.

	Total thickness (μm)	
Region	Sample 1	Sample 2
Top contact		
p^+-GaAs	0.091	0.103
p-GaAs	0.820	0.923
Load diode		
p-SL	0.369	0.424
i-SL	0.738	0.847
n-SL	0.369	0.424
Tunnel junction		
n^+-SL	0.074	0.085
p^+-SL	0.074	0.085
Quantum well diode		
p-SL	0.369	0.424
i-buffer SL	0.183	...
i-quantum wells	0.976	1.537
i-buffer SL	0.183	...
n-SL	0.738	0.847
Etch stop		
n-high Al SL	0.932	1.103

doped tunnel junction grown next effectively makes an ohmic contact between the two p-i-n diodes, converting holes to electrons and inhibiting minority-carrier diffusion, a technique employed in stacking solar cells.[13–15] Next, the load p-i-n diode is grown. Finally, GaAs contact layers are grown. The thick layer spaces the contact from the diode junction to minimize dopant spiking. The highly doped layer facilitates good contacts.

To fabricate the devices, we first form the gold contacts for the top of the mesas by conventional photolithography and evaporation followed by thermal alloying of the contacts. This leaves a two-dimensional array of contacts, each $120 \times 50 \mu$m, spaced on 400 μm centers. The GaAs top layer is removed everywhere else by masking and selective etching. Using further masking and etching, mesas approximately 200 μm square, also on 400 μm centers, are etched to a carefully controlled depth, so that the etching stops just within the bottom n-doped layers of the structure. The contacts lie toward one side of each mesa as shown in Fig. 1. Then an insulating layer of polyimide is spin deposited on the surface, and it is photolithographically patterned and etched by reactive ion etching to leave cleared squares on top of the desired mesas (see Fig. 1). Gold is then deposited on the top surface to connect all the contact regions, with $120 \times 50 \mu$m rectangular optical windows being defined photolithographically and subsequently cleared by a lift-off step. The material is then cleaved to give separate 2×2 arrays of devices. A wire is epoxied to the top gold surface of an array, the array is stuck top down to sapphire with transparent epoxy, and the substrate is removed below the mesas of interest with a selective etch. Another contact wire is taken off the remaining substrate. This fabrication leaves a 2×2 array of devices with common top and bottom contacts but otherwise insulated. Finally, a silicon oxide antireflection coating is deposited on this exposed layer. We made several 2×2 arrays from each sample, with similar performance from all arrays from a given sample.

In operation, a voltage supply reverse biases the array through the two wires, and the light beams are incident on the mesa side of the devices (although the infrared light can be shone from either direction). We used a Styryl 9 dye laser for the infrared source for most experiments, although the devices were also tested using a commercial cw GaAs diode

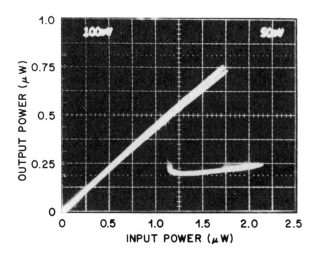

FIG. 2. Input/output characteristics for the four devices in an array superimposed. Sample 2. HeNe control power 1.95 μW (at 632.8 nm). Dye laser wavelength 857 nm. Supply voltage 28 V.

laser. HeNe and infrared beams were focused on the same mesa (they need not be coincident) using a 25 mm focal length lens, with a spot diameter of ~ 20 μm. Powers were measured with Si photodiodes. To test for bistability, the infrared power was ramped up and down using an acousto-optic modulator. A typical input/output characteristic for the infrared beam is shown in Fig. 2. This picture shows the characteristics for the *four devices in an array superimposed.*

By varying the HeNe control beam power, we were able to vary the switching power and (reciprocally) the switching speed over a very large range. (By switching power we mean the total infrared input power at the point where the output power switches from high to low transmission). With sample 1 (sample 2) we observed bistability with switching powers as low as ~ 40 pW (~ 80 pW) and switching time of ~ 10 s (~ 15 s) at ~ 40 pW (~ 100 pW) HeNe control power; slightly lower power switching may be possible. At the high power extreme, we observed bistability up to a switching power of 6.1 μW (470 μW) at 100 μs (2 μs) with 6.3 μW (610 μW) HeNe power; for higher powers, the bistable loop progressively narrowed and disappeared. Bistability could be observed between 834 and 859 nm (846 and 869 nm), for voltages greater than 4 V (9 V). The optimum wavelength for large contrast ratio between the transmission in the two states was ~ 851 nm (857 nm), with ratios of 2.8:1 (3.5:1) obtained over a range of ~ 2 nm in both cases, at operating voltages of 25 V (28 V). The transmission in the high-transmission state under these conditions is $\sim 53\%$ (40%). Lower operating voltages gave smaller contrast ratio.

The lower limit on switching power will probably be set by leakage current, which we measure to be < 20 pA for the whole array. We believe that the upper limit on switching speed is set by the limited forward tunneling current of the tunnel junctions. Preliminary data on samples with improved tunnel junctions show switching speeds less than 1 μs, currently limited by laser power. The switching time at a given power is consistent with charging the device capacitance (~ 9 pF) to the supply voltage with the photocurrent.

In conclusion, we have demonstrated integrated arrays of SEED's with excellent uniformity of bistable switching characteristics. Optical control allows bistability over nearly seven orders of magnitude in switching power and speed. We expect that these devices can be scaled to much smaller lateral dimensions, with proportionate reduction in switching power, and possibly to larger arrays. More generally, this device demonstrates the real potential for sophisticated and novel optoelectronic devices operating under practical and flexible conditions using quantum well technology.

[1] H. M. Gibbs, *Optical Bistability: Controlling Light with Light* (Academic, New York, 1985).
[2] See, for example, D. J. Hagan, H. A. MacKenzie, H. A. Al Attar, and W. J. Firth, Opt. Lett. **10**, 187 (1985); D. J. Hagan, I. Galbraith, H. A. MacKenzie, W. J. Firth, A. C. Walker, J. Young, and S. D. Smith, in *Optical Bistability III*, edited by H. M. Gibbs, P. Mandel, N. Peyghambarian, and S. D. Smith (Springer, Berlin, 1986), p. 189.
[3] J. L. Jewell, Y. H. Lee, J. F. Duffy, A. C. Gossard, and W. Wiegmann, Appl. Phys. Lett. **48**, 1342 (1986).
[4] T. Venkatesan, B. Wilkens, Y. H. Lee, M. Warren, G. Olbright, N. Peyghambarian, J. S. Smith, and A. Yariv, Appl. Phys. Lett. **48**, 145 (1986).
[5] D. A. B. Miller, D. S. Chemla, T. C. Damen, A. C. Gossard, W. Wiegmann, T. H. Wood, and C. A. Burrus, Appl. Phys. Lett. **45**, 13 (1984).
[6] D. A. B. Miller, D. S. Chemla, T. C. Damen, T. H. Wood, C. A. Burrus, A. C. Gossard, and W. Wiegmann, Opt. Lett. **9**, 567 (1984).
[7] D. A. B. Miller, D. S. Chemla, T. C. Damen, T. H. Wood, C. A. Burrus, A. C. Gossard, and W. Wiegmann, IEEE J. Quantum Electron. **QE-21**, 1462 (1985).
[8] J. S. Weiner, D. A. B. Miller, D. S. Chemla, T. C. Damen, C. A. Burrus, T. H. Wood, A. C. Gossard, and W. Wiegmann, Appl. Phys. Lett. **47**, 1148 (1985).
[9] D. A. B. Miller, J. S. Wiener, and D. S. Chemla, IEEE J. Quantum Electron. **QE-22**, 1816 (1986).
[10] D. A. B. Miller, D. S. Chemla, T. C. Damen, A. C. Gossard, W. Wiegmann, T. H. Wood, and C. A. Burrus, Phys. Rev. Lett. **53**, 2173 (1984).
[11] D. A. B. Miller, D. S. Chemla, T. C. Damen, A. C. Gossard, W. Wiegmann, T. H. Wood, and C. A. Burrus, Phys. Rev. B **32**, 1043 (1985).
[12] P. M. Petroff, R. C. Miller, A. C. Gossard, and W. Wiegmann, Appl. Phys. Lett. **44**, 217 (1984); H. Morkoç in *Technology and Physics of Molecular Epitaxy*, edited by E. H. C. Parker (Plenum, New York, 1985), Chap. 7, p. 213.
[13] M. Ilegems, B. Schwartz, L. A. Koszi, and R. C. Miller, Appl. Phys. Lett. **33**, 629 (1978).
[14] P. Bouchaib, H. P. Contour, F. Raymond, C. Verie, and F. Arnaud D'Avitaya, J. Vac. Sci. Technol. **19**, 145 (1981).
[15] D. L. Miller, S. W. Zehr, and J. S. Harris, J. Appl. Phys. **53**, 744 (1982).

Symmetric self-electro-optic effect device: Optical set-reset latch

A. L. Lentine and H. S. Hinton
AT&T Bell Laboratories, Naperville, Illinois 60566

D. A. B. Miller, J. E. Henry, and J. E. Cunningham
AT&T Bell Laboratories, Holmdel, New Jersey 07733

L. M. F. Chirovsky
AT&T Bell Laboratories, Murray Hill, New Jersey 07974

We demonstrate an integrated symmetric self-electro-optic effect device consisting of two quantum well *p-i-n* diodes electrically connected in series. The device acts as a bistable optical memory element with individual set (S) and reset (R) inputs and complementary outputs (optical S-R latch). The switching point is determined by the *ratio* of the two inputs, making the device insensitive to optical power supply fluctuations when both power beams are derived from the same source. The device also shows time-sequential gain, in that the state can be set using low-power beams and read out with subsequent high-power beams. The device showed bistability for voltages greater than 3 V, incident optical switching energy densities of ~ 16 fJ/μm^2, and was tested to a switching time of 40 ns.

The potential use of optics in telecommunications switching and computing has generated considerable interest in recent years. One approach is to use planar two-dimensional arrays of logic or memory devices interconnected with free-space optics. One proposed class of devices uses Fabry–Perot étalons containing semiconductor nonlinear refractive materials.[1–4] Uniform operation over different areas of the resonator[3] and fast switching times[4] have been demonstrated. Another device is the quantum well self-electro-optic effect device (SEED).[5–8] SEED's rely on changes in the optical absorption that can be induced by changes in an electric field perpendicular to the thin semiconductor layers in quantum well material. This effect has been called the quantum confined Stark effect (QCSE).[9,10] Combining the QCSE with optical detection within the same structure can cause optoelectronic feedback and bistability. Uniform arrays of bistable integrated diode-biased SEED's (D-SEED's)[7] as large as 6×6 (Ref. 8) have been demonstrated.

A problem with optically bistable systems is that, to obtain the gain necessary for cascadable logic, the devices need to be biased very close to their switching transition. Such critical biasing imposes very strict tolerances on the uniformity of the devices and the bias beam powers. One solution is to operate an étalon (without bistability) with two wavelengths.[3,11] This requires two different devices, one to down-convert and the other to up-convert wavelengths. The device that we demonstrate here, the symmetric SEED (S-SEED) is optically bistable, but is bistable in the *ratio* of two beam powers, rather than any absolute power. When both beams are derived from the same source, source power fluctuations are, therefore, unimportant. Furthermore, the device does not need to be biased close to a switching point to give gain, so that uniformity of bias beams is not critical. It can show time-sequential gain by being switched at low power and read out at high power. Such time-sequential operation also gives good input/output isolation, because a small reflection back onto the device will not change its state. Hence this device offers a potential solution to some of the major problems of optical bistability for logic.

The D-SEED previously demonstrated[7,8] consists of a *p-i-n* diode with quantum wells in the intrinsic region vertically integrated in series with another photodiode that is used for the load. The load photodiode is transparent to the wavelengths used for the quantum wells (~ 850 nm) but absorbs all of the incident light for wavelengths less than 750 nm. When a He-Ne laser is used to "bias" the photodiode load, the array behaves as a set of nearly identical bistable devices whose switching power and speed can be set over many decades by adjusting the intensity of the He-Ne laser. In addition, the array can operate as a dynamic memory where the state of the device is preserved when both the bias and the infrared beams are removed. The stored state may be read out sometime later (up to 30 s) by simultaneously increasing the two beams.[8] In the symmetric SEED, the load is a second *p-i-n* diode with multiple quantum wells in the intrinsic region. Since the two diodes are identical, either diode can be considered as the load of the other. The basic physical mechanisms in the D-SEED and the S-SEED are therefore similar, but the ability to work with two complementary beams at the same wavelength greatly increases the usefulness of the S-SEED.

The schematic diagram and physical layout of the device are shown in Fig. 1. The material was grown by molecular beam epitaxy on a Si-doped *n*-type GaAs substrate. The *p* region was grown as a fine-period GaAs/AlGaAs superlattice as in the integrated D-SEED to improve the quality of the material. The structure, which is *n-i-p-i-n*, results in two back-to-back diodes between the substrate and the top *n* regions. The multiple quantum well *p-i-n* diodes (on top) are connected "horizontally" by an external connection rather than "vertically" during growth as in the D-SEED, circumventing the problem that limited the speed of the D-SEED, namely, the limited current that could be carried by the internal ohmic contact in that structure. The isolation diodes (bottom), comprised of the AlGaAs *i* and *n* and superlattice *p* regions, ensure electrical isolation between the *p* regions of the two quantum well diodes. In operation, both of the isolation diodes are always reverse biased by connecting the substrate to a positive voltage. The quantum well diodes are made by etching separate mesas and electrically connected

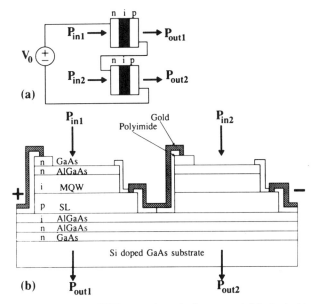

FIG. 1. Symmetric SEED: (a) schematic diagram and (b) physical layout (not to scale). The Al mole fraction in $Al_xGa_{1-x}As$ is 0.4. Epitaxial layer thicknesses and dopings (from bottom to top): n-GaAs (0.17 μm, $n = 10^{18}$ cm^{-3}); i-AlGaAs (1.92 μm); p-superlattice (SL) (250 periods of alternate 25 Å AlGaAs and 21 Å GaAs, $p = 10^{18}$ cm^{-3}); i-multiple quantum wells (MQW) (63 periods of alternate 80 Å AlGaAs and 105 Å GaAs layers); n-AlGaAs (0.64 μm, $n = 10^{18}$ cm^{-3}); and n-GaAs (0.105 μm, $n = 10^{18}$ cm^{-3}). Both the GaAs substrate and the GaAs buffer layer are removed by a selective etch under the device active area.

using ohmic contacts and evaporated gold over a polyimide insulator. The mesas are $\sim$200 μm$\times$200 μm on 400 μm centers, and the optical window is $\sim$100 μm$\times$200 μm. The substrate is cleaved into separate 1$\times$2 arrays that are packaged individually. The electrical connections to the device are made using silver epoxy to hold the bond wires to the gold metallization. The GaAs substrate and buffer layer are etched away underneath the optical windows, and an antireflection coating is applied.

The operation of the device can be understood through the use of load lines as shown in Fig. 2. By solving for the voltages across the quantum well diodes as a function of the two input power levels, we can determine the optical transmission of the two diodes and hence the input/output characteristics. As shown in Ref. 6, the device is bistable when there are three intersection points, and this will only occur if we operate the device at a wavelength where there is a region of decreasing absorption (and therefore decreasing current) for increasing voltage. Furthermore, the device will only be bistable when the optical input power levels are comparable, and will have only a single state when the power in one diode greatly exceeds the power in the other. Figure 2 also shows that if we increase (or decrease) the optical input power levels into both devices by the same factor (for example doubling both input power levels) or by the same fixed amount, the voltage at which the current-voltage curves intersect will not change, and hence the state of the device will not change. This allows us to set the state of the device using low-power beams of different intensities and then read out the state of the device using higher-power beams of equal intensities, thus achieving time-sequential gain.

The measured input/output characteristics of the device shown in Fig. 3 agree with the calculated results using the method described above. The characteristics were measured using an argon-ion-laser-pumped styryl 9 dye laser, although some measurements were also made using a com-

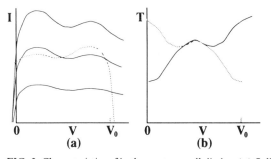

FIG. 2. Characteristics of both quantum well diodes. (a) Solid curves are the photocurrent vs voltage for a quantum well diode at three different input intensities. Dashed line is the photocurrent on the load quantum well diode vs voltage on the first quantum well diode for an input intensity on the load diode equal to that incident on the first diode in the middle curve. (b) Solid curve is the transmission vs voltage for the first quantum well diode. Dashed curve is the transmission of the load quantum well diode vs the voltage on the first diode.

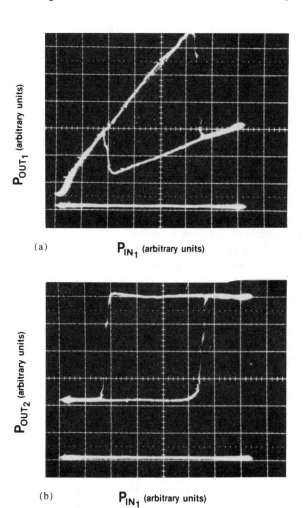

FIG. 3. Optical input/output characteristics of the S-SEED measured at a supply voltage of 15 V at 856 nm with average input power levels of $\sim$20 μW. (a) Optical power exciting first diode (P_{out_1}) vs optical power incident on first diode (P_{in_1}) with the optical power incident on second diode (P_{in_2}) held constant. (b) Optical power exciting second diode (P_{out_2}) vs P_{in_1} with P_{in_2} held constant. Optical transmission levels are $\sim$41% and 14% for the two states shown in the figure.

mercially available AlGaAs semiconductor diode laser. To measure bistability and switching speed, the light emitted from the laser was split into two paths using a polarization beamsplitter. The optical power in one side was varied using an acousto-optic modulator. The two beams are combined using another polarization beamsplitter and focused through a 5× microscope objective onto the device. The spot size of the optical beams was varied from ~7 to 50 μm in diameter as measured using a television camera looking at the light transmitted through the device. The relative intensity of the light in each path was adjusted by rotating a half-wave plate to vary the polarization of the light incident on the first beamsplitter.

The device showed bistability at 856 nm for supply voltages greater than 3 V, and showed bistability from 845 to 862 nm at 15 V. The transmission (P_{out}/P_{in}) of the quantum well diodes at 856 nm in the two states was 31% and 14% at 5 V (2.2:1 contrast ratio), increasing to 41% and 14% (2.9:1 contrast ratio) at 15 V. The devices had greater than a 2:1 contrast ratio from 855.5 to 857 nm at 5 V and from 854.7 to 857.7 at 15 V. All of the above measurements were made with optical input intensities below 100 μW. The fastest switching time measured was 40 ns when the modulation frequency of the acousto-optic modulator was 1 MHz, increasing to 60 ns for a 100 kHz sine wave input. This speed was limited by the maximum optical power available rather than any intrinsic limit of the device. For this measurement, the supply voltage was 10 V, the spot sizes were 50 μm in diameter, and the average optical input power levels into each p-i-n diode were 8 mW at 866 nm. The wavelength was adjusted for optimum contrast ratio at these power levels, and the shift in wavelength from the optimum at low-power levels was probably due to local heating of the device. The speed-power product for a 1 MHz sine wave input corresponds to a minimum optical switching energy of 640 pJ. Since the usable device area is 40 000 μm², the minimum optical switching energy density is 16 fJ/μm². This agrees with the calculated energy required to charge and discharge the capacitance of the device. Additionally, the switching energy per unit area could be reduced by a factor of 2 by growing the material on a semi-insulating substrate and eliminating the two isolation diodes, one of which needs to be charged with the photocurrent. We tested that, and just as in the D-SEED,[7] as we reduce the power, the switching speed and power scale inversely over several decades, resulting in a constant switching energy. We also observed that, for small spot sizes only, a degradation in contrast ratio was seen at power levels greater than 1 mW. We believe this is related to saturation of the quantum well material.

Since bistability was observed at optical input intensities varying from less than 10 nW to greater than 8 mW, we can have a large effective signal gain. We have demonstrated this by using the output of one device to drive another with greater than 10 dB of loss between the devices. Of course, switching at low powers takes proportionately longer, so that gain is obtained at the expense of switching speed. Another way of expressing this is to say that the device has a constant gain-bandwidth product, just like many other amplifiers. Note that this time-sequential gain mechanism gives some input-output isolation, an important attribute for a logic device. If the two incident signal levels are roughly equal, a small reflection of one of the outputs back onto the device will not switch the state of the device because of the wide bistable loop, independent of the intensity of the inputs. The reflection of a large signal never coincides in time with the application of a small signal that is used to switch the device; hence we have a device with a large gain that is insensitive to reflections back onto the device. This isolation is a classic attribute of a good "three-terminal" device such as a transistor, where fluctuations in the load are not themselves amplified by the gain of the device, but is not shared by "two-terminal" devices such as tunnel diodes or conventional optically bistable devices.

In conclusion, we have demonstrated an integrated symmetric self-electro-optic effect device that consists of two quantum well p-i-n diodes electrically connected in series. The device acts as a bistable optical memory element with individual set (S) and reset (R) capabilities and complementary outputs (optical S-R latch), and can also function as a logic amplifier because it shows time-sequential gain. In contrast to conventional bistable devices, it is insensitive to optical power supply fluctuations, does not require critical biasing, and possesses significant input/output isolation when used as an amplifier. These attributes, characteristic of a "three-terminal" rather than a "two-terminal" device, make it an attractive device for optical processing systems.

[1] H. M. Gibbs, *Optical Bistability: Controlling Light with Light* (Academic, New York, 1985).
[2] D. J. Hagan, H. A. MacKenzie, H. A. Al Attar, and W. J. Firth, Opt. Lett. **10**, 187 (1985); D. J. Hagan, I. Galbraith, H. A. MacKensie, W. J. Firth, A. C. Walker, J. Young, and S. D. Smith, in *Optical Bistability III*, edited by H. M. Gibbs, P. Mandel, N. Peyghambarian, and S. D. Smith (Springer, Berlin, 1986), p. 189.
[3] J. L. Jewell, Y. H. Lee, J. F. Duffy, A. C. Gossard, and W. Weigmann, Appl. Phys. Lett. **48**, 1342 (1986).
[4] T. Venkatesan, B. Wilkens, Y. H. Lee, M. Warren, G. Olbright, N. Peyghambarian, J. S. Smith, and A. Yariv, Appl Phys. Lett. **48**, 145 (1986).
[5] D. A. B. Miller, D. S. Chemla, T. C. Damen, A. C. Gossard, W. Weigmann, T. H. Wood, and C. A. Burrus, Appl. Phys. Lett. **45**, 13 (1984).
[6] D. A. B. Miller, D. S. Chemla, T. C. Damen, T. H. Wood, C. A. Burrus, A. C. Gossard, and W. Weighmann, IEEE J. Quantum Electron. **QE-21**, 1462 (1985).
[7] D. A. B. Miller, J. E. Henry, A. C. Gossard, and J. H. English, Appl. Phys. Lett. **49**, 821 (1986).
[8] G. Livescu, D. A. B. Miller, J. E. Henry, A. C. Gossard, and J. H. English, in *Conference on Lasers and Electro-optics* (Optical Society of America, Washington DC, 1987), pp. 283 (postdeadline papers).
[9] D. A. B. Miller, D. S. Chemla, T. C. Damen, A. C. Gossard, W. Weigmann, T. H. Wood, and C. A. Burrus, Phys. Rev. Lett. **53**, 2173 (1984).
[10] D. A. B. Miller, D. S. Chemla, T. C. Damen, A. C. Gossard, W. Weigmann, T. H. Wood, and C. A. Burrus, Phys. Rev. B **32**, 1043 (1985).
[11] J. L. Jewell, Y. H. Lee, M. Warren, H. M. Gibbs, N. Peyghambarian, A. C. Gossard, and W. Wiegmann, Appl. Phys. Lett. **46**, 918 (1985).

Hard Limiting Opto-electronic Logic Devices

P. Wheatley[1], M. Whitehead[1], P.J. Bradley[1], G. Parry[1], J.E. Midwinter[1], P. Mistry[2], M.A. Pate[2], and J.S. Roberts[2]

[1] Dept. of Electronic and Electrical Engineering,
University College London, Torrington Place, London WC1E 7JE, UK
[2] SERC Central Facility for III-V Semiconductors,
University of Sheffield, Mappin St., Sheffield, UK

We report the demonstration of two novel opto-electronic devices whose characteristics appear to make them suitable for optical logic. One of the devices may be developed into a NOR or NAND gate and the other into an OR or AND gate. The devices are based on the use of electronic gain together with optical input and output and the specific implementation uses a photo-transistor to provide photo-detection and gain and a GaAs/AlGaAs multiple-quantum-well (MQW) electro-absorption modulator [1] to enable optical output. The components are connected in such a way as to provide a characteristic suitable for logic [2]: the output is hard-limited, the input is isolated from the output and the switching time is potentially fast. Electronic nonlinearities are strong, giving the potential for low power switching, and are fundamentally fast and optical access has the potential of high data rate communication with low crosstalk.

Both devices use the structure shown in fig.1 which consists of a photo-transistor in series with an electro-absorption modulator between a constant supply voltage and ground. A beam of constant optical power, the pump power, is incident on the modulator; part of this is absorbed by the modulator to give rise to photo-current and part is transmitted to give the optical output of the device. A second optical beam, the signal power, is incident on the base of the photo-transistor and contains the information input to the device. Thus the device is an optical three terminal device since the output power is derived from a constant optical supply which does not follow the same path as the input signal. The operation of both devices is based on Kirchhoff's current law, which in this situation states that the current through the photo-transistor must equal the current through the modulator. Both of these currents are substantially optically

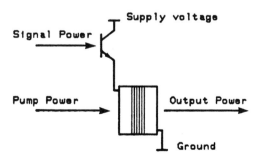

Fig. 1. Schematic diagram of device

generated and the voltage across the modulator varies in such a way as to ensure that the currents are equal. The two different responses of the device arise from the use of different wavelengths for the pump power which changes the response of the modulator.

The current generated by the photo-transistor depends on the voltage between its collector and emitter and on the optical signal power incident on its base. In general, the current is proportional to the optical power, but if the collector-emitter voltage is very small (less than about 0.5V) the current falls very sharply. The ratio of current to incident power (the effective quantum efficiency) must be significantly greater than unity to achieve optical gain. The photo-transistor could of course be replaced by other components which behave in the same way, such as a photo-diode followed by a transistor.

In the first device [3], whose characteristic is inverting, the wavelength of the pump beam should be such that the modulator photo-current increases with applied reverse bias voltage, as shown by the solid line in fig.2. This is achieved in this case by working on the low energy side of the band-gap of the MQW material. Most of the change in photo-current is due to a change in absorption in the modulator, but at very low voltages, the current becomes very small as the quantum efficiency falls. There is also a small dark current which increases with applied voltage. The power transmitted through the modulator is high with low applied voltage and falls nonlinearly as the reverse voltage is increased. In general, the photo-transistor and modulator currents are equalised by the following mechanism which was first reported in [4]. The modulator is considered as being supplied by a constant current. If that current is greater than the current generated in the modulator, then charge builds up on the modulator in such a way as to increase the voltage across it. This increases the modulator current until it is equal to the applied current. If the modulator current is the greater, then the voltage across the modulator drops, reducing its current until it again matches the applied current.

The device has two régimes of operation. Firstly there is "self-linearization" [4] in which the effective quantum efficiencies of the photo-transistor and modulator are constant and current equalisation is achieved by varying the absorption of the modulator. In this régime, the power absorbed in the modulator is proportional to the photo-transistor current

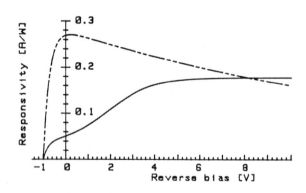

Fig. 2. Modulator responsivity vs. bias

which is itself proportional to the signal power. Thus the transmitted power falls approximately linearly as a function of optical signal power.

The second, limiting, régime is of interest in a logic device. This régime is characterised by constant modulator absorption with current equalisation being achieved by variation of effective quantum efficiency. If the signal power is very low, the photo-transistor current is low and therefore the modulator voltage is also low. In this situation, the modulator quantum efficiency varies rapidly with applied voltage, but its absorption and transmission are essentially constant. Thus for signal powers up to a certain level the output power of the device is high and constant. If the signal power is high, then the modulator current is required to be high. However, the maximum value of modulator current is determined by the product of the maximum absorption of the modulator and the value of the pump power. If the required current is greater than this, then the voltage across the modulator becomes so large that the collector-emitter voltage of the photo-transistor is small and therefore the latter's effective quantum efficiency is small. Thus for signal powers above a certain level the output power is low and constant.

This operation was demonstrated experimentally using a p-i-n structure MQW modulator described in [3]. The signal power was supplied by an LED, sealed in a light-tight box with the photo-transistor and the pump power, at a wavelength 866nm, was taken from a dye laser. A low frequency triangular wave was applied to the LED, from which the signal power could be deduced, and the transmitted power was measured. The experimental results for the input-output characteristic are shown in fig.3. The low contrast is due to the small change in transmission of the modulator, which it is believed is due to residual doping of the intrinsic region. One obvious way to improve the contrast is by the use of a waveguide modulator and contrast of 20:1 has been seen by this method [5].

In the second, non-inverting, device the modulator absorption should decrease with applied reverse bias voltage, which is achieved by working exactly at the exciton peak of the MQW material. Again the quantum efficiency of the modulator falls when the applied voltage is very small. The overall effect, shown as the broken line in fig.2, is that the responsivity is very low for low voltage, increases rapidly with increasing voltage and then falls nonlinearly. The operation of the device again

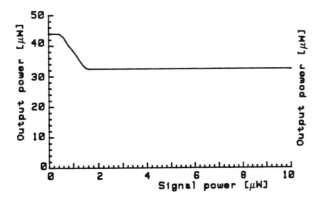

Fig. 3. Inverting characteristic

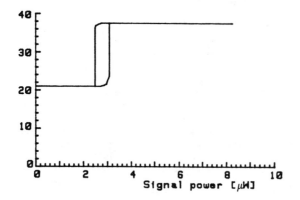

Fig. 4. Non-inverting characteristic

relies on current equalisation, but in this case there is no "self-linearization". For low signal powers, the modulator voltage is low, absorption is high and variations cause changes in quantum efficiency, thus resulting in an output which is low and constant. Above a certain level of signal power, the photo-current must increase above its maximum, causing the modulator voltage to rise which causes a further mismatch between the currents. The only stable state requires high modulator voltage and therefore low photo-transistor quantum efficiency with the result that the output switches abruptly to high power. The experimental results are shown in fig.4 for a pump wavelength of 855nm and the same experimental arrangement as before.

Multiple-input logic gates may be made by employing several photo-transistors. Using the wavelength of the first, inverting, device a NOR gate may be made if the two photo-transistors are in parallel and a NAND gate if they are in series. The second, non-inverting, device may be made into an OR or AND gate with the transistors in parallel or series respectively. These four devices have been made and successfully demonstrated using a separate LED for each transistor.

Finally, integrated versions of these devices have been fabricated by growing a hetero-junction bipolar photo-transistor directly onto MQW material using MOVPE at the University of Sheffield. By etching a mesa for the photo-transistor on top of the mesa for the modulator, devices can be electrically isolated from each other. A 6x2 array of such device has been fabricated. The compatibility of these devices with existing technology for the fabrication and processing of III-V semiconductors, already very common in opto-electronics, is undoubtably a great advantage. These devices could also be made to work with wavelengths of 1.3-1.5μm compatible with optical fibre systems by using materials based on InP.

This work was supported principally by the UK SERC and British Telecom.

References

1. Wood, T.H., et al, Appl.Phys.Lett. 44, 16 (1984)
2. Keyes, R.W., Optica Acta 32, 525 (1985)
3. Wheatley, P. et al, Electron. Lett. 23, 92 (1987)
4. Miller, D.A.B et al, Optics Lett. 9, 567 (1984)
5. Bradley, P.J. et al, Electron. Lett. 23, 213 (1987)

Section 2.3: Optical Amplifiers

Semiconductor Laser Optical Amplifiers for Use in Future Fiber Systems

M. J. O'MAHONY

(*Invited Paper*)

Abstract—This paper discusses the application of semiconductor laser amplifiers to future long wavelength optical fiber systems. The basic equations defining laser amplifier characteristics are presented together with experimental results. Linear and nonlinear modes of operation are considered; the former includes use as an optical gain block, a linear repeater, and a preamplifier, while the latter includes use as a bistable element, an electrically controlled optical switch, and an injection locked amplifier.

I. INTRODUCTION

SEMICONDUCTOR laser optical amplifiers will play an important role in future optical systems, both for optical fiber transmission and for optical data processing. Semiconductor laser amplifiers provide high gain, with low power consumption and their single-mode waveguide structure makes them particularly suitable for use with monomode fiber. There have been many studies on laser amplifiers [1]–[8], but it was not until recent years that high performance semiconductor lasers (with, for example, low threshold currents, narrow far field radiation patterns, and good antireflection coatings) became available, allowing practical amplifiers to be tested.

Semiconductor laser amplifiers can be used in both linear and nonlinear modes of operation and there are many possible applications to future optical systems. Some of the applications suggested to date, are the following.

A General Linear Optical Gain Block: For example, to compensate for losses in splitting networks etc and provide gain in particular optical configurations such as ring lasers [9] or narrow linewidth sources [10].

A Simple Nonregenerative Repeater: In optical transmission systems using single longitudinal mode lasers the effects of fiber dispersion may be small and the main limitation on repeater spacing is the signal attenuation due to fiber loss. Such systems do not necessarily require a complete regeneration of the signal at each repeater and amplification of the signal is sufficient. Thus semiconductor laser amplifiers can be used as linear optical repeaters, for intensity modulated or coherent systems, and this has been demonstrated in several experiments [11]–[15] for both single and multiple repeaters. Linear amplifier repeaters have the additional potential advantages of being bit rate independent and bidirectional, with the capability of supporting multiplex operation [16], [17].

A Receiver Preamplifier: Using a semiconductor laser amplifier to linearly amplify the optical signal prior to the photodetector in an optical receiver can increase the detection sensitivity. The improvement can be particularly significant for bit rates in excess of 1 Gbit/s [18], [19] and allows the development of sensitive wide-band receivers, which will be required for future fiber systems.

Pulse Shaping and Bistable Element: With a suitable choice of operating conditions a semiconductor laser amplifier can be made to exhibit nonlinearity and bistability [20], [21]. This type of behavior can, for example, be used to provide pulse shaping, such that a dispersed optical pulse can be reshaped into a rectangular pulse; this is part of the requirement for an all optical regenerative repeater. Bistable elements are required for future optical processing systems.

Optical Routing Switch: If the amplifier bias current is switched off the input and output ports are isolated. The degree of isolation is good—of the order of 30 dB (optical) for a 200-μm-long device. In the "on" state there is signal gain, thus an amplifier will allow lossless optical fiber routing networks to be constructed [22].

Injection Locked Amplifier: If the amplifier is biased above threshold, i.e., lasing, the output frequency and phase will under certain conditions lock to the frequency and phase of a weak monochromatic signal incident on the input facet. In this mode the device can be used as a linear amplifier of angle modulated signals [23].

The aim of this paper is to review the role of semiconductor laser amplifiers in future optical fiber systems and we shall do this by concentrating on the applications discussed above. The theory describing amplifier behavior, in both linear and nonlinear modes of operation, has been developed and refined by many authors and in Section II we give some of the more important equations governing the performance of amplifiers and discuss their implications; Fabry–Perot (FP), traveling wave (TW), and near traveling wave (NTW) amplifiers are considered. The nonlinearity associated with the variation of material refractive index with input power is also included. Section III discusses a number of the applications listed above, considering both the current state of laboratory experi-

Manuscript received October 10, 1986.
The author is with British Telecom Research Laboratories, Martlesham Heath, Ipswich IP5 7RE, U.K.
IEEE Log Number 8718513.

ments and the future performance expectations. A brief summary is given in Section IV.

II. Theoretical Models

A. General Description

Fig. 1 is a schematic diagram of a semiconductor laser amplifier, which is based on the normal semiconductor laser structure, with active region width W, thickness d, and length L. The input and output laser facet reflectivities are denoted R_1 and R_2, respectively, and for a normal laser $R_1 = R_2 = 0.3$. In this paper we confine ourselves to index guided structures, although gain guided devices have also been investigated [24]. Antireflection coatings may be applied to the laser facets to reduce or eliminate the end reflectivities; this has the effect of increasing the amplifier bandwidth and makes the transmission characteristics less dependent on fluctuations in bias current, temperature, and input signal polarization. In practice the device is biased below the lasing threshold and light incident on one facet appears amplified at the other, together with noise. In the literature two types of amplifier are often distinguished, the Fabry–Perot (FP) amplifier and the traveling wave (TW) amplifier, both have the normal semiconductor laser structure, but differ in their facet reflectivities.

A Fabry–Perot amplifier is a resonant amplifier in which the factor $G_s \sqrt{R_1 R_2}$ is close to unity, where G_s is the single pass gain through the device. Generally this implies operation at a bias current just below the lasing threshold current (I_{th}) and we are usually considering facet reflectivities in the order of 0.01–0.3. Under these conditions resonant internal gains in the order of 25–30 dB are obtainable, where the word internal is used to denote the gain from just outside the input facet to just outside the output facet, i.e., we do not include any losses involved in coupling light to and from the amplifier. As an example, Fig. 2 shows the measured peak gain versus normalized bias current (normalized to the lasing threshold current) characteristic of a 1.5-μm DC-PBH laser with no facet coatings (i.e., $R_1 = R_2 = 0.3$, $L = 200$ μm). The lasing threshold current was 15 mA and the maximum gain was approximately 25 dB. The significant feature of this characteristic is that high gains are only obtained close to threshold and the slope of the curve indicates that the gain is very sensitive to current changes. As the FP amplifier is a highly resonant amplifier, the transmission characteristic comprises very narrow passbands (Fabry–Perot resonances—Section II-C). This inherent filtering can be useful in certain applications, but the transmission characteristics are very sensitive to fluctuations in bias current, temperature, and signal polarization. For example, at an internal gain of 25 dB the ± 3-dB resonance bandwidth of the FP amplifier shown in Fig. 2 was measured as 6 GHz, to maintain the gain within ± 1 dB the temperature had to be controlled to within ± 0.1°C [25] and the measured gain difference between orthogonal signal polarizations was 12 dB. FP amplifiers also exhibit gain saturation at a lower input power level than TW amplifiers

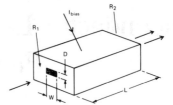

Fig. 1. Amplifier schematic.

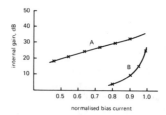

Fig. 2. Amplifier gain versus current for: A: NTW ($L = 500$ μm, $R = 8 \times 10^{-4}$, $I_{th} = 55$ mA) B: FP ($L = 200$ μm, $R = 0.3$, $I_{th} = 15$ mA). Input power = -40 dBm.

(Section II-C). Because of the resonant nature of FP amplifiers, and their high internal fields, they are used for nonlinear applications such as pulse shaping or as bistable elements.

A traveling wave amplifier has antireflection coatings applied to its facets to reduce their reflectivities. In the limiting case, an amplifier with zero reflectivity facets would be a travelling wave amplifier, with the factor $G_s \sqrt{R_1 R_2}$ equal to zero. In practice, however, even with the best antireflection coatings there is some residual facet reflectivity (the current lowest reflectivity at 1.5 μm is $R = 1 \times 10^{-4}$ [26]) and the amplifier lies between traveling wave and Fabry–Perot: the term near traveling wave (NTW) has been used in this context [27] and we shall define the term to denote an amplifier with the factor $G_s \sqrt{R_1 R_2} \leq 0.17$; when the equality holds the residual Fabry–Perot resonances cause a 3-dB peak to trough ripple across the passband (Section II-C). The main advantages of the NTW amplifier are wide bandwidth (with an attendent reduction in sensitivity to temperature and bias current fluctuations) low sensitivity to signal polarization and an improved gain saturation characteristic. Fig. 2 shows the gain versus current characteristic of a DC-PBH 1.5-μm NTW amplifier ($L = 500$ μm, $R_1 = R_2 = 8 \times 10^{-4}$). In this case a higher maximum gain is achieved (~ 33 dB) and the characteristic is almost linear. One effect of antireflection coating is to increase the lasing threshold current, which was 55 mA in this example. The ± 3-dB bandwidth, at an internal gain of 25 dB, was measured as 75 GHz; the temperature could be varied by ± 1.5° for a gain variation of ± 1 dB; the measured gain difference between orthogonal signal polarizations was approximately 2.5 dB [19]. The principal disadvantages of NTW amplifiers are an increased spontaneous noise component, and in practical terms, an increased sensitivity to reflections from outside the amplifier (as the facet reflectivities are reduced outside sources of reflection such as splices, etc., begin to affect performance).

The performance of semiconductor laser amplifiers in optical fiber systems depends upon a number of factors, such as:

a) the frequency transmission characteristic,
b) gain saturation and nonlinear effects,
c) gain stability with respect to bias current and temperature,
d) the sensitivity of the gain to signal polarization,
e) the noise introduced by the amplifier,
f) the efficiency of the coupling to the amplifier,
g) the sensitivity to reflections.

In the following discussion we consider the equations defining some of these factors and their implications. Unless otherwise stated quoted parameter values refer to InGaAsP devices operating in the 1.5-μm region; and the numbers are only given by way of example. Some topics, e.g., the sensitivity to reflections, require further theoretical analysis.

B. Amplifier Equations

Amplification takes place in the active region of the semiconductor laser and the material gain coefficient per unit length g_m is generally assumed to have a peak value proportional to the carrier density n, a parabolic gain-wavelength characteristic, and a peak gain wavelength which is a function of the carrier density. Thus [28]:

$$g_m = a(n - n_0) - a_2(\lambda - \lambda_p)^2 \quad (1)$$

where a and a_2 are gain constants (2.7×10^{-16} cm^2 and 0.15 cm$^{-1} \cdot$ nm^{-2}, respectively), n_0 is the transparency density (1.1×10^{18} cm^{-3}) and λ_p the peak gain wavelength, defined as

$$\lambda_p = \lambda_0 + a_3(n - n_0) \quad (2)$$

where $a_3 = 2.7 \times 10^{-17}$ nm $\cdot$ cm^{-3}. The ± 3-dB gain curve bandwidth is readily derived as

$$2\Delta\lambda = 2\sqrt{\frac{a(n - n_0)}{2a_2}}. \quad (3)$$

With these values, for example, the bandwidth of the gain curve, when operating close to threshold in a DC-PBH laser ($L = 350$ μm, $n_{th} = 1.8 \times 10^{18}$ cm^{-3}) is approximately 54 nm. These equations show that an increase in bias current (and therefore the carrier density) increases the peak material gain and decreases the peak wavelength. The shift to lower wavelengths with increasing carrier density presents a problem in the design of low reflectivity amplifiers. When antireflection coatings are applied to a semiconductor laser to form a TW or NTW amplifier the operating current is generally higher than the original threshold current, thus lasers designed for 1.55 μm, for example, will result in NTW amplifiers with gain peaks at considerably shortened wavelengths, a wavelength shift of 50 nm has been observed, for example, in an amplifier with residual facet reflectivities of 10^{-3}. This problem can be alleviated by using long devices, which for the same gain operate with a lower carrier density.

The net gain per unit length g is defined in terms of the material gain coefficient g_m, the optical confinement factor Γ, and the effective loss coefficient per unit length α (50 cm^{-1}):

$$g = \Gamma g_m - \alpha. \quad (4)$$

The carrier density and the optical intensity are related by the rate equation:

$$\frac{dn}{dt} = \frac{j}{ed} - R(n) - \frac{\Gamma g_m}{E}(\beta I_{sp} + I) \quad (5)$$

where $R(n)$ is the recombination rate, I the signal intensity, I_{sp} the intensity of the total spontaneous emission, j the injected current density, e the electron charge, and E the photon energy. β is the spontaneous emission coefficient, which defines the fraction of the total spontaneous emission coupled into the guided wave

$$\beta = \frac{c}{N}\frac{a\tau}{V} \quad (6)$$

where V is the active volume, c the velocity of light, τ the carrier lifetime (~ 2 ns), N the material index; for example with $V = WdL = 1.5 \times 0.15 \times 500$ μm^3, $a = 2.7 \times 10^{-16}$ cm^2; $\beta = 3.6 \times 10^{-5}$. In general the intensity and gain depend on the position along the amplifier length, however Adams et al. [29] have shown that the assumption of position independence gives accurate results when the intensities are averaged. For TW amplifiers, with high injected currents, the recombination rate should include Auger recombination terms [30], and is modeled by the expression $R(n) = An + Bn^2 + Cn^3$; where A (1×10^8 s^{-1}), B (8×10^{-17} m$^3 \cdot$ s^{-1}), C (4×10^{-41} m$^6 \cdot$ s^{-1}) are constants (values given are for 1.3 μm, from [30]). To obtain approximate analytic results, the recombination rate is often assumed to be linearly proportional to the carrier density, thus $R(n) = n/\tau$. Then in the steady state, and assuming $\alpha = 0$:

$$\frac{j}{ed} = \frac{n}{\tau} + \frac{\Gamma g_m}{E} I. \quad (7)$$

Equations (1), (4), and (7) can be combined to define the material gain coefficient in terms of the averaged signal intensity. Rewriting the results of Mukai [31] and Adams [32]:

$$g_m = \frac{g_0}{1 + I/I_s}$$

$$I_s = \frac{E}{\Gamma a \tau}, \quad g_0 = g_m|_{n=n_1} \quad (8)$$

where I_s is the saturation intensity, g_0 the unsaturated material gain coefficient in the absence of input signal and n_1 the corresponding carrier density. G_s is the single pass gain:

$$G_s = \exp[gL] = \exp\left[\left(\frac{\Gamma g_0}{1 + I/I_s} - \alpha\right)L\right] \quad (9)$$

thus the single pass gain decreases with increasing intensity, and the material gain coefficient is reduced by a factor of two when the internal intensity I is equal to the saturation intensity I_s.

The material refractive index N is a function of the carrier density and therefore the optical intensity (equation (7)). Thus, in addition to the nominal phase shift associated with a single pass of the amplifier, there will be an additional component due to the change in carrier density from the nominal density in the absence of signal. The total phase shift is [32]:

$$\phi = \phi_0 + \frac{g_0 L b}{2}\left[\frac{I}{I + I_s}\right] \quad (10)$$

where $\phi_0 = 2\pi \ln/\lambda$ is the nominal phase shift, and b is the linewidth broadening factor ($b = 5$).

Equations (9) and (10) show that the single pass gain and phase are functions of the optical intensity. For a constant intensity signal, e.g., one with frequency modulation, there is no inherent signal distortion, but with a time varying intensity it is clear that the gain and phase may vary with the signal, causing distortion. A full analysis, based on the time dependent rate equation (5) and operating in the unsaturated regime, showed that the form of the gain and phase functions depended on the factor $2\pi f_m \tau$, where f_m was the frequency of the input signal (a sinousoid was assumed). In the case where $f_m \ll 1/2\pi\tau$, the gain and phase were described as in (9) and (10), with the intensity I varying as the input signal; for $f_m \gg 1/2\pi\tau$, the time variation vanishes and the intensity I is proportional to the mean signal intensity. The absence of nonlinear distortion at high frequencies occurs because the variation in carrier density significantly decreases at frequencies much greater than the reciprocal of the carrier lifetime, which causes a reduction in the variation of refractive index. Thus intensity modulation formats will suffer some distortion of low frequencies.

The single pass gain represents the gain available in a TW amplifier. The +3-dB bandwidth for the unsaturated amplifier is obtained using (1), (4), and (9) as

$$2\Delta\lambda = 2\sqrt{\frac{\ln 2}{a_2 \Gamma L}}. \quad (11)$$

The bandwidth is independent of the absolute gain and is a function of device length, decreasing as the length increases. For example, the bandwidth of an amplifier with $L = 500\ \mu m$, $\Gamma = 0.3$ is 35 nm, which corresponds to a frequency bandwidth of 4.6×10^{12} Hz. Fig. 3 is the measured peak gain envelope versus wavelength characteristic of a 1.5-μm NTW amplifier ($L = 500\ \mu m$, $R_1 = R_2 = 8 \times 10^{-4}$); the fine structure 3-dB peak to trough ripple associated with the residual Fabry–Perot resonances (Section II-C) is shown in Fig. 4. The gain bandwidth is 45 nm; the discrepency with theory is possibly due to an overestimation of the factor Γ. Similar results have been reported by Eisenstein et al. [33].

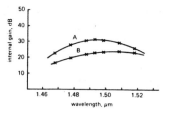

Fig. 3. Peak gain envelope versus wavelength ($L = 500\ \mu m$, $R = 8 \times 10^{-4}$). A: $I_{bias} = 0.98 \times I_{th}$; B: $I_{bias} = 0.7 \times I_{th}$. B corresponds to NTW operation with 3-dB passband ripple, bandwidth = 45 nm.

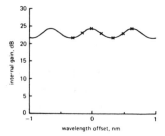

Fig. 4. Passband gain ripple as a function of offset from center wavelength (approximately 1.51 μm), with 25-dB peak gain.

The output saturation power of a TW amplifier, i.e., the power at which the single pass gain has decreased by 3 dB, can be approximated using (8) and (9) (defining the average signal intensity in the cavity as $I = I_o/gL$, where I_o is the output intensity, and setting the loss coefficient α to zero) as

$$P_{os} = \frac{WdE}{\Gamma \tau a} \ln 2. \quad (12)$$

For W, d, a as defined previously, and $\Gamma = 0.3$, the saturation power is +1.0 dBm; for a 25-dB unsaturated gain this corresponds to an input power of −24 dBm. To increase the saturation power, the carrier lifetime or the confinement factor must be decreased for a fixed active area. Gain bandwidth and saturation are important parameters in system applications and the latter two are maximimised in a TW amplifier.

C. Influence of Cavity (FP amplifiers)

In practice even antireflection coated facets will exhibit some residual reflectivity and form an optical cavity. Thus the laser amplifier transmission characteristic contains resonant peaks whose absolute wavelength and spacing depends on the cavity dimensions.

Fig. 5 shows a typical transmission characteristic for a FP amplifier, with $R_1 = R_2 = 0.3$ and a maximum gain of 25 dB. Assuming that the period of the input signal's maximum frequency component is much greater than the cavity round trip time, the intensity amplitude and phase transfer functions are defined as:

$$G = \frac{(1 - R_1)(1 - R_2) G_s}{(1 + \sqrt{R_1 R_2}\, G_s)^2 + 4\sqrt{R_1 R_2}\, G_s \sin^2 \phi}$$

$$\psi_o = \psi_{in} - \phi - \tan^{-1}\left[\frac{\sqrt{R_1 R_2}\, G_s \sin 2\phi}{1 - \sqrt{R_1 R_2}\, G_s \cos 2\phi}\right]. \quad (13)$$

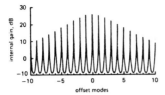

Fig. 5. FP amplifier passband. Mode 0 corresponds to peak gain wavelength; mode spacing = $\lambda^2/2NL = 1.5$ nm for $L = 200$ μm.

As G_s and ϕ are functions of the input intensity, the cavity will exhibit bistable and nonlinear characteristics at high input powers. The resonance has the effect of exaggerating any changes in G_s, thus for example the mode envelope bandwidth ($\simeq 9$ nm for the case in Fig. 5) is less than the single pass bandwidth and the cavity gain G is greater than the single pass gain G_s; the gain relationships are illustrated in Fig. 6 as a function of reflectivity, for a peak gain of 25 dB. The output saturation power of a FP amplifier is less than the TW amplifier, with the same peak gain:

$$P_{os}(FP) \simeq \frac{WdE}{\Gamma\tau a}\ln\left(\frac{2}{1+RG_s}\right)\frac{(1-R)G_s}{(1+RG_s)(G_s-1)}$$

$$R_1 = R_2 = R, \quad G_s \gg 1 \quad (14)$$

where $(1 + RG_s)/2$ is the reduction in the single pass gain required to reduce the resonant gain by 3 dB. For example with $R = 0.3$, $G = 25$ dB a decrease in the single pass gain to $0.97 \times G_s$ (where $G_s = 5$ dB from Fig. 6) reduces the gain by 3 dB; thus the output saturation power is -14.5 dBm, a decrease of 15.5 dB from the TW amplifier. Fig. 7 shows the output saturation power as a function of reflectivity for various values of unsaturated gain. For high gain combined with high saturation power, the facet reflectivities must be low.

The ± 3-dB bandwidth of a single longitudinal mode can be expressed as a function of the cavity gain and facet reflectivities:

$$2\Delta f = \frac{c}{\pi NL}\sin^{-1}\left[\frac{1}{2}\left(\frac{(1-R_1)(1-R_2)}{\sqrt{R_1R_2}\,G}\right)^{1/2}\right]. \quad (15)$$

At 25 dB, for example, with an uncoated laser ($L = 200$ μm) the bandwidth is the order of 6 GHz. The peak-trough ratio of the passband ripple is:

$$V = \left[\frac{1+\sqrt{R_1R_2}\,G_s}{1-\sqrt{R_1R_2}\,G_s}\right]^2. \quad (16)$$

The ratio is shown in Fig. 8 as a function of reflectivity and gain. When the ratio is 3 dB, $G_s\sqrt{R_1R_2} = 0.17$ and $G = 0.25/\sqrt{R_1R_2}$ ((13) and (16) with $R_1, R_2 \ll 1$).

Residual facet reflectivity introduces an additional problem associated with the use of amplifiers in optical systems, namely the effect of backward gain. The gain of the backward traveling signal is [34]:

$$G_b = \frac{(\sqrt{R_1}-\sqrt{R_2}\,G_s)^2 + 4\sqrt{R_1R_2}\,G_s\sin^2\phi}{(1-\sqrt{R_1R_2}\,G_s)^2 + 4\sqrt{R_1R_2}\,G_s\sin^2\phi}. \quad (17)$$

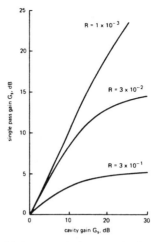

Fig. 6. Relationship between single pass gain G_s and cavity gain G.

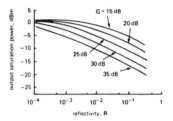

Fig. 7. Output saturation power as a function of reflectivity ($W = 1.5$ μm, $d = 0.15$ μm, $\tau = 2$ ns, $\Gamma = 0.3$).

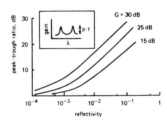

Fig. 8. Passband ripple as a function of cavity gain (G) and facet reflectivity.

For NTW amplifiers the peak forward-backward gain ratio is approximately $1/G_sR$ ($R = R_1 = R_2$), thus with a forward gain of 25 dB, a ratio of 20 dB requires a facet reflectivity $R = 3 \times 10^{-5}$. In systems with cascaded amplifiers, isolators may be necessary to avoid amplifier interaction, unless the backward wave amplitude is sufficiently small.

In general the above relations are only approximate in that the effects of spontaneous noise have been ignored. For example, calculations at low signal intensities, where the spontaneous noise contribution may be significant, can result in optimistic estimates of gain. A more exact analysis which includes the effect of spontaneous noise is given by Henning et al. [34].

D. Gain Dependence on Polarization and Temperature

Amplifier gain is dependent on the polarization of the input signal. The single pass gain is different for TE and TM polarization modes, because the confinement factors differ, i.e., $\Gamma_{TE} \neq \Gamma_{TM}$. This effect is magnified in a resonant cavity as the mode propagation constants and modal facet reflectivities are also polarization dependent. Our measurements have shown that the gain difference close to threshold in resonant devices can vary significantly from one device to another even though they have the same nominal dimensions, and this implies that polarization controllers may be necessary for using such devices. The gain difference is minimized for TW devices [35]. Fig. 9 shows the measured gain characteristics, as a function of signal polarization, for a 1.5-μm NTW amplifier ($L = 500$ μm, $R = 8 \times 10^{-4}$), and a 1.5-μm Fabry-Perot amplifier ($L = 350$ μm, $R = 0.03$); at 25-dB gain the difference is approximately 2.5 dB for the NTW and 10 dB for the FP devices.

Gain also depends on temperature. For NTW amplifiers the dominant factors are the reduction in material gain constant (a in (1)) and the increase in transparancy density (n_0) associated with an increase in temperature. At 1.5 μm the gain decreases by approximately 3 dB if the temperature is increased by 5°C (for $L = 500$ μm). A decrease in temperature increases the gain, but also increases the passband ripple in the case where there is residual reflectivity. In a FP amplifier these effects are compounded by a shift in mode wavelength of approximately 10 GHz/°C associated with the variation of refractive index with temperature. For high gain FP amplifiers, therefore, the temperature must be controlled to within 0.1°C.

E. Noise

Amplifier noise has been studied by several authors [6], [35], [36], [37]. In this paper we use the results of Mukai et al. [12]. The total mean noise power at the output facet of a FP amplifier of peak gain G is:

$$P_n = E n_{sp} m_t (G - 1) \Delta f_1 \quad (18)$$

where n_{sp} is the population inversion parameter $n/(n - n_0)$, m_t is the number of effective transverse modes, Δf_1 is the normalized spontaneous shot noise bandwidth and is a function of facet reflectivities:

$$\Delta f_1 = \frac{c}{2NL(G-1)} \cdot \sum_j \frac{(1 + R_1 G_s(j))(1 - R_2)(G_s(j) - 1)}{(1 - R_1 R_2 G_s(j)^2)} \quad (19)$$

where the summation is over the number of longitudinal modes. G is the peak cavity gain and G_s varies with wavelength ((1) and (9)) and hence mode number. The mean noise power is important for many applications; for example, it can limit the number of linear repeaters that can be cascaded (Section III) because of gain saturation. Fig. 10 is a graph of the mean noise power as a function of reflectivity, for a fixed gain of 25 dB, with $n_{sp} = 1.5$ (calculated using (18) and (19), for $L = 500$ μm). In the NTW region the noise power tends to a fixed value of -3 dBm, which is in good agreement with measurement.

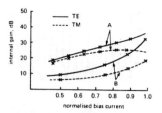

Fig. 9. Amplifier gain for TE and TM polarized signals as a function of normalized bias current. A: NTW [$L = 500$ μm, $R = 8 \times 10^{-4}$, $I_{th} = 55$ mA] B: FP [$L = 350$ μm, $R = 0.03$, $I_{th} = 35$ mA].

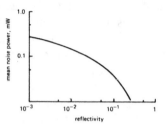

Fig. 10. Mean output noise power with $G = 25$ dB. [$L = 500$ μm, $n_{sp} = 1.5$].

The variance of the photon number, at the amplifier output, is described as:

$$\sigma^2 = \frac{GP}{E} + n_{sp} m_t (G - 1) \Delta f_1 + 2G(G - 1) n_{sp} \chi \frac{P}{E} + (G - 1)^2 m_t \Delta f_2 n_{sp}^2 \quad (20)$$

where P is the mean input power. The first two terms represent amplified signal and spontaneous emission shot noise, the second two represent signal-spontaneous beat noise and spontaneous-spontaneous beat noise, respectively. Δf_2 is the normalized beat noise bandwidth, and χ the excess noise factor defined as $X = (1 + R_1 G_s)/(1 - R_1)$; [$G, G_s \gg 1$]. For a TW amplifier Δf_1 and Δf_2 are equal to the single pass bandwidth, and the excess noise factor, which depends on facet reflectivity, approaches unity. As the noise variance depends upon optical bandwidths and input power, the noise components that dominate will depend upon amplifier application and this is discussed in Section III.

III. Application to Future Systems

In this section we consider amplifier applications, the current state of laboratory experiment and the performance expectations for future systems. We commence with a look at linear amplifiers. Most of the initial characterization was done at the shorter wavelengths, in particular 850 nm, but experimental results are now available at the longer wavelengths of 1.3-1.5 μm, and we shall concentrate on these referring to the short wavelengths results as necessary.

A. Linear Applications

The experimental characteristics of linear amplifiers show a good agreement with theory at both short and long wavelengths and this enables us to make some general comments on the limits of amplifier performance. It should be noted, however, that laser amplifiers are usually developed from unmodified semiconductor laser structures (apart from the application of antireflection coatings) and such structures may not necessarily yield the best amplifiers. The maximum amplifier gains that can be achieved are in the order of 30–35 dB, the higher values generally being achieved with TW amplifiers. In a resonant amplifier, the mode bandwidth decreases with increasing gain (15) so, at high gain the resonant amplifier may not have sufficient bandwidth for distortionless amplification of a high bit rate intensity modulated signal. The maximum output power is also achieved with a TW amplifier and this is generally in the range 0–5 dBm; a FP device can be 15 dB lower (Fig. 7). It has also been noted that FP amplifiers generate a considerable backward traveling wave and this can cause problems in many applications. For future wide-band optical systems, therefore, TW type amplifiers seem most appropriate although they do have problems associated with an increased noise bandwidth and an increased susceptibility to external reflections because of the lower facet reflectivities.

As the maximum amplification is limited, minimizing the losses incurred in coupling the light to and from the amplifier is extremely important. In the laboratory, coupling losses of 2 dB/facet at 1.55 μm have been achieved using tapered lensed fibers [39], allowing maximum net gains of 25–30 dBm, but more typical losses are in the order of 3.5 dB/facet. The minimum gain difference between TE and TM signal polarization states can be in the order of 3 dB for TW amplifiers and as much as 14 dB for FP amplifiers operating near threshold. Special laser structures may be required to significantly reduce polarization sensitivity.

B. Optical Gain Block

When amplifiers are used as a general gain optical gain block (for example, to compensate for losses in splitting networks), bandwidth, gain saturation and backward gain are of particular importance. Generally these considerations lead to the choice of a TW (or NTW) amplifier. For low input powers net gains of 20 dB have been demonstrated [40]. With high input powers gain saturation is a problem. Fig. 11 is the measured gain versus input power characteristic of the NTW amplifier and this shows that the gain has fallen to 10 dB when the input level is −10 dBm; with coupling losses this reduces the net gain to approximately 3 dB. Thus to compensate for a single branch the line input power should not exceed −7 dBm. For use in future systems, therefore, device structures with an improved saturation performance will be required. An advantage of low gain, of course, is that a very low bias current is required, operation is truly traveling wave

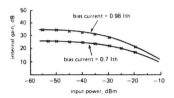

Fig. 11. Gain versus input power for NTW amplifier.

($G_s \sqrt{R_1 R_2} = 0.01$ in the above case), the backward wave is negligible (equation (17)) and the amplifier noise is very low (equation (18)). Another example of the use of an amplifier as a gain block is given by Jopson et al. [9], using the amplifier as the gain medium in an optical fiber ring laser. Use of the amplifier as a routing switch (discussed later) also falls into this category.

C. Linear Repeater

The use of a laser amplifier as a linear amplifier repeater in an optical transmission system has many attractions [41] and a number of experiments have been reported [13]–[15], using both intensity and coherent modulation. To date however all the reported experiments at long wavelengths have used amplifiers with facet reflectivities in the order of 3 percent. Fig. 12 is the schematic of a typical experiment using a single repeater in an intensity modulated optical fiber system. The two fiber sections comprise 100 km of monomode optical fiber and a fiber directional coupler is used prior to the amplifier repeater to allow monitoring of the input power. The amplifier internal gain was 27 dB and the total coupling losses were 12 dB, given a net fiber to fiber repeater gain of 15 dB. At 27-dB gain, with facet reflectivities of 0.03, the amplifier is highly resonant with a 20-dB passband ripple (Fig. 8). For this reason a tunable source was used for the transmitter and the wavelength adjusted to match the peak gain wavelength of the amplifier. With a NTW amplifiers these problems are relaxed and we have successfully used DFB lasers, and in some cases multimode Fabry–Perot lasers as sources. The second fiber section terminated in a p-i-n photodiode/FET optical receiver with a sensitivity of −44 dBm at 100 Mbit/s. The total fiber loss at 1.50 μm (the operating wavelength) was 51 dB.

The mean power launched into the fiber was −7 dBm. With a total fiber loss of 51 dB and a repeater gain of 15 dB the power at the receiver was −43 dBm, sufficient to give a 10^{-9} bit error rate. A system penalty of 1 dB was noted, but only part of this was attributed to the amplifier. Using the expression for the noise power variance (equation (20)), the SNR at the terminal receiver decision point can be expressed as:

$$\text{SNR} = \frac{4(c_1 c_2 P G \eta)^2}{\frac{E^2}{2T}\left[2\left(\frac{P c_1 G}{E} + (G-1) n_{sp} m_t \Delta f_1\right) c_2 \eta + \left(2 G(G-1) n_{sp} \chi \frac{P}{E} c_1 + (G-1)^2 n_{sp}^2 m_t \Delta f_2\right) c_2^2 \eta^2\right] + \left(\frac{S\eta}{6E}\right)^2} \quad (21)$$

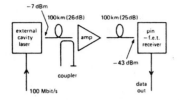

Fig. 12. Linear amplifier repeater experiment.

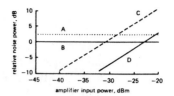

Fig. 13. Amplifier noise components as a function of signal power in input fibre tail (less 6 dB for input facet power). Powers are relative to receiver noise. A: Spontaneous emission shot noise. B: Receiver noise. C: Signal shot noise. D: Signal-spontaneous beat noise.

where P is the mean launched power, T the bit duration, S the receiver sensitivity, η the photodiode quantum efficiency, and c_1, c_2 represent the loss from the launch end to the input facet and from the output facet to the photodetector, respectively. In a fiber system there is also spatial filtering of the spontaneous emission, which is this case reduced the mean spontaneous noise by approximately 4 dB. The first two components in the denominator represent the signal and spontaneous shot noise, the second two are the bast noise components and the final term represents receiver noise. The expression shows that the beat noise terms diminish faster than the shot noise terms, with respect to output loss c_2. Fig. 13 shows the noise terms relative to the receiver noise as a function of mean signal power at the amplifier input fiber tail (input facet power is 6-dB lower); assuming a TW amplifier with a bandwidth of 45 nm. In this case the input power was -34 dBm and the graph shows that the spontaneous emission shot noise component dominates giving a penalty of approximately 2 dB. This, however, is a worst case—a TW amplifier with no optical filtering and a low bit rate. In the experiment discussed above, a FP amplifier was used with a noise bandwidth considerable less than 45 nm, and the calculated penalty was approximately 0.5 dB. As the bit rate increases beyond 100 Mbit/s the receiver noise component in Fig. 13 begins to dominate, thus 140 Mbit/s might be considered the minimum usable line rate for systems using TW amplifier repeaters (with no filtering).

Single amplifier repeater systems are attractive because of their simplicity and such a system has many of the advantages of a repeaterless system, e.g., line rate independence, it can support multiplex and bidirectional operation. With NTW amplifiers, repeater gains (allowing some margin), will probably be of the order of 20 dB and this would stretch system lengths by approximately 100 km at 1.55 μm. Single repeaters can also be combined with a terminal receiver which incorporates an optical receiver preamplifier and this is particularly attractive for bit rates in excess of 1 Gbit/s (see preamplifier section). Laser amplifiers have great potential for wide-band systems as they provide high gain over a very wide bandwidth. The above expression for the SNR shows that the amplifier detected noise power increases at a rate of 3 dB/octave; in practice receiver noise increases in the order of 6–9 dB/octave thus the effects of amplifier noise become less significant at high bit rates. Furthermore, the maximum repeater gain of 20 dB will not be far removed from that obtainable from a conventional electrooptic repeater at high bit rates and significantly easier to engineer. The above experiment illustrated some of the limitations associated with the use of amplifiers as repeaters, namely the following.

Gain Saturation: As the output power is limited to the saturation power, the maximum power that can be launched into a following fiber is unlikely to exceed 0 dBm (assuming a TW amplifier and a coupling loss of 3.5 dB).

Maximum Gain: The maximum internal gain that is likely to be achieved with a NTW amplifier is 30 dB (assuming $R = 1 \times 10^{-4}$), allowing 3.5-dB/facet loss and 3-dB margin gives a maximum net gain of 20 dB. This of course is for a device based on conventional semiconductor laser structures.

Gain Stability: Variations in gain may occur due to fluctuations in input signal polarization, ambient temperature, and bias current; minimization of these factors requires the use of a NTW amplifier. Unless a polarization controller is used, a performance margin of 2–3 dB is necessary to allow for the signal polarization drift at the end of a fiber section.

System experiments have also been conducted using two amplifier repeaters in tandem, for both intensity modulation at short [12] and long wavelengths [15] and an ASK heterodyne system [14]. For multiamplifier systems the following additional problems arise.

Gain Saturation due to Amplified Spontaneous Emission: The maximum gain of each amplifier can be significantly reduced by gain saturation caused by the mean spontaneous emission of the previous amplifier. The mean spontaneous power accumulates in proportion to the number of repeaters. Thus the mean power at the output of the kth amplifier in a chain in kP_n, where P_n is defined by (18). The right hand axis of Fig. 14 shows the accumulated noise power as a function of the number of repeaters and optical bandwidth, the latter is varied by means of an external filter at the output. A NTW amplifier is assumed, with an internal gain of 25 dB, total coupling losses of 5 dB, and an output saturation power of $+5$ dBm. The loss between each amplifier is 20 dB. The input signal power of each amplifier is -20 dBm, and gain saturation occurs when the accumulated noise power equals the signal power. Fig. 14 shows that this occurs after 10 repeaters if no optical filtering is used, assuming each amplifier has

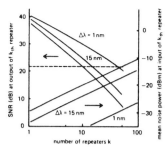

Fig. 14. Noise in multiple repeater systems. The right-hand axis shows total mean noise power at input of k_{th} repeater. The left-hand axis shows detected SNR at output of k_{th} repeater in 140-Mbit/s system. $G = 25$ dB, $R = 1 \times 10^{-3}$ Coupling loss = 2.5 dB/facet, fiber loss between amplifiers = 20 dB. Amplifier output noise power = -5 dBm. $\Delta\lambda$ = bandwidth of external optical filter.

a mean noise power of -5 dBm. With a 1-nm optical filter, the number increases to 400.

Signal to Noise Ratio Degradation: The detected signal to noise ratio is a function of the number of repeaters [12]. Fig. 14 shows the SNR at the kth repeater output (in a 140-Mbit/s system). Again without additional optical filtering the number of repeaters must not exceed 10 to maintain the SNR above 21 dB.

Interaction between Repeaters: It was shown that, with finite facet reflectivities, the backward traveling wave can be significant. In intensity modulated systems the effect of the backward wave on the previous amplifier is to cause pulsations and instability. It is essential, therefore, that its amplitude is reduced, either by providing sufficient loss, using an isolator, or reducing the amplifier gain. These effects have been noted in our experiments [15] and those of Mukai [12]. Experiments with 3-percent reflectivity amplifiers indicate that the forward to backward gain ratio at the amplifier output facet should be at least 30 dB to contain the problem. Use of an isolator, however, to overcome this problem would preclude bidirectional use of the system.

From the above discussion we can make some predictions regarding the role of amplifier repeaters in future fiber systems. It has been shown that single amplifier systems are attractive, particularly for high bit rate systems, and they will have effective gains in the order of 20 dB, for either coherent or intensity modulated systems. Multiamplifier systems need further evaluation, and it is suggested that such systems will incorporate only small numbers of amplifiers (≤ 10) because of noise problems and the effects of fiber dispersion in long systems. Multirepeater systems would seem to call for an amplifier integrated with an isolator and filter. There is also a need for a comprehensive study on the effect of near-end reflections on NTW amplifiers.

D. Optical Receiver Preamplifier

Semiconductor laser amplifiers can be used as optical preamplifiers to increase the sensitivity of the wide-band optical receivers required for future systems. A NTW laser amplifier, for example, offers a high gain over a very large bandwidth, e.g., 25 dB over 4×10^{12} Hz, which is unmatched by any photodetector (the best APD's for example, have a gain bandwidth product of 6×10^{10} Hz). Conventional high sensitivity optical receivers use either p-i-n-diode/FET or APD/FET designs. Although a p-i-n photodiode can have very large bandwidths (to 15 GHz), high sensitivity wide bandwidth p-i-n-diode/FET receivers are difficult to realize; the figure of -36 dBm at 1.2 Gbit/s is typical of the present limit [42]. APD receivers offer a wider bandwidth (a typical best sensitivity is -23 dBm at 8 Gbit/s [43]), but their performance at high bit rates is far from ideal; they also have attendent disadvantages; for example, the sensitivity is very dependent on primary dark current and consequently is very sensitive to nonzero extinction ratio or any associated CW component such as the local oscillator in a coherent detection scheme. The combination of an optical preamplifier with a simple detector (such as a p-i-n diode followed by a low noise 50-Ω amplifier) offers very real advantages for optical systems requiring wide bandwidth detection.

The performance of optical preamplifiers has been considered by a number of authors [6], [12], [18], [44]. Mukai *et al.* [12], for example, combined experimental work with a theoretical analysis of the performance of an optical preamplifier at 850 nm. Their results showed an improvement in sensitivity of 7.4 dB for a low noise silicon APD receiver (base sensitivity of -39.6 dBm) operating at 100 Mbit/s. As far as is known, there are no experimental results for long wavelength receivers (apart from those to be discussed below); however, Fye [18] has done an analysis based on the theory of Mukai *et al.* [12] to illustrate some of the main features of optical preamplifiers when used with modern long wavelength receivers.

For a practical preamplifier system we consider only TW or NTW amplifiers, because with FP amplifiers the required tolerance on signal wavelength and polarization together with bias current and temperature control make their use unlikely except in very special circumstances; in addition, the modulation format for very high bit rate systems will demand a large amplifier bandwidth. To illustrate the discussion we consider the NTW 1.5-μm amplifier discussed earlier. Fig. 15 shows the experimental receiver. The input coupling was via a tapered lensed fiber, the coupling loss being 4.5 dB. The photodetector was a commercial high-speed p-i-n-diode (rise time = 40 ps) followed by a commercial 2-GHz 50-Ω amplifier; the base sensitivity of this receiver was -22 dBm at 1 Gbit/s. The amplifier output was guided to the phodetector by a monomode fiber with an antireflection coated tapered lens at the amplifier end and a coated spherical lens at the photodetector end. An optical filter (to reduce spontaneous emission noise) could be inserted in the collimated beam section between the spherical lens and the photodetector. With no filter, the loss between the amplifier facet and photodetector was 3.5 dB. Insertion of a 2.5-nm (bandwidth) tunable Fabry–Perot filter increased the loss by 2 dB. A fiber polarization controller was also used to reduce the effect of reflections. The effectiveness of a filter depends on the preamplifier gain and the receiver noise level;

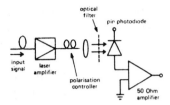

Fig. 15. Optical receiver using NTW laser preamplifier.

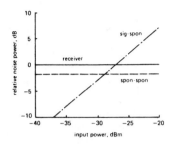

Fig. 16. Detected noise components at preamplifier receiver output as a function of optical power on input fiber tail. $G = 25$ dB, input coupling loss = 4.5 dB, output facet to photodetector loss = 3.5 dB.

unless the gain is sufficiently high to lift the amplifier noise above the receiver noise there will be little benefit. The theoretical performance can be evaluated from (21) using the appropriate values for c_1 and c_2. Fig. 16 shows the dominant noise components, with an amplifier internal gain of 25 dB and no optical filtering. As all the components have the same bit rate dependence these relationships are valid for all rates. The main components are the signal-spontaneous beat noise, the spontaneous-spontaneous beat noise, and the receiver noise. As the receiver noise dominates, optical filtering is of little benefit and the increase in the overall sensitivity almost matches the available net amplifier gain. This is illustrated in Fig. 17, which shows measured and theoretical results; sensitivities range from -41 dBm at 140 Mbit/s to -35 dBm at 2 Gbit/s. Testing at higher rates was limited by test equipment. Fig. 18 shows the eye pattern at 1 Gbit/s, measured with a 1-GHz oscilloscope. When the 2.5-nm optical filter was included, the sensitivity at 1 Gbit/s increased by 0.7 dB (to -37.7 dBm) and by 1.5 dB at 140 Mbit/s (to -42.5 dBm) and these results confirmed the dominance of receiver noise. It should be noted that in practical terms the absence of an optical filter means that this type of receiver is very simple to realize. Furthermore, as p-i-n photodiodes and low impedance amplifiers are available with bandwidths in the order of 10 GHz, very wide-band sensitive receivers are possible.

The SNR (equation (21)) shows that by choosing a suitable low noise receiver and a narrow bandwidth optical filter signal-spontaneous beat noise will dominate and the preamplifier performance can approach to within the factor n_{sp}/c_1 of the quantum limit:

$$\text{SNR} = \frac{4PT}{E}\left(\frac{c_1}{n_{sp}}\right) \quad (22)$$

assuming a TW amplifier. This expression illustrates the importance of minimizing the input coupling loss. In our case n_{sp} was estimated (by measurement) as 1.5 and our lowest coupling loss as 2 dB ($c_1 = 0.63$); thus in theory the preamplifier can attain a sensitivity within 3.8 dB of ideal PSK homodyne detection. At 2 Gbit/s, for example, this corresponds to a preamplifier sensitivity of -52.2 dBm, and it is instructive to see how close a practical receiver might come to this limit. We assume an amplifier internal gain of 25 dB, with coupling losses of 2 dB/facet. Tunable 1-nm filters are possible with insertion losses of 2 dB. Fig. 19 shows the preamplifier performance as a function of bit rate for receiver sensitivities of -25 dBm and -30 dBm at 1 Gbit/s; a bit rate dependence of 1.5

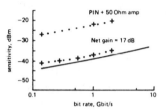

Fig. 17. Experimental and theoretical sensitivities. exp.: +++, theory: ---.

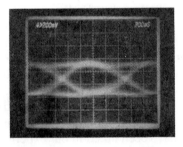

Fig. 18. Eye pattern at receiver output with $2^{15} - 1$ prs at 1 Gbit/s.

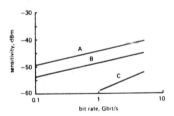

Fig. 19. Performance of optimized preamplifier. Coupling loss = 2 dB/facet; 1-nm filter (2-dB insertion loss). A: receiver sensitivity = -25 dBm at 1 Gbit/s. B: receiver sensitivity = -30 dBm at 1 Gbit/s. C: Quantum limit.

dB/octave is assumed. At 2 Gbit/s, with -30-dBm receiver sensitivity, the overall sensitivity is -47 dBm, which is 9 dB worse than PSK homodyne detection and within 3 dB of FSK detection. Optical preamplifiers, therefore, offer an interesting alternative to high bit rate coherent systems.

Optical preamplifiers look very promising for future wide-band optical systems. The advantage is most evident at high bit rates where the performance of conventional receivers is degraded and in this regime optical filtering may not be necessary. Preamplifier receivers may provide the very large bandwidth required by FSK system detectors [45].

E. Nonlinear Applications

The nonlinear behavior of amplifiers has become of great interest recently [13], [14], [33]. Devices exhibiting nonlinearity and optical bistability are required for a variety of future systems, such as optical processing networks and all optical regenerative repeaters [46]. Optical amplifiers have the particular advantage that they provide gain in association with the nonlinearity.

Uncoated FP amplifiers are generally used in nonlinear applications and are normally biased just below threshold, thus the amplifier is highly resonant and the mode bandwidth is narrow. Fig. 20 illustrates the form of the nonlinear transfer characteristic. At a particular input power the output increases rapidly and hysteresis may be present. Operation is as follows. As the input power increases, the number of photons in the cavity increases and the carrier density decreases (equation (7)). The refractive index is a function of the carrier density and increases with a decrease in the number of carriers. The change in index alters the optical length of the cavity shifting the resonance to longer wavelengths. If the signal wavelength is initially offset to the longer wavelength side of the resonance, then as the input power increases the resonance moves towards the input signal wavelength, increasing the signal gain and generating additional photons; these in turn cause the resonance to move even closer to the signal. This behavior is implicit in (13) which gives the Fabry-Perot gain as a function of the single pass gain and phase, which in turn are functions of the input intensity.

The attraction of the nonlinear amplifier is that nonlinearity occurs at low input powers; for example, bistability is evident at input powers of -30 dBm and this level is suitable for many fiber system applications. Fig. 21 shows how the nonlinear behavior can be used for reshaping optical pulses. In the photographs a triangular pulse at the amplifier input is reshaped to give a square output pulse, ie optical regeneration, and this is part of the requirement for an all optical regenerator. Some means of optical clock recovery and retiming is also required. Fig. 21 also shows the optical hysteresis loop associated with the bistable operation of the amplifier. As the nonlinearity is dependent on a variation of carrier density, the degree of nonlinearity is dependent on the carrier recombination time, which for long wavelength devices is in the order of 2 ns. Thus although the rise and fall transitions are fast (in the order of 10 ps), we have found that bistability vanishes for bit rates above 500 Mbit/s [47]. A particular disadvantage of bistable amplifiers is that they are inherently narrow-band, as they are based on resonant Fabry-Perot amplifiers. Typical bandwidths would be in the order of 5 GHz and this imposes stringent stability requirements on the source. In addition, apart from the limitation of carrier recombination time, intensity modulated signals would be limited by this bandwidth to modulation frequencies in the order of 1 GHz.

Another application is the use of a laser amplifier as an optical gate or switch and this has been investigated in some detail [22]. Optical switches will have an important

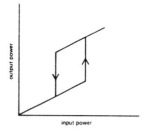

Fig. 20. Nonlinear transfer characteristic showing hysteresis.

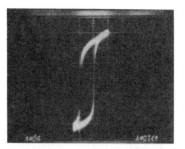

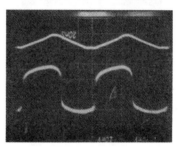

Fig. 21. Pulse shaping and associated hysteresis loop using FP amplifier ($L = 200$ μm, $R = 0.3$) biased at 95 percent of threshold current.

role in future optical networks, and the amplifier switch has particular advantages. By switching the amplifier bias current an optical signal may be suitably routed, as shown in Fig. 22; i.e., the amplifier is an electrically controlled optical routing switch. As well as providing good isolation between input and output in the off state, gain is provided in the on state. This makes lossless switching networks possible. When operated as switches the input power will generally be high, greater than -20 dBm for example, and under these conditions the optimum operating bias current will be considerably less than the threshold current (50-60 percent). With a NTW amplifier, therefore, the switch will have a wide bandwidth, low noise and low sensitivity to fluctuations in signal polarization, temperature and bias current; backward gain will also be low. The performance of the switch shown in Fig. 22 was as follows. Fiber tailed packaged amplifiers were used (Fig. 23); the amplifiers were 1.5-μm DCPBH lasers ($L = 200$ μm), with facet reflectivities of 0.03. For each amplifier the total coupling loss was approximately 10 dB. The switch input power was -17 dBm and the amplifiers were operated at 75 percent of the lasing threshold current; at these values a fiber to fiber gain of 3 dB was achieved, which compensated for the splitter loss. The contrast ratio at the gate output was 32 dB (optical) which is similar to that obtained by Ikeda [22]. The gain

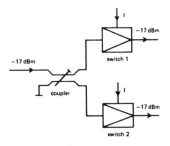

Fig. 22. Amplifiers as electrically controlled optical switches in simple routing network.

Fig. 23. Amplifier fiber tailed package.

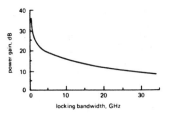

Fig. 24. Gain versus locking bandwidth characteristic of injection locked 1.5-μm amplifier. Output power = 1 mW; $I_{bias} = 1.5 \times I_{th}$.

difference between orthogonal signal polarizations was 3 dB, and the 3-dB gain bandwidth was measured as 60 GHz. As an experiment the gates were connected in tandem and a contrast ratio of 60 dB was measured, thus isolation can be increased by the tandem connection of switches. In operation there was no noticable interaction between devices, even when both devices were on simultaneously on. As discussed above these results will improve with the use of NTW amplifiers.

Semiconductor lasers can be used in an injection locked mode to amplify angle modulated signals. In this case the device is biased above threshold and under the right conditions the free running frequency and phase of the laser will lock to that of a weak incident signal. The locking bandwidth (i.e., the maximum difference between the free running frequency and the input frequency that still enables locking) is related to the amplifier gain. Fig. 24 shows the calculated relationship [23]. For gains in the order of 25 dB, the locking bandwidth is the order of 1 GHz; thus extreme source stability is required and this would appear to preclude the tandem connection of these devices. Injection locked amplifiers may prove useful in narrow-band FSK systems as preamplifiers prior to noncoherent discriminator detection.

IV. Summary and Conclusions

In this paper we have considered the performance and application of semiconductor laser amplifiers to future optical systems. The amplifier is a monomode structure with a wide bandwidth and hence is well suited to monomode fibre systems. Furthermore amplifiers are based on conventional laser diode structures, which are reasonably well understood, the only additional technology is the development of good stable antireflection coatings at long wavelengths. For linear system applications TW or NTW amplifiers are of most interest for the reasons discussed namely: wide bandwidth, tolerance to temperature and bias current fluctuations, low polarization sensitivity, low backward wave amplitude, and high output saturation power. A rough definition of the boundary of FP and NTW amplifiers was defined by the factor $G_s \sqrt{R_1 R_2}$; for NTW operation $G_s \sqrt{R_1 R_2} \leq 0.17$; at a value of 0.17 the peak–peak passband ripple is 3 dB. Therefore, for a minimum useful gain of 25 dB this requires facet reflectivities not exceeding 0.1 percent. The gain envelope bandwidth in such cases will exceed 30 nm, the exact bandwidth depending on the length and material constants. The maximum output saturation power, i.e., the output power at which the gain has fallen by 3 dB, was shown to be in the order of 0 to 5 dBm for NTW amplifiers. The noise at the amplifier output comprises a number of components and their significance depends on the amplifier application. For example when used as a repeater the spontaneous shot noise dominates, as a preamplifier the beat noise components were the most significant; deciding factors are the input power level, the loss between amplifier and photodetector, and the degree of optical filtering. Intrinsically the noise performance of the amplifier is good and for high input powers approaches within 4 dB of the direct detection limit.

The application of linear amplifiers to future systems was discussed with particular reference to use as a gain block, linear repeater, and optical preamplifier. As a linear repeater the maximum effective net gain will be in the order of 20 dB for a NTW device and this has been demonstrated in the laboratory. Single repeater systems present no special problems and are particularly attractive at high bit rates. In multiamplifier systems isolators may be necessary to avoid interaction problems caused by backward traveling waves. The maximum number of repeaters that may be cascaded is likely to be limited by the gain saturation and signal-to-noise degradation caused by amplified spontaneous emission. Without optical filtering and assuming NTW repeaters the number that may be cascaded is approximately 10. Reflections from near end splices will degrade performance and this is an area requiring further study. Preamplifiers were also discussed. The very large gain bandwidth associated with laser amplifiers is used most effectively in the wide-band amplifier optical receiver. The simple combination of an amplifier, p-i-n photodiode and low impedance amplifier, will enable high sensitivities at high bit rates, exceeding the performance of APD or p-i-n diode/FET receivers; such a combination has given a sensitivity of −37 dBm at 1 Gbit/s, without any optical filtering. Practical direct detection preamplifiers receivers should achieve sensitivities close to that of FSK receivers, at bit rates exceeding 1 Gbit/s.

FP amplifiers are preferred for nonlinear applications, because the high internal fields increase the nonlinearity. Bistable behavior is of particular interest at present and bistability has been observed at input powers of 1 μW. Amplifier bistable devices have the attraction of providing gain as well as hysteresis. Particular problems associated with amplifier bistables are that they are necessarily narrow-band devices (bandwidths in the order of 5 GHz) and hence source stability is a problem; also as the nonlinearity depends on changing carrier concentrations the hysteresis vanishes at frequencies corresponding to the carrier recombination time (2 nS).

Using the amplifier as an electrically controlled switch is an important application for future systems. Operating at low gain (and low bias current) the amplifier has many desirable characteristics and provides a good isolation when the bias is removed. Integration of many switches should be possible to provide a switch matrix for signal routing.

ACKNOWLEDGMENT

The author would like to thank his colleagues in BTRL for their support, particularly M. Adams and his colleagues in the Devices Division for helpful discussions, theory, and lasers. Thanks are also due to STL Harlow for their excellent NTW amplifiers and to the Directors of Research and Technology Applications for permission to publish this work.

REFERENCES

[1] H. A. Steinberg, "The use of a laser amplifier in a laser communication system," *Proc. IEEE*, vol. 51, p. 943, 1963.
[2] S. D. Personick, "Applications for quantum amplifiers in simple digital optical communication systems," *Bell. Syst. Tech. J.*, vol. 52, no. 1, pp. 117-133, Jan. 1973.
[3] G. Zeidler and D. Schicketanz, "Use of laser amplifiers in a glass fiber communications system," *Siemens Forsch.-u. Entwickl.*, pp. 227-234, 1973.
[4] Y. Yamamoto, "Characteristics of AlGaAs Fabry-Perot cavity type laser amplifiers," *IEEE J. Quantum Electron.*, vol. QE-16, no. 10, pp. 1047-1052, 1980.
[5] Y. Yamamoto and T. Kimura, "Coherent optical fiber transmission systems," *IEEE J. Quantum Electron.*, vol. QE-17, pp. 897-905, 1981.
[6] Y. Yamamoto, "Noise and error rate performance of semiconductor laser amplifiers in PCM-IM optical transmission systems," *IEEE J. Quantum Electron.*, vol. QE-16, no. 10, pp. 1073-1081, Oct. 1980.
[7] T. Mukai and Y. Yamamoto, "Gain, frequency bandwidth, and saturation output power of AlGaAs DH laser amplifiers," *IEEE J. Quantum Electron.*, vol. QE-17, no. 6, pp. 1028-1034, June 1981.
[8] Y. Yamamoto, "Characteristics of AlGaAs Fabry-Perot cavity type laser amplifiers," *IEEE J. Quantum Electron.*, vol. QE-16, no. 10, pp. 1047-1052, Oct. 1980.
[9] R. M. Jopson, G. Eisenstein, M. S. Whalen, K. L. Hall, U. Koren, and J. R. Simpson, "A 1.55-μm semiconductor-optical fiber ring laser," *Appl. Phys. Lett.*, vol. 48, no. 3, pp. 204-206, Jan. 1986.
[10] F. Favre, D. LE Guen, J. C. Simon, and B. Landousies, "External cavity semiconductor laser with 15 nm continuous tuning range," *Electron. Lett.*, vol. 22, no. 15, p. 796, July 1986.
[11] J. C. Simon, I. Joindot, D. Hui Bon Hoa, and J. Charil, "On line GaAs(GaAl)As laser diode amplifier for single mode fiber communi-cation systems," presented at IOOC'81, San Francisco, CA, 1981, pap. MH2.
[12] T. Mukai, Y. Yamamoto, and T. Kimura, "*S/N* and error rate performance in AlGaAs semiconductor laser preamplifier and linear repeater systems," *IEEE Trans. Microwave Theory Tech.*, vol. MTT-30, no. 10, pp. 1548-1556, 1982.
[13] M. J. O'Mahony, I. W. Marshall, H. J. Westlake, W. J. Devlin, and J. C. Regnault, "A 200-km 1.5-μm optical transmission experiment using a semiconductor laser amplifier repeater," presented at OFC 1986, Atlanta, GA, pap. WE5.
[14] N. A. Olsson, "ASK heterodyne receiver sensitivity measurements with two in-line 1.5 μm optical amplifiers," *Electron. Lett.*, vol. 21, no. 23, p. 1086, Nov. 1985.
[15] I. W. Marshall, M. J. O'Mahony, and P. Constantine, "Optical system with two packaged 1.5 μm semiconductor laser amplifier repeaters," *Electron. Lett.*, vol. 22, no. 5, p. 253, Feb. 1986.
[16] G. Grosskopf, R. Ludwig, and H. G. Weber, "Crosstalk in optical amplifiers for two channel transmission," *Electron. Lett.*, vol. 22, no. 17, p. 900, 1986.
[17] R. P. Webb, "Evaluation of a semiconductor laser amplifier for multiplexed coherent systems," presented at 2nd Int. Tech. Symp. on Optical and Electrooptical Applied Science and Engineering, Cannes, France, Nov. 1985.
[18] D. M. Fye, "Practical limitations on optical amplifier performance," *J. Lightwave Technol.*, vol. LT-2, no. 4, pp. 403-406, Aug. 1984.
[19] M. J. O'Mahony, I. W. Marshall, H. J. Westlake, and W. G. Stallard, "Wide-band optical receiver using traveling wave laser amplifier," *Electron. Lett.*, 1986.
[20] W. F. Sharfin and M. Dagenais, "Room-temperature optical bistability in InGaAsP/InP amplifiers and implications for passive devices," *Appl. Phys. Lett.*, vol. 46, pp. 819-821, 1985.
[21] H. J. Westlake, M. J. Adams, and M. J. O'Mahony, "Measurement of optical bistability in an InGaAsP laser amplifier at 1.5 μm," *Electron. Lett.*, vol. 21, no. 21, pp. 992-993, Oct. 1985.
[22] M. Ikeda, "Tandem switching characteristics for laser diode optical switches," *Electron. Lett.*, vol. 21, no. 6, p. 252, Mar. 1985.
[23] G. N. Brown, "A study of the static locking properties of injection locked laser amplifiers," *Brit. Telecom Tech. J.*, vol. 4, no. 1, Jan. 1986.
[24] H. A. Haus and S. Kawakami, "On the excess spontaneous emission factor in gain-guided laser amplifiers," *IEEE J. Quantum Electron.*, vol. QE-21, no. 1, Jan. 1985.
[25] H. J. Westlake and M. J. O'Mahony, "Gain characteristics of a 1.5-μm DCPBH InGaAsP resonant optical amplifier," *Electron. Lett.*, vol. 21, no. 1, pp. 33-34, 1985.
[26] T. Saitoh, T. Mukai, and O. Mikami, "Theoretical analysis and fabrication of antireflection coatings on laser diode facets," *J. Lightwave Technol.*, vol. LT-3, no. 2, pp. 288-293, Apr. 1985.
[27] G. Eisenstein and R. M. Jopson, "Measurements of the gain spectrum of near-traveling-wave and Fabry-Perot semiconductor optical amplifiers at 1.5 μm," *Int. J. Electron.*, vol. 60, no. 1, pp. 113-121, 1986.
[28] L. D. Westbrook, "Measurements of dg/dN and dn/dN and their dependence on photon energy in 1.5 μm InGaAsP laser diodes," *Proc. Inst. Elec. Eng.*, vol. 133, no. 2, pp. 135-141.
[29] M. J. Adams, J. V. Collins, and I. D. Henning, "Analysis of semiconductor laser optical amplifiers," *Proc. Inst. Elec. Eng.*, vol. 132, no. 1, pp. 58-63, Feb. 1985.
[30] J. W. Wang, H. Olesen, and K. E. Stubkjaer, "Recombination, gain, and bandwidth characteristics of 1.3-μm semiconductor laser amplifiers," in *Proc. IOOC-ECOC'85*, pp. 157-160.
[31] T. Mukai, Y. Yamamoto, and T. Kimura, "Optical direct amplification for fiber transmission," *Rev. Elec. Commun. Lab.*, vol. 31, no. 3, p. 340, 1983.
[32] M. J. Adams, H. J. Westlake, M. J. O'Mahony, and I. D. Henning, "A comparison of active and passive bistability in semiconductors," *IEEE J. Quantum Electron.*, vol. QE-21, no. 9, Sept. 1985.
[33] G. Eisenstein, R. M. Jopson, R. A. Linke, C. A. Burrus, U. Koren, M. S. Whalen, and K. L. Hall, "Gain measurements of InGaAsP 1.5 μm optical amplifiers," *Electron. Lett.*, vol. 21, no. 23, p. 1076, Nov. 1985.
[34] I. D. Henning, M. J. Adams, and J. V. Collins, "Performance predictions from a new optical amplifier model," *IEEE J. Quantum Electron.*, vol. QE-21, pp. 609-613, 1985.
[35] J. C. Simon, "Semiconductor laser amplifier for single-mode optical fiber communications," *J. Optical Commun.*, vol. 4, no. 2, pp. 51-62, 1983.

[36] J. C. Simon and J. L. Favennec, "Comparison of noise characteristics of Fabry-Perot type and traveling wave type semiconductor laser amplifiers," *Electron. Lett.*, vol. 19, no. 8, pp. 288-289, 1983.

[37] C. H. Henry, "Theory of spontaneous emission noise in open resonators and its application to lasers and optical amplifiers," *J. Lightwave Technol.*, vol. LT-4, no. 3, pp. 288-297, Mar. 1986.

[38] N. A. Olsson, "Heterodyne gain and noise measurement of a 1.5-μm resonant semiconductor laser amplifier," *IEEE J. Quantum Electron.*, vol. QE-22, no. 5, pp. 671-676, May 1986.

[39] I. W. Marshall, "Low loss coupling between semiconductor lasers and single mode fiber using tapered lensed fibers," *Brit. Telecom Tech. J.*, vol. 4, no. 2, Apr. 1986.

[40] M. J. O'Mahony, I. W. Marshall, W. J. Devlin, and J. C. Regnault, "Low reflectivity semiconductor laser amplifier with 20 dB fiber to fiber gain at 1500 nm," *Electron. Lett.*, vol. 21, no. 11, pp. 501-502, May 1985.

[41] M. J. O'Mahony, "Semiconductor laser amplifiers as repeaters," presented at IOOC-ECOC 85, Venice, Italy.

[42] M. C. Brain, P. P. Smyth, D. R. Smith, B. R. White, and P. J. Chidgey, "p-i-n-FET hybrid optical receivers for 1.2 Gbit/s transmission systems operating at 1.3 and 1.55 μm wavelength," *Electron. Lett.*, vol. 20, no. 21, pp. 894-895, Oct. 1984.

[43] B. L. Kasper, "Sensitivity limits on direct detection receivers," presented at OFC86, Atlanta, GA, 1986, pap. WJ4.

[44] J. C. Simon, "Light amplifiers in optical communication systems," *Proc. NATO Advanced Study Inst.*, vol. 1, 1984.

[45] M. C. Brain and D. W. Smith, "Coherent optical communication systems," in *Proc. EFOC-LAN '86* (Amsterdam, The Netherlands), p. 51-56.

[46] R. P. Webb, "Experimental demonstration of an all optical regenerator," presented at CLEO '86, San Francisco, CA, pap. ThU8.

[47] H. J. Westlake, M. J. Adams, and M. J. O'Mahony, "Assessment of switching speed of optical bistability in semiconductor laser amplifiers," *Electron. Lett.*, vol. 22, no. 10, pp. 541-543, May 1986.

All-optical repeater

Y. Silberberg

Bell Communications Research, Inc., Holmdel, New Jersey 07733

Received November 29, 1985; accepted March 31, 1986

An all-optical device containing saturable gain, saturable loss, and unsaturable loss is shown to transform weak, distorted optical pulses into uniform standard-shape pulses. The proposed device performs thresholding, amplification, and pulse shaping as required from an optical repeater. It is shown that such a device could be realized by existing semiconductor technology.

A repeater in an optical communication link receives the optical-pulses train that was attenuated and distorted by the transmission medium and transmits an amplified, reshaped version of the same pulses. Such a repeater is usually constructed from an optical detector, followed by an electronic system that performs the thresholding and timing required for the regeneration of the pulses, which are then reemitted by a laser diode. Repeaters are necessary not only in long-haul transmission systems but also in many local-area network schemes, in which attenuation is caused by multiple tappings or by distribution of the same signal into many channels.

In this Letter an all-optical repeater is proposed that can perform the thresholding, amplification, and reshaping without transforming the optical pulses into electrical ones. The proposed device combines saturable optical gain and loss as well as unsaturable loss. It is conceivable that such a device could be fabricated by standard semiconductor technology. The advantages of such a device, if it can be realized, are many. It is simple in structure and thus can be relatively cheap. It can be fast, since it is not limited by the speed of electronics. An all-optical repeater will not destroy the coherence of the optical signal, which can be useful for various communications schemes. Other wavelengths (in the semiconductor case, longer wavelengths) can propagate through the repeater without interacting with it, another useful option for network design. Note, however, that an all-optical repeater does not retime the optical pulses, and thus it lacks one important function of the electronic repeater. In addition, proper operation of the device requires the pulse duration to be much shorter than the interval between pulses.

All-optical devices performing nonlinear optical functions, such as limiters, differential amplifiers, and logic gates, have been proposed before, usually in the context of optical bistability.[1] These devices are usually based on a nonlinear index-of-refraction change. At the present stage these devices still require extensive material research in order to prove useful. Our proposed device, by using optical gain and loss, avoids most of the problems associated with bistable devices. The operation principle behind this device is similar to that of passively mode-locked lasers—a balance in the saturation of the two species. The succcessful operation of mode-locked lasers in general, and in particular the recent achievements in mode locking of semiconductor diode lasers,[2] may indicate that realization of all-optical repeater is quite possible. It was recently brought to the author's attention that a transmission line consisting of laser amplifiers and saturable absorbers was proposed by Zeidler.[3] While his proposal is for a discrete element system, he pointed to the same regenerative nature of the combination.

The analysis presented here follows that of a mode-locked laser with a slow saturable absorber.[4] Assume a medium of length L that contains a homogeneous mixture of gain and loss. The analysis will still be valid if the gain and loss are alternated in separate sections, provided that the optical density of each section is small. We use the rate-equation approximation for the gain and loss and assume that the incoming pulse length is much shorter than the response times of both media. We further assume a plane-wave geometry, which is also a good approximation for a waveguide geometry. Let N_g and N_a be the densities of the inverted population of the gain medium and the ground-state population of the loss medium, respectively. The rates of change of the two species are

$$dN_g/dt = -N_g \sigma_g I/h\nu, \quad (1)$$

$$dN_a/dt = -N_a \sigma_a I/h\nu, \quad (2)$$

where $I(z, t)$ is the light intensity and σ_g and σ_a are the emission and absorption cross sections, respectively. Equations (1) and (2) are trivially solved by

$$N_i = N_i^0 \exp\left(-\sigma_i \int_{-\infty}^{t} I dt/h\nu\right), \quad i = g, a, \quad (3)$$

where N_i^0 are the initial values of N_i, $i = g, a$.

Consider a section of thickness dz. Let E be the pulse energy entering this section. The pulse emerging from this section gains $(N_g^0 - N_g^f)dz$ photons from stimulated emission and loses $(N_a^0 - N_a^f)dz$ photons by absorption, where $N_i^f, i = a, g$ is the value of N_i after the passage of the pulse. The incremental gain to the pulse energy $E = \int_{-\infty}^{\infty} I dt/h\nu$ is then

$$dE/dz = N_g^0[1 - \exp(-\sigma_g E)] - N_a^0[1 - \exp(-\sigma_a E)] - \alpha_0 E, \quad (4)$$

where α_0 is the background unsaturable loss. A low-energy pulse will be amplified or attenuated according to

$$(1/E)dE/dz = N_g^0 \sigma_g - N_a^0 \sigma_a - \alpha_0, \quad (5)$$

while for very-high-energy pulses the loss approaches the unsaturable absorption α_0.

If the system is to be useful as a repeater, we should require loss for low-energy pulses below a certain threshold and gain for pulses above this threshold. However, in order to ensure that all the amplified pulses reach the same level, we should require also that the gain stop once the pulse energy reaches a certain final value. The first requirement can be met if the low signal amplification [Eq. (5)] is made negative (i.e., loss) and the absorber is saturating faster than the gain, so that stronger pulses see gain. This can be realized only if $\sigma_a > \sigma_g$. The second condition requires a finite unsaturable loss, i.e., $\alpha_0 > 0$. Figure 1 depicts the incremental gain $(1/E)dE/dz$ from Eq. (4) as a function of input energy. There are two energy values for pulses that are neither amplified nor attenuated as they propagate through such a medium. These energies are the steady-state solutions, and they are given by the intersections of the gain line with the energy axis. The lower of these values is unstable: any perturbation in the energy value will cause it to move further away from the steady-state value. This is the threshold energy, $E_{\rm th}$. Any input pulse with energy below it will be attenuated as it propagates along the medium, while pulses with $E > E_{\rm th}$ will be amplified. If the medium is long, the propagating pulse energy will asymptotically approach the other steady-state value, E_f.

In order to be able to predict the shape of the final pulse, further assumptions are necessary. In particular, a bandwidth-limiting process must be invoked in order to prevent the propagating pulse from becoming infinitely narrow.[4] Note that the effect of a bandwidth-limiting element is to increase the loss for some frequency components of the pulse, and hence it can be expected that the energy values predicted above will be altered. The following analysis is based on the approach developed by Haus,[4] as can be easily adapted to a traveling-wave situation.[5,6] Using the slowly varying amplitude approximation, the wave equation for the field amplitude A of a pulse traveling in the z direction can be written as

$$\Delta \partial A/\partial s + \partial A/\partial z = D\partial^2 A/\partial s^2 + (N_g \sigma_g - N_a \sigma_a - \alpha_0)A/2, \quad (6)$$

where $s = t - z/v$ is the local time in the pulse frame and $\Delta = 1/v - n/c$ is the difference in inverse velocities, with v being the pulse group velocity and c/n the velocity of light in the medium. D is the bandwidth-limiting term. If bandwidth limitation is coming from the shape of a Lorenzian gain, then $D = G/\omega_g$, where G is the peak gain and ω_g is the FWHM of the gain function.[4]

To find the steady-state pulse that propagates along z without attenuation or distortion, we try the solution $A = A_0 \,{\rm sech}(s/\tau)$. The energy in this pulse is $E = 2A_0^2 \tau$. Substituting this pulse shape into Eq. (3) leads to

$$\begin{aligned}N_i &= N_i^0 \exp\{-\sigma_i E[\tanh(s/\tau) + 1]/2\} \\ &= N_i^0 \exp(-\sigma_i E/2)\{1 - \sigma_i E \tanh(s/\tau)/2 \\ &\quad + 1/2[\sigma_i E \tanh(s/\tau)/2]^2 \ldots\}. \end{aligned} \quad (7)$$

Keeping the expansion in Eq. (7) to second order and substituting in Eq. (6) lead to three algebraic equations for terms with different functional forms:

$$D/\tau^2 = 1/2[\sigma_g N_g^0 \exp(-\sigma_g E/2) - \sigma_a N_a^0 \times \exp(-\sigma_a E/2)], \quad (8)$$

$$-D/\tau^2 = 1/8[\sigma_g N_g^0 \exp(-\sigma_g E/2)(\sigma_g E/2)^2 - \sigma_a N_a^0 \exp(-\sigma_a E/2)(\sigma_a E/2)^2], \quad (9)$$

$$\Delta/\tau = 1/2[\sigma_g N_g^0 \exp(-\sigma_g E/2)\sigma_g E/2 - \sigma_a N_a^0 \times \exp(-\sigma_a E/2)\sigma_a E/2]. \quad (10)$$

Equations (8) and (9) can be used to solve for E and τ, and Eq. (10) is then used to find the pulse delay Δ. Note that an equation for the pulse energy can easily be derived from Eqs. (8) and (9), which will be independent of the bandwidth term D. There are two solutions for this equation, corresponding to E_f and $E_{\rm th}$. These energy values are lower then those predicted by Eq. (3) because of the lower effective gain that is due to the finite bandwidth. The stable steady-state value E_f' is shown in Fig. 1. The new threshold value is very close to the original one. The final pulse width is a function of the spectral bandwidth. The delay term reflects the fact that the pulse can be pushed forward or backward in time by the combined effect of saturable absorption and gain.

We have performed simulations of pulse propagation in a medium described by Eqs. (1), (2), and (7). Figure 2 shows the results of these simulations for several values of input energy. Note the attenuation of a low-energy pulse and the amplification of a higher-energy pulse. The parameters for the simulations were the same as those used in Fig. 1. The final energy

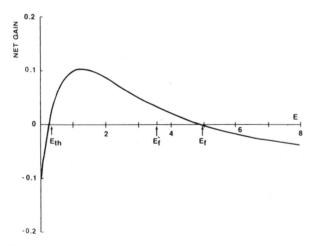

Fig. 1. The net gain as a function of pulse energy. The energy is in units of $h\nu/\sigma_g$. The parameters are $N_g^0 \sigma_g = N_a^0 \sigma_a = 1$, $\alpha_0 = 0.1$, and $\sigma_a/\sigma_g = 2$. The arrow marks the final energy value in the case of a finite-gain bandwidth.

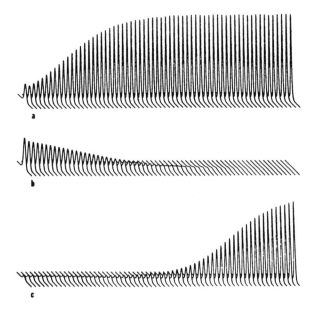

Fig. 2. Simulations of pulse propagation through a medium with parameters as in Fig. 1. The pulse propagates from left to right, and its shape is shown at intervals of $z = 1/N_g\sigma_g$ along the medium. a, Input pulse energy $E = 1.4$, above the threshold value. b, Input pulse energy $E = 0.25$, below threshold. The vertical scale is 10× compared with a and c. c, input pulse energy $E = 0.35$, above but close to threshold.

value obtained is $E_f = 3.65\sigma_g^{-1}$, as predicted by Eqs. (8) and (9).

A question of practical importance is: What is the total length required for sufficient pulse shaping? There is no single answer to this question. The closer the input energy is to E_{th}, the slower the pulse evolution will be. This behavior is demonstrated in Fig. 2c, where the input energy is just slightly above the threshold energy. This is equivalent to the ambiguity of determining the value of pulses close to the threshold value in any kind of digital device. Proper design of the system should minimize the occurrence of such pulses. Note, however, that even for higher-energy pulses the actual gain is the net effect of the gain and loss, which may be of a small magnitude and therefore may require a long propagation distance. From Fig. 1 it can be estimated that the average net gain for that set of parameters is less then 5% of the value of the gain alone. It can be concluded then that long propagation distances, equivalent to many optical densities of small-signal gain, are required for pulse standardization to be obtained.

One system that can supply enough gain is a semiconductor-laser medium. Obviously this is also the most practically important system for communications applications. The device can be fabricated from an antireflection-coated diode laser, which is modified by the introduction of saturable loss along the gain. This loss can be generated, for example, by proton bombardment.[7] To prevent the proton bombardment from affecting the gain mechanism, in may be advantageous to separate gain and loss in different regions in space. This can be done by bombarding short stripes, with widths of a few tens of micrometers, across the gain axis. If those regions are kept short enough that the gain and loss in each section is small, the device will behave as if the gain and loss were distributed. Moreover, by varying the relative density and width of these sections one can have an additional degree of freedom for the design of a practical device.

The semiconductor repeater should be much longer than a common laser diode. This will help in obtaining the high small-signal gain needed for proper operation. A centimeter-long semiconductor device can exhibit a small-signal gain coefficient exceeding 50. Our simulations show that this is enough to produce substantial pulse amplification and shaping. Note that a long device will not be plagued with high amplified spontaneous-emission noise because of the net small-signal loss, which will prevent the buildup of this noise. Moreover, the small-signal loss also relaxes the requirements on the quality of the antireflection coatings and eliminates the need for isolators between stages. Generally, reflections from all surfaces should be kept smaller than E_{th}/E_f to ensure that they will be attenuated by the loss.

The temporal characteristics of the all-optical repeater are determined by the lifetime of the two dynamic species of the loss and the gain. One disadvantage of this device is that the pulse rate should be kept low enough that the gain and loss can recover to their unsaturated values between the pulses. On the other hand, each pulse must be much shorter than these lifetimes. The result is that the bit rate should be much lower than the maximal rate determined by the pulse width. It should be noted, however, that the pulse rate can be still higher than that achievable with electronic repeaters. In the semiconductor device described before it can be expected that the recovery times of the loss and gain can be in the 100-psec range, which means that a few gigabit rates are possible. The width of the standard pulse emitted by the repeater will probably be in the 1–10-psec range. The width is expected to be limited by group-velocity dispersion in the medium and not only by the bandwidth of the system.

References

1. C. M. Bowden, H. M. Gibbs, and S. M. McCall, eds., *Optical Bistability 2* (Plenum, New York, 1984).
2. P. W. Smith, Y. Silberberg, and D. A. B. Miller, J. Opt. Soc. Am. B **2**, 1228 (1985).
3. G. Zeidler, Siemens Forsch. Entwicklungsber. 2, 235 (1973).
4. H. A. Haus, IEEE J. Quantum Electron. **QE-11**, 736 (1975).
5. H. A. Haus, *Fields and Waves in Optical Electronics* (Prentice-Hall, Englewood Cliffs, N.J., 1984), pp. 283–287.
6. D. Haas, J. Wurl, J. Mclean, and T. K. Gustafson, Opt. Lett. **9**, 445 (1984).
7. J. P. van der Ziel, W. T. Tsang, R. A. Logan, R. M. Mikulyak, and W. M. Augustyniak, Appl. Phys. Lett. **39**, 525 (1981).

Guided-Wave Optical Gate Matrix Switch

AKIRA HIMENO, HIROSHI TERUI, AND MORIO KOBAYASHI

Abstract — An optical gate matrix switch, which is made by integrating InGaAsP laser diode gates with high-silica guided-wave splitter and combiner circuits in a hybrid fashion, is proposed and demonstrated. A preliminary experiment for a 4 × 4 matrix switch shows that the switch is capable of more than 400-Mbit/s bandwidth signal switching.

I. INTRODUCTION

MANY types of optical switches have been investigated in an attempt to construct flexible optical transmission networks. These switches include space-division switches, such as an 8 × 8 LiNbO$_3$ guided-wave switch [1], and time-division switches employing optical memories [2]. Since space-division switches offer the potential of broad-band switching functions, they are expected to be applied optical video-communications networks. Such networks require that space-division switches be applied in matrix switch configurations. In addition, point-to-multipoint switching functions are needed to simultaneously distribute video signals from a central station to multiple subscribers. A bulk-type optical gate matrix switch capable of meeting these requirements has already been proposed and demonstrated [3]. However, the bulk-type switch is difficult to handle because of its bulky structure. The development of a compact guided-wave type gate switch is necessary.

This paper describes a compact guided-wave optical gate matrix switch composed of high-silica glass guided-wave optical circuits and semiconductor diode-laser optical gates.

II. SWITCH CONFIGURATION

A general $N \times N$ optical gate switch configuration is shown in Fig. 1. The switch is composed of $1 \times N$ optical splitters, optical gates, interconnections, $N \times 1$ optical combiners, and optical amplifiers.

Any input-output line connection can be performed by opening the appropriate optical gate. When multiple gates connected to a common input line are opened at the same time, a point-to-multipoint connection can be made. Such a point-to-multipoint connection function is readily obtained, because the optical gate switch is free of serious problems (such as impedance matching and increment in load capacitor) that often occur in electronic switching circuits. This type of switch, however, suffers from inherent optical loss due to optical power splitting. To restore the optical signal level due to such a loss, optical

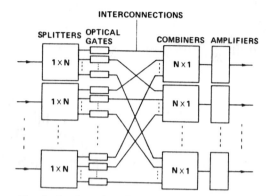

Fig. 1. Configuration of optical gate matrix switch.

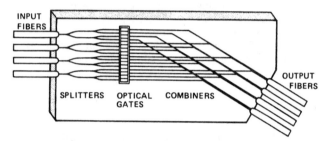

Fig. 2. 4 × 4 guide-wave optical gate matrix switch.

amplifiers are placed after the combiner circuits. In cases where systems using this switch have a sufficiently large loss margin, amplifiers are not necessary.

An experimental guided-wave 4 × 4 switch configuration is shown in Fig. 2 [4]. The optical splitter and combiner circuits including interconnections were formed using high-silica channel waveguides on a common silicon substrate. Specially designed InGaAsP LD gates were used as the optical gates. One LD gate was assembled with the high-silica guided-wave splitter and combiner circuits to determine the basic switching characteristics. Here, the amplifiers are eliminated to evaluate the basic performance of the switch circuit. The operating wavelength was 1.35 μm. The substrate size was 10 × 25 mm. Nonpolarized light was used in the experiment.

III. FABRICATION

A. High-Silica Guide-Wave Optical Circuits

1. High-Silica Channel Waveguide: High-silica waveguides fabricated using a flame hydrolysis deposition have low-loss fiber compatible dimensions, and mechanical stability [5]-[6]. The cross-sectional structure of a channel waveguide is shown in Fig. 3. In the waveguide fabrication process [7], a planar high-silica waveguide with

Manuscript received December 18, 1986; revised April 21, 1987.
The authors are with NTT Electrical Communications Laboratories, Tokai, Ibaraki 319-11, Japan.
IEEE Log Number 8717053.

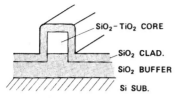

Fig. 3. Cross-sectional structure of high-silica single-mode channel waveguides.

a SiO_2–TiO_2 core/SiO_2 buffer/Si substrate structure was formed by flame-hydrolysis deposition. Next, a ridged-core circuit was formed by the photolithographic process using reactive ion etching. CVD–SiO_2 cladding was then deposited on the upper and side walls of the etched core-ridges. The core size for the splitter circuits was 5 μm wide and 5 μm deep. The core width for the combiner circuits was wider than 5 μm, which will be described later. The cladding thickness for both circuits was 3 μm. The core-cladding refractive index difference was 0.7 percent. The measured propagation loss in the straight channel waveguide was 0.5 dB/cm.

2. Splitter Circuits: To split input optical power uniformly, each splitter circuit was constructed by connecting two 1 × 2 branch circuits in a tree fashion, as shown in Fig. 2. Each branching waveguide was constructed by the combination of two curved waveguides, as shown in Fig. 4(a).

The splitter circuits were made to support a single-mode state in both the vertical and lateral directions in order to match the single-mode operation of the optical gates. Single-mode operation in the splitter circuits was confirmed by observing the output near field pattern for various offset excitation at a given input port. If the circuit has a multimode structure, the near field pattern varies with the offset excitation. On the other hand, for a single-mode circuit, the near field pattern remains constant. It was confirmed by the above-mentioned method that the experimental splitter circuits operated in the single-mode state.

Measured and calculated excess losses for 1 × 4 optical splitter circuits are shown as a function of the curved waveguide radius in Fig. 4(b). The dotted line is the calculated excess loss based on the optical field mismatch in the branching region [8]. Here, the loss in the curved waveguide after branching was ignored. This is because the calculated bending loss was less than 0.1 dB at a radius of greater than 2 mm [9]. The calculated results show that the excess loss decreases as the radius of the curved waveguide increases. However, the loss reduction becomes small at a radius of greater than 20 mm. An excess loss of 2.6 dB was obtained experimentally at a radius of 30 mm. The measured and calculated loss show a similar tendency. The discrepancy between these losses is caused by fabrication errors such as branching point sharpness and curved waveguides. Splitter circuits with a radius of 30 mm were used in the experimental gate matrix switch.

3. Combiner Circuits: The combiner circuits including interconnections were composed of reflection combining and intersecting circuits, as shown in Fig. 5. To elim-

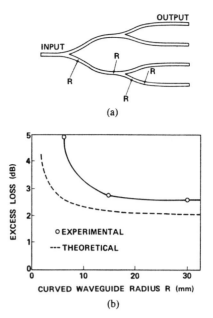

Fig. 4. Optical power splitters. (a) Structure, (b) excess loss for 1 × 4 optical splitters as a function of curved waveguide radius.

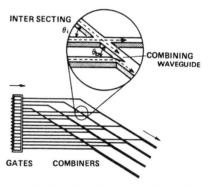

Fig. 5. Configuration of optical combiner circuits.

inate the inherent combining loss of 6 dB in the single-mode 4 × 1 combiner, the combiner circuits were made from waveguides with a multimode state in the lateral direction, while maintaining a single-mode state in the vertical direction. The combining waveguide core width was made wider by 17 μm at each combining point, as shown in Fig. 6(a). The waveguide core width before the first combining point was 17 μm.

The measured combining region losses are shown in Fig. 6(b) for the 4 × 1 combining circuit illustrated in Fig. 6(a). The excess loss originated from imperfect reflectivity and imperfect perpendicularity at the reflecting facet, and light scattering due to the roughness of the reflecting region. The reason for the angle-dependent excess loss is that these causes are a function of the angle [10].

When the bending angle is less than 30°, more than 70 percent of the input power can be combined.

The simple intersecting circuit structure shown in Fig. 7(a) was adopted for ease of circuit fabrication. The measured excess loss versus the intersecting angle is shown with a parameter of combining waveguide width in Fig.

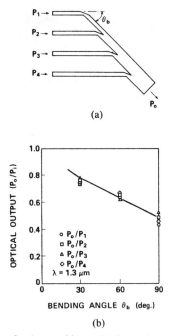

Fig. 6. Optical reflection combiners. (a) Structure (waveguide core width $d_1 = 17$ μm), (b) losses for 4 × 1 optical reflection combiners as a function of reflection bending angle.

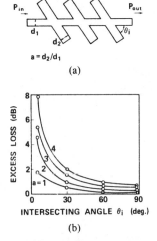

Fig. 7. Intersecting waveguides. (a) Structure, (b) excess loss for intersecting waveguides as a function of intersecting angle.

7(b). Here, the most difficult case is treated in which a waveguide transverses three combining waveguides. When the intersecting angle was 30°, the intersecting excess loss was less than 2 dB. For a 30° angle, the measured crosstalk was −35 dB. Despite the simplicity of the intersecting circuit structure, the crosstalk was very low. These results mean that the waveguide has relatively lower order modes whose light widths in the lateral direction do not spread much in the intersecting region.

In order to obtain low insertion loss in the combining circuits, both intersecting and reflection bending angles were set at 30° after taking the above-mentioned experimental results into account.

The SEM photographs of fabricated high-silica guided-wave optical circuits are shown in Fig. 8.

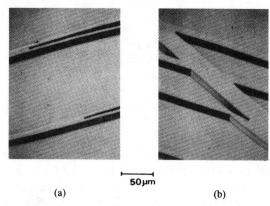

Fig. 8. SEM photographs of fabricated optical circuit, (a) branching region in splitter circuit, (b) combining circuit.

B. Optical Gates

Laser diode (LD) gate was used as the optical gate because of its high on–off extinction ratio and optical amplifier function [11]. A specially designed SCH-LD gate was fabricated for hybrid integration with the guided-wave optical circuits. The structure and SEM photograph of the SCH-LD gate are shown in Fig. 9. The gate structure was a p-InGaAsP cap layer/p-InP cladding layer/p-InGaAsP guiding layer (bandgap: 1.02 eV, thickness: 1 μm)/p-InP carrier confinement layer (1.35 eV, 0.1 μm)/InGaAsP active layer (0.92 eV, 0.2 μm)/n-InP carrier confinement layer (1.35 eV, 0.1 μm)/n-InGaAsP guiding layer (1.02 eV, 1 μm)/n-InP cladding layer. The gate was fabricated by the LPE method. Both facets of the gate received a SiO antireflection coating.

Calculated optical field profiles for the present SCH-LD gate and standard LD gate are shown in Fig. 10. Injection current for both gates was assumed to be 60 mA. It can be seen that, in the SCH-LD gate, both InGaAsP guiding layers are effective in extending the field. The field width is around twice as wide as that in the standard BH-LD gate. This field extension improves the optical field matching between the LD gate and the optical circuit waveguide, which results in reducing the optical coupling loss in the waveguide-gate-waveguide structure.

The input–output power ratio versus injection current in the waveguide-gate-waveguide configuration is shown in Fig. 11. The waveguides were butt-coupled to the gate. Input power was fixed at −8.2 dBm (150 μW). Maximum output power was obtained at an injection current of 60 mA. The input–output power ratio was reduced when the injection current was more than 60 mA. The reason is that the optical field profile in the gate is distorted by a reduction in the refractive index of the active layer due to large injection current.

Optical characteristics were measured at an injection current of 60 mA. The insertion loss was 6.3 dB. The insertion loss was improved by 3.4 dB compared with that of a standard BH-LD gate. The gate on/off extinction ratio was 18 dB. Spontaneous emission power was −29.6 dBm.

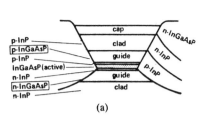
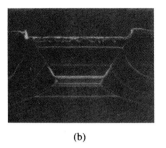

Fig. 9. SCH-LD gate. (a) Structure, (b) SEM photograph of the fabricated SCH-LD gate.

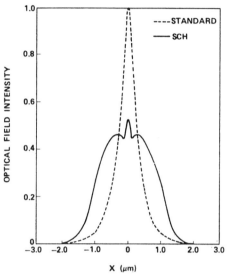

Fig. 10. Calculated optical field profiles in the SCH-LD gate.

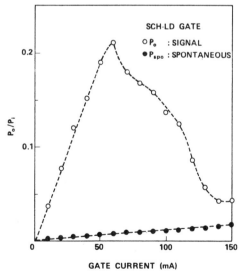

Fig. 11. Input-output power ratio versus injection current in the waveguide-LD gate-waveguide configuration.

C. LD-Gate Integration

The LD-gate, bonded on a Si heat-sink plate, was fixed in the gate-positioning groove, as shown in Fig. 12. The alignment between the LD-gate and the waveguides was carried out by maximizing the light intensity launched from the waveguide ends when the LD-gate was operated

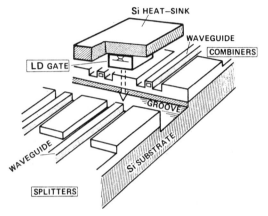

Fig. 12. Hybrid-integration of LD gate.

in the LED mode. The alignment accuracy was ±1 μm. The minimum insertion loss was 8 dB. Comparing to the above-mentioned butt-coupled experiment, the additional 1.7-dB loss was due to the misalignment between the LD-gate and the waveguides.

The spontaneous emission power to the signal ratio was −13 dB for a typical optical power of 0 dBm in an input fiber.

IV. Switching Characteristics and Discussion

Switching characteristics of the 4 × 4 experimental switch, which had one integrated LD gate, were evaluated at a 1.35-μm wavelength. The switch was operated at 60-mA injection current.

The total throughput loss was 23 dB. The loss budget is summarized in Table I. Splitter loss was 8.6 dB including an inherent loss of 6 dB due to power splitting. The combiner loss was between 1.5 and 3 dB. The loss deviation was caused by differences in both the number of intersecting points and the combining waveguide width in intersecting.

Crosstalk arose mainly from the power leaking from the LD gates in the off-state. Crosstalk degradation in the combiner circuits can be ignored because of the low crosstalk which was less than −35 dB, as mentioned above. Since the LD-gate had an extinction ratio of 18 dB, crosstalk in the matrix switch was estimated to be less than −13 dB.

The bandwidth of a switch is determined by its insertion loss, crosstalk, and spontaneous emission power. For 400-

TABLE I
LOSS BUDGET IN EXPERIMENTAL 4 × 4 MATRIX SWITCH

Fiber connections	1.0 dB
Splitters	
Power splitting	6.0 dB
Excess	2.6 dB
Combiners	
Reflection combining	1.5 dB
Intersecting	0-1.5 dB
Optical gates	
Integration	8.0 dB
Others	
Propagation loss, etc.	2.0 dB
Total	max. 22.6 dB

Mbits/s NRZ optical signal transmission, the practical minimum optical power required is −30 dBm. In the worst possible case that both crosstalk and spontaneous emission are white Gaussian noise, the power penalty is less than 1 dB, when the input optical power is 0 dBm [3]. Therefore, the experimental switch is capable of 400-Mbit/s NRZ signal transmission with an optical power margin of about 6 dB. When the multimode fibers are connected to the switch, a 2-km transmission distance is possible with an optical power margin of 4 dB.

Here, matrix size is discussed. Matrix size is determined by insertion loss, crosstalk, spontaneous emission power, and required bandwidth. For a large matrix switch, insertion loss, crosstalk, and spontaneous emission power were estimated based on the experimental results. Based on the estimations, it is possible to construct an 8 × 8 matrix for 400-Mbit/s switch bandwidth. The most serious factor limiting the matrix size is insertion loss. Especially matrix-size dependent losses such as inherent splitting loss in the splitter and intersecting region loss in the combiner were strict.

In order to construct a switch larger than 8 × 8, a multistage arrangement of switch units is necessary. In this case, optical amplification using optical/electrical conversion, an electrical signal amplifier, and electrical/optical conversion is indispensable to restore optical reduction and to convert multimode optical signals to single-mode ones.

V. CONCLUSION

An optical gate matrix switch composed of high-silica guided-wave optical splitter and combiner circuits and InGaAsP SCH-LD gates was proposed and demonstrated. The switch has a point-to-multipoint switching function. The use of the high-silica guide-wave circuits and semiconductor optical gates makes it possible to realize a compact optical switch. The size of the experimental switch was 10 × 25 mm.

Insertion loss for the switch was 23 dB while crosstalk was −13 dB. This level of performance is sufficient to switch 400-Mbit/s optical signals with an optical power margin of 6 dB.

The improved switch element characteristics and a multiple integration of LD gates are now under study.

This type of optical gate switch is expected to have a wide number of applications, especially as an optical video-signal exchanger for optical subscriber systems and as an optical switch for dynamic optical interconnections between computer circuit boards.

ACKNOWLEDGMENT

The authors wish to thank T. Miyashita and M. Kawachi for their helpful discussions. They are also grateful to Y. Yamada and M. Yasu for their technical support.

REFERENCES

[1] P. Granestrand, B. Stoltz, L. Thylen, K. Bergvall, W. Doldissen, H. Heinrich, and D. Hoffmann, "Strictly nonblocking 8 × 8 integrated optical switch matrix," *Electron. Lett.*, vol. 22, pp. 816–818, 1986.
[2] M. Sakaguchi and H. Goto, "High speed optical time-division and space-division switching," in *IOOC/ECOC '85 Tech. Dig.* (Venezia, Italy), vol. 2, 1985, pp. 81–88.
[3] A. Himeno and M. Kobayashi, "4 × 4 optical gate matrix switch," *J. Lightwave Technol.*, vol. LT-3, pp. 230–235, 1985.
[4] M. Kobayashi, A. Himeno, and H. Terui, "Guided-wave optical gate matrix switch," in *IOOC/ECOC '85 Tech. Dig.* (Venezia, Italy), vol. 3, 1985, pp. 73–76.
[5] A. Himeno, M. Kobayashi, and H. Terui, "High-silica optical reflection bending and intersecting waveguides," *Electron. Lett.*, vol. 21, pp. 1020–1021, 1985.
[6] Y. Yamada and M. Kobayashi, "Single-mode fiber connection to high-silica waveguide with fiber-guiding grooves," in *1986 Optoelectronics Conf. Tech. Dig.* (Tokyo, Japan), 1986, pp. 60–61.
[7] Y. Yamada, M. Kawachi, M. Yasu, and M. Kobayashi, "Optical fiber coupling to high-silica channel waveguides with fiberguiding grooves," *Electron. Lett.*, vol. 20, pp. 313–314, 1984.
[8] K. Mitsunaga, M. Masuda, and J. Koyama, "Characteristics of an optical branching waveguide in LiNbO$_3$," *Trans. IECE Japan*, vol. J63-C, pp. 178–185, 1980.
[9] E. A. J. Marcatili, "Bends in optical dielectric guides," *Bell Syst. Tech. J.*, vol. 48, 1969, pp. 2103–2132.
[10] A. Himeno, H. Terui, and M. Kobayashi, "Loss measurement and analysis of high-silica reflection bending optical waveguides," *J. Lightwave Technol.*, to be published.
[11] M. Ikeda, "Laser diode switch," *Electron. Lett.*, vol. 17, pp. 899–900, 1981.

Section 2.4: Hybrid Switching Devices

Hybrid optoelectronic integrated circuit

R. I. MacDonald, D. K. W. Lam, and B. A. Syrett

The distribution of optical signals to a monolithic array of GaAs photoconductors by means of ion-exchanged glass optical waveguides is demonstrated. In this hybrid technique both optical and electronic interconnections of semiconductor elements are achieved through the use of a metallic interconnect layer deposited on the surface of a glass substrate which has a mating waveguide pattern. The low optical loss, ease of fabrication, and low material cost of diffused glass waveguides with such layers permit relatively large optoelectronic circuit boards to be made, in which numerous semiconductor active optoelectronic devices can be included. The device reported here serves as the signal distribution and cross-point switching section of an optoelectronic switch matrix.

I. Introduction

The hybrid integration of optical and semiconductor optoelectronic elements can be advantageous for several reasons. Fabrication of both optical waveguide and optoelectronic active elements in monolithic semiconductors requires processes that are not necessarily compatible. Furthermore, the expense of III-V compound semiconductors makes them inconvenient substrates for large-scale distribution of optical signals, where long waveguides may be necessary. By using separate substrates for passive waveguides and active optoelectronic elements, the fabrication processes for each can be independently optimized. The choice of glass as the waveguide material permits easy fabrication of a high-quality multimode or single-mode optical waveguide. For many applications the multimode waveguide may be preferable to the single mode because modal dispersion is negligible and input coupling is facilitated. Since optical losses are low in glass waveguides, relatively large glass substrates can in principle be used to provide optical interconnections among numerous optoelectronic devices. Furthermore, by depositing a metal interconnection layer on the surface of the waveguide substrate, both electrical and optical interconnection can be made simultaneously to form a kind of optoelectronic circuit board. Such a device is shown diagramatically in Fig. 1. The work reported here was undertaken to develop an integrated signal distribution network for an optoelectronic switch matrix (OSM). Such matrices employ optical transmission methods to distribute broadband signals to cross-points that consist of switched photodetectors.[1] Very high isolation[2] and very fast switching[3] have been demonstrated in monolithic arrays of GaAs photodetectors at frequencies above 1 GHz, and applications in broadband analog switching[4] and the large-scale switching of high-rate digital signals[5] are foreseen for matrices of this type. Full realization of their potential, however, requires integrated optical signal distribution by methods such as we report.

II. Hybrid Integration of Planar Waveguide with GaAs Photoconductive Detector Arrays

A. Photoconductive Detector Arrays

The photoconductive detector array (PCDA) used in these experiments consisted of interdigitated photoconductive elements fabricated in 3-μm thick undoped epitaxial layers of GaAs grown by organometallic pyrolysis on semi-insulating liquid-encapsulated Czochralski grown substrates. Thin highly doped cap layers were used to facilitate ohmic contacts to the channel. The contacts were made with alloyed Au-Ni-Ge. The photoconductors were mesa-isolated, and the cap layer was removed in the light-sensitive region. Each detector had an overall dimension of 100- × 300-μm.

The central 100- × 100-μm active area was contacted by interdigitations 5 μm wide and separated by a channel length of 10 μm. The remaining two 100- × 100-μm metallized portions served as contact pads to which the electrical connections on the glass waveguide substrate were subsequently bonded. The monolithic GaAs PCDA consisted of five photoconductors placed side by side in a center spacing of 200-μm. These arrays were cleaved from the wafer as chips ~1- × 2-mm in dimension.

B. Planar Optical Waveguide

The planar optical waveguides were made by electric field aided ionic replacement of sodium with silver in soda-lime glass. The substrates were 2.54-cm (1-in.) square sections of microscope slides. Waveguide pat-

---

B. A. Syrett is with Carleton University, Electronics Department, Ottawa, Ontario K1S 5B6. When this work was done both of the other authors were with Department of Communications, Communications Research Centre, P. O. Box 11490, Station H, Ottawa, Ontario K2H 8S2; D. K. W. Lam is now with Northern Telecom, Transmission Networks Division, P. O. Box 3511C, Ottawa, Ontario K1Y 4H7; and R. I. MacDonald is now with University of Regina, Faculty of Engineering, Regina, Saskatchewan S4S 0A2.

Received 5 December 1986.
0003-6935/87/050842-03$02.00/0.

terns were formed on them by the electrolysis of ~2000 Å of evaporated silver[6] through an aluminum mask. Graphite electrodes were used for both anode and cathode, and the ionic exchange was carried out at an electric field of 0.3 kV/cm and a temperature of 250°C for 2 h. The high-index silver-doped glass layers so formed were ~10 μm deep, as determined from mode-coupling angles using the reverse WKB method.[6] A postdiffusion of a half hour at 480°C was carried out at a later stage to increase the waveguide depth to 60 μm and reduce the N.A. The waveguide patterns used in the experiment were straight lines 100- or 50-μm wide, and the resulting waveguides had a propagation loss of 1 dB/cm as measured with scattered light at λ = 632.8 nm using a TV camera. The edges of the slides were polished, and the waveguides were excited by butt-coupling with 50-μm core graded-index fibers held in micromanipulators. The total insertion loss of a 2.54-cm long section of planar waveguide which accounted for the input and output coupling efficiency and the propagation loss was ≈10 dB. The major loss occurred at the input coupling due to a mismatch between the shapes and sizes of the fiber core and the diffused waveguide. This can be improved with better matching.

C. Hybrid Optoelectronic Integrated Circuit

To construct the hybrid optoelectronic integrated circuit shown in Fig. 1 it is necessary to couple a portion of the optical power in the waveguide into each detector. Since the detectors are relatively large and can be located very close to the surface of the waveguiding substrate, the output coupling need not be very directional. Indeed simple isotropic scattering may be suitable if the loss of half of the radiation into the substrate is tolerable. For optical switches designed for digital signals, such a 3-dB loss is in many cases acceptable.[4] Glass substrates with scattering couplers are simpler to produce than structures using directional etching to obtain facets in semiconductors for use as mirrors[7,8] and are particularly suitable when only small proportions of the propagating power are to be coupled out, as in a distribution network.

Scattering perturbations are difficult to produce in glass by normal photolithographic processes, which usually involve some form of etching. Glass tends to undercut badly in wet etches, leaving edges with slopes too shallow to decouple light from the waveguide, particularly since the modes are strongly bound at the high-index difference air–glass interface. Dry etching of several microns of bulk glass requires special processes and is slow. We have produced perturbations in the waveguides used for the present demonstration by a masked abrasion technique. The mask was a thick composite structure to withstand the abrasive. To produce it the silver-coated aluminum mask remaining after the ion exchange step was recovered with aluminum, and 100-μm square openings were made in this layer in the regions where perturbations were required. (These openings exposed the surface of the glass because the electrolysis of silver in the waveguide

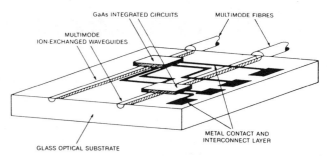

Fig. 1. Diagram of a hybrid optoelectronic integrated circuit.

Fig. 2. Assembled hybrid optoelectronic integrated circuit. This view is through the glass substrate and shows the perturbed areas with the photoconductors bonded over them and the metallized conductive pattern on the glass surface to which the photoconductors are bonded.

regions was complete.) The resulting composite metal mask was then covered with a thick layer of positive photoresist (AZ1375) and exposed from the rear (i.e., through the glass substrate) to produce a polymer mask over the metal mask after development. The polymer was hardened at 150°C for a half hour. The abrasion step was then carried out by sandblasting the masked substrate using 0.3-μm aluminum polishing grit. The resulting pitting provided the required waveguide perturbation. To facilitate the alignment of the semiconductor chips, the abraded areas were first delineated with a light HF etch. The HF bath also smoothed out some of the damages.

Electrical connections on the glass substrate were made by etching a sputtered film of Au–Ti. The GaAs photoconductor arrays were aligned to contact pads on this metallization layer and bonded to it by means of conductive epoxy. A view through the substrate of the final assembly is shown in Fig. 2. The photoconductors, perturbations, and electrical connections can be seen. The waveguide is also faintly visible.

The efficiency of the perturbed multimode waveguide in distributing optical power to the PCDA was determined in two steps. Prior to bonding the PCDA chip to the waveguide surface, a 100-μm core optical fiber was used to monitor the power that was scattered from the perturbations. With a 1-mW input power, three perturbations at an interspacing of 300-μm registered an output power of 2.5, 2.5, and 2.0 μW. It is estimated that the coupling efficiency from the perturbation to the fiber is ~80%, or −1-dB coupling loss is incurred. The waveguide was earlier measured to have a 10-dB total insertion loss and a propagation loss of 1 dB/cm. The perturbations are near the center of the 2.54-cm square glass slide. Therefore, there is ~1-cm travel from the edge to the perturbations. The actual coupling efficiency of the propagating signal in the waveguide to the perturbation is calculated as follows:

Total loss from input to coupling point:

$$10 \log(2.5 \, \mu W/1mW) = -26 \text{ dB}.$$

Various losses incurred:
(1) Coupling from monitor fiber = 1 dB.
(2) Waveguide insertion loss with 1-cm travel: From total insertion loss with 2.5-cm travel = 10 dB and propagation loss = 1 dB/cm, the waveguide insertion loss with 1-cm travel = $10 - (2.5 - 1) \times 1 = 8.5$ dB.

Coupling loss from the waveguide to the perturbation is $(-26 + 1 + 8.5)$ dB = -16.5 dB.

The uniformity in the coupling efficiency is within 20% from the three measurements of 2.5, 2.5, and 2.0 μW.

After the PCDA chip was bonded to the waveguide surface, the coupling efficiency was again measured by comparing the direct responsivity when coupling the signal through the backside of the substrate with that when coupling from the waveguide. Direct responsivity was measured at 0.3 A/W. Responsivity via coupling from the waveguide was measured at $0.5 \times 10E-4$ A/W or a -28-dB total loss from the input to the coupling point. The extra 2-dB loss (recall -26-dB loss when using a monitor fiber) is attributed to the alignment between the PCDA and the waveguide and the physical separation between them. The monitor fiber used was end-butted into the perturbations. The frequency response of the PCDA via the indiffused waveguide is shown in Fig. 3. The response is less smooth than when the PCDA is mounted on a microstripline circuit. This is because the metallic contact pattern used on the waveguide substrate does not have a 50-Ω characteristic impedance. Figure 3 indicates a bandwidth of close to 1 GHz even in this condition.

The weak coupling of -16.5 dB achieved in the experiment here is suitable for an optoelectronic distribution network (ODN) in a very large-scale optoelectronic switch matrix (OSM), where there are a large number of coupling points. Improvement on the throughput coupling efficiency can be made by better matching at the input. Better coupling from the waveguide perturbation to the PCDA can be achieved by finding a suitable chemical etchant that does not undercut badly in glasses. Chemical wet etching produces a much smoother perturbed surface than what mechanical abrasion can produce. The coupling efficiency for chemical etching can thus be expected to be higher.

III. Conclusion

The experimental results here demonstrate the feasibility of the integrated optics approach to the ODN in a large-scale OSM. The approach can also be ap-

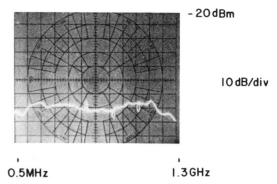

Fig. 3. Frequency response of the PCDA via the optical waveguide coupling.

plied to other optoelectronic integrated circuit (OEIC) assemblies where both optical and electrical interconnections are necessary. The advantage of this hybrid approach is that the processing steps for the dielectric, and the semiconductor substrates are separated from each other. No conflicting steps were encountered in the experiment. This approach also allows the attachment of several semiconductor chips on a single dielectric substrate. The dielectric substrate thus acts as an OEIC breadboard. This eliminates the need for a large semiconductor substrate which is much more expensive than a dielectric substrate. Finally, electrooptic and acoustooptic effects can be introduced as well by using the appropriate substrate, say lithium niobate or quartz. Complex and modular OEICs as in an optical repeater, an optical modulator and optoelectronic signal processors can be realized.

References

1. R. I. MacDonald and D. K. W. Lam, "Optoelectronic Switching— Recent Developments," Proc. Soc. Photo-Opt. Instrum. Eng. 24, 220 (1985).
2. D. K. W. Lam and R. I. MacDonald, "Crosstalk Measurements of Monolithic GaAs Photoconductive Detector Arrays in the GHz Region," Appl. Phys. Lett. 46, 1143 (1985).
3. D. K. W. Lam and R. I. MacDonald, "Fast Optoelectronics Crosspoint Electrical Switching of GaAs Photoconductors," IEEE Electron Dev. Lett. **EDL-5**, 1 (1984).
4. R. I. MacDonald, R. H. Hum, R. Kuley, D. K. W. Lam, and J. Noad, "Experimental Demonstration of an Optoelectronic Switch Matrix for Satellite-Switched TDMA Signals in the 0.3–4.0 GHz Band," CRC Tech. Note 717, Department of Communications, Ottawa (June 1984).
5. R. I. MacDonald, "Optoelectronic Switching in Digital Networks," IEEE J. Selected Areas Commun. **SAC-3**, 336 (1985).
6. J. M. White and P. F. Heidrich, "Optical Waveguide Refractive Index Profiles Determined from Measurement of Mode Indices: A Simple Analysis," Appl. Opt. **15**, 151 (1976).
7. Y. Kokubun, T. Baba, and K. Iga, "Silicon Optical Printed Circuit Board for Three-Dimensional Integrated Optics," Electron. Lett. **21**, 508 (1985).
8. R. Trommer, "Monolithic InGaAs Photodiode Array Illuminated Through an Integrated Waveguide," Electron. Lett. **21**, 382 (1985).

4 × 4 OEIC Switch Module Using GaAs Substrate

T. IWAMA, T. HORIMATSU, MEMBER, IEEE, Y. OIKAWA, K. YAMAGUCHI, MEMBER, IEEE, M. SASAKI, MEMBER, IEEE, T. TOUGE, MEMBER, IEEE, M. MAKIUCHI, H. HAMAGUCHI, AND O. WADA, MEMBER, IEEE

Abstract—A compact 4 × 4 optical switch module consisting of a monolithic 4-channel OEIC receiver chip, a 4 × 4 GaAs IC chip, and a 4-channel OEIC transmitter chip has been developed for the first time. Our module offers good performance, without an optical loss, a bandwidth of more than 600 MHz, and a crosstalk between neighboring channels of less than −20 dB. It has a good switching and distributive performance for high speed optical input signals of 560 Mbit/s. The switch module is attractive for use in high data-rate optical communication systems, particularly in local area networks, CATV systems, and intra-office links.

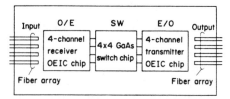

Fig. 1. Block diagram of optoelectronic switch.

I. INTRODUCTION

FIBER-optic communication's technologies are progressing mainly for applications in trunk lines, such as 400-Mbit/s and 1.6-Gbit/s systems. Applications to non-trunk-line systems, such as local area networks and CATV systems, have been expanding. Of these applications, wide-band optical networks for relaying high-definition images are attracting attention. Multichannel optical switches are key devices in such networks. From this point of view, we have proposed a 4 × 4 optical switch composed of optoelectronic integrated circuits (OEIC's) [1]. OEIC's are integrated circuits in which optoelectronic elements such as lasers and detectors, and electronic circuits are monolithically integrated on a single semiconductor substrate. Many research groups have been developing OEIC technologies because OEIC's have potential advantages for giving high performance, particularly in terms of high speed and low noise, and for constructing compact and reliable systems that can be assembled simply at low cost [2]. Several technologies that are fundamental and unique to optoelectronic integration had to be developed for fabricating OEIC's. A number of OEIC's using both GaAs and InP material systems [3]–[11] have already been demonstrated. We have also demonstrated GaAs-based photodiode/preamplifier OEIC [13], laser/driver OEIC [14], and multichannel integration of these OEICs [15], [16].

This paper describes the design of the OEIC switch module in Section II, and the construction of the whole module, its basic characteristics, and the results of the transmission experiments in Section III.

Manuscript received October 2, 1987; revised December 15, 1987.
The authors are with the Optoelectronic Systems Laboratory, Fujitsu Laboratories Ltd., Nakahara-ku, Kawasaki 211, Japan.
IEEE Log Number 8719350.

II. 4 × 4 OEIC SWITCH MODULE DESIGN

A. Outline of Switch

The block diagram of the optical switch discussed in this study is shown in Fig. 1. We have designed a 4 × 4 OEIC switch composed of three different chips: a receiver OEIC array (O/E), an electronic switch circuit (SW), and a transmitter OEIC array (E/O) using GaAs-based material systems. In this OEIC switch, incident optical signals are converted to electric signals and amplified in the O/E section. Signals are electrically switched in the SW section, and these signals are then converted to optical signals in the E/O section and transmitted through the fiber array. The use of a GaAs system in SW section has the advantage of using high-speed GaAs IC technology. This type of OEIC switch has the advantages of good yield of device, flexibility of construction (can be set amplifier between O/E and E/O for example), and elimination of the effect of DC-level drift. So this type of switch is a very attractive component for use in complex optical-fiber networks. Using this type of switch the handling of multichannel signals from and to different remote locations can be possible with a relatively compact system construction.

B. Characteristics of the Three Kind of Chips

1. Transmitter OEIC Chip: The circuit diagram of a one-channel transmitter OEIC is shown in Fig. 2 [16]. A quantum-well laser, a photodiode for monitoring the laser power, a driver circuit composed of three MESFET's, and a 50-Ω matching resistor are included in each channel. The laser and the monitoring photodiode are embedded in a moat formed on a semi-insulating GaAs substrate. The laser has one microcleaved facet [16], and one substrate-cleaved facet. The driver circuit is formed directly on the semi-insulating substrate. FET Q3 works as a constant current source, and FET's Q1 and Q2 constitute a differential circuit to modulate the laser current. The bias voltage supply and ground lines are interconnected in com-

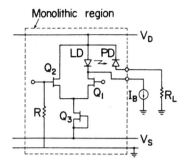

Fig. 2. Circuit diagram of one channel of a LASER/MONITOR/DRIVER transmitter OEIC.

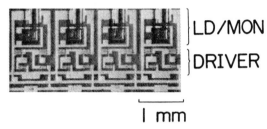

Fig. 3. Surface photomicrograph of four-channel LASER/MONITOR/DRIVER transmitter OEIC. The chip measures 4 × 2 mm.

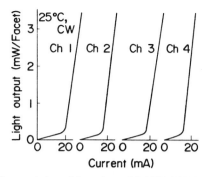

Fig. 4. DC-characteristics of four-channel LASER/MONITOR/DRIVER transmitter OEIC.

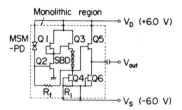

Fig. 5. Circuit diagram of one channel of an MSM-PD/AMP receiver OEIC.

mon through four channels. Fig. 3 shows a surface photomicrograph of the four-channel transmitter OEIC chip. The lasers are 1 mm apart and the chip measures 4 mm × 2 mm.

Fig. 4 shows the characteristics of laser light output versus current measured under CW operation of the transmitter OEIC. A low-threshold current (15 to 21 mA) and high quantum efficiency (50-60 percent/facet) are obtained for each laser. The laser emission wavelength is 0.83 μm. The efficiencies of the monitor current range from 1.8 to 3.0 μA/mW among the four sample photodiodes. The FET characteristics are very uniform over the wafer, and the transconductance is typically 26 mS ±1 mS for a 500-μm channel width. The conversion ratio of the input voltage to output power exceeds 6 mW/V. This conversion ratio agrees fairly well with the measured values of the laser quantum efficiency and the FET transconductance. A good eye opening has been observed with the application of 1.5-Gbit/s pseudorandom input signals [14]. The crosstalk between neighboring channels is −25 dB at 600 MHz [14]

2. Receiver OEIC Chip: The circuit diagram of the receiver OEIC for a single channel is shown in Fig. 5. A metal-semiconductor-metal (MSM) photodiode and a feedback amplifier composed of six FET's, four Schottky diodes, and two resistors, are included in each channel. We have proposed the use of an MSM photodiode in the receiver OEIC [15]. The GaAs MSM photodiode not only has a simple, planar structure, but full processing compatibility with GaAs MESFET's. The MSM photodiode capacitance is less than 0.2 pF. The transconductance of the FET is typically 73 mS/mm, and the pinchoff voltage is about 2.5 V. The dark current of the MSM photodiode is well below 1 μA in bias regions below 10 V. The sensitivity estimated from measurements of the photocurrent is 0.20-0.25 A/W at with a 6-10 V bias. A good eye opening is observed when 2-Gbit/s NRZ pseudorandom optical pulses are fed into the MSM photodiodes of the receiver circuits [15]. The results of transimpedance and equivalent input noise current density for the frequency range from 0.1 MHz to 1.5 GHz are shown in Fig. 6. The transimpedance is 53.2 dB · Ω and the minimum value of the equivalent input noise current density is 4.5 pA/$\sqrt{Hz}$. The crosstalk between neighboring channels is −40 dB at 600 MHz. A surface photomicrograph of the fabricated four-channel receiver OEIC chip is shown in Fig. 7. Fifty-two elements are integrated in one chip. Four MSM photodiodes are distributed at a constant interval of 1 mm, the chip size measures 4 × 1.2 mm. The average circuit sensitivity for the four channels at a 50 μW light power is 100 mV/mW, and the uniformity among channels is estimated to be as small as ±9 percent.

3. GaAs IC Chip: The GaAs IC switch consists of a matrix switch, an address and decoder, an output buffer, and bias circuits [18]. The conventional technique of implanting ions into a semi-insulating GaAs substrate is used to fabricate the FET's. To achieve good isolation characteristics, the circuit design of the unit switching cell shown in Fig. 8 was adopted. The cell consists of three depletion-mode FET's: Q1, Q2, and Q3. The chip measures 2.3 × 2.7 mm and incorporates 788 elements, including MESFET's, Schottky diodes, and resistors. The maximum data rate available with present data generators is 2 Gbit/s [17]. The crosstalk between neighboring channels is −40 dB at 600 MHz [17].

C. Multichannel Optical Coupling and Package Scheme

In packaging the OEIC chips, simple connection of the fiber feeds to the chip using a small optical coupling struc-

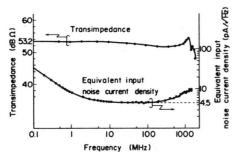

Fig. 6. Transimpedance and equivalent input noise current density of a MSM/AMP chip.

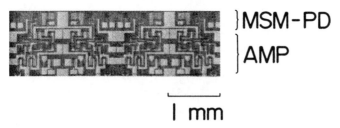

Fig. 7. Surface photomicrograph of four-channel MSM-PD/AMP receiver OEIC. The chip measures 4 × 1.2 mm.

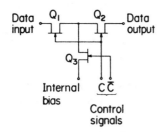

Fig. 8. Schematic of GaAs IC switch cell.

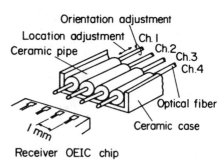

Fig. 9. Coupling scheme of four-channel fiber array for receiver OEIC chip side.

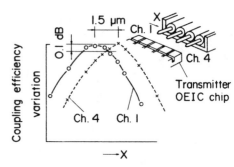

Fig. 10. Coupling efficiency characteristics of the fiber array and the transmitter OEIC chip.

TABLE I
OPTICAL COUPLING CHARACTERISTICS OF THE TRANSMITTER OEIC CHIP AND THE RECEIVER OEIC CHIP. (Loss increase is less than 0.5 dB.)

Item		Tapered hemispherical fiber (Transmitter)	Slanted-end fiber (Receiver)
Coupling efficiency		>50 %	>90 %
Positioning tolerance	X axis	8.6 μm	56 μm
	Z axis	19 μm	46 μm
Curvature of the fiber end		30 ~ 40 μm	—
Slanted-end angle		—	38 degree
Fiber		50/125 μm G.I	

ture is essential for reducing size and cost. The tapered hemispherical-end fibers are coupled to the transmitter OEIC chip and the slanted-end fibers are coupled to the receiver OEIC chips. Both coupling methods are simple and have low coupling losses, thereby providing the most suitable coupling for compact modules.

Optical coupling characteristics of the transmitter OEIC chip and the receiver OEIC chip are shown in Table I. A 50-μm-core graded-index fiber is used. In the case of tapered hemispherical-end fibers, the coupling loss is relatively unaffected by curvature of the fiber end in the range 10–40 μm, and is a minimum at around 20 μm [18].

The distance between the tapered-hemispherical-end fibers and the OEIC chip is about 50 μm, and the positioning tolerances from the optimum position are 8.6 μm in the transverse and 19 μm along the optical axis. The loss change is less than 0.5 dB within these displacements. In the case of slanted-end fibers, the slanted angle is 38° for total internal reflection at the polished surface [19]. The positioning tolerance is large, and displacements of 40–60 μm increase the coupling loss by only 0.5 dB. This large tolerance simplifies the coupling procedure. A minimum coupling loss of 1.2 dB is obtained for the transmitter side and 0.2 dB for the receiver side.

A 4-channel fiber array for the 4 × 4 OEIC switch module is shown in Fig. 9. Using a microscope, we can adjust the location and orientation of each fiber in the ceramic pipe array. Using the ceramic pipe array enables us to locate and orient each fiber. The diameter of the ceramic pipe is 1 mm. The tolerances of the diameter of the ceramic pipe and the tolerances in the eccentricity of the hole to insert the fiber and the diameter of the ceramic pipe are within 1 μm. The coupling efficiency of the fiber array and the transmitter OEIC chip are shown in Fig. 10. These characteristics were obtained by scanning the fiber array along the X axis. Misalignments between channels 1 to 4 for an interval of each LD is only 1.5 μm. The coupling efficiency degradation is only 0.1 dB, so we achieved a coupling efficiency of more than 50 percent. These figures confirm that the proposed fiber array fully satisfies our requirements. After adjusting the optical coupling between optical elements and fibers using this fiber array, we fixed the fiber array to the package with an adhesive.

III. Construction of the Whole Module and its Characteristics

A. The Whole Module

Electrical connection structure of 4 × 4 GaAs OEIC switch module is shown in Fig. 11. Three chips are connected through coupling capacitor to eliminate the effects of dc-level drift. A 200-Ω load resistor was used at the output of the GaAs switch chip to get the enough input voltage and enough bandwidth for the transmitter OEIC. A schematic figure of the 4 × 4 OEIC switch is shown in Fig. 12. The three chips are bonded separately onto the carriers. The dc bias line of each chip is connected to bypass chip capacitor on the same stem. In this OEIC switch, incident optical signals are converted to electric signals and amplified in the 4-channel MSM/AMP array chip. Signals are electrically switched in the 4 × 4 GaAs switch chip, and these signals are then converted to optical signals in 4-channel LD/MONITOR/DRIVER array chip and transmitted through the fiber array.

The OEIC switch module is shown with its cover off in Fig. 13. The module measures 68 × 79 × 21 mm, and is almost one order of magnitude smaller than the module using conventional discrete devices [20]. The pins are for the power supply and switching control signals. There is a bias circuit beside the OEIC chips in the module.

B. Module Characteristics and Results of Transmission Experiments

The frequency response and crosstalk between neighboring channels are shown in Fig. 14. In this trace, the bandwidth is 600 MHz, and the crosstalk between neighboring channels is less than −20 dB. This bandwidth is limited by the interface of the GaAs IC switch and the LD/driver chip. It is wide enough for 560-Mbit/s transmission, and the crosstalk figure of less than −20 dB is good enough for an optical switch. This switch module has no loss. Eye diagrams of the 560-Mbit/s NRZ optical output of the module, whose switching function is a distribution mode, (Input: channel 1, Output: channels 1 to 4) are shown in Fig. 15. A good eye opening is obtained with minimized intersymbol interference. The ununiformity of the optical power output results from variations of the conversion ratio of the input voltage to output power of the transmitter OEIC chip. In Fig. 16, good switching characteristics are confirmed by using 560-Mbit/s NRZ signals and clock signals. In this experiment, both the input and output ports are channel 2 and channel 3.

The transmission characteristics with and without the crosstalk of the switch module are shown in Fig. 17. These characteristics were measured to evaluate the effects of crosstalk. First, 560-Mbit/s signals are applied to the input of channel 1, and then pass to the output. Then, 560-Mbit/s signals are also applied to the input of channel 2, and also pass to the output. The channel 1

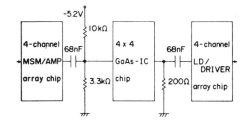

Fig. 11. Electrical connecting structure of 4 × 4 OEIC switch module.

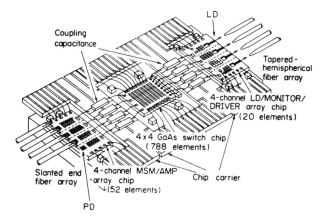

Fig. 12. Structure of 4 × 4 OEIC switch module with four-channel array chips.

Fig. 13. 4 × 4 OEIC switch module.

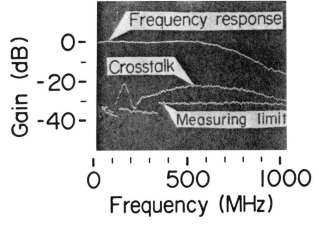

Fig. 14. Frequency response and crosstalk between neighboring channel.

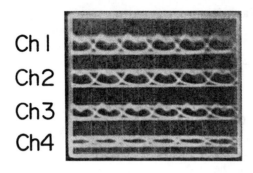

Fig. 15. Eye diagrams of 560-Mbit/s optical output of module. The switching function is a distribution mode (Input: channel 1, Output: channels 1 to 4).

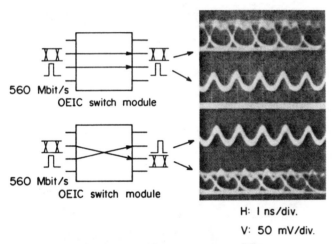

Fig. 16. Switching characteristics of module.

TABLE II
CHARACTERISTICS OF 4 × 4 OEIC SWITCH MODULE

Wavelength	0.83 μm
Bandwidth	600 MHz
Crosstalk	< −20 dB
Optical loss	0 dB
Power consumption	3.7 W
Module size	68 × 79 × 21 mm

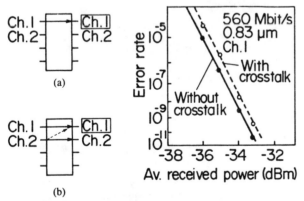

Fig. 17. Transmission characteristics of the module (a) without and (b) with crosstalk.

crosstalk is caused by the channel 2 signals. The solid line corresponds to signals without crosstalk, and the dotted line to signals with crosstalk. In this graph, the degradation due to crosstalk from the neighboring channel is less than 0.5 dB. This confirms that the crosstalk influence is very small. The characteristics of the 4 × 4 GaAs OEIC switch module are listed in Table II. By obtaining a greater transimpedance of the receiver OEIC chip and the conversion ratio of the input voltage to output power, an optical gain can be expected. The feasibility of optical gain is an advantage characteristic to the OEIC approach of constructing not only optical switches but other advanced optical components. The power consumption is 3.7 W, almost half that of the module using conventional discrete devices.

IV. CONCLUSIONS

A compact 4 × 4 OEIC switch module consisting of two types of 4-channel monolithic OEIC chips and a 4 × 4 GaAs IC chip, has been developed. Very precisely aligned 4-channel fiber array has been achieved by using a ceramic pipe array. With this fiber array, high coupling efficiency of each LD and MSM-PD has been accomplished. The module offers a good performance, a bandwidth of more than 600 MHz, crosstalk between neighboring channels of less than −20 dB, and shows good switching performances for optical input signals at 560 Mbit/s without a insertion loss. Because optoelectronic integration makes the module compact and offers high speed, the fields of application for optical transmission systems are expected to widen further through this approach.

ACKNOWLEDGMENT

The authors would like to thank Dr. Takanashi, Dr. Nakagami, and Dr. Sakurai for their continued encouragement. They would also like to thank Mr. Sanada, and Mr. Miura for OEIC processing and Mr. Ohtsuka for useful suggestions on module structure.

References

[1] T. Iwama, Y. Oikawa, K. Yamaguchi, T. Horimatsu, M. Makiuchi, and H. Hamaguchi, "4 × 4 GaAs OEIC switch module," in *Tech. Dig. OFC/IOOC '87* (Reno, NV), Jan. 1987, p. 161.

[2] O. Wada, T. Sakurai, and T. Nakagami, "Recent progress in optoelectronic integrated circuits," *IEEE J. Quantum Electron.*, vol. QE-22, no. 6, pp. 805–821, June 1986.

[3] C. P. Lee, S. Margalit, I. Ury, and A. Yariv, "Integration of an injection laser with a Gunn oscillator on a semi-insulating GaAs substrate," *Appl. Phys. Lett.*, vol. 32, no. 12, pp. 806–807, June 1978.

[4] R. F. Leheny, R. E. Nahory, M. A. Pollack, A. A. Ballman, E. D. Beebe, J. C. Dewinter, and R. J. Martin, "Integrated $In_{0.53}Ga_{0.47}As$ p-i-n FET photoreceiver," *Electron. Lett.*, vol. 16, no. 10, pp. 353–355, May 1980.

[5] H. Matsueda, S. Sasaki, and M. Nakamura, "GaAs optoelectronic integrated light sources," *J. Lightwave Technol.*, vol. LT-1, no. 1, pp. 261–269, Mar. 1983.

[6] J. K. Carney, M. J. Helix, and R. M. Kolbas, "Gigabit optoelectronic transmitters," in *Tech. Dig. GaAs IC Symp.* (Phoenix, AZ), Oct. 1983, pp. 48–51.

[7] K. Ohtsuka, H. Sugimoto, Y. Abe, and T. Matsui, "Monolithic integration of InGaAs/InP p-i-n PD with MISFET on stepless surface," in *Tech. Dig. 1st Optoelectron. Conf.* (Tokyo, Japan), July 1986, pp. 44–45.

[8] K. Kasahara, T. Terakado, A. Suzuki, and S. Murata, "Monolithically integrated high speed light source using 1.3 μm wavelength DFB-DC-PBH laser," in *Tech. Dig. 5th IOOC/11th ECOC* (Venezia, Italy), Oct. 1985, pp. 295–298.

[9] S. Hata, M. Ikeda, Y. Noguchi, and S. Kondo, "Monolithic integration of an InGaAs PIN photodiode, two InGaAs column Gate FETs and an InGaAsP laser for optical regeneration," in *Proc. 17th Conf. Solid State Devices and Materials* (Tokyo, Japan), Aug. 1985, pp. 69–82.

[10] A. Yariv, "The beginning of integrated optoelectronic circuits," *IEEE Trans. Electron. Devices*, vol. ED-31, no. 11, Nov. 1984, pp. 1656–1661.

[11] C. S. Hong, D. Kasemset, M. E. Kim, and R. A. Malano, "Integrated quantum-well-laser transmitter compatible with ion-implanted circuits," *Electron. Lett.*, vol. 20, no. 18, pp. 733–735, Aug. 1984.

[12] O. Wada, H. Hamaguchi, S. Miura, M. Makiuchi, K. Nakai, T. Horimatsu, and T. Sakurai, "AlGaAs/GaAs p-i-n photodiode/preamplifier monlithic photoreceiver integrated on a semi-insulating GaAs substrate," *Appl. Phys. Lett.*, vol. 46, no. 10, pp. 981–983, May 1985.

[13] T. Sanada, S. Yamakoshi, H. Hamaguchi, O. Wada, T. Fujii, T. Horimatsu, and T. Sakurai, "Monolithic integration of a low threshold current quantum well laser and a drive circuit on a GaAs substrate," *Appl. Phys. Lett.*, vol. 46, no. 3, pp. 226–228, Feb. 1985.

[14] K. Kuno, T. Sanada, H. Nobuhara, M. Makiuchi, T. Fujii, O. Wada, and T. Sakurai, "Four-channel AlGaAs/GaAs optoelectronic integrated transmitter array," *Appl. Phys. Lett.*, vol. 49, no. 23, pp. 1575–1577, Dec. 1986.

[15] O. Wada, H. Hamaguchui, M. Makiuchi, T. Kumai, M. Ito, K. Nakai, T. Horimatsu, T. Sakurai, "Monolithic four-channel photodiode/amplifier receiver array integrated on a GaAs substrate," *J. Lightwave Technol.*, vol. LT-4, no. 11, pp. 1694–1703, Nov. 1986.

[16] H. Nobuhara, M. Kuno, M. Makiuchi, T. Fujii, and O. Wada, "Optoelectronic integrated transmitter with a microcleaved faced AlGaAs/GaAs quantum well laser," in *Tech. Dig. IEDM* (Washington, DC), Dec. 1985, pp. 650–653.

[17] Y. Nakayama, T. Outsuka, H. Shimizu, S. Yokogawa, K. Kameo, and H. Nishi, "A GaAs data switching IC for a gigabits per second communication system," *IEEE Solid State Circuits*, vol. SC-21, no. 1, pp. 157–161, Feb. 1986.

[18] T. Horimatsu, T. Iwama, Y. Oikawa, T. Touge, M. Makiuchi, O. Wada, and T. Nakagami, "Compact transmitter and receiver modules with optoelectronic-integrated circuits for optical LANs," *J. Lightwave Technol.*, vol. LT-4, no. 6, pp. 680–688, June 1986.

[19] T. Horimatsu, M. Sasaki, H. Yamasita, T. Okiyama, T. Ohtsuka, K. Iguchi, H. Hamaguchi, and T. Nakagami, "High-speed photoreceiver front-end module with a monolithic p-i-n/FET and a GaAs amplifier," in *Tech. Dig. 10th ECOC* (Stuttgart, W. Germany) Sept. 1984, pp. 222–223.

[20] M. Sasaki, Y. Oikawa, T. Ohtsuka, H. Yamashita, and T. Horimatsu, "High speed 4 × 4 optical switch," in *1984 Nat. Conv. Rec. IECE Japan*, 1984, p. 391 (in Japanese.)

Section 2.5: Spatial Light Modulators

Two-dimensional magneto-optic spatial light modulator for signal processing

William E. Ross
Litton Data Systems
8000 Woodley Avenue
Van Nuys, California 91409

Demetri Psaltis
California Institute of Technology
1201 E. California Blvd.
Pasadena, California 91125

Robert H. Anderson
Litton Data Systems
8000 Woodley Avenue
Van Nuys, California 91409

Abstract. This electrically alterable magneto-optic device can be used as a two-dimensional spatial light modulator in an optical image processor or a display system. A thin magnetic garnet film is epitaxially deposited on a transparent nonmagnetic garnet substrate, in the manner of magnetic bubble memory films. Semiconductor photolithographic techniques are used to etch the film into a two-dimensional array of mesas and to deposit X-Y drive lines for matrix-addressed current switching of the mesa magnetization. Electromagnetic switching provides higher speed switching than a previously reported thermal switching method. The axis of polarization of polarized light transmitted through the film is rotated by the Faraday effect in opposite directions where opposite magnetic states have been written, and a polarization analyzer converts this effect into image brightness modulation. The resulting high speed random access light modulator is applicable to all three planes of the classic coherent three-plane correlator. Although the basic effect is binary, there are at least four possible configurations which achieve gray scale rendition.

Keywords: spatial light modulator; optical image processing; optical signal processing; magneto-optics.

Optical Engineering 22(4), 485-490 (July/August 1983).

CONTENTS
1. Introduction
2. Magneto-optic device
3. Optical signal processing applications
4. Gray scale considerations
5. Summary
6. Acknowledgments
7. References

1. INTRODUCTION

A salient need in optical signal processing is the provision of an electrically alterable high speed two-dimensional spatial light modulator. Such a modulator is needed as an electrical-optical interface to address optical images into the object plane, to address spatial filters into the Fourier transform plane, and, in some cases, to modulate or scan the processed output image adjacent to the output or detector plane. While there are several candidate devices for this role[1-3] they are at present typically limited in their applications either by high price, high weight, high power, large volume, poor uniformity, volatile storage, or long write or erase times. Improved high speed reusable spatial light modulators are essential for obtaining the advantage of high throughput which is inherent in optical image processing. The need is especially pressing in optical signal processing, since no fast, simple method previously existed for producing transparencies from input signals.[4] The device described here is a high speed electrically addressed reusable transparency having non-volatile image storage and a relatively modest projected cost expected to be comparable to that of integrated circuits of the same size. The principal present limitation being overcome is that the gray scale methods described here will not be implemented until after higher resolution binary chips are further developed.

2. MAGNETO-OPTIC DEVICE

The LIGHT-MOD (Litton iron garnet H-triggered magneto-optic device) consists of a bismuth substituted iron garnet film grown on a nonmagnetic substrate and having its direction of uniaxial anisotropy oriented perpendicular to the plane of the film and of magnitude greater than the saturation magnetization of the film. The film is structured into isolated mesas, as can be seen in the scanning electron microscope (SEM) picture shown in Fig. 1. The drive conductors are deposited and structured using conventional semiconductor metals, dielectrics, and photolithography. A picture of one of the 48 × 48 test arrays is shown in Fig. 2.

The use of the LIGHT-MOD as a light valve is depicted in Fig. 3. Vertically polarized light exits from the polarizer. As the light passes through the film the polarization direction is rotated clockwise or counterclockwise depending upon the sense of magnetization of the film. The amount of rotation depends upon the Faraday constant of the material, θ_F, and upon the thickness of the film. The analyzer has been set so that it blocks the light for the mesas magnetized with the north pole into the film, as viewed in the figure, and transmits light for the mesas with the opposite direction of magnetization. For simplicity of understanding, the total rotation is depicted as being ±45 degrees in Fig. 3. Practical devices do not exhibit this total rotation. Therefore the analyzer polarization does not fully pass the "on" state as depicted in Fig. 3.

To change the direction of magnetization of a mesa, current is passed through the two adjoining conductors intersecting at the

Paper 1890 received May 24, 1982; revised manuscript received Nov. 29, 1982; accepted for publication Dec. 16, 1982; received by Managing Editor Jan. 6, 1983. This paper is a revision of Paper 341-25 which was presented at the SPIE conference on Real Time Signal Processing V, May 4-7, 1982, Arlington, Virginia. The paper presented there appears (unrefereed) in SPIE Proceedings Vol 341.
© 1983 Society of Photo-Optical Instrumentation Engineers.

Fig. 1. SEM of structured film.

Fig. 2. 48×48 LIGHT-MOD.

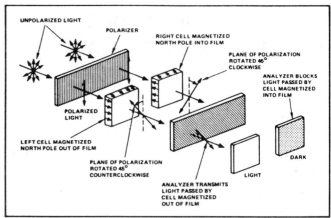

Fig. 3. Operation of LIGHT-MOD as a light valve.

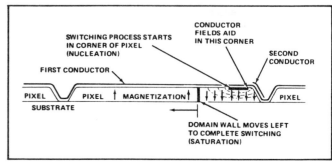

Fig. 4. Pixel switching process.

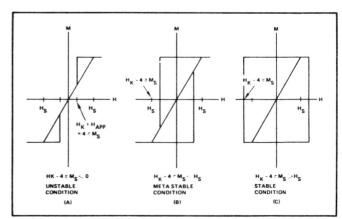

Fig. 5. Magnetic stability of material.

selected mesa. The combined magnetic fields will switch the state of the selected mesa only. The magnetic field generated by current flowing in a single conductor is insufficient to change the state of a mesa. Switching takes place in two steps as depicted in Fig. 4. The switching threshold of the applied field required for nucleation is determined by the difference between the anisotropy field H_k and the demagnetization field $4\pi M_s$. Switching occurs as coherent rotation of a volume of magnetic dipoles in the corner adjacent to the selected conductor intersection. This establishes a magnetic domain which is enlarged by the propagation of the domain wall through the thickness of the film. When the domain wall reaches the bottom of the film, the mesa has been nucleated. Removal of the drive current at that time would result in the mesa being demagnetized. By maintaining drive currents until the domain wall has propagated to the opposite corner, the mesa will complete switching and will then be saturated in the opposite direction of magnetization from that prior to switching. Propagation of the domain wall across the entire mesa requires that the magnetic field at the farthest mesa corner from the drive line intersection exceeds the saturation field H_s. When the mesa has switched, it will remain in the new state until it is again nucleated in the opposite direction by drive currents.

The LIGHT-MOD achieves pixel stability by meeting the condition $H_k - 4\pi M_s > H_s$ as described by Pulliam, et al.[5] When the above condition is satisfied, the mesas can only be in the saturated single-domain states, unless drive currents are terminated prior to completion of switching. However, there are, in general, three different possible types of stability of the various films that have been considered for this device, which result from the three cases in which the term $H_k - 4\pi M_s$ is greater than the saturation field H_s, less than H_s, or less than zero.

In Fig. 5(a), the magnetization curve M versus H is shown for the case in which $H_k - 4\pi M_s < 0$. When a field greater than the saturation field H_s is applied to the film, a single domain is formed, which is stable until the applied field is reduced below the point where $H_k + H_{app} = 4\pi M_s$. However, reversion to a multiple-domain condition occurs below this point, and if the applied field is reduced to zero, the film demagnetizes. The domain structure then has the characteristic serpentine pattern associated with equilibrium energy.

In Fig. 5(b), the second case is shown, in which $H_k - 4\pi M_s < H_s$. In this case, a single domain formed by the application of a field greater than H_s remains stable if the applied field is reduced to zero, but multiple domains form if the applied field is increased, in the reversed direction, to the point that $H_{app} + 4\pi M_s > H_k$. If the applied field is then reduced directly to zero, the demagnetized equilibrium domain pattern is formed, but if the applied field is first taken above H_s in the reversed direction and then reduced to zero, a single domain of reversed saturation is formed and remains stable when the applied field reaches zero.

In the third case [Fig. 5(c)], $H_k - 4\pi M_s > H_s$, and the mesa film is completely bistable.

The pixel switching is quite fast. Wall velocities as high as 900 m/sec have been observed in these expitaxial garnets under high drive fields and with field components normal to the wall.[6] Both conditions are achieved in this device. Switching times of about 1 μs are observed for 100 μm pixels. Average wall velocities are, thus, approximately 100 m/sec. Speed of operation is a function of cell size and mode of operation. The smaller the cell size the faster the cells switch, assuming the same wall velocity. Therefore, for serial addressing of a given array area, the total switching time can be approximately constant whether it is structured into a 128×128 array or a 512×512 array. Since the image is nonvolatile, refresh is not required, and only those pixels which are to be changed need to be switched. This results in very low power consumption and effective bandwidth compression during image transmission.

The LIGHT-MOD can operate at a much higher speed than that of the devices utilizing the temperature-compensation writing technique reported by Krumme et al.[7] and Hill and Schmidt.[8] Hill and Schmidt used bismuth-doped garnets showing a compensation temperature near room temperature. Switching was achieved by elevating the temperature of the addressed pixel while applying a reversal field. The lack of magnetism at the compensation temperature assures pixel stability in their device after switching. The principal differences between their device and the LIGHT-MOD are that Hill and Schmidt use temperature triggering rather than the H field (magnetic field) triggering used here, and that their device requires operation at the compensation temperature, while the LIGHT-MOD can be operated over a wide temperature range up to the Curie temperature.

The processing steps for producing this device are relatively simple in comparison to semiconductor or bubble memory devices, and utilize photolithographic equipment which has high throughput capability. Arrays having pixels from 10 μm size to 400 μm have been evaluated. Pixel size and shape are not critical. The maximum pixel size that can be saturated with the drive line is in the 250 to 400 μm range. External field coils can be utilized to advantage to complete switching for some applications.

The optical operation of the device is influenced by optical absorption and Faraday rotation. Figure 6 shows how the overall transmission is influenced by differences in Faraday rotation for a given absorption coefficient. These curves are for a configuration in which the polarization analyzer sheet axis is crossed at 90° to the "off" pixel image component polarization axis, so it is 2 θ_F from the "on" pixel polarization axis. The transmission equation used to generate the curves is therefore $I/I_o = e^{-\alpha t} K_1 \sin^2(2\theta_F t)$, where θ_F is Faraday rotation in degrees per micrometer, t is thickness in micrometers, α is the absorption coefficient in reciprocal micrometers, k_1 is the absorption factor of 0.89 for the analyzer, and I/I_o is the light transmission ratio. It can be seen that the light transmission rotation. Work is therefore in progress to increase the Faraday rotation of the film.

At the present state of development of this device, the physical characteristics which are known or clearly predictable are as follows: about 1000:1 contrast for signal processing applications; about 10:1 to 50:1 contrast for white light display applications; 1 W power requirement at high frame rates; 1% transmission uniformity, switching times 50 to 1000 μsec for a 512×512 array, depending on refresh requirements; pixel size 10 μm and up; array size 1 to (5 cm)^2 and mosaics possibly up to (10 cm)^2; array configuration M×N, where M≥1 and N≤2048; and cost near that of an integrated circuit of an equivalent size. Characteristics to be determined by future experimental work include noise figure and diffraction

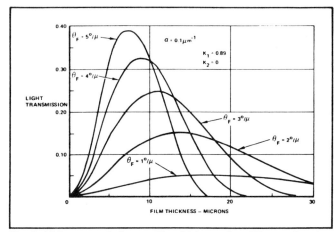

Fig. 6. Light transmission vs film thickness for different Faraday rotations.

effects, a more accurate determination of contrast and spatial uniformity, and the identification of the most suitable analog/digital (A/D) signal conversion and driver circuits for the chip.

3. OPTICAL SIGNAL PROCESSING APPLICATIONS

The LIGHT-MOD can be used in optical signal processing systems in several capacities. The most obvious use is as a spatial light modulator. The data to be processed by the optical computer is written on the device through the electrode structure. If a raster scan format is used, each pixel can be addressed sequentially by activating the appropriate crossed electrodes, or each line can be written in parallel by activating one of the horizontal electrodes and simultaneously applying the signal corresponding to a raster line to the vertical electrodes. Each pixel or line can be switched in approximately 1 μsec, which allows a 512×512 frame to be written in 0.256 sec serially and 0.512 msec in parallel. The parallel addressing scheme is attractive because of its speed, but the driving electronics is more complex in such a configuration. The scanning format is not restricted to the conventional raster; any desired addressing sequence can be implemented with digital control. The size of the arrays is projected to be 512×512 for the next generation of the devices, with each square pixel having an active area approximately (50 to 100 μm)^2 and separation of 6 to 12 μm.

The array can also be used as a random access optical scanner by allowing one (or several) pixels to be transmissive at a time. Therefore only a selected portion of the incident field is allowed to propagate through the plane of the magneto-optic device. If the device is interfaced with a high quality photodetector, a two-dimensional random access image sensor could be constructed that combines in principle the excellent detector properties of the photodetector and the resolution and speed of the magneto-optic device.

It may be possible, for example, to generate the autocorrelation of an image placed on an optical scanner chip by scanning with two pixels at a time (or alternately) and multiplying their respective signal outputs. Blurred or coded images may be restored by scanning with the appropriate multipixel deconvolving scan pattern. Although these concepts are not fully developed, multiple pixel scanning is expected to become a generic concept. Other uses of the device include logic operation on two-dimensional binary data fields and use as a programmable scanning or variable spatial filter.

Garnet films can be grown with good optical flatness, which will allow them to be used in coherent optical systems. The electrode structure will have to be masked off to minimize scatter as well as the background light level. Another important consideration is the efficiency of the magneto-optical device, measured as the percentage of the incident light that is transmitted as spatially modulated light. There are three primary factors that determine the efficiency. The active post area A_1, relative to the light blocking electrode area A_2, results in an attenuation of the incident beam by $A_1/(A_1 + A_2)$. The second factor is the optical absorpton of the garnet films. The attenuation constant of the material that is presently used is 2000/cm. Finally, the analyzer will absorb a portion of the light. The contrast of the light output is defined by the following equation:

$$CR = \frac{I_{max} - I_{min}}{I_{max} + I_{min}} = \frac{\cos^2(\phi - \theta) - \cos^2(\phi + \theta)}{\cos^2(\phi - \theta) + \cos^2(\phi + \theta)}, \quad (1)$$

where ϕ is the angle of the analyzer with respect to the polarization angle of the incident light and $+\theta$ is the angle with respect to the incident light polarization by which the polarization of the light is rotated when it propagates through the magnetized film. Obviously the attenuation is minimized if $\theta = 45°$. At the present time, however, θ is approximately 10°. The rotary power of the magnetized film is proportional to its thickness. Therefore it would appear that one might improve the contrast by growing thicker films. The optical attenuation, however, grows with the thickness, and, therefore, for a given material there is an optimum thickness at which the films are grown. There are other factors that also affect the efficiency, such as polarizer coefficients of about 0.8 to 0.9, and reflections at the surface, which could be eliminated with antireflective coatings. Indications are that the research that is presently being conducted at Litton will result in improvements in the transmittance and Faraday rotation of the magnetic film, which will result in significantly more efficient devices.

The present devices are binary spatial light modulators; the output light intensity amplitude at each pixel can take only one of two possible states depending on the direction in which the pixel is magnetized. Therefore these initial devices can only be used in applications where the input data to the optical processor is binary. Text processing and certain robotics applications are important examples where the input scenes are indeed binary. In order to extend the applicability of the LIGHT-MOD to a broader class of problems, however, it is important to be able to represent many gray shades. In the following section we will explore several methods for providing gray scale capability.

4. GRAY SCALE CONSIDERATIONS

Gray levels may be obtained by allocating more than one magnetic mesa to each pixel. The different mesas comprising a pixel may be on different arrays which are optically combined, or may be placed in a coplanar pixel area on the same array, or may be placed in a registered stack of serially transmissive arrays. The data at each pixel are represented by a binary word, and each bit of the word sets the magnetization state of one of the mesas that comprise the pixel.

One configuration of this type is shown in Fig. 7, which schematically represents the class of instruments which optically combine three array images in registration. The intensity of each pixel is derived from the overlaid images of three mesas on different arrays, which are combined with beam splitters. The magnetic states of corresponding mesas on arrays 1, 2, and 3 are set, respectively, by the most significant bit, second bit, and least significant bit of the switching signal. Each beam splitter is 50% transmissive, but the three image-forming beams each pass through one, two, or three beam splitters. Accordingly, written mesas contribute different pixel intensity components in the binary ratio 1:2:4, giving eight possible output intensity states, which are a linear mapping of the 3-bit binary input word. In general, N arrays can be used to produce 2^N gray levels. This system is complex, but it does have the advantages of dividing the connector requirement into smaller parts, accepting parallel addressing of the arrays, retaining the full resolution of the mesa spacing pattern, and obtaining gray scale without requiring an additional more specialized array design.

A different gray scale method using pixel area modulation on a single array is shown in Fig. 8. Four mesas comprise one pixel, and the relative areas of pixels 1, 2, 3, and 4 are 1, 2, 4, and 8 respectively. The direction of magnetization of each mesa can be independently

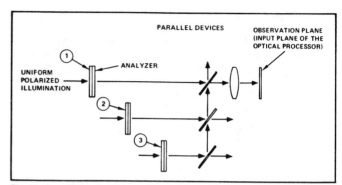

Fig. 7. Parallel devices.

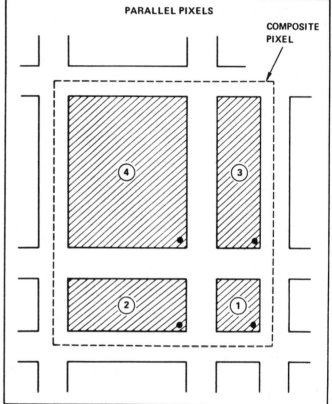

Fig. 8. Composite pixels.

set, therefore the pixel consisting of four mesa can be set in any of $2^4 = 16$ states. The gray scale is achieved at the expense of reduced resolution and addressing speed, and the need for appropriate hardware and software. Of course, it is inherent in any gray scale imaging device that the gray scale input information must be addressed into the device in some form, and complexity increases with the number of gray levels provided.

In another gray scale method, a series cascade of mesas may be used to form each pixel, as depicted in Fig. 9. Three separate devices are cascaded by either stacking them one against the other or imaging one onto the next. The cascade is illuminated with polarized light, and an analyzer is placed after all three devices. In this arrangement, the three Faraday rotations are first added in binary combinations; then the different sums of rotation angles for each pixel result in intensity modulation in proportion to the squares of the projection cosines of those angles on the analyzer polarization axis. There are $2^3 = 8$ possible states of the pixels, but four of these states are degenerate (duplicates), and only four distinct output intensity levels result.

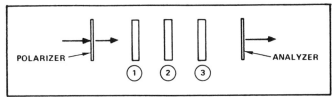

Fig. 9. Cascaded LIGHT-MODs with films of equal thicknesses.

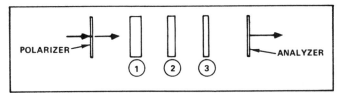

Fig. 10. Cascaded LIGHT-MODs with films of unequal thicknesses.

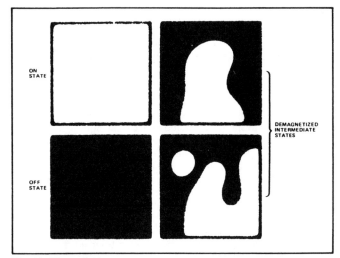

Fig. 11. Three stable states.

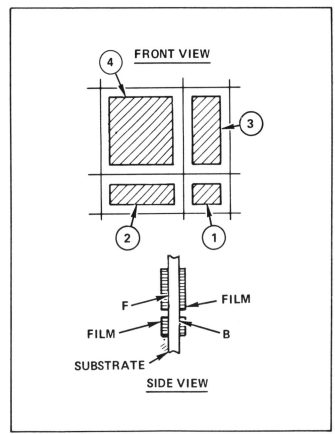

Fig. 12. A possible hybrid implementation..

In another, and similar, arrangement of three arrays shown in Fig. 10, the degenerate states can be removed, and eight gray levels obtained. Garnet films of unequal thickness are used, having thicknesses and Faraday rotations in the binary ratios 1:2:4. Rotation angles such as $\pm 2°$, $\pm 4°$, and $\pm 8°$ might, for example, result in eight pixel combination rotations from $-14°$ to $+14°$ in steps of two degrees. Since output intensity is a cosine-squared function of the relative angle of the polarization axis of light going into the polarization analyzer, the input states are not mapped linearly into output intensity in this and the preceding configuration. For some signals this is unimportant or results in a useful contrast enhancement. In other cases, the binary encoding levels of the array drive signal may be chosen to compensate for this effect and for nonlinearities of an output vidicon or photographic film or other system components.

In principle, any number N of devices can be cascaded, resulting in 2^N gray shades. In any cascaded arrangement of films, compounded light losses due to absorption in the multiple films is of concern. However, the film thicknesses in the configuration of Fig. 10 may be chosen to be in the ratio $1/2$, $1/4$, and $1/8$, compared to a relative thickness of 1 for the single film device of Fig. 3. Accordingly, the total thickness and absorption of the configuration of Fig. 10 will be less than that of Fig. 3. The same argument applies to any number of cascaded films, since any partial sum of the well-known geometric series $1/2 + 1/4 + 1/8 + \ldots$ is always less than one. In practice, N is limited by the ability to control reflective losses with antireflective coatings.

In our discussion of gray scale thus far, the mesas of the LIGHT-MOD have been treated as bistable. In fact, the demagnetized state is also stable, and it can be accessed by applying the appropriate signals to the electrodes, as previously discussed. Mesas that are magnetized in all three states, viewed through crossed polarizers are shown in Fig. 11. The individualistic meandering patterns which are characteristic of the magnetic domains of the demagnetized state are shown for two mesas. In the demagnetized state, half of the area of each mesa is magnetized in one direction and the other half is magnetized in the opposite direction. Therefore, the intensity of the light transmitted through a pixel in the demagnetized state is the average of the intensities of the two magnetized states, creating a third level of transmittance of the mesa. The resulting tristable mesa may be combined in various ways with the previously described configurations in order to select optimum combinations of advantages and limitations for particular gray scale applications. An illustrative arrangement of interest, shown in Fig. 12, uses a double array cascade of four parallel mesas, each having three stable states. A cascade of two mesas can be conveniently fabricated by growing garnet films on both sides of the same substrate wafer. The mesas on each side are addressed by separate electrode structures. Four mesas of unequal areas comprise a pixel on each side of the substrate. The eight separate mesas can be set at $3^8 = 6561$ possible combinations of magnetic states.

If two cascaded mesas have the same thickness and three stable states are used for each mesa, each cascaded mesa pair has five distinct intensity levels, and the number of output intensity levels for a four-area pixel is $5^4 = 625$ levels. The remaining 5936 magnetic states are degenerate for output intensity. However, if the film thicknesses are unequal and are in a 1:3 ratio, each cascaded mesa pair has nine distinct intensity levels, and the number of intensity levels of $9^4 = 6561$ levels. There are no degenerate states in this case.

In this illustrative example, far more gray levels are obtained than would be wanted in most applications, particularly in view of the need for high speed 13-bit A/D conversion and addressing, and the need for extreme uniformity of intensity across the LIGHT-MOD and all other optical system components, to preserve such a large gray scale. However, the design principles for obtaining gray levels from relatively simple monolithic structures are well illustrated by the examples.

5. SUMMARY

A reusable magneto-optic spatial light modulator which has the capability of high speed conversion of electrical input signals into an image on an optical transparency has been described. The transparency employs a structured iron garnet film which is X-Y matrix addressable with conductive drive lines and which has random access and nonvolatile storage. The basic effects are bistable magnetic switching of mesa-sized magnetic domains and modulation of polarized light by means of the Faraday effect. The switching properties have been described in terms of magnetic domain nucleation and enlargement of the domains by propagation of the domain wall. Several methods for obtaining gray scale have been outlined, including the use of optical superposition of binary image components, multimesa composite pixels, cascaded films, and tristable domain configurations.

Applications include input to the object plane and transform plane in optical signal processors, selected applications as a scanning detector in the output plane of optical signal processors, large screen projection displays, helmet-mounted or head-up displays, hard copy displays, and hand-held direct view displays.

6. ACKNOWLEDGMENTS

The authors are gratefully indebted to B. MacNeal, D. Cox, W. Robinson, and S. Mills for device design, system design, electronic assembly, and software; to M. Shone, K. Vermuri, and R. Belt (Airtron Division, Litton Industries) for garnet films; and to G. Pulliam for Figs. 5 and 6, and for technical editing.

7. REFERENCES

1. J. Grinberg, A. D. Jacobson, W. P. Bleha, L. Miller, L. Fraas, D. Boswell, and G. Myer, Opt. Eng. 14, 217(1975).
2. J. Feinleib and D. S. Oliver, Appl. Opt. 11, 2752(1972).
3. A. R. Tanguay, Jr., "Future Directions for Optical Information Processing," Final Report, pp. 52-76, Texas Tech. Univ. Lubbock, Texas (1980).
4. D. Casasent, ed., *Optical Data Processing*, p. 242, Springer Verlag, New York (1978).
5. G. R. Pulliam, W. E. Ross, B. E. MacNeal, and R. F. Bailey, J. Appl. Phys. 53, 2754(1982).
6. K. Vural and F. Humphrey, J. Appl. Phys. 51(1), 549(1980).
7. J. P. Krumme, J. Verwell, J. Haberkamp, W. Tolksdorf, G. Bartels, and G. P. Espinoza, Appl. Phys. Lett. 20(11), 451(1972).
8. Bernard Hill and Klaus-Peter Schmidt, Society for Information Display (SID) 1979 International Symposium, Digest of Technical Papers 10, 80(1979).

Characteristics of the deformable mirror device for optical information processing

Dennis R. Pape
Larry J. Hornbeck
Texas Instruments Incorporated
Central Research Laboratories
P.O. Box 225936/MS 134
Dallas, Texas 75265

Abstract. A new two-dimensional, fast, analog, electrically addressable, silicon-based membrane spatial light modulator (SLM) has been developed for optical information processing applications. This SLM, the deformable mirror device (DMD), consists of a 128×128 array of deformable mirror elements addressed by an underlying array of metal-oxide-semiconductor (MOS) transistors. Coherent light reflected from the mirror elements is phase modulated producing an optical Fourier transform of an analog signal input to the device. The DMD architecture and operating parameters related to this application are presented. A model is developed that describes the optical Fourier transform properties of the DMD. The calculated peak first-order diffraction efficiency of 8.2% is in good agreement with the value of 8.4% obtained from experimental optical Fourier transform measurements.

Keywords: spatial light modulator; optical information processing; optical Fourier transform; deformable mirrors.

Optical Engineering 22(6), 675-681 (November/December 1983).

CONTENTS
1. Introduction
2. Device description
3. Membrane deflection model
4. Fourier transform model
5. Experimental setup
6. Step grating optical Fourier transform results
7. DMD optical Fourier transform results
8. Future development
9. Conclusion
10. Acknowledgments
11. References

1. INTRODUCTION

An integral component of an optical information processor is a spatial light modulating device.[1] Such a device may be used to input information into the object plane of the processor or filter information in an image or Fourier plane. Present devices each have limitations affecting their use in these applications. Some of these limitations include slow response time,[2] inherent one-dimensionality,[3] and inherent digital operation.[4] This paper describes the use of a new two-dimensional, fast, analog, electrically addressable, silicon-based spatial light modulator—the deformable mirror device (DMD).

Section 2 of the paper describes the device. A more complete description is given elsewhere.[5] A model of the mirror deflection is given in Sec. 3. In Sec. 4 we present a model of the Fourier transform of an image imposed on the device. Experimental data are presented and compared with this model in Secs. 5 through 7. Finally, we outline future development of the device in Sec. 8.

2. DEVICE DESCRIPTION

A perspective view of the DMD is shown in Fig. 1. The device consists of an X-Y array of deformable mirror elements that can be addressed by an underlying array of metal-oxide-semiconductor (MOS) transistors. Each of the two devices described in this paper has an array size of 128×128 pixels with pixel elements 51 μm square. This gives a resolution of 9.8 lp/mm. Device A has an active (deformable) area of 23×36 μm and an air gap dimension of 620 nm. Device B has an active area of 28×33 μm and an air gap dimension of 800 nm.

Figure 2 shows the line-addressed organization of the DMD. Video data are fed to an off-chip serial-to-parallel converter that is connected to the 128 drain lines of the MOS transistors. These lines are charged to a potential $(\phi_d)_{nm}$. A decoder, connected to the 128 gate lines, selects one gate to be turned on. The floating sources of all the MOS transistors in that particular gate line are then charged to the potential of their respective drains. The gate is then turned off. The mirror is held at a fixed voltage V_M. An electrostatic force then exists on the mirror proportional to $V_M - (\phi_d)_{nm}$ causing the mirror element to deflect. The mechanical response time of the mirror and hence the line settling time is approximately 25 μs. Once the sources have been set, the next line of video is transferred onto the drain lines and the next gate line is selected by the decoder. The source storage time is approximately 200 ms. This line-address process continues at standard video TV line rates of 63.5 μs. Because the DMD has a line settling time less than the TV line rate, the device may be applied to taking continuous transforms at TV rates. The DMD may be operated at much higher line rates consistent with the line propagation delays estimated to be 0.5 μs.

Chip photomicrographs are shown in Fig. 3. Figure 3(a) shows a corner of the central array. Bondpads are on the left-hand side of the figure. Figure 3(b) is a magnified view of one side of the chip showing air channels etched in the polysilicon to allow air to escape when the membrane is placed on the chip.

Figure 4 shows schlieren projection images of device B generated at TV rates. A checkerboard pattern of blocks of 16 elements is shown in Fig. 4(a). Figure 4(b) shows horizontal bars consisting of alternating sections of 16 elements ON and 16 elements OFF. Figure 4(c) demonstrates the resolution of the device. The upper left quadrant of Fig. 4(c) shows all pixels ON. The upper right and lower left quadrants show alternating pixels ON. The lower right quadrant of Fig. 4(c) shows a checkerboard pattern of pixels ON. Figure 4(d) shows all elements OFF. Nonuniformities in these images are caused primarily by the projection optics. Brightness nonuniformity of the

---

Invited Paper LM-102 received Apr. 18, 1983; revised manuscript received July 15, 1983; accepted for publication July 18, 1983; received by Managing Editor July 26, 1983. This paper is a revision of Paper 388-09 which was presented at the SPIE conference on Advances in Optical Information Processing, Jan. 20-21, 1983, Los Angeles, CA. The paper presented there appears (unrefereed) in SPIE Proceedings Vol. 388.
© 1983 Society of Photo-Optical Instrumentation Engineers.

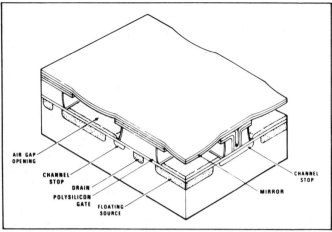

Fig. 1. Perspective view of the DMD.

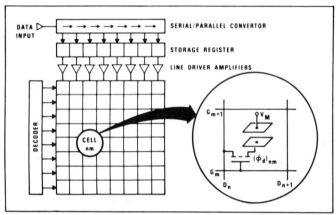

Fig. 2. Line-addressed organization of the DMD.

device is approximately ±5% and the defect level is approximately 0.4%.

3. MEMBRANE DEFLECTION MODEL

The characteristics of the deflection of the DMD membrane can be approximated using an electromechanical model for membrane modulators.[6] This model assumes the membrane has a circular support boundary of radius R and air gap t. The membrane is uniformly loaded by an electrostatic field where end effects are ignored. The membrane tension T_o is assumed constant and acts only in a tangential direction. It is also assumed that no pressure differential exists across the membrane. The numerical analysis solution of this model shows that for small peak membrane deflections Δ_o,

$$\Delta_o/t = V^2 R^2 / T_o t^3 \, , \quad (1)$$

where V is the air gap voltage. When Δ_o/t exceeds 0.44, the membrane tension can no longer counteract the electrostatic force and the membrane collapses. The voltage at which collapse occurs, V_c, is given by

$$V_c^2 \propto T_o t^3 / R^2 \, . \quad (2)$$

Typical DMD parameters are $t = 0.62\ \mu m$, $T_o = 5\ N/m$, $(\Delta_o)_{collapse} = 0.27\ \mu m$, and $V_c = 28\ V$. An effective radius is used for R:

$$R = \sqrt{(WL/\pi)} \, , \quad (3)$$

where W and L are the width and length of the air gap opening.

Figure 5(a) is a graph showing membrane deflection versus volt-

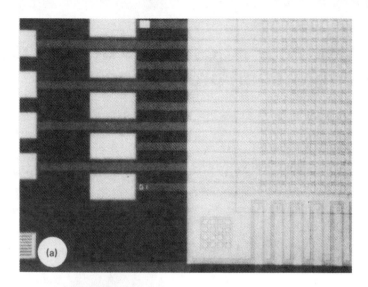

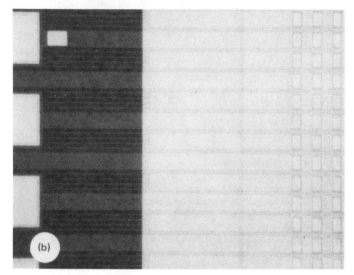

Fig. 3. Photomicrographs of the DMD: (a) corner and (b) side of the central array.

age. The extremely nonlinear behavior will be seen later in the Fourier transform data. Figure 5(b) is a graph of collapse voltage versus membrane radius.

The application of this model to the present device breaks down due to the noncircular air gap opening and the existence of a pressure differential across the membrane. A rectangular membrane deflection model is being developed but is not used in this paper.

4. FOURIER TRANSFORM MODEL

Consider a wave of complex amplitude U(x,y) and wavelength λ incident on the DMD. The field G(x,y), representing the spatially modulated and reflected wave immediately in front of the DMD, is given by

$$G(x,y) = U(x,y)\ S(x,y) \, , \quad (4)$$

where S(x,y) is the complex reflectance amplitude function describing the surface of the DMD mirror. In the following analysis we have scaled S(x,y) by the reflectance of the mirror.

The function S(x,y) is pictorially represented in Fig. 6. It consists of a unit cell convolved with an X-Y array of delta functions. The unit cell, $S_u(x,y)$, can be written as

$$S_u(x,y) = S_s(x,y) + S_M(x,y) \, , \quad (5)$$

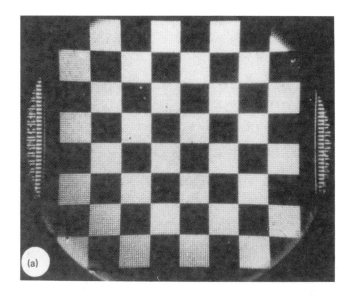

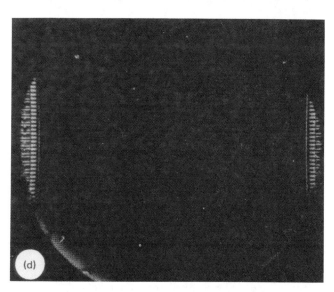

Fig. 4. Schlieren images of the DMD: (a) checkerboard, (b) horizontal bars, (c) quadrants showing all cells ON, horizontal and vertical bars, and checkerboard, and (d) all cells OFF.

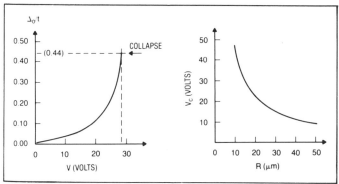

Fig. 5. Membrane deflection model: (a) normalized deflection vs voltage and (b) collapse voltage vs support radius.

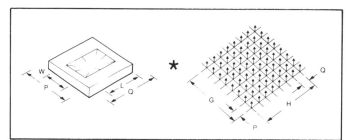

Fig. 6. A pictorial representation of the function $S(x,y) = S_u(x,y) * S_D(x,y)$ used in the Fourier transform model.

where $S_s(x,y)$ (describing a flat mirror over a polysilicon support structure of width P and length Q with a central opening of width W and length L) is given by

$$S_s(x,y) = [\text{Rect}(x/P)\text{Rect}(y/Q) - \text{Rect}(x/W)\text{Rect}(y/L)], \quad (6)$$

and $S_M(x,y)$ describes a deformed mirror of maximum central deflection Δ_0 bounded by the central rectangular opening. We assume the mirror is deformed parabolically. With the mirror deflection Δ given by

$$\Delta = \Delta_0(1 - 4x^2/W^2)(1 - 4y^2/L^2), \quad (7)$$

the second term of $S_u(x,y)$ is

$$S_M(x,y) = [\text{Rect}(x/W)\text{Rect}(y/L)]$$
$$\times \exp[-i\phi_0(1 - 4x^2/W^2)(1 - 4y^2/L^2)], \quad (8)$$

where

$$\text{Rect}(x/a) = \begin{cases} 1 & |x| < a/2 \\ 0 & |x| > a/2 \end{cases}, \quad (9)$$

and

$$\phi_0 = 4\pi\Delta_0/\lambda . \quad (10)$$

The delta function array, $S_D(x,y)$, with delta functions on centers P,Q and bounded by width G and length H is represented by the function

$$S_D(x,y)$$
$$= (1/PQ)\,\text{Comb}(x/P)\,\text{Comb}(y/Q)\,\text{Rect}(x/G)\,\text{Rect}(y/H), \quad (11)$$

where

$$\text{Comb}(x/a) = \sum_{m=-\infty}^{+\infty} \delta(x/a - m) = a \sum \delta(x - ma) . \quad (12)$$

Combining Eqs. (5), (6), (8), and (11), $S(x,y)$ is written as

$$S(x,y) = S_u(x,y) * S_D(x,y)$$
$$= \Big[\text{Rect}(x/P)\text{Rect}(y/Q) - \text{Rect}(x/W)\text{Rect}(y/L)$$
$$+ \text{Rect}(x/W)\text{Rect}(y/L)\exp\{-i\phi_0(1 - 4x^2/W^2)$$
$$\times (1 - 4y^2/L^2)\}\Big]$$
$$* \Big[(1/PQ)\text{Comb}(x/P)\text{Comb}(y/Q)\text{Rect}(x/G)\text{Rect}(y/H)\Big]. \quad (13)$$

The reflected wave now undergoes Fraunhofer diffraction. The field distribution $F(x,y)$, a distance Z from the DMD, is given by

$$F(x,y) = \frac{1}{i\lambda Z}\exp\left[i\frac{2\pi Z}{\lambda} + \frac{i\pi}{\lambda Z}(x^2 + y^2)\right]$$
$$\times \int_{-\infty}^{+\infty}\int_{-\infty}^{+\infty} G(x_1,y_1)\frac{\exp(2\pi i x x_1)}{\lambda Z}\frac{\exp(2\pi i y y_1)}{\lambda Z}dx_1 dy_1, \quad (14)$$

which we recognize, aside from multiplicative phase and amplitude factors, as the Fourier transform (F.T.) of $G(x,y)$ evaluated at the spatial frequencies $u = x/\lambda Z$ and $v = y/\lambda Z$ and denoted by $G(u,v)$. It is convenient to normalize $G(u,v)$ to the Fourier transform of the field distribution from an undeformed DMD ($\Delta = 0$) at $(u,v) = (0,0)$:

$$G_N(u,v) = G(u,v)/G_0(0,0) . \quad (15)$$

Since it is the energy spectrum and not the wave amplitude that one measures in the Fourier plane, the modulus squared of $G_N(u,v)$ is evaluated. The undeformed DMD field distribution can be written as

$$G_0(0,0) = \int_{-\infty}^{+\infty}\int_{-\infty}^{+\infty} G(x_1,y_1)dx_1 dy_1 . \quad (16)$$

Evaluating Eq. (16) and substituting the result into the modulus squared of Eq. (15) gives

$$|G_N(u,v)|^2 = |\text{F.T.}[S(x,y)]/GH|^2 . \quad (17)$$

The Fourier transform of $S(x,y)$ is the product of the Fourier transforms of $S_u(x,y)$ and $S_D(x,y)$:

$$\text{F.T.}[S(x,y)] = \text{F.T.}[S_u(x,y) * S_D(x,y)]$$
$$= \text{F.T.}[S_u(x,y)]\,\text{F.T.}[S_D(x,y)] . \quad (18)$$

We can evaluate Eq. (18) using the fact that

$$\text{F.T.}[\text{Rect}(x/a)] = a\,\text{sinc}(au) , \quad (19)$$

and

$$\text{F.T.}[\text{Comb}(x/a)] = a\sum_{m=-\infty}^{+\infty} \delta(ua - m) , \quad (20)$$

where

$$\text{sinc}(au) = \sin(\pi au)/\pi au . \quad (21)$$

The Fourier transforms of the two components of $S(x,y)$ can then be written as

$$S_u(u,v) = \text{F.T.}[S_u(x,y)]$$
$$= [PQ\,\text{sinc}(Pu)\text{sinc}(Qv) - WL\,\text{sinc}(Wu)\text{sinc}(Lv)]$$
$$+ \int_{-L/2}^{+L/2}\int_{-W/2}^{+W/2}\exp\{-i\phi_0(1 - 4x^2/W^2)(1 - 4y^2/L^2)\}$$
$$\times \exp\{2\pi i v y\}\exp\{2\pi i u x\}dx\,dy , \quad (22)$$

and

$$S_D(u,v) = \text{F.T.}[S_D(x,y)]$$
$$= (1/PQ)\left[\sum_M \delta(u - M/P)\,G\,\text{sinc}(Gu)\right]$$
$$\times \left[\sum_N \delta(v - N/Q)\,H\,\text{sinc}(Hv)\right]. \quad (23)$$

The normalized Fourier transform energy spectrum of the DMD is then

$$|G_N(u,v)|^2 = \Big|\sum_M\sum_N \Big[PQ\,\text{sinc}(M)\text{sinc}(N)$$
$$- WL\,\text{sinc}(WM/P)\text{sinc}(LN/Q)$$
$$+ \int_{-L/2}^{+L/2}\int_{-W/2}^{+W/2}\exp\{-i\phi_0(1 - 4x^2/W^2)(1 - 4y^2/L^2)\}$$
$$\times \exp\{2\pi i Ny/Q\}\exp\{2\pi i Mx/P\}dx\,dy\Big]$$
$$\times [(GH/PQ)\text{sinc}(G(u - M/P))\text{sinc}(H(v - N/Q))]\Big|$$
$$/GH\Big|^2 . \quad (24)$$

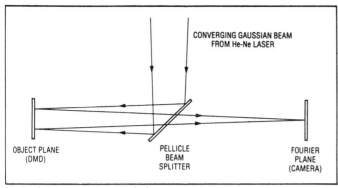

Fig. 7. Fourier transform optical setup.

The integral in the above expression must be evaluated numerically. In the following analysis we have used an adaptive Romberg method to evaluate this integral over its two-dimensional region. Finally, the energy spectrum $|G_N(u,v)|^2$ is computed for comparison with experimental data.

5. EXPERIMENTAL SETUP

A diagram of the apparatus used to measure the optical Fourier transform of the DMD is shown in Fig. 7. We have used the converging beam technique to avoid the aberrations and noise inherent in the traditional parallel beam Fourier transform lens system.[7] A converging 2.5 mW Gaussian beam from a He-Ne laser is reflected off a pellicle beam splitter onto the DMD. Much higher beam intensities may be used due to the effective light blocking layer of the floating field plate in the well of each pixel. The beam, with a diameter of approximately 5500 μm, illuminates approximately 9000 DMD pixels. The spatially modulated beam, reflected from the DMD mirror, passes through the pellicle beam splitter onto a CCD detector array located in the Fourier plane of the system. The photographs in the next section were taken from an x-y monitor connected to the video output from the CCD camera. Each video line has been offset to give the appearance of three-dimensionality to the frame of video data displayed on the x-y monitor. The energy spectrum of intensity data, compared with model calculations in the next section, were obtained with the video output from the CCD camera connected to an oscilloscope.

6. STEP GRATING OPTICAL FOURIER TRANSFORM RESULTS

To test the experimental technique and theoretical model described in the previous two sections, we fabricated 128×128 arrays of rectangular wells etched in silicon dioxide. This step grating array exactly matched the geometry of the device A polysilicon air gap openings as the photomask used to define the openings was the same in both cases. Thirteen slices of arrays were fabricated with well depths having nominally 25 nm increments from 0 to 300 nm. These oxide structures were coated with 50 nm of aluminum.

Figure 8 shows the normalized peak diffraction intensities of both the zeroth-order, I_0, (filled circles) and first-order, I_1, (open circles) diffraction peaks of these step gratings. The eight first-order peaks from the grating are not of equal intensity because the wells are not square. We have plotted the intensity of the (0,1) peak, which has the maximum intensity of the eight first-order peaks. On the same graph, in the solid lines, are shown the results of the calculation of the energy $|G_N(u,v)|^2$, from Eq. (24), for both the zeroth and first order. In this calculation we have replaced the parabolic mirror deflection expression in the two-dimensional integral with the exact expression for a well of depth Δ_0:

$$\Delta = \Delta_0 . \qquad (25)$$

The agreement between theory and experiment is quite good.

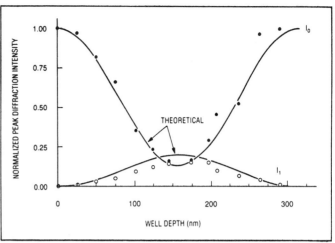

Fig. 8. Zero- and first-order diffraction intensity vs well depth for a DMD step grating.

As expected, destructive interference, leading to the minima in I_0, occurs at a well depth of $(1/4)\lambda$ or $\phi_0 = \pi$. I_0 is not zero at this point, however, because the area of the support structure is larger than that of the wells. The first-order peak diffraction efficiency is 19.8%, considerably higher than the 12.8% found when the areas are equal.

7. DMD OPTICAL FOURIER TRANSFORM RESULTS

Fourier transform images of a uniformly deflected DMD (device B) at three different drain voltage levels are shown in Fig. 9. For these measurements, the mirror was held with a fixed negative bias of −28 V, and the substrate at −4 V. Figure 9(a) shows the image with 0 V on the drains, the condition for all pixels off. Note the absence of any signal for the higher-order peaks. This demonstrates the high signal-to-noise ratio of this device, to be discussed later. Figure 9(b) shows the maxima in the first-order occurring at +16.9 V, and Fig. 9(c) shows the minima in the zeroth-order occurring at +21.8 V for this particular device.

Figure 10 shows quantitatively the zeroth-order and maximum first-order normalized peak diffraction intensities as a function of air gap voltage for device A. The nonlinearity of the mirror deflection as a function of air gap voltage, shown in Fig. 5, is responsible for the small change in I_0 and I_1 below an air gap voltage of about 16 V. The transform signal versus input signal may be linearized at the input to the serial-to-parallel converter shown in Fig. 2. Unlike the step grating case, the minima in I_0 does not coincide with the maxima in I_1. This is caused by the fact that the shape of the diffracting surface, the deformed membrane, is not flat but has an x,y dependence. It is expected that the location of the minima in I_0 and maxima in I_1 will occur at a greater central membrane deflection than that for the step grating. Mirau interferometric measurements of the DMD mirror deflection show that indeed the minima in I_0 occur at 325 nm and the maxima in I_1 occur at 275 nm. The measured first-order peak diffraction efficiency of this device is 8.4%. The model gives a value of 8.2%. These values are in contrast to the 19.8% efficiency of the step grating, the maximum possible efficiency for this geometry.

Figure 10 also shows, in solid lines, the calculated values of I_0 and I_1 from Eq. (24). We used the Mirau interferometric mirror deflection measurements to relate the amplitude of the membrane deflection to the air gap voltage for these graphs. Although the calculated intensities of I_0 and I_1 agree well with experimental values, the location of the minima and maxima in the curves does not. This indicates that the membrane deflection has less curvature than predicted by a parabolic shape.

Figure 11 shows the Fourier transform image of device B at the Nyquist frequency. Figure 11(a) is the central line of the video frame displayed in Fig. 11(b). The half-width of the zero-order peak in Fig. 11(a) is 10% of the Nyquist frequency. Therefore, the usable spatial

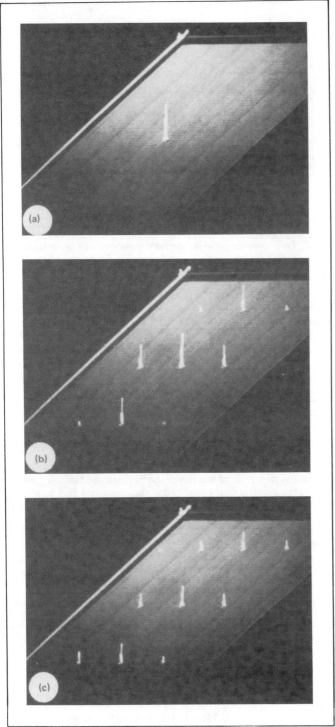

Fig. 9. Optical Fourier transform images of a DMD: (a) all pixels OFF, (b) maxima in first-order, and (c) minima in zeroth-order.

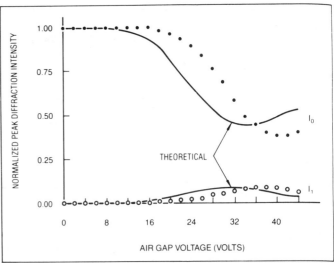

Fig. 10. Zero- and first-order diffraction intensity vs air gap voltage for a DMD with all pixels ON.

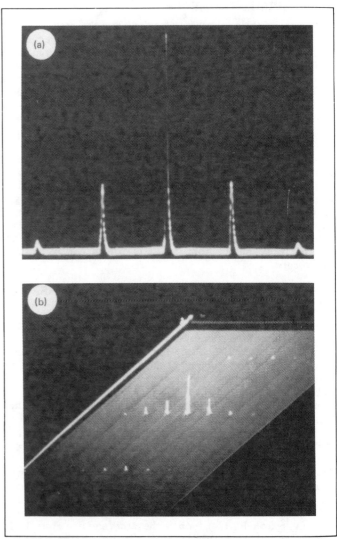

Fig. 11. Optical Fourier transform image of the DMD at the Nyquist frequency: (a) central video line and (b) full video frame.

frequency range of this device is from 10% to 100% of the Nyquist frequency.

The signal-to-noise (S/N) ratio is defined as the maximum sinusoidal signal at 1/4 Nyquist frequency (f_N) capable of being sampled before spurious frequency components appear in the Fourier plane divided by the minimum detectable signal. The present device is driven by an off-chip digital serial-to-parallel converter, limiting inputs to digital signals. Therefore, we define the S/N ratio for a square wave at f_N as the maximum first-order peak diffraction intensity at f_N divided by the noise at f_N. The noise from perodic defects at the pixel frequency is at twice f_N and is small, as can be seen in Fig. 9(a). The noise at f_N is predominantly caused by random defects in the membrane surface. This noise, however, is spread into all frequency components, and, thus, its contribution at any one

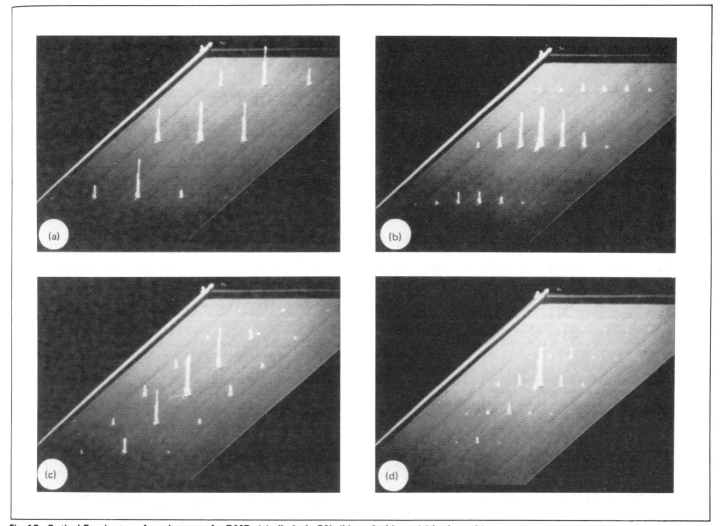

Fig. 12. Optical Fourier transform images of a DMD: (a) all pixels ON, (b) vertical bars, (c) horizontal bars, and (d) checkerboard pattern at the Nyquist frequency.

frequency is also small. For this device the S/N was measured to be in excess of 57 dB. Our measurement was limited by the temporal noise in the CCD camera electronics.

Figure 12 demonstrates optical Fourier transforms of the four image conditions shown on device B in Fig. 4(c). All photographs were taken with the zeroth-order saturated in order to enhance the amplitude of the higher-order peaks. Figure 12(a) shows all pixels ON. Figures 12(b) and 12(c) show vertical and horizontal bars, respectively, at the Nyquist frequency. Figure 12(d) shows a checkerboard pattern at the Nyquist frequency.

8. FUTURE DEVELOPMENT

A program is under way to reduce the particulate contamination between the membrane and the silicon substrate, the major cause of defects in the device. Another program has begun to develop on-chip address circuitry for the device. Because the membrane tension varies with humidity, a hermetic package is being developed. This package will also allow operation of the device in vacuum. Although gas diffusion across the membrane is negligible, air compressed in the pixel air gap increases the response time and reduces sensitivity. Finally, a new optically addressed DMD is in development.

9. CONCLUSION

The deformable mirror device is a two-dimensional, fast, analog, spatial light modulator integrated on silicon. The device's operating parameters useful for optical information processing applications have been presented. A model has been developed to characterize the Fourier transform properties of the DMD. The model was compared with step grating array data and found to be in good agreement. Optical Fourier transform data were taken using the DMD. The modeled diffraction efficiency behavior was compared with the experimental data with some differences discussed. The calculated value of the peak first-order diffraction efficiency, 8.2%, is in good agreement with the measured value of 8.4%.

A S/N ratio in excess of 57 dB was measured at the Nyquist frequency. Finally, the usable spatial frequency range of the device was measured to be from 10% to 100% of the Nyquist frequency.

10. ACKNOWLEDGMENTS

We wish to acknowledge the contributions of M. W. Cowens, G. A. Frazier, and W. E. Nelson for their development of the mirror and tail-end DMD process and G. L. Purcell for his design of the drive electronics.

11. REFERENCES

1. D. Casasent, Proc. IEEE 65(1), 143(1977).
2. M. J. Little, P. O. Braatz, U. Efron, J. Grinberg, and N. W. Goodwin, SID Digest of Technical Papers, p. 250 (1982).
3. R. A. Sprague, W. D. Turner, and L. N. Flores, in *Advances in Laser Scanning Technology,* Leo Beiser, ed., Proc. SPIE 299, 68(1982).
4. W. E. Ross, D. Psaltis, and R. H. Anderson, in *Real Time Signal Processing V,* Proc. SPIE 341, 191(1982).
5. L. J. Hornbeck, IEEE Trans. Electron Devices ED-30(5), 539(1983).
6. R. A. Meyer, D. G. Grant, and J. L. Queen, Technical Memorandum TG 1193A, Johns Hopkins University (1972).
7. D. Joyeux and S. Lowenthal, Appl. Opt. 21, 4368(1982).

The current status of two-dimensional spatial light modulator technology

Arthur D. Fisher and John N. Lee

Optical Sciences Division, Naval Research Laboratory, Washington, DC, 20375

Abstract

An introduction and comparative overview to the state of the art of two-dimensional spatial light modulator technology is provided, touching on the basic operation and performance of most of the more promising electronically- and optically- addressed device technologies. The fundamental functional capabilities and potential applications of these light control devices are also discussed, and some projections are offered on the future directions of spatial light modulator technology.

1. Introduction

Devices for impressing information onto one- and two-dimensional optical data fields play an essential role in configuring optical information processing systems to efficiently exploit the inherent speed, parallel-processing, and interconnection capabilities of optics. These "spatial light modulator" (SLM) devices generally modify the phase, polarization, amplitude, and/or intensity of a spatial light distribution as a function of electrical drive information or the intensity of another light distribution. An attempt is made here to provide a perspective of where two-dimensional spatial light modulator development is today and where it can be expected to be in the near future. First, SLM applications are briefly overviewed below to provide an appreciation of why SLMs were initially developed and are becoming increasingly critical to the field of optical information processing. An overview of the state of the art of many of the more promising 2-D SLM technologies which have been advanced over the last two decades follows in Section 2 below, with a large number of both electronically- and optically- addressed devices being discussed. A description of the fundamental functions that are performed by 2-D SLMs in optical processing systems follows and provides an additional perspective for a comparative discussion of the functional and performance capabilities of the various SLM devices. Finally, some comments will be made on future directions for 2-D SLM technology.

Much of the initial motivation for the development of light control devices was provided by the quest for projection and flat-screen devices for television and computer display. Other related early applications included page composers or formatters for printing purposes, and proposals to utilize the high information densities of 2-D data fields for memory applications. The next major impetuses for SLM development followed from the discovery of holography and the recognition of the the Fourier transforming properties of optical propagation and simple lenses. The invention of the laser made both of these areas practical, and the field of coherent optical information processing was born [1,2,3]. Initial proposed applications of holography included robust optical memory exploiting the distributed nature of holographic recordings and nondestructive testing utilizing the ability of a double-exposure and time-exposure holography to visualize tiny deformations and vibrations. The initial applications of Fourier principles progressed from spatial-filtering systems for image enhancement and deblurring [4,5] to the matched-filter correlator for pattern recognition [6] and to a major success in the processing of synthetic aperture radar (SAR) signals [3,7]. These application areas were first pioneered with film, but require SLMs in the roles of "real-time film" and electrical-to-optical input transducers to be truly practical. A variety of review articles [8-16] provide a more detailed glimpse of early SLM devices and their motivating applications.

From these roots, the number of potential SLM applications has increased rapidly with developments in the field of optical information processing in general, which is used here in a broad sense to encompass the interrelated areas of optical signal processing, data processing, and computing. A few of these potential SLM applications are outlined here, but a more complete overview can be found in the growing number of books, special journal issues, and conferences on optical computing and related areas, a very small subset of which are listed in references17-28. High-performance optical systems have been configured to perform such real-time operations on temporal electronic signals as [28,29]: spectrum analysis, convolution, correlation, adaptive filtering, triple-product processing, and generation of time-frequency representations (e.g., ambiguity or Wigner distributions). However, those systems have generally made much heavier use of the capabilities of 1-D acousto-optic modulators, rather than 2-D SLMs.Other recent thrusts in optical information processing which could benefit from 2-D SLMs include: enhanced- matched-filter [30,31] and non-correlation[32,33,34] approaches to pattern recognition, white-light/color image processing [35], tomographic transformations [28], solution of partial differential and

Table I   Two-Dimensional Spatial Light Modulators

No.	NAME/TYPE	MODULATING MATERIAL	ADDRESSING OPTICAL SENSOR	ADDRESSING ELECTRONIC	DEVELOPED AT
O.1	Phototitus	KD*P (KD$_2$PO$_4$)	Amorph. Se	n	LEP (France)
O.2		,,	Si photodiode	n	Lockheed
E.3	Titus	,,	n	e-beam	LEP,CMU, Sodern*
O.4	LCLV	Twisted nematic liquid crystal	CdS	n	Hughes*
O.5		,,	BSO	n	Thom.CSF
O.6		,,	,,	n	Lockheed
O.7		,,	Si photodiode	n	Hughes
E.8		,,	n	CCD	
O.9		Ferroelectric liquid crystal	BSO	n	Lockheed,NEC, U. of Co.,
E.10		,,	n	Matrix	Displaytech
O.11		Guest-host LC	n	Si Circuits	U. of Edinburgh
O.12	VGM	Nematic liquid crystal	ZnS	n	Hughes,USC, Xerox
O.13	PLZT	PLZT	Si phototrans.	n	UCSD
E.14		,,	n	Si transist.	,,
O.15	FERPIC/	,,	ZnCdS	n	Bell
O.16	FERICON/ CERAMPIC	,,	PVK	n	Sandia
E.17		,,	n	matrix	
O.18	RUTICON	Deformable Elastomer	Amorph. Se	n	Xerox
O.19		,,	PVK:TNF	n	Xerox,Harris
E.20		,,	n	e-beam	IBM
O.21	MLM	Deform. mem.	Si photodiodes	n	Perkin-Elmer
E.22		,,	n	electrodes	,,
O.23	DMD	,,	Si phototrans.	n	TI
E.24		,,	n	Si CCD,trans.	TI,RCA
E.25	Micro-mechanical	Cantilevered beams (SiO$_2$ or metal film)	n	Si Circuits	TRW,IBM, Telesensory,TI
E.26			n	e-beam	Westinghse,RCA
O.27	TP (Thermoplastic)	Thermoplastic	PVK:TNF	n	Harris,NRC*, Xerox,Honeywell Kalle-Hoechst* Fuginon*
E.28	Lumatron	,,	n	e-beam	CBS ,ERIM
O.29	PROM	BSO or BGO	BSO or BGO	n	ITEK,USC, Sumitomo*
O.30	PRIZ	,,	,,	n	USSR
O.31	Volume-holographic	Photorefractive crystals	Photoferroelec. (Holo.,TWM,FWM)	n	Hughs,others, Thom.CSF,
O.32	PICOC	BSO	BSO	n	USC,Cal.Tech.
E.33	Electro-absorption	GaAs (Franz-Keldysh)	n	GaAs CCD	Lincoln
O.34			Photogeneration	n	
O.35	MSLM	EO XTal (LiNbO$_3$,KDP, LiTaO$_3$)	Photocath&MCP	n	MIT,Optron*, Hamamatsu*
E.36			n	e-beam	MIT
O.37	PEMLM	Deformable membrane	Photocath&MCP	n	England,NRL
E.38			n	e-beam	
O.39	Liquid-film	Oil-film (Polysiloxan)	Heat absorption in plastic substr.	n	Switzerland
E.40		Ethanol	n	Resistive heat matrix	Canon
E.41	Eidophor Talaria	Oil-film	n	e-beam	Greytag*, GE*
O.42		Deformable gel	Photoconductor	n	Switzerland
O.43	VO$_2$	VO$_2$ (Vanadium dioxide)	VO$_2$ (Heat abs.)	n	Vought,USC
E.44			n	e-beam	
O.45	Librascope	Smectic liquid crystal	Liquid XTal (heat absorp.)	n	HP,Singer*
E.46	LIGHT-MOD SIGHT-MOD	YIG (Y$_3$Fe$_5$O$_{12}$) (Magneto-optic)	n	Matrix	Litton* Semetex*
E.47	LISA	,,	n	,,	Philips
E.48	Particle suspension Opt.Tun.Arr.	Anisotropic particle suspension	n	Matrix	Bell, Holotronics, VARAD
E.49	Bragg	Multichannel 2-D Bragg cell	n	Acoustic beams	Harris
E.50	TIR	LiNbO$_3$	n	Si Circuits	Xerox*
O.51	Platelet laser	GaAs	GaAs	n	ETL (Japan)

NOTES.
* = SLM is commercially available from this company.
n = not applicable (e.g., optical sensor for E-SLM)

integral equations [36], and high-resolution adaptive wavefront estimation and compensation [37,38]. There has also been considerable interest in optical numeric processing, which has resulted in a large number of analog and digital optical processing architectures, most of which could profit from high-perfomance SLMs. These are generally linear algebraic processors, based on the performance of matrix-vector and/or matrix-matrix products by a variety of schemes ranging from direct parallel mapping with anamorphic optical systems [39,40] to [18-20] parallel&serial systolic, engagement, and outer-product optical processing approaches.

Presently, there is a growing trend towards more fully exploiting the unique three-dimensional geometric capability of optics to provide arbitrary, fully populated, reconfigurable interconnections between two-dimensional processing planes. There is a major thrust is in the development of novel electronic processors which utilize SLMs in implementing parallel optical interconnections [24,41] between systems, boards, processors, integrated circuits (ICs), and/or even individual devices on an IC. Another major research direction is drawing upon an emerging body of knowledge about intrinsically parallel computational models and algorithms to guide the design of efficient parallel optical architectures. For example there have been exciting recent proposals for highly-parallel optical computing architectures which exploit SLMs to implement cellular automata [42,43,44], generalized parallel finite-state machines [45], associative and neural-network processing [24,46,47], parallel logical inference [48,49], or parallel low-level vision algorithms [50]. There is hope that these new concepts will be able to address demanding problem domains which are defying the capabilities of conventional electronic processing approaches- such as multi-sensor processing, large numeric models, image understanding, robotic manipulation, speech understanding, high-level symbolic processing, and a variety of difficult artificial intelligence problems.

2.Specific Device Technologies

The characteristics of many of the more promising SLM technologies are summarized in Tables I and II. The optically- or electronically- addressed devices are denoted by the either the letter "O" or "E", respectively, in the first column of Tables I and II, and referred to in the text as O-SLM's or E-SLM's. Not all variants or developers of each modulator type have been included, and some are mentioned only to provide a historical perspective.

Optically-Addressed Devices

Most optically-addressed spatial light modulators can be loosely grouped into one of a few classes based on their overall structure, with some subclassification by modulating

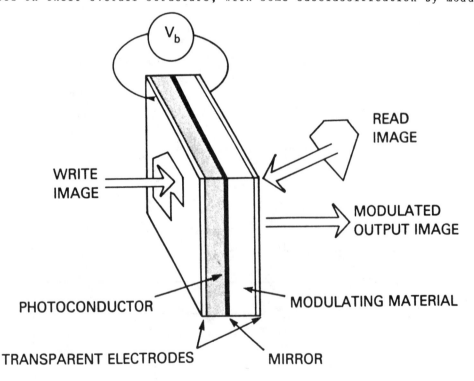

Figure 1. Basic sandwich structure of many optically-addressed 2-D spatial light modulators.

material. Many of the earliest optically-addressed spatial light modulators adopted the basic sandwich structure illustrated in Fig.1. In operation, a bias voltage applied to the sandwich is shunted within the illuminated regions of the photoconductor to a voltage-controlled phase, amplitude, and/or polarization modulating material (e.g., electrooptic). In the reflective configuration illustrated, there is a mirror and light-blocking layer at the center of the sandwich, which allows the written input information to be read out by reflection from the modulating-material side of the SLM. As seen in Table I, the modulating materials which have been employed in this configuration include electrooptic crystals (e.g., KD*P) in the Phototitus [51,52,53], liquid crystals in the Liquid Crystal Light Valve (LCLV) [54,55] and Variable Grating Mode (VGM)devices [56], and ferro-electric ceramics in the PLZT devices [65].

In the slightly modified sandwich structure of Fig.2, the photosensor transfers the bias voltage to a mechanical material which is deformed by the applied electric field. Reflecting a read beam from the deformed surface thus produces a phase modulated output beam, which can be viewed as an intensity image with suitable interferometric or phase-contrast/schlieren optics. Flexible materials which have been employed in this basic configuration include: bulk deformable elastomers in the Ruticon devices [57,58,59], plastics which deform only when heated in the thermoplastic (TP) device [60], a thin film of an insulating liquid [61] or gel [62], thin conducting membranes stretched over an array of insulating holes in the Membrane Light Modulator(MLM) [63,64], and arrays of tiny metallic or silicon dioxide diving-board-like, cantilevered-beam structures [66]. In most of these structures electrical contact is directly made to a reflecting metallic coating on the deformable surface; however, in the Beta-Ruticon [59] and Thermoplastic devices the deformable surface is contacted by an ionized gas. A conducting liquid is employed for contact in the Alpha-Ruticon [57].

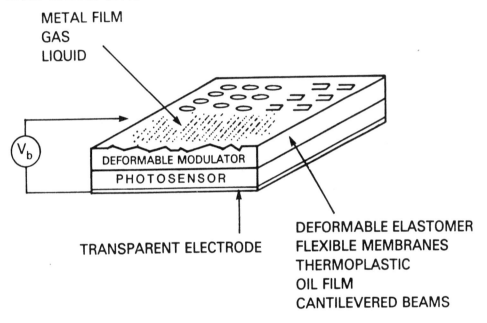

Figure 2. Spatial light modulator structure employing mechanical deformation of materials.

Materials which have commonly been employed as the photosensor layer in the sandwich-type SLMs of Figs. 1 and 2 include the photoconductors PVK:TNF, amorphous Se, CdS, ZnS, and ZnCdS. More recently, sensitivity, speed, and resolution improvements have been obtained by replacing the photoconductor in the LCLV and Phototitus sandwiches by a large planar silicon photodiode junction [53,67,68]. In a related development, new hybrid planar 2-D SLM structures are emerging which are compatible with planar silicon microcircuits such as arrays of phototransistors. For instance, one version of the Deformable Membrane Device (DMD) [69] is fabricated directly on a standard Si wafer by etching holes in a thick polysilicon layer over an array of Si phototransistors, and then covering the device with a reflecting flexible membrane. The phototransistor in each resolution cell controls the potential applied to an electrode below the membrane modulator element formed at each hole. In a more recent variant of the DMD, the membrane elements are replaced by etched-out overhangs of silicon dioxide which form flexible cantilevered beams. In another device [70], islands of recrystallized polysilicon are deposited on a PLZT substrate to produce an array of modulator cells; each Si island contains a phototransistor which controls the potential applied across the surface of an adjacent region of the PLZT. A read beam transmitted through this PLZT region, and on through the PLZT substrate, is thus modulated by the transverse electrooptic effect in the PLZT. The addition of a reflecting coating on the PLZT surface between the Si islands results in a reflective configuration, where the readout beam enters through the PLZT substrate and the output beam is reflected back out. This configuration offers lower operating voltages and better isolation between the write and read beams. Another proposed improved version of this device is built on a commercially-available, silicon-on-sapphire substrate with islands of single crystal PLZT deposited through openings etched in the Si.

By utilizing a modulating electrooptic crystal which is also photoconductive, the Pockels Readout Optical Modulator (PROM) [71,72,73] and PRIZ [74] devices can assume a very simple sandwich structure, which eliminates the separate photosensor. These devices consist merely of two transparent electrodes which are isolated from a photo-refractive crystal such as BSO ($Bi_{12}SiO_{20}$) or BGO ($Bi_{12}GeO_{20}$) by thin dielectric layers; in some instances even the dielectric layers can be eliminated. Different write-beam and read-beam wavelengths are generally employed in these devices. A short-wavelength write-beam generates charge carriers, and the read beam is at a longer wavelength where additional carriers are not generated. By properly applying an external bias voltage the photo-generated charges can be made to drift through the crystal and produce spatial electrooptic refractive index changes as a function of the input imagery. These index changes phase modulate a read beam transmitted through the crystal. Reflective readout similar to the other modulators discussed above can be also obtained, by coating one surface with a reflective layer.

Even simpler structures, with many important SLM properties, are possible with such photorefractive ferroelectric materials as BSO, BGO, Fe:$LiNbO_3$, $BaTiO_3$, KTN, or SBN. Employing one of these materials with no additional structure or only an applied bias voltage, one can perform such operations as: [75-81] image multiplication and amplification by two-wave mixing, three-wave mixing, or by degenerate four-wave mixing (FWM), real-time holography, and long-term storage of high-resolution holograms. A relatively new operating mode for these materials, called the Photorefractive Incoherent-to-Coherent Optical Converter (PICOC) [82], fills in most of the missing functions of a general SLM including wavelength conversion and incoherent-to-coherent conversion. This mode of operation involves writing a spatial photorefractive index grating in the material with two coherent light beams and selectively erasing the grating with the image information in a third beam which need not be coherent with the other beams. The modulated grating can be read out with a fourth read-beam. The volume-holographic photorefractive operations mentioned here involve separation of the photo-induced charges over the wavelength-scale distances of interference fringes. The PROM and PRIZ discussed above generally involve much larger charge separations, which restricts the number of suitable photorefractive materials and reduces the image resolution by a factor of more than one hundred, due largely to fringing electric fields. On the other hand, the PROM/PRIZ structure is generally easier to use, being a bulk modulator which does not require coherent auxillary beams or spatial carriers.

Electro-absorption, or the Franz-Keldysh effect, provides the basis for another type of single-material modulator [83] The active material, GaAs, is photo-conductive, and a device can be configured where optically-generated carriers produce fields which change the hole-electron bandgap and thereby modify the optical absorption at wavelengths near the band edge. (GaAs can also potentially be employed as a photorefractive modulator.)

Heat generated by absorption of the writing image intensity is an alternative optical sensing method which results in fairly simple device structures. The vanadium dioxide (VO_2) [84,85] and smectic liquid crystal devices consist simply of a single thin film of the active material. Optical heating of the material causes phase transitions that modify its optical properties. In some instances an auxiliary heating means is provided for active temperature control and erasure purposes. In another variant (0.39 in Table I), a reflective oil film is deformed by local heating from an underlying, light-absorbing, plastic substrate [61].

A significantly different SLM structure, where photoemission is employed as the light sensing mechanism, is illustrated in Fig. 3. In operation, an incident write image is converted by a photocathode into an electron image, which is amplified by a device called a microchannel plate (MCP). The MCP is an array of tiny tubes (~10 um diameter), each acting as a continuous-dynode electron multiplier, where electrons "bouncing" down the tubes generate additional electrons by collisons with the walls. The electrons exiting from the MCP are deposited onto a mirror on the back of a light modulating material. There is also a grid structure between the MCP and modulating material to facilitate the active removal of charge by secondary emission of electrons from the modulating material. The grid controls the energy of the primary electrons and collects the emitted secondary electrons [37]. The modulating material can be an electrooptic crystal, such as lithium niobate or lithium tantalate in the Microchannel Spatial Light Modulator (MSLM) [86,87,88], or a reflective, flexible membrane array covering the MCP in the Photo-Emitter Membrane Light Modulator (PEMLM) [89,90]. These devices offer storage of images as a net positive or negative charge distribution on the modulating material. Storage times of more than six months have been observed in some MSLM devices.

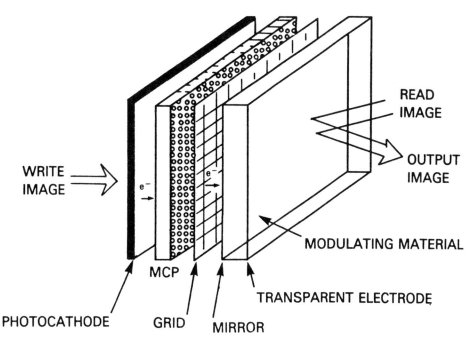

Figure 3. Spatial light modulator employing photoemissive light sensing and microchannel plate electron amplification.

Bistable optical devices (BODs), which have been receiving increasing attention in recent years, can also perform many of the functions of an optically-addressed spatial light modulator. (References 21, and 91-97 provide an overview of recent developments in this field.) Bistable optical devices often take the form of an optical etalon, such as a Fabry-Perot cavity or multilayer interference filter, containing a nonlinear material whose optical absorption, refractive index, and/or physical length are a function of the incident optical intensity. Nonlinear materials which have been utilized in the Fabry-Perot cavities include Na vapor, CS_2, nitrobenzene, CdHgTe, CdS, CuCl, ZnS, ZnSe, InAs, InSb, GaAs, and multiple quantum well (MQW) structures consisting of very thin (0.5-100 nm) layers of GaAs and $Al_{1-x}Ga_xAs$. Recent attention has focused on the last three materials and on interference color filters with layers of ZnS or ZnSe, all of which offer fairly large nonlinearities at room temperature. In operation, part of the incident intensity on the bistable device is transmitted or reflected to form an output beam. The nonlinear material provides a feedback mechanism, whereby the **intensity** in the etalon changes the effective cavity length (and/or absorption) which in turn modifies the cavity **intensity**. The net transmission or reflection follows a nonlinear transfer function of the general bistable form shown in Fig. 4, which is characterized by hysteresis with two possible stable outputs for input intensities near I_2. The feedback required for bistablity has also been

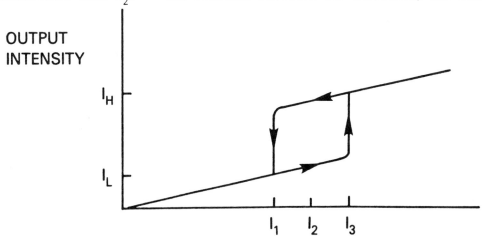

Figure 4. General nonlinear transfer function of bistable optical devices.

realized in other configurations, such as: the self-electrooptic effect device (SEED) which places an MQW structure inside a p-i-n photodiode [94,98]; hybrid structures where the output intensity from an etalon is incident on a detector which it turn drives an electrooptic (e.g., lithium niobate) modulator in the cavity[99]; laser diode cavities where the external intensity tunes the cavity resonance [97]; a conventional SLM with its output beam fed back to its input [37,96,101,102]; or reflection at an interface with a nonlinear material.

To be useful in SLM applications these bistable devices must be able to operate on 2-D information or image fields. To date, simultaneous parallel operation of only 2 to 25 elements has been demonstrated [100]. Low-contrast, with I_L large relative to $I_H - I_L$ in Fig.4, and large input-offset or holding intensity, I_2, also limit the utility of many of the bistable techonologies. High speed is a potential strong point of these devices; some of the bistable etalon devices offer picosecond switching times. Sustained operation of most of the bistable devices has generally required 1 to 10 mW of optical power per pixel (for switching times in the nsec to psec range), which implies that sustained operation of a large array may demand impractically large optical intensities. For example, a 10^6 element array constructed in 1 cm^2, would require 1 to 10 KW/cm^2 of optical intensity, with 1 KW/cm^2 or more being dissipated as heat [103]. An alternative is a low duty cycle operating mode, with occasional transient high-speed decisions being made.

A closely-related SLM device is the image-emission platelet laser, which is essentially a GaAs etalon, but it is pumped by a write beam image of one wavelength to produce a self-luminous output image at another wavelength by lasing [104]. Operation on imagery, with a resolution of 50 lp/mm has been demonstrated.

Electronically-addressed devices

Many electronically-addressed SLM's (E-SLMs) are closely related to a corresponding optically-addressed device sharing the same modulator material and hence material-dictated performance properties; in many instances both devices were developed by the same research group. This is reflected in Tables I&II, where the electronically-addressed devices are grouped among the optically-addressed SLM's, mostly according to the modulating material. An alternative classification is shown in Fig.5 according to addressing mechanism, which tends to dictate the overall physical structure. The discussion below is organized according to this addressing mechanism classification.

In the electron-beam devices, the modulating material is placed in a vacuum envelope and written on by a scanned electron beam, much like the phosphor screen in a cathode ray tube. Specific examples of electron target materials include: electrooptic crystals in the Titus [105,106] and MSLM [107], electrostatically deformable materials in the MLM [108,109], an oil film in the Talaria and Eidophor [110], or thermally modifiable materials in the vanadium dioxide [84] and thermoplastic devices. Since there is no optical sensor to be isolated from the read beam, readout by either transmission or reflection is often possible. An additional flood electron gun is sometimes included for electron erasure by secondary emission. In the PEMLM and MSLM devices, an MCP in front of the electrooptic crystal intensifies the electron beam for faster charge writing, and the incorporation of a photocathode results in a dual-function device offering both electrical and optical addressing [107].

Individual pixels are adddressed in the electrode matrix devices at the intersections of two perpendicularly-crossing linear arrays of electrodes. The crossed electrodes can either be on opposite faces of the modulating material, as in the optical tunnel array (OTA) device, or in a single plane, as in the LIGHTMOD/SIGHTMOD [111] or LISA magnetooptic devices. The OTA device contains a dielectric with pores or "tunnels" filled with a particle-in-liquid suspension. Changes in opacity are produces by reorienting the particles under the influence of a voltage applied between the opposing transparent electrodes. The SIGHTMOD/LIGHTMOD and LISA devices consists of a 2-D array of small mesas (76 x 76 um in the S/ LIGHTMOD) of the magnetooptical material yttrium iron garnet (YIG). Each mesa is a single magnetic domain which can be in one of two opposite magnetic orientations; the two orientations produce different Faraday rotation in the polarization of a transmitted readout beam. Operation between crossed polarizers thus yields two binary amplitude modulation levels. The conjunction of currents in both crossing electrodes at the ion-implanted corner of a mesa induces switching in the SIGHTMOD/LIGHTMOD. In the LISA there is a small resistive element at each mesa corner to thermally induce domain flipping. The Canon liquid-film device also employs local heating by a resistive element at each matrix pixel [112].

Simple electrode matrix addressing schemes are generally restricted to modulating mechanisms with a threshold-like or rectifying nonlinear drive characteristic that allows discrimination of the drive signal due to two intersecting electrodes from that of one electrode. An alternative is to add an active device such as a thin film transistor at each

ELECTRONICALLY-ADDRESSED SLMs

Electron Beam

NAME	MOD. MATERIAL	DEVELOPER
Titus	E-O	LEP, CMU
MSLM	E-O	MIT, Hamamatsu
PEMLM	membrane	NRL, England
Ruticon	Elastomer	IBM
Micro-mech.	Cant. beams	Westinghse, RCA
Eidophor, Talaria	Oil film	GE, Swiss
Lumatron	Thermoplas.	CBS, ERIM
VO_2	VO_2	Vought, USC

Electrode Matrix

NAME	MOD. MATERIAL	DEVELOPER
PLZT	PLZT	Bell, Sandia
MLM	Membrane	Perkin Elmer
L/SIGHTMOD, LISA	YIG (Magnetoopt.)	Litton Philips
Liq. Film	Thermal I^2R	Canon
Opt. Tun. Arr.	Anisotropic part. sus.	Holotronics

Semiconductor

NAME	MOD./ADDR. MATERIAL	DEVELOPER
LCLV	Liq. Xtal/ Si CCD	Hughes
PLZT	PLZT/ Si Transist.	UCSD
DMD	Membrane/ Si CCD, Transist.	TI, RCA
Micro-mech.	Cant. Beams/ Si Circuits	TI, Telesens. TRW, IBM
TIR	$LiNbO_3$/Si Circuits	XEROX
Electro-abs.	GaAs/GaAs CCD	Lincoln

Other

Multichan. Bragg	Acoustooptic	Harris
Display Dev.	Misc.	Many

Figure 5. Electronically-addressed spatial light modulators classified by addressing mechanism.

intersection, which places the device in the semiconductor-addressed category and can greatly increase the device's complexity, fabrication/yield difficulty, and expense. Matrix addressing often offers the option of loading an entire row of information simultaneously to obtain a frame update time equal to the material response time multiplied by the number of rows. However in a device with storage, such as the LIGHTMOD/SIGHTMOD, random pixel-by-pixel addressing can be advantageous for some applications.

Many of the newer additions to the repertoire of E-SLMs employ semiconductor addressing. As mentioned above, some of these employ semiconductor devices at the junctions of a crossed matrix to provide nonlinearity. Others implement specialized serial addressing structures such as shift registers or strings of charge coupled devices (CCDs). For example, the CCD-addressed LCLV [67,68] contains an array of parallel 1-D analog CCD shift registers. In operation, imagery is first formatted into lines by a master serial-input CCD register, and is then shifted line-by-line into the array via parallel operation of all the other CCD shift registers. After N shifts a full NxN frame has been loaded, and the charge image is then dumped through the silicon substrate to the mirror on the back of a liquid crystal modulator cell. Another silicon-addressed liquid crystal modulator is under development, with each pixel containing an nMOS memory element [113]. In an electronically-addressed variant of the DMD [114], the pixels are arranged in an x-y matrix, with each pixel containing a transistor and storage element. Each line of imagery is serial-to-parallel converted in a master shift register and then latched and parallel transfered into a line of the 2-D matrix under the control of an address decoder. Other proposed variants of the DMD employ full CCD addressing, similar to the LCLV, and more recent versions of the DMD deflect cantilevered overhangs of silicon dioxide instead of the flexible membrane elements mentioned above. The Total Internal Reflection device (TIR) [115] also places an electrooptic material (lithium niobate) in direct contact with silicon addressing circuitry. Closely-spaced line electrodes metalized on the Si substrate produce fringing electric fields which electrooptically induce refractive index changes in the lithium niobate . A phase-grating results that diffracts and modulates a 1-D readout beam entering into the side of the lithium niobate crystal and then reflected by total internal reflection off the bottom (electrode-side) surface of the crystal. To date, the TIR device has has been fabricated in only a 1-D format. However, with a suitably large crystal, it should be possible to apply a full read beam with multiple rows of imagery at the required oblique angle for total internal reflection. The GaAs electroabsorption device has also been implemented using direct electronic access, with the modulating fields being provided by charge packets in a high-speed GaAs CCD [83]. The silicon+PLZT [70] techonology discussed above is also amenable to implementing silicon addressing circuitry.

In a significant departure from the above electronic-addressing schemes, the multichannel Bragg cell [116,117] addresses a 2-D format by stacking an array of 1-D acoustooptic Bragg cells. In practice, a single slab of acoustooptic material is employed, with a linear array of transducers launching parallel acoustic beams accross the readout beam aperture. The acoustic beams produce traveling refractive index gratings which can diffract a readout image to produce a modulated output beam. The multichannel Bragg approach can potentially offer a larger, more spatially uniform, dynamic range of modulation than most other SLM technologies, and promises a respectable number of resolution elements, particularly along the direction of propagation. However, lack of storage is a severe limitation, unless multichannel high-bandwidth electronic data happens to be available as part of the application. The Bragg cells often are operated with very low duty-cycle illumination, using only a very short flash of light at the instant all the information has propagated into the cell. Therefore, in such cases, the optical processor sits dormant most of the time, and its capabilities are thus not fully utilized. In addition the short light pulses must be very intense to provide sufficient energy to a subsequent integrating detector, and optical power is wasted if the pulses are created by a shutter, with the source remaining continuously on.

The rapidly developing field of display devices can also provide some potential candidates for an electronically-addressed SLM. (References 8, 118, and 148-151 provide an overview of display technology. The yearly digests of the Society for Information Display International Symposia (SID) and regular Proceedings of the Society of Photo-Optical Instrumentation Engineers (SPIE) devoted to display devices are also excellent sources for additional information.) Many of the addressing schemes employed in E-SLMs were first pioneered in a display context, e.g., scanned electron beams, electrode matrices, and more recently, arrays of thin film transistors in amorphous Si and direct Si addressing. In fact, some of the SLMs in Table I were initially developed as display devices. Most display devices have tended to lack the cosmetic quality and flatness required for use as an E-SLM with coherent light. Furthermore, many of the display technologies are self-luminous, which, as will be discussed in the next section, restricts their potential uses. Some of the emissive display technologies currently being most actively pursued are electroluminescent, vacuum fluorescent, gas plasma panel, LED and, of course, CRT phosphor. Recently, there have been some very promising developments in such non-luminous display technologies as electrophoretic, electrochromic, and, in particular, liquid crystal. In fact, good-quality, low-cost ($\approx$$100.), nematic liquid crystal television displays with up to 240 x 240 resolution elements are now commercially available [118] and are beginning to be exploited by optical researchers [119,120].

3.Spatial light modulator functional capabilities

An appreciation of the fundamental functions performed by a spatial light modulator provides an understanding the issues which have motivated their development and set their performance requirements, and offers a perspective for evaluating and comparing the large variety of alternative devices discussed above. The most basic functions of an optically-addressed SLM can be seen by looking back to Fig.1 as a somewhat generic view of a spatial light modulator. In a typical amplitude-modulation configuration [121], each point in a "read" image incident on the right side in Fig.1 is multiplied by an **effective reflectivity** which is a function of the intensity at a corresponding point in a controlling "write" image incident on the left side, to produce a third reflected "output" image. Since reflectivity is a multiplicative process, the "output" image is essentially the product of the " write" and "read" images, and this device is fundamentally a real-time parallel multiplier of 2-D information patterns. With suitable optics, this basic image multiplication operation can be exploited for applications ranging from actual analog numeric multiplication of vectors and matrices [17,18,39,40] to matched filtering [6,30,31], wavefront conjugation [37,38], reconfigurable interconnect between 2-D fields [24,41], and programmable template-masking operations.

The generic SLM of Fig. 1 is a three-port configuration which can also be loosely viewed as a two-dimensional "optical transistor". Amplification is implemented by applying an intense spatially-uniform read beam; the output image can then become an amplified version of a weak write beam. The desirable feature of decoupling or isolation of the optical system producing the weak write-beam from the amplified output-beam is usually also obtained. Furthermore, when the uniform read beam has different properties than the write beam, the SLM can perform very useful conversion functions, such as from incoherent to coherent light or from one wavelength to another.

The electronically-addressed SLMs generally serve as interface transducers between electronic and optical processing subsystems. Many implement a parallel image multiplication function analogous to the O-SLMs, with an incident read-image being multiplied by an effective reflectivity which is a function of the electronically-written input information. This multiplication operation can be applied to such applications as analog multiplication by: matrices, matched filters, or masks stored or generated by a digital computer; or implementing control or programmable interconnects under the direction of an electronic sequencer. The E-SLMs also serve to input and/or format data into an optical processor, for example, from such non-optical sources as multi-sensor arrays or the results of digital manipulations in a hybrid electro-optical processor. The formatting role depends on the specific addressing mechanism, with sequential-electronic to the 2-D parallel-optical conversion being common. The input and format functions usually employ a uniform read-beam and hence do not demand the multiplication capability. Once the transduction to optics is accomplished it is appealing to remain in the optical processing domain by exploiting the functional capabilities of optically-addressed SLMs, which are emphasized in much of the remainder of this section.

Storage is another basic function provided by SLMs, with the output image becoming the product of a previously-stored write image and the current read image. SLMs with storage can perform memory or information latching functions, and optically-adddressed SLMs with storage are in general a reusable replacement for photographic film. Storage is particularly important in the electronically-addressed devices to avoid continuous updating of a 2-D data array, which places huge information bandwidth demands on the driving electronic processor. In optically-addressed SLMs storage capabilities are often accompanied by the ability to detect very low light level input imagery by integration over an extended period. Modulators in Tables I a&b which can store images for periods on the order of hours, and in some instances much longer, include the Phototitus, Titus, Smectic or Ferroelectric Liquid Crystal, PLZT, PROM/PRIZ, MSLM, PEMLM, Thermoplastic, Photorefractive, PICOC, LIGHTMOD/ SIGHTMOD, LISA, Bistable, and Vanadium dioxide devices.

SLMs exhibiting storage can usually also perform numeric addition between successively-stored images. Virtually all SLMs can also perform addition between simultaneously-applied incoherent image intensities. Some SLMs can also directly perform image subtraction. For example, the previously-mentioned ability of the MSLM and PEMLM devices to either add or remove electrons to/from the stored-charge image distribution on the modulating material can implement algebraic subtraction between the stored image and a new image written with opposite polarity [86,87,122,123]. A few other SLM's, such as the Phototitus can also be operated in an analogous subtraction mode. With the inclusion of additional optics, the arithmetic operations of parallel subtraction [124] and division [125] of image fields can also be implemented with most SLMs.

Another major additional class of SLM functions is in the performance of nonlinear operations on each point in the input write-image intensity. In practice, most modulators are intrinsically nonlinear; for example, many modulators are used in conjunction with crossed-polarizer or interferometric readout, which produces an output beam amplitude proportional to the sine of the write-beam intensity (or output intensity proportional to the sine-squared of the input intensity). Modulation approximately linearly proportional to the input intensity can be obtained by operation near a $\sin 2n\pi$ point of the modulation characteristic. For example the PEMLM and MSLM are easily biased by erasure to a uniform stored-charge distribution corresponding to the $\sin 2n\pi$ point. On the other hand, it is possible to obtain reversed-contrast modulation by operation about a $\sin(2n-1)\pi$ point. The contrast reversal operation can often be implemented on a previously-stored image by manipulating crossed polarizers or schlieren optics in the readout system. A few SLMs can also perform contrast reversal by other means, for instance, in the MSLM or PEMLM by switching between the electron accumulation and removal modes. Other examples of intrinsic nonlinearities are output intensity proportional to $\sin^2(\text{input-intensity}^2)$ in some configurations of the mechanical modulators (e.g., Cantilevered beam, PEMLM, MLM, or DMD), or exp(input intensity) in the electroabsorption devices. (A thin electroabsorption device can offer an output amplitude approximately proportional to to the input intensity.) Other modulation mechanisms, such as photo/cathodochromic materials can also provide output amplitude directly proportional to input intensity.

Another important category of SLM nonlinear operations is image thresholding, where there is little or no change in the output intensity until the input write-beam intensity exceeds a specific threshold level. In analog thresholding, after the input intensity has exceeded the threshold, the output intensity varies as a function of the input intensity. In binary or "hard-clip" thresholding there are two fixed output levels, corresponding to input intensities below or above threshold, respectively. Thresholding behavior can be obtained by employing specific modulator materials, such as liquid crystals, YIG, PLZT, or vanadium dioxide or by using particular aspects of the device physics, as in the MSLM and PEMLM, the platelet laser, and most of the bistable optical devices. The MSLM and PEMLM devices have a very useful intrinsic hardclipping mode where all three critical parameters: input threshold level, below-threshold output level, and above-threshold output level are fully adjustable, including inversion with the above-threshold output less intense than the below-threshold output. Hardclip image thresholding allows decisions (e.g., yes/no, true/false, exists/doesn't-exist) to be made in an optical processor in terms of the two output intensity levels and can also provide a "Schmitt trigger" function to regenerate binary optical signals in cascaded systems. In addition, Boolean logic operations can be performed between multiple input beams; for example setting the input threshold below the logical '1' intensity level produces the OR operation between two input beams, and setting the input threshold at 3/2 of the logical '1' level produces an AND gate between two input beams. With inverted contrast, the NOT, NAND, and NOR operations can be implemented. Hardclipping can also be utilized to build an analog-to-digital converter for optical image fields. Furthermore, hardclipping can form the nucleus of an optical half-tone system which can implement virtually arbitrary nonlinear intensity functions [126,152]; particularly useful nonlinear functions include Log, for transforming multiplicative signals and/or noise into additive signals, and square-root, for obtaining output amplitude modulation directly proportional to the write-beam amplitude.

Some SLMs also have intrinsic abilities to directly perform more advanced processing functions. For example, the variable grating mode (VGM) liquid crystal device produces an output beam modulated by spatial gratings proportional to the local intensity of the input write-beam [56]. Almost any nonlinear function of the input intensity can be implemented by passing the output beam through an appropriate spatial filter in a subsequent Fourier plane. The programmable hardclipping mode of the MSLM and PEMLM devices can be used to implement edge detection beween regions of the input image having intensities above and below the threshold level as follows [87,122,123]. Essentially, both the above- and below-threshold output levels are set to produce the same output intensity (e.g., white), but separated by one full cycle of the device's sinusiodal modulation characteristic. Due to finite device resolution, a line of the intermediate half-cycle level (e.g., black) is produced along portions of the input image divided by the threshold. Other devices such as the PRIZ [72,74]; and photorefractive configurations [127] also offer intrinsic spatial differentiation or edge detection capabilities. The MSLM and PEMLM devices can also intrinsically implement the more exotic function of operating as a pixel array of lock-in amplifiers or synchronous detectors [122,123,128]. This function is implemented by oscillating the device between its charge deposition and removal modes at a rapid rate. Portions of a write image which are flashing on in synchronism with, for example the charge-deposition cycles, result in net charge accumulation over many cycles which eventually integrates up to a visible modulation level. However, portions of the write image which are at constant intensity or oscillating at a slightly different frequency result in no net charge accumulation. Applications of this lock-in amplifier array include discrimination from the ambient background of a structured light pattern projected into a scene for robot vision, removal of the DC bias from a heterodyne interferometric image, and target designation.

Actual experimental demonstrations of some of these more advanced processing functions are illustrated in Fig. 6. These demonstrations were implemented with an MSLM at the Naval Research Lab [122,123,129]. Part 1 b to f of Fig. 6 shows real-time hardclip thresholding of the 6-step grey scale in part 1a, at five different adjustable-threshold levels. Part 2a is an image of two people; and 2b shows the corresponding MSLM output in its normal, non-threshold mode. Reverse-contrast hardclip thresholding of this image at two different threshold levels is illustrated in parts 2c and 2e. Parts 2d and 2f demonstrate real-time edge detection, with the below- and above-threshold output levels separated by one full modulation cycle. (The object on the left of 2e and 2f is the darker leg of the left person.) Part 3 depicts algebraic addition and subtraction by charge deposition and charge removal, respectively. The intermediate results of the following series of matrix additions and subtractions are shown.

$$\begin{matrix} 000 \\ 000 \\ 000 \end{matrix} + \begin{matrix} 100 \\ 100 \\ 100 \end{matrix} - \begin{matrix} 001 \\ 001 \\ 001 \end{matrix} + \begin{matrix} 111 \\ 000 \\ 000 \end{matrix} - \begin{matrix} 000 \\ 000 \\ 111 \end{matrix} = \begin{matrix} +2 & +1 & 0 \\ +1 & 0 & -1 \\ 0 & -1 & -2 \end{matrix}$$

The lock-in amplifier or synchronous detection function is demonstrated in part 4, where 4a-c show the output response elicited by three LEDs simultaneously pulsing at 19, 20, and 21 Hz, as the MSLM is successively locked to each frequency. The normal, noncycling response of the MSLM to the simultaneous presentation of a C-clamp image oscillating at 18

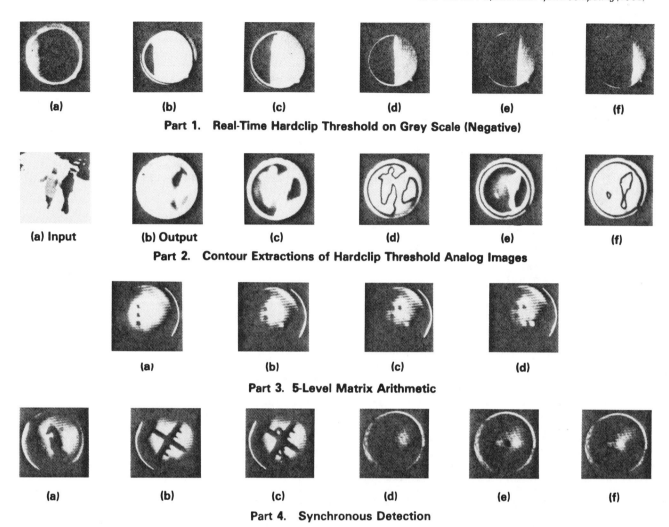

Figure 6. Experimental demonstrations of four advanced intrinsic processing functions of the Microchannel Spatial Light Modulator.

Hz superimposed with an airplane image at 20 Hz is shown in part 4d. In parts 4e and 4f, the cycling MSLM is synchronized to detect only one image despite the simultaneous presence of the other image oscillating at a nearby frequency. This synchronous detection mode can easily operate with kilohertz cycling rates, and discrimination bandwidths as small as 0.01 Hz have been demonstrated.

Some of the technologies mentioned in Section 2, for example, the bistable, self-luminous, and volume-holographic devices, do not completely fall into the basic SLM functional framework outlined at the start of this section and are probably more accurately classed as "SLM-related" devices. For example, most of the bistable optical devices are fundamentally two-terminal devices, with only a write beam and an output beam, and thus do not directly offer the basic real-time image multiplication and image-transistor functions outlined above. They are more analogous to electronic diodes in that their output intensity is a nonlinear interaction function of the beam intensities applied to their input port. The general bistablity characteristic of Fig.4 offers switching and/or memory functions; however, the bistable devices require the continuous application of illumination to store information, unlike many SLMs which have a passive storage mode requiring no power. Thresholding can be implemented by operation near I_3 in Fig.4; and it is possible to configure devices with only threshold and no hysteresis, i.e., $I_1=I_3$ in Fig.4. Some of the SLM three-port operations can be approximated with bistable devices by superimposing two beams at the input, a weak "write" beam accompanied by a stronger "read" beam which provides power for the "output" beam, for example the weak beam can be amplified when the stronger beam biases the device to just below threshold. Similarly, switching between output hysteresis states can can be induced by a weak beam accompanied by a bias beam. The threshold behavior also provides a basis for performing Boolean logic between multiple input beams. The relatively flat regions below and above the hysteresis region provide an optical limiting function, for example, to supress intensity fluctuations in a noisy beam. As mentioned previously, a major potential strength of the bistable device technology is speed - offering nanosecond to picosecond switching times, as opposed to the millisecond operation projected for most SLMs.

The fundamental SLM function of multiplying the read-beam image by the input write information requires a phase, amplitude, and/or polarization modulating material, and hence is not provided by such self-luminous technologies as emissive display devices, arrays of LEDs, or laser diode arrays. Self-luminous devices, however, can still be useful for input and formatting functions. The lack of versatility of emissive devices may explain why there are very few examples of self-luminous optically-addressed modulators. However, there was some work in the 1960's and early 1970's on photoconductor-accessed LED arrays [2] and electroluminescent devices [130,131]. Currently, there is interest in platelet lasers [104], and semiconductor integrated circuits incorporating photodiode-activated LEDs or surface-emitting 2-D planar arrays of injection lasers [41,132,133].

The photorefractive volume-holographic devices can in principle perform most SLM functions, including multiplication and amplification, but their use is generally subject to a variety of additional constraints [76-81]. For example, in four-wave mixing the write (or "pump"), read (or "probe") and output (or "conjugate") beams must have specific geometric, polarization, and intensity relations to each other and must be mutually coherent. The coherence condition generally means that the write and read beams must come from a common laser source, which precludes wavelength conversion or incoherent-to-coherent conversion. The coherence constraints are relaxed in the previously-mentioned PICOC configuration [82]. The four-wave and two-wave mixing technologies also offer additional functions, such as direct conjugate wave generation [135].

4. Spatial Light Modulator Performance Issues

There is a large number of parameters for quantitatively evaluating and comparing how well the various SLM technologies perform their basic functions. Due to space limitations, only spatial resolution and/or total number of pixels, optical-addressing sensitivity, and time response characteristics are given in Table II. Readout dynamic range, e.g., in phase, amplitude, and/or polarization, is another critical parameter which is coupled to such additional issues as: contrast, signal-to-noise ratio, input-output nonlinearity, spatial variations in response uniformity, and the practical consideration of the required operating voltages. Optical quality is also extremely important, involving such issues as scattering, cosmetic defects, spatial uniformity, and optical flatness. The total spatial-temporal throughput, i.e., area x resolution/frame-time, is sometimes used to provide a single measure for crude comparison of the various technologies. Other important parameters include fabrication difficulty, lifetime, reproducibility of behavior, specialized capabilities, and fundamental performance limitations.

Assigning realistic, comparable numeric values to SLM performance is difficult. Performance numbers are critically dependent on measurement conditions, but the conditions are often not given in the literature or are not consistently related to actual device applications. For instance, resolution, sensitivity, and speed parameters can be made arbitrarily good by making the measurement with an arbitrarily small depth of modulation. As another example, dynamic range is closely linked with spatial uniformity in these 2-D format devices. While the measured dynamic range at a single point may be very large, this is useless for most applications if a given analog level is not the same from point to point across the device. In fact, very few 2-D devices can claim a "spatial dynamic range" better than 20 levels (5% uniformity). Performance for new device technologies is also sometimes confused with overly optimistic projections based on perceived limitations. In addition, research groups often concentrate on one or a few performance characteristics at the expense of others.

The performance parameters in Table II should be interpreted with care; they are generally best-reported values or near-term, low-risk projections. They are sometimes not all simultaneously achievable in one version of a given device. The sensitivity and speed parameters in Table II are generally given for a response of 90% of maximum-contrast amplitude modulation and/or 90% of half-wave phase modulation (i.e., 0.9π radians). The time-response data is generally for updating a full frame in the optically-addressed devices, and is footnoted as to whether it is for a full frame or an individual pixel response in the electronically-addressed devices. In the latter case, the full-frame write time is usually the pixel response time mutiplied by the total number of pixels, or, for line-by-line addressing, by the total number of lines. The resolution data is footnoted as to whether it is for 50% or 10% contrast expressed in line-pairs/mm, or for pixel density of discrete-array devices expressed in pixels/mm. When available, the contrast numbers are derived from the modulation transfer function (MTF), which is essentially a plot of contrast of the output beam as a function of the spatial frequency of a sinusoidal fringe pattern applied to the write image; the contrast being measured as (max amplitude - min amplitude) divided by (max amplitude + min amplitude). A footnote in Table II also designates that

Table II   Two-Dimensional SLM Performance

No.	NAME/TYPE	RESOLUTION lp/mm (No. Pixels)	SENSITIVITY[g] ($\mu J/cm^2$)	TIME RESPONSE[g] WRITE (msec)	ERASE (msec)	STORE
O.1	Phototitus	15^a, 30^b	10	0.01	0.03	1 hr
O.2		10^a	2	2	<0.5	5 sec
E.3	Titus	20^a, 30^b	n	30^i	5^i	1 hr
O.4	LCLV	30^a, 40^b	6	10	15	15 msec
O.5		10^a, 40^b	20	15	15	20 msec
O.6		10^a	1	1	1	20 msec
O.7		15^a	1	5	10	10 msec
E.8		$(256 \times 256)^d$		5^{hi}	10^{hi}	10 msec
O.9		10^i	1	1	1	secs
E.10		$(32 \times 32)^d$	n	1	1	secs
O.11		$(16 \times 16)^d$	n	200	400	hrs^f
O.12	VGM	5^a	15	10^3	$<10^3$	$secs^f$
O.13	PLZT	10^d	$2^k, 10^{-3 k}$	$<0.01^e$	$<0.01^e$	$secs^f$
E.14		10^d	n	$<0.01^{h,e}$	$<0.01^{h,e}$	
O.15	FERPIC/	20^a, 40^b	5	<1	10	hrs
O.16	FERICON/ CERAMPIC	20^a, 40^b	5	10	1000	hrs
E.17		$(128 \times 128)^d$	n	0.01^h	0.01^h	10 msec
O.18	RUTICON	$40-120^{a,c,k}$ $10-45^{a,c,k}$	30	5	4	15 min
O.19		..	5	30	70	15 min
E.20		$15^{a,c}$	n	20^i	20^i	mins
O.21	MLM	10^d	2	<0.001	10	10 msec
E.22		$(100 \times 100)^d$	n	$<0.001^h$	$<0.001^h$	$\mu secs$
O.23	DMD	$(128 \times 128)^d$	2	0.025	0.04	200 msec
E.24		$(128 \times 128)^d$	n	0.025^h	0.025^h	$200 msec^f$
E.25	Micro-mechanical	$(18 \times 2)^d$	n	$<0.01^h$	$<0.01^h$	0.01^f
E.26		40^d	n	30^i	30^i	months
O.27	TP	$200-1600^{a,c}$	5	10	100	years
E.28	Lumatron	$70^{a,c}$	n	$<1^h$	1000^i	years
O.29	PROM	6^a, 12^b	5	<0.1	<0.1	<2 hrs
O.30	PRIZ	10^a, 20^b	5	<0.1	<0.1	<2 hrs
O.31	Volume-holographic	$>1500^{a,c}$	1 to 10^4	<30	<30	msecs to
O.32	PICOC	50^a	1	30	10	hours
E.33	Electro-absorption	$(16 \times 1)^d$	n	$10^{-6 h}$	$10^{-6 h}$	msecs
O.34						
O.35	MSLM	10^a, 20^b	$3 \times 10^{-6 e,j}$	10	20	days to months
E.36		3^a	n	$1^e, 200^i$	$1^e, 50^i$	..
O.37	PEMLM	$80^d, 40^a$	$3 \times 10^{-4 e,j}$	0.01^e	0.01^e	days to
E.38			n	$0.01^{e,h}$	$0.01^{e,h}$	months
O.39	Liquid-film	4^a	100	100	<100	<0.1 sec
E.40		$(256 \times 1)^d$	n	1	1	1 msec
E.41	Eidophor	$50^{a,c}$	n	15^i	15^i	0.3 sec
	Talaria	(1023×1023)	n			
O.42		10^a	3	10	10	10 msec
O.43	VO_2	150^a	2×10^4	$3 \times 10^{-5 h}$	<1	years
E.44		20^a	n	$10^{-3 h}$	$10^{-4 i}$	years
O.45	Librascope	40^a	10^4	0.005^h	0.001^h	months
E.46	LIGHT-MOD SIGHT-MOD	$(128 \times 128)^d$ $(512 \times 512)^{d,e}$	n	$10^{-3 h}$	$10^{-3 h}$	years
E.47	LISA	$(256 \times 128)^{d,e}$	n	0.015^h	0.015^h	years
E.48	Particle suspension Opt.Tun.Arr.	$(16 \times 16)^d$	n	1	1	15 msec
E.49	Bragg	$(32^d \times 120)$	n	0.002^i	0.002^i	10 nsec
E.50	TIR	$(5000 \times 1)^d$	n	$<0.001^h$	$<0.001^h$	<1 μsec
O.51	Platelet laser	50^a	10^3	10^{-6}	10^{-6}	nsecs

Notes:
a) 50 % MTF,   b) 10 % MTF,   c) Spatial bandpass (limits or bandwidth),   d) Discrete pixels: pix/mm or ( )=No. of pix
e) Expected performance in near future,   f) Potential exists for Si memory circuits,
g) 90 % of half-wave or full contrast, unless noted otherwise,   h) Time for one pixel (or line),   i) Time for full frame,
j) Quantum-limited (S/N=10, at given resolution),   k) For different implementations of device.
n=Not applicable

some of the devices have a spatial-bandpass MTF, which falls off at both low and high spatial frequencies; either the upper and lower 50% points or the total range of the 50% passband is given. For device comparison purposes, the 50% resolution of a discrete-pixel device can be taken as approximately 1/2 of the actual pixel density. In applications, the total number of available pixels is often of greater import than the actual resolution, particularly with discrete-pixel devices.

The improvement of one SLM performance parameter at the expense of another is often unavoidable as a result of fundamental physical tradeoffs. In many electrooptic modulators, for example, a thinner slab of electrooptic material offers higher resolution, but at the same time also increases the device capacitance, which in turn decreases the device speed. A similar tradeoff is also seen with some mechanical modulators; as the modulator elements become smaller more charge is required to deflect them and the device is slowed [90]. There is also a fundamental tradeoff between resolution and sensitivity. As the resolution elements become smaller more optical energy per unit area (i.e., $\mu J/cm^2$) must be collected to obtain enough photons in each pixel to maintain a given quantum-limited signal-to-noise ratio (S/N). The MCP-intensified PEMLM and MSLM modulators can be operated in this quantum-limited regime, and it is often desirable to optimize the MCP gain so that the number of photons required to drive the device to a usable modulation depth also provides enough photons to obtain the desired S/N [37].

Many aspects of SLM performance arise from an interplay between limits imposed by material properties and device geometries. As an example, consider time response. The typical response times of a variety of modulating materials are given in Table III; however, most SLMs operate considerably slower than these ultimate limits. For instance, the maximum speed of the MCP and silicon-addressed devices is generally limited by thermal dissipation constraints on the maximum current available to charge the modulating material. However, this is impacted by the amount of charge required to obtain full depth modulation (e.g., π phase shift), which is a function of both modulation material properties and the geometric configuration of the modulator (for instance, the thickness and orientation of the crystal in an electrooptic modulator). The time response of photoconductive modulators tends to be limited by the material properties of the photoconductor, particularly carrier decay lifetimes and carrier mobility.

Table III   RESPONSE TIMES OF MODULATOR MATERIALS

Material	Response Time
Electrooptic crystals	1 picosec
GaAs (In MQW or Etalon)	1 picosec
PLZT (Single Crystal)	1 nanosec
Deformable membranes	0.5 microsec
Cantilevered beams	1 microsec
Magnetooptic switching	1 microsec
Liquid crystal (Ferroelectric)	1 microsec
Acoustooptic (Bragg cell)	1 microsec
PLZT (Polycrystalline)	10 microsec
Particle suspensions	1 millisec
Deformable elastomers	1 millisec
Liquid Crystal (Nematic)	10 millisec

5. Current Status and Future Directions

It may be surprising to note that despite all the potential applications, device types, and functional capabilities of SLMs discussed in Sections 1-3, very few, if any, applications of two-dimensional SLMs have progressed to the point of routine use, mass production, and/or commercialization. In fact, the majority of the 2-D spatial light modulators listed in Table I exist only as laboratory prototypes and some are not even being actively developed anymore. The difficulty has been a chicken and egg problem of sorts. The development of high-performance, versatile SLMs has proven to be more difficult than initially anticipated, and in the absence of such a device, researchers have historically been reluctant to propose and attempt to implement sophisticated applications. On the other hand, in the absence of these technologically competitive and/or profitable applications, corporations and funding agencies have not been motivated to provide the levels of support required to tackle the formidable challenges inherent in SLM development. In addition, competing electronic computing and information processing approaches have made rapid advances, particularly in the two decades since the advent of integrated circuits.

There is reason for optimism, however, as recent developments in materials and fabrication techniques begin to produce a new generation of high performance 2-D SLMs [134-139,147]. A variety of specific SLM devices have recently emerged from laboratories into manufacture by commercial sources, these commercially-available devices include the:

LCLV, MSLM, Librascope, PROM, and Thermoplastic optically-addressed devices and LIGHTMOD/SIGHTMOD, Talaria/Eidophor, and Titus electronically-addressed devices. In addition, the DMD, Optical Tunnel Array, PLZT, Phototitus, PEMLM, Ruticon, and/or Vanadium dioxide modulators could potentially become available within the next 5 to 10 years, depending on corporate and funding agency decisions. Also, as mentioned above, nonemissive display devices, such as the liquid-crystal television, have recently become commercially available at low cost and are beginning to offer adequate performance for some E-SLM applications.

Furthermore, it is becoming possible to prototype simple cases of quite a few "unrealizable" sophisticated applications using some of the currently-available SLMs, and very exciting optical information processing concepts are beginning to emerge. Some of these concepts, such as a few of the previously-mentioned architectures implementing intrinsically-parallel computational models and algorithms, offer the potential of significantly exceeding the projected capabilities of conventional electronic computing and processing approaches. The initial prototyping activity can serve an important function in identifying critical parameters and defining directions for future device research.

The presently-achievable performance level is on the order of 100 x 100 resolution elements, 10 Hz framing rates, 1 sec storage, less than 50 $\mu J/cm^2$ sensitivity (O-SLM), 5 level dynamic range, a few wavelengths flatness, and 10 % spatial uniformity. In practice the actual performance levels required are application specific, and all of these features are seldom required simultaneously.

There are a number of promising, but less developed, modulator technologies which may have an important impact in the future; some of these are new organic crystal and polymer electrooptic materials [140,141], ferroelectric liquid crystals [142,143], 2-D arrays of diode lasers [132,133], and stacks of integrated-optical modulators. Other modulation mechanisms which are also currently being explored include photochromic, cathodochromic, electrophoretic, photodichroic, electrowetting, electrocapillary [144], and various magneto-optical effects. Some of the newer semiconductor-compatible technologies offer the exciting prospect of bypassing many of the limitations imposed by material properties and the simple sandwich-type device geometries. One can conceive of a hybrid 2-D SLM built on a semiconductor substrate with simple electronics between a phototransistor and modulator in each resolution cell. A reprogrammable SLM could result which accurately implements such useful nonlinear intensity functions as log, square-root, inversion, and hardclip thresholding, as well as optical logic, image storage, switching functions, and general arithmetic operations (i.e., +, _, x, /). It should also be possible to include connectivity between SLM pixels [145,146], for example to implement arithmetic and/or logic operations between adjacent pixels, spatial image shifting operations, or structures for more sophisticated algorithms. Such a device could play an important role in optical interconnect and SIMD (single instruction, multiple data) parallel processing applications. Candidate silicon-addressed technologies include cantilevered beam, deformable membrane, PLZT, liquid crystal, and electrooptic crystal devices, and also the electroabsoption and surface-emitting injection laser devices constructed with GaAs.

Current SLM research is moving in the direction of a device with better than: 1000 x 1000 resolution elements, kilohertz framing rates, quantum-limited sensitiviy (O-SLM), hour storage time, 100 levels of dynamic range, 1/5 wavelength flatness, and 1% spatial uniformity. There are also a variety of practical issues which should guide future SLM development, these include: ease of use, reproducibility and reliability of operation, minimal and unsophisticated support electronics and other equipment, simple low-loss optical readout system, small footprint on the optical table, and non-intimidating cost (< $5K).

The future of SLM development can be expected to be marked by a continuation of the interplay between applications and devices. Application prototyping efforts set device performance requirements, provide incentives for device development, and point out directions for future device research. On the other hand, optical architectures built today must by necessity be designed to exploit the strengths of existing devices while deemphasizing their weaknesses. One goal of this overview has been to attempt to aid this process by providing systems developers with a perspective of the device technologies and functional capabilities which are currently available.

References

1. Pollock, D. K., Koester, C. J., and Tippett, J. T. (1963). Eds, "Optical Processing of Information." Spartan Books, Baltimore.
2. Tippett, J. T., Berkowitz, D. A., Clapp, L. C., Koester, C. J., and Vanderburgh, A., Jr., eds. (1965). "Optical and Electro-optical Information Processing." MIT Press, Cambridge, MA.
3. Goodman, J. W. (1968). "Introduction to Fourier Optics." McGraw-Hill, New York.
4. O'Neill, E. L.(1956). Trans IEE IT-2, p. 56.
5. Tsujiuchi, J.(1962). In "Progress in Optics", Vol. 2, (E. Wolf, ed), p.133. North Holland Publ. Co., Amsterdam.
6. Vanderlugt, A. B.(1964). Trans IEEE IT-10, p. 139.
7. Cutrona, L. J., Leith, E. N., Porcello, L. J., and Vivian, W. E.(1966). Proc. IEEE 54, p. 1026.
8. Heynick, L. N., Reingold, I., Sobel, A. (1973). Guest eds., Special issue on display devices. IEEE Trans. on Elec. Dev. ED-20.
9. Flannery, J. B. (1973). IEEE Trans. on Elec. Dev. ED-20, p.941.
10. Thompson, B. J. (1977). Proc. IEEE 65, p.62.
11. Casasent, D.(1977). Proc. IEEE 65, p.143.
12. Casasent, D. (1978). Guest ed., Special issue on spatial light modulators. Opt. Eng. 17, p.307.
13. Casasent, D. (1978). In "Applied Optics and Optical Engineering" (R. Kingslake, ed.), Vol. 6, p.143. Academic Press, New York.
14. Bartolini, R. A. (1977). Proc. of the SPIE 123, p.2.
15. Lipson, S. (1979). In "Advances in Holography" (N. Farhat, ed). Marcel Dekker, New York.
16. Knight, G. R. (1981) In "Optical Information Processing" (S. Lee, ed.), p.111. Springer Verlag, New York.
17. Goodman, G. W.(1982). Jour. of Electrical and Electronics Engineering Austrailia 2, p. 139.
18. Caulfield, H. J., Horvitz, S. Tricoles, G. P., and VonWinkle, W. A.(1984). Guest eds, Special issue on optical computing, Proc. IEEE 72, p.755.
19. Psaltis, D.(1984). Guest ed, Special issue on optical computing, Opt. Eng. 23, p.1.
20. (1985). Technical Digest of the OSA '85 Topical Meeting on Optical Computing.
21. Dove, B. L. (1985). Ed., Digital Optical Circuit Technology. AGARD Conference Proceedings No. 362. NTIS, Springfield, VA.
22. Neff, J. A.(1985).Guest ed, Special issue on optical computing. Opt. eng. 24, p.1.
23. Neff, J. A.(1984).Guest ed, Special issue on optical computing. Proc. of the SPIE 456.
24. Horvitz, S. and Neff, J. A.(1985).Guest ed, Special issue on optical computing. Proc. of the SPIE 625.
25. Bell, T. E.(1986). IEEE Spectrum 23, p. 34.
26. Proc. of the SPIE, regular special issues on optical signal processing, information processing, computing, and related topics.
27. Digests of International Commission for Optics (ICO) sponsored meetings, e.g.: 10th IOCC, Boston, 1983; ICO-13,Sapporo, 1984; Image Science '85, Helsinki, 1985; IOCC Jerusalem, 1986.
28. Horner, J. L.(1987). Ed., "Optical Signal Processing." Academic Press, New York.
29. Berg, N. J., and Lee, J. N., eds. (1983). "Acousto-optic Signal Processing: Theory and Implementation." Marcel Dekker, New York.
30. Casasent, D. and Psaltis, D.(1977). Proc. IEEE 65, p.770.
31. Casasent, D.(1985). Opt. Eng. 24, p 27.
32. Leger, J. and Lee, S. H.(1982). J. Opt. Soc. Am. 72, p.556.
33. Guo, Z. H. and Lee, S. H.(1984). Appl. Opt. 23, p. 822.
34. Guo, Z. H. and Lee, S. H.(1984). Opt. Eng. 23, p.723.
35. Yu, F. T.S.(1983)."Optical Information Processing." John Wiley, New York.
36. Lee, S. H.(1985). Opt. Eng. 24, p.41.
37. Fisher, A. D. (1981). "Techniques and Devices for High-Resolution Adaptive Optics." Ph.D. Thesis. M.I.T., Cambridge, MA.
38. Fisher, A. D. (1985). Proc. of the SPIE 551, p.102.
39. Goodman, J. W., Dias, A., and Woody, L. M. (1978). Opt. Lett. 2, p. 1.
40. Tamura, P. N., and Wyant, J. C. (1976). Proc. of the SPIE 83, p.97.
41. Goodman, J. W., Leonberger, F. J., Kung, S. Y., and Athale, R. A. (1984). Proc. IEEE 72, p. 850.
42. Sawchuk, A. A. and Jenkins, B. K. (1984). Technical Digest of the OSA '85 Topical Meeting on Optical Computing, p.TuA2.
43. Jenkins, B. K. and Sawchuk, A. A.(1985). Digest of the IEEE Conference on Computer Architectures for Pattern Analysis and Image Database Management (CAPAIDM), p.61.
44. Yatagai, T. (1986). Proc. of the SPIE 625, p.54.
45. Huang, A. (1985). Technical Digest of the OSA '85 Topical Meeting on Optical Computing, p.WA2.
46. Psaltis, D., and Farhat, N. (1985). Opt. Lett. 10, p. 98.
47. Fisher, A. D. and Giles, C. L. (1985). Proc. of the IEEE 1985 COMPCON Spring Meeting

CH135-2/85, p. 342.
48. Warde, C. and Kottas, J.(1984). Appl. Opt. 25, p.940.
49. Eichmann, G. and Caulfield, H. J.(1985). Appl. Opt. 24, p. 2051.
50. Ichioka, Y. and Ono, S.(1985). Proc. Image Science '85, Helsinki, p. 241.
51. Donjon, J., Dumont, F., Grenot, M., Hazan, J.-P., Marie, G., and Pergrale, J.(1973). IEEE Trans. Elec. Dev. ED-20, p.1037.
52. Casasent, D.(1978). Opt. Eng. 17, p. 365.
53. Armitage, D., Anderson, W. W., and Karr, T. J. (1985). IEEE J. Quant. Electron. QE-21, p. 1241.
54. Bleha, W. P., Lipton, L.T., Wiener-Avnear, E., Grinberg, J., Reif, P. G., Casasent, D., Brown, H. B., and Markevitch, B. V. (1978). Opt. Eng. 17, p. 371.
55. Aubourg, P., Huignard, J. P., Hareng, M., and Mullen, R. A. (1982). Appl. Opt. 21, p. 3706.
56. Tanguay, A. R., Wu, C. S., Chaval, P., Strand, T. C., Sawchuck, A. A., and Soffer, B. H. (1983). Opt. Eng. 22, p. 687.
57. Sheridon, N. K. (1972). IEEE Trans.on Elec. Dev. ED-19, p.1003.
58. Lakatos, A. I. (1974). J. Appl. Phys. 15, p.4857.
59. Ralston, L. M. and McDaniel, R. V.(1979). Proc. SPIE 185, p. 86.
60. Colburn, W. S., and Chang, B. J. (1978). Opt. Eng. 17, p.334.
61. Schneeberger, B., Laeri, F., Tschudi, T., and Mast, F. (1979). Opt. Comm. 31, p.13.
62. Hess, K. and Danliker, R. (1985). Digest of the Conference on Lasers and Electro-Optics (CLEO), p.44.
63. Preston, K. P. (1969). Optica Acta 16, p. 579.
64. Reizman, F. (1969). Proc. Electrooptic System Design, p.225.
65. Land, C. E. (1978). Opt. Eng. 17, p.317.
66. Brooks, R. E. (1985). Opt. Eng. 24, p.101.
67. Efron, U., Bratz, P.O., Little, M.J., and Schwartz, R. N. (1983). Opt. Eng. 22, p.682.
68. Efron, U., Grinberg, J., Braatz, P. O., Little, M.J., Reif, P. G., and Schwartz, R N.(1985). J. Appl. Phys. 57, p.1356.
69. Pape, D. R. (1985). Opt. Eng. 24, p.107.
70. Lee, S. H., Esener, S. C., Title, M. A., and Drabik, T. J. (1986). Opt. Eng. 25, p.250.
71. Horwitz. B. A. and Corbett, F. J. (1978). Opt. Eng. 17, p.353.
72. Owechko, Y., and Tanguay, A. R. (1982). Opt. Lett. 7, p.587.
73. Casasent, D., Caimi, F., and Khomenko, A.(1981). Appl. Opt. 20, p.4215.
74. Petrov, M. P. (1981). In "Current Trends in Optics" (F. T. Arecchi and F. R. Aussenegg, eds.), p.161. Taylor and Francis, London.
75. Burke, W. J., Staebler, D. L., Phillips, W., and Alphonse, G. A.(1982). Opt. Eng. 17, p.52.
76. Pepper, D. M. (1982). Guest ed., Special issue on nonlinear optical phase conjugation. Opt. Eng. 21, p.156.
77. Huignard, J. P., Rajbenbach, H., Refregier, Ph., and Solymar, L. (1985). Opt. Eng. 24, p. 586.
78. Pepper, D. M., AuYeung, J., Fekete, D. and Yariv, A.(1978). Opt. Lett.3, p.7.
79. Huignard, J. P., Herriau, J. P., Augborg, P., and Spitz, E.(1979). Opt. Lett.4, p.21.
80. Huignard, J. P. and Marrakchi, A.(1981). Opt. Comm. 38, p. 249.
81. White, J. O. and Yariv, A.(1980). Appl. Phys. Lett.37, p.249.
82. Marrakchi, A., Tanguay, A. R., Yu, J., and Psaltis, D. (1985). Opt. Eng. 24, p. 124.
83. Kingston, R. H., Burke, B. E., Nichols, K. B., Leonberger, F. J. (1984). Proc. SPIE 465, p. 9.
84. Eden, D. D. (1979). Proc. SPIE 185, p.97.
85. Strome, D. H. (1984). Proc. SPIE 465, p.192.
86. Warde, C., Weiss, A. M., Fisher, A. D. and Thackara, J. I. (1981). Appl. Opt. 20, p. 2066.
87. Warde, C., and Thackara, J. I. (1983). Opt. Eng. 22, p. 695.
88. Hara,T., Sugiyama, M., and Suzuki, Y. (1985). In "Advances in Electronics and Electron Physics", (P. W. Hawkes, ed.), Vol. 64B , p.637. Academic Press, London.
89. Somers, L.E. (1972). In "Advances in Electronics and Electron Physics" Vol.33A, p.493. Academic Press, New York.
90. Fisher, A. D., Ling, L.-C., Lee, J. N., and Fukuda, R. C. (1986). Opt. Eng. 25, p.261.
91. Gibbs, H. M. (1985). "Optical Bistability: Controlling Light with Light." Academic Press, New York.
92. Peyghambarian, N. and Gibbs, H. M. (1985). Opt. Eng. 24., p. 68.
93. Smith, S. D., Janossy, I., MacKenzie, H. A., Mathew, J. G. H., Reid, J. J. E., Taghizadeh, M. R., Tooley, F. P. A., and Walker, A. C. (1985). Opt. Eng. 24, p.569.
94. Chelma, D. S., Miller, D. A. B., Smith, P. W.(1985). Opt. Eng. 24, p. 556.
95. Gibbs, H. M., Mandel, P., Peyghambarian, N., and Smith, S. D. (1986)."Optical Bistability III." Springer Verlag, Heidelberg.
96. Collins, S. A. (1980)., Guest ed, Special issue on feedback in optics. Opt. Eng. 19, p.441.
97. Dagenais, M., and Sharfin, W. F. (1986). Opt. Eng. 25, p.219.
98. Miller, D. A. B., Chelma, D. S., Damen, T. C., Gossard, A. C., and Wiegmann, W., Wood, T. H., and Burrus, C. A. (1984). Appl. Phys. Lett 45, p. 13.

99. Smith, P. W., and Turner, E. H. (1977). Appl. Phys. Lett. 30, p. 280.
100. Private communications with H. M. Gibbs and S. D. Smith.
101. Garmire, E., Marburger, J. H., and Allen, S. D.(1978). Appl. Phys. Lett. 32, p. 320.
102. Akins, R. P., Athale, R. A., and Lee, S.H.(1980). Opt. Eng. 19, p. 347.
103. Gibbs, H. M., Jewell, J.L., Lee, Y. H., Macleod, A., Olbright, G., Ovadia, S., Peyghambarian, N., Rushford, M. C., Warren, M., Weingerber, D. A., and Venkatesan, T.(1985). Digital Optical Circuit Technology. AGARD Conference Proceedings No. 362., NTIS, Springfield, VA, p. 8-1.
104. Seko, A. and Nishikata, M. (1977). Appl. Opt. 16, p.1272.
105. Groth, G., and Marie, G. (1970). Opt. Comm. 2, p. 133.
106. Casasent, D. (1978). Opt. Eng. 17, p.344.
107. Schwartz, A., Wang, X.-Y., and Warde, C. (1985). Opt. Eng. 24, p.119.
108. Van Raalte, J. A. (1970). Appl. Opt. 9, p. 2225.
109. Thomas, R. N., Guldberg, J., Nathanson, H,. C. , and Malmberg, P. R.(1975). IEEE Trans. on Elec. Dev. ED-22, p. 765.
110. Mol, J.(1974). Digest of Internat. Optical Computing Conf., p.34.
111. Ross, W. E., Psaltis, D., and Anderson, R. H. (1983). Opt. Eng. 22, p.485.
112. Minoura, K., Usui, M., Matsouka, K., Baba, T., Suzuki, M., and Asai, A. (1984). Digest of the 13th Conference of the ICO ICO-13, p.154.
113. Underwood, I., Sillitto, R. M., and Vass, D. G., (1985). IEE Colloq. on Optical Techniques in Image and SIgnal Processing.
114. Hornbeck, L. J. (1983). IEEE Trans. on Elec. Dev. ED-30, p. 539.
115. Johnson, R. V., Hecht, D. L. , Sprague, R. A., Flores, L. N., Steinmetz, D. L., and Turner, W. D. (1983). Opt. Eng. 22, p. 665.
116. VanderLugt, A., Moore, G.S., and Mathe, S. S. (1983). Appl. Opt. 22, p. 3906.
117. Pape, D. R. (1985).Technical Digest of the OSA '85 Topical Meeting on Optical Computing, p. TuC6.
118. Blechman, F. (1986). Radio Electronics 57, July, p.39 and August, p.47.
119. Lui, H.-K., Davis, J. A., Lilly, R. A. (1985). Opt. Lett. 10, p. 635.
120. McEwan, J. A., Fisher, A. D., Rolsma, P. B., and Lee, J. N. (1985). J. Opt. Soc. Am. A 2, p. 8.
121. Yariv, A.(1971). "Intoduction to Optical Electronics." Holt, Rinehart, and Winston, New York.
122. McEwan, J. A., Fisher, A. D., and Lee. J. N. (1985). Digest of the Conference on Lasers and Electro-Optics (CLEO), p. PD-1.
123. McEwan, J. A.(1985). Optical Information Processing with a Microchannel Spatial Light Modulator. M.S. Thesis. EE Dept. Univ. of Maryland, College Park, MD.
124. Marom, E. (1986). Opt. Eng. 25, p.274.
125. Efron, U., Marom, E., and Soffer, B. H. (1985). Technical Digest of the OSA '85 Topical Meeting on Optical Computing, p. TuF2.
126. Dashiell, S. R., and Goodman, J. W. (1975). Appl. Opt. 14, p.1813.
127. Feinberg, J. (1980). Opt. Lett. 5, p. 330.
128. Barrett, H. H., Gmitro, A. F., and Chiu, M. Y.(1981). Opt. Lett.6, p.1.
129. The MSLM used was provided by Hamamatsu Photonics HPK/Hamamatsu Corp.
130. Shaefer, D. H., and Strong, J. P. (1975). Tse Computers. NASA Report X-943-75-14. Goddard SFC, Greenbelt, MD.
131. Bray, T. E.(1963). In "Optical Processing of Information." (D. K.Pollock, C. J. Koester, and J. T. Tippett,eds.),p.216.Spartan Books, Baltimore.
132. Uchiyama, S. and Iga, K. (1985). Digest of the Conference on Lasers and Electro-Optics (CLEO), p.44.
133. Liau, Z. L., Walpole, J. N., and Tsang, D Z. (1984). Technical Digest of the 7th Topical Meeting on Integrated and Guided-Wave Optics. OSA, Washington, DC., p. TuC5.
134. Tanguay, A. R. (1985). Opt. Eng. 24, p.2.
135. Yariv, A. and Pepper, D. M.(1980). Opt. Lett. 1, p.16.
136. Efron, U. (1984). Ed.,Issue on spatial light modulators and applications, Proc. SPIE 465.
137. Tanguay, A. R., and Warde, C. (1985). Guest eds., Special issue on optical information processing components. Opt. Eng. 24, p.91.
138. Fisher, A. D.(1985)., Technical Digest of the OSA '85 Topical Meeting on Optical Computing, p.TuC1.
139. Warde, C., and Efron, U. (1986). Guest eds., Special issue on Materials and devices for optical information processing. Opt. Eng. 25, p. 197.
140. Williams, D. J. (1983). Ed., "Nonlinear Optical Properties of Organic and Polymeric Materials." American Chemical Society, Washington, D.C.
141. Garito, A. F., and K. D. Singer.(1982). Laser Focus, February, p. 59.
142. Clark, N. A., Handschy, M. A., and Lagerwall, S. T. (1983). Mol. Cryst. Liq. Cryst. 94, p. 213.
143. Armitage, D., Thackara, J. I., Clark, N. A., and Handschy, M. A. (1986). Digest of the Conference on Lasers and Electro-Optics (CLEO), p.366.
144. Lea, M. (1984). Proc. SPIE 465, p. 12.
145. McAulay, A. D. (1983). Proc. SPIE 431, p.215.
146. McAulay, A. D. (1985). Digital Optical Circuit Technology. AGARD Conference Proceedings

No. 362. NTIS, Springfield, VA.,p.15-1.
147. Tanguay, A. R. (1983). Guest ed., Special issue on SLMs: Critical Issues. Opt Eng. 22, p.663.
148. Hirshon, R.(1984). Electronic Imaging, January, p.40.
149. Perry, T. S. (1985). IEEE Spectrum 22, July, p.53.
150. Apt, C. M. (1985). IEEE Spectrum 22, July, p.60.
151. Kmetz, A., and Von Willisen, F. K. (1976). Eds, "Non-emissive Electro-optic Displays." Plenum, New York.
152. Armand, A., Strand, T. C., Sawchuk, A. A., and Soffer, B. H.(1982). Opt. Lett. 7, p. 451.

Section 2.6: Ultrafast Devices (Subpicosecond)

Femotosecond switching in a dual-core-fiber nonlinear coupler

S. R. Friberg,* A. M. Weiner, Y. Silberberg, B. G. Sfez,† and P. S. Smith

Bell Communications Research, 331 Newman Springs Road, Red Bank, New Jersey, 07701-7020

Received April 25, 1988; accepted June 22, 1988

We report all-optical switching of 100-fsec pulses in a fused-quartz dual-core-fiber directional coupler. The length of the device is 0.5 cm, and the switching power is 32 kW. Pulses are routed to either of two separate fiber guides, depending on the input power. Measurements of pulse reshaping by the nonlinear coupler provide compelling evidence of the device's ability to response on a femotosecond time scale.

One reason for the current upsurge of interest in photonic switching has been the realization that all-optical devices have the capability to switch at rates much higher than those possible with electronics technology. Such high rates are likely to be necessary in future high-speed communications and computing systems.[1]

Optical glasses have a number of advantages as nonlinear materials for all-optical devices. Their high transparency permits long interaction lengths in guided-wave structures and essentially eliminates the thermal heating problems that have affected the performance of all-optical devices made from semiconductor and nonlinear organic materials.[2] Certain optical glasses have the highest figure of merit of any nonlinear-optical material.[2]

In the past few years, a number of glass optical-fiber switching devices have been demonstrated, including the birefringent-fiber polarization switch,[3] the optical-fiber Kerr gate,[4-6] and the two-core-fiber directional coupler.[7,8] For the coupler, a signal is routed between two waveguides; for the polarization switch and the Kerr gate, a signal is transferred between polarization modes in a single waveguide. In this Letter we report substantially complete switching of 100-fsec optical pulses by a glass-fiber nonlinear coupler. We believe this to be the fastest switching time ever measured in a guided-wave all-optical device.

A nonlinear coupler, shown schematically in Fig. 1(A), consists of two closely spaced, parallel, single-mode waveguides in a material with an intensity-dependent index of refraction. At low light levels, the device behaves as a linear directional coupler. Because of evanescent coupling, signals introduced into guide (1) transfer completely to guide (2) in one coupling length L_c. Higher intensities induce changes in the refractive index and detune the coupler. Coupling is inhibited for input powers above the critical power $P_c = A\lambda/n_2 L_c$, where A is the effective mode area, λ is the vacuum wavelength, and n_2 is the nonlinear index.[9] A nonlinear coupler of length L_c exhibits particularly useful switching characteristics. The solid curves in Fig. 1(B) show the calculated fractional power emerging from each of waveguides (1) and (2) as a function of input power, for a constant-intensity input signal.[9] Low-power signals introduced into guide (1) emerge from guide (2), whereas high-power signals ($P \gg P_c$) emerge from guide (1).

Our nonlinear coupler consists of a 5.0-mm length of dual-core fused-quartz optical fiber. The fiber contains two 2.8-μm-diameter, Ge-doped cores with 8.4 μm between core centers and with a core–cladding index difference of 0.003. Each fiber core is single mode for wavelengths longer than 500 nm. We determined the coupling length by using a low-power white-light measurement technique, in which a tunable monochromatic source is focused into a single fiber core. The relative output of each fiber core was measured over a wavelength range of 500–800 nm, and the

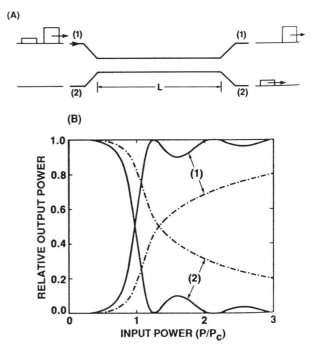

Fig. 1. (A) Schematic of a nonlinear coupler. (B) Calculated fractional output power emerging from waveguides (1) and (2) as a function of input power, for a coupler of length L_c. The input power is normalized to P_c. The curves were computed from Eq. (26) of Ref. 8. Solid curves: constant-intensity input signal. Dashed–dotted curves: coupler response integrated over a sech$^2(t)$ pulse intensity profile.

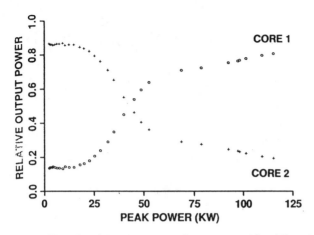

Fig. 2. Fractional output power from waveguides (1) and (2) for the 5-mm, dual-core-fiber nonlinear coupler. These data are the response for 100-fsec input pulses.

results were fitted to a theoretical model of the coupler. From this procedure, we estimate a coupling length of 4.7 mm at 620 nm. As part of the low-power characterization, we verified that the coupling length was independent of polarization and that the coupler maintained the polarization state of the input light.

Switching experiments were performed using pulses from a colliding-pulse mode-locked dye laser and a copper-vapor-laser pumped dye amplifier system.[10] The laser produced 100-fsec pulses at a wavelength of 620 nm, which were amplified at an 8.6-kHz repetition rate to 100 nJ. When desired, longer pulses could be obtained by adjustment of the dispersion-compensating prisms within the laser. Amplified pulses were focused by a 20× microscope objective into one fiber core, which we denote guide (1); the other input core was carefully blocked by the edge of a razor blade. An attenuator wheel was used to control the input intensity to the fiber. The output from each core was focused by a 40× microscope objective onto a separate power meter, and the average power emerging from each core was measured as a function of the input power. The output from each core could be directed to a real-time autocorrelator to monitor the pulse shape or to a 0.32-m spectrometer and photodiode array to observe the power spectrum.

The fraction of the average output power emerging from each guide is plotted in Fig. 2 as a function of peak power. Similar data were obtained for several pieces of dual-core fiber, for various laser pulse widths in the 100–200-fsec range, and for various polarization angles. Because the power meters respond slowly compared to the pulse duration, we expect to measure not the cw response curves shown in Fig. 1(B) but rather a response integrated over the pulse's intensity profile. At high power the data are in good agreement with the theoretical integrated response function [dashed–dotted curves in Fig. 1(B)], calculated for a coupler of length L_c and a sech^2 intensity profile. From the data we determine a critical power $P_c = 32$ kW, about a factor of 2 lower than the 60-kW value obtained from the formula $P_c = A\lambda/n_2 L_c$, using the known nonlinear coefficient for silica ($n_2 = 3.2 \times 10^{-16}$ cm^2/W) and using 15 μm^2 as a rough estimate of the effective mode area. Because of the difficulty in accurately measuring the peak power of femtosecond pulses and because of the uncertainty in the effective mode area, we consider that this discrepancy is within our experimental limits. At low power the data depart from theory and indicate incomplete energy transfer from one guide to the other. We attribute this result to partial excitation of radiation modes that are not completely extinguished in the 5-mm length of fiber.

Output pulses from the nonlinear coupler can be strongly reshaped.[11] For peak powers above P_c, the intense central portion of the input pulse emerges from guide (1), whereas the output of guide (2) is a pulse doublet corresponding to the low-intensity wings of the input. This is confirmed by autocorrelation measurements of pulses emerging from the nonlinear coupler. At low powers, the pulses emerging from guides (1) and (2) are identical. Autocorrelations obtained for 200-fsec pulses at a power of $\simeq 2P_c$ are plotted in Figs. 3(A) and 3(B), respectively, for guides (1) and (2). While the pulses from guide (1) are similar to those at low power, the pulses from guide (2) are strongly reshaped. The triply peaked autocorrelation trace corresponds to a doubly peaked intensity profile, and the 340-fsec peak separation is consistent with the duration of the input pulse. Further, we note that the individual peaks are significantly narrower than the input. These data show that for input powers $>P_c$ the coupler slices out the central part of the input pulse and directs it to the other output port. Our data show that both on and off switching times are <100 fsec.

In addition to nonlinear reshaping, intense pulses propagating in a nonlinear coupler are spectrally and temporally broadened by self-phase modulation and group-velocity dispersion. For peak input powers below $0.5P_c$, we observe that the spectra are unbroadened, with a width of 10 nm FWHM. For higher powers the spectra broaden continuously, doubling to 20 nm at $\simeq 1.3P_c$. As the power is raised further, spectra corresponding to pulses emerging from guide (2) become increasingly complex, whereas the spectra

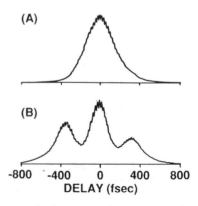

Fig. 3. Autocorrelation traces for output pulses from the coupler. The input pulses were 200 fsec in duration. (A) Guide (1), peak power $P \simeq 2P_c$. (B) Guide (2), peak power $P \simeq 2P_c$. The asymmetry in the trace results from a slight misalignment of the autocorrelator.

of pulses emerging from guide (1) remain relatively unstructured. The temporal broadening of pulses emerging from guide (1) can easily be estimated if we neglect nonlinear reshaping effects. In that case we must consider pulse propagation under the combined influence of self-phase modulation and group-velocity dispersion; this topic has been analyzed in detail in the literature.[12] For a dispersion of 0.265 psec/nm-m at our 620-nm wavelength, a 100-fsec pulse with peak power P_c would broaden by $\simeq 50\%$ at the fiber output; a 200-fsec pulse at P_c would broaden by $\simeq 25\%$. Temporal broadening within the nonlinear coupler would normally be expected to limit the maximum bit rate attainable with the device. At a wavelength of 1.3 μm, the group-velocity dispersion is zero; other factors, such as the wavelength dependence of the coupling length, would determine the maximum bit rate.

During the course of our experiments, we observed that prolonged exposure to high-intensity illumination modified the coupler operation. Specifically, after exposure to peak powers of $\simeq 100$ kW (i.e., $\simeq 3P_c$) for a period of a few minutes, the fraction of light coupled at low power from guide (1) to guide (2) decreased, typically by several percent, although the switching behavior was not noticeably affected. Significantly longer exposures caused a decrease by as much as 25% in the low-power coupling ratio and an increase in switching power. Annealing the coupler for 4 h at 600°C restored the original response. Our observations are consistent with previous observations of persistent photoinduced refractive-index changes in Ge-doped fibers[13,14]; a persistent index change could cause a mismatch between the two fiber cores and inhibit coupling. We expect that photosensitivity effects may be reduced by using fibers with pure fused-quartz cores[14] and by operating at longer wavelengths.[15]

The performance of the nonlinear coupler could be enhanced in several ways. The switching power could be reduced by using a longer device fabricated from a more highly nonlinear material. For example, a 20-cm device fabricated from SF-59 glass (which has a nonlinear refractive index 25 times higher than that of fused quartz[2]) should switch at $\simeq 30$ W. Decreasing the fiber-core size would reduce the switching power further and bring it closer to the range accessible to mode-locked diode lasers. Operation of the coupler with ultrashort square pulses[16] would lead to further enhancements: measurements would demonstrate sharper switching transitions and would avoid incomplete energy transfer owing to pulse fragmentation.

In conclusion, we have demonstrated switching of 100-fsec pulses in a 0.5-cm dual-core-fiber nonlinear coupler. To our knowledge this is the fastest switching ever observed in a waveguide all-optical switching device.

We thank E. Snitzer of Polaroid Corporation for the dual-core fiber. We gratefully acknowledge technical assistance by D. E. Leaird and enlightening discussions with M. Saifi and W. J. Tomlinson.

* Present address, NTT Basic Research Laboratories, Nippon Telegraph and Telephone Corporation, 9-11 Midori-cho, 3-Chome, Musashino-shi, Tokyo 180 Japan.

† Present address, Ecole Nationale Supérieure des Télécommunications, Paris, France.

References

1. P. W. Smith, Philos. Trans. R. Soc. London Ser. A **313**, 349 (1984).
2. S. R. Friberg and P. W. Smith, IEEE J. Quantum Electron. **QE-23**, 2089 (1987).
3. S. Trillo, S. Wabnitz, R. H. Stolen, G. Assanto, C. T. Seaton, and G. I. Stegeman, Appl. Phys. Lett. **49**, 1224 (1986).
4. N. J. Halas, D. Krokel, and D. Grischkowsky, Appl. Phys. Lett. **50**, 886 (1987).
5. K. Kitayama, Y. Kimura, and S. Seikai, Appl. Phys. Lett. **46**, 317 (1985).
6. T. Morioka, M. Saruwatari, and A. Takeda, Electron. Lett. **23**, 453 (1987).
7. S. R. Friberg, Y. Silberberg, M. K. Oliver, M. J. Andrejco, M. A. Saifi, and P. W. Smith, Appl. Phys. Lett. **52**, 1135 (1987).
8. D. D. Gusovskii, E. M. Dianov, A. A. Maier, V. B. Neustruev, V. V. Osiko, A. M. Prokhorov, K. Yu. Sitarskii, and I. A. Scherbakov, Sov. J. Quantum Electron. **17**, 724 (1987); A. A. Maier, Yu. N. Serdyuchenko, K. Yu. Sitarskii, M. Ya. Shchelev, and I. A. Scherbakov, Sov. J. Quantum Electron. **17**, 735 (1987).
9. S. M. Jensen, IEEE J. Quantum Electron. **QE-18**, 1580 (1982).
10. W. H. Knox, M. C. Downer, R. L. Fork, and C. V. Shank, Opt. Lett. **9**, 552 (1984).
11. K. Kitayama and S. Wang, Appl. Phys. Lett. **43**, 17 (1983).
12. W. J. Tomlinson, R. H. Stolen, and C. V. Shank, J. Opt. Soc. Am. B **1**, 139 (1984).
13. K. O. Hill, Y. Fujii, D. C. Johnson, and B. S. Kawasaki, Appl. Phys. Lett. **32**, 647 (1978).
14. J. Stone, J. Appl. Phys. **62**, 4371 (1987).
15. G. Meltz, J. R. Dunphy, W. H. Glenn, J. D. Farina, and F. J. Leonberger, Proc. Soc. Photo-Opt. Instrum. Eng. **798**, 104 (1987).
16. A. M. Weiner, J. P. Heritage and R. N. Thurston, Opt. Lett. **11**, 153 (1986); A. M. Weiner, J. P. Heritage, and E. M. Kirschner, J. Opt. Soc. Am. B (to be published).

Section 2.7: Optical Interconnects

Attributes

Comparison between optical and electrical interconnects based on power and speed considerations

Michael R. Feldman, Sadik C. Esener, Clark C. Guest, and Sing H. Lee

Conditions are determined for which optical interconnects can transmit information at a higher data rate and consume less power than the equivalent electrical interconnections. The analysis is performed for free-space optical intrachip communication links. Effects of scaling circuit dimensions, presence of signal fan-out, and the use of light modulators as optical signal transmitters are also discussed.

I. Introduction

Advances in VLSI technology have dramatically increased device densities and speeds. However, interconnection of devices, relying on aluminum or polysilicon lines, has not significantly changed. As a result, performance of current VLSI systems is often limited by the power dissipation, delay time, and surface area required by existing interconnect technology.[1-3]

Recently, attention has been given to using optical technology to improve connection of devices on a single chip or wafer and between chips or modules.[1,4-6] For circuits requiring complex interconnection, optical interconnections will occupy less area than their electronic counterparts.[7] This paper will address the other important issues of power dissipation and delay time. It will be shown that for a range of practical situations optical interconnects can transmit information at a higher data rate and consume less power than the corresponding electrical interconnections.

Although a brief analysis of this type was performed in Ref. 6, the dependence of switching energy on signal rise time was not included in that analysis. In this paper, the energy dependence on rise time is included, and a more extensive analysis is performed, including the effects of smaller minimum IC dimensions, of increased signal fan-out, and of increased optical link efficiency. Also, the implementation of optical interconnects with light modulators (rather than lasers) as optical signal transmitters is discussed.

In particular, this paper will focus on replacing intrachip electronic interconnects with free-space optical interconnects. The derived equations can be applied to chip-to-chip interconnects with minor modifications as described in Sec. V.

The modeling of optical interconnects that employ semiconductor lasers to transmit optical signals is described in Sec. II and electrical interconnect modeling in Sec. III. Section IV contains a comparison between electrical and optical interconnects with a single signal receiver, based on 3-μm CMOS design rules. In Sec. V the comparison is extended to include cases with signal fan-out. Effects of scaling device and interconnect dimensions are discussed in Sec. VI, and effects of improved optical link parameters in Sec. VII. The use of light modulators as optical signal transmitters is discussed in Sec. VIII.

II. Modeling of Optical Interconnects

A system for performing optical interconnects is shown in Fig. 1. Electronic signal lines are replaced with integrated optical signal transmitters, detectors, and a hologram. The optical transmitters can be GaAs lasers, light modulators, or LEDs. Several methods of creating hybrid systems combining GaAs and silicon are under development.[8,9] We shall first analyze a system consisting of a semiconductor laser transmitting a signal to a single photodetector (fan-out of 1).

The optical detector circuits, illustrated schematically in Fig. 2, are CMOS compatible optical gates[5] consisting of a photodiode and a load transistor. The optical gate is required to drive a standard CMOS inverter gate, also shown in Fig. 2. Although the use of

---

The authors are with University of California, San Diego, Department of Electrical & Computer Engineering, La Jolla, California 92093.

Received 27 July 1987.

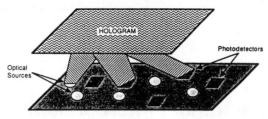

Fig. 1. System configuration for implementation of free-space optical interconnects.

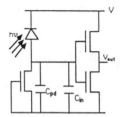

Fig. 2. Schematic diagram of a photodetector circuit.

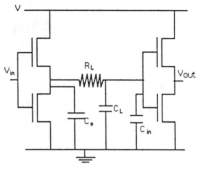

Fig. 3. Schematic diagram of electrical interconnection of two CMOS gates.

a sophisticated amplifier circuit following the detector would decrease optical power requirements, it is impractical to fabricate complex amplifiers near every photodetector in a high density interconnection scheme. In this section, expressions for the switching energy, defined as the total energy required to switch a receiving inverter from one state to the other and back, and the rise time of an optical interconnect will be derived. We neglect the switching energy and rise time associated with the driver of the laser diode and light propagation delay. The validity of these assumptions is discussed in Sec. IX.

The switching energy of the optical interconnect illustrated in Figs. 1 and 2 contains two components: E_1, the electrical energy supplied by the power supply of the detector circuit, and E_2, the electrical energy required to drive the optical emitter. E_1 is given by

$$E_1 = (2 \cdot 2 \cdot 0.5)(C_{PD} + C_{in})V^2, \quad (1)$$

where C_{PD} is the photodiode capacitance, C_{in} is the input capacitance of a minimum size CMOS gate, and V is the power supply voltage. E_1 includes the energy dissipated in the photodiode and the load transistor during the charging of the receiving gate and the energy dissipated in the load transistor while the gate is discharging. The first factor of 2 in Eq. (1) is due to our definition of switching energy which accounts for both the charging and discharging of the gate. The second factor of 2 results from the presence of the resistive load during the charging of the gate.

The second component, E_2, can be determined by noting that the photocurrent, I_P, generated within the detector is given by

$$I_P = 2q(P_L - P_{th})\eta/(h\nu), \quad (2)$$

where h is Planck's constant, ν is the optical frequency, q is the electronic charge, P_{th} is the electrical power required to bias the laser at threshold, and P_L is the average total electrical power required to drive the laser. The total optical link efficiency, η, is defined as

$$\eta = \eta_L \eta_H \eta_D, \quad (3)$$

where η_L is the external differential efficiency of the laser (incremental optical power out divided by incremental electrical power in) and η_H is the efficiency of the hologram. The photodetector quantum efficiency, η_D, is approximately equal to $[1 - \exp(-\alpha d)]$ (where α is the absorption coefficient, and d is the thickness of the detector active region) if surface reflections and near surface absorption are neglected. The switching energy E of an interconnect is related to the average power dissipation P by

$$E = 2\tau P, \quad (4)$$

where τ is the interconnect rise time, defined as the time needed for the receiving gate's input voltage to rise from 10% to 90% of its final value. For an optical interconnect, $\tau \approx 2R(C_{PD} + C_{in})$ where R is the resistance of the load transistor. Setting R to a value of V/I_P allows the signal to rise to the power supply voltage, and using Eqs. (2) and (4) yields the following expression for E_2:

$$E_2 = 2V(C_{PD} + C_{in})\frac{h\nu}{\eta q} + 2\tau P_{th}. \quad (5)$$

Thus the total switching energy of an optical interconnect E_0, given by the sum of E_1 and E_2, is

$$E_0 = 2V(C_{PD} + C_{in})\left[\frac{h\nu}{\eta q} + V\right] + 2\tau P_{th}. \quad (6)$$

Since in many cases $h\nu/(\eta q) \gg V$, Eq. (6) simplifies to

$$E_0 \approx 2V(C_{PD} + C_{in})\frac{h\nu}{\eta q} + 2\tau P_{th}. \quad (7)$$

Note that for small rise times the switching energy is directly proportional to the detector circuit capacitance $C_{PD} + C_{in}$. For long rise times the switching energy is proportional to τ.

III. Modeling of Electrical Interconnects

In this section a model for CMOS VLSI interconnections is developed for a driving gate sending a signal to a single receiving gate (a fan-out of 1). The model[10] is based on the circuit diagram in Fig. 3. The use of additional repeater inverter gates placed periodically along the line to restore the signal rise time is not accounted for in this model. Also, the switching energy and delay time of the additional inverter stages required to drive large line-driving gates are neglected. (Correspondingly, the energy and time delay associated with the drivers of the optical sources were neglected in the previous section.) R_L is the resistance per square of the transmission line. The three capacitances in the figure are the input capacitance of the receiving gate C_{in}, the output capacitance of the CMOS driving gate C_O, and the line capacitance C_L. C_{in} includes the gate capacitances of the two transistors comprising the receiving gate. (The Miller capacitance is neglected.) The output capacitance C_O can be divided into two capacitances as follows:

$$C_O = MC_{OA} + C_{OB}, \quad M \geq 1, \tag{8}$$

where M is the ratio of the gate width of the driving inverter to the gate width of the minimum size inverter allowed by the process design rules. C_{OA} is the junction capacitance of a minimum size inverter gate. C_{OB} is the additional capacitance due to the area needed to form a contact to the drain regions.

The line capacitance, C_L, is given by

$$C_L = LWC_{LA} + LC_{LB}, \tag{9}$$

where L and W are the length and width of the line, C_{LA} is the parallel plate line capacitance per unit area and C_{LB} is the fringing capacitance per unit length. The total capacitance of an electrical interconnect line, C_T, is given by

$$C_T = C_O + C_L + C_{in}, \tag{10}$$

and the switching energy of an electrical interconnect by

$$E_E = C_T V^2 = (MC_{OA} + C_{OB} + C_{in} + C_{LA}LW + C_{LB}L)V^2. \tag{11}$$

The interconnect rise time can also be estimated from the model of Fig. 3. By approximating the dynamic resistance of a CMOS inverter as $V/(2I_0)$, where I_0 is the maximum current that can be supplied by a minimum size CMOS inverter gate (which can be determined from SPICE[11] simulations), the total transmission line rise time can be calculated as

$$\tau \approx R_L C_{LA} L^2 + R_L C_{LB} L^2/W + 2C_{in} R_L L/W + [V/(MI_0)](C_T). \tag{12}$$

The terms were found by multiplying each capacitance with the sum of the resistances that occur before it on the transmission line.[12] The first two terms are due to the distributed RC time delay of the transmission line itself. The third term is the RC time delay associated with the line charging the receiving gate. The fourth term is due to the driving gate charging all three capacitances.

A computer simulation program was developed to determine the validity of this interconnection model. Employing the SPICE circuit simulation program as a subroutine, the computer program can determine rise time and switching energy from user inputs of line length and process parameters. The program was used to model polysilicon interconnect lines for the MOS implementation service (MOSIS)[13] 3-μm CMOS

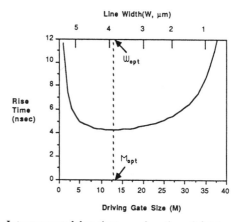

Fig. 4. Interconnect delay time as a function of driving gate size and linewidth for constant energy loss. $E = 11$ pJ, $L = 1.5$ mm for a polysilicon interconnect line

process. The simulation results agree to within 15% of the analytic estimations over a wide range of signal line properties.

For fixed switching energy and line length, M and W should be chosen to minimize the interconnect rise time. For example, a plot of delay time vs M and W using Eqs. (11) and (12), a switching energy of 11 pJ, and a line length of 1.5 mm for polysilicon lines fabricated by the MOSIS 3-μm CMOS process is shown in Fig. 4. [Note that, since the switching energy is held constant, M and W are related by Eq. (11) and thus there is only one independent variable.] The figure clearly indicates that the delay time is minimized for a particular value of M. This optimum value of M is denoted by M_{opt}, and the corresponding value of linewidth by W_{opt}.

In general, W_{opt} and M_{opt} are the values of that minimize the rise time for given values of switching energy and line length. Thus, W_{opt} and M_{opt} can be calculated by employing LaGrange multipliers; Eq. (12) is the equation to be minimized and Eq. (11) is the equation of constraint. This results in

$$M_{opt} = \frac{\frac{E_E}{V^2} - (C_{OB} + C_{in} + C_{LB}L)}{C_{OA} + C_{LA}L\sqrt{B}}, \tag{13}$$

$$W_{opt} = \frac{E_E/V^2 - (C_{OB} + C_{in} + C_{LB}L)}{C_{OA}/\sqrt{B} + C_{LA}L}, \tag{14}$$

where $B = (R_L C_{LB} L + 2R_L C_{in})I_0 V C_{OA}/(E_E C_{LA})$.

These two equations are valid if they yield values of M and W that are physically realizable; that is, if $M_{opt} \geq 1$ and $W_{opt} \geq W_{min}$, where W_{min} is the minimum linewidth allowed by the process design rules. Equations (13) and (14) allow the optimization of linewidth and driving gate dimensions in the design of electrical interconnects.

IV. Comparison of Electrical and Optical Interconnects for a Fan-Out of One Based on 3-μm CMOS Design Rules

The switching energies of optical and electrical interconnects can be compared by examining the differences between Eq. (11) describing the switching energy of an electrical interconnect E_E and Eq. (6) describing the switching energy E_O of an optical interconnect. To illustrate the behavior of these equations, both E_E and E_O are plotted as functions of rise time τ in Fig. 5. The optical interconnect switching energy curve is a plot of Eq. (6) for a 3.6-μm thick, 10-μm square detector size, yielding a photodiode capacitance of 5.3 fF (including sidewall capacitances). The detector quantum efficiency, equal to $\sim[1-\exp(-\alpha d)]$, is $\sim$30% for an assumed semiconductor laser wavelength of 0.8 μm ($\alpha \approx 1000$ cm^{-1}). Also, a total optical link efficiency η of 9.0% (e.g., laser external efficiency = 40%, hologram efficiency = 75%, detector efficiency = 30%) and a laser threshold power P_{th} of 1.0 mW were assumed. (Values of $V = 5$ V and $C_{in} = 17$ fF were based on 3-μm MOSIS CMOS specifications.) For small rise times (less than $\sim$0.3 ns in this case), the second term in Eq. (6) is negligible, and the first term yields an optical switching energy of $\sim$5 pJ which is independent of τ. For long rise times the second term dominates and E_O is proportional to τ.

The electrical switching energy curve is a plot of Eq. (11), where again 3-μm MOSIS CMOS parameters were employed, this time for a 1-mm long first level aluminum line. The values of M and W in this equation were chosen by one of two different methods. We define τ_0 as the rise time given by Eq. (12) with $M = 1$ and $W = W_{\min}$. For rise times longer than τ_0, M and W are each set to their minimum values. In this region data rates are slow enough that there is no need to increase M or W to reduce resistance and hence E_E can be kept at its minimum value. However for rise times smaller than τ_0, M and/or W must be increased to reduce the rise time. In this region, values of M and W are given by Eqs. (13) and (14) to minimize E_E for each value of τ. [If one of the equations yields a physically unrealizable result, the corresponding parameter is set to its minimum value and the other parameter value is determined from Eq. (11).] Since E_O is an increasing function of rise time and E_E is a decreasing function, E_O will be less than E_E for high data rate applications.

The ratio of the energy dissipated in an optical interconnect to that dissipated in an electrical interconnect can be obtained from Eqs. (7) and (11):

$$\frac{E_O}{E_E} = 2\frac{h\nu}{qV}\frac{(C_{PD}+C_{in})}{\eta C_T} + \frac{2\tau P_{th}}{C_T V^2}. \quad (15)$$

If $\tau \ll \tau_2$ where τ_2 is defined as

$$\tau_2 = \frac{h\nu(C_{PD}+C_{in})V}{q\eta P_{th}}, \quad (16)$$

Eq. (15) reduces to

$$\frac{E_O}{E_E} = 2\frac{h\nu}{qV}\frac{(C_{PD}+C_{in})}{C_T}\frac{1}{\eta}. \quad (17)$$

The second factor in Eq. (17) is simply the ratio of energies of optical and electrical fundamental particles. The third factor is the ratio of the capacitances associated with each type of interconnect. Note that

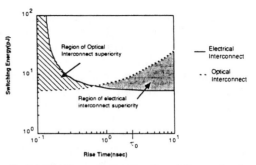

Fig. 5. Switching energy vs rise time for a 1.0-mm aluminum line and for a 9.0% efficient optical interconnect with a 1-mW laser diode threshold power.

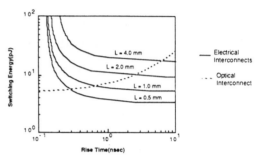

Fig. 6. Switching energy vs rise time for aluminum lines of four different lengths and for a 9.0% efficient optical interconnect with a 1-mW laser diode threshold power.

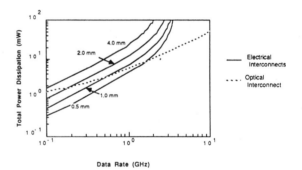

Fig. 7. Total power dissipation per interconnect vs maximum data transmission rate for the same interconnect parameters as in Fig. 6.

the capacitance associated with electrical interconnects grows with increasing line length whereas that associated with optical interconnects remains constant. This is illustrated in Fig. 6 where E_E is plotted vs τ for several values of line length L. Thus, in general, given any parameter values, for a large enough value of L or a small enough value of τ, E_O will be less than E_E.

The energy vs time plot in Fig. 6 was converted to a power vs speed plot in Fig. 7 by using the following relations:

$$1/(2\tau) = \text{maximum data transmission rate}, \quad (18a)$$

$$\text{data rate} = \frac{\text{total power dissipation}}{\text{switching energy}}. \quad (18b)$$

From Figs. 6 and 7 [or Eq. (18b)] it can be seen that,

when the switching energy of an optical interconnect is less than that of a corresponding electrical interconnect, the optical interconnect is able to transmit data at higher rates than the corresponding electrical interconnect, when subjected to the same power dissipation limits. Or conversely, the optical interconnect is able to dissipate less power if data are transmitted at the same rate.

For a given rise time, we define the line length for which $E_O/E_E = 1$ as the break-even line length, denoted by L_{be}. L_{be} is plotted vs τ in Fig. 8 using the same values of η, and process parameters as were used for Fig. 6. Note that $L_{be}(\tau)$ is given by the value of L for which an E_E curve crosses the E_O curve in Fig. 6. For interconnect lengths longer than L_{be}, optical interconnects will have a smaller switching energy. For lengths shorter than L_{be}, electrical interconnects will have a smaller switching energy. Thus for interconnect lengths and rise times corresponding to points lying in the shaded area labeled region of optical interconnect superiority in Fig. 8, optical interconnects will have a smaller switching energy than the corresponding electrical interconnects. Similarly, the region of electrical interconnect superiority indicates values of line length and rise time for which electrical interconnects have a smaller switching energy.

Although the dependence of L_{be} on τ is rather complicated, there are three regions where the curve obeys rather simple equations. For long rise times, the second term in Eq. (15) is dominant and thus

$$L_{be}^{th} \approx 2\tau P_{th}/[V^2(W_{min}C_{LA} + C_{LB})]. \qquad (19)$$

We define this length as the threshold limited break-even line length, L_{be}^{th}, and the corresponding values of τ for which $L_{be} \approx L_{be}^{th}$ as the threshold limited region. The threshold limited region, indicated in Fig. 8, occurs for $\tau > \sim 2\tau_2$, where τ_2 was defined by Eq. (16). Note that L_{be}^{th} is equal to the energy required to bias a laser diode at threshold, divided by the energy per unit length required to charge an electrical interconnect line capacitance.

On the other hand, for $\tau \ll \tau_2$, the second term in Eq. (15) can be neglected and hence,

$$L_{be} \approx \left(2\frac{h\nu}{qV}\frac{C_{PD} + C_{in}}{\eta} - MC_{OA} - C_{OB} - C_{in}\right)\Big/(WC_{LA} + C_{LB}). \qquad (20a)$$

If τ is near τ_1 [defined as the rise time given by Eq. (12) with $M = 1$, $W = W_{min}$ and $L = L_{be}^C$], M and W can be replaced with their minimum values and Eq. (20a) becomes

$$L_{be}^C \approx \left(2\frac{h\nu}{qV}\frac{C_{PD} + C_{in}}{\eta} - C_{OA} - C_{OB} - C_{in}\right)\Big/(W_{min}C_{LA} + C_{LB}). \qquad (20b)$$

This length is defined as the capacitance limited break-even line length L_{be}^C. If the capacitances associated with the transmitting and receiving gates of the electrical interconnect are neglected, L_{be}^C is proportional to the photodiode circuit capacitance divided by the electrical interconnect line capacitance per unit

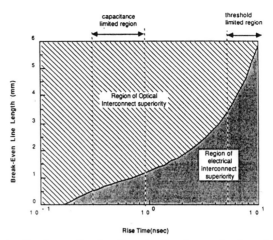

Fig. 8. Break-even line length vs rise time for an aluminum interconnect line (3-μm minimum feature size) and a 9.0% efficient optical system, with 1.0-mW threshold power lasers.

length. The capacitance limited region, defined as the values of τ for which $L_{be} \approx L_{be}^C$, occurs for $\sim \tau_1/10 < \tau < \sim \tau_2/2$. This region is also indicated in Fig. 8.

A third region of the break-even line length vs τ plot can be defined for values of τ for which the break-even line length is limited primarily by the RC delay of an electrical interconnect. From Eq. (12) one can show that

$$\tau > R_L C_{LA} L^2 + VC_{OA}/I_0, \qquad (21a)$$

and hence

$$L_{be}^{RC} < \sqrt{\frac{\tau - (VC_{OA}/I_0)}{R_L C_{LA}}}. \qquad (21b)$$

This bound, defined as the RC limited break-even line length, is an upper bound on the break-even line length that holds for all τ. Since for the case illustrated in Fig. 8 (3-μm IC minimum feature size, 9.0% efficient optical system) $L_{be}^{RC} \gg L_{be}$, a RC limited region does not exist for this case. However, it will be shown in Sec. VI that L_{be}^{RC} decreases with IC dimension scaling resulting in a RC limited region for small values of τ.

V. Fan-Out Considerations

Digital systems often require fan-out, i.e., sending of an output signal from a single gate to the inputs of several receiving gates. The rise time of an electrical transmission line performing F-fold fan-out depends on the manner in which the F gates are distributed along the interconnect line. Three types of fan-out that are representative of the majority of cases are illustrated in Fig. 9.

The first type, illustrated in Fig. 9(a), occurs when all the receiving gates are located very close to each other on the IC. For an optical interconnect to perform this type of fan-out there would be no advantage in placing more than one photodetector by the receiving gates as this would increase the switching energy without reducing line capacitance or RC delay. Thus the optical system would provide a 1:1 connection and the fan-out would be performed electrically as indicat-

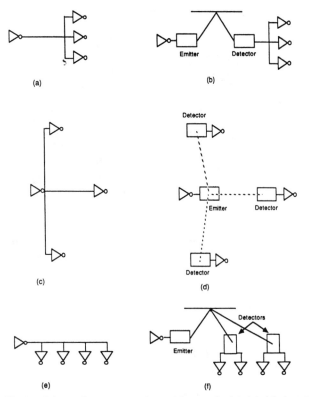

Fig. 9. Schematic representations of fan-out for (a), (c), (e) electrical interconnects and (b), (d), (f) optical interconnects.

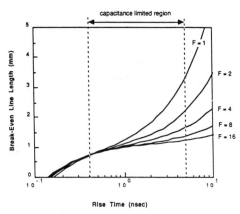

Fig. 10. Break-even line length vs rise time for five different values of remote fan-out. The efficiency of the optical system is 9.0%, the laser threshold power is 1 mW. The minimum feature size of the IC is 3 μm. The capacitance limited region is indicated for $F = 8$.

ed in Fig. 9(b). For both electrical and optical systems, better performance can be obtained by placing an additional inverter immediately prior to the receiving gates. In this case the break-even line length is identical to the case of a 1:1 gate connection.

A second type of fan-out, denoted remote fan-out, is illustrated in Fig. 9(c). Here the receiving gates are not located in the same vicinity and hence require distinct interconnect lines. For simplicity we assume that the F lines providing the F-fold fan-out are each equal in length. Electrical interconnect remote fan-out is modeled by replacing C_{in} and C_L in Eq. (10) with FC_{in} and FC_L, respectively. Optical interconnect remote fan-out can be accomplished by allowing the hologram to divide the incoming beam into F output beams, focusing each output beam onto a distinct detector as illustrated in Fig. 9(d). This can be accounted for by replacing $(C_{PD} + C_{in})$ in Eq. (7) with $F(C_{PD} + C_{in})$. Note that this results in no change in L_{be}^{RC} and has only a small effect on L_{be}^{C}, since the fan-out factor F increases the values of terms in both the numerator and denominator of Eq. (20b). However, the threshold limited break-even line length becomes

$$L_{be}^{th} \approx 2\tau P_{th}/[V^2 F(W_{min} C_{LA} + C_{LB})]. \quad (22)$$

As F increases, L_{be}^{th} decreases linearly. Also, the rise time, which determines the transition from the capacitance limited region τ_2 to the threshold limited region, becomes

$$\tau_2 = V(h\nu/q)(C_{PD} + C_{in})F/(P_{th}\eta). \quad (23)$$

As the fan-out increases, the threshold limited region is shifted to longer rise times and the width of the capacitance limited region is extended. These effects are illustrated in Fig. 10 where L_{be} is plotted vs τ for five different values of remote fan-out. Note that for $F = 8$, the capacitance limited region extends from $\tau \approx 0.4$ ns to $\tau \approx 5$ ns.

The third type of fan-out occurs when the receiving gates are evenly distributed along the interconnect line as illustrated in Fig. 9(e). In this case there is little or no reduction in line capacitance when an optical link replaces an electrical one (for a high density of gates). If optical interconnects were to be implemented by placing a photodetector next to each gate, the additional capacitance of this detector would be larger than the small amount of interconnect line that is removed. The advantage obtained in this case is only a reduction in RC delay. Depending on the delay times required, it is advisable to break the interconnect line into segments as indicated in Fig. 9(f). In this way fan-out is performed partially by the hologram and partially by the electronics.

Note that this situation is quite different from that of an electrical interconnect with repeater inverter gates placed at each detector location. Although RC delay can be reduced in the electrical interconnect case, additional delay is introduced due to the product of the new line resistance and the input capacitance of the repeaters.

The results obtained in this section are specific to intrachip communication. For communication between chips and between boards, the advantages of optical interconnects generally improve with increasing fan-out. For example, chip-to-chip interconnects can be modeled by replacing C_{in} and C_O in Eq. (10) with $C_{in} + C_B$ and $C_O + C_B$, respectively, where C_B is the bonding pad capacitance.[6] This additional capacitance (C_B) reduces the performance of electrical interconnects, especially for large fan-outs. For board-to-board interconnects line lengths are typically long enough that transmission line effects are significant. In Ref. 14 it is shown that, for terminated transmission lines, a maximum distributed fan-out exists that can be achieved for a given line length for any value of P and τ (about one 3-pF load per 2 cm for a lossless polymide strip line). Thus optical links can provide higher distributed fan-outs over a large range of data rates for interconnects between boards.

Despite these advantages, an optical interconnect system utilizing a semiconductor laser as the signal transmitter is limited in fan-out capabilities due to the limited light power that can be produced by a single laser. Denoting this maximum light power as P_{max} yields

$$P_{max} = (P_L - P_{th})\eta_L, \quad (24a)$$

where P_L, P_{th}, and η_L were determined in Sec. II. Using Eqs. (2) and (3) gives

$$P_{max} = V(C_{PD} + C_{in})h\nu F/(\eta_H \eta_D q\tau), \quad (24b)$$

and thus the maximum fan-out is given by

$$F < \eta_H \eta_D q\tau P_{max}/[V(C_{PD} + C_{in})h\nu]. \quad (24c)$$

For $P_{max} = 50$ mW,[15,8] and $\tau = 1$ ns, the maximum fan-out is ~67. We note that high power laser diodes typically have threshold powers $\gg 1$ mW.[15,8] The effects of larger laser threshold powers on break-even line length is discussed in section VII.

VI. Effects of Scaling Electronic Circuit Dimensions

Break-even line length has been calculated as a function of delay time based on typical parameters for a 3-μm CMOS process. For this process it was found that L_{be} is <2.0 mm, for τ < 2 ns, for aluminum first layer lines. To determine how L_{be} changes with current and future semiconductor device processes, a scaling rule is applied. The scaling rule chosen is that of Gardner, et al.[3] in which all dimensions scale linearly and the power supply voltage scales quasistatically. Denoting the minimum feature size of the process (the size of the minimum transistor gate length) as $\tilde{\lambda}$, gives the expressions for the process parameters listed in Table I. The form of each expression was obtained from the scaling rules. The constants were chosen so that the parameters agree with those of the 3-μm MOSIS process for $\tilde{\lambda} = 3$ μm. We assume that the optical wavelength and the detector capacitance C_{PD} remain constant. The fringing capacitance was calculated by

Table I. Integrated Circuit Process Parameters Expressed as Functions of Circuit Minimum Feature Size, $\tilde{\lambda}$ (in μm) for First-Level Aluminum Lines

Parameter	Symbol	Value	Units
Minimum linewidth	W_{min}	1.5 $\tilde{\lambda}$	μm
Gate input capacitance	C_{in}	170 $\tilde{\lambda}$ ϵ_{ox}	fF
Output capacitance	C_{OA}	0.72$\tilde{\lambda}^2$	fF
Contact output capacitance	C_{OB}	3.1$\tilde{\lambda}^2$	fF
Line capacitance	C_{LA}	$\epsilon_{ox}/(0.35 W_{min})$	fF/μm^2
Fringing line capacitance	C_{LB}	~0.061	fF/μm
Line resistance	R_L	$\rho/(0.09 \tilde{\lambda})$	Ω/sq
Power supply voltage	V	2.9 $\tilde{\lambda}^{1/2}$	V
Inverter saturation current	I_0	0.26	mA

Note: ϵ_{ox} = the permittivity of silicon dioxide = $3.9 \times 8.85 \times 10^{-3}$ fF/μm; ρ = the resistivity of aluminum = 0.0274 Ω μm.

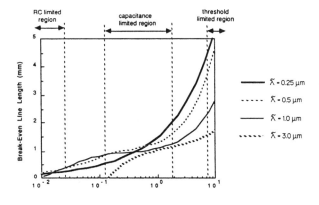

Fig. 11. Break-even line length vs rise time for a 9.0% efficient optical system with 1-mW threshold power lasers and an aluminum interconnect line, both for a fan-out of 8. The minimum feature size of the IC varies from 3 to 0.25 μm. The RC limited, capacitance limited, and threshold limited regions are indicated for a 1-μm IC minimum feature size.

modeling the line as a parallel plate with half-cylinders on each side.[16]

In Fig. 11 L_{be} is plotted vs τ for four different values of $\tilde{\lambda}$ for an aluminum interconnect line. The behavior of L_{be} with decreasing $\tilde{\lambda}$ can be described by examining the behavior of the L_{be} vs τ curve in the three regions described by Eqs. (20b), (21b), and (22). The RC limited line length can be written in terms of $\tilde{\lambda}$ as

$$L_{be}^{RC} = 0.22 \tilde{\lambda}\{\tau - (8 \text{ ps}/\mu\text{m}^{5/2}) \tilde{\lambda}^{5/2}\}^{1/2}(\rho\epsilon_{ox})^{-1/2}$$

$$\approx 0.22 \tilde{\lambda}[\tau/(\rho\epsilon_{ox})]^{1/2} \quad (\text{for } \tilde{\lambda}^{5/2} \ll \tau/8 \text{ ps}, \tilde{\lambda} \text{ in } \mu\text{m}).$$
$$(25)$$

For small $\tilde{\lambda}$, L_{be}^{RC} is directly proportional to $\tilde{\lambda}$. Thus, as feature sizes reduce, the RC limited break-even line length decreases. The RC limited region occurs for values of τ for which L_{be} approaches L_{be}^{RC}. This occurs for large values of M and W or for $\tau \ll \tau_1$. The RC limited, capacitance limited, and threshold limited regions are indicated in Fig. 11 for $\tilde{\lambda} = 1$ μm.

The situation is just the opposite in the capacitance limited and threshold limited regions. From Eq. (20b) and Table I it can be seen that for small $\tilde{\lambda}$ the capacitance limited line length approaches

$$L_{be}^C \to \frac{h\nu}{q} \frac{1}{V} \frac{1}{\eta} \frac{C_{PD}}{W_{min}C_{LA} + C_{LB}}. \quad (26)$$

In this expression, all the values are independent of $\tilde{\lambda}$, except the power supply voltage V which varies with $\tilde{\lambda}^{1/2}$. Thus L_{be}^C is proportional to $\tilde{\lambda}^{-1/2}$ (for small $\tilde{\lambda}$) and, as feature sizes scale, capacitance limited break-even line length increases. However, L_{be}^C cannot increase indefinitely. Since the minimum supply voltage V is given by $2kT/q$,[17] the maximum value of L_{be}^C is given by

$$(L_{be}^C)_{max} = \frac{h\nu}{kT} \frac{1}{\eta} \frac{C_{PD}}{W_{min}C_{LA} + C_{LB}}. \quad (27)$$

This is an upper bound on break-even line length for arbitrarily small minimum features (and for $\tau \ll \tau_2$).

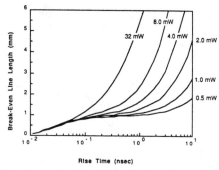

Fig. 12. Break-even line length vs rise time for varying threshold powers of the laser transmitters. Optical system efficiency is 9.0%, IC minimum feature size is 1.0 μm, and the fan-out = 8.

This bound is proportional to $h\nu/kT$, the ratio of a single quanta of light energy to a single quanta of thermal energy. It is also proportional to the ratio of the photodiode capacitance to the line capacitance per unit length.

The threshold limited break-even line length region ($\tau > \sim \tau_2$) is affected by scaling in two ways. First, the transition rise time τ_2 decreases with decreasing $\tilde{\lambda}$. For small $\tilde{\lambda}$, τ_2 scales with $\tilde{\lambda}^{1/2}$ and thus L_{be} approaches L_{be}^{th} for smaller values of τ. Second, L_{be}^{th} scales with $1/\tilde{\lambda}$ [from Eq. (22) and Table I], resulting in larger break-even line lengths in the threshold limited region with feature size scaling.

VII. Effects of Improved Optical Link Parameters

Up to this point, optical interconnects have been evaluated based on a laser threshold power of 1 mW and an optical system efficiency of 9.0%. Although GaAlAs lasers have been fabricated with threshold currents as low as 1 mA (Ref. 18) ($P_{th} \approx 1.5$ mW), the reported threshold currents of lasers fabricated as hybrids on top of silicon have been much larger.[8,9] Note that the laser threshold power affects the break-even line length only in the threshold limited region ($\tau > \sim 2\tau_2$), where $L_{be}^{th} \propto P_{th}$. However, as the threshold power increases, τ_2 decreases [Eq. (23)]. This behavior is illustrated in Fig. 12 where L_{be} is plotted vs τ for several values of P_{th} for a fan-out of 8 and 1.0-μm minimum feature size. For large values of P_{th} ($P_{th} > \sim 100$ mW in this case), since $L_{be}^{th} > L_{be}^{RC}$, the L_{be} vs τ curve is virtually equivalent to a plot of L_{be}^{RC} vs τ.

Recall that the optical system efficiency η was defined in Eq. (3) as the product of the laser differential conversion efficiency, the hologram diffraction efficiency, and the detector quantum efficiency. Although from Eqs. (21b) and (22), L_{be}^{RC} and L_{be}^{th} are independent of η, it is evident from Eq. (20b) that L_{be}^{C} is inversely proportional to η. This dependence of L_{be} on η is illustrated in Fig. 13 where L_{be} is plotted vs τ for $\tilde{\lambda} = 1.0$ μm and $F = 8$ for five different values of η. Again, since for $\eta = 0.1\%$, L_{be}^{C} is very large, this particular curve is virtually equivalent to a plot of L_{be}^{RC} vs τ. As η increases, the energy limited line length decreases. Note from both Figs. 12 and 13 that, as optical communication technology improves, the break-even line length will decrease dramatically. On the other hand,

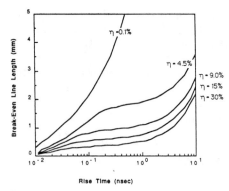

Fig. 13. Break-even line length vs rise time for varying optical system efficiency. Minimum feature size = 1.0 μm and the fan-out = 8.

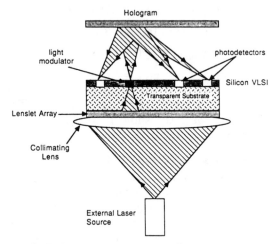

Fig. 14. Optical interconnect system with modulators as optical signal transmitters.

L_{be} is relatively insensitive to microelectronic technology improvements through minimum feature size scaling as indicated by Fig. 11.

VIII. Use of Light Modulators as Optical Signal Transmitters

As mentioned earlier, modulators are an alternative to semiconductor lasers as optical signal transmitters.[19-21] Figure 14 illustrates an optical interconnect system with light modulators as optical transmitters. (Systems utilizing reflective modulators have also been proposed.[20]) A dc biased laser illuminates the back side of a VLSI chip that has a transparent substrate (e.g., silicon on sapphire). Modulators and detectors are integrated along with the silicon circuitry on the transparent substrate. Modulators are attractive because (1) they may be easier to integrate on a VLSI chip and (2) they may dissipate less on-chip power since the electrical to optical power conversion occurs off-chip. Although several silicon based light modulator technologies have been developed,[19,22] these modulators have thus far exhibited limited switching speeds (<2 MHz). Methods for combining GaAs based multiple quantum well modulators with silicon are under development.

In this section the switching energy and delay time of an optical interconnect employing a light modulator as the optical signal transmitter are calculated. The switching energy calculation consists only of on-chip energy dissipation—energy dissipated in the off-chip laser is neglected.

An optical link with a light modulator as the signal transmitter can be modeled by replacing η, η_L, and P_L with η', η_M, and P_M, respectively, in Eqs. (2) and (3); where η', η_M, and P_M are defined as follows: $\eta' = \eta_M \eta_H \eta_D$, P_M is the optical power input to a modulator, and η_M the optical modulator efficiency, is equal to the optical power emitted by a modulator divided by P_M. Thus Eq. (5) becomes

$$E_2 = 4V(C_{PD} + C_{in})h\nu/(\eta' q). \qquad (28)$$

This is the optical energy required of an external source for illumination of each modulator. [The form of this expression is a factor of 2 larger than that of Eq. (5) because the modulator requires source illumination even when it is transmitting no light.] Electrical energy dissipated in each detector circuit is $2(C_{PD} + C_{in})V^2$, as described in Sec. II. Additional electrical power is required to switch the state of a modulator. This power is given by $C_M V_M^2/(2\tau)$, where C_M is the capacitance of the modulator and V_M is the modulator driving voltage supply. This results in a total interconnect switching energy of

$$E_0 = 2VF(C_{PD} + C_{in})[2h\nu/(\eta' q) + V] + C_M V_M^2 \qquad (29)$$

for an optical link with an on-chip light modulator transmitter and a hologram performing an F-fold fan-out.

For a multiple quantum well modulator, typical values are $\eta_M = 60\%$ and, the modulator switching energy, $C_M V_M^2 = 2$ pJ (for a 225-μm^2 device).[23] In Fig. 15 break-even line length is plotted vs rise time for $\eta' = 36\%$ (e.g., $\eta_M = 60\%$, $\eta_H = 75\%$, $\eta_D = 80\%$) for several values of remote fan-out. Since increasing F affects only one of the terms in Eq. (29), break-even line length decreases with increasing fan-out when F is small. For large values of F, the first term in Eq. (29) dominates and the break-even line length is relatively insensitive to changes in fan-out. Note that the break-even line length for modulators as optical interconnect signal transmitters is smaller than for laser diodes (Fig. 10), especially for large rise times.

IX. Conclusions

Switching energy and delay time have been expressed as functions of interconnect line length and IC process parameters for both the optical link of Figs. 1 and 2 and the electrical link of Fig. 3. Although the analysis was performed for free-space intrachip optical interconnects, a modification of the model to account for chip-to-chip interconnects was described in Sec. V. Similarly, fiber-optic interconnects could be accounted for by replacing η_H in Eq. (3) with the coupling efficiency of the fiber-optic link.

Several approximations were incorporated into the expressions developed here. For example, for electri-

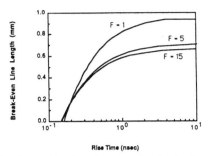

Fig. 15. Break-even line length vs rise time for implementation of optical interconnects with light modulators. Fan-out varies from 1 to 15. Optical link efficiency is 45%. Minimum feature size is 3 μm.

cal interconnects the switching energy and delay time of the additional inverters required to drive large line-driving gates were neglected. A similar approximation was made for optical interconnects where the energy and delay associated with the driving circuitry of the lasers was neglected. Both of these approximations become invalid only for small rise times, or large fan-outs when large amounts of electrical power are required (assuming laser diodes with small threshold powers are employed). For such cases, both optical and electrical interconnects would suffer from approximately equal amounts of additional energy dissipation and delay time. Other assumptions, including the neglect of the speed of light delays for optical interconnects, the neglect of transmission line effects, and not accounting for the use of repeaters for electrical interconnects are valid for the range of delay times and line lengths considered here.

The expressions for switching energy and rise time were used to plot break-even line length as a function of rise time for a 3-μm CMOS process (Fig. 8). For interconnect lengths longer than the break-even line length optical interconnects have a smaller ratio of power dissipation to data transmission rate than have corresponding electrical interconnects. Although for small rise times the switching energy for optical interconnects is approximately constant, the switching energy for electrical interconnects increases as the rise time decreases, since small rise times require large capacitative line-driving gates. Therefore the break-even line length decreases with increasing data transmission rates as indicated in Fig. 8. The break-even line length is <2 mm for rise times less than ~2.5 ns. As the remote signal fan-out [illustrated in Figs. 9(c) and (d)] is increased, Fig. 10 indicates that the break-even line length decreases, especially for rise times greater than ~1 ns. The effect of IC dimension scaling on the break-even line length was found by applying a scaling rule to the parameters in the previously developed equations. Results plotted in Fig. 11 indicate that, for a conservative estimate of total optical link efficiency of 9.0%, a laser diode threshold power of 1.0 mW, and a signal fan-out of 8, the break-even line length remains below ~2 mm for rise times < ~2 ns and IC minimum feature sizes >0.25 μm. Figure 12 indicates the large sensitivity of the break-even line length to the laser diode threshold power. Figure 15 shows that the use of light modulators rather than laser diodes as optical signal transmitters can result in smaller break-even line lengths (<1 mm for a 3-μm IC minimum feature size), especially for rise times longer than ~1 ns. These results suggest that optical interconnects may be advantageous for intrachip communication in large area VLSI circuits or wafer scale integrated circiuts, especially when high data rates and/or large fan-outs are required.

This research is supported by the Defense Advanced Research Projects Agency and administered by the Jet Propulsion Laboratory under DARPA contract 5494.

References

1. J. W. Goodman, F. I. Leonberger, S. Y. Kung, and R. A. Athale, "Optical Interconnections for VLSI Systems," Proc. IEEE **72**, 850 (1984).
2. K. C. Saraswat and F. Mohammadi, "Effect of Scaling of Interconnections on the time delay of VLSI Circuits," IEEE Trans. Electron Devices **ED-29**, 645 (1982).
3. D. S. Gardner, J. D. Meindl, and K. C. Saraswat, "Interconnection and Electromigration Scaling Theory," IEEE Trans. Electron Devices **ED-34**, 633 (1987).
4. L. A. Bergman et al., "Holographic Optical Interconnects in VLSI," Opt. Eng. **25**, 1109 (1986).
5. W. H. Wu et al., "Implementation of Optical Interconnections for VLSI," IEEE Trans Electron Devices **ED-34**, 706 (1987).
6. R. K. Kostuk, J. W. Goodman, and L. Hesselink, "Optical Imaging Applied to Microelectric Chip-to-Chip Interconnections," Appl. Opt. **24**, 2851 (1985).
7. R. Barakat and J. Reif, "Lower Bounds on the Computational Efficiency of Optical Computing Systems," Appl. Opt. **26**, 1015 (1987).
8. S. Sakai, H. Shiraishi, and M. Umeno, "AlGaAs/GaAs Stripe Laser Diodes Fabricated on Si Substrates by MOCVD," IEEE J. Quantum Electron. **QE-23**, 1080 (1987).
9. S. Sakai, X. W. Hu, and M. Umeno, "AlGaAs/GaAs Transverse Junction Stripe Lasers Fabricated on Si Substrates Using Superlattice Intermediate Layers by MOCVD," IEEE J. Quantum Electron. **QE-23**, 1085 (1987).
10. C. Mead and L. Conway, *Introduction to VLSI Systems*, (Addison-Wesley, Menlo Park, CA 1980), pp. 11–12.
11. T. Quarles, A. R. Newton, D. O. Pederson, and A. Sangiovanni-Vincentelli, *SPICE Version 3A7 User's Guide* (U. California, Berkeley, 23 Sept. 1986).
12. L. A. Glasser and D. W. Dobberpuhl, *The Design and Analysis of VLSI Circuits*, (Addison-Wesley, Menlo Park, CA, 1985), pp. 139–141.
13. "The MOSIS System (What It is and How to Use It)," Report ISI/TM-84-128, Information Sciences Institute, U. Southern California, Marina del Rey, CA 90292 (Mar. 1984).
14. P. R. Haugen, S. Rychnovsky, A. Husain, and L. D. Hutcheson, "Optical Interconnects for High Speed Computing," Opt. Eng. **25**, 1076 (1986).
15. T. Shibutani et al., "A Novel High-Power Laser Structure with Current-Blocked Regions Near Cavity Facets," IEEE J. Quantum Electron. **QE-23**, 760 (1987).
16. Ref. 12, pp. 135–136.
17. Ref. 10, pp. 341–342.
18. P. L. Derry and A. Yariv, "Ultralow-Threshold Graded-Index Separate-Confinement Single Quantum Well Buried Heterostructure (Al,Ga)As Lasers with High Reflectivity Coatings," Appl. Phys. Lett. **50**, 1773 (1987).
19. R. E. Brooks, "Micromechanical Light Modulators on Silicon," Opt. Eng. **24**, 101 (1985).
20. E. Bradley and P. K. L. Yu, "Proposed Modulator for Global VLSI Optical Interconnect Network," Jpn. J. Appl. Phys. **26**, L971 (1987).
21. G. D. Boyd, D. A. B. Miller, D. S. Chemla, S. L. McCall, A. C. Gossard, and J. H. English, "Multiple Quantum Well Reflection Modulator," Appl. Phys. Lett. **50**, 1119 (1987).
22. S. H. Lee, S. C. Esener, M. A. Title, and T. J. Drabik, "Two-Dimensional Silicon/PLZT Spatial Light Modulators: Design Considerations and Technology," Opt. Eng. **25**, 250 (1986).
23. D. A. B. Miller, D. S. Chemla, T. C. Damen, T. H. Wood, C. A. Burrus, Jr., A. C. Gossard, and W. Wiegmann, "The Quantum Well Self-Electrooptic Effect Device: Optoelectronic Bistability and Oscillation, and Self-Linearized Modulation," IEEE J. Quantum Electron. **QE-21**, 1462 (1985).

Perfect Shuffles and Their Equivalents

Optical perfect shuffle

Adolf W. Lohmann, W. Stork, and G. Stucke

Physikalisches Institut der Universität Erlangen-Nurnberg, 8520 Erlangen, Federal Republic of Germany.
Received 25 May 1985.

Recently some proposals for an optical perfect shuffle (PS) have been made.[1-4] The reason for this interest is that the PS is a useful concept in computer architecture. The main problem in multiprocessor configurations is the communication between the processing elements. An ideal interconnection network has to be fast and flexible, and the PS is the basis for such a network.

The PS performs a certain permutation of $N = 2^k$ elements (see Fig. 1). If the addresses of the elements are represented by binary numbers, ranging from 0 to $N - 1$, the PS can be described as a cyclical rotation to the left of the address bits. Mathematically the PS is defined by the following expression:

$$n' = 2n + [2n/N] \bmod N; \qquad n = 0,1\ldots N-1. \qquad (1)$$

The brackets [ ] indicate the largest integer less than or equal to the arguments.

This permutation can be performed very fast using classical optics, as described later. But to achieve flexibility the PS has to be supplemented by so-called "exchange boxes."[5] These boxes offer the facility to exchange two adjacent elements, or not to exchange them, which is usually called bypass. With the PS and $N/2$ exchange boxes it is possible to generate every arbitrary connection permutation in only a few steps, the number of steps being of the order of $\log N$.[6] Some examples of PS applications like FFT and sorting algorithms are described in Refs. 5 and 7.

In a digital optical computer the PS can be implemented in several ways. One could use optical fibers or integrated optics similar to a conventional electronic computer. But that approach would require a larger number of material connections. Optical shuffles with free space propagation can be implemented by holograms or by classical optics.

The basic principle of our setup, which uses lenses and prisms, is shown in Fig. 2. The first operation is to divide the input elements into upper and lower halves. These two halves have to be stretched in one direction to the size of the original input. An appropriate mask guarantees that there is no overlap of adjacent elements. The final step is the recombination of the two halves. This interlace operation is equivalent to the PS.

Figure 3 shows a simple setup, suitable for collimated illumination. The input object, indicated by the numbers 0–7, may be a spatial light modulator. The first pair of prisms separates the two halves, so that different shifts can be applied to the upper and lower halves of the input. These shifts are caused by a second pair of prisms in the Fourier plane. In the output plane the shuffled version of the input object appears with an overall reversed sequence. This reversal can be compensated if necessary by standard optical means using lenses, mirrors, prisms, or other standard components. Instead of masking out the unwanted light, smaller pixels are used in the input plane which do not overlap after stretching. The smaller pixels may be produced by a 2-D lenslet array.

The output size depends on the ratio of the two focal lengths f_2/f_1. A magnification of 2 generates an output of the same size as the input in the shuffle direction and of double the size in the orthogonal direction. To avoid this an anamorphic setup can be used. No anamorphic lens system is needed if the PS is generalized to two dimensions. In that case the four quadrants of the input array are to be interlaced.

Fig. 1. Perfect shuffle permutation for $N = 16$ elements. The small boxes represent processing elements (PE), numbered from 0 to 15. As an example the output of PE 2 is connected to the input of PE 4.

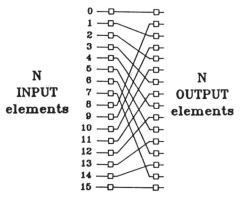

Fig. 2. Basic concept of a PS setup. The PS permutation is achieved by interlacing the upper and lower halves of the input.

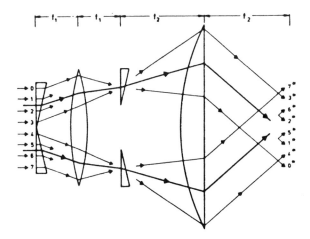

Fig. 3. Simple PS setup with two prism pairs.

Reprinted with permission from *Appl. Opt.*, vol. 25, no. 10, pp. 1530–1531, May 15, 1986.
Copyright © 1986, Optical Society of America.

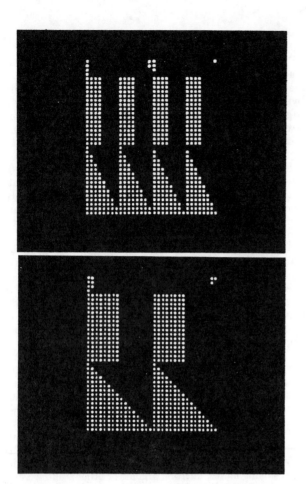

Fig. 4. Input object (upper) and its shuffled version (lower). The columns may represent numbers shuffled in a horizontal direction.

Figure 4 presents a special input object and its optically shuffled version.

This Letter is based on a paper presented at the OSA Topical Meeting on Optical Computing, 18–20 Mar. 1985.

References
1. J. W. Goodman, J. F. Leonberger, S. Y. Kung, and R. A. Athale, "Optical Interconnections for VLSI Systems," Proc. IEEE **72**, 850 (1984).
2. M. E. Marchic, "Combinatorial Star Couplers for Single-Mode Optical Fibers," FOC/LAN 84, pp. 175–179.
3. A. W. Lohmann, W. Stork, and G. Stucke, "Optical Implementation of the Perfect Shuffle," in *Technical Digest of Topical Meeting on Optical Computing* (Optical Society of America, Washington, DC, 1985).
4. K.-H. Brenner and A. Huang, "Optical Implementation of the Perfect Shuffle Interconnection," **to be submitted to Appl. Opt.**
5. H. S. Stone, "Parallel Processing with the Perfect Shuffle," IEEE Trans. Comput. **C-20** (Feb. 1971).
6. C. Wu and T. Feng, "The University of the Shuffle/Exchange-Network," IEEE Trans. Comput. **C-30** (May 1981).
7. D. S. Parker, Jr., "Notes on Shuffle/Exchange Type Switching Networks," IEEE Trans. Comput. **C-29** (Mar. 1980).

Compact optical generalized perfect shuffle

George Eichmann and Yao Li

CUNY-City College, Department of Electrical Engineering, New York, New York 10031.
Received 7 October 1986.

For the solutions of many scientific and engineering problems, parallel processing has been shown to be a fast way to process information. To distribute and to interconnect massive amounts of data between stages of parallel processing elements, fast and efficient interconnection networks are needed. It has been indicated that, for some applications, shuffle-exchange networks[1-3] are very effective in handling such data interconnections. Shuffle-exchange networks are implemented using repeated stages of the so-called perfect shuffle (PS) together with arrays of exchange boxes that can independently either exchange or bypass the adjacent lines. Different combinations of PS and exchange box arrays have found applications[1-3] in evaluating polynomials, in sorting data, in transposing matrices, as well as in computing the fast Fourier transform.

Given the inherent parallelism of optics, interest has been focused on developing parallel optical computing architectures, and in particular, on the implementation of optical shuffle-exchange networks. Goodman et al.[4] and Marhic[5] proposed the use of optical fibers or waveguides for an OPS. However, for large data arrays, large bundles of fibers are needed. To take full advantage of the free-space propagation property of optical waves, Marhic,[5] Lohmann et al.[6,7] and Brenner and Huang[8] suggested the use of unguided implementation approaches. An unguided OPS consists of either a hologram or a suitable lens and prism combination. While the holographic OPS requires monochromatic light inputs, the lens/prism-based counterpart can also be used with white light illumination. In this Letter, additional and more compact unguided OPS geometries are suggested. New transmissive and reflective OPSs are described. Finally, an implementation of an optical generalized PS (OGPS) is also discussed.

The PS $P_N(i)$ is defined as[2]

$$P_N(i) = (2i + [2i/N]) \bmod N \quad 0 \leq i \leq N-1, \quad (1)$$

where $N = 2^j$, i and j are integers and $[2i/N]$ represents the largest integer that is $\leq 2i/N$. When binary symbols are used as input line addresses, after a PS permutation, the binary addresses of the output lines represent a right shift operation. Using this PS cyclic shift permutation property together with arrays of exchange boxes, any address configuration can be permuted into any other configuration of the order of $(\log N)^2$ steps.[3]

In the stretch-mask-add approach,[7] the unguided OPS consists of four prism wedges and two positive spherical lenses with focal lengths f_1 and f_2, respectively. Correspondingly, the total length of the system is $2f_1 + 2f_2$. To maintain the same output channel spacing as that of the input, the length f_2 must be twice the length of f_1 leading to a total optical system length of $6f_1$.

A more compact OPS, using a new unguided OPS implementation, is suggested here. A PS requires that half of the inputs diverge by a factor of 2 while they interlace with those from the second half inputs. To obtain this divergence, a negative cylindrical lens may be employed. In Fig. 1(a), a negative cylindrical lens-based OPS is shown. Here, side by side, two identical aperture (D) and focal length (f) negative lenses are used. For simplicity, the sketch shows plano-concave negative lenses, where the unused portions of the lenses are not shown. Collimated input beams illuminate the plane of the lenses, where the input mask is located. The output beams, at the back focal plane of the lenses, represent the shuffled result. For an N-bit input, using geometric optics, the bit or channel period (d) and spot size (a) are determined by

$$d = \frac{D}{N-1}, \quad (2)$$

$$a \leq \frac{D}{2(N-1)}. \quad (3)$$

Because the output spot size is magnified by a factor of 2, the input spot size [Eq. (3)] is constrained to one-half of the input bit period d. For example, if the input bit spot size a and their spacing d is 0.1 and 0.25 mm, respectively, a 50- ×

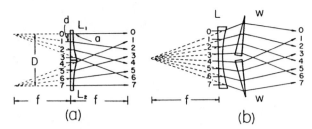

Fig. 1. Schematic diagrams of an OPS where D is the system aperture, a is the input channel size, and d is the channel spacing. (a) A lens-based system: L_1 and L_2 are two identical focal length (f) negative cylindrical lenses. (b) An alternative lens-based system where a single negative cylindrical lens together with two identical prism wedges are used.

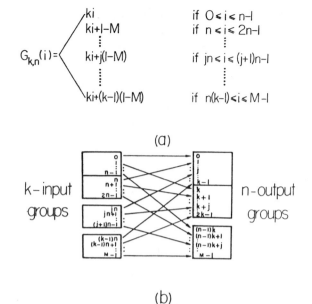

Fig. 2. Generalized PS permutation interconnection. (a) A more detailed input and output relation. (b) A graphic example showing the input and output relation. Note that both the input and output are each divided into k and n groups.

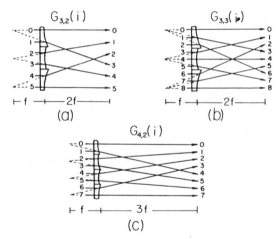

Fig. 3. Three lens-based OGPS implementation examples: (a) $G_{3,2}(i)$; (b) $G_{3,3}(i)$; (c) $G_{4,2}(i)$.

50-mm² aperture OPS can optimally handle as many as 40,000 light channels. Compared to the stretch-mask-add approach, this system is more compact since it has only two optical elements while its length is reduced by a factor of 6. With this method, the size of the output bit is identical to the stretch-mask-add approach spot size. However, because this OPS generates a divergent output, in a shuffle-exchange network the exchange boxes must be able to recollimate the optical beam. When a two-port optical waveguide switch is used as the exchange box, by proper front and back lens adjustments the beams can be demagnified to their original sizes. This compact OPS can also be implemented with large aperture reflective optical elements, i.e., two identical, side by side, radius R convex cylindrical reflective surfaces. If necessary, the output spots can be separated out by a beam splitter. In either case, the divergence operations are performed by two identical components (lens or mirror).

It is also possible to generate the required operations with only a single negative cylindrical optical element (lens or a mirror). In Fig. 1(b) the use of a single cylindrical lens-based OPS is illustrated. First, using a single negative lens, the two divergence operations are performed. Second, using two prism wedges, each half of the diverging results is stretched and interlaced together to generate the final OPS output.

Next, an optical implementation of a generalized PS (GPS)[9] is described. The interest in GPS stems from the fact that in many applications, instead of using $N = 2^j$ inputs, the use of other composite integer (M) inputs is required. A GPS [$G_{k,n}(i)$] characterized by the two integers k and n ($k \geq 2, n \geq 2$) such that the total number of inputs M ($M = kn$) is defined as[9]

$$G_{k,n}(i) = ki + [i/n](1 - M). \quad (4)$$

In Fig. 2(a), some details of the GPS permutation formula are given, while in Fig. 2(b) a corresponding permutation example is illustrated. Here, both the input and output ports are divided into k and n groups, respectively. In each of the k input groups, for example, in the jth group, there are n input lines that are to be distributed to a fixed place (the jth line as in the example) in each of the n output groups. Note that a PS $P_N(i)$ is a special case of GPS, i.e., $G_{2,N/2}(i)$. Since for each of the k input groups an identical magnification divergence operation is performed, for an OGPS implementation, k pieces of either transmissive or reflective optical elements, cut from either identical focal length negative cylindrical lenses or identical radius cylindrical reflective surfaces, can be utilized. In Fig. 3, using transmissive optical elements (a negative cylindrical lenslet array), three OGPS cases, $G_{3,2}(i)$, $G_{3,3}(i)$, and $G_{4,2}(i)$, are illustrated. In general, for each of k identical size elements the aperture A is

$$A = \frac{2D}{M-1}. \quad (5)$$

Because the OGPS output is collected at a distance $(k-1)f$ measured from the input plane, compared to input, the output diverges by a factor of $k - 1$. Thus, the input bit or channel size must be chosen as

$$a \leq \frac{D}{(k-1)(M-1)}. \quad (6)$$

While this method can be used for arbitrary k and n, because of the beam divergence it is only practical for relatively small k.

New unguided OPS geometries have been proposed. Using either a pair of negative cylindrical lenses or convex reflective surfaces, either transmissive or reflective OPSs can be implemented. Compared to the stretch-mask-add approach, this method uses fewer optical elements and a more compact geometry. The method can also be generalized to implement an OGPS.

This work was supported in part by grants 84-0144 and 85-0212 from the U.S. Air Force Office of Scientific Research.

References

1. H. S. Stone, "Parallel Processing with the Perfect Shuffle," IEEE Trans. Comput. **C-20,** 153 (1971).
2. D. S. Parker, Jr., "Notes on Shuffle/Exchange Type Switching Networks," IEEE Trans. Comput. **C-29,** 213 (1980).
3. C. L. Wu and T. Y. Feng, "The University of the Shuffle-Exchange Network," IEEE Trans. Comput. **C-30,** 324 (1981).
4. J. W. Goodman, J. F. Leonberger, S. Y. Kung, and R. A. Athale, "Optical Interconnections for VLSI Systems," Proc. IEEE **72,** 850 (1984).
5. M. E. Marhic, "Combinatorial Star Couplers for Single-Mode Optical Fibers," FOC/LAN **84,** 175 (1984).
6. A. W. Lohmann, "What Classical Optics Can Do for the Digital Optical Computer," Appl. Opt. **25,** 1543 (1986).
7. A. W. Lohmann, W. Stork, and G. Stucke, "Optical Perfect Shuffle," Appl. Opt. **25,** 1530 (1986).
8. K.-H. Brenner and A. Huang, "Optical Implementations of Symbolic Substitution," J. Opt. Soc. Am. A **1,** 1292 (1984).
9. J. Mikloško, "Correlation of Algorithms, Software and Hardware of Parallel Computers," in *Algorithms, Software and Hardware of Parallel Computers*, J. Mikloško and V. E. Kotov, Eds. (Springer-Verlag, Berlin, 1984).

Optical implementations of the perfect shuffle interconnection

Karl-Heinz Brenner and A. Huang

The concept of a perfect shuffle is reviewed. Holographic approaches based on implementing a point spread function equivalent to the sum of two shifted delta functions are suggested. Interferometric approaches based on splitting an image and then combining them in a shifted manner are also suggested.

I. Introduction

Optical interconnections will be a major application of optics in the future. The reason optics is preferred in this field lies mainly in the fact that optics offers 2-D operation, low crosstalk, and high bandwidth. A question still discussed, however, is the trade-off between arbitrary interconnections vs regular interconnections.

Arbitrary interconnections can be made only at the expense of a high space–bandwidth product. To interconnect N inputs, N^2 degrees of freedom are necessary. The space–bandwidth product provided by holograms limits the number N to the order of 10^2.

Regular interconnections need a very low space–bandwidth product. They are easy to realize with optical components because optical systems are naturally space invariant. The use of a regular interconnect seems to be topologically restrictive. However, it is possible to construct networks which are capable of supporting an arbitrary interconnection. One of the most useful and universal regular interconnects is the perfect shuffle.[1]

In this paper we give an overview on the capabilities and applications of the perfect shuffle and then discuss holographic and interferometric methods for implementing this kind of interconnection.

II. Perfect Shuffle

The significance of the perfect shuffle was first realized by Stone,[1] who demonstrated that several signal processing algorithms such as Fourier transformation, polynomial evaluation, and sorting have interconnection patterns which can be implemented via a perfect shuffle. Such an interconnect is shown in Fig. 1. Wu and Feng[2] demonstrated that other networks such as the Benes,[3] Batcher,[4] and Delta[5] can all be recast into perfect shuffle interconnections. The elegance and power of this kind of network[6,7] have inspired approaches for a perfect shuffle-exchange network in VLSI[8,9] and also in integrated optics.[10]

The perfect shuffle gets its name from playing cards. A perfect shuffle of a deck of cards involves dividing the deck into two equal halves and interlacing the second half into the first half.

The perfect shuffle can also be expressed in a more formal manner. Given a number of inputs A_k ($k = 0 \ldots N-1$), where N is an even number and an equal number of outputs $B_{k'}$, the perfect shuffle is a permutation of the numbers k and k' so that

$$k' = \begin{cases} 2k & \text{if } 0 \le k < N/2, \\ 2k - N + 1 & \text{if } N/2 \le k < N. \end{cases} \quad (1)$$

If N is a power of two, the perfect shuffle is a cyclic left shift of the binary representation of the index k. This follows from Eq. (1), since if k is less than $N/2$, multiplication by two is a left shift of the binary representation, and if k is larger or equal to $N/2$, the leftmost bit in k (which is a 1) disappears, while the remaining bits are shifted to the left, and a one is placed in the rightmost place. These two conditionals are equivalent to a cyclic left shift.

The perfect shuffle has several magical properties. One property is that if you perfect shuffle N elements $\log_2 N$ times, the elements will return to the original order. Another property is that if you perfect shuffle N elements and then interchange or not interchange

---

When this work was done both authors were with AT&T Bell Laboratories, Holmdel, New Jersey 07733; K.-H. Brenner is now with University of Erlangen, Physics Institute, 8520 Erlangen, Federal Republic of Germany.

Received 1 April 1987.

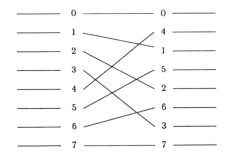

Fig. 1. Perfect shuffle permutation of eight inputs.

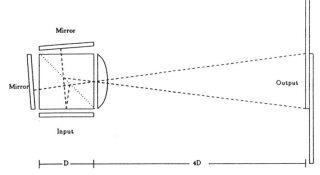

Fig. 3. Perfect shuffle implemented via a Michelson arrangement.

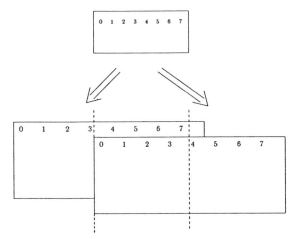

Fig. 2. Perfect shuffle implemented via interlacing two images.

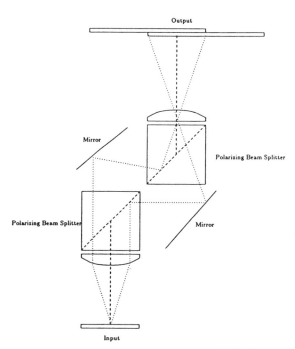

Fig. 4. Perfect shuffle implemented via a Mach-Zehnder arrangement.

neighboring pairs of elements in a prescribed manner and repeat this $3 \log_2 N$ times, you can rearrange the N elements in any order. A third property is that you can build any size perfect shuffle out of smaller perfect shuffles interconnected in a perfect shuffle manner.

III. Optical Implementation of the Perfect Shuffle

As mentioned previously, a perfect shuffle permutation can be achieved by interleaving the right half of the input with the left half. Figure 2 shows that a superimposition of shifted input images also produces a perfect shuffle permutation. The task of the optics reduces to producing a twofold magnified image, splitting the image, shifting each image, and discarding the nonoverlapping area. This can be accomplished optically with either a holographic or interferometric approach.

A. Holographic Perfect Shuffles

An optical $4f$ system is capable of producing two shifted images of an input if the point spread function of the system is the sum of two shifted delta functions. This is equivalent to a cosine-shaped transmittance function in the filter plane. In particular, if we want an absolute shift of δx in the output plane, the required cosine grating has the form

$$\tilde{p}(x) = \cos\left(2\pi \frac{\delta x}{\lambda f} x\right). \qquad (2)$$

Possible realizations include fabricated phase grat-

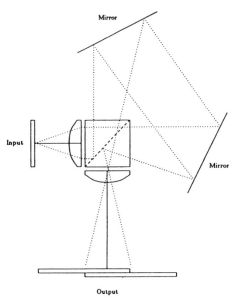

Fig. 5. Perfect shuffle implemented via a Sagnac arrangement.

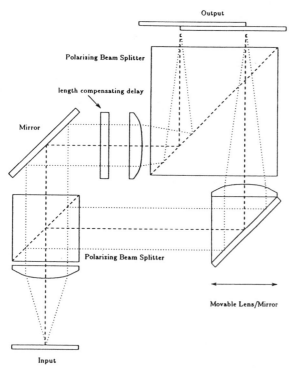

Fig. 6. Constant latency optical perfect shuffle.

ings, electrooptic phase gratings, acoustooptic modulators, volume gratings such as dichromated gelatin, etched gratings, or acoustooptic cells operating in the Bragg regime. The limited acceptance angle of volume gratings limits their space–bandwidth product. The reduced light efficiency of absorption gratings also introduces a constraint.

B. Interferometric Perfect Shuffles

Interferometers split and combine an optical beam. Interferometers can also be used to divide and interlace images. Figures 3, 4, and 5 illustrate this principle in case of a Michelson, Mach-Zehnder, and a Sagnac interferometer. Beam splitters divide the input into two equal images, and mirrors deflect these images to produce a relative shift in the output plane.

In the Michelson arrangement the mirrors mounted on the beam splitter cube allow independent deflection of the two beams and also result in a very compact device. Quarterwave plates assure that all the light reflected from the mirrors are directed toward the output plane.

In the Sagnac arrangement the mirrors are fixed, and the relative shift can be achieved by rotating the beam splitter.

In the Mach-Zehnder arrangement a shift can be achieved by tilting either the mirrors or the beam splitters. The output will appear at either of the two sides, depending on the input polarization angle (+45 or −45°). Polarization can be used to improve the light efficiency in any of the arrangements shown in Figs. 3, 4, and 5.

Figure 6 shows a configuration based on the Mach-Zehnder interferometer. However, the shift is not achieved by tilting mirrors but rather by moving the combination of mirror $M1$ and lens $L1$ horizontally. The optical path length of the two branches can be made equidistant by introducing an optical path length compensation delay D. Constant latency becomes more important as the optical signals become shorter. The configuration described also shifts the output planes without tilting the beam. This simplifies the coupling of signals into direction sensitive elements such as optical fibers.

IV. Summary

The concept of the perfect shuffle permutation is presented, and its significance is discussed. Holographic approaches based on implementing a point spread function equivalent to the sum of two shifted delta functions are suggested. Interferometric approaches based on splitting an image and then combining them in a shifted manner are also suggested.

Reference

1. H. S. Stone, "Parallel Processing with the Perfect Shuffle," IEEE Trans. Comput. **C-20,** 153 (1971).
2. C. L. Wu in T. Y. Feng, "The Universality of the Shuffle-Exchange Network," IEEE Trans. Comput. **C-30,** 324 (1981).
3. V. Benes, *Mathematical Theory of Connecting Networks* (Academic, New York, 1965).
4. K. E. Batcher "Sorting Networks, and their Applications," in *Proceedings, AFIPS 1968 SJCC*, *32*, No. 2 (AFIPS Press, Montvale, NJ, 19XX), pp. 307.
5. J. H. Patel, "Processor-Memory Interconnections for Multiprocessors," in *Proceedings, Sixth Annual Symposium on Computer Architecture*, New York, (Apr. 1979), pp. 168.
6. T. Lang and H. S. Stone, "A Shuffle-Exchange Network with Simplified Control, IEEE Trans. Comput. **C-25,** 55 (1976).
7. D. Nassimi and S. Sahni, "A Self-Routing Benes Network and Parallel Permutation Algorithms," IEEE Trans. Comput. **C-30,** 332 (1981).
8. A. Huang, S. Knauer, and J. O'Neill, "Case Study: A Self-Routing Switching Network," in *Principles of CMOS VLSI Design—A Systems Perspective*, Neil Weste, Ed. (Addison-Wesley, Reading, MA, 1985).
9. A Huang and S. Knauer, "STARLITE: a Wideband Digital Switch," in *Conference Proceedings, GLOBECOM '84*, IEEE 84CH2064-4, Vol. 1, Nov. 1984, pp. 121–125.
10. J. W. Goodman, F. I. Leonberger, S.-Y. Kung, and R. A. Athale, "Optical Interconnections for VLSI Systems," Proc. IEEE **72,** 850 (1984).

Crossover networks and their optical implementation

Jürgen Jahns and Miles J. Murdocca

Crossover networks are introduced as a new type of interconnection network for applications in optical computing, optical switching, and signal processing. Crossover networks belong to the class of multistage interconnection network. Two variations are presented, the half-crossover network and the full crossover network. An optical system which implements both networks is proposed and demonstrated. Crossover networks can be implemented using the full space–bandwidth product of the optical system with minimal loss of light. It is shown that crossover networks are isomorphic to other multistage networks such as the Banyan and perfect shuffle.

I. Free-Space Architectures for Optical Digital Computers

The interest attracted by optical digital computing is mainly stimulated by its potential to implement massively parallel architectures. This holds especially for free-space optical systems where 2-D arrays of logic elements can be connected using imaging setups. Free-space optical interconnections also offer the potential to do the communications in a computer at an extremely high temporal bandwidth without introducing problems such as clock skew or crosstalk.[1]

The use of optical imaging setups for parallel interconnects, however, limits the variety of feasible topologies to regular interconnects. The space–bandwidth product (SBP) of an optical system, i.e., the number of connections, reduces with increasing complexity of the interconnection scheme. For this reason, interest has grown in regular interconnection networks such as the perfect shuffle[2] or the Banyan.[3] The use of regular networks seems to limit the flexibility of designing a digital general purpose computer. It has been shown, however, that they can be used for general purpose computers efficiently in terms of gate count and throughput.[4]

The perfect shuffle and the Banyan both belong to the class of multistage interconnection network (MINs). For interconnecting N inputs to N outputs, perfect shuffle and Banyan require $\log_2(N)$ stages. Throughout this paper we shall assume that N is a power of 2. MINs are of great importance, for example, in digital signal processing for the design of fast algorithms[5] or in computing for the realization of sorting networks.[6] Various authors have addressed the use of multistage networks in optical data processing.[7–12]

One problem which arises for the implementation of MINs is the fact that they represent space-variant operations. Optical systems, however, offer a large SBP only for space-invariant operations. It is, therefore, necessary to find a way of implementing a specific network without losing too much of the space–bandwidth product.

In this paper, we present crossover interconnects as a new interconnection network, which is an interesting alternative to the Banyan and the perfect shuffle. Crossover networks offer the potential for a simple optical implementation. Specifically, it is possible to use the full SBP of an optical system. This means that the number of connections is limited, in principle, only by diffraction. Two versions of crossover networks will be presented in Sec. II, the half-crossover network and the full crossover network. An optical implementation for both, which is based on a Michelson setup, is proposed in Sec. III. Special optical components are used which are called prism gratings. Practical limitations for the implementation of the network arising from the use of these components are discussed in Sec. IV. In Sec. V we shall present several experimental results, and finally, in Sec. VI we show that crossover networks are isomorphic to the Banyan and the perfect shuffle.

II. Crossover Networks

This work was initially stimulated by a proposal for a VLSI interconnection network which was made by Wise.[13] This proposal was motivated by the need to have wires of exactly the same length to reduce path length differences between signals. The diagram of Wise's network is shown in Fig. 1.

The functional boxes as indicated by the shaded rectangles are not important for our discussion. Their function may vary with the specific application for which the network is used. Each box may in fact be composed of a mininetwork of its own. The light boxes are to be considered as mirrors which reflect the signal path. It is obvious then that this network is composed of signal paths of identical lengths. The name crossover network illustrates the pattern of the signal paths.

The authors are with AT&T Bell Laboratories, Holmdel, New Jersey 07733.
Received 2 November 1987.

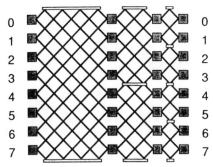

Fig. 1. Crossover network for VLSI circuits.[13]

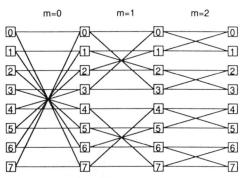

Fig. 3. Half-crossover network for eight input ports. The index m indicates the number of a specific stage.

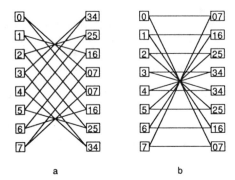

Fig. 2. (a) Bipartite graph of the first stage of the network shown in Fig. 1; (b) rearranged graph.

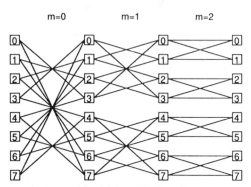

Fig. 4. Full crossover network for $N = 8$.

The diagram in Fig. 1 might lend itself to a waveguide optical implementation. The lines would directly represent the waveguides. For a free-space optical implementation it is useful to rearrange the interconnections. To this end we represent the first stage of the network shown in Fig. 1 by a bipartite graph, which shows the input ports of the stage, the output ports, and the connections between them [Fig. 2(a)]. The input ports are numbered by integers which run from 0 to 7. On the output side, the boxes are labeled according to where the connections originate.

Figure 2(b) shows the rearranged version of the graph. On the output side still the same combinations of numbers appear. However, they now appear in a different order. We note that the graph in Fig. 2(b) is very regularly structured. The interconnection pattern consists of straight-through connections and crossover connections. The straight-through connections can be implemented optically by a simple imaging step. The crossover can be implemented by imaging as well, but a spatial inversion with respect to the first image has to be introduced. We shall discuss a possibility to achieve this inversion in some detail in the next section.

We consider the diagram in Fig. 2(b) to be the first stage of our new network and assign to it the index $m = 0$. The higher stages are then obtained by the following rule. For stage m we subdivide the input into 2^m partitions. For each partition, we copy the input pixels to the output side. Furthermore, we apply a crossover within each partition. The whole network consists of $\log_2(N)$ stages where N is the width of the network. For $N = 8$, the complete network is shown in Fig. 3. The boxes were labeled according to their physical addresses. We will refer to the network shown in Fig. 3 as the half crossover network (HCN).

A variation of the half-crossover network is obtained in the following way. We subdivide the input ports to the mth stage into 2^{m+1} partitions. Then we replace the straight-through connections with crossover connections within each partition. This results in the full crossover network (FCN) shown in Fig. 4.

For simplicity, the networks are shown in Figs. 3 and 4 for 1-D inputs. For 2-D input data it is necessary to modify the operation of the network so that first the rows are processed in $\log_2(N)$ steps. Then columns are processed using also $\log_2(N)$ stages.

III. Optical Implementation of the Crossover Network

Figure 5 shows an optical setup which implements one stage of the crossover network. It is based on a Michelson setup. It should be noted, however, that the performance of the setup is not based on interference. The use of optical architectures which are derived from two-beam interferometers is simply a convenient way of implementing the splitting and combining of 2-D arrays of pixels. The setup shown in Fig. 5 implements the HCN. As we shall see, only a minor change has to be made to realize the connections for the FCN.

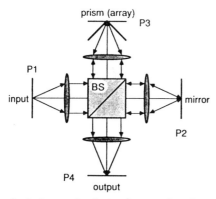

Fig. 5. Optical setup for the implementation of one stage.

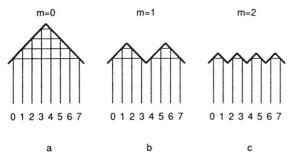

Fig. 6. Flipping of the input pixels using partitioned prisms.

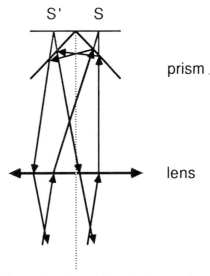

Fig. 7. Operation of the reflective 90° prism.

The beam splitter BS splits the input signal into two different paths. The planes P_1, P_2, P_3, and P_4 are related to each other by imaging steps. The path $P_1 \rightarrow P_2 \rightarrow P_4$ with a mirror in P_2 implements the straight-through connections. The path $P_1 \rightarrow P_3 \rightarrow P_4$ implements the crossover connections. This is achieved for the first stage of the network by placing a reflective 90° prism in P_3. The effect of the prism is visualized in Fig. 6(a). It illustrates the flipping of the input pixels with respect to one symmetry axis as required for the first stage. To implement the crossover connections for the higher stages, we have to partition the prism. A double prism is required for the second stage ($m = 1$), a quadruple prism for the third stage ($m = 2$), etc. [see Figs. 6(b) and (c)].

Figure 7 depicts the use of the 90° prism by showing the light paths. Light that would be focused to a position S in the back focal plane of the lens is reflected twice and is sent back into the optical system as if it would emerge from position S'. S and S' are symmetric with respect to the center position of the prism. It is important to note that, for light traveling to different positions, i.e., for different pixels, the path lengths to travel are exactly the same. Therefore, no relative time delays are introduced between different signals.

Two features should also be mentioned. First, the setup shown in Fig. 5 implements one stage of the network. For the complete network one has to use $\log_2(N)$ such setups. Second, the use of polarization optics allows us to implement each stage without losing light energy except for Fresnel losses.

The full crossover network can be realized with the same setup as shown in Fig. 5. The only change which has to be made is that the mirror in P_2 is replaced by a prism grating.

IV. Practical Limitations for the Implementation of Crossover Networks

In this section we will address a problem which occurs when using prism gratings for implementation of the higher stages ($m \neq 0$) of the network. The situation is shown in Fig. 8(a). Here the optical setup for implementing the second stage is shown with a double prism in plane P_3. For simplification the optical path $P_1 \rightarrow P_3$ is displayed in an unfolded way. As shown in the drawing, some of the light which is supposed to be focused down to a spot close to the center will hit the wrong facet of the prism grating. After reflection from this facet the light will be reflected out of the optical system and will be lost. This situation occurs when the numerical aperture of lens 3 is too large.

The situation will be most severe for the pixel closest to the optical axis. Only light traveling under an angle which is smaller than $\beta/2$ will arrive at this spot R. Assuming a pixel separation in the input array of δx and an optical setup with a 1:1 magnification, R will be at a distance $\delta x/2$ from the center. To avoid losses the numerical aperture of the light on the right-hand side of the optical setup has to be smaller than a certain maximum value N.A.$_{3,\max}$.

This value is determined by

$$\text{N.A.}_{3,\max} = \tan(\beta/2). \tag{1}$$

The value for $\beta/2$ depends on the stage number m and on the size N of the input array. To compute $\beta(m)/2$ we have to make a few geometrical considerations. The height of the prism which is used for the mth stage is

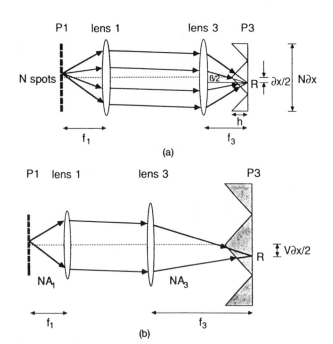

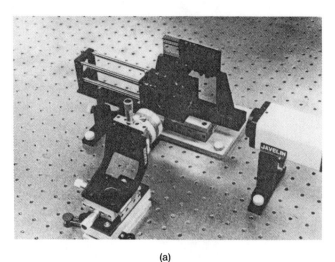

Fig. 8. (a) Light losses due to a large numerical aperture. The situation is shown for the second stage. Dark lines indicate the light path for a fully opened aperture. The dashed lines indicate the light path for a reduced aperture. The reflected light rays are not shown for simplicity. (b) Reduction of the numerical aperture using magnification.

$$h(m) = (N/2^{m+1})\delta x, \quad m = 0,\ldots,M, \tag{2}$$

where M is defined as

$$M = \log_2(N) - 1. \tag{3}$$

From this it follows that

$$\text{N.A.}_{3,\max}(m) = \frac{\delta x}{2h(m)} = \frac{2^m}{N}, \quad m = 1,\ldots,M. \tag{4}$$

Equation (4) indicates that the problem is most severe for the second stage ($m = 1$) where N.A.$_{3,\max}$ takes on very small values. For example, for an array with $N = 64$ pixels in one dimension one obtains N.A.$_{3,\max} = 1/32$. This would be a severe restriction to the use of crossover interconnects especially, since the SBP of the setup is proportional to the numerical aperture. Without taking any measures one would either lose light for the pixels close to the center or be restricted to relatively large pixels with a diameter $2\lambda/\text{N.A.}_{3,\max}$.

However, there is a simple way to circumvent this problem completely. By using magnification it is possible to reduce the numerical aperture of the light cones. Magnification is introduced by making the focal length f_3 of lens 3 larger than the focal length f_1 of lens 1 [Fig. 8(b)]. The spot pattern will then be magnified by a factor V:

$$V = \frac{f_3}{f_1}. \tag{5}$$

Accordingly the numerical aperture of the light on the right-hand side of the setup will be reduced by V.

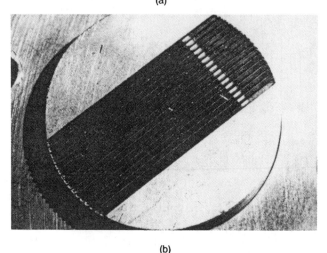

Fig. 9. Experimental setup (a) and the prism grating with a period of 640 μm (b).

$$\text{N.A.}_3 = \frac{\text{N.A.}_1}{V}. \tag{6}$$

This results in a larger value for N.A.$_{3,\max}$:

$$\text{N.A.}_{3,\max}(m) = 2^m \frac{V}{N}. \tag{7}$$

As an example we consider again the case that $N = 64$ and that on the input side we use a lens with a numerical aperture N.A.$_1 = 0.25$. This corresponds to a lens with $f/4$. Then for the second stage of the network ($m = 1$) a magnification factor $V = 8$ would be required to achieve a lossless implementation. For the higher stages the required magnification is reduced significantly.

We would like to add the remark that besides saving light energy the magnification trick has another beneficial side effect. Since the spot pattern is magnified it is possible to use relatively coarse prism gratings which are easier to fabricate than very fine ones. Therefore, magnification can be used specifically to select a certain period for the prism grating.

It is also worthwhile to note that the overall magnification between planes P_1 and P_4 is not affected by the focal length of lens 3. Instead it is strictly determined

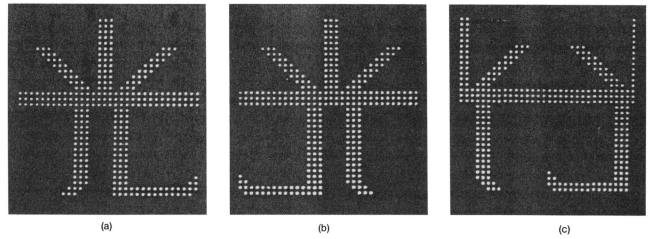

Fig. 10. Experimental results: (a) input object: Chinese character (light). Output of the first (b) and second (c) stage.

by the ratio f_4/f_1 where f_4 is the focal length of the lens near plane P_4 (Fig. 5).

V. Experimental Results

The experimental results which are presented in this section were obtained by using the Michelson setup shown in Fig. 5 and prism gratings. The photograph in Fig. 9(a) shows the actual optical system. Two different prism gratings were used for the experiments. One had a period of 640 µm, the second had a period of 10 µm. Figure 9(b) shows a photograph of the coarse grating.

Using the 640-µm grating the first and the second stages of the half-crossover network were implemented. For clarity, only the output patterns due to the crossover connections are shown in Figs. 10(b) and (c). In both cases the input pattern was the Chinese character shown in Fig. 10(a). It consists of 10-µm spots on a grid of 32 × 32 pixels. The spots were separated by 20 µm. The same grating was used for implementation of both stages. For the first stage a 1:1 magnification was used and a 2:1 magnification for the second stage.

With regard to the question how practical prism gratings are in general, a grating with a very fine period of 10 µm was tested. A regular spot pattern was used as the input. The spots were $(3 \mu m)^2$ in size. They were separated by 10 µm in the x direction and by 5 µm in the y direction. The setup was used to implement the crossover connections of the final stage of the network. After reflection from the grating the output pattern looks the same as the input pattern. The output pattern is shown in Fig. 11. Obviously, the pattern has a good contrast indicating that the use of prism gratings is feasible even in the range of a few micrometers. For this experiment, microscope objective lenses ($f/2$) were used at their diffraction limit ($\lambda = 0.633 \mu m$). Some losses occurred due to light surrounding the center spots of the Airy disks hitting the corners of the prism grating and scattering out of the optical system.

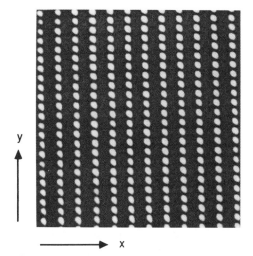

Fig. 11. Output pattern of an experiment with the 10-µm prism grating.

VI. Isomorphism of Crossover Networks with Banyan and Perfect Shuffle

Finally, we would like to point out that the crossover networks are topologically equivalent to other MINs like the Banyan and the perfect shuffle. This is important because it allows Banyan-based algorithms such as the fast Fourier transformation to be transformed to a crossover-based implementation.

First, we note that each stage of the crossover can be described mathematically in terms of two permutations. In the following we denote the binary address of the nth pixel ($0 \leq n \leq N-1$) by $\langle n \rangle$. For example, for $N = 8$ the binary address is a 3-tuple (n_2, n_1, n_0), where

$$n = 4n_2 + 2n_1 + n_0, \quad n_k = 0 \text{ or } 1. \qquad (8)$$

We put

$$\langle n \rangle = (n_2, n_1, n_0) \qquad (9)$$

and write $\langle n \rangle_k$ when we invert the kth bit; e.g.,

$$\langle n \rangle_1 = (n_2, \overline{n_1}, n_0). \qquad (10)$$

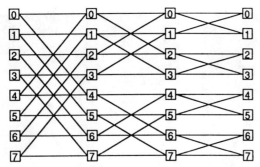

Fig. 12. Diagram for the Banyan network ($N = 8$).

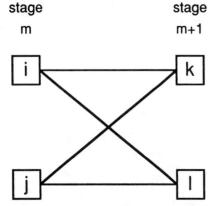

Fig. 13. Buddy property of switching nodes.

With this notation we can describe the permutations $\Pi_1^{(m)}$ and $\Pi_2^{(m)}$ for the mth stage of the HCN as follows:

$$\Pi_1^{(m)}: \quad \langle n \rangle \rightarrow \langle n \rangle; \tag{11}$$

$$\Pi_2^{(m)}: \quad \langle n \rangle \rightarrow \langle n \rangle_{M-m}. \tag{12}$$

For showing the isomorphism of the crossover and other MINs we make use of a proof which was given by Agrawal.[14] It says that two MINs are equivalent if two conditions are fulfilled. First, both have to be constructed out of nodes with two inputs and two outputs. This is obviously the case for the crossover networks (Figs. 3 and 4) as well as for the Banyan network which is shown in Fig. 12. Second, there has to hold a so-called buddy condition. This means that each pair of nodes of stage m is connected with only one pair of nodes of stage $m + 1$. This buddy property is represented graphically in Fig. 13.

In a formal way the buddy property can be shown for the half-crossover network by using the permutation formalism introduced earlier. Equations (11) and (12) show that the pixel with address $\langle n \rangle$ in the mth stage is connected to two pixels in the $m + 1$st stage with addresses $\langle n \rangle$ and $\langle n \rangle_{M-m}$. The same is true for the pixel with the address $\langle n \rangle_{M-m}$ in stage m because

$$\Pi_1^{(m)}: \quad \langle n \rangle_{M-m} \rightarrow \langle n \rangle_{M-m}, \tag{13}$$

$$\Pi_2^{(m)}: \quad \langle n \rangle_{M-m} \rightarrow \langle n \rangle. \tag{14}$$

From Eqs. (13) and (14) we can conclude that the crossover network falls into the class of MIN satisfying both conditions as given above. Similarly the buddy property can be shown to hold for the Banyan. Therefore, the equivalence between the HCN and the Banyan is proved according to Ref. 14. We would like to add the remark that a formal proof for the isomorphism between the half-crossover and the Banyan was given by Cloonan.[15]

VII. Conclusion

The crossover network was introduced as a new multistage interconnection network with applications for optical computing and photonic switching. We have presented two variations of crossover networks, the half- and the full crossover. Both are isomorphic to the more familiar Banyan and perfect shuffle. The crossover network was designed specifically for a free-space optical implementation. An optical system for the implementation of crossover interconnections was proposed and demonstrated. It makes use of special optical components which are called prism gratings. The use of these prism gratings is possible down to very small feature sizes in the micron range. It is possible to implement the crossover network with a large space–bandwidth product and a high light efficiency.

The authors would like to thank Vijay Kumar for bringing their attention to the work of D.S. Wise and Alan Huang for helpful comments on this subject.

References

1. A. Huang, "Architectural Considerations Involved in the Design of an Optical Digital Computer," Proc. IEEE **72**, 780 (1984).
2. H. S. Stone, "Parallel Processing with the Perfect Shuffle," IEEE Trans. Comput. **C-20**, No. 2, 153 (1971).
3. L. Goke and G. Lipovski, "Banyan Networks for Partitioning Multiprocessor Systems," in *Proceedings, First Annual Computing Architecture Conference* (1973), pp. 21–28.
4. M. J. Murdocca, A. Huang, J. Jahns, and N. Streibl, "Optical Design of Programmable Logic Arrays," Appl. Opt. **27**, 1651 (1988).
5. E. O. Brigham, *The Fast Fourier Transform* (Prentice-Hall, Englewood Cliffs, NJ, 1974).
6. D. J. Kuck, *The Structure of Computers and Computations* (Wiley, New York, 1978).
7. J. Jahns, "Efficient Hadamard Transformation for Large Images," Signal Process. **5**, 75 (1983).
8. A. W. Lohmann, W. Storck, and G. Stucke, "Optical Perfect Shuffle," Appl. Opt. **25**, 1530 (1986).
9. G. Eichmann and Y. Li, "Compact Optical Generalized Perfect Shuffle," Appl. Opt. **26**, 1167 (1987).
10. G. E. Lohman and A. W. Lohmann, "Shuffle Communication Component for Optical Parallel Processing," J. Opt. Soc. Am. A **4**(13), P106 (1987).
11. K.-H. Brenner and A. Huang, "Optical Implementation of the Perfect Shuffle," Appl. Opt. **27**, 135 (1988).
12. J. Jahns, "Optical Implementation of the Banyan Network," to be published.
13. D. S. Wise, "Compact Layouts of Banyan/FFT Networks," in *VLSI Systems and Computations*, H. T. Kung, B. Sproull, and G. Steele, Eds. (Computer Science Press, Rockville, MD, 1981), pp. 186–195.
14. D. P. Agrawal, "Graph Theoretical Analysis and Design of Multistage Interconnection Networks," IEEE Trans. Comput. **C-32**, 637 (1983).
15. T. J. Cloonan, "Topological Equivalence of Simple Crossover and Banyan Networks," to be published.

Hierarchic and combinatorial star couplers

M. E. Marhic

Department of Electrical Engineering and Computer Science, Northwestern University, Evanston, Illinois 60201

Received March 20, 1984; accepted May 12, 1984

An interconnection method for transmissive stars is presented that permits the fabrication of arbitrarily large stars from elementary ones. Being free from multiple-path interference, this synthesis is particularly well suited for single-mode stars. The method can be implemented with any of a number of techniques; in particular, the use of recently developed high-quality 3-dB couplers should result in the synthesis of large single-mode stars with negligible excess loss.

Stars are desirable components for fiber-optic local-area networks (FOLAN's), since they evenly divide the input signals among the receivers, thereby making possible the interconnection of many identical stations.[1] Stars have been developed for multimode FOLAN's in either transmissive or reflective versions.[2,3] The methods utilized to construct the latter, however, are not suitable to make high-quality single-mode stars. To date, single-mode stars with two to ten terminal pairs, made by fiber etching and twisting, have been demonstrated,[4,5] but it is doubtful that the technique used to fabricate them could be directly extended to much larger devices. Methods based on the interconnection of Y's cannot provide low-loss single-mode stars because of the inevitable loss associated with backward propagation through single-mode Y's[6] (however, such techniques can be used for multimode systems[7]). Single-mode stars can also be made by using lenses[8] or diffraction gratings,[9] but these methods have problems with uniformity of the outputs and overall efficiency. Thus it is generally acknowledged that none of these existing techniques allows one to make large single-mode stars. In this Letter a new approach is proposed to synthesize large stars from smaller ones, and this approach is in particular shown to be uniquely suited to fabricate high-quality single-mode stars from simple components that have already been developed, namely, 3-dB couplers.

An $N \times N$ unidirectional coupler, or N-port coupler, is a device with N (>1) inputs and N outputs. If this device evenly distributes the input power among all outputs, it is called a transmissive star, and we refer to it as an N-star to simplify the language in the rest of the Letter. In practice, N stations can be interconnected through such an N-star, with the N transmitters Ti connected to the inputs and the corresponding N receivers Ri connected to the outputs. If Pi is the power of Ti, and $P'j$ the power received by Rj, then the transmissive N-star is characterized by the following expressions:

$$\text{if } Pi \neq 0, \quad Pj = 0 \quad \text{for } j \neq i; \quad (1)$$
$$\text{then } P'j = Pi/N \quad \text{for } j = 1, 2, \ldots, N, \quad (2)$$

i.e., the output power from one station is evenly divided among the receivers of all other stations. Hence all receivers can operate at the minimum power required for a specified bit-error rate and no higher. When signals are transmitted simultaneously into all inputs, each output carries the average of all inputs (provided that there are no interference effects resulting from spatial and/or temporal coherence).

Let us assume that L-stars and M-stars exist. They can be interconnected to form an $(L \times M)$-star in the following manner (Fig. 1):

(1) Consider L M-stars and M L-stars.
(2) Take one output from each M-star and connect these L outputs into the inputs of one L-star.
(3) Repeat this operation until all outputs of the L M-stars are connected to all the inputs of M L-stars.

There are then $N = L \times M$ inputs, outputs, and interconnections. It can be seen that each input-to-output connection through the whole arrangement is unique, so that the received signals will be free of the interference problems that might result from multipath propagation. Consequently, it can readily be ascertained that the whole arrangement always divides any input power into $N = L \times M$ equal outputs, i.e., it constitutes an N-star. The above interconnection principle, for $L = M$, has essentially been described elsewhere.[10]

Single-path propagation is a requirement in single-mode systems since multipath interference effects could make such systems unpredictable and unreliable. On the other hand, averaging the effects of speckle patterns in multimode systems renders the latter less susceptible to multipath interference effects. Thus the above method is uniquely suited to making single-mode stars.

This method thus enables one to construct $(L^l \times M^m)$-stars, where l and m are arbitrary integers, by repeating it. If many types of star are available, a correspondingly greater variety can be obtained. On the other hand, however, at least one type of star must exist for this method to be meaningful.

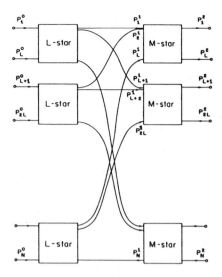

Fig. 1. An $L \times M$-star formed by interconnection of L M-stars and M L-stars.

This method can be called hierarchic (H), as it groups elementary stars into stars of progressively higher orders from the inputs to the outputs. Furthermore, although a star constructed in this manner can be viewed as a combination of elementary stars in several other ways, its design and operation are easily understood in terms of the above hierarchic-interconnection principle.

If the elementary stars are reciprocal, so is the entire arrangement, and one can then perform the hierarchic synthesis by starting from either the inputs or the outputs. There is no reason to conclude, however, that the whole structure can be viewed as H in both directions, although this may happen. We will refer to a H structure used backward as being reverse-hierarchic (RH).

This systematic approach to the synthesis of large stars is not to be confused with other interconnection methods for stars.[5,7,11] These do not provide means of making arbitrarily large stars, free of multiple-path interference as required for single-mode systems.

Assuming that M-stars are available, one can form M^n-stars from them, as shown in Fig. 2. The small squares represent M-stars. The dotted lines pass through n identical layers of M^{n-1} M-stars; each such layer can be viewed as an $(M^n \times M^n)$ port. These are connected through $n - 1$ intermediate layers of connections, each effecting a different arrangement of outputs to inputs. Since there are n layers of M-stars, the total number of required M-stars is

$$n \times M^{n-1}. \qquad (3)$$

H and RH structures have been defined above as structures in which one can view elementary stars being aggregated into progressively larger stars, as one adds new layers from the input or output, respectively. These simple structures provide adequate solutions to the single-mode star problem, but it is interesting to ask whether other star-interconnecting schemes can also perform similar functions. It is shown here by an example that the answer is yes, but no attempt is made to find all other such schemes.

Consider for simplicity elementary 2-stars. Such single-mode components are typically reversible, and one can thus make reversible 2^n-stars from them by the hierarchic method. It can be shown that 4-stars come in only one form, which is reversible, i.e., is both H and RH. This is no longer true of 8-stars, for which the following types exist: H and RH, H but not RH, RH but not H. The diversity grows further for 16-stars: In addition to the three simple types just listed, one can form 16-stars that are neither H nor RH. To see this, let us construct such a star from one H and one RH 8-star, interconnected by 8 2-stars. Inspection shows that the overall result is neither H nor RH; hence there exist other interconnection schemes besides H or RH for $N > 8$. Such other schemes cannot reduce the required number of elementary stars, but they might help in other respects, such as achieving planar, or at least less tangled, structures.

The general guiding principle for the synthesis of these stars is simply that all inputs to any elementary star should be independent, i.e., that any two inputs to such a star be the averages of two nonoverlapping sets of input signals (to the whole network). This rule clearly avoids multipath propagation through the whole network, and this is essential for single-mode systems. To realize a complete 2^n-star one must have just enough elementary 2-stars to ensure that all network outputs can be the average of all network inputs [expression (3)] and find suitable permutations of connections ($I_1, \ldots, I_{n-1}$ in Fig. 2) that do not violate the nonoverlap condition. Stars synthesized in this manner can be called combinatorial (C) stars. H (or RH) stars are only particular members of that general class; by an example we have shown that there are C stars that are neither H nor RH, but it is beyond the scope of this Letter fully to discuss and classify C stars in general.

The general interconnection principle(s) enunciated above can be implemented in practice by means of whatever technology can provide (at least) one type of star. In particular, one could use all the existing stars mentioned at the beginning of this Letter. The following is a partial list of what can be done with existing technologies; all are potentially compatible with single-mode operation.

A 50% beam splitter can be used as a 2-star. In turn,

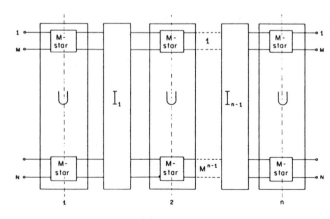

Fig. 2. M^n-star formed by interconnection of n layers of M^{n-1} M-stars. U represents the same layer of M-stars repeated n times; $I_1, I_2, \ldots, I_{n-1}$ represent $n - 1$ interconnection layers.

such elements can be used to make higher-order stars, generally with the help of mirrors. A 4-star can be made of just four beam splitters. An 8-star requires the use of mirrors. To use this method with single-mode fibers, lenses must also be introduced to couple into and out of the fibers. Alignment considerations will determine whether this approach can be useful in practice.

Guided waves can be divided among waveguides by means of evanescent-wave coupling or similar mechanisms. In particular, coupling between two waveguides is functionally equivalent to the beam-splitting operation considered above. Such coupling can be implemented in either (planar) integrated-optic or fiber-optic form.

Couplers with a 3-dB coupling ratio can thus be used as 2-stars. When manufactured by twisting and fusing optical fibers, they can exhibit insertion losses as low as 0.05 dB,[12] i.e., virtually negligible compared with the useful coupling loss. One possible way to fabricate a 2^n-star is then to take commercially available 2-stars and to interconnect them with, for instance, fusion splices that may exhibit average losses as low as 0.11 dB.[13] An interesting alternative, however, resides in the possibility of starting with 2^n lengths of single-mode fiber. The two ends of each fiber would be one input and one output for the total 2^n-star, which would be realized by making 3-dB fused couplers from pairs of fibers wherever required. This method would do away with the splices and their associated losses. The actual spatial configuration of the fibers and couplers is not unique, and many equivalent arrangements can be used. However, it does not appear that this method can be implemented in planar form, with no crossovers, except for 2-stars and 4-stars (8-stars can be made planar, but the inputs and outputs are not all conveniently located at the network periphery). This seems to rule out the possibility of using standard integrated-optics techniques to implement the method for arbitrarily large stars. Eventually, buried-channel technology[14] might make a monolithic structure possible, but this is not practical today. Hence the fiber-based method(s) described above appear to offer the best possibility for rapid fabrication of large single-mode stars. Fused-fiber technology should yield stable devices suitable for use in FOLAN's.

Although the fused-fiber technology described above may provide predictable and reliable 2^n-stars, it may be possible to simplify the fabrication process and the resulting structure by using other techniques. For instance, the polished-fiber approach to making couplers[15] in principle offers the possibility of making the required $n \times 2^{n-1}$ couplers in n steps, or even in a single step, and simultaneously doing away with the need for splices.

In summary, hierarchic and combinatorial methods have been presented that synthesize large transmissive stars from smaller ones. For single-mode fiber systems, these approaches can in particular be implemented by using 3-dB couplers, either as discrete components connected by splices or as directly fabricated between interwoven lengths of fiber; alternatively, beam-splitter-based systems can also be designed. If gigabit-per-second operation of local area networks is required in the future, single-mode fiber technology, including single-mode stars, will have to be used. The hierarchic or combinatorial methods presented in this Letter make possible for the first time to the author's knowledge the design of large, high-quality, single-mode stars suitable for this purpose.

References

1. A. F. Milton and A. B. Lee, Appl. Opt. **15**, 244 (1976).
2. M. C. Hudson and F. L. Thiel, Appl. Opt. **15**, 2540 (1974); E. G. Rawson and M. D. Bailey, Electron. Lett. **15**, 432 (1979).
3. E. G. Rawson and R. M. Metcalfe, IEEE Trans. Commun. **COM-26**, 983 (1978); M. K. Barnoski, *Fundamentals of Optical Fiber Communications* (Academic, New York, 1981), Chap. 7.
4. S. K. Sheem and T. G. Giallorenzi, Opt. Lett. **4**, 29 (1979).
5. S. K. Sheem and T. G. Giallorenzi, Appl. Phys. Lett. **35**, 131 (1979).
6. D. H. McMahon, J. Opt. Soc. Am. **65**, 1479 (1975).
7. T. Tamura, M. Nakamura, S. Ohshima, T. Ito, and T. Ozeki, J. Lightwave Technol. **LT-2**, 61 (1984).
8. A. F. Milton, "Star coupler for single mode fiber communication systems," U.S. Patent 3,937,557 (1976).
9. U. Killat, G. Rabe, and W. Rave, Fiber Integr. Opt. **4**, 159 (1982).
10. R. W. Uhlhorn and S. L. Storozum, Proc. Soc. Photo-Opt. Instrum. Eng. **296**, 141 (1981).
11. A. Yoshida, R. G. Lamont, and D. C. Johnson, Appl. Opt. **20**, 2340 (1981).
12. J. D. Beasley, D. R. Moore, and D. W. Stowe, in *Digest of Sixth Topical Meeting on Optical Fiber Communication* (Optical Society of America, Washington, D.C., 1983), paper ML5.
13. J. S. Leach, G. J. Gannel, A. J. Robertson, and P. Gurton, Electron. Lett. **18**, 697 (1982).
14. W. J. Tomlinson, I. P. Kaminow, E. A. Chandross, R. L. Fork, and W. T. Silfvast, Appl. Phys. Lett. **16**, 486 (1970).
15. R. A. Bergh, G. Kotler, and H. J. Shaw, Electron. Lett. **16**, 260 (1980).

COMBINATORIAL STAR COUPLERS FOR SINGLE-MODE OPTICAL FIBERS

M.E. Marhic

Northwestern University
Dept. of Electrical Engineering
Evanston, Illinois U.S.A.

SUMMARY

The feasibility of fabricating large single-mode optical stars by aggregating smaller optical components is studied. Various possible interconnection schemes are examined, and their effects on overall topology and design are discussed. The basic issues of output uniformity, efficiency and stability are considered. Other issues such as polarization and wavelength dependence, and planar designs are also considered. Finally, a number of possible implementations are discussed, using either beam-splitters and free-space beams, or beam-splitters and fibers, or integrated optics, or fiber couplers.

INTRODUCTION

Let us define a (transmissive) N-star as an optical device with N inputs and N outputs, performing the operation

$$P'_j = \frac{1}{N} \sum_{i=1}^{N} P_i , \quad (1)$$

where P_i is the input power into port i, and P'_j the output power out of port j; in other words, the power at each output is the average of all input powers. More generally, one may say that the N-star performs a linear transformation upon the vector of the input powers to yield the output powers. It has been shown in Reference 1 that the transformation defined by Eq. (1) can be implemented by means of a network of smaller stars, each handling a small number of inputs. This procedure is analogous to the approach followed in the electronic implementation of large parallel computations, wherein many elementary processors are connected into large arrays, capable of performing massive computations, such as matrix x vector multiplication. Since Eq. (1) is itself an example of the latter operation, albeit a simple and analog one, we expect by analogy that processor architectures developed for processor arrays should be similar or identical to those required to form a large star from smaller ones. In the following we show that some standard processor architectures are indeed suitable to make large stars, and we then discuss practical considerations concerning the physical realization of such structures.

SOME ARCHITECTURES FOR COMBINATORIAL STAR NETWORKS

Let us restrict the discussion to elementary 2-stars. By using n layers of 2^{n-1} such elements, one can form a 2^n-star [1]. There are many possible ways to do this, distinguished by the permutations of outputs to inputs between the star layers; the only restriction on these is that the two inputs to any 2-star should come from two nonoverlapping sets of inputs to the whole network. Some networks can be viewed as sets of progressively larger stars from inputs to outputs, and are termed hierarchic (H); used backwards, they are reverse-hierarchic (RH). The practical advantage of H or RH structures is that they can be used only partially, and function as smaller stars; this is generally not true of stars that are neither H nor RH. Of course, small stars can always be used to make large ones, regardless of their hierarchic properties[1].

Figure 1 shows a suitable 2-star network based on an architecture related to the data manipulator[2]. The result is RH, but not H. Figure 2 shows a star network based on the cube connection scheme [3]; Figure 3 shows a 16-star which is neither H nor RH, as stated in Reference 1.

The examples of Figures 1-3 are such that the interconnection layers are all different. However, there could be advantages to having structures where these layers are all the same, the main reason being to try to reuse the same layer a number of times, thus saving components and reducing complexity. A well-known example is given by the perfect shuffle architecture[4] shown on Figure 4. The structure is H, but not RH. There may be other architectures where the interconnection layers are identical, but they have not been identified at this time (except for repetitions of the perfect shuffle).

Generally speaking, the problem of finding and classifying all possible star-interconnection methods has not been solved. The above examples are just a few representative ones, shown with a 2D layout. Of course optics allows one to work in the third dimension if desired, or required, so that these 2D schemes could be distorted and rearranged in many physically different ways, and yet perform the same function.

PRIMARY DESIGN CHARACTERISTICS

Some basic considerations apply to all star couplers, be they combinatorial or made by some other method.

Uniformity of Outputs

The ideal star coupler should have N equal outputs, regardless of the distribution of the inputs. Combinatorial star couplers can in principle provide this. In practice, however, the elementary couplers may not have exactly even outputs due to imperfect coupling ratios and/or excess loss. Furthermore, the connections between elementary couplers may also introduce uneven losses. The uniformity is expected to deteriorate as the number of layers is increased, especially if no particular care is exercised to select or match components. An important feature of combinatorial star couplers is that they can be made of independent elements, which can be characterized separately. The detailed information about the elements can then be used to find the best way to arrange them to maximize uniformity. Since the initial reason to use a star is to have all receivers operating at the same power level, to minimize dynamic range problems, a suitable criterion to perform this optimization might be to minimize the difference between the largest and the smallest output powers. Alternatively, a criterion based upon maximizing the smallest output may also be suitable. Criteria based on rms deviation from average should be used with care, as they would not detect possibly unacceptable, extreme individual values.

Efficiency

The overall efficiency of the star, η, is the ratio of total output power to total input power. For an ideal star, η is always equal to unity. In practical devices excess losses of couplers or connections will reduce η, and consequently, the signal-to-noise ratios of the receivers. Except for lossless stars, η will generally depend on the distribution of input powers. Thus to define a criterion to optimize overall efficiency, one first has to define standard input conditions, such as $P_1 = P_2 = \ldots = P_N$, and then find the arrangement of couplers which maximizes η.

Trade-off between uniformity and efficiency

It is clear that if the outputs of a star are uneven, they can be made uniform by attenuating them to the level of the weakest. Of course the extra loss so introduced will reduce η. Thus one can, to some extent, trade efficiency for uniformity. This is also evidenced by the fact that the optimization procedures for uniformity or efficiency, described above, are unlikely to yield the same structures; hence a choice will generally have to be made as to which parameter is more important, and what are acceptable ranges for their values.

Stability

Once the star is designed and fabricated according to the above criteria, one would like it to retain its characteristics over time. This may not be easy to achieve, for instance, if focusing with lenses into and out of single-mode fibers is required. On the other hand, all-guided-wave structures should be more stable.

SECONDARY DESIGN CHARACTERISTICS

Related to the preceding considerations are the polarization and wavelength characteristics of the components and light sources used, as these could greatly alter the design and performance of the overall system if they are not properly characterized and taken into account. For instance, if the elementary couplers are polarization-dependent the system may have to be designed to accept and preserve only one polarization. Likewise, if the elements are wavelength-dependent, one should ensure that the light sources all fall within the bandwidth of satisfactory performance of the whole system.

Other important considerations are the possible reduction of the number of components by multiple use, and the compatibility with planar designs. These aspects are greatly affected by the technology used.

POSSIBLE PHYSICAL IMPLEMENTATIONS

Free-space beams, beam-splitters and mirrors

Since a 50% beam-splitter can be viewed as a 2-star, such elements can be used in conjunction with mirrors to construct large stars. An example is shown in Figure 5; note the small number of components, due to multiple use of some of them. The minimum overall size of such a device will eventually be set by diffraction.

Although this is more difficult to visualize, it can be shown that an 8-star can simply be provided by three mutually orthogonal beam-splitting planes.

Another possible approach, utilizing a recycling geometry based on the perfect shuffle, is shown in Figure 6. Each ray enters above one of the small mirrors, traveling at a small angle with respect to the plane perpendicular to all mirrors, passes through (or is reflected by) the beam-splitter three times, and finally exits below one of the small mirrors. Although only one beam-splitter is used, the presence of the small mirrors and the tilted rays render this approach less appealing than that of Figure 5.

Hybrid systems: Free-space beams and guided waves

Alternatively one may conceive of a system as shown in Figure 7, where rows of fibers are imaged onto each other via a beam-splitter and lenses. The configuration of the fiber bundles is related to the particular connection scheme chosen; in particular, for the perfect shuffle, all rows of the bundle could perform the same permutation, thus yielding a simple structure. An interesting feature of this approach is that it is actually 3D, so that the fiber ends can actually be redistributed in a plane, a fact which might be utilized to bring out other characteristics of the particular connection scheme used, such as the sub-cyclic permutations of the perfect shuffle, or the cyclic permutations of the data manipulator.

Integrated optics

There are many obstacles to the fabrication of large stars with this method. First is the fact that for $N \geq 16$, planar configurations do not seem to exist.[1] Hence, if 2D confining guides are used, one is faced with the problem of overlaps; these may introduce excess loss and/or unwanted coupling. Laser-writing of guides in three dimensions may alleviate this.[5] The problem of bending losses, however, will still be present, and it will dictate the minimum size of bend radii, and hence of the whole structure, which may become quite sizeable for large N. On the other hand, one may simply use 1D confinement in a slab guide, and propagation of beams, with coupling and reflections provided by grating-like structures, to make structures similar to those of Figures 5 and 6. With integrated optics an inevitable loss will also have to be suffered upon coupling to fibers. The primary advantage of this technology would be in the stability of the resulting monolithic devices.

Fiber couplers

Perhaps the most straightforward way to fabricate large single-mode combinatorial stars today is to use the high-quality single-mode fiber-optic couplers which have already been developed. Fused couplers have the best reported performance, and offer the best prospect for long-term stability.[6] However, they are usually fabricated individually, and would have to be spliced together. An interesting alternative resides in the use of polished couplers[7], as a number of them can be fabricated simultaneously[8]. This implies that one might be able to make sub-units, or even the whole array, without resorting to splices, so that the physical structure might look as in Figure 8.

Distributed stars

The combinatorial method lends itself well to the fabrication of stars that could be physically spread out over a large area, thereby avoiding the potential problem of single-point failure.

CONCLUSION

There are a number of potentially interesting interconnection methods and technological approaches for the fabrication of large single-mode combinatorial stars. The identification of the best candidates will require examination of the issues described above, and appropriate experimentation.

Finally, it might be worthwhile to conclude by examining the general feature of combinatorial stars brought out in the Introduction: they can be viewed as arrays of processors, performing a simple, analog, linear transformation of the input powers. If we now recall that the N-star is itself generally used to interconnect computers so that they can process data in parallel, we see that the communication medium itself has taken on a structure reminiscent of the entire system. It would be far-fetched to suggest that this analog processor might thus take over some of the computational functions of the whole system, but the analogy is nevertheless interesting, and suggests that the combinatorial synthesis of star couplers may indeed be a very natural way to approach the interconnection network in parallel processing and possibly in supercomputers.

ACKNOWLEDGMENT

I would like to thank D. Nassimi for informative discussions about processor arrays, and for supplying References 2-4.

REFERENCES

1. M.E. Marhic, "Hierarchic and Combinatorial Star Couplers," Opt. Lett. 9, (August 1984).
2. T.-Y. Feng, "A Survey of Interconnection Networks," Computer 14, 12 (December 1981).
3. H.J. Siegel and R.J. McMillen, "The Multistage Cube: A Versatile Interconnection Network," Computer 14, 65 (December 1981).
4. H.S. Stone, "Parallel Processing with the Perfect Shuffle," IEEE Trans. Comp. C-20, 153 (1971).
5. W.J. Tomlinson, I.P. Kaminow, E.A. Chandross, R.L. Fork and W.T. Silfvast, "Photoinduced Refractive Index Increase in Poly(methylmethacrylate) and its Applications," Appl. Phys. Lett. 16, 486 (1970).
6. J.D. Beasley, D.R. Moore and D.W. Stowe, "Evanescent Wave Fiber Optic Couplers: Three Methods," Sixth Topical Meeting on Optical Fiber Communication, paper ML5, New Orleans, Louisiana (1983).
7. R.A. Bergh, G. Kotler and H.J. Shaw, "Single Mode Fiber Optic Directional Coupler," Electron. Lett. 16, 260 (1980).
8. S.A. Newton, J.E. Bowers, G. Kotler and H.J. Shaw, "Single-Mode-Fiber 1 X N Directional Coupler," Opt. Lett. 8, 60 (1983).

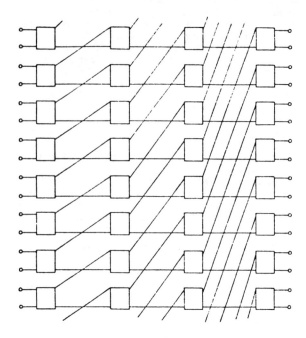

Figure 1. Star network based on data manipulator architecture.

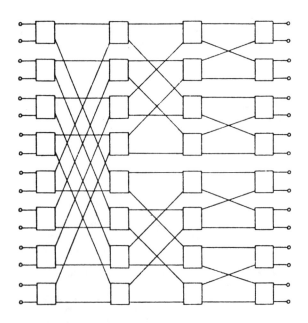

Figure 2. Star network based on cube connection scheme.

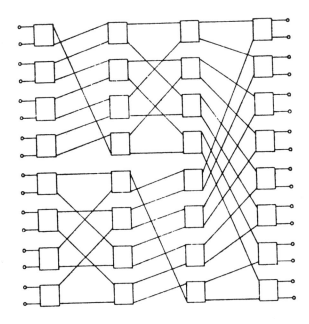

Figure 3. A 16-star which is neither H nor RH.

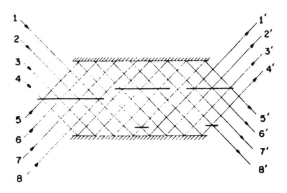

Figure 5. Possible configuration for an 8-star using free-space beams, beam-splitters and mirrors.

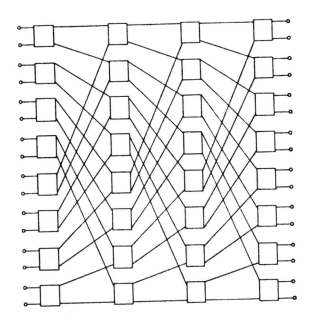

Figure 4. Star network based on perfect shuffle architecture.

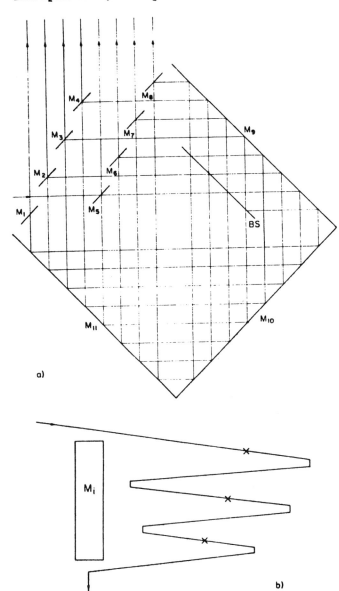

Figure 6. An 8-star based on the perfect shuffle, using a single beam-splitter, BS; M_1 to M_{11} are mirrors. a) top view; b) side view of a particular ray path.

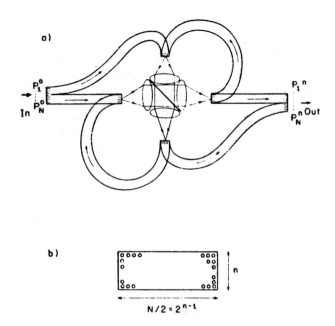

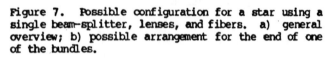

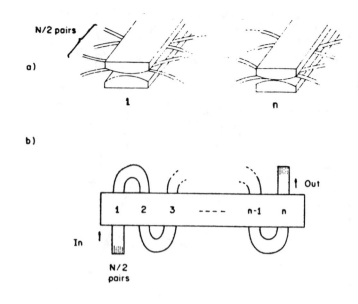

Figure 7. Possible configuration for a star using a single beam-splitter, lenses, and fibers. a) general overview; b) possible arrangement for the end of one of the bundles.

Figure 8. Star networks using blocks of polished couplers. a) n blocks; b) a single block.

Small Loss-Deviation Tapered Fiber Star Coupler for LAN

SHIGERU OHSHIMA, TAKAO ITO, KEN-ICHI DONUMA, HISAYOSHI SUGIYAMA, AND YOHJI FUJII

Abstract — A biconical tapered-fiber star coupler with a mixer rod was proposed to achieve small loss-deviation and low-excess loss. Design and fabrication techniques for this star coupler were minutely discussed, and a 100 × 100 star coupler was experimentally fabricated. The loss-deviation and average-excess loss for this star coupler was 0.37 and 3.2 dB, respectively.

I. INTRODUCTION

AN OPTICAL star network is one of the most promising approaches for optical local area networks. Some star network applications require 100-, or more- port optical star couplers [1]. The star coupler loss-deviation is required to be so small that the receiver dynamic range is reduced [2]. Especially in the star network system based on the carrier sense multiple access with collision detection (CSMA/CD) acess protocol, the loss-deviation is required to be fairly small (for example, within ± 1 dB [1]).

Two kinds of 100-port star couplers have been reported. One is the structure using a low-loss slab waveguide and fiber arrays [3]. The other is the biconical tapered fiber star coupler [4]. The latter achieved low-excess loss, but the loss-deviation was over 8 dB. The rotation-splice structure for the tapered-fiber star coupler [5] was developed in order to reduce the loss-deviation. However, this structure is not suitable for fabricating star couplers with more than 10-odd ports.

This paper proposes a tapered-fiber star coupler with a mixer rod to achieve small loss-deviation. Design and fabrication techniques for the star coupler are described. Loss characteristics for the experimentally fabricated 100-port star coupler are also described.

II. STRUCTURE

Former biconical tapered-fiber star couplers had large loss-deviation. This is because some modes of light in an input fiber are not converted into cladding modes of light

Manuscript received November 5, 1984.
S. Ohshima and T. Ito are with Electron Devices Laboratory, Toshiba Research and Development Center, Toshiba Corporation, Kawasaki, 210, Japan.
K. Donuma is with Solid State Device Engineering First Department, Electron Tube and Device Division, Toshiba Corporation, Kawasaki, 210, Japan.
H. Sugiyama is with Yokosuka Electrical Communication Laboratory, Nippon Telegraph and Telephone Public Corporation, Kanagawa, 238-03, Japan.
Y. Fujii is with Musashino Electrical Communication Laboratory, Nippon Telegraph and Telephone Public Corporation, Tokyo, 180, Japan.

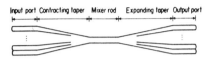

Fig. 1. Structure of the star coupler.

in the contracting taper section, and they couple strongly to a certain output fiber core [6]. Therefore, the loss-deviation is reduced if these nonconverted modes of light couple into the cladding at the input plane of the expanding taper.

The new star-coupler structure is schematically shown in Fig. 1. This star coupler is composed of contracting and expanding taper sections and a cylindrical mixer-rod section. The mixer rod is inserted between the tapers to couple the nonconverted modes of light into the cladding of the expanding taper.

The light incident on one of the input ports propagates through the contracting taper section, and some modes of light are converted into cladding modes of light. The other modes of light remain as core modes.

The converted modes of light are mixed effectively as they propagate through the contracting taper section and are mixed uniformly in the mixer rod. The light is reconverted into guided modes of light as it propagates through the expanding taper-fiber section and couple into every output fiber with equal power.

The nonconverted modes of light (remaining as core modes of the contracting taper) are spread in the mixer rod, and coupled into the expanding taper by two different ways. Some light of the modes couples into the cladding at the input plane of the expanding taper fiber section. Then, the light is mixed uniformly and reconverted into the core modes of light as it propagates through the expanding tapered-fiber section.

The other light of the modes couples directly into the cores of the expanding taper. The light may increase loss-deviation when light mixing is not sufficient at the mixer-rod output face. However, the optical power of this kind of light is a very small portion of the total power (~ 0.2 percent) as discussed in Section III-A. Therefore, smaller loss-deviation is achieved.

III. DESIGN

A. Transmission Coefficient

The transmission coefficient C_{ij} from port i to port j is described as follows:

$$C_{ij} = \frac{1}{N} K_R K_T + \frac{1}{N}(1-K_R)(1-Ck^2)K_T$$
$$+ S_{(i,j)}(1-K_R)Ck^2 \quad (1)$$

where N is the number of ports, K_R is the power coefficient of the converting core modes into the cladding modes in the contracting taper, K_T is the power coefficient of the reconverting cladding modes into the core modes in the expanding taper, Ck^2 is the power coefficient coupled into the cores at the input of the expanding taper-fiber section, k is the ratio of the core diameter to the outer diameter, and C is the coupling constant. $S_{(i,j)}$ shows the power distribution at the mixer output face.

Now let us estimate $(1-k_R)Ck^2$, because the deviation of C_{ij} is dependent on only the third term of (1). We assumed that the standard graded-index fiber (50-μm core diameter, 125-μm outer diameter, 0.21 numerical aperture) is used, and the modal power distribution in the input fiber is uniform. Under this condition, the parameters k and C are 0.4 and 2/3, respectively. K_R is calculated by the following equation:

$$K_R = 1 - 1/R^2 \quad (2)$$

where R is the taper ratio [7], [8].

When the taper ratio is 7, for example $(1-K_R)Ck^2$ is calculated to be 2×10^{-3}. This value is negligibly small. Therefore, small loss-deviation is achieved even if $S_{(i,j)}$ deviates from uniformity.

B. Taper Ratio

The taper ratio is an important parameter of the biconical taper-fiber star coupler. From (1) and (2), it is understood that a larger taper ratio gives smaller loss-deviation. However, excess losses are increased due to converting some cladding modes of light into radiation modes in the contracting taper section and mixer rod when the taper ratio exceeds a certain boundary value. The boundary taper ratio is discussed in this section.

The ray for the cladding modes of light must be traced to discuss the boundary taper ratio. Since the ray trajectories for cladding modes of light were very complicated, only the meridional ray incident on the center of the tapered fiber core was traced by the numerical method. The core of a conical graded-index taper fiber was divided into multilayers. Each layer was a cone concentric with the z axis and its index was uniform. The propagation angle θ_1 at the input is equal to the maximum propagation angle in the core modes. We defined the changing angle efficiency to be $\sin\theta_2/(R\sin\theta_1)$, where θ_1 and θ_2 are propagation angles at the input and output of the contracting tapered fiber. R is the taper ratio which is given by

$$R = a_1/a_2 \quad (3)$$

where a_1 and a_2 are the outer radii of the conical tapered fiber at the input and output, respectively (see Fig. 2). The ray behavior for the cladding modes of light approaches the ray behavior for core modes in a step-index fiber as

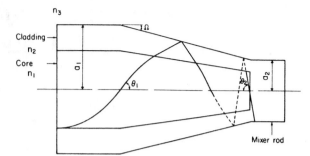

Fig. 2. Ray trajectory in tapered fiber.

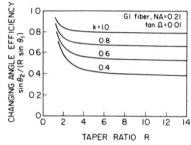

Fig. 3. Changing angle efficiency $\sin\theta_2/(R\sin\theta_1)$ versus taper ratio R.

the ray propagates through the contracting tapered fiber. For core modes in a step-index fiber, $\sin\theta_2/(R\sin\theta_1)$ is unity. Therefore, we defined the changing angle efficiency to be $\sin\theta_2/(R\sin\theta_1)$ and calculated this efficiency. Fig. 3 shows the changing angle efficiency versus the taper ratio for $\tan\Omega = 0.01$, where Ω is the taper angle (see Fig. 2).

Almost the same calculated results as above were obtained when taper angles were 1° or less. The changing angle efficiency is expressed by

$$\frac{\sin\theta_2}{R\sin\theta_1} \simeq k(1-0.2\,k) \quad (4)$$

when the taper ratio is a large value, for example, more than 6. In comparison with that for a step-index tapered fiber [9], the changing angle efficiency was smaller by $0.2\,k^2$.

The propagation angle θ_2 which does not lead to radiation modes is given as follows:

$$n_1 \sin\theta_2 < \sqrt{n_2^2 - n_3^2} \quad (5)$$

where n_1, n_2, and n_3 are, respectively, the refractive indexes of the core, cladding, and its surrounding media. The taper ratio R, which does not cause radiation loss, is obtained from (4) and (5).

$$R < \frac{\sqrt{n_2^2 - n_3^2}}{(NA)\,k\,(1-0.2\,k)} \quad (6)$$

where (NA) is the numerical aperture of the fiber.

We considered the lossless requirement even if water is stuck around the tapered fiber and mixer rod section. We obtained $R < 7.8$ when the practical parameter of $k = 0.4$, $NA = 0.21$, $n_2 = 1.46$, and $n_3 = 1.33$ were applied. Therefore, the taper ratio was determined to be 7.

As for a tapered-fiber star coupler, the taper ratio R is given by

$$R = \frac{a\sqrt{q}}{d} \quad (7)$$

where a is the outer diameter of the fiber, q is the number of fibers, and d is the diameter of the mixer rod. The mixer rod diameter d is obtained from (7). For a 100-port star coupler, parameters of $a = 125$, μm, $R = 7$, and $q = 121$ (as discussed in Section III-C) yield $d = 200$ μm.

C. Redundancy Optimization of the Number of Ports

Redundancy of ports is very effective to reduce loss-deviation and increase the yield rate, though it increases the excess losses for the coupler. In this section, the effectiveness of the redundancy of ports for the reduction of loss-deviation is discussed with the following three assumptions having been made.

i) When the jth input port is excited, the loss distribution in the output ports is a normal distribution with a mean value m_j and a standard deviation σ_j.

ii) Whether the loss for a certain output port is larger or smaller than another output port is independent of which input port was excited.

iii) The distribution of the mean values m_j is a normal distribution with a standard deviation σ_m.

The standard deviation (s.d.) σ_q for the whole distribution of the $q \times q$ star-coupler losses is expressed by

$$\sigma_q = \sqrt{\frac{\sum_{j=1}^{q} \sigma_j^2}{q} + \sigma_m^2}. \quad (8)$$

The s.d. σ_j is reduced to σ_j' by selecting the best p ports out of the q output ports. σ_j' is expressed by

$$\sigma_j' = \sigma_j \int_{-b}^{b} \frac{t^2}{\sqrt{2\pi}} e^{-t^2/2} dt \equiv \sigma_j F(b) \quad (9)$$

where b is given by the following equation:

$$\frac{P}{q} \int_{-b}^{b} \frac{1}{\sqrt{2\pi}} e^{-t^2/2} dt. \quad (10)$$

Similarly, the s.d. σ_m is reduced to σ_m' by selecting the best p ports out of the q input ports. Thus the standard deviation σ_p is $F(b) \sigma_q$. Fig. 4 shows the reduction factor $F(b)$ as a function of p/q. When p/q is larger than 0.8, the s.d. σ_p is reduced effectively. Setting p/q to be less than 0.8 is not expected to reduce the s.d. effectively but cause large optical loss. When p/q is equal to 0.8, for example, the s.d. σ_p is about half of the s.d. σ_q. For this case, the increase of excess loss is 1 dB.

IV. Fabrication

The fabrication process of this star coupler is as follows. First, fibers are bundled together. The bundled fibers are heated with an oxyhydrogen flame and pulled into a biconical taper shape. Next, the biconical taper is cut at the

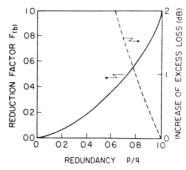

Fig. 4. Reduction factor $R(b)$ as a function of redundancy p/q.

Fig. 5. Cross-sectional view of the tapered-fibers waist.

Fig. 6. Exterior view of the star coupler.

waist and a cylindrical mixer rod is inserted between the tapers by using the fusion-splice technique. The fibers must not be twisted through the taper fabrication to decrease loss-deviation. This is because twisting the fibers puts the fiber arrangement out of order and causes microscopic bending on the fibers, and the bending leads to needless mode conversion increasing loss-deviation.

A 100×100 star coupler was experimentally fabricated using standard graded-index silica-glass fibers (50-μm core diameter 125-μm outer diameter, 0.21 numerical aperture). The total number of fibers used in this star coupler was selected to be 121, based on the result discussed in Section III-C. The taper ratio R was selected to be 7 and the taper waist diameter was nearly equal to 200 μm. The cross-sectional view of the fabricated tapered fiber waist is shown in Fig. 5. The mixer-rod diameter was set equal to the taper-waist diameter. The mixer-rod length was set to be 50 times that of the diameter.

Fig. 6 shows the exterior of the star coupler. It is possible to connect the transmission lines to the star coupler by an FC type connector [10].

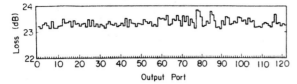

Fig. 7. Examples of loss variations in output ports.

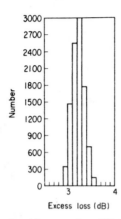

Fig. 8. Excess loss histogram for a 100-port star coupler.

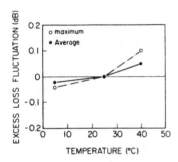

Fig. 9. Temperature dependence of excess loss.

V. CHARACTERISTICS

The excess losses of the 121 × 121 matrix were measured automatically by computer-controlled equipment. A 0.85 μm wavelength light, radiated from an LED, was passed through a mode scrambler and coupled into the star coupler. An index matching liquid was used at both input and output connections. Typical measured insertion losses for an input port are shown in Fig. 7. The insertion losses for the other input ports were almost the same as this figure. The best 100-port pairs were selected out of the 121-port pairs on the basis of the measured 121 × 121 matrix excess losses. Fig. 8 shows a histogram for the 100 × 100 excess losses. The average excess loss was 3.2 dB. Maximum loss deviation was +0.37 dB.

Temperature dependence of excess loss is shown in Fig. 9. The maximum loss fluctuation was 0.15 dB within the temperature range of 5°–40°.

VI. CONCLUSION

A tapered-fiber star coupler with a mixer rod has been proposed. The transmission coefficient, the taper ratio, and the redundancy optimization of the number of ports have been minutely discussed. The 100 × 100 star coupler has been fabricated based on these discussions. The average excess loss for the coupler was 3.2 dB. The maximum excess loss was 3.6 dB and the loss-deviation was +0.37/−0.34 dB. The fluctuation of excess loss was only 0.15 dB for the 5°–40° temperature change.

ACKNOWLEDGMENT

The authors wish to thank Dr. M. Koyama and Dr. J. Minowa of Yokosuka Electrical Communication Laboratory, NTT, Japan, for their helpful guidance and continuous encouragement, and A. Yoshinaga, Sakakibara, and other colleagues of Toshiba Corporation for their valuable discussions and measurements.

REFERENCES

[1] Y. Hakamada and K. Oguchi, "32 Mb/s star configured optical local area network design and performance," to be published.
[2] J. C. Williams, S. E. Goodman, and R. L. Coon, "Fiber-optic subsystem considerations of multimode star coupler performance," in *Proc. Conf. Opt. Fiber Commun.* (New Orleans, LA), Jan. 23-25, 1984, Pap. WC6.
[3] Y. Fujii, N. Suzuki, and J. Minowa, "A 100 input/output-port star coupler composed of low-loss slab-waveguide," in *Tech. Dig. 4th Int. Conf. Integrated Opt. Opt. Fiber Commun.* (Tokyo, Japan), June 27-30, 1983, Pap. 29C2-4.
[4] E. G. Rawson and M. D. Bailey, "Bitaper star couplers with up to 100 fiber channels," *Electron. Lett.*, vol. 15, pp. 432-433, 1979.
[5] S. Ohshima and T. Ozeki, "Rotation-splice tapered fiber star coupler," in *Proc. Opt. Commun. Conf.* (Amsterdam, The Netherlands), Sept. 1979.
[6] T. Ozeki and B. S. Kawasaki, "New star coupler compatible with single multimode-fiber data links," *Electron. Lett.*, vol. 12, pp. 151-152, 1976.
[7] ——, "Mode behaviour in a tapered multimode fiber," *Electron. Lett.*, vol. 12, pp. 407-408, 1976.
[8] T. Ito, M. Itoh, and T. Ozeki, "Bidirectional tapered fiber star couplers," in *Proc. 4th Eur. Conf. Opt. Commun.* (Geneva, Italy), Sept. 1978, pp. 318-322.
[9] Y. Uematsu, T. Ozeki, and Y. Unno, "Efficient power coupling between an MH LED and a taper-ended multimode fiber," *IEEE J. Quantum Electron.*, vol. QE-15, pp. 86-92, 1979.
[10] K. Nawata, "Multimode and single-mode fiber connectors technology," *IEEE J. Quantum Electron.*, vol. QE-16, pp. 618-627, 1980.

Reflective Single-Mode Fiber-Optic Passive Star Couplers

A. A. M. SALEH, FELLOW, IEEE, AND H. KOGELNIK, FELLOW, IEEE

Abstract—A key component in many architectures of high-speed optical local-area networks (LAN's) is a single-mode-fiber passive star coupler. Techniques are known in the literature for constructing transmissive $n \times n$ star couplers, e.g., for n equaling an arbitrary power of two, using 3-dB couplers. Such an $n \times n$ star can serve as the central node in an n-user LAN, where each user would be connected by two fibers—one for transmission to the input side of the star, and the other for reception from the output side of the star. The amount of fiber and components required for such an implementation is quite large. One obvious way to reduce the amount of fiber by a factor of two is to use a reflective n-star coupler. For this n-port structure, power entering any one of the ports is divided equally and "reflected back" out of all the ports. In this case, each of the various users would then be connected to the star by a single fiber.

In this paper we present arrangements for constructing such single-mode fiber-optic passive star couplers using simple components, e.g., 3-dB couplers and mirrors. Besides the saving of fibers, the number of components needed to construct a reflective star is shown to be half of that needed for an equivalent transmissive star. To avoid unpredictable interference effects, we only consider structures that are free of multipaths.

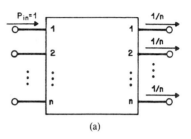

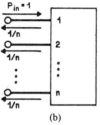

Fig. 1. (a) Passive transmissive $n \times n$ star coupler. (b) Passive reflective n-star coupler.

I. Introduction

A KEY COMPONENT in many architectures of high-speed optical local-area networks (LAN's) is a single-mode-fiber passive star coupler. Ideally, a *transmissive* $n \times n$ star coupler (Fig. 1(a)) or simply an $n \times n$ star, divides the power entering any of its n input ports equally among its n output ports. Of course there will be some excess power loss due to absorption and scattering, which is not considered in Fig. 1. The $n \times n$ star can serve as the central node in an n-user LAN, where each user would be connected by two fibers—one for transmission to the input side of the star, and the other for reception form the output side of the star. This creates a broadcast-type LAN, where a message transmitted by any user can be received by all users. Different messages can be transmitted and received simultaneously by the various users through the application of any of a number of protocols such as wavelength-division multiple access (WDMA) and time-division multiple access (TDMA).

The main advantage of the star architecture over other LAN architectures such as the *bus*, where the various users are coupled in succession to a single fiber, is its small excess loss. For example, as will be made clear shortly, the excess loss of an n-user star increases only logarithmically with n, while that of an n-user bus increases linearly with n. The difference in loss can be quite significant, especially for large values of n. On the other hand, the amount of fiber and components required to implement a transmissive-star architecture is quite large. One obvious way to reduce the amount of fiber (and, as it turns out, also the number of components) by a factor of two is to use a *reflective* n-star coupler as shown in Fig. 1(b). In this case, power entering any one of the n ports is divided equally and "reflected back" out of all ports. The various users would then separate their transmitted and received signals using diplexers at their own locations.

The purpose of this paper is to show architectures for constructing reflective single-mode fiber-optic n-stars out of ordinary directional couplers and other simple components. In Section II, we briefly review known architectures for constructing single-mode-fiber transmissive stars, which are needed as building blocks in our development. Reflective n-stars are presented in Section III for n equaling a perfect square, e.g., an even power of two, and in Section IV for n equaling twice a perfect square, e.g., an odd power of two. Finally, in Section V, the presentation is generalized to n equaling an arbitrary multiple to a perfect square.

Manuscript received March 9, 1987.
The authors are with AT&T Bell Laboratories, Crawford Hill Laboratory, Holmdel, NJ 07733.
IEEE Log Number 8717614.

Reprinted from *J. Lightwave Tech.*, vol. 6, no. 3, pp. 392-398, Mar. 1988.

II. Transmissive $n \times n$ Star Couplers

The simplest nontrivial single-mode-fiber transmissive star is the familiar evanescent-field 3-dB coupler depicted in Fig. 2, which, in fact, is a 2×2 star. It can be manufactured by bringing the cores of two single-mode fibers sufficiently close together over an appropriate coupling length L_c. Various such structures have been built through the use of etching [1], grinding and polishing [2], or fusion [3]. Passive integrated-optical 3-dB couplers are also possible, but they tend to have more excess loss than the aforementioned fiber-based designs. A free-space 3-dB beam splitter, oriented at oblique incidence, also acts as a transmissive 2×2 star. Most of our presentation is also applicable to such a structure.

A 2×2 star can be used as a building block to construct larger n-stars, with n equaling an arbitrary power of two [4], [5]. For example, Fig. 3 shows a transmissive 4×4 star made of four 2×2 stars. More subtle examples are given by the transmissive 16×16 stars of Figs. 4 and 5 [4]. The two figures are identical from a topological point of view. They each contain 32 2×2 stars. In general, a transmissive $n \times n$ star, with n equaling a power of two requires $\log_2(n)$ stages, each containing $(n/2)$ 2×2 stars, for a total of $(n/2) \log_2(n)$ 2×2 stars. Since the excess loss suffered by a signal propagating through the structure is proportional to the number of stages that is traversed, the previously mentioned logarithmic behavior of the excess loss follows.

The connection scheme of Fig. 4, which resembles that of the standard fast Fourier transform (FFT) algorithm [6], is more formally known as the cube algorithm [4], [7]. The different spacings between the various stages in Fig. 4 were chosen to yield the same slope for all inclined connections, which may be advantageous in an integrated-optics implementation. The connection scheme of Fig. 5, which is attractive because it produces identical interconnection stages, is formally known as the perfect-shuffle algorithm [4], [8]. Various other layouts are also possible [4].

A critical requirement for the proper construction of transmissive stars or, for that matter, reflective stars, is that each pair of input and output ports must be connected by only one path. Having more than one path between two ports would almost certainly produce unpredictable and unstable results because of destructive and constructive multipath interference. While the absence of multipaths is somewhat hard to verify in Fig. 5, the hierarchical nature of the layout of Fig. 4 makes this task more systematic. For example Fig. 4 can be considered as the cascade of two separate 8×8 stars (the top and the bottom halves of the left three stages) and eight 2×2 stars (the right stage). Since the two inputs of each of these 2×2 stars come from different 8×8 stars, no multipaths exist in the last interconnection stage. In turn, each 8×8 star can be considered as the cascade of two 4×4 stars (the top and the bottom halves of the left two stages of the 8×8 star) and four 2×2 stars (the right stage of the 8×8

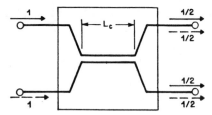

Fig. 2. A single-mode evanescent-wave 3-dB coupler used as a passive transmissive 2×2 star.

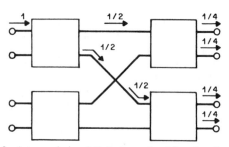

Fig. 3. A transmissive 4×4 star made of four 2×2 stars.

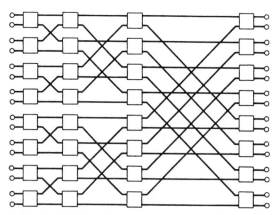

Fig. 4. A transmissive 16×16 star made of 32 2×2 stars using a cube, or a standard-FFT, connection algorithm.

star). Again, this interconnection stage has no multipaths, for the same reason given above. Proceeding in this manner, one can indeed verify that the structure is free of multipaths. In general, one can systematically construct a transmissive $mn \times mn$ star, free of multipaths, by cascading a stage of m $n \times n$ stars and a stage of n $m \times m$ stars with the inputs of each of the $m \times m$ stars connected to the outputs of different $n \times n$ stars [9]. Later, we will give similar interconnection rules for constructing multipath-free reflective stars.

Thus far we have only considered transmissive stars that are based on the 2×2 star coupler. This restricted the number of ports to a power of two. However, based on realizations of transmissive single-mode-fiber 3×3 stars [10], [11], transmissive stars whose number of ports is a power of 3 are also possible. For example, transmissive 9×9 stars have been reported in [10] and [12]. In fact, using the hierarchical cascading algorithm of [9], one can construct transmissive stars whose number of ports is the product of arbitrary powers of two and three. It is inter-

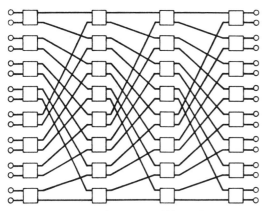

Fig. 5. A transmissive 16 × 16 star made of 32 2 × 2 star using a perfect-shuffle, or identical-stages, connection algorithm.

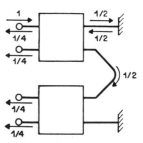

Fig. 6. A reflective 4-star constructed by vertically bisecting, or folding, the symmetric transmissive 4 × 4 star structure of Fig. 3.

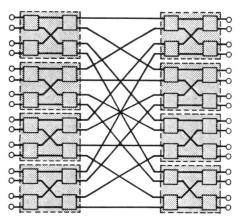

Fig. 7. Rearranging the positions of the various 2 × 2 stars in each of the four columns of Fig. 4 or 5 to produce a symmetric transmissive 16 × 16 star structure that is suitable for vertical bisection, or folding.

esting (and challenging), to note that, to the authors' best knowledge, it is not yet known how to construct single-mode transmissive stars whose number of ports is a prime number other than two or three.

III. REFLECTIVE n-STARS WITH $n = m^2$

Here we show how to construct reflective n-stars, where n is a perfect square, e.g., 4, 16, 64, 256, or 1024. As will be made clear shortly, if $n = m^2$, the construction is based on the availability of transmissive $m \times m$ stars. We first note that simply adding mirrors to the output ports of a single-mode-fiber transmissive, $n \times n$ star does *not* produce a reflective n-star. The reason is that a multipath situation would occur in this case, which, as previously mentioned, would lead to unpredictable and unstable outputs. More specifically, there would be exactly n different paths for the signal entering any input port and ultimately "reflected back" out of any port.

A multipath-free reflective 4-star made of two transmissive 2 × 2 stars and two mirrors is shown in Fig. 6. Tracing the power flow through the structure, as is indicated on the figure, one can indeed verify that a signal entering any port will emerge equally divided among the four ports with a single path to each port.

To be able to generalize this structure to any other $n = m^2$, we need to follow the systematic procedure used for its construction. This is done by starting with the *vertically symmetric* transmissive 4 × 4 star of Fig. 3, bissecting it along its vertical axis of symmetry, removing the right half, and finally inserting a vertical mirror at the bisection line. In effect, the insertion of this single mirror is equivalent to the termination of each of the two fibers that are cut at a right angle by its own mirror, and the splicing together of the two fibers that are cut at an oblique angle, which directly leads to Fig. 6. Another way of obtaining Fig. 6 is by folding the right-hand side of Fig. 3 about the vertical axis of symmetry onto the left-hand side and conceptually, "uniting" together the two folded halves. With either approach, the equally divided signals that used to emerge from the ports on the right-hand side of Fig. 3, now are "reflected back" from the ports on the left-hand side.

To construct a reflective 16-star using the above procedure, we need to start with a vertically symmetric transmissive 16 × 16 star. We note that neither of the 16 × 16-star structures of Figs. 4 and 5 satisfy this symmetry requirement. However, by properly rearranging the positions of the various 2 × 2 stars in either figure, one can obtain the desired symmetric 16 × 16 star of Fig. 7. Note that eight transmissive 4 × 4 stars (shaded boxes) can be segregated. Applying to this structure the concept of vertical bisection and insertion of a mirror, or the concept of folding about the vertical axis of symmetry, one obtain that sought reflective 16-star of Fig. 8. It consists of four transmissive 4 × 4 stars (or 16 transmissive 2 × 2 stars) and four mirrors.

Note that there are half as many stages of 2 × 2 stars in the above reflective n-stars compared to their transmissive counterparts. This factor-of-two saving in components is another advantage of reflective stars over transmissive stars, besides the previously mentioned factor of two saving in the required amount of fiber needed for a LAN implementation. Note also that the signal has to traverse the structure twice in the reflective case. Thus, the excess loss of reflective and transmissive stars with the same number of ports are identical.

The arrangement of an arbitrary reflective n-star, with $n = m^2$, which is a natural generalization of Figs. 6 ($n = 4$) and 8 ($n = 16$), is given in Fig. 9. The structure contains m transmissive $m \times m$ stars and m mirrors. The connection algorithm of the ports on the right-hand side of

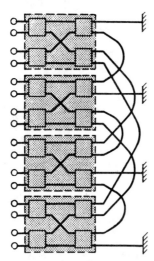

Fig. 8. A reflective 16-star constructed by vertically bisecting, or folding, the symmetric 16 × 16 star structure of Fig. 7.

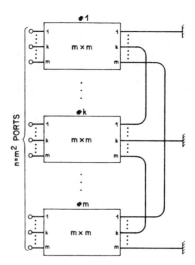

Fig. 9. A general arrangement of a reflective n-star, where $n = m^2$. Note that m mirrors and m transmissive $m \times m$ stars are needed.

the figure is quite simple: 1) Port i of $m \times m$ star #i, $i = 1, 2, \cdots, m$, is terminated in a mirror. 2) Port i of $m \times m$ star #j, $i, j = 1, 2, \cdots, m$, $i \neq j$, is connected to port j of $m \times m$ star #i.

The arrangement of Fig. 9 is also applicable for all values of n of the form $n = m^2$, not just those that are powers of two. The only restriction is that it must be possible to construct transmissive $m \times m$ stars. For example, using the transmissive 3 × 3 stars reported in [10]–[12], one can use Fig. 9 to construct a reflective 9-star ($m = 3$). In fact, upon the elimination of the missing ports in Fig. 9, which are indicated by the dots, that figure itself becomes a representation of a reflective 9-star.

IV. Reflective n-Stars With $n = 2m^2$

Here we show how to construct reflective n-stars, where n is twice a perfect square, e.g., 8, 32, 128, or 512. As will be made clear shortly, if $n = 2m^2$, the construction is based on the availability of transmissive $m \times m$ stars. In addition, as we had to introduce mirrors in the previous section to realize n-stars with $n = m^2$, here we have to introduce reflective 2-stars.

Fig. 10 shows three possible realizations of reflective 2-stars. The realization of Fig. 10(a), which we will use as the representative of reflective 2-stars, simply employs a half-reflective mirror within the fiber. This can be made, for example, by polishing the ends of two fibers, depositing an appropriate half-reflective coating on one of the ends, and finally splicing it to the other end. Various other realizations also exist. The realization of Fig. 10(b) is obtained by vertically bisecting the evanescent-wave 3-dB coupler of Fig. 2 and terminating half of the structures with a mirror. It can also be verified that the mirror can be replaced by a fiber loop of arbitrary length as depicted in Fig. 10(c).

We now show how to construct a reflective 8-star. The construction is more subtle than that used in the previous section to construct reflective 4- and 16-stars because one cannot construct a *physically symmetric* transmissive 8 ×

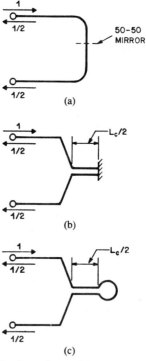

Fig. 10. A reflective 2 × 2 star formed by (a) using a half-reflecting mirror within the fiber, (b) vertically bisecting the 3-dB coupler of Fig. 2 and terminating by a mirror, or (c) replacing the mirror by a fiber loop of arbitrary length.

8 star out of 2 × 2 stars. However, as shown in Fig. 11, one can construct a *functionally symmetric* transmissive 8 × 8 star by treating the 8-port shaded box in the middle of the figure as a single structure. We now proceed to conceptually bisect Fig. 11 along its vertical line of symmetry and to introduce a mirror at that line. Alternatively, we could be thinking of folding the right-hand side of the figure on its left-hand side and "uniting" together the two folded halves. The resulting structure, which is the desired reflective 8-star, is shown in Fig. 12, which consists of five transmissive 2 × 2 stars and two reflective 2-stars.

The bisection of the top and bottom transmissive 2 ×

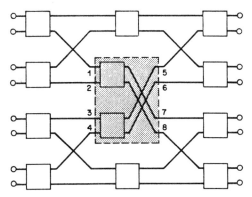

Fig. 11. A transmissive 8 × 8 star with its component 2 × 2 stars arranged to produce a functionally symmetric structure that is suitable for vertical bisection, or folding.

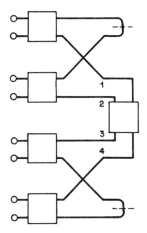

Fig. 12. A reflective 8-star constructed by vertically bisecting, or folding the "symmetric" 8 × 8 star structure of Fig. 11.

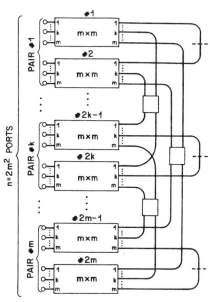

Fig. 13. A general arrangement of a reflective n-star, where $n = 2m^2$. Note that m reflective 2-stars, $m(m-1)/2$ transmissive 2 × 2 stars and $2m$ transmissive $m \times m$ stars are needed.

2 stars in the middle of Fig. 11 produces the two reflective 2-stars at the top and bottom right of Fig. 12. The vertically oriented transmissive 2 × 2 star at the middle of the right-hand side of Fig. 12 is, functionally, a vertical bisection (or folding) of the shaded box in Fig. 11. This can be verified by noting that an input signal, for example, at port 1 or 2 of the shaded box is divided equally between ports 7 and 8, which, upon folding, coincide with ports 3 and 4. Thus, in the resulting folded structure, an input signal at port 1 or 2 should be equally divided and directed to ports 3 and 4. This is exactly the action produced by the vertically oriented 2 × 2 star in Fig. 12. Of course, an analogous operation is performed for an input signal at port 3 or 4.

Note that the two crossovers in Fig. 12 can be eliminated by horizontally flipping the top and bottom 2 × 2 stars on the left-hand side. As it stands, however, Fig. 12 is a natural consequence of our development process that is more suitable for generalization to other values of n, which we now present.

Fig. 13 shows the general arrangement of a reflective n-star, where $n = 2m^2$ and m is arbitrary. This arrangement requires $2m$ transmissive $m \times m$ stars (which are grouped in m pairs), m reflective 2-stars (each joining one pair of the $m \times m$ stars), and $m(m-1)/2$ vertically oriented transmissive 2 × 2 stars (which connect together the different pairs of $m \times m$ stars). The connection algorithm of the ports on the right-hand side of the figure is as follows: 1) Ports i of the top and the bottom $m \times m$ stars of pair # i, $i = 1, 2, \cdots, m$, are joined together via a reflective 2-star. 2) Ports i of the top and the bottom $m \times m$ stars of pair # j, $i, j = 1, 2, \cdots, m, i \neq j$, are connected via a vertically oriented transmissive 2 × 2 star to ports j of the top and the bottom $m \times m$ stars of pair # i.

The arrangement of Fig. 13 is also applicable for all values of n of the form $n = 2m^2$, not just those that are powers of two. The only restriction is that it must be possible to construct transmissive $m \times m$ stars. For example, using the transmissive 3 × 3 stars reported in [10]–[12], one can use Fig. 13 to construct a reflective 18-star ($m = 3$). In fact, upon the elimination of the missing ports in Fig. 13, which are indicated by the dots, that figure itself becomes a representation of a reflective 18-star.

V. CONCLUSIONS AND FURTHER GENERALIZATION TO $n = lm^2$

Thus far we have shown how to construct reflective single-mode fiber-optic n-star couplers for n equaling either a perfect square ($n = m^2$) or twice a perfect square ($n = 2m^2$). The constructions are possible only when transmissive $m \times m$ stars can be built. This is known to be the case, for example, when m is a power of two [1]–[5], a power of three [9]–[12], or, more generally [9], a product of a power of two and a power of three. The constructions for the case of $n = m^2$ (Fig. 9) require the introduction of mirrors, while those for $n = 2m^2$ (Fig. 13) require the

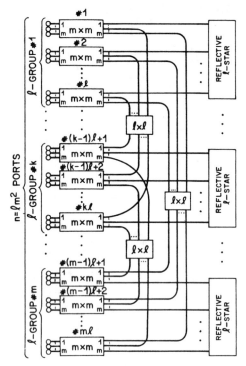

Fig. 14. A natural generalization of the arrangements of Fig. 9 and 13 yielding a reflective n-star, where $n = lm^2$. Note that m reflective l-stars, $m(m - 1)/2$ transmissive $l \times l$ stars and ml transmissive $m \times m$-stars are needed.

introduction of reflective 2-stars, which were also described.

Actually, the general arrangements of Figs. 9 and 13 can be further generalized to construct reflective n-stars with n equaling an arbitrary multiple of a perfect square, i.e., $n = lm^2$. Such a generalization, which is presented in Fig. 14, requires the use of ml transmissive $m \times m$ stars, $m(m - 1)/2$ transmissive $l \times l$ stars, and m reflective l-stars. The connection algorithm of the ports on the right-hand side of the figure is as follows: 1) Ports i of the l transmissive $m \times m$ stars of l-group i, $i = 1, 2, \cdots, m$, are joined together via a reflective l-star. 2) Ports i of the l transmissive $m \times m$ stars of l-group # j, $i, j = 1, 2, \cdots, m$, $i \neq j$, are connected via a vertically oriented transmissive $l \times l$ star to ports j of the l transmissive $m \times m$ stars of l-group # i.

It is clear that Fig. 13 is a special case of Fig. 14 for $l = 2$. Moreover, upon the realizations that a mirror can be considered as a reflective 1-star and that a straight-through fiber connection can be considered as a transmissive 1×1 star, it also becomes clear that Fig. 9 is indeed a special case of Fig. 14 for $l = 1$.

If l is a perfect square, then the reflective l-stars and the transmissive $l \times l$ stars in Fig. 14 can be absorbed in the transmissive $m \times m$-star groups, which, in fact, would make the resulting structure topologically equivalent to that of Fig. 9 (with m replaced by $m\sqrt{l}$). Similarly, if l is twice a perfect square, then the structure of Fig. 14 becomes topologically equivalent to that of Fig. 13 (with m replaced by $m\sqrt{l/2}$). In fact, the only cases where Fig. 14 can represent a reflective n-star that cannot be represented by either Fig. 9 or 13 is if the factorization of l contains odd powers of any prime other than two.

As an example, let us consider the case for $l = 3$. We know from [10]–[12] that transmissive 3×3 stars can be constructed. Moreover, applying to these structures the same bisection procedure used to obtain Fig. 10(b) from Fig. 2, we can construct reflective 3-stars. Using this value of l, and, say, also picking $m = 3$, we can use Fig. 14 to construct a reflective 27-star. In fact, upon the elimination of the missing ports in Fig. 14, which are indicated by the dots, that figure itself becomes a representation of a reflective 27-star.

In conclusion, we have shown how to construct reflective single-mode fiber-optic, n-star couplers, for n equaling an arbitrary multiple of a perfect square ($n = lm^2$). More specifically, because of restrictions on the component transmissive and reflective stars that are known to be realizable in practice, namely those based on radical-two and three constructions, the value of m is restricted to be a product of arbitrary powers of two and three, while the value of l, and hence also n, is restricted to be one, two, or three times the product of even powers of two and three. Unless constructions based on prime radicals other than two and three becomes known, the above restrictions, for example, would limit us to reflective n-stars with n = 2, 3, 4, 8, 9, 12, 16, 18, 27, 32, 36, 48, 64, 72, 81, 108, 128, 144, 162, 192, 243, 256, 288, 324, 432, 512, 576, 648, 729, 768, 972, 1024, $\cdots$, etc. which, in fact, are not that limiting!

References

[1] S. K. Sheem and T. G. Giallorenzi, "Single-mode fiber-optical power divider: Encapsulated etching technique," *Opt. Lett.*, vol. 4, no. 1, pp. 29-31, Jan. 1979.

[2] R. A. Bergh, G. Kotler, and H. J. Shaw, "Single-mode fiber optic directional coupler," *Electron. Lett.*, vol. 16, no. 7, pp. 260-261, Mar. 27, 1980.

[3] B. S. Kawasaki, K. O. Hill, and R. G. Lamont, "Biconical taper single-mode fiber coupler," *Opt. Lett.*, vol. 6, no. 7, pp. 327-328, July 1981.

[4] M. E. Marhic, "Combinatorial star couplers for single-mode optical fibers," in *Proc. Eighth Int. Fiber-Optics Communications and Local Area Network Exposition FOC/LAN 84*, Sept. 17-21, 1984, pp. 175-179.

[5] D. B. Mortimore, "Low-loss 8 × 8 single-mode star coupler," *Electron. Lett.*, vol. 21, no. 11, pp. 502-504, May 23, 1985.

[6] A. V. Oppenheim and R. W. Schafer, *Digital Signal Processing*. Englewood Cliffs, NJ: Prentice-Hall, 1975, p. 294, Fig. 6.7.

[7] H. J. Siegel and R. J. McMillen, "The multistage cube: A versatile interconnection Network," *Computer*, vol. 14, no. 12, pp. 65-76, Dec. 1981.

[8] H. S. Stone, "Parallel processing with the perfect shuffle," *IEEE Trans. Computers*, vol. C-20, no. 2, pp. 153-161, Feb. 1971.

[9] M. E. Marhic, "Hierarchic and combinatorial star couplers," *Opt. Lett.*, vol. 9, no. 8, pp. 368-370, Aug. 1984.

[10] S. K. Sheem and T. G. Giallorenzi, "Single-Mode fiber multiterminal star directional coupler," *Appl. Phys. Lett.*, vol. 35, no. 2, pp. 131-133, July 15, 1979.

[11] W. K. Burns, R. P. Moeller and C. A. Villarruel, "Observation of low noise in a passive fiber gyroscope," *Electron. Lett.*, vol. 18, no. 15, July 22, 1982.

[12] C. C. Wang, W. K. Burns, and C. A. Villarruel, "9 × 9 single-mode fiber-optic star couplers," *Opt. Lett.*, vol. 10, no. 1, pp. 49-51, Jan. 1985.

Efficient N × N Star Couplers Using Fourier Optics

C. DRAGONE, SENIOR MEMBER, IEEE

Abstract—A technique for constructing an efficient $N \times N$ star coupler with large N at optical frequencies is described. The coupler is realized in free-space using two arrays, each connected to N single-mode fibers. The highest efficiencies are obtained using a planar arrangement of two linear arrays separated by a dielectric slab serving as free-space region. The coupler is suitable for mass production in integrated form, with efficiencies exceeding 35 percent.

I. INTRODUCTION

EFFICIENT $N \times N$ star couplers, with N as large as 100, are needed for lightwave distribution in high-capacity local-area networks. A well known technique, which can be used to realize arbitrarily large N, involves the interconnection of a large number of elementary couplers [1], [2], but the complexity of such a coupler increases logarithmically with N. Here we consider a different approach, which is attractive for large N. Its theoretical efficiency for large N is independent of N, and it approaches 100 percent under ideal conditions. In practice, efficiencies exceeding 35 percent are obtainable at optical frequencies, with a simple arrangement of two linear arrays.

We construct the coupler in free space using two arrays, each composed of N elements. We determine the conditions that maximize power transfer between the two arrays. More precisely, we maximize the coupler efficiency β defined here as

$$\beta = NT_m \quad (1)$$

where T_m is the lowest value assumed by the power transmission coefficient T between two elements.

The coupler can be realized in two or three dimensions. In the former case, one obtains a geometry analogous to that of a Rotman lens [3], used at microwaves to form multiple beams in antenna arrays. This arrangement has two advantages. First, it is more efficient than a three dimensional arrangement. Second, it can be realized at optical frequencies in integrated form, on a glass substrate, by means of two arrays of strip waveguides separated by a free-space region formed by a slab waveguide as in Fig. 1. Each strip waveguide can be coupled efficiently to a single-mode fiber as in [4], [5].

Such a coupler operates as follows. Power entering any of the $2N$ ports excites the dominant mode ψ of one of the strip waveguides. The resulting radiation pattern in the free-space region is determined as shown in Section II by

Manuscript received January 25, 1988; revised May 4, 1988.
The author is with AT&T Bell Laboratories, Crawford Hill Laboratory, Holmdel, NJ 07733.
IEEE Log Number 8823583.

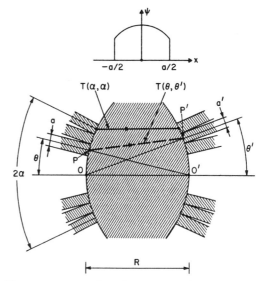

Fig. 1. Planar arrangement of two linear arrays related by a Fourier transformation.

the Fourier transform Φ of ψ. One finds that a fraction μ of the radiated power is intercepted by the receiving array, and the remaining power is lost because of spill-over. The value of μ is determined by the angular aperture 2α of the receiving array, and it will be shown to be close to 60 percent, under optimized conditions. However, the efficiency β is substantially lower than μ, since the dominant mode of a receiving element only accepts, in general, a fraction η of the incident power illuminating the element aperture. Thus, in this article we determine the conditions that maximize the efficiency β. Clearly, the value of β depends on the properties of the element distribution ψ. If no restriction is imposed on ψ, then it is shown in Section II that it is possible to obtain $\beta = 1$, and the coupler is then characterized by $T = T(\theta, \theta') = 1/N$, for all values of the input and output coordinates (θ and θ') locating the array elements on right and left of Fig. 1. In this article, however, ψ will be required to vanish outside the element aperture. Because of this restriction (then the array elements have nonoverlapping distributions) we obtain the following results: a) the highest possible β using *uniform* arrays is $\beta = 0.34$, b) the efficiency of such a coupler is approximately wavelength independent, c) it is possible, by using *nonuniform* arrays, to reduce the nonuniformity of $T(\theta, \theta')$ thus increasing β, d) the highest possible β obtainable with nonuniform arrays is 0.438. The planar coupler described in the following sections is the only known geometry that is realizable (efficiently) for large N in integrated form suitable for use with single-mode fibers. The coupler performs approximately a finite

Fourier transformation and, therefore, it has numerous applications including scanning antennas. The above results contradict the common notion that power transfer between two arrays in free space is necessarily inefficient [2].

It will be assumed throughout the article that N is large

$$N \gg 1$$

and the analysis will be based on the paraxial approximation of Fresnel's diffraction formula as in [6]. In all cases to be considered, the receiving elements will be in the far field of the transmitting elements. One can show that it is impossible, for $N \gg 1$, to construct an efficient coupler without this condition.

II. Preliminaries

In this section consideration is restricted to a plane. The plane may represent for instance a planar structure, such as a dielectric slab, or one of the two principal planes of a separable structure, as the square coupler of Section VII. Throughout Sections II–VI consideration will be restricted to the planar geometry of Fig. 1 involving two arrays, each consisting of N elements located on a circle of radius R. Each circle contains the center of the other circle, as required according to [6] to obtain a Fourier transformation, produced between the field distributions of the two circles as pointed out in Appendix E.

Power transfer between two elements P and P' of the two arrays is given by Friis transmission formula [7]

$$T = \frac{aa'}{\lambda R} \eta \eta' \qquad (2)$$

where the apertures of the two elements have widths a, a' and efficiencies η, η'. The product $a'\eta'$ is the absorption width of the receiving element. Thus η' is the ratio between the received power (i.e., the power accepted by the dominant mode ψ of the receiving waveguide) and the total power incident on the receiving aperture. A similar interpretation can be given to η, provided the sense of transmission is reversed. The element P then becomes a receiving element, with absorption width $a\eta$ determined by the dominant mode ψ of the receiving waveguide corresponding to P. To determine η assume that the dominant mode of the waveguide corresponding to P produces

$$\psi = \psi(u), \quad \text{with } u = \frac{2x}{a} \qquad (3)$$

where x denotes the distance from the center P of the element aperture and a is the aperture width, as shown in Fig. 2. Then one can show as in Appendix E that [6], [7]:

$$\eta = \eta(w') \simeq 2 \frac{|\Phi(w')|^2}{\int |\psi|^2 \, du} \qquad (4)$$

where $\Phi(w')$ is the Fourier transform of $\psi(u)$

$$\Phi(w') = \tfrac{1}{2} \int_{-1}^{1} \psi(u) e^{jw'u} \, du \qquad (5)$$

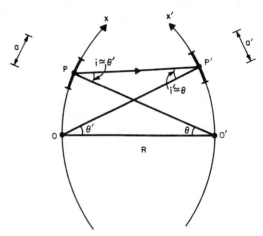

Fig. 2. Geometry of two circles related by a Fourier transformation.

and w' is specified by the angular displacement $i \simeq \theta'$ of P' from the axis PO' of the radiating element

$$w' = \frac{\pi a \sin \theta'}{\lambda}. \qquad (6)$$

Similarly, for the receiving element, by assuming on the circle through O' the distribution

$$\psi' = \psi'(u')$$

where $u' = x'/a'$ is determined by the distance x' from P', one finds that

$$\eta' = \eta'(w) = 2 \frac{|\Phi'(w)|^2}{\int |\psi'|^2 \, du'} \qquad (7)$$

where now w is a function of the angle of incidence $i' \simeq \theta$

$$w = \frac{\pi a' \sin \theta}{\lambda} \qquad (8)$$

and Φ' denotes the Fourier transform of ψ'. Notice that in Fig. 2 the angle of incidence i' at P' coincides, within the paraxial approximation, with the coordinate θ of P, and a similar property applies to i. As pointed out in Appendix E, the above expressions can be justified as follows. Over the aperture of the receiving element P' the incident wave is approximately a circular wave originating from P. This wave produces, on the circle containing P', phase distribution approximately given in the vicinity of P' by the phase factor

$$e^{-jwu'}. \qquad (9)$$

Thus, the coupling coefficient between $\psi'(u')$ and the incident distribution is approximately the Fourier transform of $\psi'(u')$ and one obtains (7).

From (2):

$$T(\theta, \theta') = \frac{aa'}{\lambda R} \eta'(w) \eta(w'). \qquad (10)$$

As pointed out in the introduction, the above derivation,

based on the paraxial approximation of [6], assumes that the angles θ, θ' are small and

$$a, a' \gg \lambda. \tag{11}$$

Furthermore, in order for the phase factor (9) to be produced accurately over the entire receiving aperture, the receiving element must be in the far field of $\psi(u)$.

Efficiency β

We are now ready to determine the coupler efficiency β. It is convenient to first consider the simplest configuration, involving two identical arrays of (nonoverlapping) elements arranged along the two circles as in Fig. 1. Initially, let the elements have all the same distribution $\psi(u)$, so that

$$\psi(u) = \psi'(u) \quad \eta(w) = \eta'(w) \tag{12}$$

and let them be uniformly spaced by a so that their total number is

$$2N = 2\frac{2R\alpha}{a}. \tag{13}$$

The properties of such a coupler are entirely determined by two factors: the distribution $\psi(u)$ and the parameter

$$w_\alpha = \frac{\pi a \sin \alpha}{\lambda} \tag{14}$$

representing the space-bandwidth product determined by the element width a and the array angular width 2α. Ideally, one would like the power radiated by any element of either array to be divided equally, and without loss, among the N elements of the other array so that

$$T = \frac{1}{N} \tag{15}$$

giving $\beta = 1$. Such an ideal coupler can be shown to require

$$|\Phi(w)| = A \text{ rect}\left(\frac{2w}{w_\alpha}\right) \tag{16}$$

where A is an arbitrary constant and rect $(2w/w_\alpha)$ is zero outside the interval

$$|w| < w_\alpha. \tag{17}$$

From (16), by taking the Fourier transform of $|\Phi|$ multiplied by an arbitrary phase factor, one can derive a variety of distributions $\psi(u)$ satisfying condition (16). All such $\psi(u)$ for

$$w_\alpha = \frac{\pi}{2} \tag{18}$$

are characterized according to (7), (10) by $T = 1/N$ and, furthermore, they satisfy, for any nonzero integer p, the orthogonality condition

$$\int_{-\infty}^{\infty} \psi(u - p) \psi^*(u) \, du = 0$$

which must be satisfied because $\psi(u - p)$ and $\psi(u)$ are the distributions obtained by exciting different (orthogonal) ports corresponding to elements spaced by pa. However, all such $\psi(u)$ assume nonzero values outside the interval $|u| \leq 1$ corresponding to the element aperture. For instance, by taking the Fourier transform of (16) for $A = 1$

$$\psi(u) = \frac{\sin\left(\frac{\pi}{2} u\right)}{\frac{\pi}{2} u}, \quad \text{for } w_\alpha = \frac{\pi}{2}$$

which does not vanish for $|u| > 1$. Here, instead, it will be assumed that

$$\psi(u) = 0, \quad \text{for } |u| \geq 1. \tag{19}$$

Then, it is impossible to cause Φ to vanish everywhere outside the interval $|w| < w_\alpha$ as required by (16). As a consequence, only a fraction μ of the power radiated by an element of either array will be intercepted by the other array. Furthermore, each receiving element will in general accept only a fraction η of the incident power illuminating the element aperture. Of importance, then, is the problem of maximizing the coupler efficiency defined in the introduction as

$$\beta = NT_m$$

where T_m denotes the lowest value assumed by $T(\theta, \theta')$. In all cases to be considered we will see that

$$T_m = T(\alpha, \alpha) \tag{20}$$

i.e., $T(\theta, \theta')$ will attain the lowest value between the marginal elements, corresponding to the edges of the arrays. From (7), (10), (13), (20) one obtains

$$\beta = \frac{2w_\alpha}{\pi} |\Phi(w_\alpha)|^2 \tag{21}$$

if $\psi(u)$ is normalized so that

$$\frac{1}{2} \int_{-1}^{+1} |\psi(u)|^2 \, du = 1. \tag{22}$$

The optimum $\psi(u)$ that maximizes $\Phi(\omega_\alpha)$ for a given w_α, under the condition (19), is shown in Section III to be

$$\psi(u) = A \cos(w_\alpha u) \tag{23}$$

and the value of β then becomes a function of only one parameter w_α. The optimum value of w_α is readily found to be 0.9, resulting in

$$\frac{0.34}{N} < T(\theta, \theta') < \frac{0.55}{N}.$$

An important property of such a coupler is that β is stationary with respect to λ and, therefore, β is approximately frequency independent. The above coupler, constructed with uniform arrays, can be further improved by using nonuniform arrays, by reducing the nonuniformity of $T(\theta, \theta')$. Thus in Section V we obtain $\beta \simeq 0.41$ with a geometry satisfying the simple condition

$$\beta \simeq \mu\eta$$

with μ, η calculated for $\theta = \theta' = \alpha$. The problem of determining the optimum $\psi(u)$ that maximizes the product $\mu\eta$ is described by an integral equation which cannot be solved exactly, in closed form. Thus in the following section we derive $\psi(u)$ by an approximate technique applicable in general to a variety of variational problems. It will be assumed, throughout this article, that $\psi(u)$ satisfies the condition

$$\psi(u) = \psi(-u) = \psi^*(u) \qquad (24)$$

which can be shown to be required for maximum efficiency, because of the symmetry of the problem for $N \gg 1$, $a \gg \lambda$. Because of the above condition, $\eta(w) = \eta(-w)$ and from (10)

$$T(\theta, \theta') = T(-\theta, \theta') = T(\theta, -\theta').$$

Therefore consideration will be restricted to

$$\theta, \theta' > 0.$$

III. Optimization of $\psi(u)$, w_α

Consider a transmitting element. Its total radiated power is proportional to the integral of $|\Phi(w)|^2$ for $|w| \le \infty$ and, clearly, it must be equal to the integral of $|\psi(u)|^2$ over the element aperture

$$\int_{-1}^{1} |\psi(u)|^2 \, du = \frac{2}{\pi} \int_{-\infty}^{\infty} |\Phi(w)|^2 \, dw$$

which is known as Parseval's theorem. It follows, taking into account (4) that

$$\int_{-\infty}^{\infty} \eta(w) \, dw = \pi. \qquad (25)$$

Now, of the total radiated power, only a fraction μ_α is radiated in the region $-\alpha < \theta < \alpha$ occupied by the receiving elements. The above region corresponds to the interval

$$-w_\alpha < w < w_\alpha$$

determined by the parameter w_α given by (14). Taking into account (25):

$$\mu_\alpha = \frac{2}{\pi} \int_0^{w_\alpha} \eta(w) \, dw \qquad (26)$$

and we will see that the optimum $\psi(u)$ that maximizes μ_α is given by

$$\psi(u) \simeq A \cos\left(\frac{w_\alpha}{\sqrt{3}} u\right) \qquad (27)$$

to a good approximation, for

$$w_\alpha < 1.$$

For the application considered here, however, a quantity that is more important than μ_α is the product

$$\mu_\alpha \eta(w_\alpha)$$

which will be shown in Sections IV–VI to give accurately the efficiency β of a coupler optimized using nonuniform arrays. The optimum $\psi(u)$ that maximizes the above product can be determined to any desired accuracy using the following theorem, derived in Appendix B.

Let $w_1, w_2, \cdots, w_r$ be r particular values of w and let $\eta_1, \eta_2, \cdots, \eta_r$ denote the corresponding values of $\eta(w)$. Suppose the problem is to maximize a given real-valued function $F(\eta_1, \eta_2, \cdots, \eta_r)$ satisfying the condition

$$\sum \left|\frac{\partial F}{\partial \eta_i}\right|^2 > 0, \quad \text{for } \sum |\eta_i|^2 > 0.$$

Then, in general, one finds that the optimum $\psi(u)$ under the constraint (24) can be written in the form

$$\psi(u) = \sum_{i=1}^{r} A_i \cos(w_i u) \qquad (28)$$

where the r coefficients A_i are determined by the partial derivatives of F. The ith term in (28) can be interpreted as the optimum distribution that maximizes $\eta(w_i)$. It is recalled, in fact, that the efficiency $\eta(w_i)$ is determined by the coupling coefficient between $\psi(u)$ and a plane wave

$$e^{\pm jw_i u} \qquad (29)$$

and therefore the optimum $\psi(u)$ that maximizes $\eta(w_i)$ under the constraint (24) is simply the even part of the above distribution.

Now consider $\mu_\alpha \eta(w_\alpha)$. By replacing the integral in (26) with a discrete sum, μ_α can be expressed accurately in terms of the values assumed by $\eta(w)$ at r discrete points in the interval $(0, w_\alpha)$. Then letting

$$\mu_\alpha \eta(w_\alpha) = F(\eta_1, \eta_2, \cdots, \eta_r) \qquad (30)$$

the optimum $\psi(u)$ according to the above theorem is obtained in the form (28).

Clearly the above procedure can be very accurate if r is large. For our present purpose, however, excellent accuracy is obtained by simply letting $r = 1$. In fact, for

$$w_\alpha < 1.2$$

one finds that $\eta(w)$ can be approximated very accurately for $|w| \le w_\alpha$ by a quadratic function of w. As a consequence

$$\mu_\alpha = \frac{2}{\pi} \int_0^{w_\alpha} \eta(w) \, dw \simeq \frac{2w_\alpha}{\pi} \eta\left(\frac{w_\alpha}{\sqrt{3}}\right) \qquad (31)$$

and

$$\mu_\alpha \eta(w_\alpha) \simeq \frac{2w_\alpha}{\pi} \eta\left(\frac{w_\alpha}{\sqrt{3}}\right) \eta(w_\alpha) \simeq \frac{2}{\pi} w_\alpha \eta\left(\sqrt{\frac{2}{3}} w_\alpha\right), \qquad (32)$$

as one can verify by expanding μ_α and $\mu_\alpha \eta(w_\alpha)$ in powers of w_α and neglecting terms of power 4 in w_α. It immediately follows that the optimum distributions that maximize μ_α and $\mu_\alpha \eta(w_\alpha)$ are given by

$$\psi(u) = A \cos(\gamma u) \qquad (33)$$

with $\gamma = w_\alpha / \sqrt{3}$ and

$$\gamma = \sqrt{\tfrac{2}{3}} w_\alpha \qquad (34)$$

respectively. In the former case the optimum distribution according to [8] is exactly a prolate spheroidal function, and therefore (27) is an approximation to this function for small w_α.

The Fourier transform of the distribution (33) can be determined straightforwardly, and one finds that

$$\eta(w) = P(\gamma, w)$$
$$= \frac{2}{1 + \frac{\sin 2\gamma}{2\gamma}} \left[\frac{w \sin w \cos \gamma - \gamma \sin \gamma \cos w}{w^2 - \gamma^2} \right]^2 \qquad (35)$$

which for $\gamma = w$ gives

$$\eta(\gamma) = P(\gamma, \gamma) = \frac{1}{2}\left[1 + \frac{\sin 2\gamma}{2\gamma} \right] \qquad (36)$$

needed in Section IV. Using (26) and (35) one finds that the largest possible value of $\mu_\alpha \eta(w_\alpha)$ is

$$\mu_\alpha \eta(w_\alpha) \simeq 0.417 \qquad (37)$$

and it is attained for $w_\alpha \simeq 1.1$ and

$$\psi \simeq \cos(0.9u). \qquad (38)$$

The dependence of μ_α and $\mu_\alpha \eta(w_\alpha)$ upon w_α for the above distribution can be determined accurately (with error < 0.6 percent for $w_\alpha < 1.1$) using the approximation (31) and it is illustrated in Fig. 3.

The optimization of NT_m in the following sections is carried out in three stages. First, in Section IV, all the elements are assumed to have the same $\psi(u)$ and the same width a. Then, in Section V, we let a vary as a function of the element location, specified by θ or θ', and obtain $NT_m \simeq 0.438$ after optimizing $a(\theta)$ so that

$$T(\theta, \theta) = T(\alpha, \alpha) \qquad (39)$$

for all $|\theta| < \alpha$. Finally, in Section VI, we let both a and $\psi(u)$ vary with θ (or θ') and improve NT_m from 0.438 to 0.450. In each case we show that the highest possible efficiency is attained by maximizing $NT(\alpha, \alpha)$ under the constraint

$$T(\theta, \theta') \geq T(\alpha, \alpha). \qquad (40)$$

The value of N is given accurately by

$$N \simeq R \int_{-\alpha}^{\alpha} \frac{d\theta}{a(\theta)} \qquad (41)$$

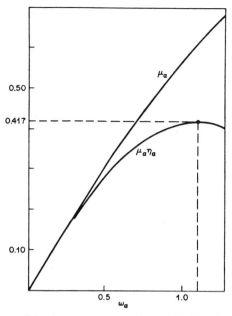

Fig. 3. Curves giving the exact and approximate behavior of μ_α and $\mu_\alpha \eta_\alpha$ for the distribution (38).

where $1/a(\theta)$ is the number of elements per unit length and the discrete variable θ, which can only assume N values, can be treated in (41) as a continuous variable because N is assumed to be large.

Each stage of the optimization will produce a different value of w_α. More precisely, the following three conditions will be obtained, respectively:

$$\frac{d[w\eta(w)]}{dw} = 0, \quad \text{for } w = w_\alpha \qquad (42)$$

$$\frac{d[w\eta^2(w)]}{dw} = 0, \quad \text{for } w = w_\alpha \qquad (43)$$

and

$$w_\alpha = \frac{\pi}{2} \qquad (44)$$

which is condition (18), whose significance is pointed out in Appendix C.

IV. Uniform Arrays

Using the same width a for all elements one obtains from (41):

$$N \simeq \frac{2R \sin \alpha}{a}. \qquad (45)$$

Then (10)–(13) give, for the marginal elements with $\theta = \theta' = \alpha$:

$$NT(\alpha, \alpha) = \frac{2}{\pi} w_\alpha \eta^2(w_\alpha). \qquad (46)$$

As pointed out in Section III, the optimum $\psi(u)$ that maximizes $\eta(w_\alpha)$ is given (exactly) by (33) with $\gamma = w_\alpha$. The optimum w_α that maximizes $T(\alpha, \alpha)$ is determined by

(43), which causes $T(\alpha, \alpha)$ to become stationary with respect to λ:

$$\frac{\partial T}{\partial \lambda} = 0, \quad \text{for } \theta = \theta' = \alpha \quad (47)$$

as can be verified from (10) and (12). From (43) and (35) for $\gamma = w_\alpha$ one obtains $w_\alpha \simeq 0.9$. Then from (10), (12), (35), (36)

$$\frac{0.554}{N} = T(0, 0) \geq T(\theta, \theta') \geq T(\alpha, \alpha') = \frac{0.340}{N}. \quad (48)$$

Such a coupler has the largest efficiency NT_m obtainable with uniform arrays. In fact, we have just shown that 0.34 is the largest value that can be attained by $NT(\alpha, \alpha)$. Thus, for a coupler with arbitrary w_α and $\psi(u)$:

$$NT_m \leq NT(\alpha, \alpha) \leq 0.34$$

and therefore the above coupler, characterized by $NT_m = 0.34$, has the largest possible NT_m.

V. Nonuniform Arrays With $\psi(u)$ Independent of θ, θ'

Next we improve NT_m by a factor $\gamma_\alpha \simeq 1.28$ by letting

$$a = a(\theta), \quad a' = a(\theta). \quad (49)$$

From (10), (12), and (41):

$$NT(\alpha, \alpha) \simeq \frac{2}{\pi} w_\alpha \eta^2(w_\alpha) \gamma_\alpha \quad (50)$$

where

$$w_\alpha = \frac{\pi a(\alpha) \sin \alpha}{\lambda} \quad (51)$$

and

$$\gamma_\alpha = \frac{1}{\alpha} \int_0^\alpha \frac{a(\alpha)}{a(\theta)} d\theta. \quad (52)$$

which can be interpreted according to (41) as the ratio

$$\gamma_\alpha = \frac{N}{N_\alpha}$$

between N and the number of elements N_α for a uniform array characterized by $a = a(\alpha)$:

$$N_\alpha = \frac{2R \sin \alpha}{a(\alpha)}. \quad (53)$$

It is shown in Appendix A that γ_α is maximized, under the constraint (40), by choosing $a(\theta)$ as required by condition (39). From (10) and (12):

$$T(\theta, \theta') = \frac{a(\theta) a(\theta') \eta(w) \eta(w')}{a^2(\alpha) \eta^2(w_\alpha)} T(\alpha, \alpha) \quad (54)$$

where now w is a function of both coordinates θ, θ'

$$w = \frac{\pi a(\theta') \sin \theta}{\lambda} \quad (55)$$

and therefore $T(\theta, \theta')$ is not separable. In order to satisfy condition (39), one must choose $a(\theta)$ so that

$$\frac{a(\alpha)}{a(\theta)} = \frac{\eta\left[\dfrac{\pi a(\theta) \sin \theta}{\lambda}\right]}{\eta(w_\alpha)} \quad (56)$$

which can be shown to admit a solution for $a(\theta)$ only if w_α does not exceed the (smallest) value specified by condition (42). The highest possible efficiency NT_m is found to be 0.438, and it is obtained for $w_\alpha \simeq 1.22$ and

$$\psi \simeq A \cos (0.9u).$$

It is now interesting to consider, instead of (56), the distribution

$$a(\theta) = \frac{\eta(w_\alpha) a(\alpha)}{\eta\left[\dfrac{\pi a(\alpha) \sin \theta}{\lambda}\right]} \quad (57)$$

which is not too different from (56) for $a(\alpha)/a(\theta) \simeq 1$. Both distributions can be shown to satisfy condition (40). According to (50) and (52) the latter distribution causes

$$NT(\alpha, \alpha) = \frac{2}{\pi} \eta(w_\alpha) \int_0^{w_\alpha} \eta(w) \, dw = \mu_\alpha \eta(w_\alpha) \quad (58)$$

and therefore a simple relation is obtained between NT_m and μ_α. As pointed out in Section III, the largest efficiency $\mu_\alpha \eta_\alpha$ is 0.417, which is not too different from the value 0.438 obtained from (56). Notice in both cases (56) and (57) the optimum $\psi(u)$ is given (approximately) by the same distribution (38) obtained in Section III with uniform spacing.

An important consideration, for some applications, is the ratio T_o/T_m between the maximum and minimum values, T_o and T_m, assumed by $T(\theta, \theta')$. In general, for a coupler characterized by a well behaved $\psi(u)$ and the optimum $a(\theta)$ of (56), one can show that $T_o/T_m - 1$ behaves as w_α^4 for small w_α. For instance, for the optimum distribution (38):

$$\frac{T_o}{T_m} \simeq 1 + 0.16 w_\alpha^4, \quad \text{for } w_\alpha < 0.9.$$

Furthermore, from (58) and (31)

$$NT_m \simeq \frac{2}{\pi} w_\alpha \eta\left(\frac{w_\alpha}{\sqrt{3}}\right) \eta(w_\alpha), \quad \text{for } w_\alpha < 0.9. \quad (59)$$

By choosing for instance $w_\alpha \simeq 0.7$, one obtains $T_o/T_m \simeq 1.04$ and $NT_m \simeq 0.35$.

VI. Nonuniform Arrays With Both a and $\psi(u)$ Independent of θ, θ'

So far, we have assumed the same $\psi(u)$ for all array elements. We now show that there is little to be gained by removing this restriction. Let $a(\theta)$ be chosen as required by condition (39) and let the aperture distribution $\psi(u)$ for an element of coordinate θ be optimized so as to maximize $T(\theta, \theta)$. Then

$$\psi = \cos\left(\frac{w_\alpha \sin\theta}{\sin\alpha} u\right) \qquad (60)$$

and one finds that the constraint (40) requires

$$w_\alpha \leq \frac{\pi}{2}.$$

By choosing $w_\alpha = \pi/2$, so that for the marginal elements

$$\psi(u) = A \cos\left(\frac{\pi}{2} u\right), \quad \text{for } \theta = \alpha \qquad (61)$$

one finds that for these elements

$$\eta(w_\alpha) = \tfrac{1}{2}$$

resulting in

$$NT_m = NT(\alpha, \alpha) = \frac{\gamma_\alpha}{4} \simeq 0.45 \qquad (62)$$

which can be shown to be the highest efficiency attainable with Fourier optics.

VII. Power Transfer Between Circular Arrays

Next consider the geometry of Fig. 4, involving two circular arrays with elements having diameters a and a' and coordinates θ, ϕ and θ', ϕ'. The aperture of either array is only partially filled by the circular apertures of the various elements, and the filling factor is easily shown to be

$$\tau = \frac{\pi}{2\sqrt{3}}.$$

Friis transmission formula [7] now gives, instead of (2):

$$T = \frac{AA'}{\lambda^2 R^2} \eta \eta' \qquad (63)$$

where A, A' are the aperture areas. The optimization of T assuming all elements have the same ψ is entirely analogous to that of Sections IV–VI and therefore only the main results will be pointed out.

The optimum element distribution $\psi(u)$ has circular symmetry:

$$\psi = \psi(u)$$

with

$$u = \frac{2\rho}{a}$$

ρ being the radial distance from the center of the element. From (63) one obtains, instead of (10):

$$T = T(\theta, \theta') = \tau \left(\frac{\pi}{4}\right)^2 \left(\frac{aa'}{\lambda R}\right)^2 \eta(w)\eta(w') \qquad (64)$$

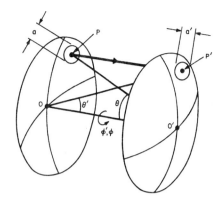

Fig. 4. Arrangement of two circular arrays. Notice the area of a circular aperture circumscribing a square aperture of area A_α is $\pi A_\alpha/2$.

where

$$\eta(w) = \eta(0)\left|\frac{\Phi(w)}{\Phi(0)}\right|^2$$

with $\eta(0)$ obtained from (4) and (5) for $w = 0$ by replacing du with $u\,du$. The function $\Phi(w)$ now denotes the Fourier–Bessel transform of $\psi(u)$:

$$\Phi(w) = \int_0^1 u\psi(u) J_0(uw) \, du \qquad (65)$$

with w given by (8). For a uniform array:

$$N = \left(\frac{2}{\pi}\frac{w_\alpha}{a}\right)^2 (\lambda R)^2 \qquad (66)$$

with w_α given by (14), and from (64) one obtains

$$NT(\alpha, \alpha) = \frac{\tau}{4} w_\alpha^2 \eta^2(w_\alpha) \qquad (67)$$

which should be compared to (46). The optimum $\psi(u)$ that maximizes $\eta(w_\alpha)$ is now found to be given by the Bessel function of order zero:

$$\psi(u) = AJ_0(uw_\alpha) \qquad (68)$$

and $\eta(w)$ is given in Appendix D. The optimum value of w_α that maximizes $NT(\alpha, \alpha)$ is found to be $w_\alpha \simeq 1.4$, resulting in

$$NT(\alpha, \alpha) \simeq 0.184\tau \qquad (69)$$

which should be compared to 0.34, obtained with planar geometry.

The above value can be increased substantially by choosing $a = a(\theta)$ so that $T(\theta, \theta) = T(\alpha, \alpha)$. As in the planar case, one finds that this will increase $NT(\alpha, \alpha)$ by a factor $\gamma_\alpha = N/N_\alpha$, where now

$$\gamma_\alpha = \frac{N}{N_\alpha} = \frac{1}{1 - \cos \alpha} \int_0^\alpha \frac{a^2(\alpha)}{a^2(\theta)} \sin \theta \, d\theta. \quad (70)$$

The optimum w_α is found to be 1.6, resulting in

$$NT(\alpha, \alpha) = 0.249\tau \quad (71)$$

obtained choosing $a(\theta)$ so that

$$NT(\theta, \theta) = NT(\alpha, \alpha). \quad (72)$$

As in Section V, the above condition is approximately satisfied for

$$a^2(\theta) = a^2(\alpha) \, \eta(w_\alpha)/\eta(w_\alpha \sin \theta / \sin \alpha)$$

resulting in $NT_m = \mu_\alpha \eta(w_\alpha)$, which can be maximized using a theorem analogous to that of Section III, with (28) replaced by

$$\psi = \sum A_i J_o(w_i u).$$

For all the above circular arrays, the value of $NT(\alpha, \alpha)$ is substantially lower than the corresponding value obtained with linear arrays. The reason is simple. Consider a coupler constructed using square arrays with separable distributions and square elements. One finds that T is then separable into a product of two factors, each describing the coupler in one of its two principal planes. Each factor is given by the same expression obtained in Sections IV–VI by considering two linear arrays. As a consequence, NT_m for a square coupler is *exactly* the square of the value obtained in Sections IV–VI for linear arrays. In particular, using uniform spacing

$$NT_m = (0.34)^2.$$

Now let such a square coupler be modified by introducing additional elements, so as to form a circular aperture with its boundary through the vertexes of the original aperture, as shown in Fig. 4. This will increase N by a factor $\pi/2$ resulting in

$$NT \simeq (0.34)^2 \frac{\pi}{2} = 0.182$$

which is not too different from the value 0.185 given by (69).

VIII. Discussion

We have seen that the efficiency of a coupler varies depending on its geometry. The highest efficiencies are obtained using the planar arrangements of Sections IV–VI. Then, the optimum NT_m varies between 0.34 and 0.45, depending on the coupler complexity. The simplest coupler, obtained using uniform arrays, has two important properties. First, once it is optimized at a particular wavelength λ_o,

$$\frac{\partial T_m}{\partial \lambda} = 0, \quad \text{for } \lambda = \lambda_o \quad (73)$$

and, therefore, it is wavelength-independent in the vicinity of λ_o. In fact, it can be shown that T_m will exceed 0.30 over the entire interval $0.7 < \lambda/\lambda_o < 1.35$, which is almost an octave. A second property of the above coupler is that its efficiency is not affected by small deviations of its parameters from their optimum values, i.e.

$$\delta(NT_m) \simeq 0$$

for small deviations. As a consequence, one can let $\psi(u)$ depart considerably from the optimum distribution without substantial sacrifice in efficiency. This is an important result, since the conditions assumed in Sections IV–VI will never be exactly satisfied, in practice. In particular, the dominant mode $\psi(u)$ of a strip waveguide in Fig. 1 will not in general vanish for $|u| \geq 1$, as required by condition (19). An approximate estimate of the coupler efficiency can nevertheless be made neglecting the values of ψ outside the interval $|u| \leq 1$. The optimization of w_α for any particular γ can then be carried out straightforwardly as in Sections IV and V. One then finds that β does not strongly depend on the value of γ, as illustrated by the following example for $\gamma = \pi/2$.

Suppose that in Fig. 1:

$$\psi(u) \simeq \cos\left(\frac{\pi}{2} u\right) \quad (74)$$

which is the distribution approached by the dominant mode of a strip waveguide for $a \gg \lambda$. Then optimizing $NT(\alpha, \alpha)$ as in Section V one finds that

$$T(\theta, \theta') \geq \frac{0.35}{N} \quad (75)$$

obtained using nonuniform arrays designed in accordance with (57) with $w_\alpha = 1.4$. The above efficiency, $NT_m = 0.35$, is not bad, particularly since the coupler is suitable for mass production.

Throughout this article consideration was restricted for simplicity to symmetrical couplers using two identical arrays, but one can verify using (10) that the transmission matrix of a coupler is not affected if the scale of one array is increased by a constant factor M, provided the other array is reduced by the same factor, so that

$$a \to Ma, \qquad a' \to \frac{a'}{M}$$

without changing R, ψ, ψ'. Finally, our main assumption, that $\psi(u) = 0$ for $|u| \geq 1$, will be removed in a future article. Then we will see that it is possible to realize efficiencies largely exceeding the values calculated in this article.

Appendix A
Optimization of $a(\theta)$ for $\psi(u)$ Independent of θ

Consider a coupler of the type considered in Section V, and let T'_m denote the lowest value of $T(\theta, \theta)$ for $|\theta| \leq \alpha$. We first show that the optimum $a(\theta)$ that maximizes NT'_m is specified by condition (39). In fact, suppose that for some values of θ

$$T(\theta, \theta) > T'_m. \quad (76)$$

Then, clearly, by reducing the corresponding values of $a(\theta)$ so as to obtain

$$T(\theta, \theta) = T'_m \quad (77)$$

N will be increased, without affecting T'_m, thus increasing NT'_m. We conclude that NT'_m is maximized by condition (39), giving condition (56).

For such a coupler, satisfying condition (77) for all $|\theta| < \alpha$, NT'_m is given by

$$NT(\alpha, \alpha) = NT(0, 0) \quad (78)$$
$$= \frac{2}{\pi} \eta^2(0) \frac{\pi a(0)}{\lambda} \int_0^\alpha \frac{a(0)}{a(\theta)} d\theta$$

which is a monotonic function of α and, therefore, it is maximized by simply choosing for α the largest value allowed by condition (56), which can be written in the form

$$\frac{\pi a(0) \sin \theta}{\lambda} = \frac{w \eta(w)}{\eta(0)} \quad (79)$$

with

$$w = \frac{\pi a(\theta) \sin \theta}{\lambda}.$$

Thus, the largest

$$w_\alpha = \frac{\pi a(\alpha) \sin \alpha}{\lambda}$$

is the value that maximizes $w\eta(w)$. One thus obtains condition (42), which according to (10) has the following significance: It specifies, for a given α, the optimum $a(\alpha)$ that maximizes the transmission coefficient $T(\alpha, \alpha)$ between two marginal elements in Fig. 1.

Conditions (56) and (42) specify the optimum values of $a(\theta)/a(0)$ and w_α that maximize NT'_m for a given $\eta(w)$. To determine $\eta(w)$, it is recalled from Section III that the optimum $\psi(u)$ is approximately a cosine distribution

$$\psi(u) \simeq A \cos(\gamma u)$$

and one can verify using (35) and (36) that the optimum γ is 0.9. From (42), one then obtains $w_\alpha \simeq 1.22$, resulting in $NT'_m \simeq 0.438$.

One can verify, by determining $T(\theta, \theta')$ of $\theta' \neq \theta$, that $T(\theta, \theta') \geq T(\theta, \theta)$ and therefore

$$NT_m = NT'_m \simeq 0.438.$$

This is the largest possible NT_m. In fact, since 0.438 is the largest possible NT'_m, and NT_m cannot exceed NT'_m, clearly NT_m cannot exceed 0.438.

APPENDIX B
DERIVATION OF (28)

Let $w_1, \cdots, w_r$ be r given values of w, let

$$M_i + jN_i = \Phi(w_i) = \tfrac{1}{2} \int_{-1}^1 \psi(u) e^{jw_i u} du \quad (80)$$

and let F be a real-valued function of M_i, N_i;

$$F = F(M_1, \cdots, M_r; N_1, \cdots, N_r) \quad (81)$$

satisfying the condition

$$\sum_{i=1}^r \left[\left(\frac{\partial F}{\partial M_i} \right)^2 + \left(\frac{\partial F}{\partial N_i} \right)^2 \right] > 0$$

for all values of M_i, N_i of interest. We wish to determine $\psi(u)$ so as to maximize F under the constraint

$$2 \int_{-1}^1 |\psi(u)|^2 du = 1. \quad (82)$$

We thus introduce a Lagrangian multiplier Λ and consider the problem

$$E = F - \Lambda \int_{-1}^1 |\psi(u)|^2 du = \text{EXTREMUM}. \quad (83)$$

In order to determine the variation δE caused by a perturbation $\delta \psi$ applied to ψ, notice, since F is a real-valued function, that

$$\delta F = \text{Re} \sum_{i=1}^r \left(\frac{\partial F}{\partial M_i} + j \frac{\partial F}{\partial N_i} \right) (\delta M_i - j \delta N_i) \quad (84)$$

and therefore by requiring $\delta E = 0$ one obtains from (81), (83), and (84)

$$0 = \text{Re} \int_{-1}^1 \left\{ \sum_{i=1}^r \left(\frac{\partial F}{\partial M_i} + j \frac{\partial F}{\partial N_i} \right) e^{-jw_i u} - 4\Lambda \psi \right\} \delta \psi^* du$$

giving the condition

$$4\Lambda \psi(u) = \sum_{i=1}^r \left(\frac{\partial F}{\partial M_i} + j \frac{\partial F}{\partial N_i} \right) e^{-jw_i u} \quad (85)$$

since $\delta \psi^*$ is arbitrary. Notice, if $\psi(u)$ is subject to the constraint (22), which requires $\psi(u) = \psi(-u) = \psi^*(u)$, then $N_i = 0$ and letting $F = F(M_1, \cdots, M_r)$ one obtains

$$4\Lambda \psi(u) = \sum_{i=1}^r \frac{\partial F}{\partial M_i} \cos(w_i u) \quad (86)$$

taking into account that both ψ and $\delta \psi^*$ are real-valued, even functions of u and therefore from (85), omitting the terms $\partial F / \partial N_i$, one obtains condition (86). In Section IV,

$$F = F(\eta_1, \cdots, \eta_r) \quad (87)$$

with $\eta_i = 4M_i^2$. Then from (86) and (87):

$$\Lambda \psi(u) = \sum_1^r 2 \frac{\partial F}{\partial \eta_i} \Phi(w_i) \cos(w_i u) \quad (88)$$

which implies (28). Notice from (80) and (88) the coefficients A_i in (28) satisfy the set of equations

$$\Lambda A_i = \frac{\partial F}{\partial \eta_i} \sum_{s=-r}^r A_{|s|} \frac{\sin(w_i - w_s)}{w_i - w_s} \quad (89)$$

which admit a solution for A_i only if their determinant vanishes. One thus obtains in general r eigenvalues for Λ. In the particular case where F is the product $\mu_\alpha \eta(w_\alpha)$, by letting $r \to \infty$ in (89) one can derive an integral equation for $\psi(u)$. If instead $F = \mu_\alpha$, then letting $r \to \infty$ one obtains the integral equation treated in [8].

Appendix C
Significance of Conditions (42 and 43)

Consider a transmitting element with aperture distribution $A\psi(u)$ and choose $A = 1/\sqrt{a}$, so that the total radiated power

$$\frac{1}{a}\int_{-a/2}^{a/2} |\psi|^2 \, dx = \frac{1}{2}\int_{-1}^{1} |\psi|^2 \, du$$

is independent of a. Then the power radiated in a particular direction is proportional to

$$w\eta^2(w), \quad \left(w = \frac{\pi a \sin \theta'}{\lambda}\right)$$

and therefore, by choosing the element width a so as to maximize the radiated power for $\theta' = \alpha$, one obtains (42). This, however, does not cause $T(\alpha, \alpha)$ to become stationary with respect to λ. In order to obtain $\partial T/\partial \lambda = 0$, one must require condition (43), which is obtained by maximizing $NT(\alpha, \alpha)$. One can verify that (44) maximizes the ratio

$$\frac{\eta(\gamma)}{\eta(0)} = \frac{P(\gamma, \gamma)}{P(\gamma, 0)}.$$

Appendix D
Aperture Efficiency $\eta(w)$ of the Distribution (68)

For the distribution (68):

$$\eta(w) = P(w, \gamma)$$

$$= \frac{4}{J_0^2(\gamma) + J_1^2(\gamma)}$$

$$\cdot \left\{\frac{wJ_0(\gamma) J_1(w) - \gamma J_0(w) J_1(\gamma)}{w^2 - \gamma^2}\right\}^2$$

which for $w = \gamma$ gives

$$\eta(\gamma) = P(\gamma, \gamma) = J_0^2(\gamma) + J_1^2(\gamma).$$

Appendix E
Arrangements Producing a Fourier Transformation

Consider the geometry of Fig. 2 involving two circles of radius R, each of which passes through the center of the other, and consider on the two circles, two points P and P' with coordinates θ and θ', respectively. For small θ, θ' the distance between the two points is a linear function of θ, θ', given by

$$R(1 + \sin \theta \sin \theta') \simeq R(1 + \theta\theta')$$

neglecting terms of power ≥ 4 in θ, θ'. As a consequence, by placing at P a point source of unit amplitude, the resulting response obtained at P' is [6]

$$U = \frac{e^{jkR}}{\sqrt{j\lambda R}} e^{jkR \sin \theta \sin \theta'}$$

within Fresnel's approximation. More generally, by considering on the circle L through 0 an arbitrary distribution of sources producing on L amplitude distribution ψ, the response produced on the other circle is obtained by convolving ψ with the impulse response U. By then expressing the distributions of L and L' in terms of $\sin \theta$ and $\sin \theta'$, a Fourier transformation is found between the two distributions. This is evident from the form of the impulse response U, whose variation with $\sin \theta$, $\sin \theta'$ is given by the factor

$$e^{jkR \sin \theta \sin \theta'}.$$

Now let a transmitting element be centered at P with aperture distribution $\psi(u)$. As pointed out earlier the resulting illumination of L' is obtained by convolving $\psi(u)$ with the impulse response U. One thus finds that a receiving element located at θ' is illuminated in Fig. 2 by a wave with angle of incidence θ and amplitude[1]

$$\frac{a}{\sqrt{\lambda R}} \Phi(w' + v')e^{jwu'},$$

where

$$u' = \frac{2x'}{a'}, \quad v' = \pi \frac{aa'}{\lambda R} \frac{x'}{a'}$$

x' being the distance from the center of the element of coordinate θ'. Here we are assuming that N is large, i.e., that a is small. Then neglecting the variation of Φ over the element aperture:

$$\Phi(w' + v') \simeq \Phi(w'), \quad \text{for } |x'| < \frac{a'}{2}.$$

The element is then illuminated by a uniform wave width phase distribution given by (9),

$$\frac{a}{\sqrt{\lambda R}} \Phi(w')e^{jwu'}$$

whose coupling coefficient with $\psi(u')$ is proportional to the Fourier transform of $\psi(u')$ multiplied by $\Phi(w')$. One thus obtains, for the power transfer between two elements of coordinates θ, θ', the two-dimensional Friis transmission formula (10).

Acknowledgment

The author is grateful to C. H. Henry, A. A. M. Saleh, M. J. Gans, T. Li, E. A. J. Marcatili, and I. P. Kaminow for discussions and suggestions. He is indebted to J. M. Fernandes for her invaluable assistance.

References

[1] A. A. M. Saleh and H. Kogelnik, "Reflective single-mode fiber-optic passive star couplers," to be published in *J. Lightwave Technol.*
[2] M. E. Marhic, "Hierarchic and combinatorial star couplers," *Opt. Lett.*, vol. 9, no. 8, pp. 368-370, Aug. 1984.
[3] W. Rotman and R. F. Turner, "Wide-angle lens for line-source applications," *IEEE Trans. Antennas Prop.*, vol. AP-11, pp. 623-632, 1963.

---

[1] Ignore a constant factor of unit amplitude.

[4] B. H. Verbeek, C. H. Henry, N. A. Olsson, K. J. Orlowsky, R. F. Kazarinov, and B. H. Johnson, "Integrated four-channel multi/demultiplexer fabricated with phosphorous doped SiO_2 waveguides on Si," submitted for publication.
[5] N. Takato, M. Yasu, and M. Kawachi, "Low loss high silica single mode channel waveguides," *Electron. Lett.*, vol. 22, pp. 321–322, 1986.
[6] J. W. Goodman, *Introduction to Fourier Optics*. New York: McGraw-Hill, 1968.
[7] J. D. Kraus, *Antennas*. New York: McGraw-Hill, 1950.
[8] D. Slepian and H. O. Pollak, "Prolate spheriodal wave functions, Fourier analysis and uncertainty—I," *Bell Syst. Tech. J.*, vol. 40, pp. 43–64, Jan. 1961.

Section 2.8: Beam-Combination

DESIGN OF AN OPTICAL DIGITAL COMPUTER

M.E. PRISE, M.M. DOWNS, F.B. McCORMICK, S.J. WALKER and
N. STREIBL

*AT and T Bell Laboratories, Crawfords Corner Road, Holmdel,
NJ 07739, U.S.A.*

Abstract A possible implementation of the design of a digital optical computer is presented. A general technique for *space-multiplexing* arrays of beams is described

Introduction
In this paper we describe a method for the implementation of a digital optical computer based on the architecture suggested by Murdocca [1]. A general discussion of the optical system is given by Prise et al [2]. The basic system configuration is that the signals in the computer are passed around as arrays of beams using free-space (as opposed to guided wave) optics. The logic is done by planar arrays of optical logic gates placed in the image plane of the optical system. Here the array of beams becomes an array of spots. Each spot corresponds to a logical bit.
Several regular interconnections between arrays of gates have been proposed. There are several networks which are all topologically equivalent to a perfect shuffle, whose optical implementation has been discussed. See for instance Jahns [3]. A simpler interconnection to implement optically is a split and shift. Murdocca [4] has shown how to implement arbitrary logic functions using this interconnect. The functionality of the system is added by blocking off some of the interconnection paths.

Devices
Before discussing the optical system we first have to make some assumptions about the devices. We assume the devices have the following properties. The signal output consists of the reflected power supply beam. We refer to these devices as reflection mode devices. The devices have two logical inputs. The two input signals do not have to be coincident with the power supply input. We wish to avoid the possibility of two signals being spatially coincident unless they are in opposite polarization since, unless all the path lengths in the system are held to interferometric precision, interference may result in a spurious logic signal. We anticipate devices such as these being produced using the SEED effect [5] and GaAs integrated circuit fabrication technology.

Optical Requirements
The purpose of the optical system connecting the arrays of gates is the following. An array of optical beams must be produced from a single power supply laser each having the same intensity. We will call this the power supply array. Our initial solution is to use binary phase gratings [6],[7]. This array of beams must be imaged onto the optical logic gate array. The reflected output must then be taken, put through an interconnect such as the split and shift with masks and then recombined on the next array of logic gates along with the power supply array for this device. We call these arrays the signal arrays since they contain the logic information.
The problem we concentrate on is having the array of logic gates in the focal plane of a single lens and feeding our array of power supply beams plus our arrays of signal beams onto the array of devices without losing power or sacrificing the resolution of the lens (which would mean we require larger devices and hence more power [2]). Each individual signal or power supply must use, as near as possible, the whole aperture of the lens so that the resolution is not reduced. Since we are using reflective devices we also have the problem of extracting the reflected power supply array which provides the input signals to the next set of devices. The method we use may have general applications for free-space opto-electronic interconnects.

Split and Shift Implementation
Before describing how we do the above, we will briefly state how a simple split and shift interconnect can be implemented. Either put a birefringent plate in the output from the previous optical logic gate array, or simply use a polarization beamsplitter to split the output array, into two seperate beam images (beam arrays), shift one with respect to the other, and recombine with a polarization beamsplitter. Several different shifts are desirable between logic arrays [4]. The architecture requires that some of the paths are blocked. This can be done by re-imaging in two separate paths and placing a mask in the image plane. An alternative masking technique is to replace selected reflectors with absorbers in the image combinatation setup described in the next section. This second method is preferable since it will lead to a more compact system and it is the only one which can be used when a birefringent plate is used to perform the split and shift. Notice that the output of this interconnect is two signal arrays of opposite polarization but spatially coincident. The problem now is to combine these signal arrays with the power supply array as discussed above.

Beam combination using patterned reflectors
One means of combine two arrays onto the same polarization without either power loss or resolution loss, is to use polarization and patterned reflectors placed in the image plane of the system. The basic principle is shown in Figure 1, for combining the power

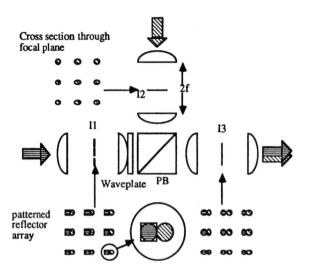

Figure 1: Method for space-multiplexing arrays of beams with no power or resolution loss.

supply array with a single signal array. The critical elements is the patterned mirror I1. These consist of arrays of reflectors. Each of the arrows represents an array of beams which corresponds to an array of spots in the image planes. In Figure 1 we show how to combine two arrays of beams losslessly without losing any of the optical system resolution. The combination is done by interleaving two counterpropagating arrays of beams in an image plane where the arrays of beams are discrete spots. The patterned reflector is arranged such that one set of beams is reflected and the other transmitted. It consists of an array of reflectors, each of which has the same dimensions as the spot size, with the same spacing as the spots in an individual array. We now have the two interleaved arrays of beams propagating in the same direction. In order to avoid these arrays propagating back along the same path as that taken by the reflected array, we use the polarization beamsplitter and a $\lambda/4$ plate. The collimated beams are incident on a polarization beamsplitter with the polarization such that they are completely reflected. They are then focussed down into the plane of the patterned reflector. They are aligned such that the individual spots are all reflected. In Figure 1 we have shown what the image looks like at the focal planes of the different lenses.

The $\lambda/4$ plate is placed between the polarization beamsplitter and the lens. The beams, on reflection back through the lens, are collimated and, this time, are transmitted through the beamsplitter. The array of beams transmitted through the patterned reflectors is arranged to have a polarization such that it is transmitted through the beamsplitter. Thus we have now obtained two arrays of beams which, when focussed down, give us two interleaved arrays of spots. The interesting thing is that both these arrays now have the same polarization and the size of each spot is limited by the full aperture of the system. We can take these arrays and combine with another array of signals by using the same system again. We can continue doing this until we have filled up the entire field of the lens. This ability to "space-multiplex" arrays of beams to an arbitrary extent may prove very useful for all types of free space optical interconnections.

We still have the other polarization channel which we can use to losslessly combine the output another array of beams (in this case the other signal beam array).

Input/Output Module for an Optical Computer

In Figure 2 we describe how this principle can be used to build a compact optical input/output system which could be used as a module in an optical computer. The power supply beams are provided by a single laser and a Dammann grating. Since we do not have two arrays of working devices, our input signal beam is provided by another laser and binary phase grating. In a real system our signal would consist of two arrays in opposite polarization obtained by taking the output from the previous optical logic gate array and putting it through a split and shift interconnect as described in section 4.

In our experiments we have used patterned reflectors to simulate the devices. These masks have reflectors of dimensions $10 \mu m \times 20 \mu m$ separated by $80 \mu m$. The reflectors are on a glass substrate about 2mm thick (for shorter focal length lenses this means we would have to use lenses corrected for spherical aberration). The structure is transparent between reflectors. These patterned reflectors

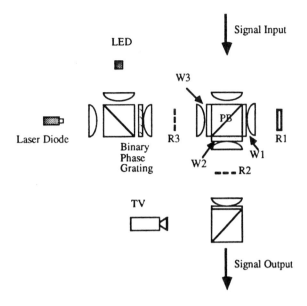

Figure 2: Experimental optical input/output module for a digital optical computer.

were fabricated using an Electron Beam Exposure System which is commonly used to fabricate the masks for VLSI processing.

The central element is the polarization beamsplitter. The power supply array comes from the laser to the right. The signal input is obtained from another laser and binary phase grating. The lasers are operating at $852nm$. Arrays consisting of 289 beams are generated from each laser diode using binary phase gratings [7]. These gratings have a period of $520\mu m$. All the lenses shown have a focal length of $5cm$. The only thing not shown is that we use a zoom lens system after the binary phase grating. This allows us to slightly magnify the binary phase grating to match it exactly to our device array size.

The light emitting diode is used to illuminate both the devices and the arrays of reflectors in the different image planes. A light emitting diode operating at the same wavelength as the laser diode was used to avoid any problems due to dispersion between the illumination light and the signal or power supply light. The illumination light is fed into the system using a polarization beamsplitter. The alignment is carried out as follows. The reflector array representing the devices (R1) is positioned $5cm$ from lens L1. The image of this reflector array is focussed on the TV camera. This reflector array is now replaced with a mirror. The input power supply array of beams is now focussed onto R1. Then by replacing the patterned array of reflectors, and using the zoom lens, the spacing between the spots in this array can be adjusted so it exactly matches the reflector array spacing.

Patterned reflector R2 is then brought into focus. If R1 and R2 are not exactly one focal length away from the respective lenses or the focal lengths are not exactly equal, these arrays of reflectors will appear to have different spacings. In this case this was not a problem. Further study is necessary to see whether this becomes a serious problem with different focal length lenses. Reflector R3 and R4 are also brought into focus. All of the reflectors arrays can be translated in both a horizontal and vertical direction with micrometers. We found the focus could be adjusted simply by manually sliding the reflector holder back and forward along the rods. The rotation of everything can also be set by manually sliding the reflectors in their mounts. We found it straightforward to align all the reflector arrays and the arrays of beams so that they would appear coincident on the video camera.

To set up the desired signal paths, we want one polarization of the split and shifted signal input to go straight through the beamsplitter, be reflected off R2, go back through W2 and be reflected onto the device array. The other polarization must be reflected through W3, back off R3 and transmitted through the beamsplitter onto the device array. In order to block off some of the signal paths which we wish to do for our split and shift interconnect, we have another fabrication stage where some of the reflectors are made absorbing, perhaps by depositing polysilicon.

The power supply input must be transmitted through R3 and W3, then transmitted through the beamsplitter and onto the device array. The reflected power supply which is the output must then be reflected back through W1, reflected off the beamsplitter, through W2 and R2, and is the output signal array.

The above was acheived by first aligning all the reflector arrays so they appeared coincident on the cameras. We then shift arrays R2 and R3 so they are half overlapping with array R1. By adjusting the position of the input signal array, which is split into two at the beamsplitter, so that it is incident on this overlapping area both polarizations now take the required paths. This appears on the camera as if these arrays (which are coincident) dissappear. The alignment of the signal beams reflected of R3 can be checked simply by moving R2. The only way to check on the alignment of the other signal beam is to place another camera behind R1 and shift R1 slightly to see if all the beams are coincident in the plane of the devices. We did this initailly but we found this check was unnecessary.

The power supply array was lined up onto the half of the R1 reflectors which were not overlapping with the reflectors R3 and R2. So on the television camera we can see the reflected power supply which is the output.

So that we could use the total output power to put into an interconnection stage, we use a waveplate / polarization beamsplitter combination to vary the portion of the light directed to the camera. (By rotating the waveplate the reflectivity of the beamsplitter can be varied).

We found the alignment straight forward. The main part of the system is very compact. In this case the beamsplitter is only $1cm^3$. Ideally we would like to use smaller focal length lenses.

The problem we have is that the spacing between devices is large, and we are limited by fabrication to binary phase gratings with a certain minimum period size. The innaccuracies inherent in the fabrication of the binary phase grating [7] mean that as we make the period size smaller the differences between the intensities of the different beams become larger. We found we could make a grating to produce a 17×17 array of beams with a period of $520\mu m$ with sufficient accuracy to obtain beams with a maximum intensity spread of 30% of the minimum intensity. The intensity spread was found both experimentally and by numerical simulation to go approximately inversely with the period size and by the square root of number of beams one trys to generate. We limit ourselves to a minimum grating period of $520\mu m$.

This period gives a spot spacing, in the image plane, of about $80\mu m$ with a $f = 5cm$ lens. We can get around this problem by demagnifying the grating but this means we require a larger grating and hence larger illuminating optics. At some point this will become impractictable. In this experiment we use a $1cm$ aperture optical system. If we have a Gaussian input beam, in order to lose less than 1% of the light at the aperture, the gaussian beam waist parameter must be less than $3.33mm$. When these beams are focused down onto our device $\sim 5\%$ of the power will miss the device. This is undesirable, but can only be changed by having a smaller period grating and a shorter focal length lens (which is a problem because of the limitations of binary phase gratings), or demagnifying a binary phase grating illuminated with a larger beam (which is a problem because we require larger aperture optics which will take up more space and will cost more) or, best of all, simply making the devices closer together.

Conclusions

In conclusion we have described a method for combining arrays of beams without losing power or optical resolution which may have useful applications in optical computing and for opto-electronic interconnections. We have experimentally demonstrated a particular application of this method and pointed out some of the system tradeoffs which have to be made.

References

[1] M. J. Murdocca, A. Huang, J. Jahns and N. Streibl, "Optical Design of Programmable Logic Arrays", to be published in *Applied Optics*, May 1988

[2] M.E.Prise,N. Streibl and M. M. Downs, "Optical Considerations in the Design of a Digital Optical Computer",*Optical and Quantum Electronics*,20, 49-77, 1988

[3] J.Jahns and M. J. Murdoccca, "Crossover Networks and their Optical Implementation", submitted to *Applied Optics*,1988

[4] M. J. Murdocca, unpublished

[5] D. A. B. Miller, "Quantum Wells for Information Processing", *Optical Engineering*,26 (3),368-372,1987

[6] H. Dammann and K. Gortler, "High Efficiency in-line multiple imaging by means of multiple phase holograms", *Optics Communications*,3,312-315,1971

[7] J. Jahns, M.E. Prise, M.M. Downs, S.J. Walker, N. Streibl, "Dammann gratings as array generators", Optical Society of America Annual Meeting, Rochester 1987, paper WJ3.

Section 2.9: Spot-Array-Generation

Array illuminator based on phase contrast

Adolf W. Lohmann, Johannes Schwider, Norbert Streibl, and James Thomas

An array illuminator converts a uniformly wide beam losslessly into an array of bright spots. These spots provide the necessary illumination for microcomponents such as optical logic gates or bistable elements. Such elements may serve as devices in a 2-D discrete parallel processor. We propose an array illuminator with a phase grating on its front end. The phase grating is illuminated uniformly and then converted into an amplitude image by means of a phase contrast setup. The bright spots of the amplitude image are to be used for illuminating the array of microdevices of a digital optical computer.

I. Definitions, Motivation, and Plan

An array illuminator is a component which may be useful in the context of a digital optical computer (DOC). The DOC will most likely be parallel in two dimensions. A certain subsystem of a DOC may consist of an array of 100 × 100 microdevices, supported by a common substrate. A typical microdevice is an optical logic gate or a bistable element. The array illuminator provides each device in an array with a separate optical power supply beam or a holding beam, in the case of a bistable element. The array illuminator should not waste light; given the energy dissipation of current optical devices and a switching rate compatible with fiber communications, we might otherwise run out of laser power quite soon. The array illuminator may consist of black boxes, which concentrate the incoming light rays and deliver them in parallel or in convergent bundles. These bundles of rays hit the microdevices, which may have a quadratic, circular, or any other shape. The compression factor C is the ratio of illuminated areas in front of and behind the array illuminator:

$$C = (d/b)^2, \qquad (1)$$

where d is the distance between adjacent devices and b is the diameter of a single device. There are several optical means for implementing an array illuminator, such as lenslet arrays or gratings.[1-3] These components may be set up in an image forming geometry or in a diffraction geometry, Fraunhofer or Fresnel version. In this study the black box consists of a macroscopic image forming system, with a phase grating as the object and a phase contrast filter in the Fourier domain. The image of the grating is the output of the array illuminator. Both object and filter should be nonabsorbing for optimum light efficiency, and both should be piecewise constant phase specimens, which is desirable for ease of manufacturing.

In this study we present a qualitative explanation (Sec. III); a more quantitative theory, including tolerance requirements, follows in Sec. III; some experimental results are shown in Sec. IV. Finally, we evaluate our array illuminator and compare it with alternate approaches. Such features as light efficiency, maximum compression ratio, volume, extensibility, coherence requirements, and error sensitivity are considered.

II. Phase Contrast Setup

The object (*OBJ* in Fig. 1) consists of a two-step phase grating. The phase contrast platelet at the center of the filter plane shifts the phase of the zero–zero diffraction order. The structure of the phase grating and the phase shifts of grating and phase contrast platelet should be selected so that the image consists of bright islands surrounded by darkness. Object phase $\Phi(x)$ and desirable image intensity $|v(x)|^2$ are shown in Fig. 2, for simplicity in one dimension only.

For our qualitative explanation we want to use the graphic representation of complex numbers (complex amplitudes). Before doing so, we have to define a few terms:

complex object amplitude: $u(x) = \exp[i\Phi(x)] = \begin{cases} 1, \\ \exp(i\Phi_0) \end{cases}$,

object phase [see Fig. 2(a)]: $\Phi(x) = 0$ or Φ_0,

The authors are with University of Erlangen-Nuremberg, Physics Institute, 8520 Erlangen, Federal Republic of Germany.
Received 20 November 1987.

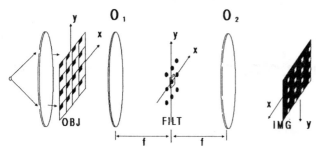

Fig. 1. Scheme of the phase contrast illuminator: *OBJ*, phase mask due to the envisioned compression ratio; *FILT*, plane where the phase shifter is positioned so that the zero–zero order is retarded; *IMG*, output plane; bright spots are the locations of the optical switches; 01,02, Fourier transform lenses.

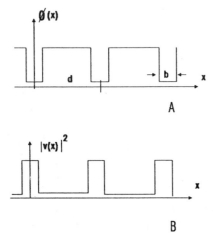

Fig. 2. (a) Phase distribution in the input plane *OBJ* and (b) intensity distribution in the output plane.

average object amplitude: $\bar{u} = p \cdot 1 + (1-p)\exp(i\Phi_0),$ (2)

relative slit width [see Fig. 2(a)]: $p = b/d,$

object variation: $\Delta u(x) = u(x) - \bar{u} = 1 - \bar{u} = u_1,$
$u_2 = \exp(i\Phi_0) - \bar{u}$ (3)

phase factor of phase contrast platelet: $\exp(i\alpha).$

The complex amplitude of the object $u(x)$ assumes one of two values on the unit circle in the complex plane [Fig. 3(a)]. The mean value $\bar{u}$ of $u(x)$ lies somewhere in a straight line between 1 and $\exp(i\Phi_0)$.

The exact location depends on the relative slit width p. The variable part $\Delta u(x)$ of the object is shown in Fig. 3(c). The mean value $\bar{u}$ will appear as the zeroth diffraction order in the filter plane (Fig. 1). The complex amplitude $\bar{u}$ will be phase shifted at the center of the filter plane by the phase platelet (Fig. 1). Hence, the bias term in the image plane will be a rotated version (by an angle α) of the former $\bar{u}$, now as in Fig. 3(d). To this rotated $\bar{u}$ we must add the variable portion $\Delta u(x)$ of the object to obtain the two values of the image amplitude $v(x)$, as in Fig. 3(e). The associated image intensity values $|v(x)|^2$ are apparently al-

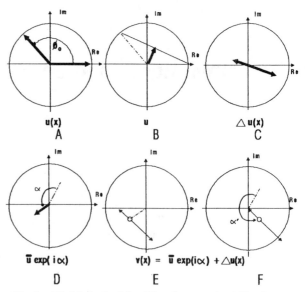

Fig. 3. Working principle of the phase contrast illuminator.

most equal in this case, since both arrowheads are about equidistant from the origin. This is not what we want. A larger phase angle α' [Fig. 3(f)] is better suited to achieve our goal [Fig. 2(b)]:

$$|v(x)|^2 \approx C = 1/p^2 \text{ or } 0. \quad (4)$$

A quantitative derivation of the suitable phase angles and relative slit widths follows in Sec. III, together with tolerances and some limitations.

III. Quantitative Approach, Limitations, and Tolerances

In a first approach the conditions to be met by the phase screens in object and filter planes (Fig. 1) are derived from 1-D screens. This procedure is justified since the 2-D case leads to similar expressions. Since both phase screens have a binary structure the conditions for Φ_0 and α are derived as a function of the ratio $p = b/d$. As outlined in Sec. II the aim of the setup is the transfer of the energy located in the region $b/2 \leq x \leq d - b/2 \pmod{d}$ to the region $-b/2 \leq x \leq b/2 \pmod{d}$ in the exit plane.

To obtain cancellation for the region $b/2 \leq x \leq d - b/2 \pmod{d}$ the complex average amplitude and the complex amplitude assigned to the latter region must have the same modulus but opposite phase. Using the definitions given in Eqs. (2) and (3) for $\bar{u}, u_1, u_2$, the following relations for u_1, u_2 hold:

$$u_1 = (1-p)[1 - \exp(i\Phi_0)], \quad (5a)$$

$$u_2 = -p[1 - \exp(i\Phi_0)]. \quad (5b)$$

Since p is positive definite and ≤ 1, the complex amplitudes u_1, u_2 are in phase opposition. Due to factorization the arguments in Eqs. 5(a) and 5(b) are equal apart from π due to phase opposition.

Modulus and argument of the average amplitude are

$$|\bar{u}| = \sqrt{p^2 + 2p(1-p)\cos\Phi_0 + (1-p)^2}, \quad (6a)$$

$$\arg\bar{u} = \arctan\{(1-p)\sin\Phi_0/[p + (1-p)\cos\Phi_0]\}. \quad (6b)$$

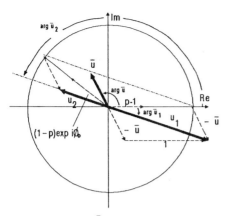

Fig. 4. Complex amplitudes $\bar{u}, u_1, u_2$ and their relative orientations. For an explanation see text.

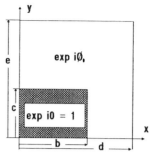

Fig. 5. One elementary cell of the OBJ plane mask.

The condition for equal moduli of $\bar{u}$ and u_2 results in

$$\sqrt{p^2 + 2p(1-p)\cos\Phi_0 + (1-p)^2} = 2p\sin\Phi_0/2.$$

This condition can be simplified into

$$\cos\Phi_0 = 1 - \frac{1}{2p}. \quad (7)$$

From Eq. (7) it follows that p can only be chosen within the range: $1/4 \leq p \leq 1$, if $\cos\Phi_0$ remains within allowed limits. The corresponding range for Φ_0 is then $\pi \leq \Phi_0 \leq \pi/3$.

In a next step the phase shift α of the phase platelet in the filter plane is derived from the arguments of $\bar{u}$ and u_2. For a cancellation of the shifted complex amplitudes $\bar{u}$ against u_2 the argument of $\bar{u}$ has to be increased by α due to the following relation (see Fig. 4):

$$\arg\bar{u} + \alpha = \arg u_2 + \pi, \quad (8)$$

where the π summand is necessary to attain phase opposition for $\bar{u}$ and u_2.

The argument of u_2 is, due to Eq. (5b),

$$\arg u_2 = \pi - \arctan\{\sin\Phi_0/(1-\cos\Phi_0)\}. \quad (9)$$

By substituting Eq. (9) into Eq. (8) α becomes (see Appendix A)

$$\alpha = 2\pi - \Phi_0 \quad (10)$$

or simply $\alpha = -\Phi_0$.

The partition of the energy between regions 1 and 2 can be derived from the complex amplitudes:

$$v_1 = (\bar{u}\exp i\alpha + u_1), \quad v_2 = (\bar{u}\exp i\alpha + u_2). \quad (11)$$

From this the following relation for I_1 is attained:

$$I_1 = 4(p^2 - p)\cos^2\Phi_0 + 4(1-p)(2p-1)\cos\Phi_0 + 4(1-p)^2 + 1 = 1/p \quad (12)$$

by using Eq. (7) for $\cos\Phi_0$. In a similar manner it can be shown that within the same range $I_2 = 0$.

The requirement for a total cancellation of the intensity in the surrounding of the bright spots can be relaxed to some extent if the dark areas do not coincide with the positions of the switches of the array. Since the maximum phase shift for Φ_0 and α is π, these phase

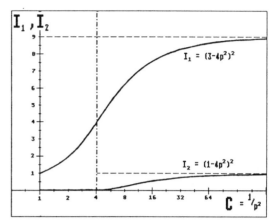

Fig. 6. Intensities I_1, I_2 in the output as a function of the compression ratio C. Note: to the left of $C = 4$ the Φ_0 is due to Eq. (7) and to the right, $\Phi_0 = \pi$ const. Here I_1: intensity in the bright spots; I_2 = intensity of the surrounding area; $C = 1/p^2$ (or $1/pq$) = compression ratio.

shifts are maintained while the ratio p is diminished beyond the $1/4$ limit. In this case the intensities I_1, I_2 are

$$I_1 = (3 - 4p)^2, \quad I_2 = (1 - 4p)^2. \quad (13)$$

The limiting case $p = 1/4$ is also described correctly by Eq. (13) and for decreasing p the intensities I_1, I_2 converge to values 9 and 1, respectively (see Fig. 6).

The contrast ratio I_1/I_2 could be increased by an absorbing phase shifter at the cost of the number of photons available in the bright regions. If the amplitude absorption constant is indicated by τ, the two intensities have the form $I_1 = [(2 + \tau) - (2 + 2\tau)p]^2$ and $I_2 = [\tau - (2 + 2\tau)p]^2$. For $p \Rightarrow 0$ the contrast ratio becomes $\lim I_1/I_2 = [(2 + \tau)/\tau]^2$, which is, e.g., $I_1/I_2 = 25$ if τ is assumed to be 0.5, but the energy efficiency drops from 9 to 6.25. So, in general, absorbing phase shifters are not quite so effective for this problem but might be useful for other applications where contrast is the major goal.

An extension to a 2-D object phase screen can be carried out by the calculation of the zero–zero diffraction order amplitude for such a screen. Since the object has a predominant periodic character it is sufficient[4] to calculate the Fourier transform of one elementary cell. To simplify matters the notation is given in Fig. 5. The phase must be zero in the small rectangular area, having extensions b and c within the elementary cell of extensions d and e, and be Φ_0 in the remaining field.

From the Fourier transform calculation we can derive the zero–zero Fourier amplitude $\bar{u}$ as

$$\bar{u} = de\{pq + [1 - pq]\exp i\Phi_0\}. \tag{14}$$

This equation is similar to Eq. (2) for the 1-D case but with the difference that p has to be replaced by pq or, for a quadratic geometry, by p^2 (with $p^2 = 1/4$ being the limiting case with total cancellation, meaning that the energy located in the whole cell in the input plane is concentrated to one-quarter of the cell area in the output). The equation for the 1-D case can be considered as a special case of Eq. (14), i.e., for $q = 1$.

In the same way other geometries can also be dealt with, e.g., circular patches in a Cartesian structure. Only minor differences can be expected in the expression for $\bar{u}$. The result for circular geometry is given without proof:

$$u = \pi q^2 + [1 - \pi q^2]\exp i\Phi_0, \tag{15}$$

with $q = R/d$, where R is the diameter of the phase shifting patches and d is their distance. As has been shown, the efficiency of a phase contrast illuminator is limited to a compression ratio $C = 1/p^2 = 4$ or with arbitrary compression ratios the intensity gain is limited to values of ~9 in comparison to the surrounding field. Since the object plane is imaged sharply onto the image plane, the optical tolerances of the object phase screen are not critical in a global sense. Commonly, the compression ratio is >4 so the phase shifts Φ_0 and α are set to π. The intensities in the two areas of one elementary cell are, therefore, the result of constructive or destructive interference only. The cos-type intensity distribution is taken in the maximum/minimum region only and is accordingly insensitive to phase errors of the phase screens. This alleviates the mask generating and control process.

In the extreme region the following holds:

$$|\Delta I| \approx V|\Delta\Phi_0|^2 \quad (\text{or } |\Delta\alpha|^2), \tag{16}$$

where V is the visibility after Michelson[5] and $\Delta\Phi_0$ and $\Delta\alpha$ are the phase errors of the phase screens. The visibility depends on the amplitude balance, i.e., on pq (where $pq \leq 1/4$ has been assumed):

$$V = \{4 - 8pq\}/\{5 - 16pq + 16(pq)^2\}, \tag{17}$$

which is of the order of $V = 4/5$ for small pq values. For a numerical estimate, the visibility can be considered independent of the compression ratio, being of the order of 1. Therefore, the intensity error due to phase errors is equal to about the square of the phase error in radians. On the one hand, there is no self-healing capability due to the sharp imaging of the phase screen from the input to the output plane, but on the other hand, the method is immune to weak phase gradients of the mask support glass for the same reason. This remains true as long as the phase variations remain in the low spatial frequency region. In general, the phase factors of the mask and the glass support multiply and, therefore, one gets in the Fourier plane a convolution of the Fourier transforms of the mask amplitude with the transform of the support. If the transform of the glass is strongly broadened crosstalk between adjacent diffraction orders will occur, and this would impair the working principle of the phase contrast illuminator.

The relative insensitivity of the method against glass thickness variations does not apply to the filter plane, since in this plane the plane wave spectrum of the mask is generated. If each spatial harmonic of the input undergoes different phase delays due to imperfections of the phase shifter glass support, the intensity distribution (being the interference pattern of the plane wave spectrum) will deviate from the distribution aimed for. Also aberrations caused by the finite thickness t of the phase shifter in the Fourier plane can impair the correct interference of the waves in the output plane if the devices to be illuminated become rather small (e.g., a few wavelengths of the illuminating laser). In this case the off-axis aberrations, coma, and astigmatism introduce additional phase shifts depending on the field variable and therefore on the diffraction order number linearly or quadratically (third-order theory). In this way different orders are progressively phase mismatched.[6] Especially in the case of small densely spaced switching elements care has to be taken that the whole optical system is corrected for such aberrations and that the thickness variations of the phase shifter can be kept within the limits estimated with the help of Eq. (16). Some comfort can be gained from the fact that the Fourier amplitudes of the higher spatial frequencies fall off with the inverse of the diffraction order number. In our experiments such effects can be neglected thanks to the coarse phase mask used.

The phase contrast illuminator does not presuppose total coherence of the light source. If an extended monochromatic and incoherent source is used, the dimensions of the source image in the FILT-plane should be smaller than the distance between adjacent diffraction orders. This might be helpful if multimode lasers (e.g., excimer lasers) or thermal light sources are used.

IV. Experimental Verification

The basic elements of the phase contrast illuminator are a phase mask and a phase platelet producing in certain areas phase lags of Φ_0 and α, respectively. As has been shown for the ratios $1/pq$ (or for quadratic cell geometries $1/p^2$) >4, the phase lag for both masks is π.

Now, production and measurement of the phase shift of the masks are discussed. Two approaches for the production of the phase masks have been tried, i.e., bleaching binary photographic masks and contact printing of photographic masks on a photoresist film of suitable thickness.

For our experiments a rather coarse phase mask was chosen with a pitch length of $d = 1$ mm to reduce the requirements on the glass support quality. In the Fourier plane of the phase contrast illuminator, therefore, the diffraction orders are separated by the dis-

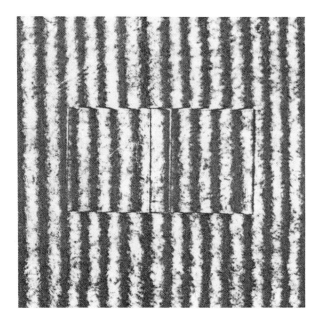

Fig. 7. Shearing interferogram of the phase platelet used as a phase shifter. Fringe number adjusted by a suitable amount of defocus of the wave entering a Michelson shearing interferometer.

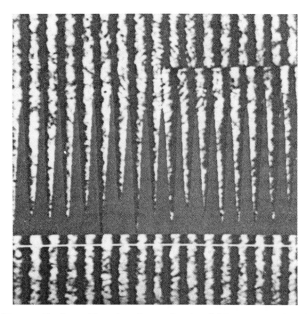

Fig. 8. Similar to Fig. 7 but detected with a CCD camera and read into a buffer memory for further processing. The evaluation has been carried out in ten cross sections (only one is shown). The overlaid picture is the first derivative of the intensity along the indicated section.

tance $a = f'\lambda/d$, where f' is the back focal length of objective O_1 in Fig. 1. For $f' = 500$ mm the width of the phase shifting region on the phase platelet becomes $a = 0.3$ mm.

The measurement of phase shifts was carried out with a Michelson type shearing interference microscope.[7] The fringe orientation is coupled to the shear direction. The number of fringes can be adjusted by choosing a suitable amount of defocus. A typical measuring result for the phase platelet with a phase shifting region of 0.3×0.3 mm is shown in Fig. 7. With the help of a CCD camera (512×512 pixels) these interferograms have been read into an image memory and evaluated by a program using a low-pass filter combined with a differentiation process to find the extrema of the cos-type interference fringes.

Figure 8 shows an interferogram with fringe adjustment to measure the phase shift introduced by the photoresist island. The low pass filtered derivative of the intensity pattern was obtained using a method due to Snyder[8] and is shown overlaid on top of the interference pattern seen by the CCD camera. From the zeros of this function the phase lag has been determined as $\alpha = (1.03 \pm 0.05) \cdot \pi$. The illuminator experiments were carried out with photoresist copies of demagnified computer plots, where the phase retarding region is opaque and the remainder of the elementary cell of the phase mask is transparent, i.e., we used a binary amplitude mask to create the binary phase illuminator plate. The measurement of the phase retardation gave similar results for all masks produced in this way as have been obtained for the phase shifting platelet (see Fig. 7).

If a phase mask is placed in the OBJ plane (see Fig. 1) the intensity distribution in the output plane without phase shift is shown in Fig. 9, which is a usual

Fig. 9. Output plane of the phase contrast illuminator when the phase shifter is strongly maladjusted. The result is a normal bright-field picture of the input phase mask.

bright-field pattern of a binary phase object. By placing the phase shifter at the position of the zero–zero order in the $FILT$ plane (Fig. 1) one gets a picture as given in Fig. 10, emphasizing the intensity compression effect.

To obtain a more quantitative insight, we performed experiments using a phase retardation of only about π, thus improving the immunity to small phase inaccuracies. This is justified since in most practical cases the compression factor C will be >4. Here only two measuring results are presented. Figure 10 shows the

Fig. 10. Direct photograph of the output plane of the PCI with the phase mask having compression ratios of $C = 100/9$ (left) and $C\ 100/16$ (right).

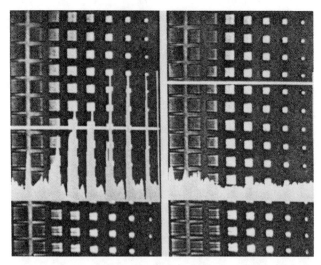

Fig. 12. Experiment with a variable phase mask. Aspect ratio p^2 varies between 4/100 and 81/100. Phases $\Phi_0 = \alpha = \pi$ have been used. Left: Scan through the bright patches of the output plane. Note: with decreasing p^2 the brightness I_1 of the patches increases as predicted (Fig. 8). Right: Scan through the surrounding showing with decreasing p^2 an increase in the background intensity I_2. For $p^2 > 25/100$ the phases Φ_0 and α show the mismatch by failure to concentrate the energy in the regions desired.

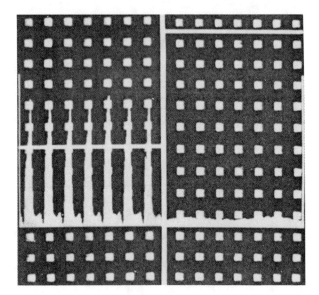

Fig. 11. Same as Fig. 10 but scanned with a CCD camera: left, scan through the bright spots; right, scan through the dark surrounding. The bright bar to the left corresponds to the dynamic range of the CCD.

output plane of the phase illuminator with a split phase mask having in the left part a compression ratio $(C = 1/p^2)$ of $C = 100/9$ and in the right part $C = 100/16$. The output plane was scanned with a CCD camera, and the data were read into a digital image memory. Two scans across the intensity distribution have been evaluated: one through the bright spots and one as reference through the surrounding field to get an impression of the background intensity I_2 [see Eq. (13)]. These are shown in Fig. 11.

As stated before, for compression ratios of $C > 4$ an appreciable energy concentration can also be achieved for $C \to \infty$. Therefore, in Fig. 12 the aspect ratio p^2 or the compression ratio $C = 1/p^2$ was varied within wide limits in one phase mask. The elementary cell size was chosen as 1, and p varies from 2/10 to 9/10 in steps of 1/10; consequently, p^2 changes from 4/100 to 81/100. As in the previous experiment a scan through the bright spots (left-hand side) and one through the dark surrounding show sufficient agreement between theory (Fig. 6) and the height of the bright spots. In the right-hand part of Fig. 12 the increase in the background intensity I_2 for $p^2 \to 4/100$ becomes obvious. In the left-hand part of Fig. 12 the increase in the intensity I_1 for $C = 4$ ($p^2 = 25/100$) to I_1 for $C = 25$ ($p^2 = 4/100$) by a factor of ~ 2 is in good agreement with the estimate of Eq. (13).

The experiment of Fig. 13 is described in addition as a curiosity. Since the phase contrast illuminator decomposes the plane wave spectrum in the FILT plane (Fig. 1) and brings the single Fourier components with the correct phase to interference in the output plane, a phase shift of any of the plane wave components will lead to image distortions. Hitherto, only the zero–zero order has been shifted resulting in the intensity distribution desired. But, if, e.g., the phase shifter is placed at the location of one of the first diffraction orders, interference fringes appear having a pitch identical to the pitch of the phase mask in the entrance plane. By choosing other orders the fringe orientation as well as the pitch can be varied. Since the higher orders have lower Fourier amplitudes the contrast of the fringes will fade out.

V. Conclusions

It has been shown, both theoretically and experimentally, that the light energy of an equally illuminated beam can be concentrated in a periodic array of bright spots by a phase contrast method. The compression ratio can be adjusted to 4:1 for the ideal case of a dark surrounding of the bright spots. If the condition for a totally dark surrounding is violated, the maximum possible light concentration is ~ 9. For

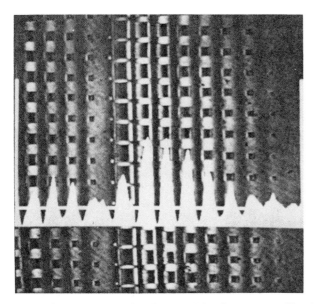

Fig. 13. Output intensity distribution of the phase contrast illuminator with the phase shifter at the position of one first order (0,+1). Due to the mismatch in this case a fringe system appears in the image plane.

compression ratios >4 the phase shift of the input phase screen and of the phase contrast phase shifter is π. Experiments have demonstrated that the phase contrast illuminator is insensitive to the majority of the possible errors.

Appendix

Substituting Eq. (9) into Eq. (8), α becomes

$$\alpha = 2\pi - \arctan\{\sin\Phi_0/(1 - \cos\Phi_0)\} + \arctan\{(1-p)\sin\Phi_0/[p + (1-p)\cos\Phi_0]\},$$

and by using the relation

$$\arctan A + \arctan B = \arctan(A+B)/(1-AB) \quad \text{for } AB \leq 1,$$

α becomes

$$\alpha = 2\pi - \arctan\{\sin\Phi_0/(1-2p)\cos\Phi_0\}. \quad (A1)$$

Substitution of Eq. (7) into Eq. (A1) yields

$$\alpha = 2\pi - \arctan\sqrt{(4p-1)/(1-2p)^2}. \quad (A2)$$

If now the identity $\arccos A = \arctan\sqrt{(1-A^2)}/A$ is used, relation (10) is valid.

A similar result for the phase relations has been attained for an intensity corrector[9] for phase-locked laser arrays.

References

1. H. Dammann and K. Gortler, "High Efficiency In-Line Multiple Imaging by Means of Multiple Phase Holograms," Opt. Commun. **3**, 312 (1971); U. Killat, G. Rabe, and W. Rave, "Binary Phase Gratings for Star Couplers with High Splitting Ratio," Fiber Int. Opt. **4**, 159 (1982).
2. M. E. Prise, N. Streibl, and M. M. Downs, "Computational Properties of Nonlinear Optical Devices," in *Technical Digest of Topical Meeting on Photonic Switching* (Optical Society of America, Washington, DC, 1987), paper FB4.
3. A. W. Lohmann, "An Array Illuminator Based on the Talbot Effect," Optik **79**, 41 (1988).
4. A. Papoulis, *Systems and Transforms in Optics* (McGraw-Hill, New York, 1968).
5. M. Born and E. Wolf, *Principles of Optics* (Pergamon, Oxford, 1980).
6. W. Smith, *Modern Optical Engineering* (McGraw-Hill, New York, 1966).
7. D. Malacara, *Optical Shop Testing* (Wiley, New York, 1978).
8. J. J. Snyder, "Algorithm for Fast Digital Analysis of Interference Fringes," Appl. Opt. **19**, 1223 (1980).
9. G. J. Swanson, J. R. Leger, and M. Holz, "Aperture Filling of Phase-Locked Laser Arrays," Opt. Lett. **12**, 245 (1987).

Part 3
Networks and Systems

We have already commented in our introduction that a key aspect of studies in "photonic switching" is the relationship between the capability offered by optical technology and requirements placed upon the switch by the problems of network design, control, and implementation. This part deals in turn with the different approaches being pursued, including switching networks based on TDMA, CDMA, WDMA, and FDMA (Section 3.1); space and time division, and code and wavelength division (Section 3.2); packet switching (Section 3.3); and general systems issues (Section 3.4). Only here do we try to resolve the really difficult questions and to identify a role for optics in switching. Each paper presents an independent view, and we leave it to the reader to form his own conclusions. One prediction is safely made—we are dealing with a rapidly changing field, and it is in this area of systems that we will see the major developments in the next few years.

Section 3.1: Switching Networks

TDMA and CDMA

Ultrafast All-Optical Synchronous Multiple Access Fiber Networks

PAUL R. PRUCNAL, MARIO A. SANTORO, STUDENT MEMBER, IEEE, AND SANJAY K. SEHGAL, STUDENT MEMBER, IEEE

Abstract—Two synchronous multiple access schemes, TDMA and CDMA, are proposed for fiber optic networks using optical signal processing. Network synchronization is achieved by using a central mode-locked laser which also serves as the source for each station. The data are converted into a high-bandwidth optical signal using electrooptic modulators. The accessing schemes use optical fiber delay lines. The feasibility of these schemes is discussed.

I. INTRODUCTION

THE importance of optical fibers in the context of long-haul digital communications has already been demonstrated. However, their potential in networks has yet to be fully realized. Various schemes have been proposed in order to exploit the high bandwidth–distance product of optical fibers. Using an optical fiber as a high-bandwidth channel is, by itself, insufficient to implement high-capacity networks. Ultimately, the capacity of the channel is limited by the processing speed of the associated electronic circuitry. The state of the art in electronic technology yields maximum processing speeds of approximately 1 GHz [1]. On the other hand, optical processing methods offer much higher processing speeds. For example, optical fiber delay line devices have bandwidths in excess of 100 GHz.

A fiber optic network utilizing high-bandwidth optical signal processing was previously demonstrated [2]. That network utilized an asynchronous, collision-free multiple access scheme with zero delay.

In general, synchronous accessing methods where transmissions are perfectly scheduled provide higher throughput (more successful transmissions) than asynchronous methods where network access is random and collisions occur [3]. In this paper, two synchronous accessing schemes using optical signal processing are investigated: fixed assignment TDMA and synchronous CDMA. The feasibility of both schemes is characterized. Fixed assignment TDMA is presented in Section II. Synchronous CDMA is discussed in Section III. Power requirements of both access schemes are explained in Section IV. In Section V, the feasibility of these synchronous schemes is discussed. Conclusions and future experiments are detailed in Section VI.

II. FIXED ASSIGNMENT TDMA

Local area networks (LAN's) were originally proposed to interconnect computer facilities where the channel utilization was relatively low [3]. Multiple access protocols were therefore designed to accommodate the low traffic demand in such networks in the most efficient way. Consequently, asynchronous contention protocols, such as token passing and carrier sense multiple access, are well suited to LAN's with low traffic demand. However, these asynchronous protocols suffer from cumulative delay as the traffic intensity increases.

Synchronous multiple access schemes, such as time division multiple access (TDMA), can accommodate higher traffic demands and do not suffer from cumulative delay. In fixed assignment TDMA, each user is assigned a fixed time slot during which that user can receive a data packet or a bit of information. In order to accommodate as many users as possible, the NRZ data are first converted into a low duty cycle signal by sampling them at the data rate. The low duty cycle signal is then transmitted in the selected destination time slot. This time division multiplexing process must be performed at high speed. The cost of high-speed electronics restricts the use of TDMA to networks where the traffic demand is heavy most of the time, in which case the high expenditure is justified. In any event, the speed limitation of electronics will ultimately restrict the minimum duty cycle of the sampled signal (width of sample or TDMA slot) and hence the number of users that can be accommodated.

The use of fiber optics for the transmission channel alone does not reduce the bandwidth requirements of the electronic interfaces, nor does it make a synchronous system more economical for network applications. It is proposed that the bandwidth requirements in the electronic interfaces for TDMA systems be reduced by performing the high bandwidth processing of the medium access protocols optically, and confining the slow electronic processing to relatively low-rate data handling.

A block diagram of such a TDMA system with optical signal processing is shown in Fig. 1. The system consists of an optical clock source, M stations with optical transmitters and optical receivers, and a passive optical star

---

Manuscript received June 13, 1986.

The authors are with the Lightwave Communications Research Laboratory, Center for Telecommunications Research, Columbia University, New York, NY 10027.

IEEE Log Number 8610780.

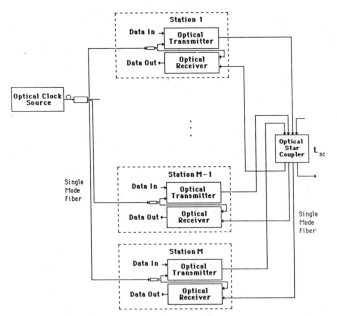

Fig. 1. Block diagram of a synchronous fiber optic network with M stations in a star topology and a central optical clock. L_{SC} is the star coupler excess loss. Single-mode fiber is used to interconnect the stations.

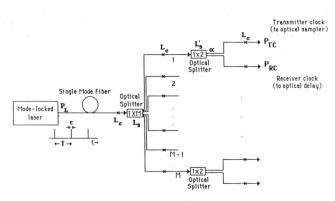

Fig. 2. Optical clock signal distribution. $1/T$ is the optical pulse repetition rate, τ is the optical pulse width, P_L is the laser peak power output, α is the splitting ratio of input to transmitter-clock-output of the 1×2 splitter, L_C is the connector/splice loss, L_S is the $1 \times M$ optical splitter excess loss, and L_s^r is the 1×2 splitter excess loss. The distance of any station to the mode-locked laser is $k_i Tc/n$ where k_i is an integer and c/n is the velocity of light in the fiber.

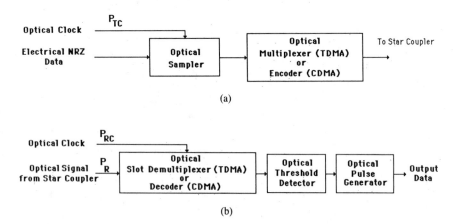

Fig. 3. (a) Block diagram of the optical transmitter. The optical clock is used to sample the data. The optical data are then either moved to the appropriate destination time slot by the optical multiplexer (TDMA) or are converted to the appropriate optical CDMA sequence corresponding to the destination address. (b) Block diagram of the optical receiver. The optical signal from the star coupler is either demultiplexed (TDMA) or decoded (CDMA) using the optical clock. The composite optical signal is then threshold detected and formatted into an electrical NRZ signal.

coupler. All interconnections utilize single-mode optical fibers. Although other configurations are possible, a star topology is chosen here to permit the maximum number of users given a fixed power budget [4].

Using fixed assignment TDMA, the optical signal at each station must be located in the correct time slot that corresponds to its destination. To synchronize all the transmitters, a centralized clock signal must be available to all stations. The clock signal can be distributed optically from a central mode-locked laser, as described in Fig. 2. The mode-locked laser generates a train of very narrow optical pulses at the data rate $1/T$. The time between pulses T and the width of the pulses τ will determine the maximum number of time slots T/τ that can be assigned. Using a mode-locked laser, low duty cycle, high peak power pulses can be routinely generated. Consequently, many time slots can be assigned. The distance of any station from the mode-locked laser must be an integral multiple of Tc/n where c/n is the velocity of light in the fiber core.

The block diagram of a station transmitter is shown in Fig. 3(a). At each transmitting station, the data must be optically sampled and multiplexed into the selected destination time slot. As seen in Fig. 4, an optical sampler can be implemented by using the optical clock pulses in combination with an electrooptic modulator [5]. The op-

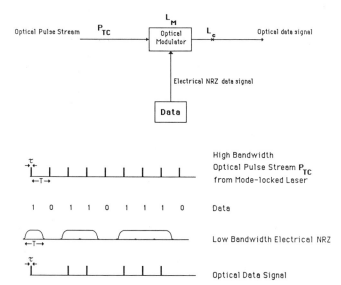

Fig. 4. Optical sampler. A stream of narrow optical pulses is gated by low bandwidth data using an electrooptic modulator. Both data and pulse stream have the same signal rate of $1/T$. L_M is the modulator total loss.

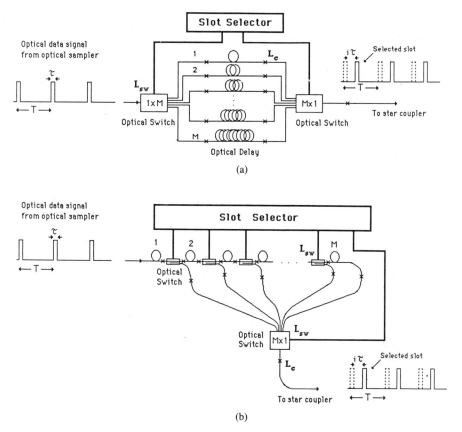

Fig. 5. Optical multiplexer. (a) Parallel multiplexer. Different fiber optic delay lines with length a multiple of $\tau c/n$ are attached to $1 \times M$ and $M \times 1$ electrooptic switches. The slot selector selects the desired fiber optic delay path by appropriately activating the $1 \times M$ and $M \times 1$ switches. (b) Serial multiplexer. Fiber delay lines of length $\tau c/n$ are connected serially by 1×2 electrooptic switches or taps. The second output of the switch is connected to an $M \times 1$ switch. The slot selector activates as many 1×2 switches as required to obtain the appropriate delay.

tical pulses are gated by the electrical data signal at the electrooptic modulator. The original electrical NRZ data signal is thus converted into a low duty cycle optical signal with the same data rate. By this procedure, conversion from a low bandwidth electrical signal into a high bandwidth optical signal is accomplished without resorting to high bandwidth electronics. The slow rise/fall time of the electrical data compared to the narrow optical pulses could result in a reduction in optical pulse amplitude if the optical pulses are gated during these transition times. To avoid this, some type of synchronization between the electrical signal and the optical pulses must be provided. This subject will be discussed in detail in Section V.

The optical data signal can be multiplexed into the correct destination time slot simply by using optical fiber delay lines. Two possible schemes for a time division multiplexer (TDM) are shown in Fig. 5. In Fig. 5(a), M delay lines of length $i\tau c/n$, with $i = 1, \cdots, M$, are connected in parallel between two active electrooptic switches [6]. To select the ith time slot, the first $1 \times M$ switch selects the fiber delay line of length $i\tau c/n$. The selected delay line is then switched into the output port using the $M \times 1$ electrooptic switch. The selections are made by the electrical slot selector unit. Fig. 5(b) shows M fixed delay lines of length $\tau c/n$ connected serially with 1×2 electrooptic switches. The slot selector unit selects the desired time delay $i\tau$ and routes the signal through the $M \times 1$ electrooptic switch. The latter scheme requires less fiber length.

The block diagram of the receiver is shown in Fig. 3(b). The receiving station must be able to identify its own time slot within the TDMA frame. To accomplish this, the station uses the slot demultiplexer (Fig. 6). In order to recognize its own time slot, the ith receiver delays the clock signal by $i\tau$ and adds it to the incoming signal. The addition of the TDMA signal and the delayed clock results in the correct signal riding atop the clock signal. The signal can then be threshold detected. It is necessary that the TDMA frames arrive at the receiver in synchronism with the optical clock, so that the clock signal can be synchronized to the position of the receiver slot in the TDMA frame. Therefore, the length of fiber from each station to the star coupler must be an integral multiple of $L = Tc/n$. The required precision in fiber lengths will be discussed in Section V.

The transmitted data are ultimately recovered by the threshold detector and pulse generator [see Fig. 3(b)]. The output of the slot demultiplexer has three well-defined signal levels: a low level corresponding to no data or a data 0 transmitted in another user's slot, a middle level corresponding to a data 1 transmitted in another user's slots or a data 0 transmitted in the receiver's own slot, and a high level corresponding to a data 1 transmitted in the receiver's own slot. The threshold detector level is set between the middle and high level. Whenever the optical signal crosses the threshold, a pulse of length T is emitted by the optical pulse generator. This optical pulse is sub-

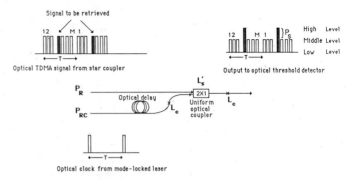

Fig. 6. Slot demultiplexer. The optical TDMA signal is combined with a delayed optical clock. In this way, information arriving in the correct slot is raised on the clock pedestal and is made easily distinguishable. P_R is the received peak power. L'_s is the coupler loss. Three signal levels are defined: a low level corresponding to no data or a data bit zero in other users' slots or a data 0 transmitted in the receiver's own slot, a middle level corresponding to a data bit one in other users' slots, and a high level corresponding to a data bit one in the receiver's own slot. P_S is the signal power.

sequently converted by the photodetector into an electrical NRZ data signal of rate $1/T$. Possible implementations for the optical threshold detector and pulse generator are discussed in Section V.

III. Synchronous CDMA

As explained in the last section, TDMA is an efficient multiple access protocol in networks with heavy traffic demands. However, in situations where the channel is sparsely used, TDMA is inefficient [3]. Contention protocols generally proposed for low traffic demands are not suitable if traffic delay is a major issue, e.g., in networks where information must be transmitted simultaneously.

A multiple access protocol which is both efficient with low traffic and has zero access delay is code division multiple access (CDMA). In CDMA, each user is assigned a code sequence which serves as its address. The transmitter encodes each information bit with the code sequence address of its intended destination, i.e., a "1" bit is transmitted as a sequence of N chips and a "0" bit is sent as an all-zero sequence.

An asynchronous all-optical LAN using CDMA was demonstrated by Prucnal *et al.* [2], [7]. Since that network was asynchronous, the number of pseudoorthogonal (distinguishable) code sequence was limited, and therefore, so was the total number of possible subscribing users. However, with synchronous CDMA, all possible time-shifted versions of a code sequence can be used. A given code sequence can be time shifted $N - 1$ times. The cross-correlation peak between two time-shifted versions of the code sequence is as high as the autocorrelation peak, but always occurs either delayed or ahead of the autocorrelation peak. Since the receiver is synchronized to the expected position of the autocorrelation peak, it is easily distinguished from adjacent cross-correlation peaks. The number of possible subscribing users is therefore $N - 1$ times larger than in asynchronous CDMA.

The network layout shown in Fig. 1, optical clock distribution scheme shown in Fig. 2, and transmitter and re-

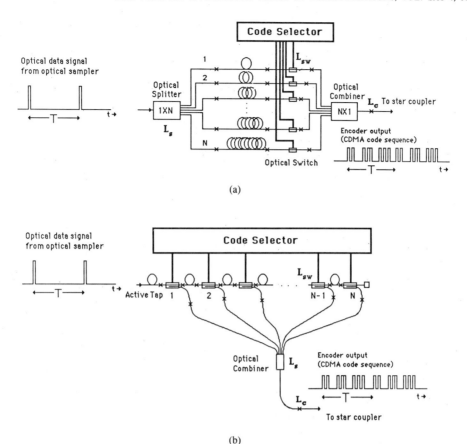

Fig. 7. Optical encoder. (a) Parallel encoder. The optical data signal is split into N different delay paths. The code selector selects the paths corresponding to the chosen code sequence. The paths are recombined at the optical combiner. L_{SW} is the switch excess loss. (b) Serial encoder. N active optical taps with variable coupling coefficients are placed at equal distance of $\tau c/n$ on a single-mode fiber. The tap outputs are combined in an optical combiner. The code selector activates the ϵ taps corresponding to the code sequence chosen in such a way that the optical CDMA sequence has equal power levels. ϵ is the number of ones in the code sequence.

ceiver configurations shown in Fig. 3 also apply to synchronous CDMA. The optically sampled data must be encoded at the transmitter with the intended receiver's code sequence. Fig. 7 illustrates two possible optical CDMA encoder configurations. In Fig. 7(a), the signal is split using a $1 \times N$ optical splitter. The code selector choses the ϵ delay paths that correspond to the positions of the ones in the intended receiver's code sequence. In Fig. 7(b), a fiber is tapped serially at intervals $i\tau$. The coupling coefficient between the two outgoing paths of each active tap is set by the code selector according to the codeword sequence. These coupling coefficients are chosen so that all of the output pulses are of equal intensity.

The receiving station must correlate the incoming optical CDMA sequence with its own address. The correlator consists of fiber delay lines as shown in Fig. 8. Two possible constructions for a correlator are given. Fig. 8(a) is a parallel correlator and Fig. 8(b) is a serial correlator. If the incoming signal is encoded with the correct address, the output of the correlator will yield an autocorrelation peak of amplitude ϵ. The details of the CDMA code and the correlation process are discussed in [2] and [7]. Transmitter and receiver synchronization is achieved by adding to the correlated signal the receiver clock signal delayed by the proper amount. The delay must be such that the peak of the autocorrelation function coincides with the optical clock pulse. The same considerations discussed earlier regarding transmitter–receiver synchronization and fiber length apply. The data are retrieved by threshold detecting the autocorrelation peaks which ride atop the clock signal. The threshold level is set below the cumulative peak of the autocorrelation and the clock, but above the cumulative peak of the cross-correlation and the clock.

IV. Power Requirements

The synchronous networks presented in Sections II and III require a mode-locked laser with sufficient peak optical power. The mode-locked laser must be able to drive the whole network, leaving a power margin for future expansions and network reliability. The power available for a station in an M-station network is divided between a

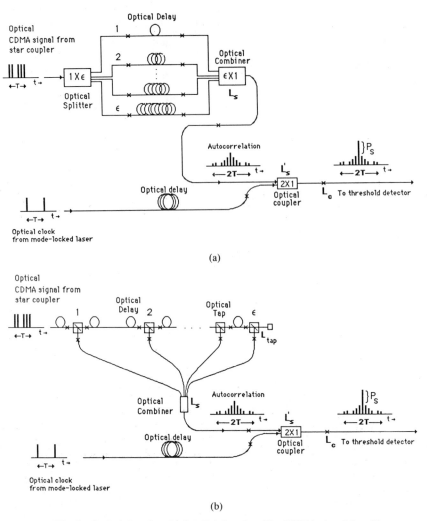

Fig. 8. Optical decoder. (a) Parallel decoder. The CDMA signal is split into ϵ different delay paths corresponding to the position of the ones in the time inverse version of the receiver address. The paths are then combined in the $\epsilon \times 1$ optical combiner. The output of the combiner is added to the appropriately delayed clock. (b) Serial decoder. ϵ passive taps with variable coupling coefficients are placed on a single-mode fiber at distances corresponding to the time inverse of the receiver code sequence (address). The taps are combined in an $\epsilon \times 1$ optical combiner. The output of the combiner is added to the clock which is delayed by the appropriate amount.

transmitter clock and a receiver clock signal (see Fig. 2). The signal intensity P_{TC} of transmitter clock is given by

$$P_{TC} = P_L - 10 \log M - L_S$$
$$- 3L_C - 10 \log \alpha - L'_S \quad (1)$$

where P_L is the laser peak power output into a fiber pigtail, L_S is the $(1 \times M)$ splitter excess loss, L_C is the connector (splice) loss, L'_S is the (1×2) splitter excess loss, and α is the splitting ratio of input to transmitter-clock-output of the (1×2) splitter. The optimal choice of this ratio will be discussed later. It is assumed for simplicity that the loss in the fiber is negligible and the power is equally divided among M output fibers.

The power at the input to the receiver is given by (see Figs. 1, 4, and 5)

$$P_R = P_{TC} - L_M - L_{SS} - 2L_C - L_{SC} - 10 \log M \quad (2)$$

with L_M the modulator excess loss, L_{SS} the multiplexer (or encoder and decoder) total loss, and L_{SC} the star coupler excess loss.

At the optical threshold detector, the signal power P_S for TDMA corresponds to the difference between the middle and high levels in Fig. 6. For CDMA, the signal power corresponds to the difference between the peak autocorrelation level and the maximum cross-correlation level. To maximize the signal power and reduce the effects of adjacent pulse interference, the clock pedestal should be at least as large as the surrounding interference, i.e., the middle level in TDMA or the maximum cross-correlation peak in CDMA. To minimize the source power required, the amplitude of the clock pedestal is set equal to the maximum interference level in both cases. This is achieved by adjusting α to equalize the losses in the clock channel and signal channel. Setting the losses in the clock channel equal to the losses in the signal channel and solving for α yields

$$\alpha = 1 + 1/\{M \log^{-1} ((L_M + L_{SS} + L_C + L_{SC})/10)\}. \quad (3)$$

With this choice of α, the signal power P_S at the optical threshold detector is (see Fig. 6)

$$P_S = P_R - L'_S - L_C - 3 \quad (4)$$

where the splitting loss in the (2×1) combiner is assumed to be 3 dB. Substituting (1) and (2) in (4), the power required at the mode-locked laser to yield a given signal level at the input to the optical threshold detector is given by

$$P_L = P_S + 20 \log M + L_S + 2L'_S + L_M$$
$$+ L_{SS} + 6L_C + L_{SC} + 10 \log \alpha + 3. \quad (5)$$

For a TDMA system with a parallel line multiplexer [Fig. 5(a)], the multiplexer loss is

$$L_{SS} = 2L_{SW} + 3L_c \quad (6a)$$

where L_{SW} is the switch loss. For a serial multiplexer [Fig. 5(b)], the total loss is

$$L_{SS} = M(3L_c + L_{SW}). \quad (6b)$$

In a synchronous CDMA system, the encoder total loss is given by

$$L_{Pe} = 2L_S + 20 \log N + 4L_C + L_{SW} \quad (7a)$$

for the parallel encoder [Fig. 7(a)] and by

$$L_{Se} = (3N - 1) L_C + NL_{SW} + 10 \log (\epsilon N) + L_S \quad (7b)$$

for the serial encoder [Fig. 7(b)], where ϵ is the number of ones in the code sequence. Additional power loss is contributed by the decoder. The total loss in a parallel decoder (L_{Pd}) is given by [see Fig. 8(a)]

$$L_{Pd} = 3L_C + 2L_S + 10 \log \epsilon. \quad (8a)$$

For a serial decoder, the total loss (L_{Sd}) is given by [see Fig. 8(b)]

$$L_{Sd} = (3\epsilon - 1) L_c + \epsilon L_{tap} + L_S + 10 \log \epsilon \quad (8b)$$

where L_{tap} is the tap excess loss. The excess loss L_S in the splitter and the combiner is assumed to be the same. The total encoder and decoder loss for the CDMA case is then

$$L_{SS} = L_{Pe} + L_{Pd} \quad (9a)$$

for a parallel implementation and

$$L_{SS} = L_{Se} + L_{Sd} \quad (9b)$$

for a serial implementation.

Figs. 9 and 10 show the minimum laser pulse energy $P_L\tau$ required versus total number of stations M in a TDMA and a synchronous CDMA fiber optic network, respectively. Two detector sensitivities and the quantum limit for a bit-error rate (BER) of 10^{-9} are considered. The de-

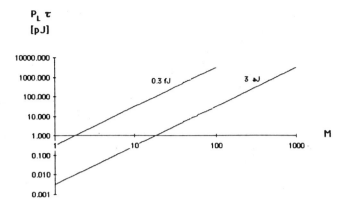

Fig. 9. Minimum laser pulse energy ($P_L\tau$) required versus total number of stations in a TDMA fiber optic network using parallel multiplexer for a p-i-n diode (0.32 fJ), and APD (15.8 aJ), and the quantum limit (3.2 aJ) with 10^{-9} BER. A total device loss of 30 dB is assumed, which allows 8 dB for system margin.

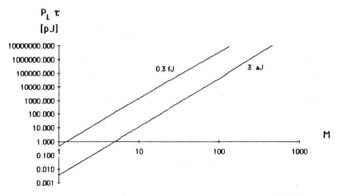

Fig. 10. Minimum laser pulse energy ($P_L\tau$) required versus total number of stations in a synchronous CDMA fiber optic network with parallel encoder and decoder for a p-i-n diode (0.32 fJ), an APD (15.8 aJ), and the quantum limit (3.2 aJ) with 10^{-9} BER.

tector sensitivities are thermal noise limited. The increase in BER due to false crossing of the threshold level by signals in adjacent slots has been neglected. The level of the clock pedestal can be adjusted to eliminate this increase in BER by appropriately selecting the splitting ratio α. The increase in power requirement is marginal.

In Fig. 9, the slope of the plot displays the quadratic relation between P_L and M as expected from (5) for the TDMA case. A parallel multiplexer is chosen since this introduces less loss than a serial multiplexer. The curve labeled 0.32 fJ corresponds to a p-i-n diode and the curve labeled 15.8 aJ to an avalanche photodiode (APD). Also shown is a curve labeled 3.2 aJ corresponding to a receiver operating at the quantum limit for a 1.3 μm wavelength. The connector/splice loss (L_C) is taken to be 0.1 dB (which is the typical loss for currently available splicers), the excess loss due to the star coupler (L_{SC}) is 2 dB, and the excess loss for the optical switch (L_{SW}) and optical splitter (L_S) are 1 dB (which represent loss values for available components). The modulator total loss (L_M) is measured to be 7 dB (this is a conservative estimate based on experimental measurements on commercially available, unpigtailed modulators using end-fire coupling). The multiplexer total loss (L_{SS}) is 6.3 dB. With these values, the total device insertion loss is 30 dB, with an 8 dB power margin allowed.

A commercially available solid-state Nd:YAG laser can be mode locked to produce a train of pulses 140 ps in duration with peak pulse powers of 150 W. The pulse energy is 21 nJ, and the network could accommodate up to 1151 TDMA users at a 6.2 Mbit/s data rate with an APD. In comparison, mode-locked semiconductor laser diodes can output a train of pulses 48 ps in duration with peak pulse powers of 15 mW [8]. The pulse energy is 0.72 pJ and can accommodate up to 7 TDMA users. The data rate for this case is about 2.97 Gbits/s.

Fig. 10 shows the minimum laser energy requirements $P_L\tau$ versus the maximum number of subscribing users M for a synchronous CDMA fiber optic network with parallel encoder/decoder. Prime code sequences are used [7]; hence, $M = \sqrt{N}(N-1)$ and $\epsilon = \sqrt{N}$. This plot does not display a quadratic relation between P_L and M as in the TDMA case. From (5) and (9a), and using the above dependence of M and ϵ on N, the slope of the plot can be verified to be 11/3. For a solid-state laser producing a train of pulses with pulse energies of 21 nJ, the synchronous CDMA system can accommodate up to 55 users. An extra loss of about $10 \log (\epsilon N)$ dB is introduced by the fiber optic combiners at the encoder and decoder. This loss could be reduced if lenses are used instead to combine the beams from the various delay paths at the encoder and decoder. The number of users will increase accordingly.

V. Feasibility Analysis

In this section, the appropriateness of the previous assumptions are discussed. Different configurations for the detection process are presented. Fiber length tolerances are specified. Finally, ways to avoid errors arising from the imprecise gating of the electrooptic modulator by a low bandwidth signal are given.

In the preceding section, equal output power distribution from the optical splitters and star couplers was assumed. Equal power distribution is important because the slot synchronizing process (TDMA) and the code sequence recognition process (CDMA) employ threshold detection. If the power distribution is unequal, false detection could result when interfering transmissions are strong enough to cross the threshold level. Although most commercially available optical splitters and passive star couplers have unequal output power distribution, recently reported was a 100-node star coupler with low excess loss and approximately equal output power distribution [9].

For the network configurations given above, $1 \times M$ optical splitters and $M \times M$ star couplers are required. Such devices are available for M up to 100. To implement a network with a larger number of users ($M > 100$), these devices could be implemented by cascading existing units of smaller size. One way of cascading these units with low loss is to interconnect $\sqrt{M} \times \sqrt{M}$ sized units in two columns of $\sqrt{M}$ units each.

Nonlinear optical effects become nonnegligible in a single-mode fiber when high power optical clock sources (above 100 mW) are used [14]. To avoid these effects, the clock distribution can be done by coupling the output of the optical clock source into a bundle of M single-mode fibers. Thus, the power per fiber is reduced by a factor of M. Coupling efficiency can be maximized by reducing the fiber cladding thickness to just a few micrometers.

Since the coherence time of the optical clock source is of the order of τ, random interference could occur when pulses are added at the optical slot demultiplexer or decoder. To avoid this, separate optical clocks of different wavelengths might be used for the transmitter and the receiver. These clocks must be driven by the same electrical synchronization signal.

As explained in Section II, the narrow peaks must be threshold detected and converted into an NRZ signal of appropriate rate. This process can be done either electronically or optically. With electronic processing, a photodetector first converts the optical signal into an electrical signal. Photodetectors are reported to have response time of less than 100 ps [10]. In the future, electronic threshold detection and data regeneration could perhaps be done up to 10 Gbits/s. In this case, however, the number of slots (TDMA) or chips (CDMA) will be limited, as will the maximum number of subscribers.

The optical threshold detection process employing an optical bistable device [11] is shown in Fig. 11. Optical bistable devices are not yet commercially available [11]. Candidates for optical threshold detectors are switches activated by photorefractive materials [12]. Switching is accomplished when the photorefractive medium has the appropriate index of refraction yielding resonance in the Fabry–Perot cavity. Each medium requires a certain minimum optical energy to change its index of refraction. For example, some organic materials have a nonlinear index of refraction of 10^{-6} [MW/cm^2]$^{-1}$ [13]. The activation time for these materials is in the subpicosecond regime. When very narrow pulses are utilized in the synchronous systems described above, the available optical energy at the receiver might be insufficient to activate such a material. With the use of optical amplifiers just before the threshold detector, the intensity could be increased to the level required to activate the switch [14]. Reference [15] examines the feasibility of using optical amplifiers.

In order to maintain network synchronism, the distance of the stations from the mode-locked laser must be an integral multiple of TDMA frame lengths (see Section II). The permissible fiber length tolerance is $\pm 0.2\delta\tau$ m, where δ is the allowed percent of slot overlap and τ is the pulse width in nanoseconds. A propagation velocity of 0.2 m/ns is assumed (this corresponds to the index of refraction of commercially available fibers operating in the near-infrared region).

At the slot demultiplexer, the delayed clock reference and the received signal must also be synchronized. The permissible tolerance in fiber delay lengths in this case is $0.2jT \pm 0.2\delta\tau$, where j is any integer. For picosecond

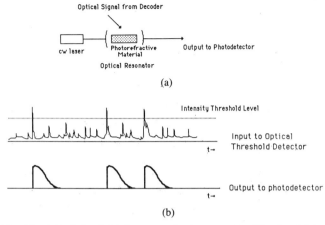

Fig. 11. Optical threshold detector and pulse generator. The signal intensity induces a change in the refractive index of the photorefractive material. When resonance conditions are achieved in the Fabry–Perot resonator, i.e., when the impinging intensity is enough to produce the appropriate refractive index change, an optical pulse of width T will be emitted.

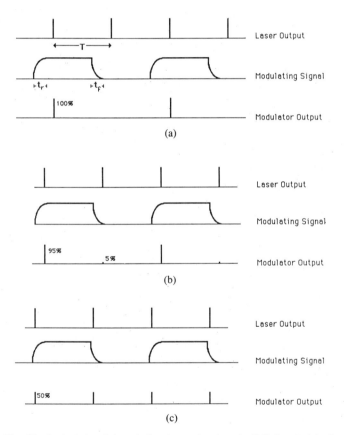

Fig. 12. Optical signal degradation due to the slow rise/fall time (t_r/t_f) of the electrical signal actuating the electrooptic modulator. (a) Proper signal gating, 100 percent pulse energy available. (b), (c) Signal gating during the transition periods. Pulse energy available for transmission is reduced.

pulses, fiber length tolerances would have to be kept within hundreds of microns. Such precision in fiber lengths would not be practical. More accurate tolerances can be achieved by using integrated waveguides with adjustable optical path length. Using the electrooptic effect, the optical path length is controlled by an applied electric field.

A final point to be addressed is the timing between optical pulses and electrical gating signal at the electrooptic modulator. As shown in Fig. 12, if the optical pulses are gated during the transition times of the electrical signal, a reduction in pulse amplitude will result. There are several ways to obviate this effect. The RF signal that mode locks the laser could also be used to synchronize the electrical signal. This would require every station to have a coaxial line carrying the RF signal from the mode-locked laser. Another solution is to send a training sequence and monitor the output of the modulator. If the average power detected is below a specified threshold, then the electrical signal will be delayed by twice the rise/fall time of the electronics. This assures that the optical pulses are not gated during the rise or fall time of the electrooptic modulator, but during the steady-state regime.

VI. Conclusion

The advantages of optical fibers over coaxial cable justify their use in networks, especially in highly noisy environments, in crowded buildings, or where high bandwidth is required. Fiber optics in combination with high-speed optical signal processing permits the implementation of some medium access protocols which would otherwise be either too costly or highly inefficient. Fiber optic delay line signal processors are easy to implement and perform sampling and multiplexing at very high rates (>100 Gbits/s).

Network synchronization can be achieved using a central mode-locked laser as a clock for the whole network. Mode locking produces very high peak power pulses, and is therefore more suitable than switching a CW laser. Mode locking of a solid-state laser can produce picosecond pulses with several nanojoules of energy. Using a central mode-locked laser would not pose a network reliability problem if a redundant laser could be provided. An optical power sensitive switch could activate the standby laser when the output power of the primary laser goes below the minimum required level for acceptable network performance.

Threshold detection requires optical gates which are not yet commercially available. Most optical gates require high optical energy to be activated, although optical materials are being investigated which will operate at high signal rates without requiring large optical power [16]. If an optical gate is not used, the system requires a fast receiver and electrical threshold detector.

As can be gathered from the two schemes presented, similarities exist in network configuration and device requirements. For high throughput requirement such as in voice applications, the TDMA scheme is most efficient. On the other hand, in the case of data transfer where traffic tends to be bursty rather than continuous, CDMA can be used for contention-free, zero delay access. This suggests the possibility of integrating voice and data on the same network. Currently, at the Columbia University Lightwave Communications Research Laboratory, an accessing scheme incorporating the above idea is under consideration, the results of which will be reported later.

Acknowledgment

The authors wish to thank Dr. P. E. Green, Jr., Dr. P. S. Henry, and Dr. I. P. Kaminow for helpful suggestions.

References

[1] K. P. Jackson et al., "Optical fiber delay-line signal processing," *IEEE Trans. Microwave Theory Tech.*, vol. MTT-33, pp. 193–209, Mar. 1985.
[2] P. R. Prucnal and M. A. Santoro, "Spread spectrum fiber optic local area network using optical processing," in *Digital Communications*, E. Biglieri and G. Prati, ed. Amsterdam: Elsevier, 1986.
[3] P. W. Davies et al., *Computer Networks and Protocols*. New York: Wiley, 1981, ch. 5 and references therein.
[4] R. V. Schmidt, "Fibernet II: A fiber optic Ethernet," *IEEE J. Select. Areas Commun.*, vol. SAC-1, pp. 701–711, Nov. 1983.
[5] M. Thewalt, "Time domain multiplexing of signals on an optical fiber using mode-locked laser pulses," *IBM Tech. Disc. Bull.*, vol. 24, pp. 2473–2476, Oct. 1981.
[6] H. Goto et al., "Optical time-division digital switching: An experiment," presented at the OSA/IEEE 6th Topical Meet. Opt. Fiber Commun., New Orleans, LA, Feb. 28–Mar. 1, 1983.
[7] P. R. Prucnal, M. A. Santoro, and T. R. Fan, "Spread spectrum fiber optic local area network using optical processing," *J. Lightwave Technol.*, vol. LT-4, pp. 547–554, May 1986.
[8] R. S. Tucker, G. Eisenstein, and I. P. Kaminow, "10 GHz active mode-locking of a 1.3 nm ridge-waveguide laser in an optical-fibre cavity," *Electron. Lett.*, vol. 19, pp. 552–553, July 7, 1983.
[9] S. Oshima et al., "Small loss-deviation tapered fiber star coupler for LAN," *J. Lightwave Technol.*, vol. LT-3, pp. 556–558, June 1985.
[10] T. P. Lee et al., "Very high speed back illuminated InGaAsP/InP PIN punch through photodiodes," *Electron. Lett.*, vol. 17, pp. 431–432, 1981.
[11] H. A. Elion and V. N. Morozov, *Optoelectronic Switching Systems in Telecommunications and Computers*. New York: Marcel Dekker, 1984.
[12] P. W. Smith, "On the physical limits of digital optical switching and logic elements," *Bell Lab. Tech. J.*, vol. 61, pp. 1975–1993, Oct. 1982.
[13] H. M. Gibbs, *Optical Bistability: Controlling Light with Light*. New York: Academic, 1985.
[14] L. B. Jeunhomme, *Single-Mode Fiber Optics Principles and Applications*. New York: Marcel Dekker, 1983.
[15] D. Fye, "Practical limitations on optical amplifier performance," *J. Lightwave Technol.*, vol. LT-2, pp. 403–406, Aug. 1984.
[16] G. M. Carter, Y. J. Chen, and S. K. Tripathy, "Intensity dependent index of refraction in organic materials," *Opt. Eng.*, vol. 24, pp. 609–612, July-Aug. 1985.

Spread Spectrum Fiber-Optic Local Area Network Using Optical Processing

PAUL R. PRUCNAL, MARIO A. SANTORO, STUDENT MEMBER, IEEE, AND TING RUI FAN

Abstract—Spread spectrum code division multiple access (CDMA) allows asynchronous multiple access to a local area network (LAN) with no waiting. The additional bandwidth required by spread spectrum can be accommodated by using a fiber-optic channel and incoherent optical signal processing. New CDMA sequences are designed specifically for optical processing. It is shown that increasing the number of chips per bit, by using optical processing, allows an increase in capacity of a CDMA LAN. An experiment is performed demonstrating the performance of an optical CDMA LAN, operating at 100 Mbd with three users.

I. INTRODUCTION

SPREAD spectrum multiplexing techniques, which have been investigated extensively in the context of satellite and mobile radio communications [1], [2], offer several potential advantages in local area networks (LAN's) as well.

First, spread spectrum makes efficient use of the channel by providing asynchronous access to each user. Since the traffic in LAN's is typically bursty, asynchronous multiplexing schemes, which allow multiple users to share the entire channel, are more suitable than synchronous multiplexing schemes (e.g., TDMA, FDMA, WDM), which dedicate a portion of the channel to each bursty user. For spread spectrum multiplexing, system performance depends on the number of simultaneous users, which may be much smaller than the number of users actually subscribing to the network. Other benefits of asynchronous access are that no scheduling is required and that new users can be easily added to the network [3].

Second, spread spectrum permits multiple users to simultaneously access the channel with no waiting time, in contrast to other asynchronous techniques (e.g., carrier sense multiple access with collision detection, or CSMA/CD), where each user must wait for the channel to become idle before gaining access. With CSMA/CD, for example, as the traffic intensity increases, so do the incidence of collisions between packets and the accumulated delay. The waiting time with CSMA/CD increases with transmission distance, which is not the case with spread spectrum.

Manuscript received August 5, 1985; revised November 13, 1985. This work was supported by NASA under contract number NAS5-29138 and by NETEK, Inc.
P. R. Prucnal and M. A. Santoro are with the Department of Electrical Engineering, Center for Telecommunications Research, Columbia University, New York, NY 10027.
T. R. Fan is with the Peking Institute of Post and Telecommunications, Peking, China.
IEEE Log Number 8607477.

However, spread spectrum (more properly referred to here as code division multiple access, or CDMA) is based upon the assignment of orthogonal codes to the address of each user, which substantially increases the bandwidth occupied by the transmitted signal. A CDMA LAN therefore requires a wide bandwidth channel such as an optical fiber; coaxial cable and twisted pairs are not suitable transmission media. In addition, CDMA requires wide bandwidth signal processing at the receiver. Conventional CDMA receivers [4] perform signal processing with electronic logic or with surface acoustic wave (SAW) devices at frequencies up to several hundred megahertz. To fully realize the potential of CDMA, higher bandwidth signal processing is required. New signal processing technologies operating at frequencies above 1 GHz include magnetostatic wave (MSW) devices and superconducting delay line (SDL) filters [5]. Optical fiber delay-line signal processors using single-mode fiber have bandwidth-distance products in excess of 100 GHz · km and losses less than 0.5 dB/km [5].

Therefore, with the purpose of exploiting the wide bandwidth offered by optical fiber channels and optical signal processing, a spread spectrum CDMA LAN is proposed and demonstrated which employs optical fiber delay-line signal procesing. The performance of new CDMA sequences designed specifically for incoherent optical correlation is analyzed and compared to conventional CDMA. Measurements of the performance of an experimental optical CDMA system operating at 100 MBd is reported. Finally, a promising topology is proposed for an all-optical CDMA LAN, involving a mode-locked laser, optical modulator, fiber-optic delay-line encoder, fiber-optic channel, fiber-optic delay-line decoder, and monostable or bistable optical switch used as a threshold detector.

II. CONVENTIONAL CDMA SEQUENCES

Each bit in a CDMA system is encoded into a waveform $s(t)$ that corresponds to a code sequence of N chips representing the destination address of that bit. Each receiver correlates its own address $f(t)$ with the received signal $s(t)$. The receiver output $r(t)$ is

$$r(t) = \int_{-\infty}^{+\infty} s(z) f(z - t) \, dz. \tag{1}$$

If the signal has arrived at the correct destination, then

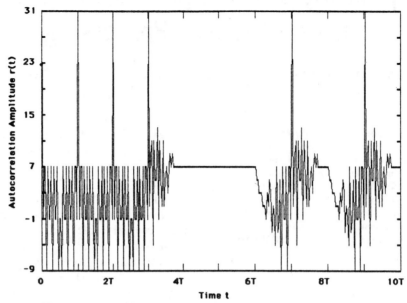

Fig. 1. Autocorrelation function for a Gold code sequence of length $N = 31$ using conventional processing. The data sequence is 1111000101.

$s(t) = f(t)$, and (1) represents an autocorrelation function. If the signal has arrived at an incorrect destination, then $s(t) \neq f(t)$, and (1) represents a cross-correlation function. At each receiver, to maximize the discrimination between the correct (destination) signal and interference (all other signals), it is necessary to maximize the autocorrelation function and minimize the cross-correlation function. This is accomplished by selecting a set of orthogonal code sequences.

Sets of orthogonal code sequences, such as maximal length codes and Gold codes, have received considerable attention in spread spectrum communications [6]. Maximal length sequences are shift register sequences of length (or period) $N = 2^n - 1$, which is the longest sequence length an n-stage binary linear feedback shift register can generate. An important characteristic of a maximal length code is its two-valued discrete autocorrelation function given by [7]:

$$r(t) = \begin{cases} N, & \text{for } t = 0 \\ -1, & \text{for } t = i\frac{T}{N}, \quad i = \pm 1; \pm 2; \cdots \end{cases} \quad (2)$$

where T is the data bit width. This characteristic makes a maximal length sequence easy to distinguish from time shifted versions of itself, which is ideal for radar or ranging applications. However, the peak of the cross-correlation function is not less than $-1 + 2^{(n+1)/2}$. This implies that certain maximal length sequences will interfere very strongly with one another. Those code sequences exhibiting the minimum possible cross-correlation peak (equal to $-1 + 2^{(n+1)/2}$) are referred to as preferred maximal length sequences. Since the set of preferred maximal length sequences is few in number [7], maximal length codes are not suitable for CDMA, where a large number of assignable addresses is required.

Gold codes are attractive for code division multiplexing because of the large number of orthogonal sequences available. Gold code sequences are generated by combining a pair of preferred maximal length sequences using modulo-2 addition [7]. A Gold code is then a set of $N + 2$ code sequences given by $\{a, b, a + b, a + \tau b, \cdots, a + \tau^{N-1} b\}$, where $\{a, b\}$ is the preferred maximal length sequence pair, and τ^i denotes an operator that shifts a sequence cyclically i chip intervals to the left. The autocorrelation function of a Gold code exhibits side lobes which depend on the pair of code sequences considered. The cross-correlation function, on the other hand, is bounded. The maximum value of the peak of the cross-correlation is equal to that of the generating preferred maximal length sequence pair [7].

An example of the autocorrelation function for a Gold code is shown in Fig. 1. The transmitted data is 1111000101. Each data bit "1" of duration T s is encoded into a waveform $s(t)$ consisting of a Gold code sequence of $N = 31$ chips of amplitude ± 1 and duration T/N s. Each data bit "0" is encoded into 31 chips of amplitude -1. In Fig. 1, the amplitude of $r(t)$ is plotted versus t in discrete chip intervals T/N. As expected the peak $= 31$.

When K users are transmitting simultaneously, the total interference at a given receiver is the superposition of $K - 1$ different cross-correlation functions. If the $K - 1$ interferors are uncorrelated, then the variance of the total interference is equal to the sum of the variances of the $K - 1$ cross-correlation functions, which are assumed to be identical. The signal-to-noise ratio (SNR) is represented as the ratio of the square of the peak of the autocorrelation function to the variance of the amplitude of the interference. Assuming synchronization between transmitter and receiver, and considering other users' interference to be the dominant source of noise in the system, the SNR for Gold codes has been shown to be [8]

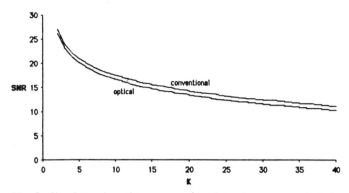

Fig. 2. Signal-to-noise ratio versus number of simultaneous users both for conventional processing using Gold code sequences of length $N = 127$ (upper curve) and for optical processing using prime code sequences of length $N = 121$ (lower curve).

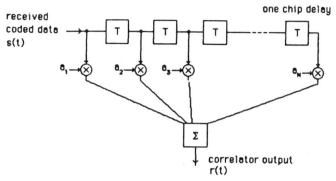

Fig. 3. Block diagram of the correlation process for either conventional processing or optical processing. The coefficients a_i take the values $+1$ or -1 for conventional processing and the values 1 or 0 for optical processing.

$$\text{SNR}_{\text{Conventional}} = 4 \left[\frac{N^3}{(K-1)(N^2 + N - 1)} \right]. \quad (3)$$

Unfortunately, the variance of the amplitude of the cross-correlation function (the denominator of the SNR) increases with both the number of users K and the number of chips N. The SNR for Gold codes is plotted versus K in Fig. 2 (labeled "conventional"), for $N = 127$. It is clear that for a given number of chips per bit, the SNR decreases gradually as the number of users increases. A decrease in SNR implies an increase in the probability of error, i.e., the more the users accessing the network at a given time, the poorer the performance of the system. It is apparent then that to increase the number of simultaneous users, the number of chips per bit must be increased.

A conventional CDMA system using Gold codes, electronic processing, and transmission over an optical channel was recently reported [8]. Although that system optically multiplexed the signals from the various users, it was nevertheless limited by the processing speed of the electronics.

The CDMA system discussed below uses optical processing and new CDMA sequences designed specifically for optical processing. The new optical CDMA system therefore has the potential for increased number of chips per bit, and increased capacity over systems using electronic processing.

III. Optical CDMA Sequence Design

It is apparent that increasing the processing speed, by using optical processing, allows an increase in N and therefore an increase in capacity of a CDMA LAN. However, a fundamental difference exists between optical processing and conventional processing, using, for example, tapped delay lines [5]. Conventional delay lines coherently combine tapped signals. Optical fiber delay lines, on the other hand, incoherently combine tapped signals, resulting simply in the summation of optical power. The resulting difference between coherent and incoherent combining of signals is illustrated in Fig. 3. The taps on the delay line a_j equal the code sequence that corresponds to the appropriate destination address (the waveform $f(t)$ in (1)). The taps equal $+1$ or -1 for conventional processing, and 1 or 0 for incoherent optical processing. Though coherent optical processing is possible in principle, it is not practical at the present time, due to the high frequency of the optical carrier.

Using incoherent optical processing, the levels of the transmitted optical code sequence correspond to light "ON" or light "OFF". The taps on the optical correlator shown in Fig. 3 are set to 0 at positions corresponding to light "OFF" in the correct code sequence, and set to 1 at positions corresponding to light "ON" in the correct code sequence. For this correlator, the peak of the autocorrelation function equals the number of 1's in the code sequence. The peak of the cross-correlation function equals the maximum number of coincidences of 1's in all shifted versions of the two code sequences. For Gold codes, the number of 1's in each code sequence varies and, therefore, so does the peak of the autocorrelation function. The number of coincidences of 1's between shifted versions of two code sequences can be large and, therefore, so can the peak of the cross-correlation function. Shown in Fig. 4(a) is the autocorrelation function of a Gold code sequence with $N = 31$ and twenty 1's in the code sequence, for the data 1111000101. The peak is equal to 20 as expected. Shown in Fig. 4(b) is the cross-correlation of the code sequence in Fig. 4(a) with a different code sequence having twelve 1's. The peak, which is equal to ten, is unacceptably large for one interfering user. Gold codes are therefore unacceptable for the type of incoherent optical processing considered.

For the above reason, a new set of code sequences with fewer coincidences of 1's is needed for optical processing. A promising candidate is the set of prime codes of length $N = P^2$ which are derived from prime sequences of length P obtained from a Galois field $GF(P)$, where P is a prime number [9]. The codes are generated as follows: starting with $GF(P) = \{0, 1, \cdots, j, \cdots, P-1\}$, a prime sequence $S_x = (s_{x0}, s_{x1}, \cdots, s_{xj}, \cdots, s_{x(P-1)})$ is constructed by multiplying every element j of $GF(P)$ by an element x of $GF(P)$ modulo P. P distinct prime sequences can thus be obtained. An example for P

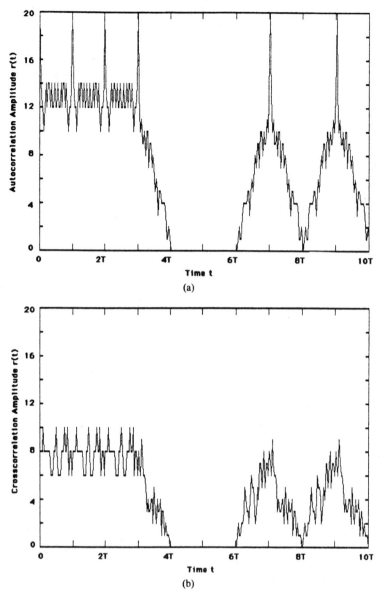

Fig. 4. Optical processing performed on a Gold code sequence of length $N = 31$. The data sequence is 1111000101. a) Autocorrelation function. b) Cross-correlation function.

TABLE I
PRIME SEQUENCES $S_x = (s_{x0}, s_{x1}, \cdots, s_{xj}, \cdots, s_{x(P-1)})$
Each element s_{xj} in the table is obtained by multiplying the corresponding x and j modulo 5.

x \ j	0	1	2	3	4
0	0	0	0	0	0
1	0	1	2	3	4
2	0	2	4	1	3
3	0	3	1	4	2
4	0	4	3	2	1

$= 5$ is given in Table I. Each element s_{xj} in the Table is the product of the corresponding x and j modulo 5. The prime sequences are then mapped into a binary code sequence $C_x = (c_{x0}, c_{x1}, \cdots, c_{xi}, \cdots, c_{x(N-1)})$, by assigning ones in positions $i = s_{xj} + jP$ for $j = 0, 1, \cdots, P - 1$ and zeros in all the other positions, i.e.,:

$$c_{xi} = \begin{cases} 1, & \text{for } i = s_{xj} + jP, \quad j = 0, 1, \cdots, P - 1 \\ 0, & \text{otherwise}. \end{cases}$$

(4)

For example, prime sequence S_3 of Table I would be mapped into the code sequence $C_3 =$ (1000000010010000000100100).

For prime codes, the number of coincidences of 1's for all shifted versions of any two code sequences is only one or two. The peak of the cross-correlation function is therefore at most two, which is much less than the peak for optically processed Gold codes. However, the number of 1's per code sequence and the autocorrelation peak both equal P, which is much lower than for Gold codes. Shown in Fig. 5(a) is the autocorrelation function of code sequence C_3 for the data 0111001010. (Note that 7 zeros

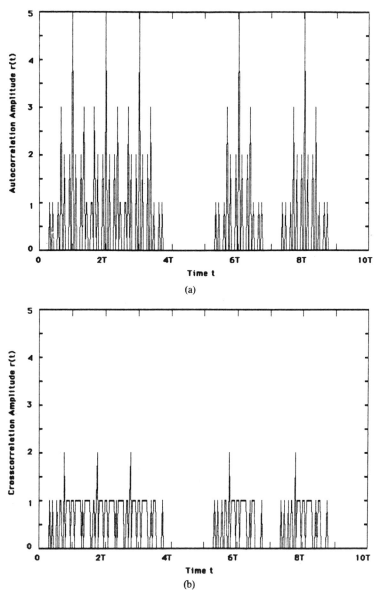

Fig. 5. Optical processing performed on a prime code sequence with $P = 5$. $N = 32$ (7 padding zeros are used). The data sequence is 0111001010.
a) Autocorrelation function. b) Cross-correlation function.

are added to the code sequence to make $N = 32$, which is a more convenient sequence length to use in the experimental setup described later.) The peak is equal to five as expected. Shown in Fig. 5(b) is the cross-correlation of the code sequence C_3 with code sequence C_2. The peak is equal to two. Computation of the average variance of the cross-correlation amplitude was carried out using all possible code sequences and several values of P. The value of the variance obtained was approximately 0.29. This value is independent of N since the number of coincidences of 1's is independent of P. Thus, these prime codes appear to be quite suitable for optical processing.

The SNR for the prime codes corresponds to the ratio of the square of the peak amplitude of the autocorrelation function $N = P^2$ to the variance of the amplitude of the cross-correlation for $K - 1$ uncorrelated interferors

$$\text{SNR}_{\text{Optical}} \approx \frac{1}{0.29}\left[\frac{N}{(K-1)}\right]. \quad (5)$$

The SNR for prime codes with $N = 121$ is plotted versus K in Fig. 2 (labeled "optical"). The SNR's for both prime codes using optical processing, and Gold codes using conventional processing, are comparable. However, the real advantage of the optical processing is that many more chips per bit can be used than in conventional systems. The SNR using optical processing increases in proportion to the increase in the number of chips per bit, which translates into more supportable users for a fiber-optic system.

The theoretical probabilities of error for prime codes with optical processing ($N = 25, 121, 529,$ and 961) and Gold codes with conventional processing ($N = 31, 127, 511,$ and 1023) over a fiber-optic channel are shown ver-

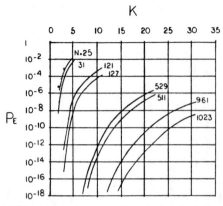

Fig. 6. Probability of error versus number of users both for optical (N = 25, 121, 529, and 961) and conventional (N = 31, 127, 511, and 1023) signal correlation. The dots correspond to the measured error rate for the optical processing experiment.

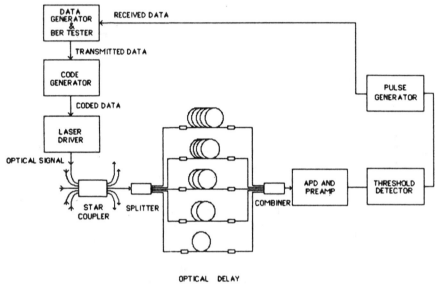

Fig. 7. Block diagram of the experimental setup.

sus K in Fig. 6. Equivalent length sequences are compared, keeping in mind that, for prime codes, N is the square of a prime number and, for Gold codes, N is a power of two, minus 1. The probability of error is computed from the appropriate error function of the SNR. This approximation is valid for large values of K, where, by the central limit theorem, the interference component approaches a Gaussian distribution. For small values of K, the approximation is not necessarily accurate, but is used here to permit a comparison to the experimental data later. As an example, consider conventional processing with N = 31, and a comparable optical system accommodating about 30 times as many chips per bit, say N = 961. For a probability of error of 10^{-8}, conventional CDMA accommodates 2 users, whereas optical CDMA accommodates 26 users. Thus, even for a conservative estimate of the increase in processing speed due to optical processing, the number of users accommodated by the optical system is substantially larger than for the conventional system. Note that the total number of subscribers can be much larger than the number of simultaneous users.

IV. EXPERIMENT

The experimental set-up shown in Fig. 7 employs a P = 5 prime code, on a 100-Mbd fiber-optic link, using 32 chips per bit. This number of chips was chosen to conform to the requirements of the code generator used, and was raised from 25 to 32 by padding extra zeros at the end of the code sequence. The additional zeros added do not influence the theoretical calculations performed, and, in fact, may help to reduce the timing jitter and the power of the interference. The data rate is 3.125 Mbps.

The system is configured as follows. A bit error rate tester generates the data pattern. The data is then encoded by the code generator with the appropriate prime code sequence. The coded data is injected into a multimode fiber using a laser diode. At the receiver, the incoming signal is split into P = 5 signals. The optical correlator then selectively delays the signals before recombining them. This processed signal impinges on an avalanche photodiode. An ac-coupled amplifier removes the dc variations and avoids baseline wander as the number of simultaneous users changes. A threshold detector triggers the

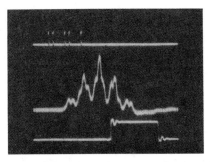

Fig. 8. Autocorrelation function (middle trace) for the prime code sequence $C_3 = (100000001001000000100100)$ ($P = 5$) (upper trace) and recovered signal (lower trace) in the absence of interference. Note that the code sequence is time inverted as required by (1).

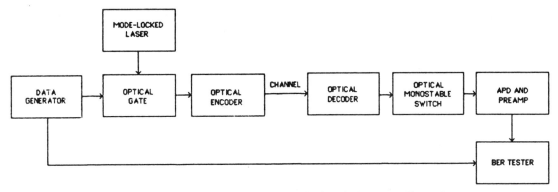

Fig. 9. Block diagram of a system using both optical data encoding and decoding.

pulse generator whenever the autocorrelation peak is present. No synchronization is used in this system. Finally, the reconstructed signal is monitored for errors.

The first experiment is performed in the absence of interfering users, to determine the effects of thermal noise (when high gain is used at the preamp), shot noise (when low gain is used at the preamp), and timing jitter on system performance. Measurements indicate that the bit error rate (BER) in the thermal noise limit is 10^{-9}, and in the shot noise limit is 10^{-11}. The prime code sequence, autocorrelation and recovered signal are shown in Fig. 8.

In the second experiment, the correct signal plus two interfering users are simulated. The data transmitted is a periodic 1010 pattern. For each binary one, the code generator produces the prime code sequence C_3 followed by 7 padding zeros, as discussed above. Interference from two additional users is simulated by superposing various combinations of prime code sequences C_0, C_1, C_2, and C_4 in the code generator output. System performance is monitored for the various interfering combinations, and found to depend on the particular interfering code sequences used. This is consistent with the computer simulations performed, in which the variance of the cross-correlation function varied with the particular interfering code sequences present. The measurements, shown as dots in Fig. 6, represent an average of data for all possible interfering code sequences. As shown in Fig. 6, the data is in good agreement with the theory.

V. Conclusion

Spread spectrum can potentially allow asynchronous multiple access to a LAN with no waiting. The additional bandwidth required by spread spectrum can be accommodated using a fiber-optic channel and optical correlation. Using incoherent optical signal processing and the appropriate optical CDMA code sequence, the number of users can be increased over a conventional CDMA system. This increase in the number of users comes about for two reasons: 1) using the new prime code sequences, the SNR of the optical CDMA system increases in direct proportion to the increase in the number of chips (Gold codes are not suitable for optical CDMA systems because they exhibit a large cross-correlation variance. The SNR is therefore smaller compared to that of an optical CDMA system where prime codes are used); 2) the larger bandwidth of the optical CDMA system allows many more chips per bit compared to a conventional CDMA system.

In the future it will be desirable to take full advantage of the speed of optical processing, by making as much of the "chip processing" as possible optical. A possible system using all optical chip processing is depicted in Fig. 9. A mode-locked laser produces a low duty cycle, high intensity pulse stream at the data rate. This sequence of pulses is modulated by an optical gate, such as a directional coupler switch, which is driven by the information waveform. Using single-mode optical fiber delay lines, each short laser pulse generates the appropriate code sequence. The optical fiber delay lines are configured so that P taps can be selected from any of N positions, according to the address of the desired receiver. At the receiver, correlation is performed by optical fiber delay lines in the way described in this paper. In order to reduce the bandwidth requirements of the detector, the narrow autocorrelation peak is used to trigger a bistable or monostable optical switch, with decay time equal to the bit width. The slowly decaying signal is detected and processed at the rate of the original data. With advances in electrooptics, this all-optical system can provide substantial improvements in the capacity of CDMA systems.

Acknowledgment

The authors wish to thank F. Levinson and T. Lockwood for valuable assistance.

References

[1] G. R. Cooper and R. W. Nettleton, "A spread spectrum technique for high capacity mobile communications," *IEEE Trans. Vehicular Technol.*, vol. 27, pp. 264–275, Nov. 1978.
[2] H. J. Kochevar, "Spread spectrum multiple access communication experiment through a satellite," *IEEE Trans. Commun.*, vol. 27, pp. 853–856, Aug. 1979.
[3] J. Y. N. Hui, "Throughput analysis for code division multiple accessing of spread spectrum channel," *J. Selective Areas Commun.*, vol. 2, pp. 482–486, Sept. 1984.
[4] P. Freret, "Wireless terminal communication using spread spectrum radio," in *Proc. IEEE Compcon '80*, pp. 244–248.
[5] K. P. Jackson *et al.*, "Optical fiber delay-line signal processing," *IEEE Trans. Microwave Theory Tech.*, vol. 33, pp. 193–209, Mar. 1985.
[6] R. C. Dixon, *Spread Spectrum Systems.* New York: Wiley, 1984.
[7] D. V. Sarwate and M. B. Pursley, "Crosscorrelation properties of pseudorandom and related sequences," *Proc. IEEE*, vol. 68, pp. 593–619, May 1980.
[8] S. Tamura, S. Nakano, and K. Akazaki, "Optical code-multiplex transmission by Gold sequences," *J. Lightwave Technol.*, vol. 3, pp. 121–127, Feb. 1985.
[9] A. A. Shaar and P. A. Davies, "Prime sequences: Quasi-optimal sequences for Or channel code division multiplexing," *Electron. Lett.*, vol. 19, pp. 888–889, 1983.

Encoding and decoding of femtosecond pulses

A. M. Weiner, J. P. Heritage, and J. A. Salehi

Bell Communications Research, 331 Newman Springs Road, Red Bank, New Jersey 07701-7020

We demonstrate the spreading of femtosecond optical pulses into picosecond-duration pseudonoise bursts. Spreading is accomplished by encoding pseudorandom binary phase codes onto the optical frequency spectrum. Subsequent decoding of the spectral phases restores the original pulse. We propose that frequency-domain encoding and decoding of coherent ultrashort pulses could form the basis for a rapidly reconfigurable, code-division multiple-access optical telecommunications network.

Diverse communication and signal-processing technologies utilize specially coded signal formats in order to achieve desirable capabilities such as error correction, interference rejection, and secrecy. The ability to encode and decode signals according to a specified format is a crucial component of such systems.

In this Letter we demonstrate frequency-domain encoding and decoding (in the time domain, spreading and despreading) of femtosecond optical pulses. Previously we described the generation of arbitrarily shaped picosecond optical pulses, by spectral amplitude and phase filtering in a fiber and grating pulse compressor.[1-3] Here we utilize a special dispersion-free grating apparatus to manipulate the phase spectra of femtosecond pulses. Specifically, by encoding pseudorandom binary phase codes onto the optical-frequency spectrum, we spread femtosecond pulses into picosecond-duration pseudonoise bursts. Subsequent decoding of the spectral phases restores the original pulse.

We propose that spectral encoding and decoding of coherent ultrashort pulses could form the basis for a rapidly reconfigurable, code-division multiple-access[4] (CDMA) optical telecommunications network. The system would provide tens to hundreds of users with asynchronously multiplexed, random access to a common fiber or free-space channel.

Our experiments utilize 0.62-μm, 75-fsec pulses from a balanced, colliding-pulse mode-locked (CPM) ring dye laser,[5] which are shaped by using the special dispersion-free grating apparatus. This apparatus consists of a pair of 1700-line/mm gratings placed at the focal planes of a unit magnification confocal-lens pair.[6] The grating separation is 60 cm, and the lenses are achromats with focal lengths of 15 cm. Spatially patterned amplitude and phase masks are inserted midway between the lenses at the point where the optical spectral components experience maximal spatial separation. The pulse shape at the output of the grating is the Fourier transform of the pattern transferred by the mask onto the frequency spectrum. The shaped pulses are measured by cross correlation, using femtosecond pulses directly out of the CPM laser as the reference.

We have verified the nondispersive nature of our grating apparatus by performing autocorrelation measurements of pulses incident upon and emerging from the apparatus. Figure 1 shows autocorrelations of 48-fsec pulses with no mask present. The pulse widths are identical. Without the lens pair, the temporal dispersion arising from the 60-cm grating separation would broaden the pulses by more than 3 orders of magnitude. By moving the gratings either closer to or farther from the lens, respectively, either positive or negative dispersion is obtained.[7] The reduction in the wings of the output pulse may be attributed to cancellation of some small chirp on the pulses from the CPM laser.

By using simple amplitude and phase masks, we have generated trains of femtosecond pulses, femtosecond odd pulses, and other shaped femtosecond pulses. These results are an extension of pulse-shaping experiments previously performed on the picosecond time scale and are reported elsewhere.[8] Here the emphasis is on spectral phase coding of femtosecond pulses. The coding work will serve as an example of the high degree of complexity that can be incorporated into the shaped waveforms.

An example of frequency-domain phase coding is

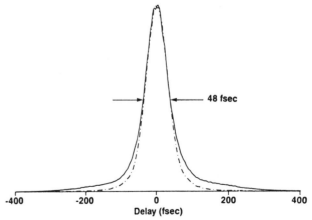

Fig. 1. Autocorrelation traces of 48-fsec pulses from the CPM laser, measured before (solid line) and after (dotted–dashed line) the grating apparatus. With no mask present, the pulse widths are identical.

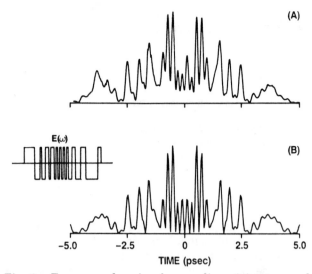

Fig. 2. Frequency-domain phase coding of femtosecond pulses. (A) Cross-correlation measurement of the coded waveform. The inset shows the 44-element phase code; the individual phases are either zero or π. (B) Calculated intensity profile.

shown in Fig. 2. A 44-element pseudorandom binary phase mask, shown in the inset, is used to scramble (encode) the spectral phases. The mask consists of a 2.23-mm clear aperture, corresponding to an optical bandwidth of 5.86 THz (7.52 nm). The spectrum of the incident CPM laser pulses is nearly flat over this bandwidth. The 2.23-mm window is divided into 44 equal pixels, each of which corresponds to a phase shift of zero or π. The masks are fabricated on fused silica by standard microlithography; the π phase shifts are obtained by reactive ion etching to a depth $D = \lambda/2(n - 1) \simeq 0.68\ \mu m$, where n is the refractive index of the fused silica.

The intensity cross-correlation measurement of the encoded waveform is shown in Fig. 2(A). As seen, spectral encoding spreads the incident 75-fsec pulse into a complicated pseudonoise burst within a 10-psec temporal envelope. The peak intensity is reduced to $\simeq 8\%$ compared with an uncoded pulse spectrally windowed[9] to the same bandwidth. For comparison, we show in Fig. 2(B) the theoretical intensity profile, obtained by squaring the Fourier transform of the spectral phase mask. The calculation includes no adjustable parameters. The excellent agreement between theory and experiment underscores the high degree of precision available with our technique.

Autocorrelation measurements of uncoded, coded, and decoded pulses are shown in Fig. 3. Figure 3(A) depicts the autocorrelation of the incident, uncoded pulses together with that of the encoded pulses of Fig. 2. The contrast ratio of $\simeq 25:1$ illustrates the dramatic reduction in intensity that accompanies encoding. In order to reconstitute the original femtosecond pulse, we place a second, phase-conjugate mask adjacent to the first mask. This phase-conjugate mask decodes (or unscrambles) the spectral phases scrambled by the first mask, thus restoring the initial pulse. On the other hand, if the second mask does not match the first, the spectral phases are rearranged but not

unscrambled. In that case the waveform remains a spread, low-intensity pseudonoise burst. Autocorrelations of such successfully and unsuccessfully decoded pulses are shown in Fig. 3(B).

For ultrashort-pulse CDMA (see below), it is desirable to choose codes that spread the incident pulses as widely and as uniformly as possible. One type of code that appears suitable is the so-called maximal length sequence (or M sequence). These binary sequences have been utilized widely in spread-spectrum communications, and their properties are well known.[4] Figure 4 shows the autocorrelation of a pseudonoise burst spread to a duration exceeding 10 psec by spectral phase coding with a 127-element M-sequence code. The autocorrelation contains a coherence spike riding upon a broad pedestal, as expected for a noise burst.[10] The observed contrast ratio is $\approx 1.4:1$, in agreement with calculation and significantly lower than the 2:1 contrast ratio expected when the field is a Gaussian random variable. This observation indicates that this pseudonoise burst spread by spectral M-sequence coding is smoother and more nearly uniform than a Gaussian noise burst, obtained, for example, as the output of an imperfectly mode-locked laser.[10]

We propose that frequency-domain phase coding of coherent ultrashort pulses could form the basis for an

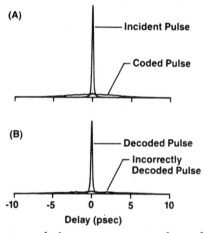

Fig. 3. Autocorrelation measurements of uncoded, coded, and decoded pulses. (A) Uncoded and coded pulses. (B) Successfully and unsuccessfully decoded pulses.

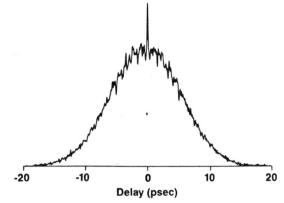

Fig. 4. Autocorrelation of 127-element M-sequence code.

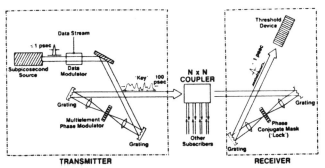

Fig. 5. Proposed ultrashort pulse, CDMA communications network.

ultrahigh-speed, CDMA optical communications network. CDMA is a type of spread-spectrum communication in which multiple pairs of subscribers, each assigned different, minimally interfering code sequences, communicate simultaneously and asynchronously over a common fiber channel.[4] Other CDMA systems have traditionally been based on time-domain encoding, and optical implementations have relied on incoherent processing.[11,12]

The proposed ultrashort-pulse CDMA system could be configured as shown in Fig. 5. N subscriber stations are connected via a passive $N \times N$ directional coupler. Each transmitter station is equipped with a picosecond or femtosecond source and an encoder that contains a multielement phase modulator; each receiver is equipped with a decoder, which contains a similar phase modulator, and an optical threshold device. Each receiving station is assigned a unique lock, or access code, which it imposes on its phase modulator. A given subscriber pair can communicate only if the transmitter encodes by using a key that is phase conjugate to the receiver's lock. In that case the decoder will reassemble the transmitted signal into an intense ultrashort pulse that can be detected by the thresholder. Signals encoded by using the wrong key will not be decoded and will be rejected.

The number of users N that the network could accommodate depends on the length and the properties of the code sequences. The code length in turn depends on the spectral resolution that the optical system can provide. As discussed in Ref. 3, the number of distinct spectral features that can be imposed within a given bandwidth is related to the divergence of the input beam and to the angular dispersion of the grating. In the present work, with a 3-mm input beam, we achieve a resolution sufficient to generate a code sequence of length 127; for a 1-cm beam diameter, the resolution would approach 500. Provided that the code length does not exceed the available spectral resolution, coding and decoding can successfully be achieved.

We have performed bit-error-rate (BER) calculations for the proposed femtosecond CDMA network, in which we model the codes as binary random sequences. The calculations assume an ideal thresholder and account for interference that is due to the other users but do not include power-budget considerations. Our analysis will be described in detail elsewhere. As a particular example, we estimate that a free-space network utilizing 128-element random codes could support simultaneous communication by 30 subscriber pairs at a BER of 10^{-9}, assuming 1-Gbit/sec individual data rates and 80-fsec input pulses. Three hundred subscriber pairs could be supported at a BER of 10^{-5}. For a fixed BER, the number of users who could be accommodated increases with increasing code length and with decreasing individual bit rates and shorter input pulses.

In summary, we have demonstrated encoding and decoding of femtosecond pulses and have suggested that this technology could be utilized for a high-capacity, optical CDMA communications network. Real-time encoding and decoding could be achieved by replacing the prefabricated masks used in the current work with multielement modulators. Ultrashort-pulse CDMA could then serve as a novel architecture for a rapidly reconfigurable optical crossbar switch, in which any transmitter station could connect to any receiver station. Our results should contribute to a new class of communications techniques based on frequency-domain manipulation of ultrashort light pulses.

We gratefully acknowledge fabrication of the masks by E. M. Kirschner, helpful discussions with O. E. Martinez and C. A. Brackett, and the technical assistance of D. E. Leaird.

References

1. J. P. Heritage, A. M. Weiner, and R. N. Thurston, Opt. Lett. **10**, 609 (1985).
2. A. M. Weiner, J. P. Heritage, and R. N. Thurston, Opt. Lett. **11**, 153 (1986).
3. R. N. Thurston, J. P. Heritage, A. M. Weiner, and W. J. Tomlinson, IEEE J. Quantum Electron. **QE-22**, 682 (1986).
4. R. Skaug and J. F. Hjelmstad, *Spread Spectrum in Communications* (Peregrinus, London, 1985).
5. J. A. Valdmanis, R. L. Fork, and J. P. Gordon, Opt. Lett. **10**, 131 (1985).
6. C. Froehly, B. Colombeau, and M. Vampouille, in *Progress in Optics*, E. Wolf, ed. (North-Holland, Amsterdam, 1983), pp. 115–121.
7. O. E. Martinez, IEEE J. Quantum Electron. **QE-23**, 59 (1987).
8. A. M. Weiner and J. P. Heritage, Rev. Phys. Appl. (December 1987), p. 1619.
9. J. P. Heritage, R. N. Thurston, W. J. Tomlinson, A. M. Weiner, and R. H. Stolen, Appl. Phys. Lett. **47**, 87 (1985).
10. E. P. Ippen and C. V. Shank, in *Ultrashort Light Pulses*, S. L. Shapiro, ed. (Springer-Verlag, Berlin, 1977), pp. 85–88.
11. P. R. Prucnal, M. A. Santoro, and T. R. Fan, IEEE J. Lightwave Technol. **LT-4**, 547 (1986).
12. J. A. Salehi and C. A. Brackett, "Fundamental principles of fiber optics code-division multiple-access," presented at the IEEE International Conference on Communications, Seattle, Wash., June 1987.

WDMA and FDMA

DEMONSTRATION OF HIGH CAPACITY IN THE LAMBDANET ARCHITECTURE: A MULTIWAVELENGTH OPTICAL NETWORK

H. KOBRINSKI
R. M. BULLEY
M. S. GOODMAN
M. P. VECCHI
C. A. BRACKETT

Bell Communications Research
Morristown, NJ 07960, USA

15th June 1987

L. CURTIS
J. L. GIMLETT

Bell Communications Research
Red Bank, NJ 07701, USA

Indexing terms: Optical communications, Networks, Video, Optical connector and couplers

The letter presents transmission measurements in a multi-wavelength optical network, using 18 channels spaced 2 nm around 1·55 μm. Each channel was modulated at 1·5 Gbit/s and transmitted through 57·8 km of single-mode optical fibre. Experimental results demonstrate a bandwidth-distance product of 1·56 Tbit s^{-1} km and a point-to-multipoint figure-of-merit of 21·5 Tbit s^{-1} km node.

Introduction: The LAMBDANET™ architecture[1] is a multi-wavelength optical star network intended for the distribution of broadband services. This network is fully connected, internally nonblocking and allows the integration of point-to-point and point-to-multipoint broadband services. The broadcasting inherent in the LAMBDANET design makes it especially suitable for video distribution applications.

The LAMBDANET network (Fig. 1) is composed of nodes (for example, telephone company central offices) which are linked together to form a star topology. Two salient features of the design are that each node on the star transmits its information on a unique wavelength, and broadcast routing among N nodes is achieved using a passive $N \times N$ transmissive star coupler at the hub of the network. Nodes use single-frequency laser diodes to transmit their information over single-mode optical fibres. Each node may receive information from all the other nodes, independently and simultaneously, using a wavelength demultiplexer and N receivers.

The LAMBDANET design has a number of useful features.[1,2] Since each node transmits on its unique wavelength, it effectively has a contention-free 'private line' available to each of the other nodes on the network. Information in analogue and digital form, and with different data formats and different data rates, may be simultaneously transported. The LAMBDANET approach, while providing transparency to carried traffic, full connectivity and passive routing among the nodes, places distributed processing requirements at the periphery of the star.

This letter presents results from initial transmission experiments intended to establish the feasibility and performance of optical components in the LAMBDANET design. Previous high-capacity WDM transmission experiments[3] were performed for point-to-point communications under optical bench conditions. The LAMBDANET experimental network, however, incorporates elements constructed from commercially available components and is a fully connectorised modular system.

Experimental set-up: The set-up (Fig. 2) closely resembles a video distribution application of the LAMBDANET system,[2] in which all the transmitters are located together at the hub (e.g. a CATV head-end) and the output is broadcast to remote nodes (e.g. local video switches). The hub of the experimental system is a 16 × 16 single-mode transmissive star coupler (Gould Inc.) composed of wavelength-insensitive directional couplers. Star coupler insertion losses range between 11·9 dB and 18·0 dB (14·2 dB average: 12 dB theoretical splitting loss + 2·2 dB average excess loss) at 1·5 μm. The experiment advantageously used this nonuniformity of coupler loss to increase the number of input channels from 16 to 18, by adding 2 × 2 directional couplers at two of the low-loss star coupler inputs.

For this experiment, distributed feedback (DFB) lasers (Hitachi Ltd.) were selected according to our specifications with centre wavelengths spaced 2 nm apart over the region from 1527 to 1561 nm. The lasers had threshold currents < 35 mA, and exhibited > 30 dB side-mode rejection ratio. We incorporated these lasers into modulator transmitter units containing electronic biasing, modulation, temperature control and monitoring circuitry. Precise tuning of the laser was accomplished through temperature control which stabilised the laser package to within ±0·01°C (for a 10°C shift in the ambient temperature), corresponding to wavelength variations of ⩽ 0·1 nm. These DFB lasers had 'pigtailed' fibres and

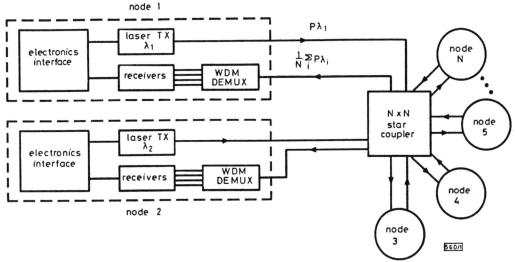

Fig. 1 *Block diagram of a LAMBDANET star*

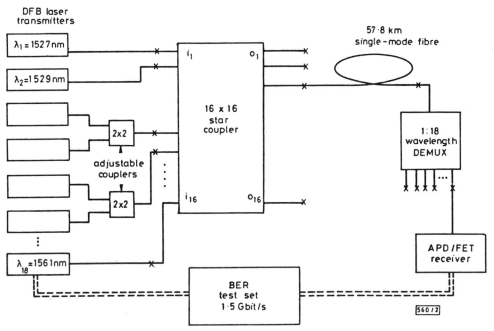

Fig. 2 *Set-up for LAMBDANET transmission experiment*

were operated without isolators. The maximum launch power ranged from -3.2 dBm to $+3.3$ dBm.

The grating wavelength demultiplexer (Instruments SA) had one single-mode fibre input and 18 multimode fibre outputs with centre wavelengths spaced $\simeq 2$ nm apart over the 1527–1561 nm range. Wavelength demultiplexer insertion losses were <3.5 dB and the demultiplexer 3 dB bandwidth of the channels was <0.85 nm.

The high-impedance receiver with a GaAs FET front-end was constructed using a commercial InGaAs APD (Fujitsu Inc.) with a multimode fibre pigtail. The receiver sensitivity for 10^{-9} BER at 1.5 Gbit/s was better than -34.5 dBm.

Results: The bandwidth-distance product was maximised by a procedure in which the laser transmitters (based on launch power minus receiver sensitivity through a long fibre) were matched with star coupler inputs (based on insertion losses at a selected output). Fig. 3 shows the spectral distribution observed at the selected output of the star coupler. Under these conditions, the link length attainable for that output with 1.5 Gbit/s modulation was 57.8 km of standard single-mode fibre. The fibre loss (including six splices and three biconic connectors) was 14.5 dB, corresponding to $\simeq 0.25$ dB/km. The system was fully 'connectorised' using biconic connectors, and the component losses stated include the connector losses.

Bit-error-rate (BER) measurements were sequentially taken for each laser, with NRZ $(2^{15}-1)$ pseudorandom word modulation. Fig. 4 shows BER curves for the channels with the best and worst receiver sensitivities. These results indicate required power levels between -28.1 dBm and -32.5 dBm for BER $=10^{-9}$. The system power penalty (due to reflections, fibre dispersion, laser extinction ratio and laser intensity noise) for different lasers ranged from $\simeq 2$ to 6 dB.

The effect of crosstalk on the BER at the output of the demultiplexer was measured in two different ways. First, the cumulative effect of crosstalk from all other channels on each channel was measured to be < -27 dB. Secondly, we measured the nearest-neighbour crosstalk, taking into account variations in the receiver power as well as the channel separation. A channel at λ_i was set for BER $=10^{-9}$ at 1.5 Gbit/s, and neighbouring channels at $\lambda_{i\pm 1}$ were modulated independently. The received powers for each of the interfering channels were adjusted to be 14 dB higher than that of the selected

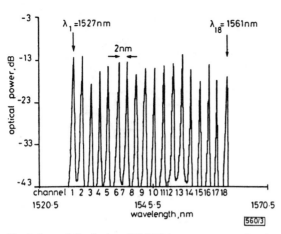

Fig. 3 *Spectral distribution of 18 DFB lasers as seen at selected output of star coupler*

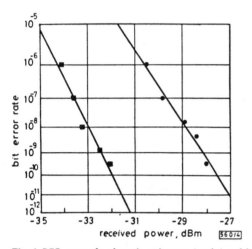

Fig. 4 *BER curves for channels with worst (circles) and best (squares) receiver sensitivities, measured after 57·8 km of fibre at 1·5 Gbit/s*

channel. No BER degradation was measured. Then, the interfering channels were purposely detuned in wavelength (by temperature control) to estimate the minimum channel separation. As the wavelength separation was reduced to 1 nm, the BER increased to 10^{-8}.

The bandwidth-distance product measured under these conditions (18 wavelength-multiplexed channels, 57·8 km and 1·5 Gbit/s/channel) was 1·56 Tbit s^{-1} km. Similar capacity was transmitted on each of the other 15 output ports of the star coupler; however, different output ports have different insertion losses, and these losses result in varying maximum link lengths at the corresponding output ports. Accounting for these variations, the network capacity for broadcasting was determined to be $\simeq 21\cdot5$ Tbit s^{-1} km node.

Summary: The initial LAMBDANET experiment using 18 wavelengths multiplexed through a single 57·8 km-long optical fibre has demonstrated a capacity of 1·56 Tbit s^{-1} km (the largest multiwavelength bandwidth-distance product demonstrated to date), corresponding to more than 24 million voice channels/km. For broadcast applications, a more appropriate figure-of-merit is the bandwidth-distance-node product, since the same information is transmitted to each of the receiving nodes. This experiment demonstrated a broadcast figure-of-merit of 21·5 Tbit s^{-1} km node. Used for a video broadcasting application, for example, such a system would be able to distribute >10000 video channels/km (at $\simeq 140$ Mbit/s/channel) to 16 nodes.

Acknowledgments: We thank Dr. H. Matsumura and Dr. Y. Koga of Hitachi Ltd. for assistance in obtaining the DFB lasers, and we also acknowledge the contributions of S. S. Cheng, J. Georges, T. P. Lee, C. N. Lo, K. W. Loh and P. W. Shumate to this work. LAMBDANET is a trade-mark of Bell Communications Research.

References

1 GOODMAN, M. S., KOBRINSKI, H., and LOH, K. W.: 'Application of wavelength division multiplexing to communication network architectures'. ICC '86 conference record, vol. 2, Toronto, 1986, p. 931

2 GOODMAN, M. S., BRACKETT, C. A., BULLEY, R. M., LO, C. N., KOBRINSKI, H., and VECCHI, M. P.: 'Design and demonstration of the LAMBDANET system: a multiwavelength optical network'. To appear in Proc. IEEE global telecommunications conf. '87, Tokyo, 1987

3 OLSSON, N. A., HEGARTY, J., LOGAN, R. A., JOHNSON, L. F., WALKER, K. L., COHEN, L. G., KASPER, B. L., and CAMPBELL, J. C.: '68·3 km transmission with 1·37 Tbit km/s capacity using wavelength division multiplexing of ten single-frequency lasers at 1·5 μm', *Electron. Lett.*, 1985, 21, pp. 105–106

WDM Coherent Optical Star Network

BERNARD S. GLANCE, FELLOW, IEEE, K. POLLOCK, CHARLES A. BURRUS, FELLOW, IEEE, BRYON L. KASPER, MEMBER, IEEE, GADI EISENSTEIN, MEMBER, IEEE, AND LAWRENCE W. STULZ

Abstract—This paper reports the results obtained with a fiber optical star network using densely-spaced wavelength division multiplexing (WDM) and heterodyne detection techniques. The system consists of three lasers transmitting at optical frequencies around 234 000 GHz spaced by a frequency interval of 300 MHz. The lasers are frequency shift key (FSK) modulated at 45 Mbit/s. A 4 × 4 optical star coupler combines the three optical signals. The WDM signals received from one of the 4 outputs of the star coupler are demultiplexed by a heterodyne receiver consisting of a tunable local oscillator (LO) laser and a balanced mixer receiver followed by a frequency-discriminator centered at an intermediary frequency (IF) of 225 MHz.

The minimum received optical power needed to obtain a bit-error rate (BER) of 10^{-9} is -61 dBm or 113 photon/bit, which is 4.5 dB from the shot noise limit. This is the best receiver sensitivity reported to date for optical FSK modulation. The degradation caused by co-channel interference was measured and found to be negligible when the channels, modulated at 45 Mbit/s, are spaced by more than 130 MHz in the IF domain. These results indicate that a WDM coherent optical star network of this type has a potential throughput of 4500 Gbit/s.

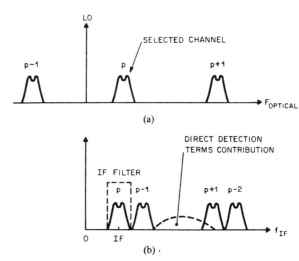

Fig. 1. Schematic showing (a) channel selection in the optical domain, and (b) the resulting power spectrum in the IF domain and the interference due to the detection process.

I. INTRODUCTION

TO DEMONSTRATE the capability of an optical fiber network, we selected the WDM coherent optical star approach [1]–[3] because: WDM allows the use of the wide available optical bandwidth [4], coherent detection yields high-receiver sensitivity and high-frequency selectivity, and a star configuration efficiently distributes [5] the optical power through the network. Our goal was to demonstrate that, by combining these three features, such an optical network can provide a large throughput (number of users × bit rate).

For a given bit rate, the system throughput depends on the number of channels which can be multiplexed in the available optical bandwidth. However, this bandwidth is much larger than any feasible IF and therefore the LO frequency must be positioned near the desired channel [6]. In addition, an optical mixer is equally sensitive to inputs above and below the LO frequency[1] (one input is the image frequency of the other). As a result the channel spacing must be large enough to avoid interference from the image frequency. Such a heterodyne process interleaves in the IF domain the channels on the low-frequency side of the LO signal with the channels on the high-frequency side of this signal. Consequently, the channels are spaced closer in the IF domain (Fig. 1). The minimum achievable channel spacing varies proportionally with the value of the IF frequency when the lowest IF channel is selected by the IF filter. As a result, a large channel spacing is needed since the IF frequency must be well above the baseband to avoid interference between the IF selected signal and the resulting demodulated signal. These two problems can be resolved conjointly by selecting the second lowest [8], [9] IF channel (Fig. 2). The result is obtained by tuning the LO frequency at the edge of the optical channel adjacent to the selected channel. But in this case, the selected IF channel is interfered by intermodulation components arising from the frequency beat between adjacent optical channels.[2] This interference can be significant in a star network, where the optical power of the received WDM signals can be comparable to that of the LO signal. Fortunately, this interference can be suppressed[3] by using a balanced mixer receiver [6], [10]. The minimum channel spacing, and thus the system throughput, is then determined by the channel spacing preventing co-channel interference in the IF domain [11].

To demonstrate the throughput capability of the proposed system, we built a WDM coherent optical star network using three optical channels spaced by the minimum frequency interval satisfying the above conditions. The circuit used for this demonstration, and the results obtained, are described in the following sections.[4]

Manuscript received June 5, 1987; revised July 24, 1987.
The authors are with AT&T Bell Laboratories, Crawford Hill Laboratory, Holmdel, NJ 07733.
IEEE Log Number 8717518.
[1]The situation is improved through the use of an image rejection mixer [7] but this requires a more complicated receiver.

[2]For FSK modulation, interference from direct terms arises mainly from frequency beat between adjacent optical channels.
[3]The amount of suppression depends on the degree of balancing.
[4]A summary of these results has been published in [17].

Reprinted from *J. Lightwave Tech.*, vol. 6, no. 1, pp. 67–72, Jan. 1988.

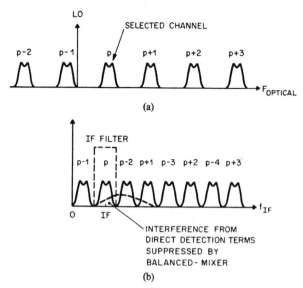

Fig. 2. (a), (b) Same as Fig. 1 for the channel configuration providing maximum throughput.

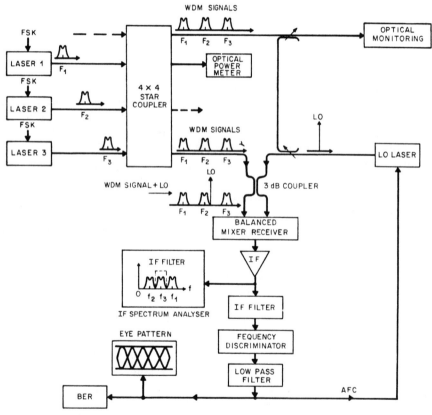

Fig. 3. Schematic of the WDM network used for making the measurements.

II. Circuit Description

The optical sources, used for transmission, are fast frequency-tunable external cavity lasers [12]. These lasers provide a narrow-line (≈ 100 kHz) single-frequency signal at 1.28 μm, which can be frequency-tuned over about 4000 GHz as well as frequency-modulated up to 100 MHz. Modulation is done by FSK at 45 Mbit/s with a modulation index of about 1 (equal to $(F_{\text{space}} - F_{\text{mark}})/\text{bit}$ rate), by a random NRZ bit stream with $2^{15} - 1$ pattern length. The three sources transmit (≈ 0 dBm each) at optical frequencies spaced by a frequency interval of 300 MHz.

The three optical signals are combined by a 4×4 fiber star coupler (Fig. 3). Each of the 4 output fibers of the coupler carries the three WDM signals. The signal from one of these fibers is combined, by means of a 3-dB fiber coupler, with the LO signal provided by a conventional external-cavity laser [13]. The polarization of the transmitted signals is manually adjusted by means of fiber polarization compensators to match the polarization state of the LO signal. The combined signals from the two outputs of the coupler feed a balanced mixer receiver [14] which heterodynes the received signal to an IF frequency of 225 MHz. The optical power of the LO signal, at the photodetectors, is about 0.7 dBm. As a result, the shot noise due to the LO signal dominates the thermal noise of the receiver. The use of a balanced mixer is essential in a WDM star network. This type of mixer eliminates the interference arising from the direct detection terms. It also allows a more efficient use of the optical power of the LO source and thus reduces the degradation caused by the shot noise originating from the received signal. (In a WDM star network, the shot noise caused by the received WDM signal is equal to the shot noise due to one transmitting optical source minus the excess loss of the network).

The IF signal is amplified and then filtered by an IF filter 60 MHz wide centered at 225 MHz. Demodulation is accomplished by a delay type frequency discriminator, and the resulting baseband signal is filtered by a low-pass filter having a 3-dB cutoff of 35 MHz. Selection of the desired channel is achieved by tuning the LO frequency to the value that centers the wanted channel within the bandwidth of the IF filter. The resulting IF is maintained by an automatic frequency control (AFC) circuit controlling the optical frequency of the LO signal (Fig. 3).

The signal from one of the three remaining output fibers from the star coupler is used, after combining with a fraction of the LO signal, to monitor the four optical signals (the three WDM signals and the LO signal). This is done using a spectrometer and a scanning Fabry–Perot etalon. Another output fiber is utilized to measure the received signal. The measurement takes into account the slight difference of received signal between this fiber and the fiber connected to the receiver.

III. IF Channel Selection

In order to operate at an IF frequency above the frequency of the baseband signal, the desired optical channel is heterodyned by an LO signal which is tuned to the edge of an adjacent optical channel. Thus, the channel selected by the IF filter is the second lowest IF channel. This arrangement minimizes the frequency spacing needed between optical channels and separates the demodulated baseband signal from its IF signal. For a modulation rate of 45 Mbit/s, the minimum optical frequency spacing preventing co-channel interference in the IF domain is about 260 MHz. To provide a protective margin, we selected a channel spacing of 300 MHz. The resulting frequency spacing between the interleaved IF channels is 150 MHz.

These results are illustrated in Fig. 4, which shows the IF power spectra of the three channels, heterodyned from an optical frequency of about 234 000 GHz, and selected successively by the IF filter centered at 225 MHz. The three first cases show the selection, respectively, of chan-

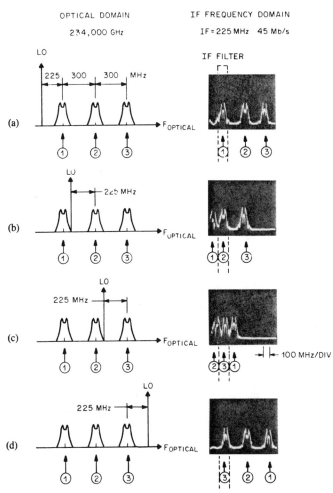

Fig. 4. Power spectra of the three IF channels corresponding to the LO frequency tuning illustrated opposite to the measured IF spectra. The first three cases show the selection by the IF filter, centered at 225 MHz of, respectively: channel 1, channel 2, and channel 3. The fourth case shows the selection of channel 3 for a LO frequency tuning yielding the reversed IF channel distribution as compared to the example illustrated in the first case.

nels 1, 2, and 3. In this case, channel 3 is interleaved between two adjacent channels spaced at 150 MHz as would occur in a system having a large number of channels. The fourth case shows the selection of channel 3 for a LO frequency tuning yielding the reversed IF channel distribution as compared to the distribution given by the selection of channel 1. Opposite to each IF channel spectrum distribution is illustrated the corresponding power spectra of the optical signals and the frequency position of the LO signal yielding the resulting IF spectra. With more than 3 channels, there would be no further channel interference since only the two channels adjacent to the selected IF channel give rise to co-channel interference.

IV. Receiver Sensitivity Measurement

For one channel, the minimum received optical power needed to obtain a BER of 10^{-8} is -61.6 dBm (Fig. 5). Extrapolating this result to a BER of 10^{-9} (direct measurement requires too long a counting time at 45 Mbit/s) yielding a received optical power of -61 dBm. This re-

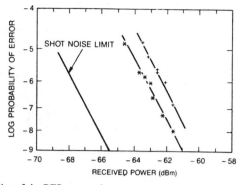

Fig. 5. Plot of the BER versus the received optical power for one channel alone. The same result is given for the same channel in presence of an interfering channel spaced by 100 MHz in the IF domain, received with the same optical power and modulated at the same bit rate of 45 Mbit/s * channel 1 alone. + channel 1 with channel 2 spaced at 100 MHz from channel 1.

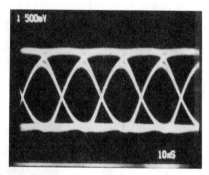

Fig. 6. Eye pattern of the received signal, in absence of noise, showing the eye closure due to mismatch of the modulating port of the laser. (45 Mbit/s, $P = -40$ dBm.)

sult corresponds to the actual optical power received at the output of the 4 × 4 optical star coupler. It does not include the additional excess loss due to the 3-dB coupler and the fiber connectors, which amounts to about 0.2 dB. Expressed in terms of the averaged number of photons per bit, the receiver sensitivity is 113 photon/bit, which is 4.5 dB from the shot noise limit of an ideal FSK optical receiver. This is the best receiver sensitivity measured to date for FSK modulation. The 4.5-dB degradation includes a 1-dB penalty due to the narrow bandwidth of the laser modulation port, which was designed for dc bias only. This limitation causes a 20-percent closure of the eye pattern in absence of system noise (Fig. 6).

V. Co-Channel Interference Measurement

We deliberately reduced channel spacing in order to determine co-channel interference. The power penalty caused by co-channel interference is 1 dB when the selected channel is interfered in the IF domain by a second channel spaced by 100 MHz, received with the same optical power and modulated at the same bit rate of 45 Mbit/s but with a different random bit stream. This result is shown in Fig. 5, which displays the measured BER versus the received power for one channel alone, and the similar result obtained when the channel is interfered by a second channel spaced by 100 MHz in the IF domain.

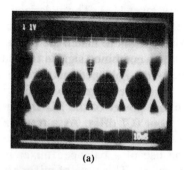

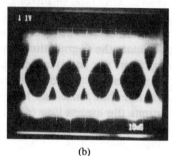

Fig. 7. (a) Eye pattern for one channel received with an optical power of -61.6 dBm. Eye pattern corresponding to the case where the channel is interfered by a second-channel spaced by 100 MHz in the IF domain, received with the same optical power and modulated at the same bit rate 45 Mbit/s. ($P = -61.6$ dm.)

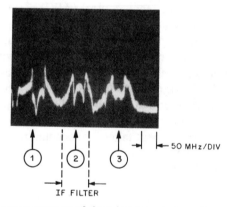

Fig. 8. IF power spectrum of three interleaved channels, each received with -61.6 dBm, modulated at 45 Mbit/s. (IF = 225 MHz.)

The eye patterns measured for these two cases are shown in Fig. 7 for a received optical power of -61.6 dBm.

The co-channel interference increases very fast when the frequency spacing is reduced to 90 MHz. It completely disappears when the frequency spacing is increased to 130 MHz. Similarly, no co-channel degradation is observed when the selected channel is interleaved by two adjacent channels spaced by 150 MHz. Fig. 8 shows the measured IF spectrum for this case, where each of the three channels is received at a level of -61.6 dBm and is modulated at 45 Mbit/s.

VI. System Throughput

The system throughput can be estimated from the above results for a network operating at the wavelength of 1.55 μm. For example, let us assume that a user transmits 0 dBm of optical power into his fiber connected to a star coupler interconnecting 2^{17} subscribers. The excess loss suffered by a signal propagating through the star coupler is proportional [4] to the 17 stages of 3-dB couplers constituting the star. Since commercially available 3-dB couplers have about 0.1-dB excess loss, such a star coupler will introduce an excess loss of about 2 dB. Adding 4 dB of fiber loss for a network having a 10-km radius yields -57.2 dBm of optical power received by each subscriber from the signal transmitted by the above user. This received power is 3.8 dB larger than that measured to obtain a BER of 10^{-9} for a modulation rate of 45 Mbit/s. Therefore, the above results indicate that 100 000 users transmitting at this bit rate could be interconnected within a radius of 10 km by a WDM coherent star network, provided that the lasers can be tuned over the frequency range filled by the WDM channels. The resulting throughput would be 4500 Gbit/s.

The most important requirement for realization of this potential throughput is a practical laser providing a single-frequency signal tunable over a large range. Other important problems, such as polarization control and confining the optical sources to a comb of frequencies, are being studied and solutions are envisioned [15], [16].

VII. Conclusion

We have demonstrated WDM of three optical channels spaced by 300 MHz, including selection of the desired channel by heterodyne detection, for optical signals FSK modulated at 45 Mbit/s. The minimum optical power needed to obtain a BER of 10^{-9} in -61 dBm or 113 photon/bit. This result, which is 4.5 dB from the shot noise limit, represents the best receiver sensitivity yet measured for FSK modulation. The degradation caused by co-channel interference is negligible when the heterodyned channels are spaced by more than 130 MHz in the IF domain for a modulation rate of 45 Mbit/s.

The above results indicate that 100 000 users transmitting at 45 Mbit/s could be interconnected within a radius of 10 km, by a WDM coherent star network. A throughput comparable to this value (4500-Gbit/s user) is expected to be achievable for a smaller number of users operating at a higher bit rate.

Acknowledgment

The authors are grateful to L. J. Greenstein and R. A. Linke, who pointed out the possibility of reducing the spacing of the optical channels by IF filtering the second lowest IF channel. They are also thankful to R. A. Linke for his helpful suggestions and advice, and to K. C. Reichman, K. Hall, G. Raybon, and E. Desurvire for help provided during this project.

References

[1] E. J. Bachus, R. P. Braun, C. Caspar, E. Grossmann, H. Foised, K. Heimes, H. Lamping, B. Strebel, and F. J. Westphal, "Ten-channel coherent optical fiber transmission," *Electron. Lett.*, vol. 22, no. 19, pp. 1002–1003, Sept. 1986.

[2] A. R. Chraplyvy and R. W. Tkach, "Narrow-band tunable optical filter for channel selection in densely packed WDM systems," *Electron. Lett.*, vol. 22, no., pp. 1084–1085, Sept. 1986.

[3] R. A. Linke, "Beyond gigabit-per-second transmission rates," in *OFC Sixth Int. Conf. Tech. Dig.*, (Reno, NV), Jan. 19-22, 1987, pp. 184.

[4] P. S. Henry, R. A. Linke, and A. A. Gnauck, "Introduction to lightwave systems," *Optical Fiber Telecommunications II*, S. E. Miller and I. P. Kaminow, Eds. New York: Academic, 1987, ch. 21.

[5] A. A. M. Saleh and H. Kogelnik, "Reflective single-mode fiber-optic passive star couplers," to be published in *J. Lightwave Technol.*,

[6] L. G. Kazovsky, "Multichannel coherent optical communication systems," in *OFC Sixth Int. Conf. Tech Dig.*, (Reno, NV) Jan. 19-22, 1987, pp. 59–60.

[7] B. Glance, "An optical heterodyne mixer providing image-frequency rejection," *J. Lightwave Technol.* vol. LT-4, no. 11, pp. 1722–1725, Nov. 1986.

[8] L. J. Greenstein and R. A. Linke, private communication.

[9] Y. K. Park, unpublished.

[10] L. J. Greenstein, private communication.

[11] Y. K. Park, S. S. Bergstein, R. E. Tench, R. W. Smith, S. K. Korotky, and K. J. Burns, "Crosstalk and prefiltering in a two-channel ASK heterodyne detection system without the effect of laser phsae noise," presented at OFC Sixth Int. Conf., Reno, NV, Jan. 19-22, 1987, post deadline pap. PDT-13.

[12] B. Glance, C. A. Burrus, and L. W. Stulz, "Fast frequency-tunable external-cavity laser," *Electron Lett.*, vol. 23, no. 3, pp. 98–99, Jan. 1987.

[13] R. Wyatt and W. J. Delvin, Jr., "10-kHz linewidth 1.5-μm InGaAsP external cavity laser with 55-mm tuning range," *Electron. Lett.*, vol. 19, no. 3, pp. 110–112, Feb. 1983.

[14] B. L. Kasper, C. A. Burrus, J. R. Talman, and K. L. Hall, "Balanced dual-detector receiver for optical heterodyne communication at Gbit/s rates," *Electron. Lett.*, vol. 22, no. 8, pp. 413–415, Apr. 1986.

[15] B. Glance, "Polarization independent coherent optical receiver," *J. Lightwave Technol.*, vol. 5, no. 2, pp. 274–276, Feb. 1987.

[16] B. Glance, P. J. Fitzgerald, K. J. Pollock, J. Stone, C. A. Burrus, and L. W. Stulz, "Frequency stabilization of FDM optical signals," *Electron. Lett.*, vol. 23, no. 14, pp. 750–752, July 1987.

[17] B. Glance, K. J. Pollock, C. A. Burrus, B. L. Kasper, G. Eisenstein, and L. W. Stulz, "Densely spaced WDM coherent optical star network," *Electron. Lett.*, vol. 23, no. 17, pp. 875–876, Aug. 1987.

MULTIWAVELENGTH OPTICAL CROSSCONNECT FOR PARALLEL-PROCESSING COMPUTERS

E. ARTHURS
J. M. COOPER
M. S. GOODMAN
H. KOBRINSKI
M. TUR*
M. P. VECCHI

4th December 1987

Bell Communications Research
435 South Street
Morristown, NJ 07960, USA

* Permanent address: School of Engineering, Tel-Aviv University, Israel 69978

Indexing terms: Computers, Parallel processing, Optical connectors

A multiwavelength optical interconnection architecture for parallel-processing computers is presented. This fully connected, internally nonblocking network, using tunable lasers, requires only two optical fibres per node. Wavelength switching measurements (access times < 20 ns) and calculated throughput are presented. Critical target parameters are established for optical device development.

Introduction: A fundamental problem in the design of parallel-processing computers is the interconnection between processors and external memory. Fully connected and internally nonblocking interconnection networks minimise the restrictions that crossconnects place on the achievable throughput of parallel-processing computers. Electronic interconnection systems utilise either parallel-bus structures or switching networks. Bus structures require a large number of interconnection points and electrical wires, while electronic switching crossconnects in general introduce significant blocking. In this letter an architecture is presented for a multiwavelength optical interconnection network (referred to as FOX, for fast optical crossconnect), which is internally nonblocking and fully connected with only two optical fibres per node. Experimental results and performance analysis are used to establish target optical parameters for the devices required by the FOX architecture.

System description: The basic structure of the FOX is shown schematically in Fig. 1. The computer is assumed to have N processor nodes ($4 \le N \le 256$) with local cache memory (CP), and N external memory nodes (MEM). FOX uses two separate interconnection networks:[1] one for transmission of data from CP to MEM, the other for the return from MEM to CP. The CP and MEM nodes each have separate optical transmission and receiver modules, with a unique, fixed-wavelength filter associated with each receiver, and a tunable-wavelength source associated with each transmitter. The N processor transmitters are connected through optical fibres to a forward $N \times N$ optical star coupler whose outputs are connected to the memory receiver nodes. Similarly, the N memory transmitters are connected to a reverse $N \times N$ star coupler. Thus the FOX supports bidirectional communication between any CP/MEM pair, to obtain a fully connected, internally nonblocking interconnection network.

The pattern of external memory references imposes an upper bound on the concurrency achievable in a parallel processor. The FOX is internally nonblocking, but if a memory node is simultaneously addressed by more than one processor, then external collisions will occur. The FOX can also support efficient conflict–resolution algorithms for external collisions. The normalised throughput of each processor is shown in Fig. 2 as a function of the fraction of memory references that are external (cache miss-rate), assuming uniform random addressing in the limit of large N. The top curve refers to the intrinsic (ideal) performance, such that for every external collision that occurs, one of the requests is satisfied. The lower curve refers to an example of a simple real conflict–resolution algorithm, based on collision detection and binary exponential backoff retransmission.[2]

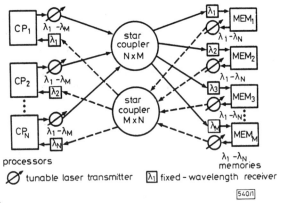

Fig. 1 *Schematic diagram of fast optical crossconnect (FOX)*

Two optical star couplers interconnect bidirectionally the processors (CP) and the external memories (MEM). FOX could support N processors and M memory nodes, including the possibility of multiple-processor nodes. For simplicity, in the text we consider the case $N = M$

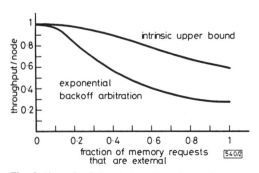

Fig. 2 *Normalised FOX throughput for each node, as a function of fraction of memory requests that are external*

These results are for large N limit, assuming uniform random addressing

For a typical case of $\simeq 10\%$ cache miss-rate, Fig. 2 shows that for both the intrinsic (ideal) performance and the binary exponential backoff algorithm the throughput for each of the N processors operating in parallel is $\simeq 98\%$ of the single-processor value. Further, for the extreme case of no cache memory (i.e. cache miss-rate = 100%), the intrinsic (ideal) effective throughput for each of the N processors operating in parallel is $\simeq 58\%$ of the single-processor value, while for the binary exponential backoff algorithm it is $\simeq 25\%$ of the single-processor value.

Experimental results: To support an equivalent 48-bit-wide bus at $\simeq 20$ MHz, optical transmission rates of $\simeq 1$ Gbit/s are required for each processor. Assuming frame-synchronous serial transmission with a fixed packet size of $\simeq 100$ bits, the laser diode (LD) sources need to be switched between any two wavelengths in a time $\ll 100$ ns. This requires electronic carrier-density tuning.[3,4] To demonstrate the feasibility of the required wavelength tuning rate, an experiment was performed as shown in Fig. 3. A 1·5 μm DFB-LD was switched between two distinct wavelengths λ_1 and λ_2 by current injection ($\simeq 0.006$ nm/mA at 45 MHz), followed by an external LiNbO$_3$ intensity modulator. A Fabry–Perot filter provided the required wavelength selectivity at the receiver node. The optical receiver used a GaInAs APD detector, followed by a high-impedance GaAs FET stage. Fig. 3a, before the optical filter, shows the combined effect of the LD injection-current wavelength tuning (a square wave, 10 mA peak-to-peak, at 45 MHz) and the external data modulation (pseudorandom sequence at 1 Gbit/s). The wavelength tuning also introduces an undesirable amplitude modulation. Fig. 3b shows the signal after filtering by the Fabry–Perot cavity tuned to wavelength λ_1, and Fig. 3c shows the corresponding signal with tuning to wavelength λ_2. The results demonstrate the fast (<20 ns access time) wavelength switching of an LD between two channels separated by 0·06 nm.

This experiment did not use LDs specifically designed for wavelength tuning. Two- and three-section DBR- and DFB-LDs are being developed as wavelength-tunable sources. In particular, DBR-LDs have been demonstrated with DC electronic tuning ranges of $\simeq 5.8$ nm.[4] We are currently investigating the fast wavelength-switching characteristics of such devices. Crosstalk measurements[1] of two closely spaced optical channels have demonstrated that a direct-detection channel spacing of < 0.16 nm (for a bit error rate of 10^{-9} at 1 Gbit/s) can be easily obtained. Therefore, current technology indicates the feasibility of multichannel operation of the FOX architecture.

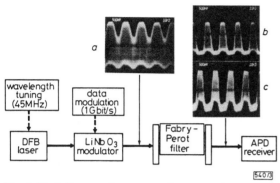

Fig. 3 *Block diagram of experimental set-up to demonstrate combined effect of laser current modulation (wavelength tuning with 45 MHz pattern) and external modulation (data at 1 Gbit/s)*

Signal detected without optical filter (inset a), after optical filter tuned to λ_1 (inset b) and after optical filter tuned to λ_2 (inset c)

Conclusions: The performance analysis and the initial experiments indicate that the FOX is an attractive new architecture for the interconnection network of parallel-processing computers. Present results also provide the basis for a realistic definition of important target parameters for future development. As a guide for component development, Table 1 presents both current and future expected values for the critical device parameters in the FOX architecture.

Table 1 TARGET PARAMETER SPECIFICATIONS

	Current	Target
Star coupler: (single-mode)		
Size ($N \times N$)	32×32	256×256
Uniformity	±6 dB	±3 dB
Tunable source: (Single-mode laser)		
Channels	32	256
Power	0 dBm	0 dBm
λ switching time	20 ns	5 ns
Modulation rate	1 Gbit/s	2 Gbit/s
λ registration	±0·1 Å	±0·1 Å
Tuning range	5·8 nm*	25 nm
Filter and receiver: (direct detection)		
λ filter BW	1·5 Å	1·0 Å
Insertion loss	10 dB	10 dB
Sensitivity	−36 dBm	−36 dBm

* Tuning range not measured under fast wavelength-switching conditions

References

1 ARTHURS, E., BULLEY, R., GOODMAN, M. S., KOBRINSKI, H., VECCHI, M. P., and GIMLETT, J. L.: 'A fast optical crossconnect for parallel processing computers'. 13th European conf. on optical comm., Helsinki, 1987, pp. 279–282
2 HUI, J., and ARTHURS, E.: 'A broadband packet switch for integrated transport', *IEEE J. Sel. Areas Commun.*, 1987, **SAC-5**, p. 1264
3 OLSSON, N. A., and TSANG, W. T.: 'An optical switching and routing system using frequency tunable cleaved-coupled-cavity semiconductor lasers', *IEEE J. Quantum Electron.*, 1984, **QE-20**, p. 332
4 MURATA, S., MITO, I., and KOBAYASHI, K.: 'Over 720 GHz (5·8 nm) frequency tuning by a 1·5 μm DBR laser with phase and Bragg wavelength control regions', *Electron. Lett.*, 1987, **23**, pp. 403–405

Demonstration of Fast Wavelength Tuning For a High Performance Packet Switch

M. S. Goodman, E. Arthurs, J. M. Cooper, H. Kobrinski, and M. P. Vecchi
Bell Communications Research
Morristown, NJ 07960-1961, USA

Future broadband networks will use packet switching to achieve uniform switching and multiplexing of multiple-bit-rate data streams. A critical element of such broadband networks is a large capacity ($\approx$ 50 Gbit/s) packet switch. We present a high performance packet switching system (HYPASS) based on a parallel, multiwavelength hybrid opto-electronic interconnection, employing wavelength selective transmitters and receivers. HYPASS represents a departure from conventional packet switch designs, and it addresses both the complex switching, and the internal transport and control functions. We describe the overall system architecture including a brief description of the control structure, and present experimental results demonstrating the feasibility of the required laser wavelength tuning for such an architecture.

HYPASS receives packets on optical fiber input trunks, and switches them to the appropriate optical fiber output trunks. HYPASS is formed from two tunable multiwavelength optical networks, a transport network and a control network, that connect the input ports of the switch to the output ports (Fig 1). Addressing is achieved in the HYPASS by using tunable transmitters in the transport network, and tunable receivers in the control network. HYPASS utilizes an output controlled/input buffered protocol to take maximum advantage of the parallelism available to the optical interconnect.

Packets arriving at an input port have their destination port address decoded, and are stored temporarily in a FIFO buffer at the input until a 'poll' or 'request-to-send' is received from the appropriate output port. The polls are broadcast over the control network to all input ports. At each input port, a wavelength tunable receiver selects the control information from its desired output port. The packets are routed and transmitted to the appropriate output port by tuning the input port laser to the unique fixed 'wavelength address' corresponding to the output port. Simultaneously, packets from other inputs are being transmitted on different wavelengths to other output ports using distinct wavelengths and the same transport network. Each output port monitors its busy status, and generates the required probing information to be sent to the input nodes. This information is converted to optical signals at a unique wavelength associated with each output port and broadcast over the control network to all input ports.

The control structure of a packet switching system is of paramount importance and the HYPASS design features an integrated control and transport structure. The HYPASS architecture incorporates a single stage routing network, and its 'output control structure' allows substantial design flexibility. In this design, contention for use of output ports is resolved by sending polls (or explicit request-to-send probes) to the input ports. HYPASS features simultaneous polling of distinct input ports by different output ports taking advantage of the multiwavelength interconnect. The broadcast nature of the control structure allows a poll to simultaneously be issued to a group of input ports. This feature allows the implementation of tree polling techniques[1] to resolve output port packet collisions. Both random (any input port to any output port) and scheduled demand (fixed input port to fixed output port) packet transport may be accommodated in this design. A detailed discussion of these arbitration and control issues is presented elsewhere[2].

The critical optical devices required by HYPASS are the tunable transmitters and the tunable receivers. The required wavelength tuning characteristics of the laser transmitters are a function of the packet lengths. Assuming packet sizes of $\approx$ 1000 bits and laser modulation rates of $\approx$ 2 Gbit/s, the target laser tuning parameters are laser switching speeds which should be at most 20 ns from any wavelength to any other wavelength, and residency times (stable transmission at a given wavelength) of $\approx$ 500 ns. A 32 port HYPASS switch with these parameters would have a capacity more than an order of magnitude larger than the largest existing electronic packet switches. Due to the short interconnection distances ($\approx$ 3 m) the overall power budget is dominated by the splitting losses in the star couplers; 32 way splitting may be accomplished with available laser launch powers ($\approx$ 0 dBm) and demonstrated receiver sensitivities (-37 dBm at 1.0 Gbit/s). The requirements on the wavelengths used and their spacing need not be the same for the tunable transmitters and the tunable receivers, as they are used in different internal networks in HYPASS.

Wavelength-selectable receivers operating at high speeds have been recently reported, using DFB optical amplifiers[3]. Two signal channels were wavelength-multiplexed with data modulations at 1.0 Gbit/s, and selected by the optical filter at nanosecond switching times. Receivers with these switching characteristics and the ability to access many channels would be compatible with the requirements of HYPASS.

Recent studies with tunable lasers[4,5] support the feasibility of wavelength switchable transmitters for the HYPASS architecture. In our work, nanosecond switching between 8 wavelengths was achieved using a Hitachi double-section distributed feedback (DS-DFB) laser. The laser was modulated using 8-level current injection patterns applied to both forward and rear section electrodes. The wavelength switching was observed by passing the laser output through a scanning Fabry-Perot interferometer with a free spectral range of 5 Å and a bandwidth of 0.18 Å. The modulation patterns for both electrodes and the Fabry-Perot output spectrum are shown in Figure 2. The total tuning range was 3.1 Å (centered at 1577.7 Å), with a 0.45 Å separation between each of the eight wavelengths. In the experiment, the channel-to-channel access time was less than 5 ns, and a residency time of 500 ns per channel was demonstrated.

In order to demonstrate the ability to switch data we also performed an experiment in which 2 of the 8 selectable wavelengths were chosen for data transmission. The experimental setup is shown in Figure 3. In this experiment, the output of the DS-DFB laser was coupled directly into a lens-tipped fiber and passed through a polarization controller (PC). The optical signal was modulated at 0.5 - 1.0 Gbit/s using a commercial LiNbO$_3$ modulator with an extinction ratio of 15 dB and insertion loss of 7 dB. The Fabry-Perot was used as a filter (no scanning) to choose between the two channels for the wavelength-switched and amplitude-modulated signals. The signal was detected by an InGaAs APD in a custom designed high impedance receiver with a sensitivity of -37 dBm at 1.0 Gbit/s. The Fabry-Perot filter was tuned alternately to the 2 channels; the wavelength selection is shown in the two oscilloscope traces measured at the APD receiver (Fig. 3). The switching time between the two wavelengths was again measured to be less than 5 ns. Also shown in the figure is an eye diagram corresponding to one of the wavelength channels modulated with a pseudorandom sequence at 1.0 Gbit/s.

We have presented a new approach to high performance packet switching for the broadband network. This architecture incorporates the transport and control functions in an integrated design, using optical routing and transmission, and electronic control processing. We have demonstrated multiwavelength tunability of a double-section DFB laser source among 8 distinct wavelengths with access times less than 5 ns and residency times ≈ 500 ns.

References

1. J. Capetanakis, 'The Multiple Access Broadcast Channel: Protocol and Capacity Considerations', IEEE Trans. Info. Theory, IT-25:505 (1979)

2. E. Arthurs, M.S. Goodman, H. Kobrinski, and M.P. Vecchi, 'HYPASS: an Optoelectronic Hybrid Packet Switching System', *to be published in IEEE J. on Select Areas Commun.*

3. M.P. Vecchi, H. Kobrinski, E.L. Goldstein, and R.M. Bulley, 'Wavelength Selection with Nanosecond Switching Times Using Distributed Feedback Optical Amplifiers', submitted to ECOC'88 (this conference).

4. J.M. Cooper, J. Dixon, M.S. Goodman, H. Kobrinski, M.P. Vecchi, E. Arthurs, S.G. Menocal, M. Tur, and S. Tsujii, 'Nanosecond Switching with a Double-Section DFB Laser', CLEO'88 Proceedings, to be published.

5. I. Mito, 'Recent Advances of Frequency Tunable DFB/DBR Lasers', in Proc. OFC'88 (THK1), New Orleans, 1988.

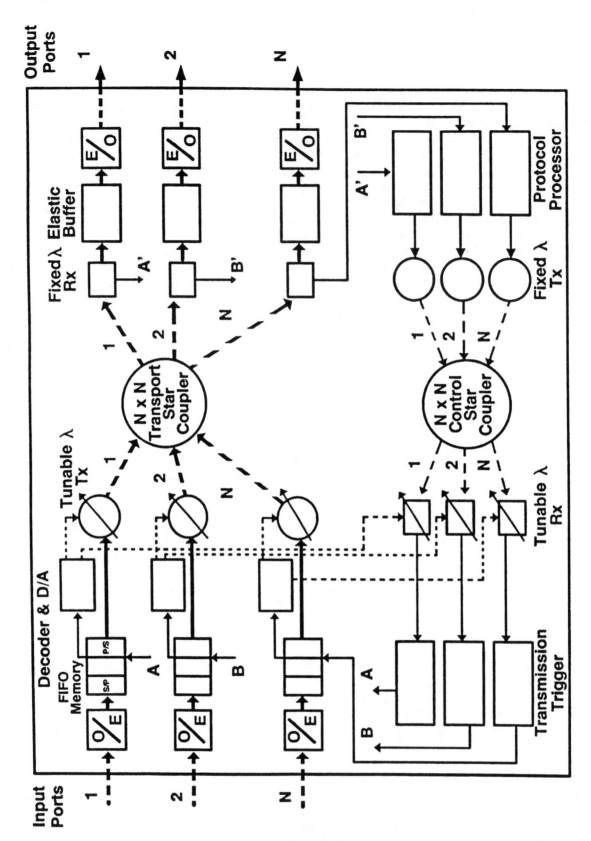

1. Schematic diagram of HYPASS architecture. Optical fiber input trunks enter at the upper left, after O-E conversion and serial to parallel (s/p) conversion the packets enter FIFO memory. Decoding of the destination address and D/A conversion takes place which results in tuning currents for the laser transmitters and control information receivers. The lower portion of the figure shows the arbitration control structure. Transport commences through the transport star coupler when requests are received from the transmission trigger (lower left). After routing, packets are stored in output buffers (upper right) before being transmitted on the output fiber trunks. Lightwave signals are indicated by dashed lines, wavelength tuning control currents by dotted lines and electronic control signals by solid lines.

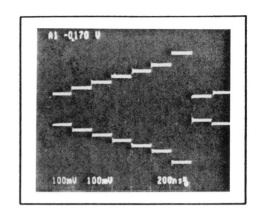
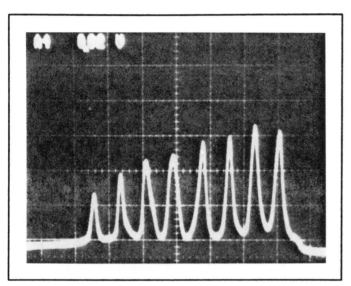

2. Scanning Fabry-Perot output indicating 8 distinct wavelengths with 3.1 Å total tuning range 0.45 Å channel separation. Also shown in the figure are the 8-level current inputs (starting with level 2) for the forward section (top trace) and rear section (lower trace) electrodes. The bias for both currents was 40 mA.

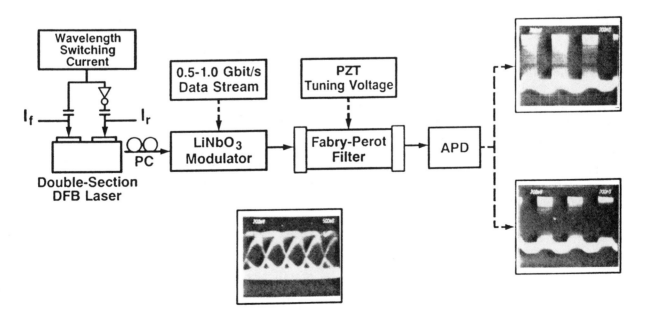

3. Experimental setup for alternating fast tuning of the DS-DFB laser to 2 of the 8 wavelengths and external modulation. Also shown is the eye diagram at 1.0 Gbit/s. The forward section current (I_f) and the rear section current (I_r) were biased at 40 mA as in figure 2.

FDMA-FSK Star Network with a Tunable Optical Filter Demultiplexer

I. P. KAMINOW, P. P. IANNONE, J. STONE, AND L. W. STULZ

Abstract—An optical frequency-division-multiple-access (FDMA) star network is analyzed and demonstrated experimentally using two 45-Mbit/s frequency-shift-keyed (FSK) laser channels at 1.5 μm. A tunable fiber Fabry–Perot (FFP) filter is used to select channels and convert FSK to intensity modulation for direct detection. The analysis predicts and experiment supports a minimum channel spacing of about 6 times bit rate B for a single FFP. The analysis predicts a channel spacing of $3B$ for a tandem FFP. These constraints are similar to those for the more complex heterodyne demultiplexing. Estimates show that a network with 1000 users, independent of bit rate, is feasible with a tandem FFP. For $B = 1$ Gbit/s per channel the network capacity would be 1 Tbit/s.

I. Introduction

A PASSIVE single-mode star coupler can provide network access to a multiplicity of users. Using frequency-division-multiplexing (FDM), each active user pair can be assigned one of a comb of optical carrier frequencies for a session to provide a frequency-division-multiple-access (FDMA) network. As with radio frequency FDM systems, we assume that the channel spacing between optical carriers is minimized in order to make efficient use of the optical spectrum. Heterodyne detection is often used for demultiplexing at the receiver. However, the network analyzed and demonstrated here makes use of a tunable optical filter—a fiber Fabry–Perot (FFP) [2], [3]—thereby avoiding the cost of a tunable local oscillator and a polarization controller, at the expense of reduced receiver sensitivity. Calculations show that channel spacing (f_c) can be as low as $6.4B$ for a single FFP and $3B$ for two FFP's in tandem, where B is the bit rate, with negligible power penalty. In addition, it is shown that the laser linewidth (f_l) is required to be $\leq 0.1B$. These constraints are comparable to those for heterodyne systems. The calculations are verified by bit-error-rate (BER) measurements on a single FFP two-channel 45-Mbit/s network with variable channel spacing. Estimates are also given for a network with 1000 users at 1 Gbit/s each for a network capacity of 1 Tbit/s.

The connectorized single-mode FFP [3] can be piezoelectrically tuned over several free spectral ranges (FSR) by increasing the optical length by several optical half-wavelengths. FFP's have been realized with finesse (F)

Manuscript received October 7, 1987; revised January 18, 1988. An abbreviated version of this work was published in *Electron. Lett.*, vol. 23, p. 1102, 1987.
The authors are with AT&T Bell Laboratories, Crawford Hill Laboratory, Holmdel, NJ 07733.
IEEE Log Number 8821128.

values up to ~200 by selecting the mirror reflectivity; and the FSR has been designed to cover a wide range, exceeding ~100 GHz, with different structures. It is expected that the small residual birefringence in present devices can be reduced to negligible levels by careful fabrication [4]. Thus, polarization control in the network would not be required. Two decoupled FFP's, with unequal lengths, in tandem can increase the effective FSR by a vernier effect [1], [5]–[7]. In addition, the transmission function of the tandem device gives a sharper filter characteristic and much larger effective finesse than the single FFP, as we show in a later section.

The basic network is illustrated schematically in Fig. 1, utilizing an $M \times M$ single-mode passive star with $M = 4$. The FDM carrier comb contains N carriers spaced by f_c and covering a band $\sim Nf_c$, where N must be less than the effective finesse. We assume M terminals; $2N$ may interact at one time. Each transmitter is assigned (permanently or dynamically) a different carrier frequency and all channels are combined in the passive $M \times M$ star coupler, where $M \geq N$. The complete FDM comb is available to each receiver, which utilizes a tunable FFP to select the appropriate channel.

Amplitude-shift-keying (ASK) could have been employed if an external modulator were used to avoid chirp. However, frequency-shift-keying (FSK) allows direct modulation of semiconductor lasers. The laser must be single frequency with an FM response, $\rho = \partial f / \partial I$, where f is the optical frequency and I the laser current, that is nearly independent of modulating frequency in order to avoid pattern effects. Alternatively, the laser drive circuit can be equalized to provide a flat FM response.

As shown in Fig. 2(a), the power spectrum of a continuous-phase FSK signal with large deviation index, $\mu = f_d/B$, with f_d the deviation frequency, consists of two peaks, spaced by f_d, of approximate width $B/2$ that decay as f^{-4} far from the peaks [8]. For large μ, the FSK power spectrum is the superposition of two independent ASK spectra each containing half the signal power. One peak corresponds to the "0"s and the other to the "1"s. Since ρ is positive and a positive increment in I increases the laser power, the "1" peak may be slightly stronger; we neglect that difference in the analysis but take advantage of it in the experiment. The FFP may be designed to have a full width at half maximum f_{FP} as small as B in order to select the "1" peak without distortion and thereby convert FSK to ASK. In a later section, we show that the

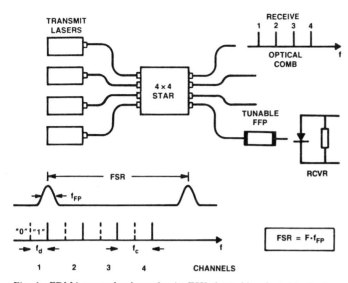

Fig. 1. FDMA network schematic. An FSK channel is selected by the tunable fiber Fabry-Perot (FFP) filter.

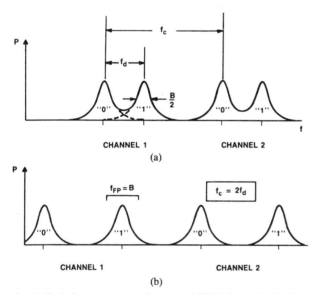

Fig. 2. (a) Optical power spectra for general FDM signal. (b) Optimized for $f_c = 2f_d$.

number of channels for given cross-talk interference is maximum when $f_c = 2f_d$ as in Fig. 2(b).

The paper is organized as follows: We begin by reviewing the transmission properties of a Fabry-Perot resonator by itself and, then, two FP's of slightly different length in tandem; the tandem FP offers the prospect of a large capacity network. We then analyze the power penalty due to the crosstalk between adjacent channels and compute the minimum channel spacing for the cases of single and tandem FFP demultiplexers; the tandem FFP network can support more than 100 times as many users as the single FFP network. Next, we consider the limitation imposed by the finite laser linewidth (f_l) and conclude that the FP passband must be at least ten times f_l, which is not a severe constraint in high-bit-rate networks. Finally, we describe a two-channel 45-Mbit/s network experiment, which supports our analysis and gives us confidence to speculate on much larger networks.

II. SINGLE AND TANDEM FABRY-PEROT FILTERS

The power transmission function T through a Fabry-Perot resonator is [9]

$$T = \left[1 + \left(\frac{2F}{\pi} \sin \delta\right)^2\right]^{-1} \quad (1)$$

where the phase increment

$$\delta = \frac{(f - f_p)}{\text{FSR}} = \frac{\pi f_{\text{FP}}}{\text{FSR}} m = \frac{\pi}{F} m \quad (2)$$

with

$$\text{FSR} = \frac{c}{2nL}, \quad (3)$$

$$f_p = p \cdot \text{FSR} \quad (p \text{ integer}). \quad (4)$$

Here, f is the optical frequency, f_{FP} the FP 3-dB passband measured as the full width at half maximum (FWHM), f_p the frequency of the pth order FP mode, L the length of the resonator, n its (group) refractive index and m the frequency shift from the mode peak measured in FP linewidths. The transmission function is plotted in Fig. 3(a); it is periodic with period FSR in f space, π in δ space and F in m space. The number of linewidths in a FSR is F. We assume throughout that $F \gg 1$. For $|m| \leq 1$:

$$T \approx [1 + 4m^2]^{-1} \quad (5a)$$

for $1 < |m| \ll F/2$:

$$T \approx (2m)^{-2} \quad (5b)$$

and for $|m| \approx F/2$:

$$T \approx \left(\frac{\pi}{2F}\right)^2. \quad (5c)$$

If two or more FP's are placed in tandem, the sharpness of the passband cut-off, the effective finesse, and the effective FSR can all be improved. For this purpose, it is necessary that each FP be independent of the other so that no standing wave is present between them. This independence can be achieved [6], [7] if the FP's are separated by either an isolator or by a small attenuator (~1 dB) and a fiber path length D such that the coherence length l_{coh} of the signal transmitted by the first FP is much shorter than the round-trip path

$$2D \gg l_{\text{coh}}. \quad (6)$$

A conservative estimate for l_{coh} is given by the laser linewidth f_l:

$$l_{\text{coh}} \approx \frac{c}{nf_l} \quad (7)$$

which is 40 m for $f_l = 5$ MHz. In practice, the effective spectral width of the modulated signal may be closer in magnitude to B, giving a shorter value for l_{coh}.

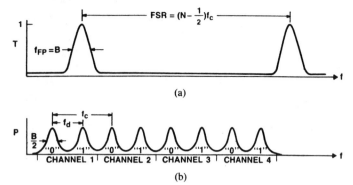

Fig. 3. (a) FFP transmission function. (b) Multichannel continuous phase FSK spectrum.

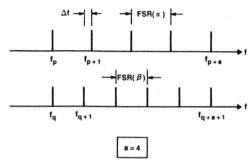

Fig. 4. Schematic of passbands of tandem vernier FP for $a = 4$.

The transmission through two independent FP's, $FP(\alpha)$ and $FP(\beta)$, is

$$T(\alpha\beta) = T(\alpha) \cdot T(\beta). \tag{8}$$

If we assume for the moment two identical FP's, with $T(\alpha\alpha) = T(\alpha)^2$, then from (5a) the ratio of FWHM linewidths for the tandem and single FP's becomes

$$\frac{f_{FP}(\alpha\alpha)}{f_{FP}(\alpha)} = (\sqrt{2} - 1)^{1/2} = 0.64. \tag{9}$$

In addition, (5b) yields

$$T(\alpha\alpha) = (2m(\alpha))^{-4}, \quad \text{for } 1 < |m(\alpha)| \ll F/2 \tag{10}$$

where $m(\alpha)$ is still measured in linewidths, $f_{FP}(\alpha)$, of the single FP. Thus, the passband of the tandem FP cuts off as f^{-4} compared with f^{-2} for the single FP.

Suppose now two FP's with approximately the same finesse $F(\alpha) \approx F(\beta)$, but with slightly different lengths $L(\alpha) \leq L(\beta)$ chosen such that the p-order mode frequency f_p for the first is the same as the q-order frequency f_q for the second. Then, from (4):

$$\frac{L(\alpha)}{L(\beta)} = \frac{p}{q} \approx 1 \quad \text{(with } p \text{ and } q \text{ integers)} \tag{11}$$

where, for example, $p \approx q \approx 4 \times 10^4$ for $\lambda = 1.5 \mu m$ and $FSR(\alpha) = 5$ GHz. The transmission function $T(\alpha\beta)$ is the product of the individual functions indicated schematically by lines in Fig. 4. The effective FSR, $FSR(\alpha\beta)$, is given by the period of overlapping modes

$$FSR(\alpha\beta) = a\, FSR(\alpha) = (a + 1)\, FSR(\beta) \tag{12a}$$

with integer a given by

$$a = \frac{L(\alpha)/L(\beta)}{1 - L(\alpha)/L(\beta)} \tag{12b}$$

in analogy with a vernier gauge. The vernier behavior hold for $1/2 < L(\alpha)/L(\beta) < 1$. In Fig. 4, $a = 4$ and $L(\alpha)/L(\beta) = 4/5$. Combining (9) and (12a), we find the effective finesse

$$F(\alpha\beta) = \frac{a\, FSR(\alpha)}{0.64 f_{FP}(\alpha)} = 1.6 a F(\alpha). \tag{13}$$

The closest approach of nonoverlapping modes is

$$\Delta f = f_{p+1} - f_{q+1} = \frac{FSR(\beta)}{a} \tag{14}$$

as shown in Fig. 4.

In order to ensure negligible added crosstalk due to the closest nonoverlapping modes of the tandem FP, we require $\Delta f \geq 5 f_{FP}(\beta)$, which limits the transmission to 10^{-2} according to (5a). Then:

$$a \leq f(\beta)/5 \tag{15}$$

$$\frac{1}{2} < L(\alpha)/L(\beta) \leq \frac{F(\beta)}{5 + F(\beta)} \tag{16}$$

and, according to (13):

$$F(\alpha\beta) \leq 0.32\, F(\alpha) F(\beta). \tag{17}$$

A tandem FP with $L(\alpha)/L(\beta) \ll 1/2$ can also be used to extend the effective finesse of individual FP's [5], but its behavior differs from that of the vernier FP. If the passband width of the shorter FP, $f_{FP}(\alpha)$, is greater than 5 times $FSR(\beta)$, then the transmission of nearest unwanted mode will be 10^{-2} and

$$F(\alpha\beta) \leq 0.2\, F(\alpha) F(\beta). \tag{18}$$

Since $f_{FP}(\alpha)$ is much larger than $f_{FP}(\beta)$, this tandem FP is easier to align than the vernier FP. However, the selectivity is not as good since the passband cuts off as f^{-2} rather than f^{-4}.

III. Demultiplexing and Detection

The simplest digital modulation format would be intensity modulation, i.e., amplitude shift keying (ASK), without any optical frequency deviation. A CW laser and external modulator could be used for this purpose, although half the laser power would be lost in the digital intensity modulator. At a bit rate B, a FFP with passband $f_{FP} = B$ could select the proper channel, without excessive distortion. Except for crosstalk from neighboring channels, the bit-error-rate (BER) would be determined by the power level at the receiver, as for ordinary direct detection. ASK might provide twice as many channels as calculated below for FSK since its power spectrum has about half the width; however, the cost of the external modulator might not be acceptable.

Frequency shift keying (FSK) can be achieved by taking advantage of the frequency deviation f_d inherent in a semiconductor laser for a change in drive current I. In a typical DFB laser, $\rho = \partial f / \partial I$ varies in the range ~ 100 MHz/mA to ~ 10 GHz/mA, depending on bit rate and laser structure. For an external-waveguide-Bragg-reflector laser [10], [11], $\rho \approx 100$ MHz/mA with a flat response between 100 kHz and 1 GHz; and, for a phase-tunable DFB laser [12], [13], $\rho \approx 10$ GHz/mA, with a flat response to ~ 500 MHz. A digital signal can be encoded with a "0" corresponding to $f_d = 0$ and a "1" corresponding to positive f_d as in Fig. 1. The optical spectrum of continuous phase digital FSK has the double-peaked form of Fig. 2(a) [8]. For $f_d \gg B$, it can be approximated as the sum of two peaks each of width $B/2$ (FWHM) separated by f_d, where the power spectrum of each peak decays as the inverse fourth power of the frequency deviation far from a peak.

In order to receive channel 1, the FP is tuned to the "1" peak. With $f_{FP} = B$ and $f_d \gg B$, the FSK signal is transformed to an intensity-modulated ASK signal, which can be received by direct detection. The bit-error-rate performance is limited by two separate mechanisms. The first mechanism may produce a false "1" when a "0" is transmitted either by the channel under consideration, channel 1, or by the adjacent higher-frequency channel, channel 2. Errors are produced by crosstalk, or leakage, of these "0"s through the FP. The other mechanism may produce a false "0" when a "1" is transmitted by the channel under consideration. These errors are produced by phase noise on the optical carrier, which is manifested by the laser linewidth f_l. For now, we assume the linewidth sufficiently small, $f_l / f_{FP} \ll 1$, as to be negligible; the effects of linewidth are considered in the next section. In the present section, we make crude but physically meaningful estimates of the maximum crosstalk from adjacent "0"s and represent it as an equivalent power penalty due to a degradation of the on–off ratio of the ASK signal at the receiver.

In a rigorous calculation of crosstalk interference, one would integrate the power spectrum of the total network over the FP transmission function $T(f)$. Since $T(f)$ decreases rapidly outside the passband for high finesse, the crosstalk is dominated by the "0" transmitted in the received channel and a possible "0" transmitted in the next-higher channel. We assume a "0" continuously transmitted by this next-higher channel but neglect contributions from other channels. (For a calculation that sums over all channels, see [14].) Since the analytic form of the "0" spectrum is not known and since f_d is several times larger than $f_{FP} = B$, we can make the following crude estimate of the overlap of "0" power spectrum and $T(f)$, where the peak of $T(f)$ and the two adjacent "0"s are separated by f_d and $(f_c - f_d)$, respectively. We represent the overlap as the sum of two components: a) the transmission through the FP of the unmodulated "0" carrier, i.e., the value of T at f_d or $(f_c - f_d)$ from the peak times $P/2$, where P is the total power in the channel, and b) the tail of the modulated "0" spectrum integrated over the FP passband, i.e., approximately the area under the function in (5a) times the value of the "0" power spectrum at the center of the passband. Consider the a) component first. From 5(b), the a) crosstalk component is approximately

$$P_I(a) = \frac{P}{2} \left\{ (2\mu_0)^{-2} + (2\mu_1)^{-2} \right\} \quad (19a)$$

for a single FP and

$$P_I(a) = \frac{P}{2} \left\{ (2\mu_0)^{-4} + (2\mu_1)^{-4} \right\} \quad (19b)$$

for the tandem FP with

$$\mu_0 = -f_d/B, \quad \mu_1 = (f_c - f_d)/B \quad (20a)$$

and $f_{FP} = B$. For the periodic spectrum of Fig. 2(b), for which $f_c = 2f_d$, we have

$$\mu_0 = -\mu_1 = \mu. \quad (20b)$$

In order to estimate the b) crosstalk component, it is necessary to assume a specific analytic form for the spectral peaks in Fig. 2. The squared-Lorentzian function provides an approximation that gives the required f^{-4} dependence far from the peak

$$G(\delta f) = \frac{4(\sqrt{2} - 1)^{1/2} P}{\pi B \left[1 + (\sqrt{2} - 1) \left(\frac{4\delta f}{B} \right)^2 \right]^2} \quad (21)$$

where δf is measured from the peak, the linewidth is $B/2$ and the power under each peak is $P/2$. (If necessary, a sharper rolloff than f^{-4} could be assured by prefiltering the baseband signal spectrum.) The crosstalk interference due to the "0" peak of the signal in channel 1 plus the interference due to the "0" peak of the signal in the next-higher-frequency channel, integrated over the FP passband, is approximately

$$P_I(b) = \frac{4(\sqrt{2} - 1)^{1/2} P}{\pi B}$$
$$\cdot \frac{\pi}{2} f_{FP} \left\{ \frac{1}{\left(1 + (\sqrt{2} - 1) \left(\frac{4 f_d}{B}\right)^2\right)^2} \right.$$
$$\left. + \frac{1}{\left(1 + (\sqrt{2} - 1) \left(\frac{4 (f_c - f_d)}{B}\right)^2\right)^2} \right\} \quad (22a)$$
$$= 2(\sqrt{2} - 1)^{1/2} P \left\{ \frac{1}{\left(1 + (\sqrt{2} - 1) (4\mu_0)^2\right)^2} \right.$$
$$\left. + \frac{1}{\left(1 + (\sqrt{2} - 1) (4\mu_1)^2\right)^2} \right\}$$
$$\approx 2(\sqrt{2} - 1)^{-3/2} P \left\{ (4\mu_0)^{-4} + (4\mu_1)^{-4} \right\},$$
$$\text{for } \mu_{0,1} > 1 \quad (22b)$$

for the single FP, where we assume $f_{FP} = B$, and

$$P_I(b) = P\left\{\frac{1}{\left(1 + (\sqrt{2} - 1)(4\mu_0)^2\right)^2} + \frac{1}{\left(1 + (\sqrt{2} - 1)(4\mu_1)^2\right)^2}\right\} \quad (23a)$$

$$\approx (\sqrt{2} - 1)^{-2} P\left\{(4\mu_0)^{-4} + (4\mu_1)^{-4}\right\},$$
$$(\mu_{0,1} > 1) \quad (23b)$$

for the tandem FP. The numerical coefficients in (22b) and (23b) are 7.5 and 5.8, respectively.

In order to maximize the number of channels for a given crosstalk interference level $P_I = P_I(a) + P_I(b)$, we calculate the minimum value of f_c as a function of f_d in (19a), (19b), (22b), and (23b). Taking derivatives, holding P_I constant, and setting $\partial f_c/\partial f_d = 0$, we find the condition $f_c = 2f_d$, or $\mu_0 = -\mu_1 = \mu$, corresponding to the periodic spectrum of Fig. 2(b). Thus, in the optimum case, the interference contributions from the "0" of the received channel and next-higher channel are equal.

The effect of crosstalk P_I on system performance can be regarded as a power penalty, since the "0" level at the receiver is P_I instead of zero. Thus, the effective off-on power increment is reduced from P to $(P - P_I)$, corresponding to an optical power penalty

$$p_I = -10 \log (1 - P_I/P). \quad (24)$$

If we consider $p_I = 0.1$ dB to be a negligible penalty, then we require $P_I/P \le 0.023$. For the single FP with $f_c = 2f_d$, from (19a) and (22b)

$$P_I/P = 0.25\mu^{-2} + 0.058\mu^{-4} \quad (25)$$

and the condition for negligible penalty is $\mu = f_d/B \ge 3.2$; the a) component is dominant due to the weak cutoff of the FP passband. For the tandem FP with $f_c = 2f_d$, from (19b) and (23b):

$$P_I/P = 0.063\mu^{-4} + 0.045\mu^{-4} = 0.108\mu^{-4} \quad (26)$$

and the condition is $f_d/B \ge 1.5$; the a) and b) components are comparable in size.

Thus, the allowable channel spacing is

$$f_c \ge 6.4B \quad (27)$$

for the single FP and

$$f_c \ge 3.0B \quad (28)$$

for the tandem FP, in order to assure $p_I < 0.1$ dB when the laser linewidth is negligible. As shown in Fig. 3, the number of channels N that can be accommodated within an effective FSR with effective finesse F is given by

$$\text{FSR} = (N - \tfrac{1}{2})f_c \quad (29a)$$

or for large N

$$N \approx \frac{\text{FSR}}{f_c} = \frac{Ff_{\text{FP}}}{f_c} = \frac{FB}{f_c}. \quad (29b)$$

Then, with (27) and (28):

$$N \approx \frac{F(\alpha)}{6.4} \quad (30a)$$

for the single FP and

$$N \approx \frac{F(\alpha\beta)}{3.0} \quad (30b)$$

for the tandem FP.

FFP's have been realized with $F = 200$ [3]. A network employing such a single FFP could support $N = 31$ channels according to (30a). A network employing a vernier tandem FFP could support

$$N = 0.11 F^2$$
$$= 4266 \text{ channels} \quad (31)$$

according to (17) and (30b). A network employing a small-ratio tandem FFP could support only a third as many channels:

$$N = 0.03 F^2$$
$$= 1250 \text{ channels} \quad (32)$$

according to (18) and (30a). In order to support a network with $N = 1000$ users, a single-FFP system would require $F = 6400$, but a vernier tandem FFP would require only $F = 95$ for the individual FFP's. If $B = 1$ Gbit/s, such a system would have a network capacity of 1 Tbit/s.

The limits in (27) and (28) take account of false "1's" due to the crosstalk by the neighboring "0" spectrum. The finite laser linewidth caused by phase fluctuations leads to false "1's" and "0's" due to the resultant intensity fluctuations, as we discuss in the next section.

IV. Effects of Laser Phase Noise

A real semiconductor laser is modulated by phase noise, which gives rise to a near-Lorentzian lineshape with linewidth f_l and side peaks at f_r due to relaxation oscillations [15]. The linewidth and the damping and frequency of the relaxation oscillations all depend upon laser structure and output power P_o. A typical value of $f_l P_o$ is 75 MHz · mW and of $f_r/\sqrt{P}$ is 3 GHz/mW$^{1/2}$. With careful control of laser structure and bias current, it should be possible to damp the relaxation peaks by more than 20 dB below the laser peak and to position them so as to cause minimal crosstalk interference. Therefore, we will neglect the relaxation peaks.

A. Effects when "0" is Transmitted

The optical power spectra represented in Figs. 2 and 3 must now be obtained as the convolution of baseband and carrier spectra. As far as crosstalk is concerned, the dominant effect is due to the asymptotic f^{-2} dependence of the Lorentzian laser carrier spectrum, since we have assumed an f^{-4} dependence [8] for the baseband FSK spectrum. The Lorentzian carrier spectrum for either a "0" or "1" peak is

$$H(\delta f) = \frac{P/\pi f_l}{1 + \left(\frac{2\delta f}{f_l}\right)^2} \quad (33)$$

where δf is measured from the frequency of the peak, f_l is the 3-dB linewidth and the power under each peak is $P/2$. Then we can proceed as in (18) and (19) to approximate the b) crosstalk component, neglecting the modulation. For the single FP

$$P_I(b) = \frac{Pf_{FP}/f_l}{1 + \left(\frac{2f_d}{f_l}\right)^2} \quad (34a)$$

$$\approx \frac{Pf_l/B}{(2\mu)^2} \quad (34b)$$

and for the tandem FP

$$P_I(b) = \frac{Pf_{FP}/2(\sqrt{2}-1)^{1/2}f_l}{1 + \left(\frac{2f_d}{f_l}\right)^2} \quad (35a)$$

$$\approx 0.78 \frac{Pf_l/B}{(2\mu)^2} \quad (35b)$$

where $f_{FP} = B$, $f_c = 2f_d$, $\mu = f_d/B$. If $f_l/B < 0.1$, then (34b) is small compared to (19a) and the phase noise does not contribute substantially to P_I for the single FP. Similarly, for $f_l/B < 0.1$, (35b) can be neglected in comparison with the sum of (19b) and (23b) for the tandem FP under spacing choices for which $p_I < 0.1$ dB.

B. Effects when "1" is Transmitted

The remaining effect of phase noise is the generation of a false "0" when a "1" is present. Energy builds up in a FP (as in a lumped resonant circuit) with a characteristic time $(\pi f_{FP})^{-1}$ (which is shorter than the bit period). However, the mean-square phase change over time t is [15] $\langle \Delta\phi^2 \rangle = 2\pi f_l t$, due to laser fluctuations. This phase change should be much less than unity over the characteristic time if the FP transmission is to permit reliable detection of a "1." That is, we require $f_l/f_{FP} = f_l/B \ll 1/2$. A numerical estimate [16] based on rigorous calculations [17] leads to the conclusion that the bit-error rate when a "1" is transmitted is (BER | "1" transmitted) $\leq 10^{-9}$ if $f_l/f_{FP} < 0.06$. Thus, these crude estimates show that the laser phase noise can be neglected if f_l/B is sufficiently small, say 0.05 to 0.1, which is similar to the linewidth constraint for FSK with heterodyne detection [17].

V. EXPERIMENT

The principle of operation of the FDM-FSK network was demonstrated with a 4 × 4 single-mode star coupler and single-FFP demultiplexer, as shown schematically in Fig. 5. Two channels were provided by packaged and pigtailed external-waveguide-Bragg-reflector lasers [10] with $f_l \approx 4$ MHz and relatively flat FM response $\rho \approx 100$ MHz/mA over 100 kHz to 1 GHz. Except for the insertion of bulk isolators, the entire network is connected with single-mode fiber. Each laser is driven at 45 Mbit/s by a pseudorandom word generator with word length ($2^{10} - 1$), limited primarily by available word lengths and the FM response of the laser. The lasers are mounted on a common heat-sink plate with separate thermoelectric coolers that maintain a controllable frequency separation that is stable over suitably long periods of time for the experiment. A frequency-lock circuit based on a fixed stabilizer-FFP [18], a balanced mixer, and a sinusoidal dither oscillation [19] of I at ~40 kHz for one laser and at ~50 kHz for the other could also be employed to provide a fixed f_c corresponding to the FSR of the stabilizer-FFP. A similar stabilizer circuit dithers the voltage applied to the demultiplexer-FFP at 10 kHz to lock it on the "1" peak of the received channel. The parameters of the demultiplexer-FFP are $f_{FP} = 115$ MHz, FSR = 5 GHz, $F = 43$, insertion loss = 1.5 dB. Tuning or scanning of the FFP's is accomplished by using a PZT piezoelectric actuator [3]. In the most recent versions of these FFP's the voltage required is below 10 V/order. The remaining output ports of the 4 × 4 single-mode star coupler are used for diagnostics, including a swept analyzer-FFP and a RF spectrum analyzer. The former is used to measure f_c and f_d; and the latter to measure f_l when two unmodulated carriers are present, and the FSK baseband spectrum when one modulated and one unmodulated carrier are present. The observed baseband spectrum, Fig. 6, was in good agreement with theoretical expectations [8] for $\mu = 3.2$.

A *differential* power penalty $\tilde{p}_I$ was measured as a function of f_c for fixed $f_d = 3.2B = 144$ MHz using the demultiplexer-FFP mentioned above, for which $\eta = f_{FP}/B = 2.6$ (rather than the minimum value of unity). The spectrum of the two channels is as illustrated in Fig. 2(a). In this measurement, a 0-dB reference was established by measuring bit-error-rate (BER) vs relative receiver power for only one channel present. Then, $\tilde{p}_I$ is the added power in the first channel required to maintain a given BER (10^{-8} for easy measurement) when a second channel separated by f_c (with power equal to the initial power in the first channel) is present. For each BER measurement, the threshold level is optimized for minimum BER. Fig. 7(a) is a plot of BER versus $\tilde{p}_I$ with f_c as a parameter.

The crosstalk interference for this experiment can be calculated from (19a) and (22b) as

$$P_I/P = \frac{(\eta/\mu)^2}{8}\left[1 + (2\sigma - 1)^{-2}\right]$$

$$+ \frac{\eta}{128(\sqrt{2}-1)^{3/2}\mu^4}\left[1 + (2\sigma - 1)^{-4}\right]$$

(36)

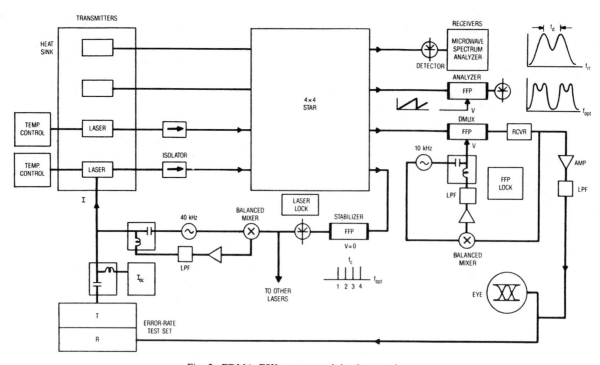

Fig. 5. FDMA-FSK star network implementation.

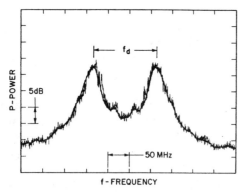

Fig. 6. Baseband power spectrum obtained by mixing an unmodulated laser with an FSK-modulated laser at a slightly higher frequency.

where $\sigma = f_c/2f_d$. Equation (36) reduces to (25) for $\eta = 1$ and $\sigma = 1$. In the experiment, the crosstalk from the "0" spectrum of the received channel is unchanged and only the crosstalk from the "0" spectrum of the adjacent channel varies with f_c. Thus, the differential crosstalk measured in the experiment[1] is the value given in (36) less its value for $\sigma \to \infty$, i.e.:

$$\tilde{P}_I/P = \frac{(\eta/\mu)^2}{8(2\sigma - 1)^{-2}}$$
$$+ \frac{\eta}{128(\sqrt{2} - 1)^{3/2} \mu^4 (2\sigma - 1)^4} \quad (37)$$

The measured $\tilde{p}_I$ is compared with the theoretical result from (36) in Fig. 7(b), which is derived from the data in Fig. 7(a) at BER = 10^{-8}. The channel separation at the

---

[1] The authors are grateful to G. L. Abbas for pointing out their failure to subtract the $\sigma \to \infty$ component in [1, Fig. 3].

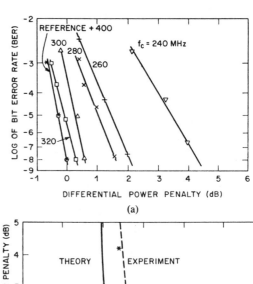

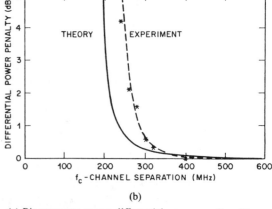

Fig. 7. (a) Bit-error-rate versus differential power penalty with respect to the single-channel case with channel spacing as a parameter. The reference measurements without the second channel are shown by asterisks. (b) Differential power penalty $\tilde{p}_I$ versus channel spacing. (Solid curve: theory, asterisks: experiment).

knee is about 50 MHz greater for the experiment. For p_I = 1 dB penalty with this nonoptimum network, theory gives f_c = 260 MHz = $5.8B$ and experiment gives f_c =

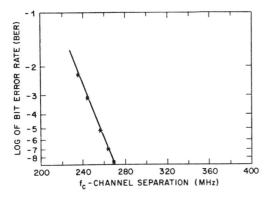

Fig. 8. Bit-error-rate in one channel as a function of f_c for one interfering channel.

285 MHz = $6.3B$. This close correspondence between theory and experiment is remarkable in view of the crude approximations in the crosstalk analysis and the sizable experimental error in the measurements. Additional sources of the discrepancy may be attributed to the finite laser linewidth, which has been neglected in (36), residual birefringence in the FFP [4], and effects of thermal noise in the receiver.

An alternative method for measuring minimum channel spacing is simply to measure BER in one channel in the presence of an interfering channel as f_c is reduced. The measurements shown in Fig. 8 indicate a minimum f_c = 270 MHz = $6B$ for BER < 10^{-9}.

VI. CONCLUSIONS

The crude network calculations made in Section IV have been shown to provide a good estimate of experimental two-channel performance in Section V. Even with non-optimum parameters (f_{FP}/B = 2.6), the experiment shows satisfactory BER performance with $f_c \approx 6B$, indicating that the criterion for negligible power penalty $p_I <$ 0.1 dB may be too conservative.

The experimental ratio of laser linewidth to FP bandwidth, f_l/f_{FP} = 0.034, was small enough to be negligible. At higher values of B and with $f_{FP} \approx B$, the external-waveguide-Bragg-reflector laser should, therefore, be suitable. Even a simple DFB laser with $f_l \approx$ 40 MHz may be satisfactory above B = 400 Mbit/s or at lower B with $f_{FP}/B > 1$. However, lasers with a flat FM response from low frequency (depending on coding constraints) up to the operating bit rate are essential for low BER. The external-waveguide-Bragg-reflector laser should be suitable up to $B \sim$ 1.5 Gbit/s.

The agreement between the analysis and the two-channel experiment encouraged us to speculate on larger capacity networks. A single-FFP network with F = 200 [3] can support N = 31 users at any bit rate B. Improved fabrication techniques may give single FFP's with $F \approx$ 1000 and $N \approx$ 156. Tuning speeds, limited by the piezoelectric transducer, may approach a few microseconds.

Vernier FFP's have been demonstrated experimentally [7], although a number of practical problems remain. However, estimates obtained here show that N = 1000 is allowed with readily fabricated individual FFP's with F = 95. With B = 1 Gbit/s, such a network offers the prospect of 1-Tbit/s capacity.

In the simplest network configuration, the equispaced laser transmitter frequencies could be fixed while the FFP in each receiver tunes over the full network bandwidth. The stabilization of a comb of 1000 carrier frequencies is a very challenging problem that is just beginning to be addressed. Ideally, each carrier would be provided by the same model of laser that can be tuned over the network band (e.g., ~3 THz or ~180 Å for the 1-Tbit/s network) or by a small set of models that can be tuned over segments of the band.

Similar channel spacing and capacity could be realized in a FSK network with heterodyne detection [20]. In that case, however, one would require a local oscillator laser capable of continuously tuning over the entire network band of ~3 THz. In addition, some means of polarization control or polarization diversity would be required. In either case, considerable progress in component and network research is required before a practical FDM network can be envisioned.

ACKNOWLEDGMENT

It is a pleasure to thank P. J. Anthony, T. E. Darcie, B. Glance, L. J. Greenstein, P. S. Henry, B. L. Kasper, J. Talman, and S. Yang for their help.

REFERENCES

[1] I. P. Kaminow, P. P. Iannone, J. Stone and L. W. Stulz, "FDM-FSK star network with a tunable optical filter demultiplexer," *Electron. Lett.*, vol. 23, p. 1102, 1987.

[2] S. R. Mallinson, "Wavelength-selective filters for single-mode fiber WDM systems using Fabry-Perot interferometers," *Appl. Opt.*, vol. 26, pp. 430-436, 1987.

[3] J. Stone and L. W. Stulz, "Pigtailed high-finesse tunable fiber Fabry-Perot interferometers with large, medium, and small free spectral ranges," *Electron. Lett.*, vol. 23, p. 781, 1987.

[4] J. Stone, "Stress-optic effects, birefringence, and reduction of birefringence by annealing fiber Fabry-Perot interferometers," *J. Lightwave Technol.*, to be published.

[5] W. Gunning, "Double-cavity electrooptic Fabry-Perot filter," *Appl. Opt.*, vol. 21, p. 3129, 1982; G. Hernandez, *Fabry-Perot Interferometers*. New York: Cambridge Univ. Press, 1986, ch. 4.

[6] A. A. M. Saleh and J. Stone, "Two-stage Fabry-Perot filters as demultiplexers in optical FDMA LAN's," *J. Lightwave Technol.*, to be published.

[7] I. P. Kaminow, P. P. Iannone, J. Stone, and L. W. Stultz, "A tunable vernier fiber Fabry-Perot filter for FDM demultiplexing and detection," presented at CLEO '88, Anaheim, CA, Apr. 25, 1988, pap. WA-2.

[8] R. W. Lucky, J. Salz, and E. J. Weldon, Jr., *Principles of Data Communication*. New York: McGraw-Hill, 1968, p. 202.

[9] M. Born and E. Wolf, *Principles of Optics*. New York: Macmillan, 1959, p. 329.

[10] N. A. Olsson, C. H. Henry, R. F. Kazarinov, H. J. Lee, R. J. Orlovsky, B. H. Johnson, R. E. Scotti, D. A. Ackerman, and P. J. Anthony, "Performance characteristics of 1.5-μm single-frequency semiconductor laser with external waveguide Bragg reflector," *Electron. Lett.* vol. 23, p. 687, 1987.

[11] N. A. Olsson, B. H. Johnson, H. J. Lee, C. H. Henry, R. F. Kazarinov, D. A. Ackerman, and P. J. Anthony, "Coherent transmission using external waveguide Bragg reflector laser with direct frequency modulation," *Electron. Lett.*, vol. 23, 687, 1987.

[12] S. Yamazaki, K. Emura, M. Shikada, M. Yamaguchi, and I. Mito, "Realization of flat FM response by directly modulating a phase tunable DFB laser diode," *Electron. Lett.*, vol. 21, pp. 283-285, 1985.

[13] S. Murata, I. Mito, and K. Kobayashi, "Frequency modulation and spectral characteristrics for a 1.5-μm phase-tunable DFB laser," *Electron. Lett.*, vol. 23, pp. 12-14, 1987.

[14] S. R. Mallinson, "Crosstalk limits of Fabry-Perot demultiplexers," *Electron Lett.*, vol. 21, pp. 759-760, 1985.

[15] C. H. Henry, "Phase noise in semiconductor lasers," *J. Lightwave Tech.*, vol. LT-4, pp. 298-311, 1986.

[16] L. J. Greenstein, private communication.

[17] G. J. Foschini, L. J. Greenstein, and G. Vannucci, "Noncoherent detection of coherent lightwave signals corrupted by phase noise," *IEEE Trans. Commun.*, vol. 36, pp. 306-314, 1988.

[18] B. Glance, P. J. Fitzgerald, K. J. Pollock, J. Stone, C. A. Burrus, G. Eisenstein, and L. W. Stulz, "Frequency stabilization of FDM optical signals," *Electron. Lett.*, vol. 23, p. 750 1987.

[19] A. Sollberger, A. Heinamaki, and H. Melchior, "Frequency stabilization of semiconductor lasers for applications in coherent communication systems," *J. Lightwave Technol.*, vol. LT-5, pp. 485-491, 1987.

[20] L. G. Kazovsky, "Multichannel coherent optical communication systems," *J. Lightwave Technol.*, vol. LT-5, pp. 1095-1102, 1987.

Section 3.2: Switches

Space-Division Switching

Architectures for Large Nonblocking Optical Space Switches

RON A. SPANKE

Abstract—This paper introduces three architectures for optical space switches that are based on a multiplicity of fiber interconnected optical components. The architectures eliminate the need for optical waveguide crossovers and reduce the complexity required in the individual elements. The architectures are strictly nonblocking and allow for easy control and routing. Architecture type 1 exhibits a low system attenuation and a high system signal-to-noise ratio for very large switch dimensions. Architectures 2 and 3 are alternatives for realizing broadcast and point-to-point architectures.

INTRODUCTION

SEVERAL architectures have been proposed for optical space switches fabricated on titanium-diffused lithium niobate (Ti:LiNbO$_3$) substrates [1]-[4]. These architectures have been constructed using various optical switching elements including directional coupler switches [5], total internal reflection (TIR) switches [6], and crossing X switches [7]. In addition, many of the classical switching architectures found in the electronic and communication domains could be implemented with photonic switching elements in the optical domain. Such architectures include the Clos, Benes, Banyan, omega, and shuffle networks to name a few.

All of these architectures could be fabricated optically; however, they all encounter difficulties when expanded to large switch dimensions where they become limited by system attenuation, system signal-to-noise ratio (SNR), and LiNbO$_3$ real estate. In addition, many of these architectures require the signal paths to cross through one another on the optical substrate between the switching elements. These passive integrated optical waveguide crossovers appear feasible [8]; however, they contribute additional attenuation and crosstalk problems for these architectures. Optimal switching architectures in the electronic domain where one attempts to minimize crosspoints are not the optimal switching architectures in the optical domain where one attempts to minimize attenuation and maximize SNR.

This paper proposes three space-division optical switching architectures based on passive splitters, passive combiners, active splitters, and active combiners. All three architectures use a multiplicity of interconnected optical components. This eliminates the need for any integrated optical waveguide crossovers and allows for large

switch dimensions without complex integration onto individual LiNbO$_3$ substrates. The three architectures have widely different attenuation and SNR characteristics, one of which exhibits excellent performance in both areas.

BASIC OPTICAL COMPONENTS

1:N passive splitters divide optical power evenly into N channels, while N:1 passive combiners combine the power of N inputs into a single output. These passive devices could be realized with guided wave devices on LiNbO$_3$ [9] or by fiber couplers [10]. Passive splitters incur at least a 3k (dB) power loss from input to each output where k represents log$_2 N$. Most present devices that perform the passive combiners function also incur a 3k (dB) power loss from any given input to the output.

1 × N active splitters and N × 1 active combiners are fabricated with 1 × 2 and 2 × 1 optical switch elements on Ti:LiNbO$_3$. These elements could be directional couplers, X switches, or any other photonic switch element. Each 1 × N splitter or N × 1 combiner is constructed from $N - 1$ switch elements arranged in a k-stage binary tree configuration (Fig. 1). To switch from the input port to the required output port, k switches (one in each stage) need to be activated. All of the elements in a given stage could possibly be electrically tied together so that only k control leads and electronic drivers would be required instead of $N - 1$. A 1 × 16 polarization-independent optical switch with 15 directional couplers in a four-stage binary tree structure has recently been demonstrated [11].

ARCHITECTURES

The three optical switching architectures described here are constructed of active splitters and active combiners (type 1), passive splitters and active combiners (type 2), or active splitters and passive combiners (type 3). A fouth type constructed of passive splitters and passive combiners is merely a novel method of implementing an $N \times N$ transmissive star. It is not switchable without an additional stage of switches on each line in the center of the switch and will not be considered in this paper.

Architectures type 1 consists of N 1 × N active splitters on the left interconnected by fiber to N N × 1 active combiners (Fig. 2) to form a nonblocking optical interconnection network. Types 2 and 3 replace the active splitters or the active combiners with their passive counterparts. An $N \times N$ switch of type 1 requires 2$N(N - 1)$ optical switch

Manuscript received October 11, 1985; revised February 4, 1986.
The author is with AT&T Bell Laboratories, Naperville, IL 60566.
IEEE Log Number 8608238.

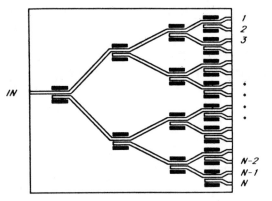

Fig. 1. 1 × N active splitter using k stages of 1 × 2 optical switch elements in a binary tree structure.

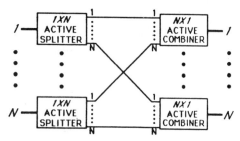

Fig. 2. N × N optical switch (type 1).

elements, while 2 and 3 require $N(N - 1)$ switch elements and $N(N - 1)$ 50/50 splitting(combining) elements.

Because optical switch types 1 and 3 use an active splitter, they function as a strictly nonblocking point-to-point switch. No rearrangement of any optical signal paths is ever required. Type 2 uses passive splitters which provide every input signal to every output combiner. The active combiner will then select the desired signal to leave the output. Architecture type 2 can therefore function as a strictly nonblocking broadcast switch. Any output can listen to any input, even if other outputs are currently listening to the same input.

ATTENUATION CHARACTERISTICS

All three architecture types have insertion losses proportional to k instead of proportional to the switch dimension N. The insertion loss in an active splitter or active combiner depends on the number of optical switch elements that the optical signal passes through and is given by

$$L_{AS} = L_{AC} = k \cdot L \text{ (dB)} \qquad (1)$$

where L represents the insertion loss in a given switch element. This L includes the straight waveguide attenuation (in dB/cm) due to scattering losses, the waveguide bending losses, and the losses due to incomplete coupling that are associated with each switch element.

The insertion loss in a passive splitter and passive combiner is given by

$$L_{PS} = L_{PC} = k \cdot (3 + E) \text{ (dB)} \qquad (2)$$

The 3 dB figure represents the 50/50 power split and E represents the excess loss in each passive power split. The insertion losses IL for the three architectures are then given by

type 1: $\qquad IL = 2 \cdot k \cdot L + 4 \cdot W$ (dB) $\qquad (3)$

types 2 and 3: $\qquad IL = k \cdot (3 + L + E) + 2 \cdot W$ (dB)
$\qquad (4)$

where W represents the waveguide-to-fiber coupling loss which is roughly 1-2 dB at each interface. The insertion losses for types 2 and 3 assume fiber devices for the passive components. If guided wave $LiNbO_3$ devices are used for the passive components, types 2 and 3 would have four waveguide/fiber transitions.

For comparison purposes, the worst case attenuation for the crossbar architecture [3] that uses switch elements having an insertion loss of L is given by

crossbar: $IL = (2N - 1) \cdot L + 2 \cdot W$ (dB). $\qquad (5)$

SNR CHARACTERISTICS

Architecture type 1 benefits from the selection of the desired signal in both the active splitter side and in the active combiner side. Every optical input that is not the desired input for a given output represents a noise signal for that output. The double selection forces every possible source of noise to be attenuated by at least two optical extinction ratios before leaking into a desired signal. This squared crosstalk isolation allows a large optical switch to have a better signal-to-noise ratio than the individual switch elements that make up the switch.

All signal and noise paths through the switch pass through the same number of devices ($2k$). The insertion loss of these devices will appear in both the signal and noise terms and will cancel out of the final SNR. The insertion losses due to the four fiber/waveguide transitions will also appear in both the signal and noise terms and will cancel out of the final SNR. The power at the input of each 1 × N active splitter is P_{IN}. The power of each of the N signals leaving the active splitter will differ based on the number of switch elements passed with an extinction ratio X (power transmittance, i.e., -20 dB $= 0.01$). The number of switch extinctions encountered will vary from zero for the desired signal channel to k for the most attenuated. The number of channels with various extinction ratios will follow a binominal distribution (Table I).

Given the fiber interconnect pattern between the splitting and combining stages, it can be shown that at the input of every $N \times 1$ combiner, the N inputs will also exhibit this binomial distribution of switch extinctions. Inside the active combiner, the desired signal will pass with no further extinction, while each of the noise terms will encounter from 1 to k additional switch extinctions. The worst case SNR will occur when the k highest power noise channels from the active splitter $P_{IN} \cdot X$ pass through the active combiner with only one additional switch extinction. The binomial distribution of worst case power levels at the $N \times 1$ active combiner output is given in Table II.

The total noise power for architecture type 1 is the sum

TABLE I
DISTRIBUTION OF POWER LEVELS LEAVING A $1 \times N$ ACTIVE SPLITTER

Splitter Extinctions	Power$_{(WATTS)}$	Number of Channels N=64	General
NONE	P_{IN}	1	1 (desired channel)
1	$P_{IN} \cdot X$	6	$\frac{k}{1!}$
2	$P_{IN} \cdot X^2$	15	$\frac{k \cdot (k-1)}{2!}$
3	$P_{IN} \cdot X^3$	20	$\frac{k \cdot (k-1) \cdot (k-2)}{3!}$
4	$P_{IN} \cdot X^4$	15	$\frac{k \cdot (k-1) \cdots (k-3)}{4!}$
5	$P_{IN} \cdot X^5$	6	$\frac{k \cdot (k-1) \cdots (k-4)}{5!}$
k	$P_{IN} \cdot X^k$	1	$\frac{k \cdot (k-1) \cdots 3 \cdot 2 \cdot 1}{k!}$

TABLE II
DISTRIBUTION OF POWER LEVELS AT $N \times 1$ ACTIVE COMBINER OUTPUT

Combiner Extinctions	Power$_{(WATTS)}$	Number of Channels N=64	General
NONE	P_{IN}	1	1 (desired channel)
1	$P_{IN} \cdot X \cdot X$	6	$\frac{k}{1!}$
2	$P_{IN} \cdot X^2 \cdot X^2$	15	$\frac{k \cdot (k-1)}{2!}$
3	$P_{IN} \cdot X^3 \cdot X^3$	20	$\frac{k \cdot (k-1) \cdot (k-2)}{3!}$
4	$P_{IN} \cdot X^4 \cdot X^4$	15	$\frac{k \cdot (k-1) \cdots (k-3)}{4!}$
5	$P_{IN} \cdot X^5 \cdot X^5$	6	$\frac{k \cdot (k-1) \cdots (k-4)}{5!}$
k	$P_{IN} \cdot X^k \cdot X^k$	1	$\frac{k \cdot (k-1) \cdots 3 \cdot 2 \cdot 1}{k!}$

of all $N - 1$ unwanted channels that find their way to the worst case output and is given by

$$P_{\text{noise}}(\text{type 1}) = P_{IN} \cdot \sum_{i=1}^{k} \frac{k!}{i! \cdot (k-i)!} \cdot X^{2i} \quad (W). \quad (6)$$

Ignoring X^4 and higher order crosstalk terms, the worst case SNR for type 1 can be approximated by

$$\text{SNR}_{\text{type 1}} = 2 \cdot X_{dB} - 10 \cdot \log k \quad (\text{dB}). \quad (7)$$

Type 2, being a broadcast architecture, does not select the desired signal in the splitter stage and does not benefit from the double crosstalk isolation in the SNR characteristics. The optical power entering the active combiner stage is the same for the signal channel and all noise channel and is represented by P_{comb}.

$$P_{\text{comb}} = P_{IN} - k \cdot (3 + E) \quad (\text{dB}). \quad (8)$$

The noise power for type 2 is the sum of the $N - 1$ unwanted signal powers entering an active combiner and is also binomial in nature.

$$P_{\text{noise}}(\text{type 2}) = P_{\text{comb}} \cdot \sum_{i=1}^{k} \frac{k!}{i! \cdot (k-i)!} \cdot X^{i} \quad (W). \quad (9)$$

A first-order approximation for the SNR yields

$$\text{SNR}_{\text{type 2}} = X_{dB} - 10 \cdot \log k \quad (\text{dB}). \quad (10)$$

Architecture type 3 experiences the same binomial distribution of extinction ratios leaving the active splitter stage as shown in Table I. Given the fiber interconnect pattern between stages, this binomial distribution of coupler extinctions is also present at the inputs to every $N:1$ passive combiner. The passive combiner combines the signal channel with these $N - 1$ channels of noise with an inherent $k \cdot (3 + E)$(dB) attenuation on all channels.

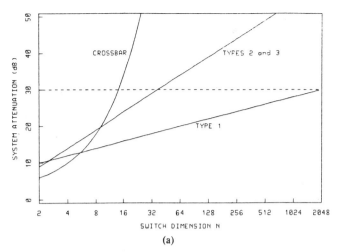

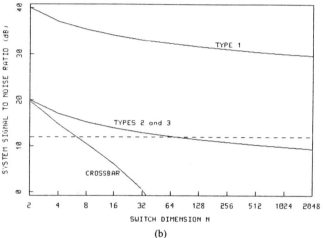

Fig. 3. System attenuation (a) and system signal-to-noise ratio (b) versus switch dimension for architecture types 1, 2, 3 and conventional crossbar switches.

Therefore, the noise power and SNR for architecture type 3 are the same as for type 2.

For comparison purposes, the worst case SNR of the crossbar [3] implemented with optical switch elements that have an extinction ratio of X_{dB} is given by

$$\text{SNR}_{\text{crossbar}} = X_{dB} - 10 \cdot \log(N - 1) \quad (\text{dB}). \quad (11)$$

SWITCH DIMENSION LIMITS

Fig. 3(a) shows the system insertion loss versus switch dimension N [(3) and (4)] for the three architecture types presented here. The values for the switch element insertion losses, excess losses, and fiber/waveguide coupling losses are assumed to be $E = L = 1$ dB and $W = 2$ dB. For this comparison, the maximum attenuation allowed from system input to output without amplification or regeneration is assumed to be 30 dB. Fig. 3(b) shows the system SNR versus switch dimension [(7) and (10)] for the three architecture types with $X_{dB} = 20$ dB. The required SNR is assumed to be greater than 11 dB to achieve a 10^{-9} bit error rate. The figures also show insertion loss and SNR characteristics [(5) and (11)] for conventional crossbar switches.

With $E = L = 1$ dB, $W = 2$ dB, and $X_{dB} = -20$ dB.

a maximum switch dimension of 32 × 32 should be possible for architecture types 2 and 3, yielding a system attenuation of 29 dB and a 13 dB SNR. With the same device characteristics, a switch dimension of 1024 × 1024 is theoretically possible with type 1 using ten-stage active splitters and ten-stage active combiners. This yields a 28 dB system attenuation with a 30 dB system SNR. Even with very large switch dimensions, SNR is not a limiting factor, and the system SNR for type 1 is considerably better than the extinction ratios of the optical switch elements themselves. In this example, we based the maximum switch dimension for architecture 1 only on the insertion loss, but in reality, fiber interconnection now becomes the limiting factor.

Larger switch dimensions could possibly be realized without complex LiNbO$_3$ integration and unreasonable fiber I/O by separating the active splitter and active combiner functions into two stages each. This increases the overall optical chip count and system attenuation, thereby reducing the maximum switch dimension, but it allows for simpler components and easier system interconnection.

Conclusion

Three space division optical switch architectures have been presented based on active and passive splitting and combining elements. Type 1 is a point-to-point switch and exhibits excellent SNR and attenuation characteristics for very large switch matrices. Type 2 is a broadcast switch exhibiting moderately good SNR and attenuation characteristics up to a 32 × 32 switch. Type 3 is a point-to-point switch exhibiting the same system characteristics as type 2 and is useful in matrices up to 32 × 32.

The type 1 and type 2 architectures appear to be well suited for large point-to-point and broadcast optical switches. These architectures offer larger switch dimensions and improved attenuation and signal-to-noise performance over other optical switching architectures proposed to date.

References

[1] M. Kondo et al., "Integrated optical switch matrix for single-mode fiber networks," *IEEE J. Quantum Electron.*, vol. QE-18, pp. 1759-1765, Oct. 1982.
[2] R. A. Spanke and V. E. Benes, "An N-stage planar optical permutation network," to be published.
[3] H. S. Hinton, "A nonblocking optical interconnection network using directional couplers," in *Proc. GLOBECOM*, 1984, pp. 26.5.1-26.5.5.
[4] L. McCaughan and G. A. Bogert, "4 × 4 strictly nonblocking integrated Ti:LiNbO$_3$ switch architecture," in *Proc. OFC/OFS '85*, San Diego, CA, Feb. 1985, p. 76.
[5] R. C. Alferness, R. V. Schmidt, and E. H. Turner, "Characteristics of Ti-diffused lithium niobate optical directional couplers," *Appl. Opt.*, vol. 18, pp. 4012-4016, Dec. 1, 1979.
[6] C. S. Tsai, B. Kim, and F. R. El-Akkari, "Optical channel waveguide switch and coupler using total internal reflection," *IEEE J. Quantum Electron.*, vol. QE-14, pp. 513-517, July 1978.
[7] A. Neyer and W. Mevenkamp, "Single-mode electrooptic X-switch for integrated optic switching networks," in *Proc. IEE 2nd European Con. Integrated Opt.*, no. 227, Oct. 17-18, 1983, pp. 136-139.
[8] E. E. Bergmann, L. McCaughan, and J. E. Watson, "Coupling of intersecting Ti:LiNbO$_3$ diffused waveguides," *Appl. Opt.*, vol. 23, pp. 3000-3003, Sept. 1, 1984.
[9] T. Findakly and B. V. Chen, "Single-mode integrated optical 1 × N star couplers," in *Tech. Dig. Topical Meet. Opt. Fiber Commun.*, 1983, paper ML-2.
[10] S. K. Sheem and T. G. Giallorenzi, "Single-mode fiber-optical power divider: Encapsulated etching technique," *Opt. Lett.*, vol. 4, p. 29, 1979.
[11] J. E. Watson, "Polarization independent 1 × 16 optical switch using Ti:LiNbO$_3$ waveguides," in *Proc. OFC/OFS '85*, paper WK-4, Feb. 1985, p. 110.

Dilated Networks for Photonic Switching

KRISHNAN PADMANABHAN AND ARUN N. NETRAVALI, FELLOW, IEEE

Abstract—We present some novel architectures for rearrangeably nonblocking multistage photonic space switches implemented using arrays of Ti:LiNbO$_3$ directional couplers. Multistage networks, studied mostly in the electronic domain, are obtained by minimizing the number of 2×2 elements needed to implement a switch. Unfortunately, straightforward extensions of these networks to the photonic domain show that the switch size has to be severely limited by the crosstalk in each of the Ti:LiNbO$_3$ 2×2 switching elements. Our networks, on the other hand, have a controllable (including almost zero) amount of crosstalk, low optical path loss, and an asymptotically optimal number of directional coupler switches for a given switch size. In addition, the switch has a simple control algorithm and its performance for light loading appears very promising. The switch is easily decomposable into smaller arrays of no more than two types, making it easy to partition the switch into chips. At the cost of a slight increase in crosstalk, the switch can be made single fault tolerant in terms of its ability to connect any input to any output.

I. INTRODUCTION

WIDE-BAND optical signals can be switched under electronic control using directional couplers between Ti:LiNbO$_3$ waveguides on a planar LiNbO$_3$ crystal [1]. The basic switching element is a directional coupler with two active inputs and two active outputs. Depending on the amount of voltage at the junction of the two waveguides which carry the two input signals, either of the two inputs can be coupled to either of the two outputs. Several architectures have been proposed to construct an $N \times N$ ($N > 2$) switch with the 2 × 2 directional coupler as the basic component [6], [12]. These architectures are essentially analogs of similar architectures for electronic switching and interconnection networks [7], [8]. However, due to the difference in characteristics of the electronic and optical switching elements, performance of the optical architectures is significantly different.

Performance of optical architectures may be characterized by the following parameters:
1) optical path loss (worst case)
2) crosstalk (worst case)
3) number of switching elements necessary to implement an $N \times N$ switch.

The attenuation of light passing through the switch has several components: 1) fiber-to-switch and switch-to-fiber coupling loss, 2) propagation loss in the medium, 3) loss at waveguide bends, and 4) loss at the couplers on the substrate. In a large switch, a substantial part of this attenuation that depends on the switch architecture is directly proportional to the number of couplers that the optical path passes through (about 0.25–0.50 dB per coupler).[1] Therefore, for the purposes of this paper, the number of couplers in an input-output path will be used to characterize the optical path loss.

Optical crosstalk results when two signal channels interact with each other. There are two ways in which optical paths can interact in a planar switching network. The channels (waveguides) carrying the signals could cross each other in order to imbed a particular topology. We call this a *channel crossover*. Alternatively, two paths sharing a switching element will experience some undesired coupling from one path to the other. We call this a *switch crossover*. Experimental results recently reported [4] show that it is possible to make the crosstalk from passive intersections of optical waveguides negligible (by keeping the intersection angles above a certain minimum amount). Thus, for the purposes of this paper, we will assume that switch crossovers are the major source of crosstalk in optical switching networks constructed out of directional couplers. Although the total crosstalk in a switch is a complex combination of the crosstalk in individual couplers, we take the number of couplers with two active inputs in the optical path as a measure of crosstalk.

The number of switching elements necessary to implement the switching network is a good measure of the cost of the network.[2] As device integration permits more and more couplers to be fabricated on a single chip, the cost measure will have to be revised somewhat. (We will consider this issue in Section VI of the paper.) However, unless the entire network can be fabricated on a single chip, partitioning arguments will ultimately bring us back to the number of switching elements needed to implement the network.

The principal contribution of this paper is a class of permutation networks, called *dilated Benes networks*, that guarantee connections between inputs and outputs with zero switch crossovers and hence negligible crosstalk. These networks are asymptotically optimal in terms of switch count ($2N \log N - N/2$) and have a simple routing algorithm.[3] In addition, the architecture provides a tradeoff between the number of switching elements and crosstalk (if a crosstalk budget is given). The network topology, control algorithm, and performance are presented in Sections III–V of the paper. In Section VI, we consider how such networks might be decomposed for fabrication if a level of integration higher than a directional coupler per chip is available. Finally, in Section VII, we illustrate how the techniques in this paper can be applied to other multistage networks, like the Omega network [10] and the Clos network [2].

II. PRIOR WORK

We begin by taking a brief look at interconnection networks that have been proposed in the literature for the purpose of permuting input lines. These are summarized in Figs. 1 and 2 and have been extracted from [7, Fig. 1(b) and (c)] and [8, Fig. 1(a)]. Fig. 1(a) is the classical crossbar implementation, in which each input is fanned out to all the outputs, and at each output, the lines from all the inputs are fanned in. If each fan-in or fan-out unit is implemented by directional couplers with

---

Paper approved by the Editor for Communication Switching of the IEEE Communications Society. Manuscript received January 23, 1987; revised April 27, 1987. This paper was presented at the Topical Meeting on Photonic Switching, Incline Village, NV, March 1987.

The authors are with the Computer Technology Research Laboratory, AT&T Bell Laboratories, Murray Hill, NJ 07974.

IEEE Log Number 8717491.

[1] Some recent experiments, e.g., [3], indicate that this loss could be decreased substantially. However, such low-loss couplers are still difficult to make.

[2] This neglects the possibility that if crosstalk is not a factor in the design of the 2 × 2 Ti:LiNbO$_3$ switching elements, then their design can be considerably simplified, resulting in lower cost for the overall switch.

[3] All logarithms in this paper are to the base 2 unless otherwise specified.

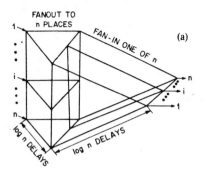

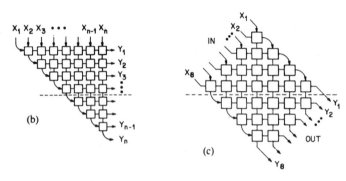

Fig. 1. Permutation network architectures. (a) Crossbar. (b) The triangular array. (c) The diamond array.

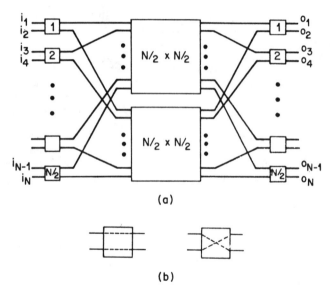

Fig. 2. Benes network construction. (a) Recursive construction. (b) Switching element and its two states (straight and cross).

not only optical path loss, but also crosstalk. If crosstalk is assumed to be from the adjacent channel in each of the 2×2 directional couplers, in a fully loaded rearrangeable array, the number of directional couplers contributing to crosstalk along any path is $(2 \log N - 1)$. It turns out that based on the budget for crosstalk in a realistic communication system and the current characteristics of directional couplers, crosstalk limits the size of the switch more than optical path loss [6], [12].

In the following sections, we present network architectures based on the rearrangeable array that will reduce the number of switch crossovers to zero (thereby reducing crosstalk to a negligible amount) while increasing the switch count to $(2N \log N - N/2)$, the channel crossovers to approximately $2N^2 - 2N \log N$, and the optical path length to $2 \log N$. Moreover, our networks have the additional flexibility of trading off crosstalk with the complexity of the network as measured by the number of couplers required to implement the network.

III. THE DILATED BENES NETWORK

A. The Benes Network

Permutation networks, defined as switching networks capable of all $N!$ permutations of the N inputs, have been extensively studied in the context of telephone switching [2], [14], [7]. A rearrangeable switching array, widely known as the *Benes network*, that achieves this function is shown in Fig. 2(a). The basic building block for such a network (the *switching element*) is a two-input two-output crossbar switch [Fig. 2(b)]. The Benes network is made up of two stages of switching elements connected to two Benes networks of half the size in the center stage. (The connections between the stages are the *inverse shuffle* and the *perfect shuffle* [13], respectively.) When the recursive construction is completed, the network consists of $(2 \log N - 1)$ stages of switching elements, with $N/2$ switches in each stage. Fig. 3 shows an 8×8 Benes network.

Not all the switches in Fig. 3 are necessary for full permutation capability. It can be shown [14] that one switch (arbitrarily chosen) in Fig. 2 can be set to a fixed state (straight or cross), leading to a total switch count of

$$S_N = N \log N - N + 1$$

for an $N \times N$ Benes network. In the limit that $N \to \infty$, $S_N \to \log (N!)$, and it is in this sense that the Benes network is considered an optimal network.

The control algorithm for the Benes network, i.e., the algorithm to derive switch settings for a particular permutation, is the *looping algorithm* presented in [11].

Let us define a 2-*active switch* in a permutation array as a switch with two active inputs, when a permutation is being realized. To support N input–output paths in a Benes network, every switch has to have two paths through it. Thus, every switch is a 2-active switch in the network. In the next section, we derive a permutation array with no 2-active switches.

B. The Dilated Benes Network—Topology

The *dilated Benes network* with N inputs and N outputs ($N > 2$) is defined recursively as follows.

1) It consists of two stages of N switching elements each, connected to two $N \times N$ subnetworks by the inverse shuffle and the perfect shuffle connections [Fig. 4(a)]. The $N \times N$ subnetworks are constructed recursively until 4×4 subnetworks are reached. In the first stage of recursion, one (fixed) input at each switch j is designated an input terminal i_j to the network, and at the last stage, one (fixed) output at each switch j is designated an output terminal o_j of the network.

2) 4×4 subnetworks, in the last stage of the recursion, are constructed with two stages of switching elements as shown in Fig. 4(b). (A 4×4 Benes network would have three stages of switches.)

one input each, the crossbar implementation requires $2N(N - 1)$ directional couplers. And since each switching element in Fig. 1(a) has only one active input (or one active output), there are no switch crossovers in a crossbar. The two cellular arrays in Fig. 1(b) and (c) require $N(N - 1)/2$ directional couplers each. The rearrangeable array in Fig. 2, which we will consider in detail later, uses $N \log N - N + 1$ couplers.

Table I lists the performance measures of interest for these networks. We see that the rearrangeable array drastically reduces switch count (and optical path length) by introducing nonadjacent cell connections (which lead to the channel crossovers). Based on the earlier discussion on performance degradation and cost associated with switch and channel crossovers, we conclude that this is an effective tradeoff to make.

Reduction of the path length (from N to $2 \log N - 1$) affects

TABLE I
PERFORMANCE MEASURES FOR SOME INTERCONNECTION NETWORKS

Network	Path Length	Channel Crossovers	Switch Crossovers	Total Crossovers	Total Couplers
Crossbar	$2 \log N$	$O(N^2 \log N)$[a]	0	$O(N^2 \log N)$	$2N(N-1)$
Triangular	$N-1$	0	$N(N-1)/2$	$N(N-1)/2$	$N(N-1)/2$
Diamond	N	0	$N(N-1)/2$	$N(N-1)/2$	$N(N-1)/2$
Rearrang.		$N(N-1)/2 -$			
Array	$2 \log N - 1$	$(N \log N - N + 1)$	$N \log N - N + 1$	$N(N-1)/2$	$N \log N - N + 1$

[a] This number has been derived by R. A. Spanke.

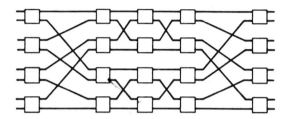

Fig. 3. An eight-input eight-output Benes network.

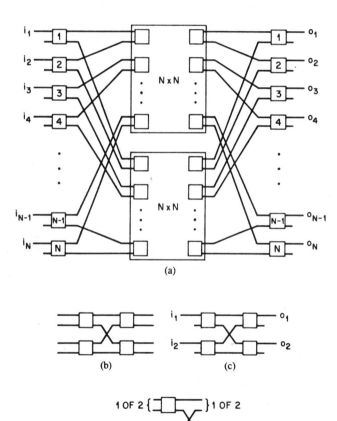

Fig. 4. Dilated Benes network construction. (a) Recursive construction. (b) Last stage of recursion. (c) A dilated Benes network of size 2. (d) The simplest switching function without 2-active switching elements.

3) A 2 × 2 dilated Benes network [Fig. 4(c)] consists of two stages of two switching elements each. A fixed input to each switch in the first stage is designated an input terminal, and a fixed output from each switch in the second stage is designated an output terminal. (By contrast, a 2 × 2 Benes network is a single switching element.)

Construction of the dilated Benes network proceeds similarly to that of the Benes network until the subnetworks are of size four. The network consists of $2 \log N$ stages, with N switching elements in each stage. An 8 × 8 network is shown in Fig. 5. It is possible to devise a control algorithm for the dilated Benes network (next section) that routes the paths without giving rise to any 2-active switches. (In essence, each subnetwork in Fig. 4(a) has only half of its inputs and outputs active, distributed one per switch. This is also the reason why the construction terminates at 4 × 4 subnetworks.) This then provides the connections with negligible crosstalk due to the absence of an active adjacent channel inside any coupler that implements the switching elements.

As in the case of the Benes network, some of the switches in Figs. 4(a) and 5 are redundant (shown in dashed boxes). The switch count for the dilated network is given by

$$S_N = 2N \log N - N/2.$$

Under the requirement that no switching element carry more than one active path, the simplest 2-(active)input/2-(active)output switch is shown in Fig. 4(d). It switches two pairs of inputs, each containing one active input, to two pairs of outputs, each containing one active output. [Note that this is a simpler switching function than that in Fig. 4(c).] This switch requires two switching elements and can be controlled by a single bit of state information [just like the switching element in Fig. 2(b)]. Thus, when no 2-active switching elements are permitted, the minimum switch count is $2 \log (N!)$, and the dilated Benes network is optimal under this condition. Its optical path length, equal to the number of switching stages, is one more than that of the Benes network, and equal to that of the fan-out/fan-in crossbar implementation [Fig. 1(a)]. (Although it may be possible to implement a 1-active coupler (in which crosstalk is not an issue) more economically than a 2-active coupler (in which crosstalk is an issue), currently the two are implemented the same way and therefore in quantifying the complexity of the switch we make no distinction between these two types of couplers.)

C. Dilated Benes Network—Control Algorithm

The routing algorithm for the dilated Benes network is a looping algorithm, similar to that for the Benes network [11]. We first present the algorithm and then show that it results in no switching element with two active inputs.

Referring to Fig. 4(a), the approach we will take is as follows. We will first apply the looping procedure to the two outer columns of switches to set these for the desired permutation. We will then show that the N active paths are directed to the two center subnetworks in such a way that the outer columns of 2 × 2 switches in these two subnetworks have only one active input (or output) per switch; thus, the two subnetworks are dilated Benes networks (of size $N/2$) and the looping algorithm can be applied to their two outer columns. This process recurses until the subnetwork become 4

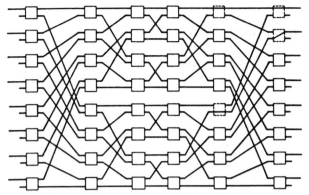

Fig. 5. An eight-input eight-output dilated Benes network.

$\times$ 4 networks, with two active inputs and two active outputs [Fig. 4(b)].

Step 1—Looping Algorithm for Setting Outer Stages: Refer to Fig. 4(a) with inputs and outputs numbered 1 through N. (We have dropped the i and o qualifiers on the input and output numbers for clarity of presentation.) The *dual* of an input (or an output) i, denoted by $\hat{i}$, is defined as

$$\hat{i} = \begin{cases} i+1 & \text{if } i \text{ is odd} \\ i-1 & \text{if } i \text{ is even.} \end{cases}$$

Let π be the permutation to be realized, i.e., input i is to be connected to output $\pi(i)$, $1 \le i, \pi(i) \le N$.

1) Choose a new input switch i which has not been *set*, i.e., its input has not been connected to center subnetwork 1 or 2. If all switches have been set, exit the algorithm.

2) Set $j = i$.

3) Set the input switch j to *up*, i.e., connect input j to center switch 1.

4) Set the input switch $\hat{j}$ to *down*, i.e., connect input $\hat{j}$ to center switch 2.

5) Set output switch $\pi(\hat{j})$ to down.

6) Set output switch $\widehat{\pi(\hat{j})}$ to up.

7) If $\pi^{-1}(\widehat{\pi(\hat{j})}) = i$, then go to 1; else set $j = \pi^{-1}(\widehat{\pi(\hat{j})})$ and go to 3.

When redundant switches are removed (as in Fig. 5), in Step 1 we will have to begin with the switch that has been removed (i.e., it has already been set).

Step 2—To Prove that the Two Center Subnetworks are Dilated Benes Networks of Size $N/2$: Consider one of the subnetworks and its N input lines. These N lines go to $N/2$ 2 $\times$ 2 switches inside the networks. At any such switch, the two input lines come from two *dual switches* in the outer stage of Fig. 4(a). The looping algorithm in Step 1 ensures that only one of these latter switches sends its input to a subnetwork. Thus, every 2 $\times$ 2 switch in the input stage of a subnetwork has only one active input. A similar argument can be made for the switches in the output column of the subnetworks. Since the subnetworks themselves are recursively constructed, this condition ensures that they are dilated Benes networks of size $N/2$.

Step 3—Setting of 4 $\times$ 4 Subnetworks: When we reach 4 $\times$ 4 subnetworks [Fig. 4(b)] by repeated application of Step 1, each switch has only one input and this has to be connected to one of the two outputs connected to the two output switches. The second input cannot request the same output switch as the first (because a permutation is guaranteed by Step 1) and so each input switch can be connected to the desired output switch.

The time complexity of the control algorithm is of the same order as that for the Benes network [11]. At a particular level in the recursion, Step 1 involves N settings (a $(j, \hat{j})$ pair) and there are log N levels in the recursion. Thus, the algorithm takes N log N time steps. The relatively long setup time of rearrangeable arrays, which results from the centralized nature of their control, generally restricts them to applications in which signal paths are used for a long time after they are set up.

IV. PARTIAL DILATION OF THE BENES NETWORK

The dilated Benes network introduced in the last section provides connections with no directional coupler carrying two signal paths. We call this condition a *zero-crosstalk connection* for purposes of discussion, bearing in mind that in reality this does involve a small but negligible amount of crosstalk. In practice, the requirement of zero-crosstalk connections is too conservative and it is desirable to permit a budgeted amount of crosstalk if this will reduce the number of devices required to implement the network. We now improve on the dilated architecture to achieve such a tradeoff between the amount of crosstalk and the number of switching elements, without changing the optical path length.

Let us define a *uniform crosstalk network* to be a permutation network in which all input–output paths have the same worst case crosstalk where crosstalk is in terms of the number of 2-active switches along the path. This worst case crosstalk is suffered by all the paths when the network is fully loaded, i.e., it has N active inputs and outputs. In the following two sections, we present two partially dilated architectures based on the Benes network, the first of which has the uniform crosstalk property, and the second does not.

A. Partial Dilation—Uniform Crosstalk

Partial dilation of the Benes network is performed by dilating the outer columns of switches in only *some* of the log N stages of recursive construction of the network. Uniform crosstalk is achieved by building dilated columns with no 2-active switches. Thus, each column of switches in the network contributes crosstalk either to all the paths or to none. As an example, consider the partially dilated Benes network in Fig. 6. Both terminals of each switching element in the first stage are designated input terminals, and thus all $N/2$ switches in that stage are 2-active switches. In the two $N \times N$ subnetworks, each switch in the first stage has only one (fixed) input connected to an external switch (by an N-way shuffle connection). Keeping the rest of the network identical to that in Fig. 4(a), we see that in a fully loaded configuration, any optical path contains only one directional coupler with two active signal paths. All the other switches in the path contain only one active input. Thus, the number of devices is reduced by $N/2$ (from that for the fully dilated network), at the expense of crosstalk in only one stage.

In general, we could have k stages ($k < \log N$) with 2-active switches. These k stages would contribute to crosstalk and we save $kN/2$ directional couplers over a fully dilated network. Thus, the strategy would be to determine k based on the amount of crosstalk permitted and then to implement k stages with $N/2$ switches per stage and the rest with N switches. The percentage saving in the number of switches is then given by

$$\text{percent saving in switches} = \frac{kN/2}{2N\log N - N/2} \times 100$$

$$= \frac{k}{4\log N - 1} \times 100 < 50 \text{ percent.}$$

For some typical values of N and k, this is shown in Table II.

The k stages containing the 2-active switches *cannot* be selected arbitrarily if permutation capability is to be retained. Consider a step in the recursive construction of the network where a $J \times J$ subnetwork is split into two $J/2 \times J/2$ subnetworks. This is shown in Fig. 7. This subnetwork has $J/2$ *active* lines coming into it. Its permutation capability

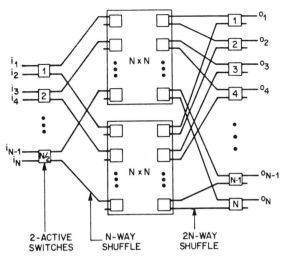

Fig. 6. A partially dilated Benes network with one stage of 2-active switches.

TABLE II
SWITCH SAVINGS FROM PARTIAL DILATION

Switch size N	32	64	128
No. of stages	10	12	14
No. of 2-active Stages k	1 2 3 4 5	1 2 4 6 8	1 2 4 6 7
Percent Saving in Switches	5 11 16 21 26	4 9 17 26 35	4 7 15 22 26

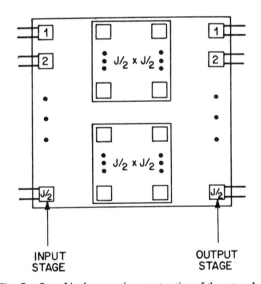

Fig. 7. Step J in the recursive construction of the network.

depends on the fact that each switch in the input stage (and each switch in the output stage) is connected to *both* subnetworks. If we now wanted to introduce 2-active switches in the next step of recursion, the subnetworks would be of size $J/4 \times J/4$ and the number of links would drop from J to $J/2$. Each switch in the input stage in Fig. 7 can then have only one output and can be connected to only one of the center subnetworks. This will result in loss of permutation capability.

Thus, 2-active switches have to be positioned in the *outermost stages* of the network, whether on the input side or on the output side. When the last two (center) stages are replaced by 2-active switches, these two stages can be combined into one, to reduce to the Benes network topology (with $2 \log N - 1$ stages).

The routing algorithm for a partially dilated network is a combination of the Opferman and Tsao-Wu algorithm and the looping algorithm specified in Section III-C. Stages with 2-active switches will be set using the former, and the remaining stages will be set using the latter algorithm.

B. Partial Dilation—Nonuniform Crosstalk

It is possible to construct dilated versions of the Benes network in which some (specified) paths can be guaranteed to be zero-crosstalk connections while the rest of the paths suffer crosstalk. The nonuniform crosstalk networks in this section have the following property. The set of inputs I is divided into two subsets I_c and $I_{\bar{c}}$, corresponding to crosstalk and no crosstalk, respectively. Similarly, the outputs are also divided into O_c and $O_{\bar{c}}$. Then 1) connections from $I_{\bar{c}}$ to $O_{\bar{c}}$ will suffer no crosstalk, 2) connections from $I_{\bar{c}}$ to O_c will suffer crosstalk only in the second half of the network, 3) connections from I_c to $O_{\bar{c}}$ will suffer crosstalk only in the first half of the network, and 4) connections from I_c to O_c will encounter crosstalk in both halves of the network. The amount of crosstalk suffered in 2), 3), and 4) depends on the sizes of $I_{\bar{c}}$ and $O_{\bar{c}}$, and will be specified below.

As an example, Fig. 8 shows two 8×8 partially dilated networks in which 25 and 50 percent of the inputs (and outputs), respectively, are designated no-crosstalk inputs. Switches numbered 2 in the figure are 2-active switches. The number of 2-active switches at a stage gets evenly distributed among the two subnetworks in the following stage. (An odd 2-active switch gets distributed as two 1-input switches in the next stage.) This ensures that no 1-active switch in an outer stage gets connected to a 2-active switch in an inner stage, and thus properties 1), 2), and 3) in the last paragraph are satisfied. If we start with k 2-active switches in the first stage, it takes $\lfloor \log k \rfloor + 1$ stages for them to become 1-active switches, and this determines the worst case crosstalk in the (first half of the) network.

The 25 percent dilation in Fig. 8(a) has increased switch count by 95 percent over an 8×8 Benes network and the 50 percent dilation [Fig. 8(b)] increases switch count by 100 percent. (By contrast, a fully dilated network requires a 120 percent increase in switch count.) These numbers improve for larger N (for $N = 64$, 25 percent dilation involves a 54 percent increase, and 50 percent dilation requires a 72 percent increase in switch count), but the nonuniform crosstalk networks are not as cost-effective as the uniform crosstalk networks of the previous section. However, if nonuniformity is part of the system specifications, the networks in this section are less expensive than the uniform crosstalk networks.

V. Performance Analysis of Dilated Benes Networks

We consider now how the uniform crosstalk dilated networks introduced in the last two sections perform (in terms of crosstalk) under different traffic conditions. In a fully loaded network, all the paths will encounter equal crosstalk (equal to the number of stages of 2-active switches in the network). In a lightly loaded network, on the other hand, the number of paths that encounter crosstalk and the amount of crosstalk each path encounters will depend on the distribution of the inputs constituting the load.

We derive an expression for the mean crosstalk per path in a uniform dilated Benes network operating under a *load l*. Load is defined as the probability, in a subpermutation, that an input is active. Thus, the expected number of active inputs at load l is Nl.

Since the network is symmetric about the center, we will do the analysis for the first $\log N$ stages of the network. Stages 1–k ($0 \leq k \leq \log N$) are constructed out of 2-active switches. (This would correspond to $2(\log N - k)$ center stages being dilated in the network.) Only the stages with 2-active switches could potentially contribute to crosstalk.

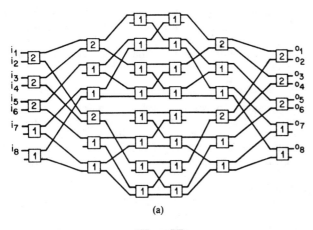

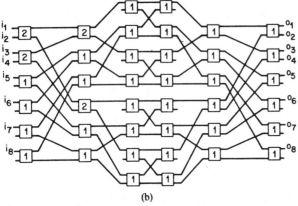

Fig. 8. Two nonuniform crosstalk partially dilated Benes networks. Network in (a) corresponds to a 25 percent dilation and network in (b) corresponds to a 50 percent dilation. Numbers inside the switching elements indicate the number of active paths through the element when a permutation is being realized.

For each stage j ($1 \leq j \leq k$), let us define

$P_2(j)$ = probability that a given switch is 2-active,

$P_1(j)$ = probability that a given switch is 1-active,

$P_0(j)$ = probability that a given switch is 0-active.

Given the definition of load, at stage 1 we get

$$P_2(j) = l^2,$$
$$P_1(j) = 2l(1-l),$$
$$P_0(j) = (1-l)^2.$$

To proceed from stage j to stage $j + 1$ ($1 \leq j \leq k - 1$), consider the two pairs of switches in the two stages that are connected, as shown in Fig. 9. We can derive the following:

$$P_2(j+1) = P_1(j)P_2(j) + P_2(j)^2$$

$$P_1(j+1) = P_0(j)P_1(j) + 2P_0(j)P_2(j)$$
$$+ P_1(j)P_2(j) + P_1(j)^2$$

$$P_0(j+1) = P_0(j)P_1(j) + P_0(j)^2.$$

By symmetry, $P_i(j) = P_i(2 \log N - j)$ ($1 \leq j \leq \log N$, $0 \leq i \leq 2$).

The expected number of 2-active switches is then given by $2(N/2)\Sigma_{j=1}^{k} P_2(j)$.

Each such switch contributes (a unit amount) to crosstalk along two of the Nl paths in the network.

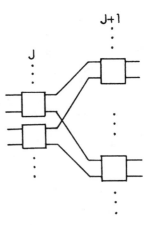

Fig. 9. Interconnection of switch pairs in two adjacent stages.

Hence, the mean crosstalk per path is given by

$$\frac{1}{Nl}(2N)\sum_{j=1}^{k} P_2(j) = \frac{2}{l}\sum_{j=1}^{k} P_2(j).$$

This expression is plotted in Fig. 10 as a function of the number of dilated stages and the network load. When no dilation is done, the networks are Benes networks, with $2 \log N - 1$ switches contributing to crosstalk when load is 1. When complete dilation is done, no paths encounter any crosstalk at any load. In between, the network load has a significant impact on the crosstalk. At very light loading, dilation has little effect. In a sense, at light loads the network has been dilated dynamically.

VI. Network Decomposition for Compact Layout

Advanced levels of integration permit us to package more than one directional coupler on a single chip [3], [5]; it is then necessary to consider how a network can be partitioned to fit on chips. Conversely, the building blocks should be designed so that larger networks can be constructed using the required number of subnetworks. Wise in [15] considers how Banyan networks can be packed into cubic space and presents a novel scheme for this purpose. Since one half of the dilated Benes network (Fig. 5) is a Banyan interconnection scheme with the first stage in the Banyan eliminated, his packing scheme can be readily extended to the dilated Benes network. This is illustrated in Fig. 11 for an 8×8 network. The network has been redrawn in Fig. 11(b) (shown by the permuted switches in the second and fifth columns) so that it now comprises of 4 $\times$ 4 subnetworks with two stages of four-way shuffles in between the networks. These shuffles can be implemented with no wire crossings in three dimensions by simply rotating the planes 90° and joining them end to end [Fig. 11(c)]. In general, the decomposition would be into $\sqrt{2N} \times \sqrt{2N}$ subnetworks. It should be pointed out that in general, the center stage subnetworks are not the same as the two outer stage subnetworks, but they can be synthesized by placing two of the latter networks back to back. Thus, if a single chip type is desired, we would have four stages of subnetworks. The connection between the second and third stages would be the identity connection; the first and last stages would involve rotation by 90° as described before. In practice, decomposition of the network into such stages will result in additional optical path loss due to the interstage interconnect. If this interconnect is accomplished by fibers, the characterization of loss is straightforward and known. We are, however, investigating direct butt-coupling of the individual stages to reduce the interconnect loss.

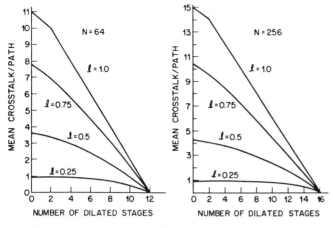

Fig. 10. Mean crosstalk/path in dilated Benes networks as a function of dilation and network load. Crosstalk is in terms of number of 2-active switches in the path.

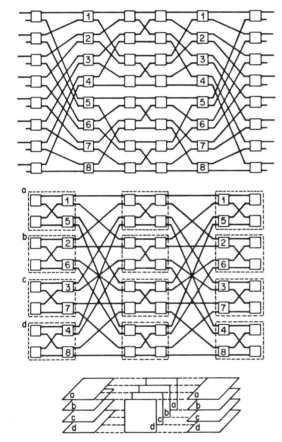

Fig. 11. Compact layout in three dimensions for an eight-input eight-output dilated Benes network. The numbers on the switches show the rearrangement to achieve the layout.

VII. Dilation of Other Multistage Networks

We have so far illustrated the concept of dilation using the rearrangeably nonblocking Benes network. In this section, we show how the technique can be applied to reduce crosstalk in other multistage networks, in particular, the blocking Omega network [10] and the strictly nonblocking Clos network [2].

Blocking networks like the Omega and other topologically equivalent structures [16] have proved popular, particularly for multiprocessor interconnections, due to their simple distributed control algorithms (in addition to their relatively low cost and delay properties). However, their blocking nature means that they cannot realize all $N!$ permutations of their inputs.

The dilated Omega networks we now define have the following properties: 1) they retain the simple distributed bit control scheme of the Omega network, 2) every permutation realizable by the Omega network can be realized by the dilated Omega network with no 2-active switching elements, i.e., with negligible crosstalk, 3) the stochastic properties of the network are superior to those of the Omega network. This last result is to be expected, however, since the hardware has been more than doubled.

An $N \times N$ dilated Omega network consists of $\log N + 1$ stages with N switching elements in each stage. Stage i is connected to stage $i + 1$ by the perfect shuffle connection ($0 \leq i \leq \log N - 1$). At the first stage, a *fixed* input at each switch is designated a network input. Similarly, at the last stage, one output of each switch is designated an output terminal. Fig. 12 shows a dilated Omega network of size 8.

To get from source $S = s_0 s_1 s_2$ to destination $D = d_0 d_1 d_2$, the source uses as a tag the destination address. Each bit d_i in the tag is used by the switch in stage i to route the message (or set up a path). (When the tag bit is 0, the message is routed to the upper output, and when the tag bit is 1, the message is routed to the lower output.) At stage $\log N$, no tag bit is required since the switch has only one output. Thus, a $\log N$-bit tag suffices, corresponding to the N destinations.

To show that the dilated Omega network will realize any Omega permutation with no switch having two paths through it, consider the path from a source $S = s_0 s_1 \cdots s_{n-1}$ to a destination $D = d_0 d_1 \cdots d_{n-1}$ where $n = \log N$. The *switch* that this path occupies at stage j ($0 \leq j \leq n - 1$) in the network is given by the n-bit window

$$s_0 \cdots s_{j-1} \boxed{s_j \cdots s_{n-1} d_0 \cdots d_{j-1}} \cdots d_n.$$

In an $N \times N$ Omega network, this window corresponds to the *terminal* at the output of stage j that the same path will occupy [10]. In an Omega permutation involving this connection, no other path can occupy this terminal. Thus, when the same permutation is considered in the corresponding dilated network, no other path can occupy the switch $s_j \cdots s_{n-1} d_0 \cdots d_{j-1}$. This is true of all switches and all paths and proves that all switches will have exactly one path passing through them.

As before, the stochastic performance of the dilated Omega network is also of interest. Other researchers [9], [17] have considered structures that have an equivalent amount of hardware, with multiple copies of a single network or replicated links between switches, for purposes of stochastic performance improvement. The idea is that network performance at a load l is better than the performance of the same network at load kl ($k > 1$). However, reduction of crosstalk has not been a concern in the multiprocessor context in which these analyses have been done.

To get an estimate for the mean crosstalk experienced by a path, we follow the technique of the analysis in Section V. A key difference here, however, is that requests need not constitute a (sub)permutation, and each request has an equal probability of being directed to any of the N outputs. Let us define l_i (the ''load'') to be the mean number of requests a link carries at the input to stage i of the network ($l_i \leq 1$). Thus, $l_0 = l$, the network load as before.

$$l_1 = l/2$$

$$l_j = \frac{1}{2} \cdot 2 \cdot l_{j-1}(1 - l_{j-1}) + \frac{3}{4} \cdot l_{j-1}^2$$

$$= l_{j-1}\left(1 - \frac{l_{j-1}}{4}\right) \qquad 2 \leq j \leq \log N$$

$$l_{\log N + 1} = 2 l_{\log N}(1 - l_{\log N}) + l_{\log N}^2 = 2 l_{\log N} - l_{\log N}^2.$$

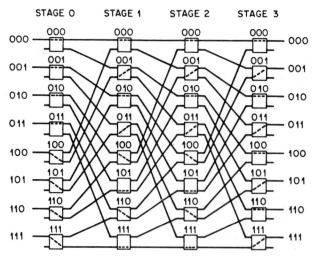

Fig. 12. An eight-input eight-output dilated Omega network. Switch settings are shown to realize the identity permutation.

TABLE III
CROSSTALK ALONG SUCCESSFUL PATHS IN AN $N \times N$ DILATED OMEGA NETWORK (NUMBER OF 2-ACTIVE SWITCHES PER PATH)

N	Load (Dilated Omega)				Load (Omega)			
	0.25	0.5	0.75	1.0	0.25	0.5	0.75	1.0
32	0.42	0.71	0.92	1.1	0.94	1.47	1.79	2.0
64	0.51	0.85	1.1	1.26	1.07	1.64	1.96	2.16
128	0.60	0.98	1.2	1.42	1.20	1.78	2.1	2.29
256	0.68	1.1	1.4	1.56	1.31	1.91	2.22	2.4

$Nl_{\log N-1}$ is thus the network bandwidth, the expected number of requests that have been accepted. As far as crosstalk is concerned, there are only this many active paths in the network. The first and last stages in the network present no crosstalk. In the intermediate stages, the probability that a switch has two active paths through it is given by $(l_{\log N+1}/2)^2$. (The probability that any one link is active is $l_{\log N+1}/2$. Note that there are $2N$ links in the intermediate stages of the network.)

Thus, the expected number of 2-active switches per path is given by

$$\frac{2(l_{\log N+1}/2)^2 N (\log N - 1)}{Nl_{\log N+1}} = l_{\log N+1} (\log N - 1)/2.$$

This expression is tabulated in Table III for several values of network size and load, along with the corresponding values for the standard Omega network.

Two points have to be kept in mind while interpreting the numbers in Table III. First, even though the numbers are small, they have to be compared to *zero* in the case of the dilated Benes network. Second, the actual number of paths existing in the network is a fraction of the product of N and the load listed in the table because many of the initial requests are not accepted. (For instance, at an input load of 1.0 in a 256 × 256 network, only 44 percent of the requests are accepted in the dilated case, and 30 percent in the undilated case.) Thus, the attrition in the number of requests caused by network blocking helps in the crosstalk properties of the successful paths, even without dilation. Dilation does improve the situation, but the crosstalk numbers are small to begin with in the stochastic operating environment of the Omega network.

The strictly nonblocking Clos network has a recursive construction like the Benes network, but uses three subnetworks in the center stage instead of two; it is this additional redundancy that results in its strict sense nonblocking characteristic. A dilated Clos network can be constructed by following a procedure analogous to dilation of the Benes network. It can be shown to require $(5N^{\log 3} - 3.5N \simeq 5N^{1.585} - 3.5N)$ couplers with 2 log N stages of 1 × 3 couplers (or 4 log N stages of 1 × 2 couplers). Distributed control algorithms for such a network can be derived. In a practical application, one will have to trade off the factors of dilation (to reduce crosstalk), redundancy (to reduce network rearrangements required), and switch complexity against one another. In some cases, it might be advantageous to construct a hybrid of Clos and Benes structures and provide dilation in the different stages appropriately.

VIII. Concluding Remarks

We have presented in this paper some novel asymptotically optimal architectures suitable for photonic switching networks. These reduce the crosstalk that input-output paths encounter without a significant increase in optical path lengths or number of switching elements. The stochastic performance of these networks is also promising.

We close this paper with a brief discussion of the fault-tolerant nature of the dilated Benes network. Considering Fig. 4(a), if there is a directional coupler failure in one of the two center subnetworks, the path through that element can always be rerouted through the other subnetwork *if* the following modification is made: the 4 × 4 subnetworks in the last stage of the recursion [Fig. 4(b)] are constructed with three stages of two switching elements each, i.e., as 4 × 4 Benes networks. The effect of this modification is to make each subnetwork in Fig. 4(a) into a Benes network (of size N), and such a subnetwork can accommodate the path that would normally take the faulty element in the other subnetwork. Note that this path will now share directional couplers with other active paths, and hence this results in a slight increase in crosstalk, in addition to the increase in switch count by N.

References

[1] R. C. Alferness, "Guided-wave devices for optical communication," *IEEE J. Quantum Electron.*, vol. QE-17, pp. 946–957, 1981.

[2] V. E. Benes, *Mathematical Theory of Connecting Networks and Telephone Traffic.* New York: Academic, 1965.

[3] G. A. Bogert, E. J. Murphy, and R. T. Ku, "A low crosstalk 4×4 Ti:LiNbO$_3$ optical switch with permanently attached polarization-maintaining fiber arrays," in *IGWO '86*, pp. 348–356.

[4] G. A. Bogert, "Ti:LiNbO$_3$ intersecting waveguides," to be published.

[5] P. Granestrad et al., "Strictly non-blocking 8×8 integrated-optic switch matrix in Ti:LiNbO$_3$," in *IGWO '86*, pp. 26–28.

[6] H. S. Hinton, "A non-blocking optical interconnection network using directional couplers," in *Proc. IEEE Global Telecommun. Conf.*, Nov. 1984, pp. 885–889.

[7] W. H. Kautz, K. N. Levitt, and A. Waksman, "Cellular interconnection arrays," *IEEE Trans. Comput.*, vol. C-17, pp. 443–451, May 1968.

[8] D. J. Kuck, *The Structure of Computers and Computations.* New York: Wiley, 1978, ch. 6.

[9] M. Kumar and J. R. Jump, "Generalized delta networks," in *Proc. 1983 Int. Conf. Parallel Processing*, Aug. 1983, pp. 10–18.

[10] D. H. Lawrie, "Access and alignment of data in an array processor," *IEEE Trans. Comput.*, vol. C-24, pp. 1145–1155, Dec. 1975.

[11] D. C. Opferman and N. T. Tsao-Wu, "On a class of rearrangeable switching networks, Part I: Control algorithm," *Bell Syst. Tech. J.*, vol. 50, pp. 1579–1600, 1971.

[12] R. A. Spanke, "Architectures for large non-blocking optical switches," *IEEE J. Quantum Electron.*, vol. QE-22, pp. 885–889, Aug. 1986.

[13] H. S. Stone, "Parallel processing with the perfect shuffle," *IEEE Trans. Comput.*, vol. C-20, pp. 153–161, Feb. 1971.

[14] A. Waksman, "A permutation network," *J. Ass. Comput. Mach.*, vol. 15, pp. 159–163, Jan. 1968.

[15] D. S. Wise, "Compact layout of Banyan/FFT networks," in *Proc. CMU Conf. VLSI Syst. Computations*, 1982, pp. 186–195.

[16] C. L. Wu and T.-Y. Feng, "On a class of multistage interconnection networks," *IEEE Trans. Comput.*, vol. C-29, pp. 694–702, Aug. 1980.

[17] P.-C. Yew, D. A. Padua, and D. H. Lawrie, "Stochastic properties of a multiple layer single-stage shuffle-exchange network in a message switching environment," *J. Digital Syst.*, vol. VI, no. 4, pp. 387–409, 1982.

Photonic Switching Modules Designed with Laser Diode Amplifiers

JOSEPH D. EVANKOW, JR. AND RICHARD A. THOMPSON, SENIOR MEMBER, IEEE

Abstract—Photonic switching elements are designed from semiconductor optical amplifiers and passive couplers with fiber-to-fiber unity gain and low crosstalk. Designs for a 2 × 2 and an asymmetric 2 × 3 element, and several designs for 4 × 4 elements, are presented. While most amplifier analyses have stressed the importance of ultra-low facet reflectivities for high-gain operation, with protection against external reflections with optical isolators, modest facet reflectivities are satisfactory for these elements. It is also shown that substantial amounts of external reflection can be tolerated. The various architectures are compared according to amplifier count, blocking characteristic, broadcast potential, noise power (amplified spontaneous emission), and fault tolerance.

I. INTRODUCTION

WHILE semiconductor optical amplifiers (SOA's) are effective linear amplifiers [1], [2], they can also function as switches [3], [4]. Although much of photonic switching has relied on lithium niobate technology, with optical amplifiers compensating for losses [5], optical switching elements can be made directly from SOA's and passive couplers [6], [7].

Described here is a series of 2 × 2, 2 × 3, and 4 × 4 unity-gain photonic switches made with fiber passive couplers and SOA's. These networks, establishing connectivity between any pair of input and output ports and not switching "by-the-bit," makes it possible to manipulate data at rates that exceed the modulation bandwidth of an individual amplifier. In the next section, device and systems considerations are used to compare the various elements. In the following sections, designs for one-stage 2 × 2 and 4 × 4 elements, two designs for multiple-stage 4 × 4 elements built from the 2 × 2 element, a 2 × 3 element, another multiple-stage 4 × 4 designed using 2 × 2 and 2 × 3 elements, and optimizations of these with some applications are presented.

A. Device Considerations

When SOA's are operated at high levels of gain (>20 dB), in the presence of even minute amounts of reflection from either the facets or any other dielectric discontinuity, periodic Fabry–Perot variations in the gain curve cause temperature instability. While most amplifier reflectivity analyses [8], [9] have stressed the importance of ultra-low facet reflectivities for high-gain operation, together with optical isolators to protect against system reflections, it is shown here that moderately high-facet reflectivities are satisfactory for the operation of each module. Although system reflections, regenerative feedback, adversely affect all SOA's, these cascaded modules distribute the gain over several SOA's, allowing low-gain operation for each amplifier and thus robustness in reflectivity for the overall module. Since the degree of Fabry–Perot variation, nonresonant gain ratio, at a given level of gain, depends on the product of the internal single-pass gain and the geometric mean of all associated reflectivities, distributed low-gain operation offers an advantage to complex cascaded structures. Since recent efforts in packaging [10]–[12] have shown that SOA's can function away from optical tables, cascaded configurations in field transmission systems is envisioned.

B. System Considerations

Photonic switching networks in lithium niobate have been built and described in the literature for sizes 4 × 4 [13]–[15] and even 8 × 8 [16]. Architectures capable of larger sizes have been proposed and analyzed [17], [18]. Two major drawbacks to lithium niobate implementations are high-insertion loss and high crosstalk (compared to even the worst electronic switches). There has been considerable research, worldwide, to produce better devices and to understand the limits of the technology. There has been a corresponding effort to invent architectural solutions, since network architectures have been designed for years around near-perfect electromechanical or solid-state devices. Clever architectural concepts have been designed around photonic device limitations yielding an overall network performance that exceeds the quality of the individual devices. An example is the proposed use of *dilation* to improve overall crosstalk in a network built from 2 × 2 switches [19]. Although the lithium niobate directional coupler is a natural 2 × 2 photonic switching element, it only operates in one of two states: *bar* and *crossed*. Broadcasting, in which one input is connected to BOTH outputs, is desirable but probably impractical. It would also be useful to disconnect inputs, when those inputs are temporarily connected to noise or other unwanted signals. Unless ports are deliberately unused, systems built from such elements tend to "conserve" noise because every input is steered to a destination.

Lithium niobate based technology is certainly well es-

Manuscript received October 21, 1987; revised April 14, 1988.
The authors are with AT&T Bell Laboratories, Murray Hill, NJ 07974.
IEEE Log Number 8822557.

tablished; however, the switching elements proposed here can easily achieve the following characteristics:
- zero net insertion loss (unity gain),
- crosstalk ≤ -60 dB, typical (dependent on cavity length).

The purpose of this paper is to convince the reader that these characteristics are not only possible, but they may border on being trivial. While temperature stabilization for the amplifiers is an unfavorable constraint, especially at higher nonresonant gain ratios, the control electronics and device drivers are similar to lithium niobate technology. During the discussion of each of the switching elements, the following system's issues will be considered:
- the number of amplifiers in an element,
- its *blocking* properties,
- its ability to broadcast or simply communicate point-to-point,
- its fault tolerance,
- and total noise power.

II. THE 2 × 2 ELEMENT

The first photonic switching element, shown in Fig. 1, is a nonblocking 2 × 2. It is made from four discrete packaged SOA's and four fiber directional couplers splitting and combining the signal equally (50–50 percent, -3 dB). It is an implementation of a traditional *beta element* [20], enabling connectivity in both the *bar* and *crossed* states, similar to a directional coupler fabricated in lithium niobate. The upper input, for example, is connected to the lower output by enabling the top SOA in the figure.

A. Device Considerations

Shown in Fig. 1 are the major losses that each SOA must overcome for unity-gain operation of the module. A major question, from a device standpoint, is whether the gain needed to overcome all the associated losses (i.e., coupling into and out of the SOA, and splitter/combiner losses) requires unrealistically stringent tolerances for the fabrication of the AR coatings [21], [22]. For example, if a TE polarized reflectivity of $R \leq 10^{-3}$ were required for a 1.30 μm laser cavity, with typical dimensions, then the thickness (h) and the refractive index (n) of the film would have the following tolerances [21]: $\Delta h \simeq \pm 50$ Å, and $\Delta n/n \simeq \pm 3$ percent. While these values are obtainable, large scale production becomes difficult. The lower the reflectivity values, the tighter the tolerances on the film's parameters. Although film tolerances for the fabrication of the AR coatings is an important consideration, it will not be analyzed here. Instead, the focus will be on the required facet reflectivity and the susceptibility to external reflection of each architectural design. Since these architectural designs do not require ultra-low reflectivities, the film tolerances will, of course, be correspondingly more relaxed.

The equations relating the gain and reflectivity are

$$G(f) = \frac{KG_o(1 - R_1)(1 - R_2)}{(1 - G_o R)^2 + 4G_o R \sin^2(\pi f/\text{FSR})} \quad (2.1)$$

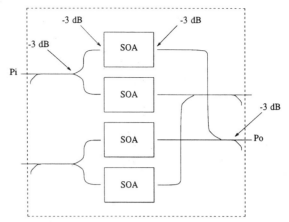

Fig. 1. 2 × 2 one-stage element with four SOA's.

where G_o is the single-pass gain ($\equiv e^{\Gamma(g-\alpha)L}$), $R_{1,2}$ are the facet reflectivities, R is the geometric mean of the facet reflectivities ($R^2 = R_1 R_2$), FSR is the free-spectral range ($c/2nL$), L is the cavity length, K is the coupling coefficient, c is the speed of light in a vacuum, and Γ is the optical confinement factor. The ratio of the minimum to the maximum gain is defined as the nonresonant gain ratio [9] $G_{\text{nr}} \equiv G_{\min}/G_{\max}$. As this value approaches zero, the gain curve is characterized by the familiar high-finesse Fabry–Perot interference fringes, resulting from a high single-pass gain–reflectivity product, while a ratio approaching one is characteristic of an ideal traveling-wave amplifier (low finesse) arising from a "reflectionless" interaction inside the gain medium,

$$G_{\text{nr}-\text{FPA}} = 0 \quad G_{\text{nr}-\text{TWA}} = 1. \quad (2.2)$$

The nonresonant gain ratio can be represented in terms of the single-pass gain and the geometric mean of the facet reflectivity as

$$G_{\text{nr}} = \frac{(1 - G_o R)^2}{(1 + G_o R)^2}. \quad (2.3)$$

The net peak value of the gain is

$$G_{p(\max)} = \frac{K(1 - G_{\text{nr}})(1 - R_1)(1 - R_2)}{4 R G_{\text{nr}}}. \quad (2.4)$$

Rearranging (2.4) and assuming $R = R_1 = R_2$ yields

$$\frac{(1 - R)^2}{R} = \frac{4 G_{p(\max)} G_{\text{nr}}}{K(1 - G_{\text{nr}})}. \quad (2.5)$$

With this last expression, it is possible to obtain the required facet reflectivities at each level of gain needed for unity-gain operation in an optical element.

Although many questions related to system requirements have not been answered, a $G_{\text{nr}} = -3.0$ dB is assumed. Since unity-gain for the circuit shown in Fig. 1 requires $G_{p(\max)} = +6$ dB and assuming a coupling coefficient of $K = -6.0$ dB, facet reflectivities of $R = 1.52$ percent cause a 3.0 dB variation in gain (G_{nr}) over the wavelength region near the center of the gain curve. While this resonant variation seems severe, the temperature can easily be stabilized over $\Delta T = \pm 0.5°C$. Earlier studies have shown [23] that net gains of $G_p = 6$ dB, with a resonant variation of slightly greater than $G_{nr} = -3$ dB, result in only a $\Delta G = 0.25$ dB over $\Delta T = \pm 0.75°C$.

This reflectivity value ($R = 1.52$ percent) is much greater than the reflectivities required for higher levels of gain. As the required gain increases, so does the need for low reflectivities. If, for example, the gain was $G_{p(\max)} = 30$ dB, then the facet reflectivity would have to be $R = 0.025$ percent, for a resonant variation of $G_{nr} = -3$ dB and perfect coupling ($K = 1$). For the same resonant gain variation ($G_{nr} = -3$ dB) with a realistic coupling coefficient ($K = -6$ dB), a net gain of $G_{p(\max)} = 30$ dB yields a required reflectivity of $R = 0.006$ percent. The previously calculated value of $R = 1.52$ percent enables higher yields in the fabrication of the AR coatings. In addition, for high levels of gain, external reflections from other amplifier facets, or even scattering in long lengths of fiber can cause the net reflectivity to quickly exceed the previously specified limits for the facets and thus significantly alter the nonresonant gain ratio. Optical isolators alleviate external reflection concerns but add to the overall cost and complexity. It will be shown that these cascaded architectural configurations, with amplifiers having nonzero facet reflectivities, alter but do not exceed previously specified system parameters. It is envisioned that optical isolators will need to be used only on the input/output ports to protect against reflections from long lengths of fiber and the lasers generating the signals.

As previously stated, an important concern is the effect of external reflection on the resonant behavior of this element. In other words, how much external reflection can each element withstand before the nonresonant gain ratio moves outside some prescribed limit?

Fig. 2 is a graph of the percentage change away from the original nonresonant gain ratio, for the optical element shown in Fig. 1, in the presence of external reflection at the input/output of the module. Since drastic changes in this ratio will cause adverse affects with the stable operation, small changes (< 10 percent) were monitored. If, for example, a nonresonant gain ratio of $G_{nr} = -5$ dB with an allowed change of $\Delta G_{nr} = 10$ percent are the system constraints, then an external reflectivity of $R_{ext} = 16.5$ percent can be tolerated.

B. System Considerations

Referring to Fig. 1, if one input is connected to either of the outputs, the remaining input can be connected to the remaining output. Thus, the architecture is completely nonblocking as is the corresponding implementation in lithium niobate.

If the top two SOA's in the figure are enabled, then data on the upper input fiber appear on both output fibers, at zero net loss. The architecture is seen to support broadcast, or one-to-two communication, as well as dual point-to-point communication. Equivalent operation by a single directional coupler in lithium niobate is difficult and prob-

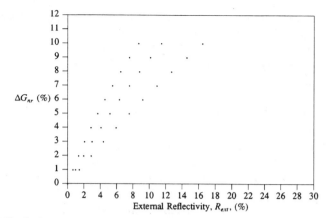

Fig. 2. Percentage change in nonresonant gain ratio for the 2 × 2 element versus external reflectivity for $G_{nr} = -1$ dB, $G_{nr} = -3$ dB, and $G_{nr} = -5$ dB, for each curve left to right, respectively.

ably impractical. An equivalent architecture could be implemented in lithium niobate by replacing each block in Fig. 1 labeled SOA by a simple on-off switch, but the corresponding loss and crosstalk characteristics can probably never be matched.

If the upper input in Fig. 1 must be connected to the lower output, the only path is via the top SOA. If it *fails*, then the connection is impossible. Thus, there is no alternate routing and the element is sensitive to faults. With no knowledge of the failure statistics of SOA's, it is difficult to conjecture the importance of this drawback, or to compare its reliability to other implementations. If the issue is critical, potential solutions include improving the reliability of an individual SOA, replicating each SOA, $n + k$ sparing the SOA's in an element, or embedding the element in a larger architecture that is tolerant of faults in its building blocks.

Since each amplifier produces spontaneous emission, successive amplification in cascaded architectures is of concern. The analysis [24], [25] for the total amplified spontaneous emission has assumed uniform coupling losses and that each successive SOA amplifies all of the previous amplifier's spontaneous emission (i.e., an ideal match between all gain curves). The noise power of each amplifier is

$$N = \frac{Bh\nu n_{sp}\chi}{\eta}(G_o - 1), \quad (2.6)$$

where

$$n_{sp} = \frac{N_e}{N_e - N_o}, \quad (2.7)$$

N_e is the carrier concentration, N_o is the carrier concentration for optical transparency, B is the filter bandwidth, $h\nu$ is the photon energy,

$$\eta = \frac{\Gamma g - \alpha}{\Gamma g}, \quad (2.8)$$

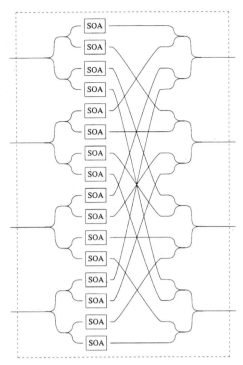

Fig. 3. 4 × 4 one-stage element with 16 SOA's.

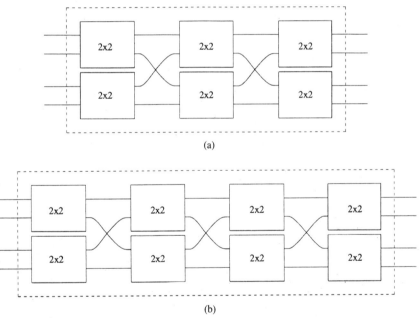

Fig. 4. 4 × 4 multistage elements with (a) 24 SOA's, and (b) 32 SOA's.

and

$$\chi = \frac{(1 + RG_o)(1 - R)}{(1 - RG_o)^2}. \qquad (2.9)$$

The simplified results for each switching element are placed in Table I (see Section VIII)—multiple values indicate a route dependent intensity.

III. A ONE-STAGE 4 × 4 ELEMENT

A 4 × 4 nonblocking optical element, shown in Fig. 3, is a generalization of the 2 × 2 element of Fig. 1. It uses a minimum number of active devices, namely 16. A similar element was described by Kobayashi *et al.* [6].

All fiber couplers are −3 dB splitters/combiners, and while the device count is minimized, it comes at the cost of a required decrease in facet reflectivity. Since there are two levels of couplers at the input and output, there is a total coupler loss of −12 dB. If the coupling coefficient is again assumed as $K = -6$ dB, then the overall chip gain must be $G = +18$ dB. This requires facet reflectivities of $R = 0.38$ percent ($G_{nr} = -3$ dB), a realistic value, but one that makes fabrication of the AR coatings much harder to obtain. This element is about as sensitive to external reflection, for $G_{nr} < -3$ dB and $\Delta G_{nr} < 5$ percent, as the 2×2 element shown in Fig. 1 because of the second tier of fiber couplers (see Section II-A). For $G_{nr} > -3$ dB and $\Delta G_{nr} > 5$ percent, this 4×4 is more tolerant of external reflections (i.e., $+2$ percent at $G_{nr} = -5$ dB, $\Delta G_{nr} = 10$ percent).

As in the 2×2 case, this 4×4 element is completely nonblocking, it supports both broadcast and point-to-point communication, and is not fault tolerant. Except for its low count of SOA's, this element does not compare well to the 4×4 elements to be described below. It has more stringent requirements on the reflectivity and the lack of alternate routes for fault tolerance. However, the reflectivity requirements are not severe and alternate routing may be obtained by using the element in a fault-tolerant super-architecture.

IV. MULTISTAGE 4×4 ELEMENTS

The 2×2 element in Fig. 1 can be used as a basic unit from which other elements can be built. Elements using six and eight such 2×2 units are shown in Fig. 4(a) and (b).

Since these two elements are built from the element in Fig. 1, the required reflectivity is similar, however, each stage affects the next, since it introduces reflections in the cascaded form that were not taken into account with the earlier calculation. While a calculated change will occur, the nonresonant gain ratio will change $\Delta G_{nr} = 4.0$ percent from its original value of $G_{nr} = -3$ dB.

The element in Fig. 4(a) will be shown to be rearrangeably nonblocking in Section VII-B. Adding a fourth stage to the element of Fig. 4(a) makes a new element, shown in Fig. 4(b) that is completely nonblocking. Since each component 2×2 unit supports broadcast, the entire element is seen to support broadcast, as well as point-to-point, communication.

To be rearrangeably nonblocking and fault tolerant, an architecture must provide multiple paths between any pair of endpoints. The failure of any single SOA does not make either element inoperable, not even for specific pairs of endpoints. For example, let the second SOA down in the upper left 2×2 unit in Fig. 4(a) fail. Then the only constraint on the 4×4 element is that the top input fiber cannot use a path through the network requiring the top fiber between the upper left unit and the upper center unit. The overall network can always be rearranged around this constraint. If, for example, the top two SOA's in the upper left unit both failed, then the upper input fiber is isolated from the rest of the network. Thus, both elements are seen to tolerate single faults, although not double faults, even though both are made from 2×2 units that are not fault tolerant.

V. OPTIMIZED MULTISTAGE 4×4 ELEMENTS

The next pair of elements are shown in Figs. 5 and 6. By using all output ports of the passive splitters, an optimum 4×4 can be made.

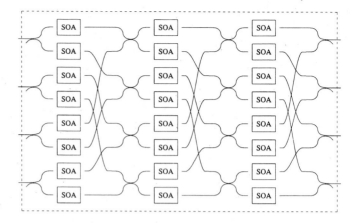

Fig. 5. Optimized 4×4 three-stage rearrangeably nonblocking element.

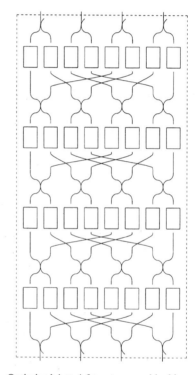

Fig. 6. Optimized 4×4 four-stage nonblocking element.

While the SOA count is the same as in the corresponding elements in Fig. 4, the total number of couplers is reduced. Starting with the three-stage configuration in Fig. 4(a), the output couplers on the left-hand units are effectively combined with the input couplers on the center units and the output couplers on the center units with the input couplers on the right-hand units. The result is the element in Fig. 5. The use of all four ports of these combined passive fiber couplers enables a more efficient use of the amplified signal, thereby reducing the constraints on the facet reflectivity.

Fig. 7 compares the robustness against changes in external reflectivity for the optimized switching elements in Figs. 5 and 6.

The blocking characteristic, support of broadcast, and fault tolerance for the elements in Figs. 5 and 6 are the

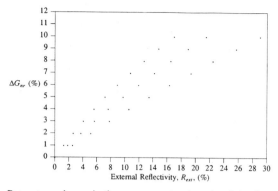

Fig. 7. Percentage change in the nonresonant gain ratio of the three- and four-stage optimized 4×4 switching elements versus external reflectivity for $G_{nr} = -1$ dB, $G_{nr} = -3$ dB, and $G_{nr} = -5$ dB, for each curve left to right, respectively.

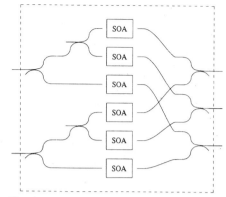

Fig. 8. 2×3 one-stage element with six SOA's.

same as for the corresponding elements in Fig. 4(a) and (b).

VI. A 2×3 Element

Another optical switching element, to be used as a basic building block for more complex elements, is shown in Fig. 8. The two stages of fiber couplers first split the input signal $1/3$–$2/3$ with the larger signal proceeding to the second tier. If the signal is then split 50–50 percent, each of the three SOA's will receive $1/3$ of the original signal.

A. Device and System Considerations

To combine the output signal onto one fiber, the output coupler adds another -3 dB loss. Thus, the total loss contributed by the couplers is -7.77 dB. If it is assumed that the coupling loss is $K = -6$ dB, the total loss that the SOA must overcome is about -14 dB. This translates into a facet reflectivity of $R = 1.02$ percent, if the nonresonant gain ratio is $G_{nr} = -3$ dB, as before. While this required facet reflectivity is lower than the 2×2 element shown in Fig. 1, it is still a reasonable value. Fig. 9 is a graph showing the susceptibility of a 2×3 element to changes in external reflection.

As with all the elements proposed in this paper, this element supports both broadcast and point-to-point communication. As with the other one-stage elements proposed in this paper, shown in Figs. 1 and 2, this element is completely nonblocking but is intolerant of even single faults (as a standalone element).

B. Nonblocking Clos Networks

Since such an asymmetric architecture is unlikely with directional couplers in lithium niobate, little architectural work has been done around such a unit. A three-stage 4×4 switching element, built with a stage of 2×2 elements (Fig. 1) in the center and stages of 2×3 elements (Fig. 8) on the edges is shown in Fig. 10. Like the other multistage elements, since the component units support broadcast, the entire element supports both broadcast and point-to-point communication. Like the other multistage elements, this element also tolerates single faults. How-

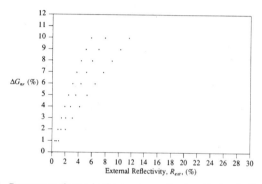

Fig. 9. Percentage change in the nonresonant gain ratio for the 2×3 element versus external reflectivity for $G_{nr} = -1$ dB, $G_{nr} = -3$ dB, and $G_{nr} = -5$ dB, for each curve left to right, respectively.

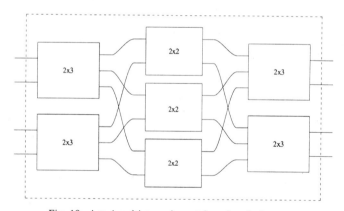

Fig. 10. 4×4 multistage element from 2×3 elements.

ever, it also tolerates failures in any two SOA's, although its blocking characteristic degrades to rearrangeably nonblocking. Unlike any other element proposed here, no double fault isolates any input or output fiber.

Clos described general symmetric three-stage networks [26]. Let the left and right stages be identical, with r distinct $n \times m$ switches and let the center stage have m distinct $r \times r$ switches, all appropriately interconnected. Thus, the overall network has rn inputs and rn outputs and each input–output pair has m different interconnecting paths, one through each distinct switch in the center stage. Two sufficient, but not necessary, conditions proven about Clos networks are

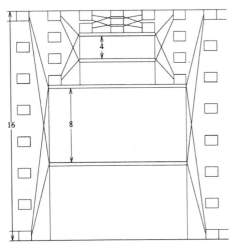

Fig. 11. $2^n \times 2^n$ recursive Clos architecture.

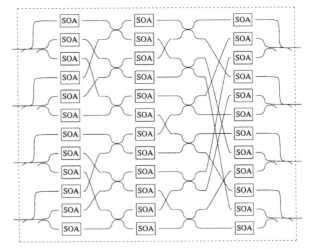

Fig. 12. Optimized multistage 4×4 Clos network.

- if $m \geq n$, then the network is rearrangeably nonblocking;
- if $m \geq 2n - 1$, then the network is completely nonblocking.

Referring to the three-stage element in Fig. 4(a) or 5, $n = m = 2$ satisfies the first condition and the element is rearrangeably nonblocking. Referring to the three-stage element in Fig. 10, $n = 2$ and $m = 3$ satisfies the second condition and the element is completely nonblocking. The element of Fig. 10 uses 36 SOA's, compared to 32 used in the only other element that is both completely nonblocking and fault tolerant (although only for single faults)—the four-stage element of Fig. 4(b) or 6.

Benes showed a recursive construction for $2^n \times 2^n$ rearrangeably nonblocking Clos networks [27]. The *beta element*, shown in Fig. 1, is the basis of the recursion, for $n = 1$. One constructs a three-stage $2^n \times 2^n$ network by using two switches with size $2^{n-1} \times 2^{n-1}$ as the center stage and by using a layer of $n/2$ *beta elements* as the left and right stages. Each *beta element* connects to a pair of ports on one side and to each switch in the center stage on the other side. Fig. 4(a) shows a 4×4 Benes network, the first step of the recursion, for $n = 2$. For every step of the recursion, $n = m = 2$, satisfying the condition for rearrangeability. If the 2×2 units support broadcast, then the overall Benes network does. All the Benes networks are single fault tolerant, possibly becoming blocking networks in the process, but never isolating any inputs or outputs.

We present a similar recursive construction for completely nonblocking Clos networks, built from 2×3 elements. One constructs a three-stage $2^n \times 2^n$ network by using three switches with size $2^{n-1} \times 2^{n-1}$ as the center stage and by using a layer of $n/2$ 2×3 elements as the left and right stages. Each 2×3 element connects to a pair of ports on one side and to each switch in the center stage on the other side. Fig. 10 shows a 4×4 recursive Clos network, the first step of the recursion, for $n = 2$. For every step of the recursion, $n = 2$ and $m = 3$, satisfying the condition for completely nonblocking networks.

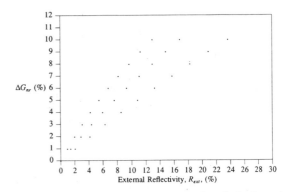

Fig. 13. Percentage change in the nonresonant gain ratio for the optimized multistage Clos network versus external reflectivity for $G_{nr} = -1$ dB, $G_{nr} = -3$ dB, and $G_{nr} = -5$ dB, for curves left to right, respectively.

If the 2×3 and 2×2 units support broadcast, then the overall recursive Clos network does. All these networks are double fault tolerant, possibly becoming blocking networks in the process, but never isolating any inputs or outputs.

Unlike the Benes construction, this recursion for Clos networks becomes inefficient for large networks (large n). For example a 16×16 network would have nine 4×4 elements, like that of Fig. 10, embedded in the center. A network with similar properties made from 4×7 elements would only require seven of them. Other variations on the 4×7 element include a 4×8, that would be fault-tolerant and still completely nonblocking, and a 4×6, that might be *virtually nonblocking* (depending on the traffic statistics). This area of architectural research is open.

VII. Optimized Multistage 4×4 Clos Network

The elimination of passive couplers in between the 2×3 and 2×2 elements reduces the transfer loss from one stage to the next. The networks in Figs. 10 and 11 simplify to the circuit shown in Fig. 12.

Fig. 13 shows the effect of external reflection on this network.

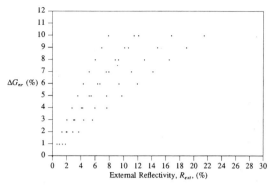

Fig. 14. Comparison of various switching element's susceptibility to changes in external reflectivity for an initial $G_{nr} = -3$ dB. The curves, from left to right, are the 2×3, 2×2, one-stage 4×4, 4×4 optimized multistage Clos, and optimized Benes multistage 4×4 network.

TABLE I
COMPARISON OF ELEMENTS AND NETWORKS

Switch Type	Device Count	Robustness[†] R_{ext}	Noise[‡] Power	$R_{facet,max}$ @$G_{nr}=-3$dB	Fault-Tolerance	Blocking Characteristic
2×3	6	7.9%	1.3N, 2N, 2.7N	1.02%	0 - Fault	non-blocking
2×2	4	11.6%	N	1.52%	0 - Fault	non-blocking
4×4	16	12.2%	2.1N	0.38%	0 - Fault	non-blocking
4×4, 3-stage	24	11.6%	3N	1.42%	1 - Fault	rearrangeably n-b
4×4, 4-stage	32	11.6%	4N	1.42%	1 - Fault	non-blocking
4×4, Clos	36	7.9%	4N, 8N	0.90%	2 - Fault	non-blocking
4×4, Clos opt.	36	16.9%	3.4N, 6.9N	2.00%, 1.52%	2 - Fault	non-blocking
4×4, 3-stage opt.	24	21.7%	3N	2.94%, 1.42%	1 - Fault	rearrangeably n-b
4×4, 4-stage opt.	32	21.7%	4N	2.94%, 1.42%	1 - Fault	non-blocking

†—$G_{nr} = -3$ dB, $\Delta G_{nr} = -10$ percent.
‡—normalized to the output intensity of the 2×2 (multiple values indicate route dependent intensity).

VIII. Summary

Fig. 14 shows a comparison of the various elements and networks with regard to susceptibility to changes in external reflectivity. Clearly, the optimized multistage Benes 4×4 network is most robust; however, there are other considerations that are also important. From the standpoint of fault tolerance, the optimized multistage Clos 4×4 is the best choice. If total amplified spontaneous emission is a major concern, then the single-stage 4×4 is the best choice among all the 4×4's.

Table I compares the advantages and disadvantages of each switching element and network. Photonic switching elements are designed from semiconductor optical amplifiers and passive couplers with fiber-to-fiber unity-gain and low crosstalk. Designs for a 2×2 and an asymmetric 2×3 element, and several designs for 4×4 elements, are presented. While most amplifier analyses have stressed the importance of ultra-low facet reflectivities for high-gain operation, with protection against external reflections with optical isolators, modest facet reflectivities are satisfactory for these elements. It is also shown that substantial amounts of external reflection can be tolerated. The various architectures are compared according to amplifier count, blocking characteristic, broadcast potential, noise power, and fault tolerance.

Acknowledgment

Special thanks to V. E. Kelly for his assistance with the computational analysis.

References

[1] Y. Yamamoto, "Characteristics of AlGaAs Fabry-Perot cavity type laser amplifiers," *IEEE J. Quantum Electron.*, vol. QE-16, pp. 1047-1052, Oct. 1980.
[2] T. Mukai and Y. Yamamoto, "S/N and error rate performance in AlGaAs semiconductor laser preamplifier and linear repeater systems," *IEEE Trans. Microwave Theory Tech.*, vol. MTT-30, pp. 1548-1556, Oct. 1982.
[3] M. Ikeda, "Switching characteristics of laser diode switch," *IEEE J. Quantum Electron.*, vol. QE-19, pp. 157-164, Feb. 1983.
[4] J. Hegarty and K. A. Jackson, "High-speed modulation and switching with gain in a GaAlAs traveling-wave optical amplifier," *Appl. Phys. Lett.*, vol. 45, pp. 1314-1316, Dec. 15, 1984.
[5] L. ThyLén, P. Granestrand, and A. Djupsjöbacka, "Optical amplification in switching networks," presented at Top. Meet. Photon. Switching, Incline Village, NV, Mar. 1987.
[6] M. Kobayashi, A. Himeno, and H. Terui, "Guided-wave optical matrix switch," in *Proc. IOOC-ECOC '85*, Venice, Italy, 1985, pp. 73-76.
[7] J. D. Evankow, Jr. and R. A. Thompson, "Photonic switching with laser diode amplifiers," unpublished.
[8] R. M. Jopson and G. Eisenstein, "Optical amplifiers for photonic switches," in *Proc. Top. Meet. Photon. Switching*, Incline Village, NV, March 18-20, 1987, FC1-1-FC1-3, pp. 116-118.
[9] G. Eisenstein and R. M. Jopson, "Measurements of the gain spectrum of near-traveling wave and Fabry-Perot semiconductor optical amplifiers at 1.5 μm," *Int. J. Electron.*, vol. 60, pp. 113-121, 1986.

[10] R. T. Ku et al. "A packaged 1.5 μm InGaAsP laser amplifier," to be published.
[11] I. W. Marshall, M. J. O'Mahony, and P. D. Constantine, "Optical system with two packaged 1.5 μm semiconductor laser amplifier repeaters," *Electron. Lett.*, vol. 22, pp. 253-255, Feb. 27, 1986.
[12] H. Kataoka and M. Ikeda, "Laser-diode optical switch module," *Electron. Lett.*, vol. 20, pp. 438-439, May 24, 1984.
[13] H. S. Hinton, "A nonblocking optical interconnection network using directional couplers," in *Proc. GLOBECOM'84*, 1984, pp. 26.5.1-26.5.5.
[14] T. Shimoe, K. Hajikano, and K. Murakami, "Path-independent insertion-loss optical space switch," in *Proc. OFC/IOOC '87*, 1987, paper WB2.
[15] G. A. Bogert, "4 × 4 Ti:LiNbO3 switch array with full broadcast capability," presented at OSA Top. Meet. Photon. Switching, 1987, paper ThD3.
[16] P. Granestrand et al., "Strictly nonblocking 8 × 8 integrated optic switch matrix in Ti:LiNbO3," presented at IGWO '86, 1986, paper WAA3.
[17] S. Suzuki, M. Kondo, K. Nagashima, K. Komatsu, and T. Mikakawa, "Thirty-two line optical space-division switching system," presented at OFC/IOOC '87, 1987, paper WB4.
[18] R. A. Spanke, "Architectures for guided-wave optical space switching systems," *IEEE Communications*, vol. 25, May 1987.
[19] K. Padmanabhan and A. Netravali, "Dilated networks for photonic switching," presented at OSA Top. Meet. Photon. Switching, 1987.
[20] A. E. Joel, "On permutation switching networks," *Bell. Syst. Tech. J.*, vol. 47, pp. 813-822, May 1968.
[21] G. Eisenstein, "Theoretical design of single-layer antireflection coatings on laser facets," *AT&T Tech. J.*, vol. 63, pp. 357-364, Feb. 1984.
[22] T. Saitoh, T. Mukai, and O. Mikami, "Theoretical analysis and fabrication of antireflection coatings on laser-diode facets," *J. Lightwave Technol.*, vol. LT-3, pp. 288-293, Apr. 1985.
[23] J. D. Evankow, Jr., N. A. Olsson, and R. T. Ku, "Performance of packaged near-traveling-wave semiconductor laser amplifier with multi-longitudinal mode input," *J. Lightwave Technol.*, to be published.
[24] C. H. Henry, "Theory of spontaneous emission noise in open resonators and its application to lasers and optical amplifiers," *J. Lightwave Technol.*, vol. LT-4, pp. 288-297, Mar. 1986.
[25] P. S. Henry, "Lightwave primer," *IEEE J. Quantum Electron.*, vol. QE-21, pp. 1862-1879, Dec. 1985.
[26] C. Clos, "A study of nonblocking switching networks," *Bell Syst. Tech. J.*, vol. 32, pp. 406-424, 1953.
[27] V. E. Benes, *Mathematical Theory of Connecting Networks and Telephone Traffic*. New York: Academic, 1965.

Time-Division Switching

An Experiment on High-Speed Optical Time-Division Switching

SYUJI SUZUKI, TOMOJI TERAKADO, KEIRO KOMATSU, KUNIO NAGASHIMA, AKIRA SUZUKI,
AND MICHIKAZU KONDO

Abstract—An experimental high-speed optical time-division switching system has been realized. The system is able to exchange digitally encoded color video signals at 256-Mbit/s highway speed.

Bistable laser diodes and directional coupler switch matrices are adopted as optical memories and optical read/write gates, respectively, in an optical time switch. The bistable laser diode operates as an optical flip-flop circuit which can be set and reset by optical and electrical signals, respectively. 256-Mbit/s highway speed has been realized with sufficient input highway operating margin using the same wavelength as that of bistable laser diodes for an electrooptical converter. Results of this experiment will be helpful data for use in constructing future optical telecommunications networks, where a variety of broad-band services need to be realized.

I. INTRODUCTION

WITH THE ADVENT of new technology and growing interest in broad-band services, requirements for broad-band telecommunications networks are increasing.

On the other hand, worldwide developments in the optical-fiber technology field have reached the stage where optical-fiber transmission systems are being introduced into telecommunications networks as essential parts to realize the integrated services digital networks (ISDN).

However, present switching systems, which are essential in telecommunications networks as well as transmission systems, are still composed of conventional electronic devices and cannot directly exchange optical signals. Therefore, electronic-optical (E/O) and optical-electronic (O/E) conversions are necessary between optical transmission systems and switching systems.

An optical switching system, which can directly exchange optical signals without E/O and O/E conversions, is expected to have advantages over conventional switching systems for use in exchanging broad-band signals. This paper describes an experimental high-speed optical time-division switching system, which can switch color video signals, using bistable laser diodes and optical switches.

Manuscript received November 15, 1985.
S. Suzuki and K. Nagashima are with the C & C Systems Research Laboratories, NEC Corporation, 4-1-1 Miyazaki, Miyamae-ku, Kawasaki, 213, Japan.
T. Terakado, A. Suzuki, K. Komatsu, and M. Kondo are with the Opto-Electronics Research Laboratories, NEC Corporation, 4-1-1 Miyazaki, Miyamae-ku, Kawasaki, 213, Japan.
IEEE Log Number 8608557.

II. SWITCHING NETWORKS CLASSIFICATION

On the analogy of conventional switching systems, optical switching systems fall into two categories, space-division and time-division switching networks. Wavelength-division switching networks also will be utilized in optical switching systems.

Space-division switching networks are suitable for small-size optical switching systems. Mechanical optical switches can be applied to the space-division switching network, because a high-speed operation is not required. Nonmechanical optical switches have the advantage of integration ease and reliability, but their loss and crosstalk are larger than those for mechanical optical switches.

Wavelength-division switching networks are very attractive because of their flexibility. To construct this switching network, a high-selectivity tunable optical filter, a wide-range variable wavelength light source, and a wavelength converter are required. Recently, a tunable optical filter and a variable wavelength laser diode have been reported. However, the wavelength converter is still not available, according to the present state of the art.

Compared with two other switching networks, time-division switching networks require high-speed control circuits. However, they are still attractive, because they are suitable for large-size optical switching systems and can easily interface with existing time-division optical-fiber transmission systems.

High-speed optical memories and write/read gates are necessary to construct time-division switching networks. An optical-fiber delay line can be used as the optical memory. Furthermore, several categories of optical bistable devices have already been developed. Nonmechanical optical switches, using the electrooptic effect, are suitable for use as optical write/read gates, because of their high-speed switching capability.

NEC reported the world's first experiment on optical time-division switching using fiber delay lines and directional coupler switch matrices [1], [2].

Furthermore, time-division switching was experimented upon, using newly developed bistable laser diodes as optical memories, which are operative at room temperature and will be integrated in small size. Directional coupler switch matrices were successively adopted as write/read gates.

Reprinted from *J. Lightwave Tech.*, vol. LT-4, no. 7, pp. 894-899, July 1986.

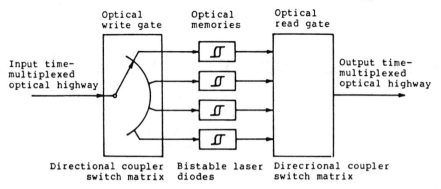

Fig. 1. Optical time-division switch.

The system was successfully used to exchange four 8-Mbit/s monochrome video signals [3].

Fig. 1 shows the optical time-division switch. The optical write gate leads the input time-multiplexed optical highway to individual optical memories in turn. Then, the optical read gate reads out optical signals, stored in optical memories, to the output highway according to controller's direction. Thus, time switching on the time-multiplexed optical highway is accomplished.

III. Optical Devices

A. Bistable Laser Diode as Optical Memory

The bistable laser diode used in this experiment is a 1.3-μm InGaAsP/InP double-channel planar-buried hetero-structure (DC-PBH) laser diode with a tandem electrode, as shown in Fig. 2 [4]. It consists of two gain sections, with a loss section between the two gain sections. The loss section acts as a saturable absorber, which realizes hysteresis characteristics.

Fig. 3 shows hysteresis characteristics as the relation among an optical input power P_{in}, output power P_{out}, and injection current I_{i1}. The bistable laser diode can operate as an optical memory, using two different methods.

Fig. 3(a) shows one method. The bistable laser diode begins laser oscillation without providing optical input power ($P_{in} = 0$), when injection current I_{i1} reaches beyond an I_1 threshold current. Laser oscillation stops when injection current I_{i1} decreases below an I_2 threshold current. Therefore, the bistable laser diode has two stable states A and C, while injection current I_{i1} is at I_b. The bistable laser diode in state A turns into an oscillating state, when more than P_0 optical power is injected into it. The bistable laser diode is able to be set ($A \rightarrow B \rightarrow C$) and reset ($C \rightarrow D \rightarrow A$) by applying an optical set pulse and a reverse injection current reset pulse, respectively.

Fig. 3(b) shows the other method. Injection current I_{i1} is set smaller than threshold current I_2, as shown in Fig. 3(b). The bistable laser diode reaches beyond a P_1 threshold power, and laser oscillation stops when optical input power P_{in} decreases below a P_2 threshold power. Therefore, the bistable laser diode has two stable states E and G, and can retain binary optical information as long as bias input power is remaining on P_b. The bistable laser diode can be set ($E \rightarrow F \rightarrow G$) and reset ($G \rightarrow H \rightarrow E$)

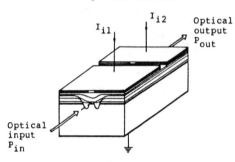

Fig. 2. Bistable laser diode structure.

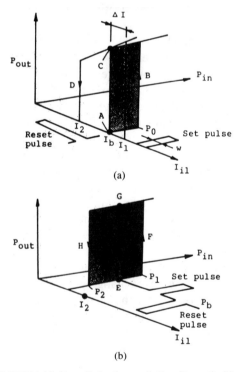

Fig. 3. (a) (b) Bistable laser diode characteristics. P_{in}: optical input power, P_{out}: optical output power, I_{i1}: injection current.

by applying an optical set pulse and reverse optical bias reset pulse, respectively.

The operation in the former method is adopted in this experiment, because it does not require the use of an additional laser diode to feed the optical bias power. A bistable laser diode with a larger noise margin, ΔI, which

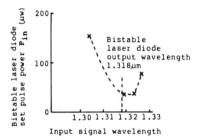

Fig. 4. Wavelength characteristic for bistable laser diode set pulse power. Set pulse width $W = 2$ ns. Noise margin $\Delta I = 0.5$ mA.

is defined as the magnitude of the difference between two currents I_1 and I_b, $I_1 - I_b$, requires a higher optical set pulse power P_{in} to be turned to its oscillating state. The optical set pulse, with a narrower pulse width W, also needs a higher optical set pulse power P_{in}, to set the bistable laser diode.

Low optical set pulse power P_{in} is desirable to obtain an optical operating margin. Large noise margin ΔI is also desirable to avoid the use of a highly regulated constant current source. However, narrow optical set pulse width W was required to realize a high-speed memory operation. Taking these relations into consideration, optical set pulses with a 2-ns pulse width W and 0.5-mA noise margin ΔI, have been applied to this experiment. As a result, a 256-Mbit/s time-multiplexed highway speed has been established.

To ensure stable operation, it is necessary to restrain I_{i1} and I_b drift caused by a temperature variation to as small a value as possible. A temperature stabilization unit, using a thermoelectric element, which can adjust the bistable laser diode module temperature at $27 \pm 0.1°C$, is used. Constant current sources, which can regulate I_b drift within ± 0.1 mA, are introduced in this system.

On the other hand, the bistable laser diode has a wavelength characteristic for optical set pulse power P_{in}, as shown in Fig. 4. Optical set pulse power P_{in} is the lowest when the input signal wavelength is nearly equal to that for bistable laser diode output. This is caused by optical gain wavelength dependency in the bistable laser diode. Consequently, bistable laser diodes set pulse power P_{in} can be reduced to -15 dBm, using the same wavelength as that of bistable laser diodes for the electronic-optical converter.

B. Optical Switch as Write/Read Gates

In this experiment, an integrated 4×4 directional coupler switch matrix, which has the advantages of low insertion loss, low switching voltage, and low crosstalk, is adopted as write and read gates. Fig. 5 shows an optical switch matrix configuration. Five directional coupler switches and waveguides connecting between them are fabricated on the LiNbO$_3$ chip by titanium diffusion [5]. When an optical input signal is injected into one of the optical wave guides, as shown in Fig. 6(a), optical output power values X and Y vary with changes in applied voltage V_e between the two pairs of electrodes, as shown in

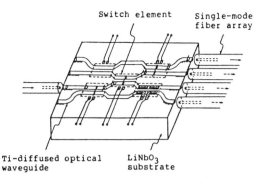

Fig. 5. 4×4 optical switch matrix configuration.

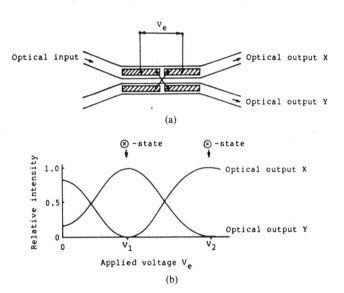

Fig. 6. (a) (b) Directional coupler switch structure and characteristic.

Fig. 6(b). Therefore, the directional coupler switch retains a crossover state, while voltage V_1 is being applied between the two pairs of electrodes. It turns to a straight-through state, when voltage V_2 is applied between them. The switch module, with optical-fiber pigtails, has an about 7-dB insertion loss, -21-db crosstalk, and 12-V switching voltage.

IV. SYSTEM CONFIGURATION

Figs. 7 and 8 show a block diagram and a timing chart of the experimental optical switching system, respectively. The optical time-division switch is constructed in a single stage time switch. A color video signal from a camera is encoded into a 64-Mbit/s digital pulse stream by a video encoder. The digital pulse stream is combined into a 256-Mbit/s pulse stream with the other pulse streams by a multiplexer and then converted to an optical signal using an electronic-optical converter.

This input time-multiplexed 256-Mbit/s optical signal is led to the input optical highway. In the optical time-division switch, each bistable laser diode is reset every signal frame by a reset current pulse in the first half of each time slot (Fig. 8(a)). Optical signals are extracted sequentially from the time-multiplexed optical signal on the input highway by the 1×4 write optical switch matrix and injected into each bistable laser diode in the latter

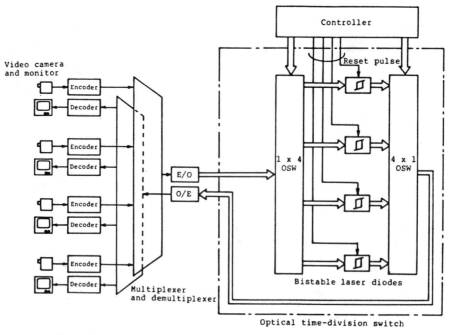

Fig. 7. Experimental optical switching system block diagram. OSW: Optical switch matrix, E/O: electronic-optical converter, O/E: optical-electronic converter.

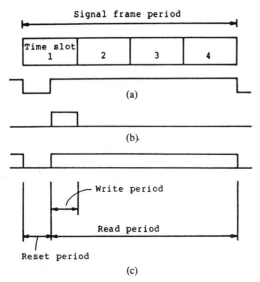

Fig. 8. Timing chart. (a) Reset pulse. (b) Bistable laser diode optical input. (c) Bistable laser diode optical output.

half of each time slot (Fig. 8(b)). Therefore, each bistable laser diode remains in the "off" state or turns to the "on" state, according to the optical signal, and memorizes the binary optical data until the next reset period (Fig. 8(c)). The 4 × 1 read optical switch matrix connects the bistable laser diode outputs to the output highway in each read period and constructs an output time-multiplexed optical signal, where time switching is accomplished. The time-multiplexed optical signal on the output highway is converted to the electric signal again, using an optical-electronic converter, and separated into four 64-Mbit/s digital pulse streams, by a demultiplexer. Finally, this digital pulse stream is decoded into the analog color video signal and applied to each video monitor.

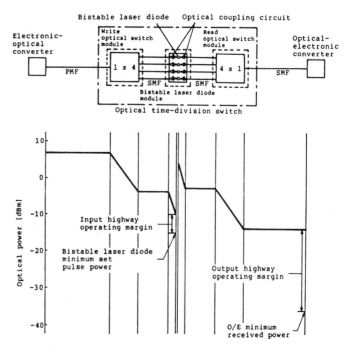

Fig. 9. Optical module configuration and optical level diagram. PMF: Polarization maintaining fiber. SMF: Single-mode fiber.

Fig. 9 shows an optical module configuration and an optical level diagram. A DC-PBH laser diode and Ge avalanche photo diode are adopted in the electronic-optical and optical-electronic converters, respectively.

A polarization-maintaining optical fiber is used as an input device to the 1 × 4 optical switch matrix module, because waveguide switch characteristics depend on incident light polarization and an unwanted polarization light could cause crosstalk.

Optical power and loss are distributed to each optical

module to obtain sufficient input and output highway operating margins, as shown in Fig. 9. The anticipated optical output power value for the electronic-optical converter is 7 dBm. The estimated optical loss values for the write optical switch matrix module and optical coupling circuit in the bistable laser diode modules are 10 dB (including splicing loss) and 6 dB, respectively. Therefore, it is expected that the optical input power value for the bistable laser diodes would be −9 dBm. Because it is required to inject a −15-dBm input power for the bistable laser diode set operation, a 6-dB input highway operating margin value would be expected. On the other hand, the anticipated optical output value for the bistable laser diode is 4 dBm. The estimated optical loss values for the optical coupling circuit in the bistable laser diode module and the read optical switch matrix module are also 5 and 10 dB (including splicing loss), respectively. Therefore, it is expected that the optical input power value for the optical-electronic converter would be −12 dBm. Because a −35-dBm minimum received power optical-electronic converter is required to satisfy the condition that the error rate is less than 10^{-10}, a 23-dB output highway operating margin value would be expected.

V. Experimental Results

Table I shows measured optical power and loss values for the experimental optical switching system components. Fig. 10 shows measured relations between input/output highway optical loss and bit-error rates. The additional input highway optical loss causes a decrease in bistable laser diode input optical power P_{in}, and bit errors occur if P_{in} is not sufficient to set the bistable laser diode. The additional output highway optical loss causes a decrease in the optical-electronic converter input optical power, also, bit errors appear, which are caused by shot-noise and thermal-noise in the optical-electronic converter. The dotted line in Fig. 10(b) shows an error rate characteristic, where output power values from individual bistable laser diodes are adjusted to channel 3 at an optical-electronic converter input. The difference between the dotted and channel-3 lines mainly depends on optical switch characteristics. From the error characteristic measurement, input and output highway optical loss margin values are 5 ~ 6 dB and 19 dB, respectively, under the condition that the error rate is less than 10^{-10}.

Figs. 11 and 12 are a photograph of experimental optical signal waveforms and a photograph of the experimental optical switching system, respectively.

VI. Conclusion

The 256-Mbit/s optical time-division switching system, which can exchange four 64-Mbit/s color video signals, has been successfully experimented on.

Bistable laser diodes and directional coupler switch matrices have been introduced into the optical time-division switching network as optical memories and optical read/write gates for them, respectively. A temperature stabilization unit and constant current sources ensure stable and

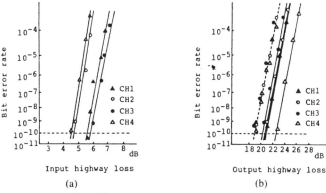

Fig. 10. (a) (b) Error-rate characteristics.

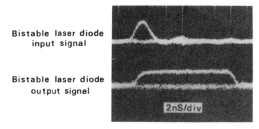

Fig. 11. Experimental optical signal waveforms.

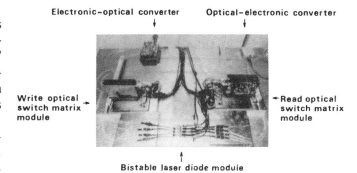

Fig. 12. Experimental optical switching system.

TABLE I
Measured Optical Power and Loss Values for the Experimental Optical Switching System Components

OPTICAL COMPONENTS	OPTICAL POWER AND LOSS VALUES
Electronic-optical converter output power	7 dBm
Write optical switch matrix module and splicing loss	11 dB
Bistable laser diode module input coupling circuit loss	4 ~ 6 dB
Bistable laser diode input power	−11 ~ −8 dBm
Bistable laser diode output power	3 ~ 4 dBm
Bistable laser diode module output coupling circuit loss	4 ~ 6 dB
Read optical switch matrix module and splicing loss	11 ~ 12 dB
Optical-electronic converter input power	−15 ~ −13 dBm

high-speed optical memory operation. Furthermore, the sufficient input highway operating margin has been obtained using the same wavelength as that of bistable laser diodes for the electronic-optical converter. This experiment will open the way for future optical telecommunications networks.

Acknowledgment

The authors gratefully acknowledge the encouragement and guidance given by Y. Kato, General Manager, C & C Systems Research Laboratories, Dr. F. Saito, General Manager, Opto-Electronics Research Laboratories, and Dr. M. Sakaguchi, Assistant General Manager, Opto-Electronics Research Laboratories. They also acknowledge the guidance given by H. Goto, S. Matusita, Dr. K. Kobayashi, Dr. T. Yamaguchi, and Y. Ohta.

References

[1] H. Goto et al., "Optical time-division digital switching: An experiment," presented at Topical Meet. on Optical-Fiber Commun. MJ6, Feb. 1983.

[2] M. Kondo et al., "High-speed optical time switch with integrated optical 1 × 4 switches and single-polarization fiber delay lines," presented at 4th Int. Conf. Integrated Optics Opt.-Fiber Commun., June 1983.

[3] H. Goto et al., "An experiment on optical time-division digital switching using bistable laser diodes and optical switches," in *Proc. Global Telecommun. Conf.*, vol. 2, Nov. 1984, pp. 880–884.

[4] Y. Odagiri et al., "Bistable laser-diode memory for optical time-division switching applications," presented at Conf. Lasers Electrooptics THJ3, June 1984.

[5] M. Kondo et al., "Integrated optical switch matrix for single-mode fiber networks," *IEEE J. Quantum Electron.*, vol. QE-18, pp. 1759–1765, Oct. 1982.

An Experimental Photonic Time-Slot Interchanger Using Optical Fibers as Reentrant Delay-Line Memories

R. A. THOMPSON AND P. P. GIORDANO

Abstract—The system interchanges data from time slots in its input to time slots in its output. Carried by single-mode optical fibers and switched by lithium-niobate directional couplers, the data remains in photonic form from input to output. The data format is a series of frames, each with a start-bit and N time slots with B bits each. An electronic controller reads the arbitrary one-to-one assignment of N input time slots to N output time slots (time-slot permutation) and orchestrates photonic directional couplers to steer the data among input, output, and N recirculating fiber-optic delay lines. While a prototype with $N = 3$, $B = 8$, and a pedestrian 90-MHz bit rate is a 3×3 switch, where each channel carries 28.8 Mbit/s, the basic design is not limited to these numbers.

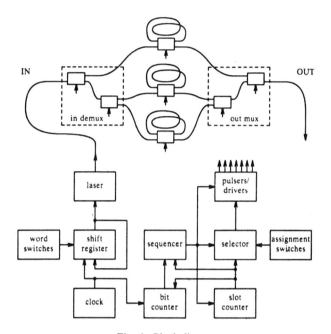

Fig. 1. Block diagram.

I. INTRODUCTION

WITH PHOTONIC switching in the space division well under investigation [1]–[4], the system described here is a basic building block for direct switching of photonic data in the time division. Time-division switching in conventional electronics technology is implemented with a VLSI circuit called an *elastic store*. Input data is stored as it is received and goes to the output by random access according to the switching assignment. Since a photonic elastic store is not expected soon, storage and synchronization of data require alternate and more primitive means. In photonics, the problem is more significant and the design is more primitive than in electronics because the technology is less mature.

A design for a time-slot interchanger is presented. The data is photonic at the input and output and remains photonic internally. The novel feature is the use of loops of optical fibers as reentrant delay-line memories. The block diagram, general timing, and principles of operation are described in Section II. The detailed design is presented in Section III and Section IV contains a discussion of future directions.

II. ARCHITECTURE

The system's function is described in Section II-A. The block diagram is shown in Fig. 1 and described in Section II-B. General timing is described in Section II-C and the principles of operation, mainly the delay-line selection algorithm, are presented in Section II-D.

Manuscript received September 1, 1986. A version of this paper was presented at the Optical Fiber Communications Conference, Atlanta, GA, February 24–26, 1986.
The authors are with AT&T Bell Laboratories, Murray Hill, NJ 07974.
IEEE Log Number 8612222.

A. Function

The input to the system is a serial binary photonic signal organized as a sequence of frames, where each frame consists of a start bit and a sequence of words in time slots. The nth time slot in each frame is assigned as a time-multiplexed digital channel for a data stream. The output from the system is similar except that the assignment of data streams to time slot may be different from the input. The purpose of the system is to read the assignment for reordering the time slots and to effect the reassignment on line as the data is throughput.

In general, let there be B bits per time slot and N time-slots per frame and let the data rate be F frames per second. Then, there are $NB + 1$ bit/frame and the data rate is $F(NB + 1)$ bits/s. Thus, the period of one bit of data is $p = 1/F(NB + 1)$ s. The inputs and outputs are organized as N data channels, each with FB bit/s and a F-bit/s *control channel*. Thus, the photonic time-slot interchanger is an $N \times N$ switch for these data channels.

Two architectures for implementing such a system have been described. One system [5] consists of $2N$ optical fibers with delay lengths of $Bp, 2Bp, \cdots, 2NBp$ s. The incoming word goes to that fiber that installs the correct amount of delay and the 2N fibers are multiplexed together at the output. This paper presents an alternate architecture, where there are N reentrant fibers, all with the same delay length Bp s. If a word must be delayed by n time slots, in the first architecture it would be steered to the appropriate delay line with delay nBp s, but in this architecture it would recirculate in any available delay line n times. As discussed in Section IV, this system is more easily extended for fault tolerance.

The second system [6] interchanges one bit at a time in a bit-interleaved format ($B = 1$). N bistable laser diode memory elements [7] store the photonic bits during each frame of N bits. The architecture presented here is a generalization to N word-sized memory elements in a word-interleaved format. An implementation using word-interleaving with bistable laser diodes would require a photonic shift register or steering at the bit level to one of $B \times N$ memory elements. Any architecture built around these devices has serious problems with high data rates.

Although the proposed design should work for many time slots, a prototype with 3 time slots/frame is deemed large enough to demonstrate the concept and the architecture. The fundamental design is practical, but it is more dependent on photonic amplification [8] in a system with practical size than the architectures previously described.

B. Block Diagram

In Fig. 1, the block diagram of the photonic time-slot interchanger shows the photonic section in the upper half and the electronic control section in the lower half. The photonic section shows, from left to right, an incoming fiber, an input demultiplexer, $N(=3)$ separate fiber-optic delay lines with transmission gates, an output multiplexer, and an outgoing fiber. The electronic control section shows, from left to right, a data generator, a clock and two counters, a sequencer circuit, a set of pulsers and corresponding drivers, and a selector circuit.

Generated by the semiconductor laser, the photonic input signal is carried by single-mode optical fiber to the input demultiplexer, where it goes to one of its three outputs, each fibered to a reentrant delay line. The outputs of the three reentrant delay lines are multiplexed and output. Steering in the input demultiplexer, output multiplexer, and in the delay-line transmission gates uses Ti:LiNbO$_3$ directional couplers [9], [10].

The electronic signal that drives the laser comes from the output stage of a reentrant 25-bit shift register. In this prototype, each channel carries an iteration of the same 8-bit word and the three words are set from manual switches and read into the shift register on RESET. A later simple change would allow the shift register to be loaded by 3 electronic input channels. The 90-MHz system clock drives the shift-register and one counter. The counters keep track of the bit position in the word and the time-slot position in the frame, respectively, of the current input bit. The sequencer produces microcontrol pulses that orchestrate the slot counter, pulsers, and the selector based on the state of the counters.

The switching assignment is input by an operator on manual switches. The selector reads and records this assignment and produces the identity of the delay line from which the current output word is read and into which the current input word is stored. This identity depends on the state of the slot counter. The pulsers produce the signals that control the switching state of the input demultiplexer, delay-line transmission gates, and output multiplexer. A driver transforms the digital output of each pulser into an electrical level appropriate for the directional-couplers.

C. Timing

1. General: Since the data in the last time slot in an input frame could go to the first time slot in the corresponding output frame:

- a delay line is provided for each of the N time slots in a frame, and
- the output frame is delayed from its corresponding input frame by one entire frame nominally, or N time slots.

The use of $N - 1$ delay lines and a throughput delay of $N - 1$ time slots presents a slight improvement, but with added complexity in the control algorithm. For simplicity, we choose to use N delay lines. Since each word may recirculate in its delay line many times, the delay time must equal the duration of a time slot. The start bit in each frame has the effect of inserting a 1-bit shift in the phase of input and output time slots relative to the phase of their respective frames. Alternate means for handling this include:

- adding a read-out transmission gate 1 bit past the read-in transmission gate on each delay-line,
- inserting a 1-bit delay in the fibering between the input demultiplexer and the output multiplexer, or
- advancing the phase of output frames by 1 bit period p relative to input frames.

Because it affects only the electronic timing logic and not the photonic portion of the system, the last solution is the simplest to implement. Then, the throughput delay is p s less than the duration of one frame or exactly the time of N time slots. The start bit of frame j is output at the same time that the last bit of the last time slot of frame j is input.

Since data in the delay lines cannot be reclocked, the delay time, $\tau = Bp$ s, is critical. Since data recirculates in the delay lines, the criticality is multiplied. In general, the greatest tolerance in the arrival of a pulse is ± 20 percent of the clock period. If the input pulses are perfectly aligned with the clock, this figure is the maximum tolerance in the worst case throughput delay. That is, $n\Delta\tau = 0.2$, where n is the most times a signal recirculates in a delay line and $\Delta\tau$ is the delay-time tolerance of the delay lines. The most times a signal recirculates occurs when data in the first input time slot goes to the last output time

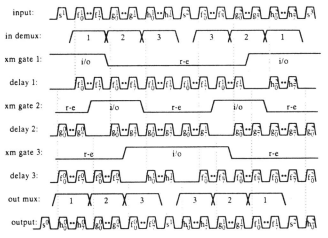

Fig. 2. Data and control waveforms.

slot and the number of recirculations is $n = 2N - 1$. Then, for the prototype, where $N = 3$, $\Delta \tau = 4$ percent, and for a system with a DS1 format, where $N = 24$, $\Delta \tau = 0.4$ percent. In the latter case, anomalies like temperature sensitivity would have to be investigated in a practical system.

2. Details: Fig. 2 illustrates a set of waveforms of the data and control signals for the photonic section of the time-slot interchanger. The duration, horizontally in the figure, is over two frames of data. The time-slot assignment illustrated is that the first and third time slots, called f and h, are interchanged and the second time slot, called g, remains second. Each bit of photonic data in the input and output streams is represented by:

- a superscript indicating the frame number,
- a letter—f, g, or h—indicating whether the bit is in the first, second, or third time slot respectively, and
- a subscript indicating the bit position within the time slot.

Examination of the input and output waveforms shows that the input is receiving frame j while the output is delivering frame $j - 1$ and that bit positions in the output are one ahead of the corresponding position in the input. The delay between input and output is 1 bit less than an entire frame.

The second waveform illustrates the control signal to the input demultiplexer. The labels 1, 2, and 3 mean that input data goes to those demultiplexer outputs that connect to delay lines numbered 1, 2, and 3, respectively. The second waveform from the bottom illustrates the control signal to the output multiplexer. The labels, 1, 2, and 3 mean that output data comes from those multiplexer inputs that connect to delay lines numbered 1, 2, and 3, respectively.

The third, fifth, and seventh waveforms are the signals that control the directional-coupler transmission gates associated with the first, second, and third delay lines, respectively. The fourth, sixth, and eighth waveforms represent the data just entering the first, second, and third delay lines, respectively.

We trace f_0^1, bit number 0 of the first time slot of frame number 1, as it moves through the time-slot interchanger. When this bit reaches the input demultiplexer, it goes to the fiber connected to delay line number 1. Since that transmission gate is in the *input/output* state, the bit enters delay line number 1. By the time it reappears, the transmission gate is in the *reentrant* state and so the bit reenters the delay line. The bit reenters the delay line in synchronism with the input of bits g_0^1, h_0^1, s^2, and f_7^2.

Note that the time that the bit is reentered is in phase with the input of a 0 bit of other time slots until the start bit of the next input frame is received and then it is in phase with the output of 0 a bit of other time slots. The next time the bit would reenter the delay line, in synchronism with the input of bit g_7^2, the transmission gate is in the *input/output* state, so the bit goes to the fiber connected to the output multiplexer. Since the output multiplexer is set to accept data from this fiber, the bit reaches the output. The bit is now bit number 0 of the *third* time slot of frame number 1.

D. Delay-Line Selection

Consider Fig. 2 again, and the first time slot in the second frame. When bits f_0^2–f_7^2 arrive at the input, they are steered to delay line 3. Notice that:

- this particular delay line is NOT the same one used for this time slot in the previous or next frames,
- since ALL other delay lines are busy, this assignment is not arbitrary,
- the ONLY available delay line is the one that is currently being emptied for the output, and
- delay-line selection is nontrivial.

The delay-line selection algorithm, implemented in the selector circuit, is a key part of the photonic time-slot interchanger. Selection must be inexpensive and, more important, quick. An example illustrates the algorithm and then a formal statement is proved.

1. Example: Consider a system with 6 time slots, labeled 0–5. An example switching assignment is shown in Table I.

As long as there is no one-to-many communication, the assignment in Table I is a permutation on the set of time slots. As a shorthand notation, the assignment may be represented as a set of *permutation cycles:*

$$(053)(1)(24).$$

The first cycle, in parentheses, represents the assignment of 0 to 5, 5 to 3, and 3 to 0. The second cycle represents the assignment of 1 to 1. The third cycle represents the assignment of 2 to 4 and 4 to 2.

Since there are also six delay lines, they are also labeled 0–5. To help clarify Table II, we use "in" to denote input time slot n, "on" to denote output time slot n, and "dn" to denote delay line n, where $0 \leq n \leq 5$. Table II represents the operation of the example as the first four frames of data are throughput.

TABLE I
EXAMPLE SWITCHING ASSIGNMENT

Input	Output
0	5
1	1
2	4
3	0
4	2
5	3

TABLE II
EXAMPLE OPERATION IN THE FIRST FOUR FRAMES

Frame #	Input Time-Slot to Delay-Line						Output Time-Slot from Delay Line					
	i0	i1	i2	i3	i4	i5	o0	o1	o2	o3	o4	o5
0	d0	d1	d2	d3	d4	d5	d3	d1	d4	d5	d2	d0
1	d3	d1	d4	d5	d2	d0	d5	d1	d2	d0	d4	d3
2	d5	d1	d2	d0	d4	d3	d0	d1	d4	d3	d2	d5
3	d0	d1	d4	d3	d2	d5	d3	d1	d2	d5	d4	d0

Since all delay lines are empty when frame 0 is input, data from the six input time slots initially go the six delay lines in order. See the left half of line 0 in Table II.

Since frame 0 is output at the same time that frame 1 is input, the right half of line 0 and the left half of line 1 are simultaneous. These twelve entries in Table II are explained in detail in this paragraph. Data in output time slot 0 of frame 0 comes from input time slot 3, now stored in delay line 3. Simultaneously, the data from time slot 0 of frame 1 is input and goes to delay line 3, the only one available. Data in output time slot 1 of frame 0 comes from input time slot 1, now stored in delay line 1. Simultaneously, the data from time slot 1 of frame 1 is input and goes to delay line 1, the only one available. Data in output time slot 2 of frame 0 comes from input time slot 4, now stored in delay line 4. Simultaneously, the data from time slot 2 of frame 1 is input and goes to delay line 4, the only one available. Data in output time slot 3 of frame 0 comes from input time slot 5, now stored in delay line 5. Simultaneously, the data from time slot 3 of frame 1 is input and goes to delay line 5, the only one available. Data in output time slot 4 of frame 0 comes from input time slot 2, now stored in delay line 2. Simultaneously, the data from time slot 4 of frame 1 is input and goes to delay line 2, the only one available. Data in output time slot 5 of frame 0 comes from input time slot 0, now stored in delay line 0. Simultaneously, the data from time slot 5 of frame 1 is input and goes to delay line 0, the only one available.

Since frame 1 is output at the same time that frame 2 is input, the right half of line 1 and the left half of line 2 are simultaneous. Six of these twelve entries in Table II are explained in detail in this paragraph. Data in output time-slot 0 of frame 1 comes from input time slot 3, now stored in delay line 5. Simultaneously, the data from time slot 0 of frame 2 is input and goes to delay line 5, the only one available. Data in output time slot 1 of frame 1 comes from input time slot 1, now stored in delay line 1. Simultaneously, the data from time slot 1 of frame 2 is input and goes to delay line 1, the only one available. Data in output time slot 2 of frame 1 comes from input time slot 4, now stored in delay line 2. Simultaneously, the data from time slot 2 of frame 2 is input and goes to delay line 2, the only one available. At this point, the reader is expected to be on board and to be able to verify the remainder of Table II.

The points of interest are the rightmost six columns in Table II. The jth entry under on is the identity of the delay line to be used for the output of time slot n of frame j. While it is being emptied to the output, this same delay line stores the contents of input time slots n of frame $j + 1$. Notice that:

Each of these columns is the expansion of the *inverse* of a permutation cycle.

The $(j + 1)$th entry in column on is a function of ONLY the jth entry in that column and the switching assignment, found from the *inverse* of Table I.

2. Generalization: The general operation is described as a thoerem and proved by induction. Let $C(\ )$ be the inverse of the switching assignment, where C stands for "comes from," as in "output time slot 5 comes from input time slot 0." Let $C^j(\ )$ be its recursive extensions. From the example:

$$C^0(5) = 5$$
$$C^1(5) = C(5) = 0,$$
$$C(C^1(5)) = C^1(C(5)) = C^2(5) = 3,$$
$$C(C^2(5)) = C^2(C(5)) = C^3(5) = 5, \text{ etc.}$$

Note that inner recursion and outer recursion are equivalent.

Theorem 1: Data in input time slot n in frame j goes to, and simultaneously data for output time slot n in frame $j - 1$ comes from, the delay line identified by $C^j(n)$.

Proof: When frame 0 is input, there is no output and, since all delay lines are empty, data in time slot n goes easily to delay line $n = C^0(n)$. The basis of the induction is thus established. Assume that when frame $j - 1$ was input that time slot n went to the delay line identified by $C^{j-1}(n)$. By the operation of the system, time slot n in output frame $j - 1$ goes at the same time that time slot n in input frame j is received. By the meaning of $C(\ )$, output time slot n of frame $j - 1$ has received the data from the input time slot identified by $C(n)$. This data now goes to the delay line identified by $C^{j-1}(C(n)) = C^j(n)$. Simultaneously, the data from time slot n of frame j goes to the delay line identified by $C^j(n)$, the only one available. The inductive step is thus established and the theorem is proved.

This operation suggests the use of two RAM's in the selector circuit in the photonic time-slot interchanger. One RAM, called CF ("comes from"), holds $C(n)$ and the other RAM, called DL ("delay line"), holds $C^J(n)$. DL identifies the active delay line when addressed by the time slot number. This delay-line identification addresses CF to get a new value, used to update the word in DL for this same time slot number in the next frame.

III. DETAILED DESIGN

The significant portions of the digital electronic design are the selector, implementing the delay-line selection algorithm described in Section II-D, and the pulsers and drivers that stimulate the directional couplers. These designs are described in Sections III-A through III-C and the remainder of the digital design is briefly summarized in Section III-D.

The electronic portion uses emitter-coupled logic (ECL), which has some peculiarities. Many gates have both the TRUE and COMPLEMENT outputs available. The standard JK flip-flop has complementary inputs $\bar{J}$ and $\bar{K}$ and changes value on the rising edge of the clock signal.

A. Selector

The selector reads the numerical position of the current time slot and determines the delay line from which the input byte is read and to which the input byte is stored. Its input signals are the time-slot assignment switches, the state of the slot counter, and control signals from the sequencer. Its output is a parallel bus that carries the identity of the active delay line. The selector, shown in Fig. 3, consists of two high-speed RAM chips, called CF and DL, a latch for the output of the DL RAM, and steering gates for the address of the CF RAM and for the data output to the DL RAM. The control signals, from the sequencer, are write-enable pulses for each RAM, a pulse to latch the DL output, and a CF RAM address steering control. The RESET synchronizing signal acts as the DL RAM data input steering control.

This circuit cycles through four operations:

- read from DL,
- read from CF,
- write to DL, and
- write to CF

during each time slot. Before the first operation, the numerical position of the imminent time slot addresses the DL RAM. This address remains through the second and third operations. The RAM's contents at that address are the identity of the delay line to be used during this time slot. This data latches and serves as the selector's output and also as the address for the CF RAM during the second and third operations. The contents of the CF RAM, at this address, is the identity of the delay line that should be used during the equivalent time slot in the next frame of data. This code must be read into the DL RAM, during the third operation, at the same address from which the current output was just read, so it will be available there in the next frame. In the fourth operation, the contents of the time-slot assignment switches are read into the CF

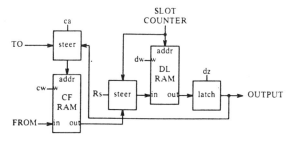

Fig. 3. The selector.

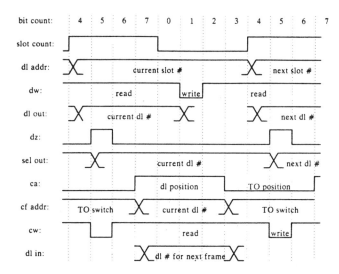

Fig. 4. Selector waveforms.

RAM. The *to* switches provide the address and the *from* switches provide the data. During this operation the time-slot identity, from the slot counter, changes and the DL RAM is ready for the first operation of the next time slot.

The relevant signals are illustrated in Fig. 4. With a 90-Mbit/s clock, each state of the bit counter, shown in the first line of Fig. 4, represents a clock period of 11 ns. The slot-counter advances at the beginning of Q_4 by the SC signal, and its new value is available shortly afterwards as the address of the DL RAM. This value remains until Q_4 of the next time-slot. Because $DW = 1$, the DL RAM is in the READ state and the delay-line identity appears at the RAM output bus 3-7 ns later. The output latch enables during Q_5 with $DZ = 1$ and the DL RAM output is gated straight through to the output bus in the early part of Q_5. This is enough time for the pulsers to be enabled before the IB and OB signals arrive. The selector is quiet during Q_6.

At the beginning of Q_7, the CA signal switches the address of the CF RAM to this identity of the active delay-line. Because $CW = 1$, the CF RAM is in the READ state and the delay-line identity for the next frame appears at the output of the CF RAM in another 3-7 ns. Except during RESET synchronization, this value is the input to the DL RAM and is available well before the beginning of Q_1. The selector is quiet during Q_0. During Q_1, $DW = 0$ causes the DL RAM to WRITE the CF RAM output into the appropriate address. Operation during RESET synchronization is described in Section III-D. The output of the DL RAM changes during the WRITE, but this change is not seen at the output of the latch. The selector is quiet during Q_2.

At the beginning of Q_3, the CA signal switches the address of the CF RAM to the value of the TO time-slot assignment switches. The output of the CF RAM changes shortly afterward, but the DL RAM WRITE is already complete. During Q_5, even after the slot counter advances to begin the next DL RAM READ, $CW = 0$ causes the CF RAM to WRITE the contents of the FROM time-slot assignment switches into the appropriate address. The address remains until Q_7.

Since the TO and FROM switches are not debounced, the switching assignment must be entered carefully or switch bounce could clobber valid data. A third TO address switch is provided and valid data is found only in locations corresponding to a ZERO value on this switch. To enter a switching assignment, the operator must first set this extra switch to the ONE state. After the FROM and remaining TO switches are configured and there is no danger of clobbering valid data, the extra switch is put back to ZERO to effect the new assignment.

B. Pulser

A pulser is a digital circuit responsible for controlling a photonic directional coupler. Its three inputs are an enable signal, a signal indicating when the output pulse should begin, and a signal indicating when the output pulse should end. Its digital output is input to a driver circuit that changes the electrical level of the pulse.

The block labeled "pulsers" on the block diagram in Fig. 1 contains a set of pulsers for the input demultiplexer, a set for the output multiplexer, and one pulser for each delay-line directional-coupler transmission-gate. This block is expanded in Fig. 5. The inputs to the block are the four timing signals IB, OB, IE, and OE, a parallel bus containing the binary code for the delay line to/from which data is steered, a clock phase for the flip-flops, and the RESET signal. Each bit line in the bus corresponds to a pulser for the input demultiplexer and for the output-multiplexer. A decoder enables one delay-line transmission-gate pulser for each valid binary combination on the bus. All enable signals are OFF during RESET. (See Fig. 6.)

The pulses for controlling the directional couplers in the input demultiplexer, the output multiplexer, and the delay-line transmission gates must be generated in different phases of the bit clock and with different widths. Furthermore, as seen in Fig. 2, they overlap. The pulses for the input demultiplexer begin slightly before the first bit of an input time slot and end with the last bit of the same time slot. The pulses for the output multiplexer begin slightly before the first bit of an output time slot (one bit ahead of the input-demultiplexer pulse) and end with the last bit of the same time slot. The pulses for the transmission gates begin when the output multiplexer pulse begins and end when the input-demultiplexer pulse ends.

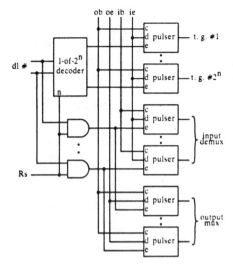

Fig. 5. The pulsers.

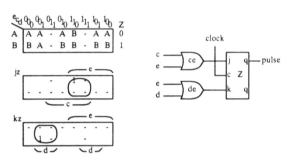

Fig. 6. Individual pulser.

Rather than centrally control these pulses, an *intelligent* pulser controls each pulse. The pulser was originally designed as an asynchronous sequential circuit, however the temporal position of the resultant output pulse was too unpredictable because there were too many logic stages between this pulse and any true synchronization.

Each pulser is a simple two-state Moore-type finite-state machine. The single flip-flop primes for setting when the particular pulser enables (E) and the bit-counter indicates the beginning of the appropriate byte (C). It primes for clearing when the enable (E) is FALSE and the bit-counter indicates the end of the appropriate byte (D). The beginning and end of the pulse synchronizes to an appropriate phase of the clock. Since the flip-flops set and reset on the same clock phase, that phase must be carefully selected.

C. Driver

The driver circuit transforms the digital signal output by a pulser into the electronic signal required by the directional coupler. Each directional coupler has its own driver circuit and they are built from discrete components onto printed wiring boards. The schematic is shown in Fig. 7.

D. Remaining Design

1. Clock: The clock rate is 90 MHz and a six-phase clock is generated with three cascaded one-input gates that

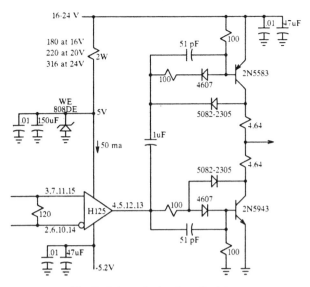

Fig. 7. Schematic drawing of a driver.

provide both outputs. The final assignment of clock phase to various points in the logic is a function of gate and wire delay and is left to experiment in the implementation phase.

2. Data Generator: The data generator is part of the breadboard, but would not be part of a real system. It is only included because a data word is needed and appropriate lab equipment is not available. The generator is a 25-bit reentrant shift register. The word is entered manually on 24 switches, that are not debounced, and parallel loaded into the shift register by the RESET pulse. The shift register advances by a clock phase when RESET is OFF. The value of the front bit of the shift register, strobed with a different clock phase ANDed with the complement of RESET, drives the laser.

3. Bit Counter: The bit counter identifies the numerical position of the current incoming bit within its time slot. Since the output phase is one bit ahead of the input, the bit counter identifies the numerical position, minus one, of the current outgoing bit within its time slot. The start bit in an incoming frame is the last bit of the last time slot of the previous frame. Therefore the counter must count from 0 to $B - 1$ in all time slots except the last and from 0 to B in the last time slot. It requires a clock phase and another input, called L, from the slot counter, that identifies the last time slot.

The bit counter is a sequential circuit with nine states. A state assignment is selected where the adjacent counter states are assigned adjacent binary codes as much as possible. This cyclic state assignment has higher cost in the counter's logic over a classical binary-coded state assignment, but it simplifies a design free of combinational hazards in the sequencer. The bit counter is not affected by RESET

4. Slot Counter: The slot counter is a classical binary counter over the number of time slots in a frame, 0 to $N - 1$. For our implementation, $N = 3$, but for a DS1 format, $N = 24$. The input is a clock signal SC from the sequencer and the output L identifies the last time slot. The slot counter is cleared by the RESET pulse.

5. Sequencer: The sequencer produces the microcontrol signals internal to the electronic control of the system. These signals are pulses with different widths that occur in different phases of the bit counter and are independent of RESET. The inputs to the sequencer are the bit-counter state and L, indicating the last time slot in a frame. Output SC serves as clock to the slot counter; outputs IB, OB, IE, and OE serve as timing signals to the pulsers; and outputs DW, CW, DZ, and CA serve as control signals to the selector.

6. Initialization: System RESET is controlled manually by a debounced toggle switch. This signal is the input to a reset circuit that generates two distinct RESET signals:

- R_p is a short RESET pulse, and
- R_s is a long RESET synchronizing signal.

Both RESET signals go ON slightly after the RESET switch is operated. The exact time is synchronized by a specific clock phase slightly before the start bit of the next frame. The duration of R_p is several bit times. The synchronizing signal R_s remains ON until the beginning of the first frame after the RESET switch has been normalized manually.

The RESET pulse R_p:

- parallel loads the 24 data switches into the reentrant shift register, and
- resets the slot counter.

The RESET synchronizing signal R_s:

- disables shifting the reentrant shift register,
- disables the laser,
- disables all enable signals to the pulser circuits, and
- controls the DL RAM data input steering gates for initialization.

From the example used in Section II-D and in the proof of Theorem 1, frame 0 must be stored in delay lines such that the data from time slot n is placed in delay line n. To effect this initial condition, the DL RAM must have data n stored in address n. Using R_s to control the DL RAM data input steering gates in the selector causes data n to be read into address n during the third of the selector's four cyclic operations. This happens in each time slot of each frame, repeatedly, for the duration of R_s.

IV. Future Directions

Future directions abound, in many dimensions, and promise to keep us busy for many months.

A. Alternate Technologies

In the initial implementation, we have restricted the photonic input to each directional coupler to the TM polarization (perpendicular to the surface of the device) by using polarization-state controllers. This constraint lowers the voltage switching requirement of the driver circuit, allowing faster operation. Alternate implementations would be to use polarization-maintaining fibers [11], [12] for all internal fibers or to use polarization-independent directional couplers, that require a higher switching voltage.

A multiplicative factor occurs with insertion loss because the signal recirculates in the delay lines. If total loss around a delay line, including the transmisison gate, is ρ_d dB, then the worst case throughput loss is $(2N - 1)\rho_d$, not including the loss in the input demultiplexer and output multiplexer. Let ρ_g be the worst-case loss of a single photonic directional coupler transmission gate, including the coupling loss to the input and output fibers. Let $\rho_d = \rho_g$ and let the losses in the demultiplexer and the multiplexer equal $(ln_2 N)\rho_g$, where $ln_2 N$ is the number of transmission gates in any path through a binary tree with N leaves. Then, the worst-case throughput loss is $(2N + 2ln_2 N - 1)\rho_g$, or $57\rho_g$ when $N = 24$. The initial implementation of the photonic time-slot interchanger will have no photonic gain in it and only provides simple demonstrations until it does. The principal reason for limiting N to 3 in the prototype is that worst-case loss is reasonable at $9\rho_g$. Even if we could get $\rho_g = 1.0$ dB, then the worst-case output signal has one-eighth the photonic power of the input signal.

We plan to install photonic amplifiers to provide enough gain to compensate for the loss in each of the reentrant loops. In addition, we plan an alternate implementation of the photonic switching with switched laser diode amplifiers as the switching elements [13] replacing the lithium niobate technology.

B. Data

Generating repetitive data in a cyclic shift register is obviously an initial implementation. We plan to provide three true switchable channels, each at 28.8 Mbit/s, by replacing the data generator with three circuits, each made from an eight-bit parallel port/buffer, shift register, and data synchronization at 3.6 Mbyte/s.

Observation of Fig. 2 shows that all the switching occurs between the last bit of a time slot and the first bit of the next one. The pulse period inside a time slot is not constrained to the switching speed of the devices but to the allowable throughput rate of the devices, a much larger number. We plan to raise the bit rate in each of the three channels enough to demonstrate digital video, switched in the time division. Later we plan to experiment with just how high we can make the bit rate in each channel and determine where the limits are.

C. Relevance

We plan to change the time-slot interchanger so that it interfaces with *real-world* transmission equipment. Some of the issues are the layers of multiplexed hierarchy, pulse stuffing, bit interleaving, and some binary signaling details. To this end, we are installing the transmission equipment illustrated in Fig. 8 in our lab and plan to make the necessary changes to the TSI, interface it to the FT3C, and switch DS0 telephone connections.

Fig. 8. Transmission equipment in lab.

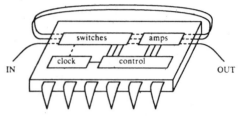

Fig. 9. Futuristic implementation.

D. Extensions

We plan changes to the control algorithm to allow for broadcast and fault tolerance. We plan to interface with space-division switching. The research is not complete until we have investigated a practical implementation, perhaps on a single carrier, shown in Fig. 9.

Fig. 9 suggests an electronic controller on a silicon chip, directional couplers on a lithium niobate chip, and amplifiers and a clock recovery circuit on a III-V chip; all on same carrier. The chips interconnect electrically and photonically. The carrier gets electronic clock signals and power from traditional pins, and has a 2-fiber silicon-V-groove port for the input and output fibers and four 12-fiber silicon-V-groove ports to which two 12-fiber ribbons attach externally to provide 24 delay lines. With electronic clocking and delay length external to the carrier, it is adaptable to various data rates and formats.

V. Conclusion

A system provides direct time-division switching of word-interleaved photonic data using optical fibers as reentrant delay-line memories. A prototype 3×3 switch has been built and many variations will be applied to it in the future.

References

[1] H. S. Hinton, "Nonblocking optical interconnection network using directional couplers," in *Proc. GlobeCom*, Nov. 1984,. pp. 26.5.1-26.5.5.
[2] R. W. Blackmore, W. J. Stewart, and I. Bennion, "An optoelectronic exchange of the future," in *Proc. Int. Switching Symp.*, May 1984, pp. 41A.4.1-41A.4.7.
[3] S. D. Personick, "Photonic switching technology and applications," in *Proc. Conf. Optical Fiber Comm.*, Feb. 1985, p. 44.
[4] R. A. Thompson, "Traffic capabilities of two rearrangeably-nonblocking photonic switching modules," *AT&T Tech. J.*, pp. 2331-2374, Dec. 1985.
[5] M. Kondo et al., "High-speed optical time switch with integrated optical 1 × 4 switches and single polarization fiber delay lines," presented at Fourth Int. Conf. Integrated Optics and Optical Fiber Communication, June 1983.
[6] S. Suzuki, Y. Odagiri, and K. Komatsu, "An experiment on optical time-division digital switching using bistable laser diodes and optical switching," in *Proc. IEEE GlobeCom*, Nov. 1984, pp. 26.4.1-5.
[7] Y. Odagiri and K. Komatsu, "Bistable laser diode memory for optical time-division switching applications," presented at Conf. Lasers and Electrooptics, June 1984.
[8] S. Kobayashi and T. Kimura, "Semiconductor optical amplifiers," *IEEE Spectrum*, pp. 26-33, May 1984.

[9] R. C. Alferness, R. V. Schmidt, and E. H. Turner, "Characteristics of Ti-diffused Lithium Niobate optical directional couplers," *Appl. Opt.*, Dec. 1979.

[10] R. C. Alferness, "Guided wave devices for optical communications," *IEEE J. Quantum Electron.*, vol. QE-17, p. 946, 1981.

[11] J. R. Simpson, "Low-loss polarization preserving fiber by preform deformation," in *Proc. Conf. Optical Fiber Comm.*, Feb. 1985, pp. pd4.1–4.

[12] R. H. Stolen, W. Pleibel, and J. R. Simpson, "High birefringence optical fibers by preform deformation," *J. Lightwave Tech.*, vol. LT-2, no. 5, pp. 639–641, 1984.

[13] M. Kobayashi, A. Himeno, and H. Terui, "Guided-wave optical gate matrix switch," presented at 5th International Conf. on Integrated Optics and Optical Fiber Comm. and 11th European Conf. on Optical Comm., Oct. 1985.

Wavelength-Division Switching

An Optical Switching and Routing System Using Frequency Tunable Cleaved-Coupled-Cavity Semiconductor Lasers

N. A. OLSSON AND W. T. TSANG

Abstract—A 1 × 4 optical switch based on a frequency tunable cleaved coupled cavity (C^3) laser and a grating demultiplexer is demonstrated. The switch, which can easily be expanded to a general $N \times N$ switch with N as high as 13 has a 2 ns access time with negligible crosstalk between the channels.

COMMUNICATION networks require some means for switching signals between different routes and channels, and with the increasing use of lightwave communication systems, a need for optical switching systems has evolved. This can be achieved indirectly by converting the optical signals to electrical signals for such functions as switching and routing. It is evident that a switching system which can manipulate optical signals directly, like integrated optic devices [1], [2] and the device presented here, have important advantages.

In this letter we propose and demonstrate a new type of optical switching and routing system based on the recently developed frequency tunable cleaved-coupled-cavity (C^3) semiconductor laser [3]. In this switching system, the route or output port an optical signal will follow is determined by the wavelength of the optical signal. By switching the lasing wavelength of the C^3 laser transmitter, different routes or output ports can be addressed. The key parameters of a switching system—the access time and the number of addressable ports—is determined in this case by how fast the C^3 laser can be wavelength switched and by how many discrete frequencies are available from the C^3 laser. Switching times as short as 1 ns have been observed [3] and up to 13 discrete frequencies are available, [4] indicating that high performance switching systems can be built with this new scheme.

A general $N \times N$ switch network is outlined in Fig. 1. It consists of N C^3 lasers that can be tuned to N discrete wavelengths, λ_1 to λ_N. The optical outputs from the lasers are combined and enter a wavelength division demultiplexer[5] (WDD). The N outputs can be combined either into a single fiber leading to a reduced throughput of approximately $1/N$ for each laser or, as indicated in Fig. 1, combined into a fiber bundle were each fiber carries the light from one laser. In the WDD the different wavelengths are separated and each wavelength is detected with a separate detector. If desirable, the separated wavelengths can be launched into lightguides for further transmission before detection. To eliminate mode partition noise (which will lead to channel crosstalk) induced by optical feedback into the lasers, reflections from the multiplexes and other optical elements should be kept below 1 percent. The frequency tunable C^3 laser has two optically coupled but electrically isolated sections. One section, the

Manuscript received August 15, 1983; revised December 5, 1983.
The authors are with Bell Laboratories, Murray Hill, NJ 07974.

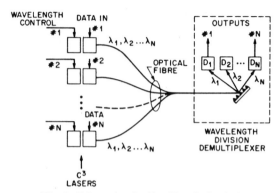

Fig. 1. Schematic of a $N \times N$ optical switch.

laser section, is operated above threshold and the optical output is taken from this side. The other section, the modulator, is below threshold and the lasing wavelength is controlled by varying the current to this section. The tuning is, in effect, a result of the shift in the Fabry-Perot modes of the modulator section, resulting from the index change which is controlled by the modulator current via the carrier density. The C^3 laser step tunes [3] between the longitudinal modes of the laser section, which are typically 20 Å apart for a 100 μm long laser operating at 1.3 μm wavelength.

In what follows we will describe an experimental realization of a 1 × 4 switching system using a C^3 laser operating a 1.3 μm wavelength. A schematic of the experimental arrangement is shown in Fig. 2. The laser current, I_L, consists of a 46 mA dc bias and a 300 Mbit/s, 64 bit, NRZ work of 20 mA peak-peak amplitude. The wavelength control current applied to the modulator section of the C^3 laser is synchronized with the data signal and is a four level step waveform (see the top trace in Fig. 3). The step amplitudes have been adjusted so that each of the current levels correspond to a different wavelength. As a result, the 64 bit data word will be transmitted in 4 subwords, each subword transmitted at a different wavelength. Using a diffraction grating, the different wavelengths can be separated and detected. The top trace in Fig. 3 shows the wavelength control signal and the lower traces are the optical outputs at $\lambda = 1.313$, 1.315, 1.317, and 1.319 μm respectively. Shown in Fig. 4(a) on an expanded time scale is the original data signal, upper trace, and, lower trace, the optical signal at 1.319 μm.

As can be seen from Figs. 3 and 4 there is negligible crosstalk between the 4 channels and the switch over from one channel to the other is very fast. A detailed recording of the switchover between two channels is shown in Fig. 4(b). The switching time is approximately 2 ns which includes the effects from the limited risetime of the wavelength control pulse (~1 ns) and the oscilloscope (~0.9 ns). When the laser is switched

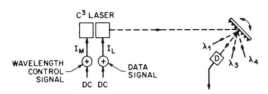

Fig. 2. Experimental setup.

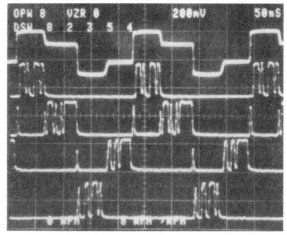

Fig. 3. From top, wavelength control current (4 mA/div), optical outputs at 1.313, 1.315, 1.317, and 1.319 μm respectively. The time scale is 50 ns/div.

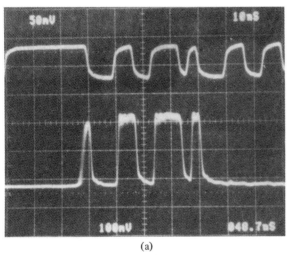
(a)

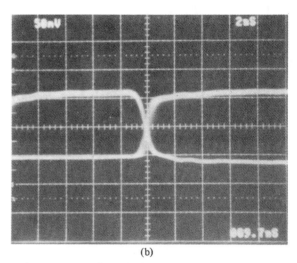
(b)

Fig. 4. (a) Top trace is the data signal and the lower trace the optical output at 1.319 μm. The time scale is 10 ns/div. (b) The crossover between two channels. The time scale is 2 ns/div.

between channels that are not adjacent to each other a short pulse or "glitch" will appear in the in-between channels. This is clearly seen in Fig. 3 for the outputs at 1315 and 1317 μm. The "glitch," however, is very short, approximately 3 ns, and does not seriously degrade the system performance.

Because the wavelength control section of the C^3 laser is operated below threshold the switching time is determined by the carrier lifetime. It is, therefore, advantageous to operate the modulator at a high current level and hence, high carrier density and shorter carrier lifetime. For the experiments presented here where only 4 channels were used the modulator current can be kept fairly high (11-19 mA) for all four channels and the switching speeds are very fast [see Fig. 4(b)]. When the number of channels is increased, however, some of them must be accessed at lower modulator currents and the access time will be longer. Some initial experiments show that access times up to 7-8 ns can be expected when the modulator current is only a few mA. The inherent crosstalk in this new optical switch is limited by the mode discrimination ratio of the C^3 laser. With the mode discrimination ratio, defined as the ratio of the power in dominant mode to the power in the next most prominent mode, ratios of 100:1 are routinely obtained.

In conclusion we have demonstrated a new kind of optical switch based upon the frequency tunable C^3 laser and a wavelength division demultiplexer. The device is characterized by a nanosecond access time and low crosstalk. In addition, it offers the advantage that the routing of the optical signal is determined at the transmitters and the channel separator at the receiver end is a simple passive device.

Acknowledgment

We wish to thank J. Spalink for suggesting this application of C^3 lasers, and we are grateful to R. A. Logan for providing us with the base lasers for forming the C^3 lasers, and to J. A. Ditzenberger and A. L. Savage for their excellent technical assistance.

References

[1] R. A. Steinberg, T. G. Gaillorenzi, and R. G. Priest, "Polarization-insensitive integrated-optical switches," *Appl. Opt.*, vol. 16, pp. 2166–1270, 1977.

[2] R. A. Soref, "Electrooptic 4 × 4 matrix switch for multimode fibre-optic systems," *Appl. Opt.*, vol. 21, pp. 1386–1389, 1982.

[3] W. T. Tsang, N. A. Olsson, and R. A. Logan, "High speed direct single-frequency modulation with large tuning rate and frequency excursion in cleaved-coupled-cavity semiconductor lasers," *Appl. Phys. Lett.*, vol. 42, pp. 650–653, 1983.

[4] W. T. Tsang, N. A. Olsson, R. A. Linke, and R. A. Logan, "1.5 μm wavelength GaInAsP C^3 lasers: Single-frequency operation and wide-band frequency tuning," *Electron. Lett.*, vol. 19, pp. 415–417, 1983.

[5] W. J. Tomlinson and C. Lin, "Optical wavelength division multiplexer for the 1–1.4 μm spectral region," *Electron. Lett.*, vol. 14, pp. 345–347, 1978.

[6] W. T. Tsang *et al.* "119 km 420 Mb/s transmission with a 1.55 μm single-frequency laser," in *Proc. OSA IEEE Conf. Opt. Fibre Commun.*, post-deadline paper, New Orleans, LA, 1983.

A COHERENT PHOTONIC WAVELENGTH-DIVISION SWITCHING SYSTEM FOR BROADBAND NETWORKS

M. FUJIWARA, N. SHIMOSAKA, *M. NISHIO, *S. SUZUKI, S. YAMAZAKI, S. MURATA and K. KAEDE

Opto-Electronics Res. Labs., *C&C Systems Res. Labs., NEC Corporation, 4-1-1 Miyazaki, Miyamae-ku, Kawasaki 213, Japan

ABSTRACT: A photonic wavelength-division (WD) switching system, utilizing a coherent wavelength switch, is proposed. Design consideration shows that over 1000 line capacity is possible, using a multi-stage switching network with 32 WD channels. The switching function was confirmed through 2 channel switching experiments, using 8GHz-spaced 280Mb/s optical FSK signals.

INTRODUCTION: Broadband networks providing various kinds of services, such as video telephony and video broadcasting, have been receiving increasing attention in recent years. For realizing such networks, optical wavelength-division-multiplexing technologies are attractive, both for transmission and switching, because of its bit-rate independency for individual channels. From this viewpoint, an optical broadband network architecture, using photonic wavelength-division (WD) switching systems and wavelength-division multiplexed (WDM) optical transmission systems, was proposed[1]. Coherent optical detection technologies can realize dense WDM optical transmission, due to their high-frequency selectivity. Moreover, coherent optical technologies are also important for achieving a large scale photonic switching system[2]. This paper proposes a novel photonic WD switching system based on the coherent wavelength switch. Results of two channel WD switching experiments, using 8GHz-spaced 280Mb/s optical FSK (Frequency Shift Keying) signals, are also reported.

COHERENT WD SWITCHING SYSTEM: The key element for the WD switching system is the wavelength switch (λ switch), which is used to accomplish wavelength interchange. Figure 1 shows the structure for the proposed λ switch, employing coherent optical detection technologies. In this system, the FSK modulation format is considered. The input WDM optical FSK signals to the λ switch are split in an optical splitter. Individual split WDM signals are detected and demodulated after being combined with tunable local oscillator (LO) light. Demodulated signals are used to create output optical FSK signals, with driving multi-wavelength laser diode array. With tuning LO wavelength, the desired signal channel can be selected at every coherent optical receiver. Through this process, wavelength interchange is accomplished. A large-capacity switching system can be constructed with multi-stage connection with λ switches. Figure 2 shows the λ^3 switching network, using wavelength multiplexers and demultiplexers (MUXs,DMUXs) in inter-stage networks[1]. The features of the proposed switching system are as follows; (1)Low crosstalk switching for dense WDM signals, due to high- frequency selectivity of coherent optical detection, (2)A large capacity network capability resulting from multiple use in WD channels. Line capacities for the λ^3 and λ^5 switching networks are expressed as n^2 and n^3, respectively, where n is the number of WD channels. On the other hand, line capacity for the WDM coherent passive star network[3]-[5] equals n. Therefore, it is clear that a large line capacity is achievable with λ^N switching networks.

DESIGN CONSIDERATION FOR BROADBAND MAN: The proposed coherent WD switching system is very attractive for realizing a broadband Metropolitan-Area-Network (MAN). The structure of MAN, utilizing the coherent WD switching system as a transit switch(TS), is illustrated in Fig. 3. User signals are multiplexed into coherent WDM signals at local switch (LS) line interfaces and transmitted to the TS. In the coherent WD TS, wavelength interchange is carried out. The feasibility of such a network has been confirmed through the consideration described below.

(1)**Number of WD channels:** The number of WD channels n is determined by LO light source tuning range and required channel separation for suppressing inter-channel crosstalk. Using 1.55μm wavelength tunable DBR LD, over 31Å(380GHz) continuous wavelength tuning has already been demonstrated[6]. Theoretical calculation and experiments have shown that required channel separation is about 10GHz, considering optical FSK dual filter detection of the 1.2Gb/s signal[7]. Therefore, 32 WD channels would be available, even for Gb/s signals. In this case, line capacity values reach 1024 and 32768 for λ^3 and λ^5 switching networks, respectively.

(2)**Optical power level diagram:** In this study, receiver sensitivity was assumed to be -45dBm. This value is possible for the 1.2Gb/s CPFSK system. Therefore, allowable loss for the switches and transmission lines (SMF) was 39dB. (fiber input power=0dBm, system margin=6dB) MUX and DMUX were assumed to consist of m=logn/log2 stage cascaded connection with 2x1(1x2) Mach-Zehnder interferometer elements[8] or 2×2 optical couplers (excess loss=0.2dB). Optical power level was examined, dividing the switching system into the following two parts.

(a)**LS line interface→coherent receiver in λ switch:** Figure 4 shows the relation between SMF length l and number of WD channels n. The parameter is MUX/DMUX element loss L_{MUX}(dB/stage). Even when commercially available optical couplers are used as the MUX in LS (L_{MUX}=3.2dB), 15km SMF transmission is possible, with utilizing 32 WD channels.

(b)**Inter-stage connection:** The inter-stage connection for 32 WD channels can be achieved within the loss budget, using MUX/DMUX elements with L_{MUX} of less than 1.5dB/stage. Moreover, as shown in Fig. 5, with a reduction in the number of MUX/DMUX element stages and with an increase in the number of inter connection lines, the inter-stage connection becomes possible, even with current technologies(L_{MUX}=3~4dB/stage[8]).

(3)**Introduction of "wavelength synchronization"[1]:** For achieving multi-stage WD switching networks, individual wavelengths for input and output WDM signals of each λ switch should be exactly the same. Wavelength locking of WDM optical sources using the "reference pulse method"[9] can be applied to achieve the wavelength synchronization.

EXPERIMENTS: Figure 6 shows the experimental switching system diagram. The "reference pulse method" was applied for wavelength synchronization. Direct output and output via Fabry-Perot resonator(F-P) from the sweep LD were distributed to wavelength multiplexer and λ switch. LD wavelengths were controlled so that the generation time for the beat pulses between the transmitter LD lights and the sweep LD light coincides with the timing for the F-P output pulses. Transmitted and switched signals are 280Mb/s optical FSK signals with frequency deviation of 1GHz. Coherent optical receivers in the λ switch consist of balanced receivers and FSK single filter detection systems with 400MHz-1.4GHz passbands. 1.55μm wavelength tunable DBR LDs were used as the transmitters, LOs and the sweep LD. Continuous wavelength tuning range for the DBR LDs was 16Å(200GHz). Beat spectral linewidths, between two DBR LDs, were around 30MHz, which is sufficiently narrow for FSK single filter detection. Frequency separation for the WDM signals was set to be 8GHz, which is sufficient to avoid inter-channel crosstalk at 280Mb/s. SMF couplers were used as MUXs and optical splitters. Optical spectra, measured using a scanning Fabry-Perot interferometer, are depicted in Fig. 7. In this case, only the LD_{i1} was modulated, where LO2 was tuned to select the channel 1 signal. Switching between synchronized WD channels is clearly observed.

CONCLUSION: A coherent photonic WD switching system for broadband networks is proposed. Design consideration has shown that a broadband MAN with a line capacity exceeding 1000, is possible with the system. The switching function was demonstrated in two channel WD switching experiments, using 8GHz-spaced 280Mb/s optical FSK signals. The coherent WD switching system will play an important role in future broadband networks.

ACKNOWLEDGEMENT: Great thanks are due to M. Sakaguchi, T. Ishiguro, K. Minemura, K. Watanabe and K. Kobayashi for their continuing encouragement.

REFERENCES:
[1] S. Suzuki and K. Nagashima; Tech. Dig. Topical Meeting on Photonic Switching, Nevada, 1987, ThA2, pp.21-23/[2] M. Fujiwara et al.; ibid. ThA4, pp.26-29/[3] M. S. Goodman et al.; ICC'86, 29.4, pp.931-933/[4] D. B. Payne and J. R. Stern; IEEE J. Lightwave Technol., LT-4, pp.864-869, 1986/[5] B. S. Glance et al.; ibid., 6, pp.67-71, 1988/[6] S. Murata et al.; Electron. Lett., 23, pp.403-405, 1987/[7] K. Emura et al.; OFC'88, WC4, p.54/[8] H. Toba et al.; Electron. Lett., 23, pp.788-789, 1987/ [9] N. Shimosaka et al.; OFC'88, THG3, p168

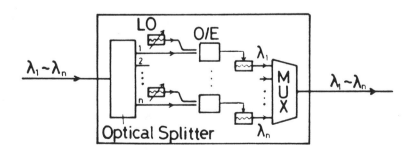

Fig. 1 Coherent wavelength switch (λ switch).

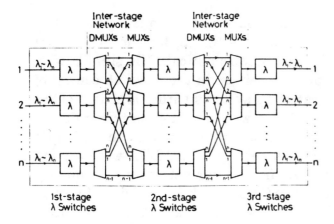

Fig. 2 λ^3 switching network.

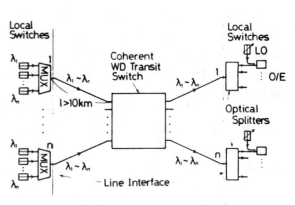

Fig. 3 MAN utilizing coherent WD transit switch.

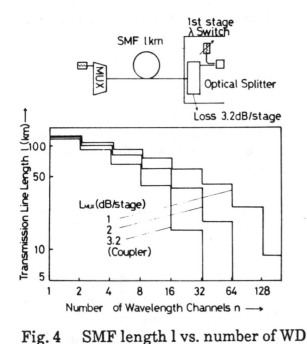

Fig. 4 SMF length l vs. number of WD channels n.

Fig. 5 The inter-stage connection with reduction in the number of MUX/DMUX element stages.

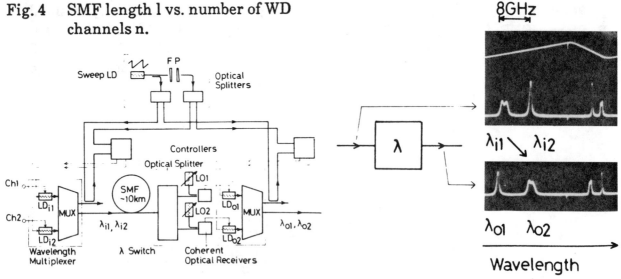

Fig. 6 Experimental coherent WD switching system.

Fig. 7 Spectra of input and output WDM signals for λ switch.

EIGHT-CHANNEL WAVELENGTH-DIVISION SWITCHING EXPERIMENT USING WIDE-TUNING-RANGE DFB LD FILTERS

M.NISHIO   T.NUMAI[*]   S.SUZUKI   M.FUJIWARA[*]   M.ITOH[**]   S.MURATA[*]

C&C Systems Res. Labs.   [*]Opto-Electronics Res. Labs.   [**]R&D Planning and Technical Service Div.

NEC Corporation, 4-1-1 Miyazaki, Miyamae-ku, Kawasaki, 213, Japan

Abstract

A photonic wavelength-division switching system using phase-shift-controllable DFB LD tunable filters was studied. Uniform eight-channel selectivity, error rate and crosstalk characteristics, measured in a switching experiment, are shown to be sufficient for achieving eight wavelength-division multiplexity with 0.8 Å separation over a 6 Å tuning range.

1. Introduction

A photonic wavelength-division (WD) switching system is expected to provide large line capacity and play an important role in the future broadband networks[1]. In the WD switching system, a tunable wavelength filter is a key component. A distributed feedback laser diode (DFB LD) tunable filter[2),3)] has the advantages of small size and integration ease. The WD multiplexity in the previously conducted WD switching experiment, using the conventional DFB LD filters, was limited to four[3], because of broad wavelength transmission bandwidth and uneven optical gain. Recently, a phase-shift-controllable DFB LD filter[4] with a wide wavelength tuning range and a narrow transmission bandwidth, whose gain and bandwidth can be constant over the tuning range, has been developed. This paper first reports results of studies made on the WD multiplexity in the switching network, using phase-shift-controllable DFB LD tunable filters, taking an optical level diagram and crosstalk characteristics into account. After that, measured eight-channel selectivity for the tunable filter, as well as the WD switching system error rate performance, are reported. Moreover, the crosstalk characteristics effect, when the tunable filter selects one wavelength signal from a 5-channel wavelength-division multiplexed (WDM) signal, is shown. These results are sufficient for achieving eight-channel WD multiplexity.

2. System consideration

Figure 1 shows the photonic WD switching system blockdiagram considered in this paper. This system consists of a wavelength multiplexer, a wavelength switch and a wavelength demultiplexer. The wavelength multiplexer consists of modulators, which intensity-modulate light carriers supplied from wavelength reference light sources according to input CH1~n signals, and an optical combiner. In the wavelength switch, an input WDM signal is split and parts are led to individual tunable wavelength filters, each of which extracts a specific wavelength signal. The output signal from the tunable wavelength filter is then converted to an electronic signal by an optical-electronic converter. A preassigned-wavelength light carrier is intensity-modulated by a modulator according to the electronic signal from the optical-electronic converter. Therefore, a specified wavelength signal, selected by the tunable wavelength filter, can be converted to a different wavelength signal. The wavelength demultiplexer consists of an optical splitter, fixed wavelength filters and optical-electronic converters.

WD multiplexity value **n** in this photonic switching network is mainly determined by the tunable filter performance, namely, optical gain, spontaneous emission, selectivity and tuning range. Tunable filter optical gain and spontaneous emission noise characteristics determine the required filter input power. Calculated filter input power values, taking the filter spontaneous emission into account, are -31.5 dBm and -38.5 dBm, to satisfy the 10^{-10} error rate for a 200-Mbps signal, on condition that tunable filter optical gain is 15 and 75, respectively. Assuming 3-dBm modulator output power and 5-dB coupling loss between optical devices and fibers, 21 dB and 28dB, respectively, for 15 and 75 optical gains, could be assigned to the combiner and splitter maximum loss, while the operation margin is 3.5 dB. Therefore, it is concluded that the **n** values can reach 8 and 16, respectively, where 0.5-dB single stage optical coupler excess loss in the combiner and splitter is counted.

Tunable filter selectivity and tuning range also bound the **n** value. The relation among required channel separation $\Delta\lambda$, wavelength tuning range **W** and the **n** value are calculated as shown in Fig. 2, permitting 1-dB power penalty due to crosstalk at a 10^{-10} error rate. The filter transmission spectrum used for the calculation assumes that the pass-band shape is Lorentzian, whose 10-dB down bandwidth is 1 Å, and that the out-band response is flat with X=16-dB attenuation from the peak gain. Therefore, **n** can reach 8, while **W** is 6 Å. When X=20 dB and W=12 Å, the **n** value can reach 16.

3. Experimental results

Figure 3 shows phase-shift-controllable DFB LD filter transmission spectra. The gain and the wavelength are controlled by active section current I_a and phase control current I_p. Constant gain (about 16) and constant 3-dB down bandwidth (0.25 Å) have been obtained over a wavelength tuning range as wide as about 6 Å, while the filter input power was -28 dBm. Bit error rate characteristics, as a function of the filter input power, measured under no-crosstalk conditions for each wavelength channel, are shown in Fig. 4. In this figure, the solid line represents the theoretical value. Eight kinds of plots represent the experimental values. They show good agreement with the theoretical value.

Figure 4 also shows error rate characteristics, when the phase-shift-controllable DFB LD filter selects the wavelength λ_4 signal from the 5-channel WDM input signal, whose wavelengths are $\lambda_2 \sim \lambda_6$. The measured power penalty is at a very low level (about 0.5 dB), which agrees with the calculated value. Therefore, no unexpected non-linear inter-channel interference, caused by four neighboring channels, was observed. Figure 5 shows a photograph of the filter output signal waveform. The experimental system is being successfully used in exchanging digitally encoded motion-video signals.

With further improvements in the optical gain, out-band attenuation, and tuning range for the filter, WD multiplexity will reach 16.

4. Conclusion

The WD switching system using phase-shift-controllable DFB LD tunable filters has been studied, taking optical level diagram and crosstalk characteristics into account. Uniform eight-channel selectivity and the error rate characteristics have been demonstrated. Moreover, the crosstalk effect has been shown. These results have shown to be sufficient for achieving eight WD multiplexity.

Reference

1) S.Suzuki et al., Topical Meeting on photonic switching 1987,ThA2,pp.21-23
2) H.Kawaguchi et al., Appl.Phys.Lett.,vol.50,pp.66-67,1987
3) T.Numai et al.,Electron. Lett.,vol.24,pp.236-237,1988
4) T.Numai et al.,to be presented at ECOC'88

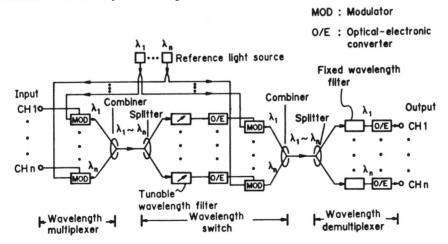

Fig.1. Wavelength-division switching system blockdiagram

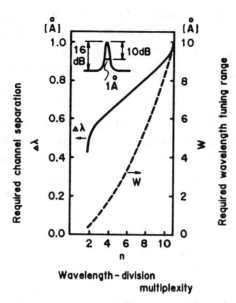

Fig. 2 Required channel separation and required wavelength tuning range vs. wavelength-division multiplexity

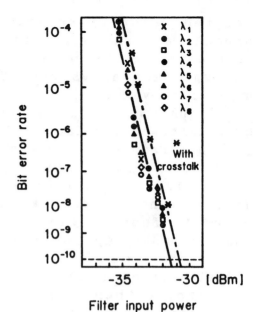

Fig. 4 Error rate characteristics

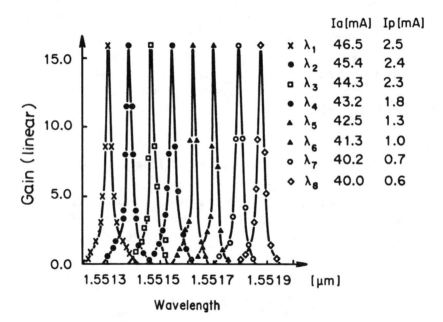

Fig. 3 Tunable filter transmission spectra

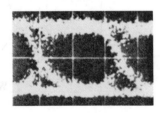

Fig. 5 Filter output signal waveform (2ns/div.)

Section 3.3: Packet Switching

DETERMINISTIC AND STATISTIC CIRCUIT ASSIGNEMENT ARCHITECTURES FOR OPTICAL SWITCHING SYSTEMS

A. de Bosio, C. De Bernardi, F. Melindo

CSELT - Centro Studi e Laboratori Telecomunicazioni, 10148 Turin Italy

The size of an optical switching network based on 2x2 optical switching elements, is limited by a series of factors: the relatively small phisical size of the substrate (no more than 100x75 mm for $LiNbO_3$), the phisical size, loss and crosstalk of the switching element, the loss introduced by the fiber-substrate coupling and the large radii (more than 1 cm) imposed to the waveguide to keep bend loss around few hundredths of dB/degree.

These limitations make it quite difficult to realize on a $LiNbO_3$ substrate a switching network, topologically organized as an N dimension array, with N greater than 10-16, if directional couplers are utilized as switching elements.

In the absence of efficient optical signal regenerators and optical memories, the improvement of an optical switching component can then be obtained, at least in a first step, by both improving the interconnections and switching element characteristics and minimizing the number of switching elements at a parity of the switching network dimension.

This approach seems particularly appropriated for those architectures, called Deterministic Circuit Assignement Architectures (DCAA), which rely on nonblocking networks operated according to synchronous fast packet switching strategies, for which the problem is to find a fast switch setting algorithm, yet keeping low the number of switching elements.

If on one side the "defects" of the optical switching elements lead to architectures based on non blocking multistage switching networks operated according to FPS schemes, on the other hand their "qualities", that is their wide bandwidth and small switch setting time (few psec), make them well suited for architectures, called Statistic Circuit Assignement Architectures (SCAA), which are based on asynchronous switching strategies and on few switching element networks whose blocking characteristics are compensated by the multiplexing capability of the optical system as a whole.

Table 1 shows some characteristics of four switching networks (numerical values are computed for N=8). The first three belong to the permutation network cathegory and are therefore utilizable for systems based on DCA architectures. The fourth belongs to the single path network cathegory, that is has exactly one pass between every arbitrary input/output pair and therefore is not able to connect its inputs to its outputs in any arbitrary way. This network and even simpler networks are well suited for systems based on SCA architectures.

Among the permutation networks, Benes network has the smallest number of stages, but the best algorithm for switch setting on a single process machine

NAME	E(n)	S(n)	CONTROL	BLOCK
BENES	$2^{n-1}(2n-1)$	$2n-1$	CONCEN- TRATED	NO (REARR.)
	20	5		
ODD-EVEN MERGE	$(n^2-n+4)2^{n-2}-1$	$n(n+1)/2$	DISTRIBUTED	NO
	19	6		
BITONIC SORTER	$(n^2+n)2^{n-2}$	$n(n+1)/2$	DISTRIBUTED	NO
	20	6		
OMEGA (DELTA)	$n\,2^{n-1}$	n	DISTRIBUTED	YES
	12	3		

$N=2^n$: network dimension
$E(n)$: number of elements
$S(n)$: number of stages
Numerical values computed with $N=8$

Tab. 1 - Networks' main features

runs in O(NlogN) time. Odd-even Merge network has a smaller number of elements than Bitonic Sorter (BS) network, but has paths with different lenghts and less regular structure. The BS network shows a good tradeoff between the number of elements and switch setting time because it works as a selfrouting network if any switching element is associated to a comparator for switch setting.

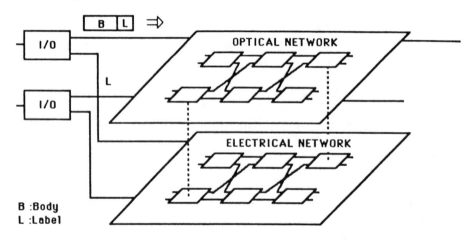

Fig. 1 - Optical switching with selfrouting electrical control

Fig. 1 shows the utilization of a BS network in an optical switching system of a DCA architecture. The input information is transformed in packets of identical size with a label and a body. The labels are sent, at the same time, to an electrical network which is topologically identical to the optical network. The autosetting of each electrical switching element causes the setting of an homologous optical switching element. When the electrical network is completely positioned, the bodies of the packetized information are sent through the optical network.

Fig. 2 shows a model based on a SCA architecture: I1,..,In and O1,..,On are respectively inputs and outputs of a switching network RC;

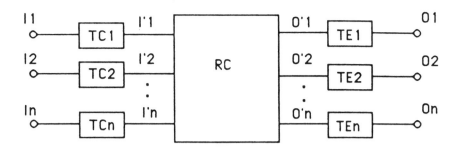

Fig. 2 - Model based on a SCA architecture

TC1,..,TCn are "time compressors" and TE1,..,TEn "time expanders". Each TC receives from an input line bursts of data of variable lenght and interarrival time; data are stored and then sent to RC at a higher bit rate, that is in a shorter time, exploiting the wide band of the optical switching elements; the inverse operation is made by each TE; the path assignement in RC is made when data are ready to be sent from any TC.

As the time compression reduce the data packets collision probability, also blocking networks, like OMEGA or DELTA, characterized by a small number of switching elements and low setting time, can be utilized for such scheme, yet providing end to end protocols for error recovery.

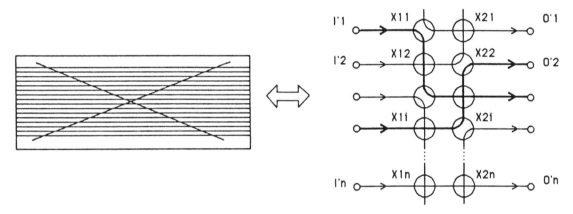

Fig. 3 - Near-neighbour mesh network

Also much simpler networks can be employed, like the ring, the bus, or the near-neighbour mesh (NNM) represented in fig. 3 which is equivalent to a linear array, but provides optical connections both from a high order input to a low order output and viceversa. Estimates made on this structure showed that on a 100x10 mm $LiNbO_3$, a 30 input/output NNM can be implemented using X junctions with a loss of 7.35 dB and a crosstalk of 25dB.

In conclusion, two switching systems based on two different switching architectures have been proposed: they both utilize electro-optical elements also suitable for semiconductor implementation. Hovever, in order to implement electro-optical systems of large dimensions, many problems are still to be solved: among them, the interconnection problem becomes remarkable, especially when logarithmical-growth networks, like those discussed in this paper, are employed.

AN OPTOELECTRONIC PACKET SWITCH UTILIZING FAST WAVELENGTH TUNING

H. Kobrinski, E. Arthurs, R. M. Bulley, J. M. Cooper
E. L. Goldstein, M. S. Goodman and M. P. Vecchi

Bell Communications Research
Morristown, New Jersey 07960-1910

Abstract

We present a new hybrid packet switching system architecture based on integrated electronic and optical subsystems. The design uses multiwavelength optical interconnects with rapidly tunable transmitters and receivers. Experimental feasibility studies of the required optical devices are reported. Transmitter tuning times < 5 ns are demonstrated using a double-section DFB laser randomly tuned among 8 channels. Receiver tuning times of ≈ 1 ns are obtained using DFB optical amplifiers switched between two wavelength-multiplexed channels separated by 0.23 nm.

1. Introduction

Future broadband networks will take advantage of packet switching to achieve uniform routing and multiplexing of multiple-bit-rate data streams. A critical element of such broadband networks is a large capacity ($\geq$ 50 Gbit/s) packet switch. We present a design for a high performance hybrid packet switching system (HYPASS) based on multiwavelength optical interconnections, employing wavelength-selective transmitters and receivers. HYPASS is an input-buffered/output-controlled packet switch using electronic components for the storage and control processing functions, and optical components for routing and transport. HYPASS represents a departure from conventional packet switch designs; its hybrid opto-electronic structure addresses both the control functions and the routing of high-bit-rate data streams. We first discuss the overall system architecture, including a brief description of the control structure and algorithms, and then present experimental results demonstrating the feasibility of the optical transmitter and receiver requirements.

2. The Packet Switch Architecture

A schematic layout of the HYPASS architecture is shown in Figure 1. Input packets are received on optical fiber trunks, and HYPASS switches them to the appropriate output optical fiber trunks. HYPASS contains two multiwavelength optical networks, a *transport* network and a *control* network, that interconnect the input and output ports. Internally, the passive transmissive star couplers in the optical networks split the light intensity from every incoming fiber and distribute it uniformly to all outgoing fibers, and hence perform the function of broadcasting the information. HYPASS is based on broadcast-and-select operations for both the data transport and the control functions. Each output port in HYPASS is identified by the unique wavelengths of the fixed-wavelength receivers (in the transport network) and the fixed-wavelength transmitters (in the control network). At the input ports, the packet destination address defines both the wavelength of the tunable transmitter (in the transport network), and the wavelength of the tunable receivers (in the control network). As shown in Figure 1, the transport network carries the packet traffic from the input ports to the output ports at the full bit-rate. The control network is used to send control information from the output ports to the input ports, and in general can operate at lower bit-rates.

HYPASS utilizes an output-controlled/input-buffered protocol to take maximum advantage of the parallelism afforded by the optical interconnect. The control structure of a packet switching system is of paramount importance and the HYPASS design features an integrated control and transport structure. A single-stage routing network is employed with an output-controlled structure that allows substantial design flexibility. In this design, contention for use of output ports is resolved by sending polls (i.e. explicit request-to-send probes) to the input ports. HYPASS features simultaneous polling of distinct input ports by different output ports taking advantage of the multiwavelength interconnect.

Packets arriving on an input optical fiber trunk are converted to word-parallel electronic signals and stored temporarily in a FIFO buffer. The destination port address is decoded, and the wavelength-tuning currents are set for both the tunable transmitter and the tunable receiver. Output port status information is broadcast over the control network to all input ports. At the input port, the wavelength-tunable receiver selects the control information from its desired output port. The control signals are processed, and a poll received from the output port triggers the transmission of the packet over the transport network. The packet is routed and transmitted to the appropriate output port, since the input port laser has been tuned to the unique fixed-wavelength-address corresponding to the output port.

HYPASS operates in a highly parallel manner. Packets from all inputs are being simultaneously transmitted on different wavelengths using the same passive transport network, while all output ports monitor their busy status and generate the required control signals to be broadcast at their unique wavelength over the control network.

A variety of different control algorithms might be implemented within the HYPASS design. Among the simplest is sequential, individual port polling from each output port. In this case, output port collisions do not occur, but the length of a transmission cycle (number of polls/packet transmission) grows linearly with the number of ports. On the other hand, the broadcast nature of the HYPASS control structure allows a poll to simultaneously be issued to a group of input ports. This feature, plus collision detection at the output port, makes it possible to implement tree polling techniques[1]. As seen from each output port, the tree polling algorithm starts by sending a poll to transmit to all the input ports. If no collision occurs (i.e. one packet or no packet sent), a new cycle starts. If a collision is detected, divide the input ports into two groups, and send a poll to all the input ports in the first group. If no collision occurs, move on to the second input port group. If a collision is again detected, divide the first group into two subgroups, and send polls to one subgroup. The iteration continues in this manner until all packets destined for this output port are cleared. The average delay for tree polling algorithms is a weak function of the number of ports, and it represents an attractive option for services that can be statistically switched.

Using a combination of individual port polling and statistical tree polling, the available bandwidth can be divided to accommodate both random demand (any input port to any output port) and prescheduled demand (fixed input port to fixed output port) packet services. A detailed discussion of these arbitration and control issues is presented elsewhere[2].

The critical optical devices required by HYPASS are the tunable transmitters and the tunable receivers. The required wavelength tuning characteristics of the laser transmitters depend on packet length. Assuming packet sizes of $\approx$ 1000 bits and data rates of $\approx$ 2 Gbit/s, the target laser transmitter parameters are switching speeds faster than $\approx$ 20 ns from any wavelength to any other wavelength, and residency times (stable transmission at a given wavelength) of $\approx$ 500 ns. The switching time should be fast enough to represent a small overhead on the packet transmission cycle. A typical packet switching cycle is shown in Table 1, with a total overhead time of $\approx$ 10% (< 100 ns).

A 32-port HYPASS switch with these parameters would have a large capacity ($\approx$ 64 Gbit/s), compatible with projected BISDN network requirements. Due to the short interconnection distances ($\approx$ 3 m) the overall power budget for the optical interconnection networks is dominated by the splitting losses in the star couplers; a 32-port switch (15 dB splitting losses plus 5 dB excess losses) may be conservatively accomplished with available laser launch powers ($\approx$ 0 dBm) and demonstrated receiver sensitivities (-32 dBm at 2.0 Gbit/s). In addition to power budget considerations, the size of HYPASS depends on the number of distinct wavelengths that can be accessed rapidly and randomly by the tunable transmitters and receivers. Note though that the requirements on the wavelengths used, and their spacing, need not be the same for the tunable transmitters and the tunable receivers, as they are used in different internal networks in HYPASS. In the following sections we describe experiments with fast tunable transmitters and receivers, and report initial results that confirm the feasibility of the required fast wavelength tuning.

3. Rapidly Tunable Transmitter Experiment

Several semiconductor laser structures have been demonstrated to exhibit electronic tuning with $\approx$ 4-10 nm tuning range[3]. Channel spacings < 0.1-0.3 nm will then be required to support several tens of ports in HYPASS. This channel spacing requirement can be met with either coherent heterodyne detection or with high-resolution optical interferometric techniques and direct detection. The lower limit on channel spacing is determined both by the baseband channel bandwidth (e.g. filter bandwidth > 2 GHz is required for an externally-modulated 2 Gbit/s ASK signal) and a sustainable power penalty due to interference from neighboring channels. For a given channel bandwidth and acceptable power penalty (typically $\leq$ 1 dB with respect to the sensitivity with no interference), the minimum channel spacing depends on the modulation schemes and filtering techniques. Using optical filtering (i.e. Fabry-Perot interferometers) and direct detection, the minimum channel spacing is "" sup $\approx$ 2-6 times the channel bandwidth [4,5].

In our work[6], nanosecond switching among 8 distinct wavelengths was achieved using a double-section distributed feedback (DS-DFB) laser along with bulk Fabry-Perot filtering and direct detection. The DS-DFB laser was tuned by current-injection, using an 8-level pattern applied to both the forward and rear section electrodes, in such a way that the total current was kept constant. The optical spectrum of the wavelength-switched DS-DFB shown in Figure 2 was obtained using a scanning Fabry-Perot interferometer with a free spectral range of 0.5 nm and a resolution (filter bandwidth) of 0.018 nm. The tuning currents for both electrodes are shown in the inset of Figure 2. The total tuning range was 0.32 nm, with $\approx$ 0.045 nm channel spacing, which is $\approx$ 2.5 times the filter bandwidth and $\approx$ 3 times a 2 GHz baseband channel bandwidth. In the experiment, the channel-to-channel switching time was less than 5 ns, and a residency time of 500 ns per channel was demonstrated.

In order to demonstrate the ability to switch data we also performed an experiment in which 2 of the 8 selectable wavelengths were chosen for data transmission. In this experiment (Fig. 3), the output of the DS-DFB laser was coupled directly into a lens-tipped fiber and passed through a polarization controller. The optical signal was externally-modulated at data rates between 0.5 and 1.0 Gbit/s using a commercial $LiNbO_3$ modulator with an extinction ratio of 15 dB and insertion loss of 7 dB. The Fabry-Perot cavity was used as a fixed filter (no scanning) to choose between the two channels for the wavelength-switched and amplitude-modulated signals. The filtered signals were detected by an InGaAs APD in a custom designed high impedance FET receiver with a sensitivity of -37 dBm at 1.0 Gbit/s. The Fabry-Perot filter was tuned alternately between each of the 2 channels, and the amplitude-modulated output of the DS-DFB at each wavelength is shown in the two oscilloscope traces measured at the APD receiver (Fig. 3). The switching time between the two wavelengths was again measured to be less than 5 ns. Also shown in the figure is an eye diagram corresponding to one of the wavelength channels modulated with a pseudorandom sequence at 1.0 Gbit/s.

4. Rapidly Tunable Receiver Experiment

Wavelength-tunable receivers operating at high speeds have been recently reported[7], using distributed-feedback optical amplifiers (DFB-OA). Two 1 Gbit/s signal channels with different wavelengths were wavelength-demultiplexed by the optical filter with nanosecond selection times.

The experimental arrangement is shown in Figure 4. The DFB-OA is a commercially available 1.5 μm BH device with a

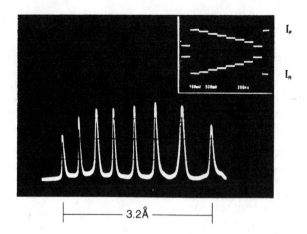

Figure 2. Optical spectrum from a tunable double-section DFB laser transmitter, as seen through a scanning Fabry-Perot filter. Notice the 8 distinct wavelengths, with 0.32 nm total tuning range and 0.045 nm channel separation. Also shown in the figure are the 8-level current inputs for the forward section electrode (top trace) and the rear section electrode (lower trace). The bias current on each electrode was 40 mA.

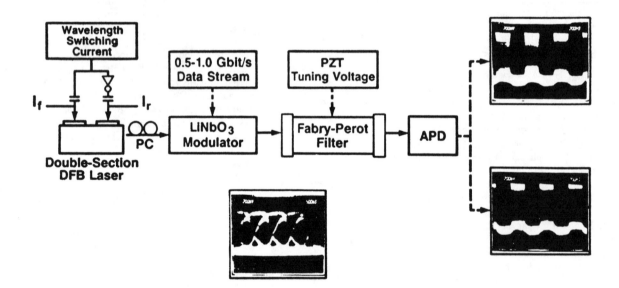

Figure 3. Experimental setup for the rapidly tuning experiment using the double-section DFB laser. The data was driving an external modulator. Two of the 8 possible wavelengths were selected alternatively by the Fabry-Perot filter, and the observed signals are shown on the two oscilloscope traces on the right. Also shown is the eye diagram at 1.0 Gbit/s. Both the forward and the rear section bias currents were 40 mA, as in Figure 2.

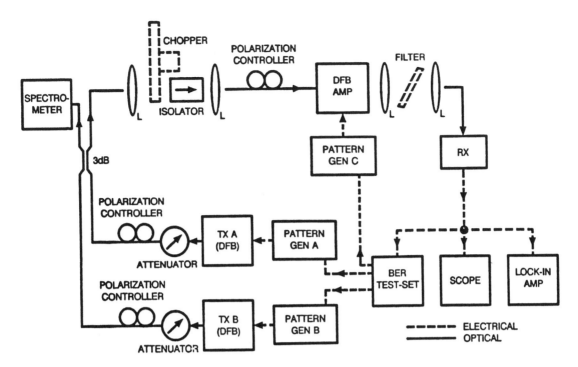

Figure 4. Experimental setup for the tunable receiver experiments based on a DFB optical amplifier. The polarization of the light incident on the AR-coated facet was matched to the TE mode of the DFB-OA. The amplified light was passed through an interferometric band-pass optical filter ($\delta\lambda \approx 2.0$ nm) in order to reduce the collected amplified stimulated emission incident on the APD receiver. A lock-in detection system was used for the dc characterization. The three Pattern Generators and the BER Test Set are synchronized by a common master clock, not shown in the Figure.

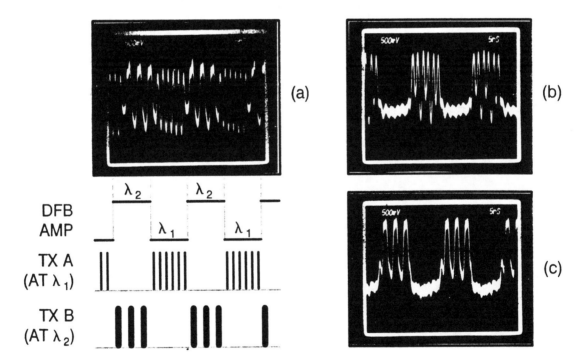

Figure 5. Tunable receiver experimental results using a DFB optical amplifier. Modulating patterns and observed waveforms for the two-wavelength packet-switching simulation experiment at 1.2 Gbit/s with $\Delta\lambda = 0.23$ nm. The pattern (a) corresponds to each packet properly placed in its corresponding time-slot. Pattern (b) corresponds to the signal channel at λ_1 with interference from λ_2, and pattern (c) corresponds to the signal channel at λ_2 with interference from λ_1.

Section 3.4: Systems Issues

Photonic Switching Technology: Component Characteristics versus Network Requirements

J. E. MIDWINTER, FELLOW, IEEE

Abstract—Components for switching in the optical domain offer substantially different characteristics to electronic switches. However, except in special cases these characteristics do not immediately map well onto the network requirements as currently perceived so that there remains great challenge in establishing a viable technology.

INTRODUCTION

IN THIS PAPER, we will use the term "photonic switching" to embrace components that either switch signals while they are in optical form or that intimately embrace the use of optoelectronic technology in the switching function. The term switching is interpreted in the telecommunications sense whereby it is synonymous with routing. This does not exclude its use in the logical switching sense (to perform logical AND, OR, etc.) but does embrace a wide range of implementations that are not suitable for implementing computing type logical interactions.

From the network point of view, we take it as axiomatic that switching will be performed electronically rather than optically unless it is either impossible to do so or is cheaper using optics. Since electronic switching is well established both for routing and multiplexing on circuits and systems at today's operating rates, it seems unlikely that optics will displace it from such applications. Thus we are forced to look for new applications where electronic technology is likely to be more seriously challenged. Broadly speaking, we propose four general areas where this might occur:

a) reconfiguring long-lines high data rate cable networks (protection and block switching),
b) ultrahigh-speed (>1 Gbit/s) multiplexing and demultiplexing,
c) routing of wide-band signals in a wide-band BB.ISDN or CATV local network,
d) routing wide-band digital data in circuit or packet format at major network nodes in the long-lines networks.

From a component point of view, the performance requirements for each of these are very distinct. Class a) applications require switches capable of transmitting very wide-band digital signals. The size of matrix required is likely to be small (16 × 16?), the time between reconfiguring the matrix long (typically measured in hours), and the time available to carry out the reconfiguration operation probably measured in micro- or milliseconds if not much longer. It is also important to note that the whole data stream is to be rerouted as a single block and that no access to the underlying frame structure of the data stream is required in this application, provided that reconfiguration occurs when the system is out of traffic. Even if it is not, data loss will occur because of different cable route lengths (and hence delays) to the destination.

In Class b), the emphasis is upon interleaving or deinterleaving very high-speed data streams. Data rates of many gigabits per second per port can be expected with interleaving complexities of 4 to 1 or 1 to 4 being typical. The challenge here is to use the optics to ease the problems of designing very high-speed electrical drive circuits. The individual switches in such a system must be driven in a fixed sequence but only the first (or last where the highest data rate is assembled) needs to switch with great precision relative to the data bit intervals. Access to the underlying timing data of the digital signal is of course fundamentally necessary in contrast to a) above.

Class c) applications typically involve very large numbers of customers with one or two fibers connected per home to provide a broad spread of new wide-band distributive and interactive services as well as telephony and digital data connections. It is axiomatic that the fiber(s) will carry many different services simultaneously and thus the means of multiplexing these together is critically linked to the means of routing them. The nature of the switching functions required can vary widely between telephony, viewphone, and entertainment TV or audio with "call duration" times spanning minutes to hours and data rates spanning kilobits per second to many megabits per second. In addition, depending upon the availability of bandwidth, the need to deliver services that are largely distributive by nature (i.e., network TV) via a switched channel may be questioned so that some form of hybrid network, part switched, part distributive, may be more appropriate.

Manuscript received February 15, 1988; revised May 25, 1988.
The authors are with University College London, Torrington Place, London WC1E 7JE, England.
IEEE Log Number 8822712.

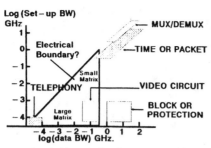

Fig. 1. Summary of switching applications presented in terms of the matrix setup time and matrix complexity.

The Class d) application almost inevitably implies operating directly on a time multiplexed data stream at very high data rate in real time. The exact details of the data formatting will vary considerably between packet type data, TDM data streams, and asynchronous time division data as favored for some future wide-band networks. In each case, the ability to reconfigure the switch very fast is essential and this must be done during time intervals that are sharply defined relative to the data flow. Exactly how much time is allowed for such a reconfiguration then depends upon the specification of the multiplex data format. Access to the timing of the multiplex signal is necessary and a typical switching operation involves moving data from one time multiplexed channel to another, implying an ability not only to time separate blocks of data but also to time shift them. Switching between many input streams and many output streams of such data presents even more complex problems.

We will now examine a variety of different photonic switching technologies against these applications, each bringing unique possibilities to the subject but also problems or limitations. The strengths of optical transmission such a huge transmission bandwidth and freedom from optical "crosstalk" generally carry over to the switching domain. However, in many cases we find that the electrical control problems of setting optical switches may be more difficult than for their electrical brothers, so that the benefit to be gained from their use is not immediately obvious. In discussing them, we will group them according to their switching characteristics rather than their optical implementations. These applications are summarized in Fig. 1 in which we attempt to relate in simple terms setup time and matrix complexity for the various categories.

Passive Pathway Switches with Electrical Control

Here, the light traverses the switch unchanged, apart from attenuation and hopefully only minor reflection and crosstalk from adjoining elements. Prime examples of such elements are those made using optical fibers as the guidance elements and those formed in a planar "integrated-optic" waveguide format in materials such as $LiNbO_3$.

Fiber switches may involve the mechanical movement of one fiber relative to another to change the coupling between them [1] or they may involve the use of a material such as a liquid crystal film sandwiched between two exposed cores electrically modified to adjust the coupling coefficient. In the former case, switching times are likely to be limited by the mechanical time constant of the movement mechanism which is likely to be measured in milliseconds. The time to switch liquid crystal films can vary over a wide range according to the material used and spans, in the extreme, microseconds to seconds. The coupling between fibers may occur in an endfire format or laterally through evanescent field coupling [2]. In the latter case, removal of the cladding is necessary to expose the core, although this might be achieved through the use of fiber having a D cross section as pulled leaving the core exposed.

In general, these fiber switches share the characteristic that they have extremely low insertion loss in the "closed" state and large isolation in the open state. If they are configured in the form of an "exchange-bypass" switch with four terminals offering either a straight-through or crossover connection, then the insertion loss in the cross and bar states is likely to be different, normally higher in the cross state. Furthermore, since both signals are simultaneously present in the same structure, crosstalk is now possible, the exact value being extremely device dependent. The devices are relatively large, typically of millimeter-length interaction regions and with fiber tails that are likely to be in centimeter dimensions. Packaging large numbers for a complex array is thus likely to involve handling large quantities of fiber. The fact that the fiber devices can offer very low insertion losses suggests it may be possible to construct large matrices using them, perhaps serving many thousands of terminations. However, the mechanical problems involved in handling such large numbers of discrete fibers in an orderly manner remain unsolved.

From the control point of view, these devices typically have two electrical connections per switch. For the fiber "reed relay" format switch, switching is bistable and the properties thus do not depend in an analog fashion on the electrical control signals. For fiber switches using precision V-groove arrays for location, the same is true. However, the switches in which the movement is controlled by a piezoelectric analogue "pusher" or by electrical modification of a liquid crystal or other material, precision control of the individual elements may be necessary. Most of the fiber devices are not polarization sensitive.

Switches formed in $LiNbO_3$, typically by the diffusion of Ti to form a planar waveguide directional coupler structure, are again electrically controlled but now through the electrooptic effect [3]. Many detailed designs exist and these have been reviewed extensively elsewhere. Their characteristics are significantly different from the fiber devices. Being produced lithographically in a planar substrate, it is now possible to form many devices simultaneously in a single substrate. Because the directional coupler is typically millimeters long yet micrometers wide, matrices of such devices tend to be very long and thin. To make better use of the substrate area and achieve a more tractable device structure, an 8 × 8 full cross point array (64 devices) has been made on a substrate measuring 68 mm in length using individual elements of 4 mm in length in a folded array, folded by reflection across the matrix diagonal [4]. Larger arrays seem likely to be fabricated by wiring together chips of this level of complexity using optical fibers, and switches of 32 × 32 Clos type have been reported using 8 × 8 building blocks [5].

Discrete switches can operate very rapidly, substantially subnanosecond [6]. However, since the capacitance of the electrode structure associated with each device is typically a few picofarads and switching voltages are of the order of 10 V, significant electrical energy is involved in switching each device ($0.5\ CV^2$ at 100 MHz rate corresponds to 5 mW). Add to this the fact that the discrete devices for a large matrix are spread over a large area of substrate and it is apparent that the electrical problem of delivering and switching signals with precise timing and no electrical crosstalk is severe. This problem is further compounded by the fact that most devices exhibit analog (nonlatching) switching characteristics and thus require close control of switching voltage. Many designs also require two separately controlled voltages per device to ease fabrication tolerance problems and variations in manufacture often mean each has to be individually adjusted. For these reasons, the construction of large arrays that must switch synchronously with a high data rate signal stream is extremely difficult if not impossible. Moreover, we note that the transit time through 2 × 70 mm of lithium niobate would approach 1 ns so that synchronizing data streams at multigigabit-per-second rates traveling via different pathways would be very difficult.

The optical characteristics of these waveguide switches share in common with the fiber devices a very large data bandwidth so that this is unlikely to be a major design consideration. However, the spectral bandwidth is finite and the switching characteristics (extinction, crosstalk, etc.) are likely to be wavelength dependent so that some control of carrier wavelength may be necessary. For systems involving multiple carrier wavelength, this may be a problem. The insertion loss is higher than for equivalent fiber devices, typically 0.4–1 dB/switch. Crosstalk tends to be more of a problem, as does incomplete extinction in the discrete device and it is these effects that currently limit the useful array size. The fiber and planar integrated optic switches could operate bidirectionally if required.

A feature not present in most fiber devices is dependence of the switching properties upon light polarization direction. The planar waveguide devices are polarization dependent unless specifically designed not to be, when other penalties accrue such as larger drive voltage and tighter fabrication tolerances [7].

Another factor that should be noted in the switching context is that both the lithium niobate and fiber-based switches tend to be used in a toggled mode, ON or OFF, exchange, or bypass. Some applications require the ability to fanout a signal to many points on a switched basis. Without some form of signal amplification, this necessarily implies attenuation of the available power by at least a factor $1/N$ where N is the fanout level. Matrices having this characteristic have been designed and built in lithium niobate using passive waveguide power splitters [8]. So also have special designs aimed at achieving extremely low levels of crosstalk [9]. A further topic of detailed study is that of crossovers between waveguides in planar circuits where NO power transfer is the design objective [10]. These are required in the construction of "wiring patterns" such as banyans or perfect shuffles which appear in the more efficient matrix designs that use fewer cross points for a given size switch.

The long term stability of $LiNbO_3$ devices is still a subject of study and some concern. The optical properties can change because of refractive-index damage effects, while the electrical properties can change because of the buildup of electrical carriers within the crystal material, leading to local space charge effects (see, for example, [11]).

Large crossbar circuit switches have also been proposed using optically written phase holograms operating in a free-space beam steering mode [12]. The concept here is to optically write a phase grating in a suitable photorefractive material and to place it normal to an incident beam. The grating then deflects the beam at some angle A to its original direction with large A corresponding to many lines per millimeter in the grating, and at some angle B in the azimuth direction set by the orientation of the grating in rotation about the beam axis as shown in Fig. 2. Thus the beam can be scanned over a circular area centred about the original beam direction. By associating a separate holographic grating with each input beam in a parallel array, perhaps derived from a square array of fibers, it is possible in principle to direct beams at a similar sized array of detectors. To couple the light directly into output fibers would then require a second grating array with matching gratings to beam steer each beam into its target fiber, and this seems likely to present further difficult implementation problems since writing the single grating array seems a challenging task in itself if it is to be done with speed and accuracy. Moreover, in a real circuit switch, it will be necessary to selectively erase single gratings in the array prior to rewriting them for connection to another output channel. The technique's major attraction appears to be that it offers some potential for scaling to very large arrays since the insertion loss appears largely independent of matrix size. However, since deflection angle is related to wavelength (and grating pitch), larger matrices will require even tighter tolerancing of these parameters. When used as an optical switch for remote sources, this could be extremely difficult to arrange.

Of the above devices, lithium niobate devices seem well suited to Class a) applications provided that the long-term stability problems can be proven solved. However, fiber devices might well be preferred on the grounds that they may offer superior reliability and long term stability and provide adequate fast switching.

The Class b) application clearly requires devices of the lithium niobate type (or another electrooptic material).

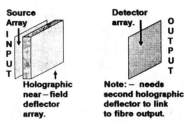

Fig. 2. Schematic layout of a holographic space switch.

Here emphasis is placed upon speed of switching coupled with modest insertion loss in a small matrix. However, it may reasonably be commented that if the electronic drive circuit can drive the modulator at the required speed, it could probably have driven an electronic modulator for the laser source directly. Some advantage is obtained in interleaving multiple random bit streams in that the interleavers are switched with a fixed frequency 10101.. signal, etc., and only the modulators, operating at a submultiple of the line rate need to operate with random data streams (see [13]). It is not obvious to this author that either the fiber or the integrated-optic devices have any substantial role in Class c) applications. The need to handle a wide variety of signals, each of which is at rates that could be switched electrically and with the transmission between the switch and the customer in a time or frequency multiplexed format suggests that other techniques may be more appropriate (see later sections) including some that are not optical. Likewise, the holographic switch, while sometimes presented as a switch for use in wide-band local networks, seems again ill suited to handling any form of multiplexed traffic so that in the application, it would be restricted to one video channel per port operation. Finally, the Class d) switches look difficult to implement using any of the above techniques since normal time multiplexed transmissions contain many time slots per frame, implying that need for a very large switch fanout and many different integer delays available on demand. If the many channels are at low data rate, then switching is more readily done electronically while if many channels of high data rate are interleaved, the line rate rapidly moves out of the electronic control range, so that only simple 1 to 4 or 4 to 1 multiplexing is conceivable. All these switches involve resetting of the switching matrix during time periods that are precisely linked to the multiplex format and which must be short relative to the byte or packet time. However, we note that some proposals for wide-band services suggest that extra time spaces be left between "packets" of data specifically for this purpose so that only these reset periods be synchronized with the switch, the underlying data being unsynchronized [14]. This would then allow the broad transmission bandwidth of the lithium niobate switch to be exploited but still allow matrix resetting times of many data clock cycles length. Note that time slot interchange switching involves not merely the breaking out of discrete packets of data destined for any given source but their reassembly in a different time sequence. Optical memory is thus a key constituent and one element that can provide fixed time delay with low attenuation is the optical fiber, offering storage delays of approximately 5 ns/m. The techniques for assembling and accessing an array of such delay lines have already been given some consideration [15].

ACTIVE PATH OPTICAL SWITCHES WITH ELECTRICAL CONTROL

These devices amplify or in other ways modify the optical data flowing through the device so that they are not optically passive like a block of glass. A prime example is the semiconductor laser amplifier (SLA) based upon a laser chip but operated with antireflection coatings to suppress laser oscillation. The device can exhibit substantial power gain, say 20 dB, under electrical control. Hence inserting the device in an optical pathway is equivalent to inserting a variable gain amplifier. Add to this the fact there is usually substantial insertion loss associated with coupling such devices to fibers or planar waveguides and we see that the potential exists for a new type of switch [16]. In Fig. 3, we show such a switch using a passive 3-dB two-way splitter with two SLA's, one in each arm. By adjusting the gain of either laser, the attenuation between them and the output port can be set to 0 dB. Turning the laser off will introduce a loss of at least 3 dB. In practice this will be much greater, probably 10–15 dB. Thus the possibility exists of building networks in which signals are fanned out and/or routed under electrical control. Combining four SLA's with four 3-dB splitters as shown in Fig. 4 then leads to an optical crossbar switch. The 3-dB bandwidth for the semiconductor laser amplifier is typically in the range 1–10 GHz, although some special laser devices respond to higher frequencies. Interconnecting discrete devices to form a switching matrix could be done with fibers although a novel and possibly better technique involves SiO_2 waveguides formed on Si substrates by chemical vapor deposition and lithographic techniques. Milling slots across an array of such guides then allows a linear array of laser devices to be inserted and electrically connected in hybrid circuit fashion to conductors on the silicon motherboard and optically connected to the optical waveguides, giving a composite electrical/optical hybrid circuit. However, a problem with this as with the $LiNbO_3$ planar waveguide circuits is the need to cater for crossing noninteracting waveguides. This can be done with optical waveguides in a single layer format by careful control of waveguide dimensions and crossing angle.

As with the lithium niobate devices, severe problems seem likely if both fast switching large matrices are to be constructed simply on the grounds of electrical crosstalk, inductance, capacitance, and delay problems. Moreover, even in the case of a slow circuit switch, the electrical power dissipated by a large number of SLA's is certain to lead to difficult problems of heat sinking. Typical SLA drive currents are likely to be 10–50 mA which, with a voltage drop of a few volts, implies 30–150 mW per amplifier and 60–300 mW per cross point. Hence a full 16 × 16 crossbar could be consuming as much as 80 W of electrical power. Thus to make a large crossbar using this

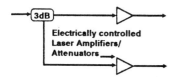

Fig. 3. Simple three-port switch based upon semiconductor laser amplifiers.

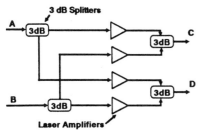

Fig. 4. Four-port switch configuration using semiconductor laser amplifiers, illustrating a minimum of 6-dB insertion loss that can be balanced by the SLA gain.

approach will require very careful attention to minimizing optical loss and maximizing SLA efficiency. (Note that at 1-mW optical per port and 6-dB attenuation per cross point, when all 256 are active, the minimum optical power loss is 0.19 W assuming 100 percent efficiency for all other elements).

OPTICALLY CONTROLLED ELECTRONIC LOGIC

Careful examination of the requirements for the Class d) fast time switch quickly leads one to question whether optics has anything to offer. Given that the data stream is time multiplexed with large dead times whenever the switch needs to be reconfigured, then the "slow circuit" switch offered by relatively large LiNbO$_3$ matrices is a potential candidate. However, if switches are required that can be operated synchronously with a wide-band (multigigabit per second data stream, this will become increasingly difficult using any of the optical devices above. An all electronic solution is also likely to be difficult, since the synchronous operation of a large digital circuit runs into severe timing problems arising from clock time skew and variable delay on different data pathways, not to mention the difficulty of feeding large quantities of wide-band real time data into and out of a large chip. Moreover, if the electronics can drive the optical switch, then it can in all probability handle the data directly. However, at such speeds, problems of timing and electrical crosstalk become extremely serious in extended circuits.

A radical approach to this problem has been proposed which seeks to exploit the wide-band interconnect properties of light, coupled with the possibility of constructing complex parallel "zero time skew" wiring patterns using imaging optical systems. This has generally been proposed in conjunction with "optical logic elements" (to be described in the next section) but could equally well be used with electronic logic given only an efficient electronic ⟨--⟩ optical interface capability. This particular problem, coupled with the type of switching matrix architecture that might become attractive using it, has been considered in some detail and ways in which a self routing

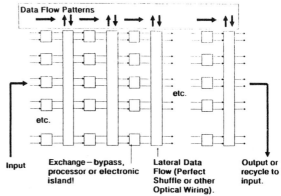

Fig. 5. Pipeline processor based upon electronic islands with optical wiring.

matrix using such a hybrid optoelectronic technology with optical "wiring" might be constructed has been proposed [17]. Closely related to this general approach are ideas of electronic islands with optical zero-time skew interconnections such as perfect shuffles allowing small electronic processors to operate at the frontiers of electronic speed and also to be connected into a large pipeline type processor closely resembling a large switching matrix [18]. Such a layout is shown schematically in Fig. 5 which highlights the proposed division between optical and electronic componentry to exploit the major strengths of each. A variety of suitable optoelectronic interfaces already exist in discrete component form. For the optical to electrical interface, one of the many types of detector is candidate. A natural approach is to think in terms of simple p-i-n detectors monolithically integrated with simple receivers. These might be formed in silicon or III-V semiconductors. For the other interface, more problems arise. It is tempting to envisage using a III-V LED or laser but these are relatively power hungry devices. Moreover, the laser is also a rather critical device whose performance is easily degraded by the presence of growth or other effects.

Another approach is to use a modulator rather than a light source, interrogating its status with light generated elsewhere. This approach has the attraction that the laser source can be located remotely and its heat dissipated well away from the logic chip. By operating as an optical power source only, it can be optimised for power efficiency. For the modulator, a variety of possibilities exist. Prime candidates are either electroabsorption or electrorefraction devices. These can be based on the Franz-Keldysh or quantum confined stark [19] effects in bulk and multiple quantum well materials, respectively. The latter look particularly promising, being very fast, low capacitance small area modulators. Monolithic integration of these devices with other electrical devices is however, in its infancy [20]. Candidate materials systems are either an all III-V semiconductor system or a hybrid combination of III-V grown on a Silicon substrate with the electronic logic.

OPTICALLY CONTROLLED OPTICAL LOGIC

Given that optical wiring has attractions, it is natural to enquire whether one cannot dispense with electronic logic altogether and substitute optical logic. Optical bistable devices which exhibit two state switching responses reminiscent of electronic devices have been studied intensively for many years and a huge literature describing their properties exists [21]. However, just as light is good for transmission since it does not interact with itself or with the guidance medium, leading to low crosstalk, attenuation, and dispersion, so the same attributes become serious limitations when interaction is positively sought as in logic circuits.

Optical nonlinear effects are well known and generally involve some form of nonlinear dielectric constant, leading to effects such as harmonic generation, parametric mixing and oscillation, frequency shifting, etc. Of particular interest for logic processing is the intensity dependent refractive index effect. Here the refractive index or optical dielectric constant of the material changes with light intensity. The effect embraces materials where the critical time constant associated with the change can be comparable to or very long compared to an optical cycle or is similar. Thus extremely fast effects can be embraced, typically in materials such as optical fibers at high intensities, or slower effects with time constants typical of thermal heating and conduction or carrier recombination in III-V or II-VI semiconductor materials [22]. To a first approximation, the product of the characteristic response time and the size of nonlinearity is constant, leading to fast high power devices or slow low power devices can be designed. Exploiting the effect involves the use of some optical positive feedback in association with the nonlinear response material to obtain switching. A typical example is the nonlinear Fabry–Perot in which a resonator filled with the nonlinear material exhibits bistable switching of transmission and reflection coefficient as a function of input light intensity. Recent results for such elements are optical switching powers of 1 mW with switching speeds of 30 μs using a 10-μm diameter spot on ZnSe material [23]. However, with a power speed product of 30 nJ it is many orders of magnitude worse than the best electronic devices (at around 100 fJ). Other devices, such as derivatives of the III-V MQW SEED, promise vastly improved performances.

Even given such advances, the problems of operating optical threshold logic devices over large areas and in very large numbers looks utterly forbidding because of the power uniformity and stray light level requirements. It is proposed that large arrays of such elements be formed in an extended planar Fabry–Perot device to provide the basis of a logical processor with all optical wiring. Even at that stage, fundamental problems remain to be overcome before truly ultrafast systems can be constructed, so that the probability of optical logic displacing electronic logic, except for in few very isolated situations, looks minimal. Nevertheless, much serious study is being devoted to matters such as the architectural design of an all optical digital processor [24], [25]. Perhaps the one clear exception is where truly ultrafast multiplexing or sampling is required, at speeds into the femtosecond region, where fiber soliton logic probably offers the only plausible approach.

WAVELENGTH ROUTING TECHNOLOGIES

In the sections above, we have identified two fundamentally different approaches to optical switching. In the cases of the passive and active path optical switches, it was assumed fundamentally that the routing information was carried by some separate channel and was available to provide electrical control signals to establish the route at the switch point. In the optically controlled electrical or optical switches, some form of synchronous logic was envisaged. That could establish routing either through the use of control data from some external source or by means of data carried within the signal, as in packet transmission where the packet header includes the destination address. Another approach that is more analogous to that used in radio communication is to code the destination through the optical carrier wavelength or frequency. Here we find three broadly different approaches as well as a number of different network formats to exploit them.

When using wavelength coding, one key parameter is the wavelength separation between channels. Until recently, this was large, measured typically in tens of nanometers, so that the number of channels that could be packed within the transmission window of an optical fiber (typically 100 nm in extent) was very limited, of order 10. This meant that the technique was seen primarily as a means of multiplexing in place of TDM for a point-to-point transmission system. Advances in laser wavelength control and stability have recently made possible very narrow-spectral-linewidth tunable semiconductor lasers. Linewidths as narrow as a few kilohertz have been reported using sophisticated external cavity sources [26] while single chip distributed-feedback lasers provide linewidths of 1–10 MHz [27]. Moreover, these devices can be turned over large spectral ranges, typically of the order 10–100 nm corresponding to 10^{12} to 10^{13} Hz in optical frequency.

It is immediately evident that such sources open up the possibility that the spectral transmission window in the fiber centered at 1300 or 1500 nm become a "guided-wave free space." To access such a space, however, narrow spectral linewidth receivers must also exist. These can be made either using fixed [28] or tunable [29] narrow-linewidth optical filters or by the use of an optical heterodyne or homodyne receiver (see, for example, [30]) with its own tunable laser local oscillator. Both approaches are being studied intensively and either offers interesting network possibilities. Note that from the numbers above, the fiber could in principle support 1000 to 10 000 channels each occupying 1 GHz of spectrum!

Assuming first that both tunable transmitters and receivers will soon be available, then we can envisage a fiber network of star format where the central node acts as a passive power splitter, splitting the power from each source equally among all receivers. Channel allocation can be done by assigning a given carrier wavelength to the transmitter and receiver(s) that wish to communicate. Routing is fully nonblocking given one wavelength per channel and wavelengths can be reassigned when released if required in addition. An alternative approach would be to assign each transmitter a unique wavelength and to instruct the receivers to tune to the appropriate wavelength to receive a chosen signal. Both approaches allow point to multipoint transmission but also offer limited security. A third approach is to assign each receiver a unique wavelength. This could be done by placing the fixed filter either at the receiver or at the central node connection. In such a network, the transmitter now selects the receiver to be addressed, affording much higher security in transmission but removing the point to multipoint option. Hybrid combinations are also possible, combining point to point with point to multipoint communications by the judicious assignment of discrete wavelengths and wavelength bands [31]. Furthermore, since very large numbers of wide-band channels are potentially available, it is suggested that "tree type" CATV networks may be possible using fibers in which broadcast and interactive wide-band communication can be combined, leading to great economy in use of fiber and fiber cable with benefit to the capital installation cost per terminal.

One major problem in implementing such a network lies in the cost of the sources, filters, and receivers necessary. At present, these are only available in research quantities and in a few laboratories. Given high volume high yield production, there seems no obvious reason why such techniques should not become cost effective. It is then interesting to note that the same technology used to provide wide-band switched and distributive network services could also be used to provide a powerful (circuit) switching capability in a large network node. The control of such a network is again assumed to be "external" and almost certainly electrical. Since the individual sources can retune very rapidly in principle, the possibility exists of establishing a very fast resetting switch, perhaps for packet handling, although the control problem looks particularly severe in this case with the digital logic based self-routing systems strongly favored.

Optimum Component and Technology Choices

From the above discussion, it will have become apparent that photonic switches can offer a variety of different attributes in the terms of switching or routing. These are summarized in broad terms against the technologies discussed below:

a) *Mechanical fiber switches:* Low insertion loss and crosstalk, potentially large matrices, slow circuit switch operation. Bidirectional, data, and broad-band or narrow-band WDM transparent according to design.

b) *Liquid crystal based devices:* Broadly similar to a) but likely to show faster setting times although not fast enough to synchronize with wide-band data. Probably higher insertion loss than a) thus limiting utlimate matrix size. Bidirectional.

c) *Electrooptic devices:* Small arrays can be extremely fast, hence suitable for high speed data interleaving, MUX, DEMUX. Larger arrays suffer insertion loss and crosstalk problems. Probably limited to 16 × 16 or 64 ×

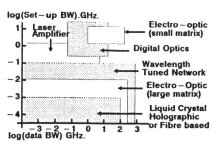

Fig. 6. Summary of photonic switch performances in terms of setup time versus routing bandwidth for comparison with Fig. 1.

64 without optical gain elements. Large arrays also severely limited in speed by electrical drive problems, hence only suitable for circuit switching or TSI switch with large time guard bands. Bidirectional data but not WDM transparent.

d) *Holographic array cross-bar switches with optically written grating deflectors may allow the operation of very large optical switches:* However, the problems in establishing suitable photorefractive materials for the hologram and in engineering the writing and beam forming optics look very formidable. The switch seems fundamentally to be of the slow circuit type. If the output is to be taken directly in optical form, then a second holographic grating is required, further increasing the engineering and control problems.

e) *Active gain switched elements:* Presence of gain offers greater extensibility. Size limited by crosstalk and noise buildup but not yet established. Heavy power consumption, unidirectional, bandwidth limited.

f) *Optically controlled electronics:* Route to very fast synchronous switches for packet or TSI. Extensible to multigigabit per second clock rates. Technology in infancy. Self routing algorithms likely to be used. Common technology base with OEIC.

g) *Optically controlled optical switches:* These appear to offer the only way to handle data in the subpicosecond regime should that be required. At slower speeds, say longer than 10 ps, they then come into direct competition with electronic logic and at present seem to offer few advantages.

In summary, we can say that while virtually all optical switching techniques offer very large transmission bandwidths, in many cases they appear to be severely limited in application either by the fact that the bandwidths actually required can already be handled electronically or because they arise from multiplexed data channels where the switch is required to operate on the underlying multiplex frame structure. In this latter case, many optical switches are excluded because of their relatively slow switching or resetting times. These conclusions are summarized in Fig. 6 against similar axes to those of Fig. 1 to highlight the overlap (or lack thereof) between the requirements and optical capability. Thus, some applications have been identified as offering real scope for optical solutions and others can be expected to emerge as the technology advances and become better quantified.

REFERENCES

[1] W. C. Young and L. Curtis, "Single-mode fiber switch with simultaneous loop-back feature," presented at IEEE/OSA Top. Meet. on Photonic Switching, Incline Village, NV, Mar. 18-20, 1987.

[2] S. R. Mallinson, J. V. Wright, and C. A. Millar, "All fiber routing switch," presented at IEEE/OSA Top. Meet. on Photonic Switching, Incline Village, NV, Mar. 18-20, 1987.

[3] R. C. Alferness and R. V. Schmidt, "Directional coupler switches, modulators, and filters using alternating delta-beta techniques," *IEEE Trans. Circuits Syst.*, vol. CAS-26, pp. 1099-1108, 1979.

[4] P. J. Duthie, M. J. Wale, and I. Bennion, "New architecture for large integrated optical switch arrays," presented at IEEE/OSA Top Meet. on Photonic Switching, Incline Village, NV, Mar. 18-20, 1987.

[5] S. Suzuki, M. Kondo, K. Nagashima, M. Mitsuhashi, H. Nishimoto, T. Miyakawa, M. Iwasaki, and Y. Ohta, "A 32-line optical space division switching system using 8×8 optical matrix switches," NEC Corp. Japan, Pub. 87, pp. 44-50, Oct. 1987.

[6] R. C. Alferness, L. L. Buhl, S. K. Korotky, and R. S. Tucker, "High speed delta-beta reversal directional coupler switch," presented at IEEE/OSA Topical meeting on Photonic Switching, Incline Village, NV, Mar. 18-20, 1987.

[7] R. C. Alferness, "Polarization independent optical directional coupler with weighted coupling," *Appl. Phys. Lett.*, vol. 35, p. 748, 1979.

[8] G. Bogert, "4×4 $TiLiNbO_3$ Switch Array with full broadcast capability," presented at IEEE/OSA Top. Meet. on Photonic Switching, Incline Vilage, NV, Mar. 18-20, 1987.

[9] G. A. Bogert, E. J. Murphy, and R. T. Ku, "A low crosstalk 4×4 $Ti:LiNbO_3$ optical switch with permanently attached polarization maintaining fiber arrays," in *Proc. Int. Guided Wave Optics Conf. IGWO'86* (Atlanta, Ga), Feb. 27, 1986.

[10] L. McCaughan and N. Agrawal, "A rigorous analysis of intersecting waveguides," presented at IEEE/OSA Top. Meet. on Photonic Switching, Incline Village, NV, Mar. 18-20, 1987.

[11] C. M. Gee, G. D. Thurmond, H. Blauvelt, H. W. Yen, "Minimizing dc drift in $LiNbO_3$ devices," *Appl. Phys. Lett.*, vol. 47, pp. 211-213, 1985.

[12] J. P. Huignard, "Wave mixing in nonlinear photorefractive materials and applications to dynamic beam switching and deflection," presented at IEEE/OSA Top. Meet. on Photonic Switching, Incline Village, NV, Mar. 18-20, 1987.

[13] R. S. Tucker, S. K. Korotky, G. Eisenstein, U. Koren, G. Raybon, J. J. Veselka, L. L. Buhl, B. L. Kasper, and R. C. Alferness, "4 Gibits/s optical time division multiplexd system experiment using $Ti:LiNbO_3$ Switch/Modulators," presented at IEEE/OSA Top. Meet. on Photonic Switching, Incline Village, NV, Mar. 18-20, 1987.

[14] T1X1.4/86-A25R1 Liason Report or R. J. Boehm, Y. C. Ching, and R. C. Sherman, "SONET (Synchronous Optical Network)," in *Proc. GLOBECOM'85* (New Orleans, LA), Dec. 1985, pp. 1443-1450.

[15] R. A. Thompson, "Optimizing photonic variable integer delay circuits," presented at IEEE/OSA Top. Meet. on Photonic Switching, Incline Village, NV, Mar. 18-20, 1987.

[16] R. M. Jopson and G. Eisenstein, "Optical amplifiers for photonic switches," presented at IEEE/OSA Top. Meet. on Photonic Switching, Incline Village, NV, Mar. 18-20, 1987.

[17] J. E. Midwinter, "A novel approach to the design of optically activated wide-band switching matrices," *Proc. Inst. Elec. Eng.*, vol. 134, pt. J, pp. 261-268, 1987.

[18] J. E. Midwinter, "Digital optics, smart interconnect or optical logic," *Phys. Technol.*, to be published, 1988.

[19] D. A. B. Miller, D. S. Chemla, T. C. Damen, T. H. Wood, C. A. Burrus, A. C. Gossard, and W. Weigmann, "The quantum well self electrooptic effect device: Optoelectronic bistability, self oscillation, and self linearized modulation," *IEEE J. Quantum Electron*, vol. QE-21, pp. 1462-1476, 1985.

[20] P. Wheatley, M. Whitehead, P. J. Bradley, G. Parry, J. E. Midwinter, P. Mistry, M. A. Pate, and J. S. Roberts, "A novel nonresonant optoelectronic logic device," *Electron. Lett.*, vol. 23, pp. 92-93, 1987.

[21] H. M. Gibbs, *Optical Bistability, Controlling Light with Light*. New York: Academic, 1985.

[22] J. E. Midwinter, "Light electronics, myth or reality?" *Proc. Inst. Elec. Eng.*, pt. J, vol. 132, pp. 371-383, 1985.

[23] S. D. Smith, I. Janossy, H. A. MacKenzie, J. G. H. Mathew, J. J. Reid, M. R. Taghizadeh, F. A. Tolley, and A. C. Walker, "Nonlinear optical circuit elements as logic gates for optical computing; the first digital optical circuits," *Opt. Eng.*, vol. 24, pp. 2-18, 1985.

[24] M. J. Murdocca and N. Streibl, "A digital design technique for optical computing," presented at Top. Meet. Optical Computing, Incline Village, NV, Mar. 16-18, 1987.

[25] A. Huang, "Parallel algorithms for optical digital computers," presented at IEEE 10th Int. Optical Computing Conf., 1983.

[26] R. Wyatt and W. J. Devlin, "10-kHz linewidth 1.5 micron InGaAsP external cavity laser with 55 nm tuning range," *Electron. Lett.*, vol. 19, pp. 110-112, 1983.

[27] L. D. Westbrook, A. W. Nelson, P. J. Fiddyment, and J. V. Collins, "Monolithic 1.5 micron hybrid DFB/DBR laser with 5 nm tuning range," *Electron. Lett.*, vol. 20, pp. 957-959, 1984.

[28] J. Stone and L. W. Stulz, "Pigtailed high-finesse tunable fiber Fabry-Perot interferometers with large, medium, and small free spectral ranges," *Electron. Lett.*, vol. 23, p. 781, 1987.

[29] R. C. Alferness and R. V. Schmidt, "Tunable optical waveguide directional coupler filter," *Appl. Phys. Lett.*, vol. 33, pp. 161, 1978.

[30] R. Wyatt, T. G. Hodgkinson, and D. W. Smith, "1.52-μm PSKJ heterodyne experiment featuring and external cavity diode laser local oscillator," *Electron. Lett.*, vol. 19, pp. 550-552, 1983.

[31] G. R. Hill, D. W. Smith, R. A. Lobbett, T. G. Hodgkinson, and R. P. Webb, "Evolutionary wavelength division multiplexed schemes for broadband networks," in *Proc. Optical Fiber Commun. Conf. OFC-77* (Reno, NV), Jan. 19-22, 1987.

Design of Lithium Niobate Based Photonic Switching Systems

W. A. Payne
H. S. Hinton

Several components required in a photonic switching system as well as the status of their development is described. The system architectural areas that either require development or resolution are then discussed, followed by an examination of the need for viable applications for photonic switching

This article discusses several system areas that require development prior to the successful implementation of the lithium niobate technology for photonic switching. The article first describes several components required in a photonic switching system as well as the status of their development. It then discusses the system architectural areas that either require development or resolution. The final area discussed is the need for viable applications for photonic switching.

Photonic Components Requiring Development

There are three major technical areas essential to eventual development of a titanium indiffused lithium niobate ($Ti:LiNbO_3$) photonic digital switching system. The first of these components is the directional coupler itself. The second factor is the further development of polarization maintaining fiber (PM fiber) which is used to interconnect different $Ti:LiNbO_3$ substrates as well as the driving lasers to the substrates. Optical amplifiers will also be required for large dimensioned photonic switches. Ideal amplifiers for this technology would be those with thresholding capability. These amplifiers would help to improve the signal-to-noise ratio of the signal passing between $Ti:LiNbO_3$ substrates.

Directional Couplers Device Design and Fabrication

The centerpiece of the $Ti:LiNbO_3$ technology is the directional coupler [1] (Fig. 1). These couplers can be interconnected to create larger photonic switching systems. The main limitation in interconnecting a large group of directional couplers to form a large switching network is the individual coupler crosstalk [2,3]. At the current time switching matrices have been fabricated with 16 directional couplers all having crosstalk less than -30 dB [4] (Fig. 2). This polarization sensitive array was fabricated with permanently attached polarization maintaining fiber [4]. In [5] an 8×8 crossbar matrix was reported, with crosstalk values of -23 dB and less, and a fiber to fiber maximum insertion loss of 13 dB. The key concept in the future as larger switch fabrics are needed will be the need to partition the photonic switching networks for the best loss and crosstalk performance. Higher integration of course should be strived for since much of the signal loss at the switch is due to fiber coupling.

Polarization Maintaining Fiber

To minimize the required drive voltages, directional couplers have been optimized to operate on a single linear polarization. In most cases this is the TM polarization. This requirement reduces the required voltage from approximately 50 volts to the 10 to 15 volt range. These lower voltages are desired to allow high speed switching of the directional couplers.

As light propagates through standard single-mode (SM) fibers its state of polarization can be changed. Thus, linearly polarized light injected into a single-mode fiber can have an elliptical polarization when it reaches a

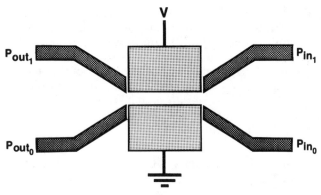

Fig. 1. *Directional Coupler.*

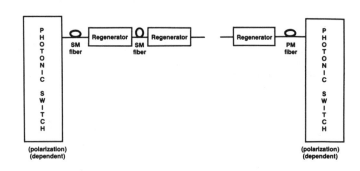

Fig. 3. *Network Architecture Assuming Polarization Dependent Lithium Niobate Photonic Switches.*

Ti:LiNbO$_3$ switching device. Another complicating factor is that the polarization effect of the fiber does not remain constant over time. To solve this problem polarization maintaining (PM) fiber is required from the laser source to the Ti:LiNbO$_3$ substrates and between substrates. Thus PM fiber would be used between the last point of regeneration and the photonic switch (Fig. 3).

Best performance values for near manufacturable PM fiber are better than −25 dB extinction ratio with loss values less than 0.3 dB/km, although typical values are −15 dB and 0.5 dB/km respectively.

In some applications the use of PM fiber might not be possible. For example, if a lithium niobate switch is to be placed in a fiber span for restoration purposes. Since the regeneration equipment is already present it could be uneconomical to introduce PM fiber between the switch and the regenerator. In this case the use of polarization independent switches or a polarization controller [6] with feedback might be required.

Optical Amplifiers

Optical amplification in a photonically switched network will be required to enhance the signal not only due to loss incurred in transmission, but also for loss suffered in the photonic switch. Signal loss across a photonic switch is proportional to the size of the matrix. Thus, there might be a need in the future to amplify the signal before it passes through the entire switch. The cost of using electronic lightwave regeneration to do this would be enormous since a regenerator would be required for most or all of the paths in the switch. The introduction of optical amplification in the photonic switch is important for the same reason that it is viewed as important for transmission facilities, namely that it should provide cheaper amplification.

Most of the optical amplifiers under investigation are linear amplifiers [7]. Generally two types of linear amplifiers exist. The Fabry-Perot (FP) type can provide very high gain (≥29 dB) but has a small bandwidth (≈ 10 GHz.) due to the cavity resonance required. Traveling wave (single pass) (TW) amplifiers have very low facet reflectivities (≥0.1 percent). These amplifiers can provide gains on the order of 20 dB and have bandwidths of around 20 A° (≈250 GHz).

Linear amplifiers have the disadvantage of amplifying the low level noise signals as well as the desired signals. Thus, the noise in the system is amplified along with the signal. With these type of optical amplifiers the SNR of a photonic switching system limits its eventual size since the crosstalk introduced to a signal is proportional to switch size.

Thresholding amplifiers would be useful because of their signal regeneration capability (Fig. 4). In the ideal case there would be no amplification for signals below a given intensity level. Once an intensity threshold has

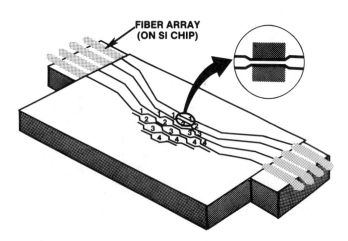

Fig. 2. *4 × 4 Photonic Crossbar Switching Matrix.*

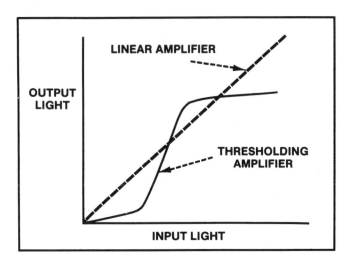

Fig. 4. *Linear and Thresholding Optical Amplifier Characteristics.*

been surpassed a large gain is desired [8]. A saturated or maximum value of output intensity is also desired. This type of thresholding amplifier is desired to both amplify and improve the SNR of the signals passing between substrates of a dimensionally large photonic switching system.

System Architectural Considerations

There are three main system level areas that need further study for the development of the lithium niobate technology. First we must find adequate switching techniques and architectures that provide good optical performance with respect to crosstalk and loss, as well as a rich set of interconnection capability. The maintenance of synchronization in a photonic network will be a very difficult task and is the second area of importance. The third area would be concerned with the incorporation of transmission formats that allow easy integration of photonic switches.

Switch Architectures And Techniques

There is a need to explore the implementation of larger switching matrices using directional coupler arrays. Most of the work has been on 4×4 arrays [4], but there has been an 8×8 array recently reported [5]. All of these have been crossbar configurations, and any larger matrices will be limited in size by the amount of crosstalk introduced to the signal.

Recently there has been a report on other switch topologies [3] which reduce the crosstalk introduced to the signal. Versions of these switches have been proposed for use in time-division switching networks [9]. These switches have reduced crosstalk due to the fact that no two paths across the switch encounter the same coupler. The penalty is the increase in the number of crosspoints required ($> n^2$).

Another method of improving system crosstalk involves the use of a multipath switching network, and carefully choosing paths such that all are isolated with respect to one another. A switch of this type is the dilated Benes network described in [10]. The switch provides comparable crosstalk performance with those reported in [3] but requires substantially less crosspoints. The reduction in crosspoints is primarily due to the fact that the switch is rearrangeably non-blocking as opposed to being strictly non-blocking.

Still another approach to reducing the crosstalk and loss in a switch matrix is described in [11]. In this scheme the lithium niobate array is operated in a bidirectional manner. Lithium niobate switches are not cognizant of the direction of information flow. The switches can also operate efficiently on different wavelengths as long as that separation is on the order of several nanometers. In this switch, originating and terminating transceivers transmit optical signals to one another on different wavelengths. This effectively reduces the size of the switch matrix by a factor of two. A folded Clos topology is also used in order to reduce the average number of crosspoints that a signal must traverse.

The reliability of a lithium niobate space matrix out in the field is currently under study. A photonic switching demonstration by AT&T has been in operation for over a year. Periodic tuning of the individual coupler voltages is required. The demonstration [12] is based on a 4×4 lithium niobate crossbar switch, and is used to switch 90 Mb/s video signals (Fig. 5). The switch is under stored program control, in fact an AT&T personal computer is used for system reconfiguration and control.

Time division multiplexed photonic switching is also currently under investigation. In [13] a lithium niobate switch is used for a DS-3 cross-connect switch. DS-3 signals are organized as time-slots and multiplexed up to 1.7 Gb/s, conforming to the FT Series G line rate. As opposed to being organized in a bit multiplexed format as is assumed in the FT Series G equipment, these signals are block multiplexed for optimal switching performance.

In [14] the lithium niobate is switched on a per-bit basis in order to provide access to the bistable laser diode storage elements. The relatively low line rate of 32 Mb/s allows this type of switching.

There is also the need to enhance the interconnection richness through the introduction of time-slot interchanging (TSI). To date there have been only a couple of approaches reported [15][14]. In [15] fiber delay lines are used to store time-slots for their rearrangement. Optical amplification would be required for implementation of a large TSI of this type. Bistable laser diodes are used in [14] but are currently rather slow devices. These and other techniques need to be further refined to insure the use of photonic switching technology for a large set of applications.

Synchronization

Some degree of synchronization must be maintained in the lithium niobate photonic switches. In order to maintain synchronization, framing will likely need to be placed over the data stream [13]. Variation in the signal phase relationship at the switch occurs due to wander and jitter on the optical fiber (temperature variations, and so on). The framing bits must be processed (electronically) and then the optical data stream could be delayed appropriately by some type of elastic store. The

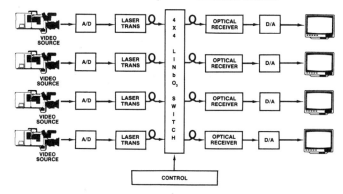

Fig. 5. 4 × 4 Photonic Space-Division Switch (AT&T Showcase)

electronic processing is performed on the order of the network frame time (currently ≈125 μs). The storage mechanism could be implemented using similar signal delay techniques to those described in [14,15]. Another alternative might be to use some sort of feedback mechanism to the optical sources of the network to insert/remove signal delay. The point is that there will need to be relatively complex synchronization electronics as well as possibly elastic storage capability at the photonic switch. The need for synchronization is not one unique to the photonic switch; the method of providing it in the photonic domain however needs more attention and study [13][16].

Another problem in the use of lithium niobate in time division switching with the absence of elastic storage is the phase discontinuities introduced in the optical data stream. The variation in phase relationship at the input to the photonic switch introduces variations in phase between time-slots on the outgoing data stream. The phase hits seen by transmission equipment downstream could cause serious problems depending upon how clock recovery is done. If a high-Q (SAW) filter is used then there will need to be a significant number of "clock recovery" bits appended to the time-slot to allow for the relatively long filter impulse response.

Transmission Formats

As the transmission bit rates increase it becomes very difficult to reconfigure the lithium niobate array such that there is no loss of information. Thus guard bands will probably need to be placed between information bursts or time-slots. These bands are considered reconfiguration overhead and should be minimized in order to get the highest transmission throughput. This fact promotes the use of a block multiplexing scheme where for example an entire DS3 frame becomes the basic switching quanta [13].

As mentioned above, there might also be a need to place a clock recovery field in the time-slot in order to allow for the acquisition of clock out of the data stream. This field would be followed by a flag pattern which delimits the data (Fig. 6).

The recommended format (SYNTRAN [17]) for multiplexing DS0 and DS1 signals to the DS3 rate provides for byte interleaving thus allowing the DS0's and DS1's to be easily identifiable. Transmission formats for signals above the DS3 rate however, are currently viewed to involve bit multiplexing (SONET [18]). Recently there has been speculation that a bit multiplexing format is not the most optimal one. Although this relieves the need for large amounts of high speed memory in the transmission equipment, it places a heavy burden on the rate of reconfiguration of the switching system. If guard bands need to be placed in the information stream then transmission formats should be reevaluated to reduce the necessary overhead.

As an example, in the FT Series G transmission systems a 1.7 Gb/s data stream contains 36 frames of DS3 information. The bits of these frames, plus overhead bits, are interleaved and mixed to the point that individual DS3 frames can not be extracted from the stream unless that stream is at least partially demultiplexed down to DS3 channels. By requiring that the DS3 frames be block multiplexed onto the high speed channel with a small guard band between them, individual extraction and insertion of DS3 frames should be possible. This also reduces the rate of reconfiguration of the switch array.

Applications/Conclusions

Perhaps the key component required to drive the Ti:LiNbO$_3$ technology into the marketplace within the next three to five years is a good application. A *good* application requires that the strengths of this new technology be used. The strength of directional couplers is their ability to control extremely high bit-rate information. The high data rates allowed by this technology eliminate the needs for highly parallel electronic switching architectures and all of the problems associated with them (demultiplexing/multiplexing, wide high speed data busses, electrical crosstalk, large amounts of circuitry, and so forth). The technology is currently limited by several factors: 1) the electronics required to control them limits their switching speed, 2) the long length of each directional coupler prevents large scale integration, 3) for large switching systems thresholding optical amplifiers will be required to clean-up and amplify the signals as they pass from substrate to substrate, and 4) some sophisticated high speed electronic processing is still required for synchronization between the transmission and switching equipment.

An initial application for this technology could possibly be its use in protection switching of optical fibers. In this scenario the lithium niobate would be held in a particular state for a long period of time. If breakage occurs in one of the cables then the switch is used for connecting the input fiber to a spare cable which bypasses the problem area. With further development of the components described, lithium niobate based photonic switching will find a wider range of applications.

References

[1] R. C. Alferness and R. V. Schmidt, "Directional coupler switches, modulators, and filters using alternating delta-beta techniques," *IEEE Trans. on Circuits and Systems*, vol. CAS-26, no. 12, pp. 1099–1108, Dec. 1979.

[2] H. S. Hinton, "A non-blocking optical interconnection network using directional couplers," *Proc. of the IEEE Global Telecommunications Conference*, vol. 2, pp. 885–889, Nov. 1984.

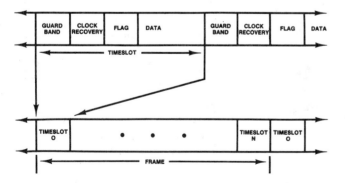

Fig. 6. Time-Multiplexed Photonic Switch Data Format.

[3] R. A. Spanke, "Architectures for large nonblocking optical space switches," *IEEE Jour. of Quantum Electronics*, vol. QE-22, no. 6, pp. 964-967, June 1986.

[4] G. A. Bogert, E. J. Murphy, and R. T. Ku, "A low crosstalk 4×4 Ti:LiNbO$_3$ optical switch with permanently attached polarization-maintaining fiber arrays," Topical Meeting on Integrated and Guided-Wave Optics, Atlanta, GA, pp. PDP 3.1-3, Feb. 1986.

[5] P. Granestrand et al., "Strictly nonblocking 8×8 integrated-optic switch matrix in Ti:LiNbO$_3$," Topical Meeting on Integrated and Guided-Wave Optics, Atlanta, GA, paper WAA3, p. 4, Feb. 1986.

[6] R. C. Alferness and l. L. Buhl, "Low loss, wavelength tunable, waveguide electro-optic polarization controller for $\lambda = 1.32$ μm," *Appl. Phys. Letters*, vol. 47, no. 11, pp. 1137-1139, Dec. 1985.

[7] S. Kobayashi and T. Kimura, "Semiconductor optical amplifiers," *IEEE Spectrum*, pp. 26-33, May 1984.

[8] Y. Silberberg, "All-optical repeater," *Optics Letters*, vol. 11, no. 6, June 1986.

[9] K. Habara and K. Kikuchi, "Optical time-division space switches using tree-structured directional couplers," *Elec. Letters*, vol. 21, no. 14, July 4, 1985.

[10] K. Padnamabhan and A. Netravali, "Dilated networks for photonic switching," *Proc. of the Topical Meeting on Photonic Switching*, March 1987.

[11] S. Suzuki, et al., "Thirty-two line optical space-division switching system," *Proc. of the Optical Fiber Conference*, p. 146, Jan. 1987.

[12] J. R. Erickson, et al., "Photonic switching demonstration display," *Proc. of the Topical Meeting on Photonic Switching*, March 1987.

[13] J. R. Erickson et al., "A 1.7 Gb/s time-multiplexed photonic switching experiment," *Proc. of the Topical Meeting on Photonic Switching*, March 1987.

[14] H. Goto et al., "An experiment on optical time-division digital switching using bistable laser diodes and optical switches," *Proc. of the IEEE Global Telecommunications Conference*, vol. 2, pp. 880-884, Nov. 1984.

[15] R. A. Thompson and P. P. Giordano, "Experimental photonic time-slot interchanger using optical fibers as reentrant delay-line memories," *Proc. of the Topical Meeting on Integrated and Guided-Wave Optics*, Atlanta, GA, paper TUB4, p. 26, Feb. 1986.

[16] W. A. Payne and H. S. Hinton, "System Considerations for Lithium Niobate Photonic Switching Technology," *Proc. of the Topical Meeting on Photonic Switching*, March 1987.

[17] G. R. Ritchie, "SYNTRAN- a new direction for digital transmission terminals," *IEEE Communications Magazine*, vol. 23, no. 11, pp. 20-25, Nov. 1985.

[18] R. Boehm, "Synchronous optical networks concept," Technical Requirements Industry Forum (TRIF), St. Louis, MO, April 1985.

Optical considerations in the design of digital optical computers

M. E. PRISE, N. STREIBL, M. M. DOWNS
AT & T Bell Laboratories, Holmdel, New Jersey 07733, USA

Received 29 May; revised and accepted 24 July 1987

The promise of digital optical computing is based on massively parallel interconnections between logic gates, which allow for novel architectures, and the possibility of ultrafast switching devices. This paper spells out the computational requirements and limitations for non-linear optical devices and optical interconnects. Relationships between the optical properties of devices (transmission and contrast) and their potential computational properties (fanin and fanout) are derived. The accuracy of the intensity levels required in the system are estimated. The requirements for a minimal device useful for digital optical computing are stated. The 'volume' of a device in phase-space limits fanin, switching energy and the degree of space variance in the interconnections. Space-invariant and space-variant interconnections are compared. Limits of random interconnects by volume holograms are discussed.

1. Introduction
1.1. Optical digital computing and computer architecture

All optical digital computers have been suggested because of the massive parallelism made possible by using optical interconnects [1], the high data rates already available in optical communications and the potentially high speed of optical logic devices [2]. To build an ultrafast computer von Neumann [3] suggested using '... substances whose dielectric constant or magnetic permeability depend on the electromagnetic field ...' at frequencies of the electromagnetic field related to circuit resonances. He illustrated the underlying principles for crystal diodes irradiated by microwaves, and thus he arrived (besides other things) at the hysteresis loops that are well-known today in optical bistability. In the past few years optical bistable devices, optical logic gates and hybrid electro-optic devices have been realized experimentally (for a recent collection of papers see [4] and references therein). In the future these devices may serve as switching elements in optics, as transistors do today in electronics and vacuum tubes did in von Neumann's days.

One approach to building a digital optical computer is to implement an architecture similar to that used in conventional electronic processors, where optical devices are connected by integrated optical waveguides [2]. This is worthwhile only if the optical devices can be clocked at much shorter cycle times than electronic gates in current computers.

Alternative architectures take advantage of the massive parallelism of imaging through free space. A 1000 × 1000 wide interconnection pays of only if a significant fraction of these lines are busy at any one instant. This requires the development of novel computer architectures that differ from the sequential architecture of many present computers. It

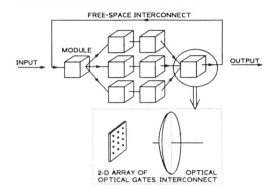

Figure 1 Schematic of a massively parallel pipelined optical processor, based on two dimensional arrays of non-linear devices (gates) and free-space interconnects.

is crucial to take advantage of the parallelism of imaging without suffering from the regularity of the interconnections provided by simple imaging set-ups. First steps in this direction are based on transformations performed with binary images [1, 5–10, Murdocca *et al.*, unpublished results].

Fig. 1 shows how we visualize such a processor; it is built out of modules, each consisting of homogeneous two-dimensional arrays of non-linear optical devices. Between each module we find optical interconnections which may consist of lenses, lenslet arrays, spatial filtering systems, holograms, beamsplitters, masks, etc. This architecture is characterized by a very wide pipeline. Naturally parallel problems, such as switching in optical communications, profit directly from this architecture. However, to implement a wide-pipeline architecture for more-general problems we must keep in mind that, as long as the pipe does not break, the throughput of the processor scales with the width of the pipeline and the latency of the non-linear optical device (not with the propogation time of the signals between the arrays). With this in mind we can adopt certain design rules, one of which is to implement state machines as large as possible. That is, we want long continuously connected circuits – to prevent registering. To put this another way, we want to break our computation into large chunks of data that we launch into a long pipeline. Registering and storing occur seldom, if at all. If fast electronics are not registered periodically, then the signals tend to get out of synchronization due to clock skews and differences in propogation delays of the gates. With optics, since the path length can be controlled accurately, constant latency architectures and pulsed logic are feasible. In fact, imaging systems can be designed such that the path lengths are constant over the whole imaged field, down to fractions of a wavelength. This corresponds to transit time differences of the order of a femtosecond.

Besides this synchronous aspect of pipelined optical parallel computers, another design rule is that to make the optical realization more feasible we want to work with constant fanin (each gate has exactly the same number of inputs) and constant fanout (each gate drives exactly the same number of outputs). All devices across the array should have the same computational and optical properties (same logic function, same fanin and fanout, etc.).

As we will discuss below, if the number of gates is large, regular interconnects are much easier to implement optically than random interconnects. It has been shown that it is possible to build arbitrary digital processors using simple regular connections with constant fanin and fanout [5, 6, 10]. Furthermore, it has also been shown that such a design can be computationally efficient despite its regularity (M. J. Murdocca *et al.*, unpublished results). Hence, although much more work is needed on these architectural and computer science topics, existing results encourage the devleopment of optical hardware.

1.2. Contents of this paper

This paper discusses the potential the the specific limits imposed by optics on computing. We hope to motivate the development of architectures based on regular interconnects with constant fanin and fanout. We also hope to motivate the development of more low-energy optical switching devices. In Section 2 we establish relationships between optical properties of non-linear devices (inherent gain, differential transmissions and contrast), system properties (system transmission and accuracies) and computational properties (cascadability, fanin and fanout). Figures of merit are derived that show what makes a good device. Also, the minimum requirements for a device useful in computing are stated. In Section 3 we consider the requirements for interfacing non-linear optical devices with free-space optical systems. Isolation is needed to prevent unwanted feedback between the different arrays in the computer. Different inputs to a device have to be in different orthogonal modes because, otherwise, interference will prevent reliable switching. In Section 4 we investigate the connectivity of free-space optics. Diffraction and aberrations limit the channel capacity of single-lens systems, which translates into the width of our pipeline. Space-invariant and space-variant optical interconnects within the pipeline are compared. The limits on the interconnects which a volume hologram can implement are discussed. The idea that a completely arbitrary interconnect can be implemented using volume holograms is shown to be unrealistic in an optical system interconnecting large numbers of small devices. In Section 5 we address power dissipation and power requirements.

2. Computational properties of non-linear optical devices

2.1. Bistable and thresholding devices

Fig. 2a to d shows various characteristics for non-inverting and inverting, bistable and thresholding devices. To use one of these devices as a logical gate the signals from previous devices are fed to the gate by an optical system. For an inverting gate (Fig. 2c), if the input power does not amount to the switching power P_{SW} the output of the device must be usable as a logical HI at the input of another identical device. If the input power exceeds the switching power, then its output power must be usable as a logical LO. Conversely, for a non-inverting gate (Fig. 2a), if the input power does not amount to the switching power P_{SW} the output of the device must be usable as a logical LO at the input of another identical device. If the input power exceeds the switching power, then its output power must be usable as a logical HI.

An external bias beam P_{BIAS}, as well as signals from previous devices, can be fed onto the device. This is essential in the case of the existing bistable devices, and more generally in the case of all devices without inherent gain.

Bistable devices can be used either as latches or as gates. If the device is used in a pulsed mode — that is, the optical power is reduced below the bistable region before each switching event — then the bistable device can be considered as a thresholding device. If this is not the case, the device is a latch — it can memorize a previous state. Here we consider only thresholding operation. Since a bistable device relies on positive feedback and the switching is achieved when the system becomes unstable, the switching window ΔP_{SW} can be considered to be zero.

The main difference between bistable devices and thresholding devices is that with the bistable device the output power must be supplied by a bias beam. This bias brings the device near the switching point, such that a small additional change in the input beams can cause switching. Therefore, everything depends critically on the biasing. This imposes tight

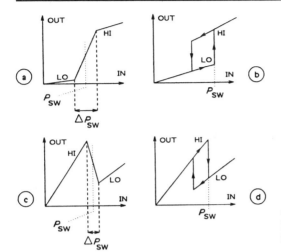

Figure 2 Characteristics of inverting (c, d) and non-inverting (a, b), bistable (b, d) and thresholding (a, c) devices.

restrictions on the accuracies of the system and of the optical power supply. In a thresholding device the bias and power supply beams for the output can be different from each other. As an example, we could think of a detector having both an input and bias beam, that drives an LED, or a modulator, such that the power supply and bias beams are independent. Also, the multiple-wavelength optical logic etalon (OLE) [11] has a supply beam decoupled from the inputs, since it does not influence the material, due to its different wavelength. Strictly, such a device with several independent input ports cannot be completely described by one simple curve such as in Fig. 2; rather we need higher-dimensional surfaces. On the other hand, we want to use all of the devices on the array in the same way (same number of inputs, same power supply beam, etc.). Hence, as long as somewhere within the device all input powers are added, and thresholding of the power supply beam depending on this sum is performed, it still has a simple characteristic curve. If there is non-linear preprocessing before the summation, the situation gets more complicated. A device where input and power supply are not decoupled from each other requires critical biasing.

It turns out that the slope of the characteristic curve of the devices, around the LO and the HI states, plays an important role: variations in input power will cause variations in output powers — that is, they cause deviations from the ideal binary LO and HI states. If we are not careful these deviations will cause erroneous switching of devices. Since bistable devices are latching, they allow an interesting trick for power normalization: after switching a device, the previous devices can be completely shut off if a 'holding power' is appropriately adjusted such that the device does not forget its state. Then the variation in output will depend only on the inaccuracies of the holding beams and not on how many preceding devices were switched LO or HI. We do not explicitly consider this adaptive biasing of bistaable devices in this paper. However, it does allow us to work with bistable devices as if the differential transmissions in the HI and LO states are zero.

2.2. Computational properties of a device
The computational properties of a gate are:

fanin — the number of inputs;
fanout — the number of outputs;
threshold — the number of HI inputs required to switch the device.

The threshold describes the logic function of the gate: for example, a NOR-gate has threshold = 1, whereas a four-input AND-gate has threshold = 4. Obviously, threshold has to be greater than zero and less than or equal to fanin.

One principal condition for the development of optical digital computing is the existence of an optical gate, which is infinitely cascadable and has a fanin and fanout of at least two. Infinite cascadability means the ability of one gate to drive other identical gates. Additionally, for digital information processing, data fork and data join operations are needed; thus, a useful gate must have a minimum fanin and fanout of two. Insufficient fanout could theoretically be remedied by optical amplifiers.

The minimum requirements for a device to be computationally useful are that one device or several devices can be made into a gate which has the following.

Infinite cascadability.
For an inverting gate: fanin = fanout = 2, threshold = 1, which corresponds to a two-input NOR-gate. As is well known, NOR gates form a logically complete set; that is, any logic function can be built out of them. For threshold = 2 we obtain a NAND-gate, which is also a complete Boolean set.
For a non-inverting gate: fanin = fanout = 2, threshold = 2, which corresponds to a two-input AND-gate. Note, that if we employ dual-rail logic [12], and if we have means to combine signals by a 'wired' OR-function [5], then we can implement any logic function with two-input AND-gates. The trick [5] is to use AND-gates for making the decisions (in dual-rail logic), and to guarantee that at a given time one and only one of the signals to be combined is HI, all the others are LO. OR-gates (threshold = 1) alone are not sufficient to build arbitrary circuits, even with dual-rail logic.

In what follows we consider one device as a logic gate. This treatment can be extended by considering the input versus output characteristics of a several devices making up a logic gate.

2.3. Optical properties of a device
A thresholding device is characterized by the following properties (Fig. 3):

P_{SW} — the switching power;
the switching window ΔP_{SW} — the difference in input power required to switch the device;
P_{ON} — the output power just after switch-on;
P_{OFF} — the output power just before switch-on;
T_{HI} — the differential transmission of the device at the switching point when the device is in the ON-state;

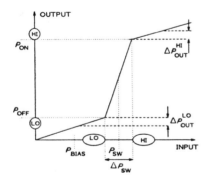

Figure 3 Characteristics of a thresholding device.

T_{LO} — the differential transmission of the device at the switching point when the device is in the OFF-state.

From this we can define a few figures of merit characterizing the device:

The switching contrast is

$$C_{SW} = \frac{P_{ON} - P_{OFF}}{P_{ON}} \qquad (1)$$

The switching transmission is

$$T_{SW} = \frac{P_{ON}}{P_{OFF}} \qquad (2)$$

The relative switching window is

$$\sigma_{SW} = \frac{\Delta P_{SW}}{P_{SW}} \qquad (3)$$

Note that $\sigma_{SW} = 0$ for a bistable device.

Whether a device is operated in transmission or in reflection is very important; Wherrett [13] pointed out that reflective operation may reduce the switching power, enhance the contrast, reduce problems with spatial hysteresis (deformation of transverse modes of a cavity due to non-linear index profile), and allow efficient cooling from the back of the devices (see also Section 5 on power problems). Our treatment is easily extended to devices that work in the reflection mode by replacing all the differential transmissions with differential reflections, and by defining the output power as the reflected power.

Note also that in some devices the switching transmission can be greater than one. We refer to these devices as having inherent or absolute gain, as opposed to differential gain.

2.4. Optical properties of the system

In addition to the device properties we have to consider the following system properties.

P_{BIAS} — an externally introduced biasing power which can be used to bring the device nearer to its switching point.

T_{SYS} which we define as the fraction of the total power available from one device to drive other devices. This contains the transmission losses of the optical system used to interconnect different devices.

Variations of the switching power in time and across the array δP_{SW}.

Variations of the bias power in time and across the array δP_{BIAS}.

Variations of the system transmission across the array δT_{SYS}.

P_{OUT}^{LO} and P_{OUT}^{HI} — the output powers of the devices in LO and HI state, respectively.

ΔP_{OUT}^{LO} and ΔP_{OUT}^{HI} — the ranges of output power which are allowed as legal LO and HI, respectively.

All of the inaccuracies intrinsic to the system can be taken into account by defining an effective switching window σ_{eff} which is bigger than σ_{SW}:

$$\sigma_{eff} = \sigma_{SW} + \sigma_{SYS} = \sigma_{SW} + \frac{\delta P_{SW}}{P_{SW}} + \frac{\delta P_{BIAS}}{P_{SW}} + \frac{\delta T_{SYS}}{T_{SYS}} \qquad (4)$$

Whether we can consider a device as being binary depends on how we operate it. It turns out that the ranges of the output powers in the LO and HI state depend on the required computational properties of the device, as well as on the system properties. Therefore, for the devices to be binary, the ranges of the output powers have to be small compared with the difference between the absolute output powers of the LO and HI state:

$$\Delta P_{\text{OUT}}^{\text{LO}} \ll P_{\text{ON}} - P_{\text{OFF}} \qquad \Delta P_{\text{OUT}}^{\text{HI}} \ll P_{\text{ON}} - P_{\text{OFF}} \qquad (5)$$

2.5. Analysis of idealized binary devices

If a device switches, its output changes by $P_{\text{OUT}}^{\text{HI}} - P_{\text{OUT}}^{\text{LO}}$. Therefore, on a following device, the input power will change by

$$\frac{T_{\text{SYS}}}{\text{fanout}}(P_{\text{OUT}}^{\text{HI}} - P_{\text{OUT}}^{\text{LO}}) \sim \frac{T_{\text{SW}} C_{\text{SW}}}{\text{fanout}} T_{\text{SYS}} P_{\text{SW}} \qquad (6)$$

The first condition for reliable switching is that this input change is bigger than the effective switching window. In terms of fanout, this means

$$\text{fanout} < T_{\text{SYS}} \frac{T_{\text{SW}} C_{\text{SW}}}{\sigma_{\text{eff}}} \qquad (7)$$

Notice that an ideal bistable device in a system with infinite accuracy can have an infinite fanout. We derive formulae describing 'real-world' devices in the Appendix. The second condition determines the logical operation that is performed by the device. P_{BIAS} has to be chosen such that:

if there are threshold $-$ 1 or less HI inputs, then the device does not switch;
if there are threshold or more HI inputs, then the device switches.

This can be written as:

$$\begin{aligned} P_{\text{SW}} + \frac{\Delta P_{\text{SW}}}{2} - \frac{\text{fanin}}{\text{fanout}} T_{\text{SYS}} P_{\text{OUT}}^{\text{LO}} - \frac{\text{threshold}}{\text{fanout}} T_{\text{SYS}}(P_{\text{OUT}}^{\text{HI}} - P_{\text{OUT}}^{\text{LO}}) < P_{\text{BIAS}} \\ < P_{\text{SW}} - \frac{\Delta P_{\text{SW}}}{2} - \frac{\text{fanin}}{\text{fanout}} T_{\text{SYS}} P_{\text{OUT}}^{\text{LO}} - \frac{\text{threshold} - 1}{\text{fanout}} T_{\text{SYS}}(P_{\text{OUT}}^{\text{HI}} - P_{\text{OUT}}^{\text{LO}}) \end{aligned} \qquad (8)$$

We would like to choose the bias power such that it is in the middle of the interval given by the inequality 8:

$$\frac{P_{\text{BIAS}}}{P_{\text{SW}}} \sim 1 - \frac{T_{\text{SYS}} T_{\text{SW}}}{\text{fanout}}[\text{fanin}(1 - C_{\text{SW}}) + (\text{threshold} - \tfrac{1}{2})C_{\text{SW}}] \qquad (9)$$

This will allow us to calculate the allowable error in the bias beam. To make physical sense, P_{BIAS} must be non-negative. This translates into the following condition for the fanin:

$$\text{fanin} < \text{fanout} \frac{1}{T_{\text{SYS}} T_{\text{SW}}(1 - C_{\text{SW}})} - (\text{threshold} - \tfrac{1}{2})\frac{C_{\text{SW}}}{1 - C_{\text{SW}}} \qquad (10)$$

Equations 7 and 10 are the basic switching conditions relating the optical and computational properties of a device with the systems properties.

2.6. Error propagation

Now we start to consider the consequences of the non-binary nature of real-world devices caused by non-zero differential transmissions T_{LO} and T_{HI}. If the differential transmissions are non-zero, then the output of any device will depend not only on whether the input is below or above threshold, but on exactly where its input is with respect to the switching point. This leads to variations of the output powers of the devices in the LO and the HI state which, in turn, leads to variations of the input powers of the next devices. In the worst case, output errors of all fanin preceding arrays $\Delta P_{\text{OUT}}^{\text{preceding}}$ will change the input in the next array which, in turn, changes its output due to non-zero differential transmissions T_{LO} and T_{HI}:

$$\Delta P_{\text{OUT}}^{\text{next}} = \frac{\text{fanin}}{\text{fanout}} T_{\text{SYS}} |T_{\text{LO/HI}}| \Delta P_{\text{OUT}}^{\text{preceding}} \quad (11)$$

The error propagation is 'graceful' (no explosive accumulation of errors), if $\Delta P_{\text{OUT}}^{\text{next}} < \Delta P_{\text{OUT}}^{\text{preceding}}$, which translates into conditions for the differential transmissions:

$$\frac{\text{fanin}}{\text{fanout}} < \frac{1}{T_{\text{SYS}} |T_{\text{LO/HI}}|} \quad (12)$$

The fanout and fanin conditions corresponding to Equations 7 and 10 for devices with non-zero differential transmission are given in the Appendix.

Studying the 'computational usefulness' of a device is complicated, since many parameters are involved: the optical properties of the device, its systems properties, and its computational parameters such as fanin, fanout and threshold. In Figs 4 to 9 we map out areas within which the switching conditions (Equations 7, 10 and A5 to A8) are fulfilled, varying one parameter at a time.

Fig. 4 shows the relationship between inherent gain (or switching transmission T_{SW}), the system transmission T_{SYS}, fanout, the switching contrast C_{SW}, and fanin for a binary (zero differential transmission) device. On the vertical axis we have the 'gain-parameter' $T_{\text{SW}} \times T_{\text{SYS}}/\text{fanout}$ and on the horizontal axis the contrast C_{SW}. Fanin is the variable parameter in this group of curves; it is increased from fanin = 2 (leftmost curve) in steps of 1 up to fanin = 150. With zero differential transmission (ideal binary device: $T_{\text{HI}} = T_{\text{LO}} = 0$) both inverting and non-inverting devices obey the same switching conditions. A device located to the right of a given curve can be used in a system with the given losses T_{SYS}, fanout, fanin, threshold = 1 and an effective switching window $\sigma_{\text{eff}} = 0.1$. Note that the effective switching window contains not only the sharpness of the device switching, but also the system inaccuracies (Equation 4). The interpretation of the curves is straightforward: a device is useless if the gain parameter is too small to drive the next

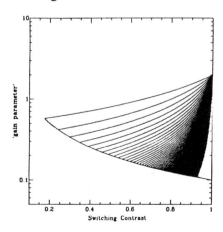

Figure 4 For a binary (zero differential transmission) device the relationship between the 'gain-parameter' (inherent gain, (or switching transmission T_{SW}) multiplied by the system transmission T_{SYS} divided by the fanout) and the switching contrast C_{SW} for a fanin of 2 to 150. A fanin of 2 gives the leftmost curve. For the given parameters operation anywhere to the right of the given curve is allowed.

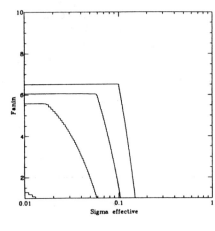

Figure 5 The relationship between fanin and the effective switching window σ_{eff} for thresholds 1 to 4. The fanout = 2 and the other optical parameters are $C_{SW} = 0.6$, $T_{SW}T_{SYS} = 0.6$, $T_{HI}T_{SYS} = 0.6$ and $T_{LO}T_{SYS} = 0.1$. The curve to the right corresponds to a threshold of 1. The allowed area is to the left of the curve.

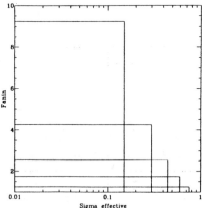

Figure 6 For an ideal binary device ($T_{HI} = T_{LO} = 0$) the fanin is plotted against the effective switching window σ_{eff} for gain parameters ($T_{SW}T_{SYS}/$fanout) of 0.25 to 1.25. The curve which extends farthest to the right corresponds to a gain parameter of 1.25. The switching contrast was chosen to be $C_{SW} = 0.6$. The area to the left of a curve is allowed.

device. If the gain parameter gets too large, the residual power of all previous LO devices combined exceeds the switching point, and again the device is useless. As we may expect, the 'feasible region' increases with increasing switching contrast and is reduced by higher fanin requirements.

In Fig. 5 we investigate the limits for fanin and the effective switching window σ_{eff} (system accuracy!) for different thresholds. We chose fanout = 2; the optical parameters are typical of the self-electro-optic effect devices (SEEDs) [14, 15] — since they are available to us in our laboratory: $C_{SW} = 0.6$, $T_{SW}T_{SYS} = 0.6$, $T_{HI}T_{SYS} = 0.6$, $T_{LO}T_{SYS} = 0.1$. It can be seen clearly that increasing the threshold decreases the possible fanin and, more importantly, requires high system accuracies.

To investigate the merits of inherent gain ($T_{SW} > 1$), Fig. 6 shows the maximum possible fanin and the limit for the effective switching window σ_{eff} for different gain parameters $T_{SW} \times T_{SYS}/$fanout for an indeal binary device ($T_{HI} = T_{LO} = 0$). The switching contrast was chosen to be $C_{SW} = 0.6$. Increasing the gain parameter gives us a larger effective switching window, yet it limits the fanin.

In Fig. 7 we study, in a similar fashion, the influence of the switching contrast C_{SW}. With increased contrast the possible fanin increases dramatically (note the different scaling) as does the required effective switching window σ_{eff}. The gain parameter for this plot was chosen to be $T_{SW} \times T_{SYS}/$fanout = 0.25.

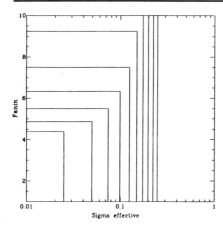

Figure 7 For an ideal binary device ($T_{HI} = T_{LO} = 0$) the fanin is plotted against the effective switching window σ_{eff} for switching contrasts C_{SW} of 0.1 to 1.0. The rightmost curve corresponds to the largest C_{SW}. The area to the left of a curve is allowed.

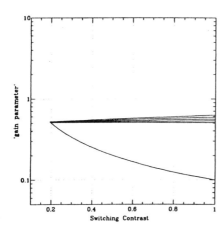

Figure 8 For a non-inverting device the gain-parameter ($T_{SW}T_{SYS}$/fanout) is plotted against the switching contrast C_{SW} for differential transmissions $T_{LO} = T_{HI} = 0.1$ to 0.5. The curve to the left has the lowest differential transmissions. The other parameters are $\sigma_{eff} = 0.1$, fanin = 2 and threshold = 1. The area to the right of the curve is allowed.

Fig. 8 shows the influence of the differential transmissions, $T_{LO} = T_{HI} \neq 0$, on the behaviour of a non-inverting device. We chose fanin = 2, threshold = 2 and $\sigma_{eff} = 0.1$. As we might guess from looking at the switching characteristic, increased differential transmissions require increased contrast. This situation corresponds to using the device as a two-input AND gate. Fig. 9 shows that the effect on the inverting device is more drastic. In this case we have a two-input NOR gate.

Summarizing these plots we find the following rules-of-thumb.

The switching transmission or the inherent gain of a device must be within a certain range, not too small and not too large. This range depends on how we use the device, on the contrast (Figs 4 and 7) and on the switching window (Fig. 6).

Avoid large thresholds — they require high system accuracies (Fig. 5).

Avoid large fanins — they also require high system accuracy (Fig. 6), as well as good contrast (Fig. 4).

Large fanouts can be realized with devices having inherent gain; otherwise the system accuracies have to be too high to be practical.

The differential transmissions hurt seriously in the case of an inverting device (Fig. 8) and, to a lesser degree, for a non-inverting device (Fig. 9).

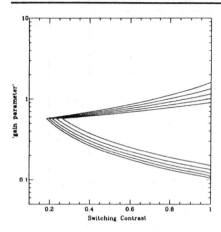

Figure 9 For an inverting device the gain parameter ($T_{SW}T_{SYS}$/fanout) is plotted against the switching contrast C_{SW} for differential transmissions $T_{LO} = T_{HI} = 0.1$ to 0.5. $T_{LO} = T_{HI} = 0.1$ is the leftmost curve. The other parameters are $\sigma_{eff} = 0.1$, fanin = 2 and threshold = 2. The area to the right of the curve is allowed.

The computational usefulness of a given device can be investigated by checking Equations A5 to A8 for fanout ≥ 2, fanin ≥ 2 and threshold ≥ 1 (inverting device) or threshold ≥ 2 (non-inverting device). With many current devices this comes down to the question of systems accuracy: how homogeneous are the bias or power supply beams and the systems transmission across the array? Clearly, tight requirements on these accuracies lead to impractical systems and defeat the idea of digital processing.

2.7. Critical slowing down

Critical slowing down occurs in devices with internal feedback where the non-linearity is dependent on the output power of the device. For example, bistable devices always exhibit critical slowing down, since they depend on positive feedback [16]. OLEs [11] should not exhibit critical slowing down, because the output beam only sees the transmission of the etalon. Ideally, it does not interact sufficiently with the material to affect its own transmission.

Critical slowing down is important when interpreting the experimental switching times of some devices, since in most experiments reported in the literature the device is overdriven; consequently, these speeds go down, if we build a system with small switching increments.

Critical slowing down can be roughly estimated by a linear approximation of the dynamics around the switching point. At the onset of switching, the temporal evolution of the output power is determined by the eigenmodes of the linearized dynamics, which are exponential in time. This exponential behaviour stems from the fact that linear differential or difference equations are solved by Laplace transformations. For an estimation of how quickly the switching is initiated, it is sufficient to consider the eigenvalue with the largest real part, corresponding to the fastest growth at the onset:

$$\delta P_{OUT}(t) \propto P_{drive}\, e^{t/\tau_{device}} \qquad (13)$$

where P_{drive} denotes the light increment used for initiating the switching and τ_{device} is a time-constant characteristic for the device. Depending on the actual device, it contains the time constants for photon absorption, carrier recombination, cavity build-up for Fabry–Perot devices and RC time constants for opto-electronic devices. For very small switching increments, P_{drive} (the actual switching time) is dominated by the time needed for this exponential onset. Thus, we estimate

$$\frac{\tau_{SW}}{\tau_{device}} \sim \text{const.} + \log\left(\frac{P_{SW}}{P_{drive}}\right) \qquad (14)$$

This estimation is not very accurate because it completely ignores the non-linear dynamics during switching. On the other hand, we get an impression of what happens without actually analysing the non-linear dynamics. A similar condition (with a different time constant τ_{sw}) should apply to the switch-off. In the set-up we are thinking about, the switching light increment P_{drive} is given by Equation 6, which yields

$$\frac{\tau_{sw}}{\tau_{device}} \sim \text{const.}' + \log\left(\frac{\text{fanout}}{T_{sw} T_{sys}}\right) \qquad (15)$$

The fanin condition of Equation 10 gives a lower bound for the fanout, which leads to the estimation

$$\frac{\tau_{sw}}{\tau_{device}} > \text{const.}' + \log\left[\text{fanin}\,(1 - C_{sw}) + (\text{threshold} - \tfrac{1}{2})C_{sw}\right] \qquad (16)$$

These relationships indicate that we have to pay for large fanout, fanin or threshold not only with high system accuracies, but also with the switching speeds due to critical slowing down. This problem becomes especially important for neural networks, associative memories and architectures relying on optical crossbars (see Section 4.5). Parallel architectures using only low fanin and fanout need not to be concerned too much about critical slowing down, since in this case we work with switching increments as large as possible. Inaccuracies in the bias beam across the array result in different switching times across the array.

3. Interfacing devices and free-space connectors
3.1. Beam combination and coherence

We expect an optical computer to be powered by a laser which produces coherent light. The reason is the high amount of optical power needed (see Section 5) and/or the fact that most devices rely on resonant effects (spectrally narrow absorption features, Fabry–Perot interferometers or resonators with discrete mode structure) which require illumination within a narrow spatial and spectral bandwidth. If we combine the signals on the device without precautions, destructive interference could 'simulate' a logical LO although there are several logical HIs. There are two ways to prevent this effect and still combine the signals with no power loss:

we may build our circuits to interferometric precision (which looks impractical to the authors, except in few special cases); or
we may combine the signal in such a way that they look mutually incoherent to the device.

For incoherent beam combination, we have to consider the interplay of (a) the modes of the electromagnetic field within the device and (b) the response of the material in the device. The modes of the field in the device are given by the boundary conditions: for example, in a cavity the electromagnetic field is a superposition of the transverse and the longitudinal modes, which are basically standing waves within the resonator. If we couple two beams into the same mode within the device, we will always suffer from the interference problem mentioned above. If we couple into different modes, then the materials characteristics come into play. Indeed, there may be interference fringes; the device will average over them under certain circumstances such that these modes (and hence the different inputs) can be considered mutually incoherent. Another way of saying this is that we wish to couple into orthogonal modes of the device.

- The inputs can be coupled into a sufficiently large device at different locations. If the non-linear material averages across all these locations, then they are mutually incoherent. Due to the diffraction limit of optical imaging, the different input ports have to be spaced by some $\lambda/N.A.$, where N.A. is the numerical aperture.
- The inputs can be coupled into a sufficiently large device from different directions. Although there will be interference fringes across the device, they will be averaged out if the active area of the material is sufficiently large. In order to have different inputs coming from different directions, each input is assigned to an individual part of the exit pupil of the imaging system. The limit of this pupil-division technique is again related to diffraction: very small devices require large numerical apertures (large areas of exit pupil) for each input, and ultimately this limits the number of inputs.
- In some devices beams can be coupled in from opposite sides of the device. Again, the device has to be sufficiently large to average out the interference fringes which occur between the different beams. This technique can only be used in direction-independent devices.
- In certain devices, inputs have different wavelengths [17]. The interference fringes formed by beams with different frequencies move with the difference frequency. If the response time of the device is too long to resolve this movement, then again we average over the fringes and the beams can be considered mutually incoherent. The response time of the device is a function of the response time of the non-linearity and the cavity build-up time for cavity-based devices. The architectural problems involved in this approach are discussed at the end of this section.
- If the response time of the device is slow compared with the optical pulse length, then again the device integrates over all pulses arriving within its response time. Therefore mutual incoherence can also be achieved by temporal stacking of different inputs.
- Finally, different polarizations can be used for the different inputs, if the device is not polarization-sensitive. This is probably the 'cleanest' approach to coupling. Unfortunately, there are only two independent polarizations.

Besides the interference problem, the losses involved in beam combination require that different inputs be in different modes of the electromagnetic field: a normal 50:50 beamsplitter combines two beams with 50% light loss. It has been suggested [18] that this effect limits fanin and fanout in an optical computer. However, if the different input beams are coupled into different modes, then we have means of influencing them separately, and of combining them losslessly (in principle). Mirrors with location-dependent reflectivity, wavelength-dependent reflectivity, pupil division, fast switchable mirrors, shutters (difficult on the timescale we are thinking about) and polarization optical beamsplitters are examples of optical elements acting differently on different modes.

Again, the key point is that all of our inputs have to be in different non-interfering (orthogonal) modes.

3.2. Isolation

It has been pointed out [19, 20] that devices such as tunnel diodes make the construction of digital circuits very difficult; the reason is that input and output of a digital device have to be 'isolated' from each other to ensure unidirectional signal propagation. It must never happen that the switching of one device influences preceding devices unintentionally.

Each device needs at least fanin + 1 input beams for the logical inputs and the bias or power supply beams. The output beam will, in all the devices we are aware of, be transmitted or reflected power supply or bias beam. Now, if all these beams were coupled into the same mode of the device, then indeed, an isolation problem may arise. For example, if a non-linear Fabry–Perot interferometer used in transmission switches from the transmitting to the reflecting state, the strongly reflected power supply beam can erroneously switch preceding devices. However, if the different beams are coupled into different othogonal modes of the device, then back-propagation can be prevented by suitable filtering. Appropriate components were described in the preceding section.

Whether putting all inputs into different modes is sufficient to prevent back-propagation depends on the devices. For example, opto-electronic devices such as the SEED can be antireflection coated. If devices are insensitive to signals entering through the output port, isolation is no problem; in a worst case, with bistable devices that switch from a transmitting to a reflecting state and that have critical biasing, it still may be necessary to use optical isolators (unidirectional devices) to prevent the reflection of input beams of a device from back-propagating. In Section 3.4 we will count the number of orthogonal modes of a device, which limits the number of inputs.

Counter-propagating beams are orthogonal modes of a device, such as the SEED (which does not rely on a Fabry–Perot cavity), if the device is longer than a few wavelengths, because the device is insensitive to the coupling or interference between the beams. Attempting to use these modes will cuase severe isolation problems unless we are using a two-wavelength device where the signal (logic) beams are highly absorbed, avoiding backward propagation through the system. In any other case, we have to use polarization isolation.

3.3. Coupling
If the non-linear optical devices are small, great care has to be taken to avoid coupling losses. In fact, since we want to couple into many devices at the same time with one single optical system, our situation is significantly more difficult than coupling into a single fibre. Our optical (imaging) systems have to be well corrected over a large field, especially for geometrical distortion, they have to be telecentric, and they have to preserve the mode structure across the field during the imaging process.

3.4. Counting the number of accessible orthogonal modes of a device
All of the considerations of the previous sections indicate that problems arise as device size is reduced to approach the diffraction limit, since we cannot fit sufficient spatial modes into the device. Similarly, for wavelength-multiplexed devices, problems arise if the response time of the device and spectral width of the non-linear interaction do not allow us to multiplex into mutually incoherent temporal modes. Generally, we may say that a device becomes difficult to operate as we approach its spatio-temporal transform limit. To optimize the switching energy and time, it is essential to scale the devices down near these limits (see Section 5). Therefore, we want to point out the minimum dimensions of a computationally useful device.

As stated above, the device must support at least fanin + 1 accessible mutually incoherent (orthogonal) modes. Since we assume free-space optics, different modes must be accessible by an optical system having a numerical aperture N.A. The diameter (approximated by the first zero) of the diffraction disc is $1.22\lambda/\text{N.A.}$ for a circular pupil. A device having a cross-sectional area A_{dev} will have, at maximum, approximately $4(\text{N.A.})^2 A_{dev}/\lambda^2$ accessible spatial modes from each side of the sample. For polarization-independent devices, this number is doubled. To achieve such a high numerical aperture ($n \simeq 3$ to 4 in semiconductors) the imaging system has to consist partially of high-index material ('semiconductor-immersion system') [17]. To estimate the number of temporal modes, we assume the non-linear effect that we are using to have a bandwidth of $\Delta\lambda$. That is, beams within this range can influence the device. On the other hand, two beams are considered incoherent if their beat frequency is larger than $1/\tau_{dev}$, where τ_{dev} is the device response time. This is the case if their wavelengths are separated by more than $\lambda^2/(c\tau_{dev})$. The number of incoherent temporal modes is estimated by $\Delta\lambda c\tau_{dev}/\lambda^2$. Therefore, we estimate that

$$\text{fanin} + 1 < 8 \frac{A_{dev}}{\lambda^2} (\text{N.A.})^2 \frac{\Delta\lambda c\tau_{dev}}{\lambda^2} \qquad (17)$$

A second limiting case occurs in quasi-monochromatic operation. If we are limited, in the time domain, by the pulse length of the laser (any pulse has a minimum wavelength spread which can be determined from its Fourier transform into frequency space). At maximum we can stack as many as τ_{dev}/τ_{pulse} pulses within the response time, which leads to

$$\text{fanin} + 1 < 8 \frac{A_{dev}}{\lambda^2} (\text{N.A.})^2 \frac{\tau_{dev}}{\tau_{pulse}} \qquad (18)$$

Notice that Equations 17 and 18 are equivalent if we equate the spectral spread of the short pulse in the second case with the spectral width the device can accept in the first case.

In some special circumstances we can access more modes by using counter-propagating beams. This is only in the case of two-wavelength devices with high absorption at the signal (logic) wavelength. This approach will probably be restricted by the engineering considerations involved in setting up an optical interconnect system with as few elements as possible and short transit times.

In principle, we can address the maximum number of modes in a non-interfering manner by using a combination of these two techniques. In practice, using spatial modes will be straightforward. We simply need to image our various beams onto different parts of our devices. This is discussed in Section 4.2. Using polarization is very simple. We use two orthogonal polarizations. Using the spectral width of the devices is a much harder problem. In the architectures we are considering, a device with more than two wavelengths would be more complex to design and use.

A system using two wavelengths and two 'flavours' of devices is feasible, and a possible technique already exists (GaAs OLE devices [11]). These devices would have a power supply beam at one wavelength and all logical inputs at another wavelength. The output would come out at the power supply wavelength. In one flavour of the devices the power supply would be at the upper wavelength and the logical inputs at the lower wavelength. In the other flavour the situation is vice versa. As mentioned previously, this may have the advantage of allowing us to use another mode by having counter-propagating (and highly absorbed) signal beams. It is possible to implement the same scheme using arrays of identical bistable devices provided they have enough bandwidth (in a SEED bistability exists over several tens of nanometres and in a Fabry–Perot resonator different resonances could be used).

Using more than two wavelengths is difficult from an architectural rather than a physical point of view. Systems with arrays of devices running at different wavelengths are possible. Yet, if we want to combine different signals from different arrays on one device, we need to be careful that all the signals come in at different wavelengths and that we use all of the available modes. Such systems are, in principle, possible, but they pose new design restrictions on possible interconnects. This technique may be useful in the future, but new system architectures will be necessary.

Using temporal stacking in a system looks feasible if the appropriate laser sources can be developed (see Section 5.3). For instance, if we use 'branching' (see Section 4.1), then differences in optical path lengths can be introduced by varying physical lengths or putting in sheets of a higher refractive index material in one 'branch' (or path).

4. Optical interconnects
4.1. Types of interconnects
Fundamentally, there are two main types of interconnects: waveguides and free-space interconnects. In this section we mainly consider parallel synchronous free-space array interconnects, as our architecture is based on these (M. J. Murdocca et al., unpublished results).

In interconnecting two arrays of devices through free space, we have two problems: (1) the optical system has to image one array of devices onto the second array of devices, and (2) we have to implement an interconnection pattern which is the point-spread function of the optical system.

For focusing or imaging we can use the following types of optical systems: single-lens imaging systems; arrays of lenses (lenslets); and holographic optical elements. We list different types of optical interconnects; the properties of these interconnects are discussed in a later section. Some operations are particularly simple to implement optically, such as the following.

Convolutions, which can be decomposed into split, shift and recombine. The reason for this is that the point-spread function of a convolution is the same across the entire image. Usually such operations are called space-invariant.

Magnification and demagnification, which may be useful in a computer architecture because they provide long-range interconnects. Magnification is a part of the optical implementation of perfect shuffles [21, K.-H. Brenner and A. Huang, unpublished results].

Image rotation and transposition, which can be performed by prism systems.

Image shearing, which can be performed by cylindrical optics.

From the point of view of conventional computer architectures, random interconnects would be desirable. A random interconnect means that each pixel has its own point-spread function, which can be different for each pixel. Such an interconnect is called space-variant.

One family of optical random interconnects are crossbars, which are implemented using optical matrix multipliers [22, 23].

Most other implementations of space-variant operations rely on some type of multiplexing, usually in space or in direction. We give some examples below.

Volume holograms have been suggested for implementing random interconnects.

A given space-variant interconnect can always be built out of several space-invariant operations. In the worst case, if all point-spread functions of all pixels are different, then we need as many space-invariant operations as there are pixels. We define the degree of the space variance of a given interconnect as the minimal number of space-invariant operations from which we can build it. Usually the degree of space variance translates directly into the degree of multiplexing needed in the optical implementation of random interconnects.

4.2. Limitations of imaging

An optical computer can consist of several parallel channels (Fig. 1). Each channel has arrays of devices interconnected by imaging systems. The computational power of this system is related to the width of the interconnets; that is, how many ports of the devices are independently addressable per channel. This is the 'channel capacity' of the interconnect. The maximum channel capacity is determined by the resolution of the imaging system and its field size.

Since the minimum resolvable spot size is diffraction limited, and small devices are necessary to minimize the switching energies, we want to have a diffraction-limited optical system. Separating the ports according to the Rayleigh criterion for the resolution of an optical system is not adequate, since too much cross-talk between adjacent signals will result. This becomes a serious consideration if we attempt to couple into different spatial modes of each device. A strong power supply beam may mess up other inputs due to diffractive spillover. Depending on the operating characteristics of the device, different input ports have to be separated by several $\lambda/N.A.$ If the optical system is not diffraction limited, then the separation has to be several times the diameter of the point-spread function.

The aberrations of a single-lens imaging system scale with the aperture and the field size. If high channel capacities are required, the design difficulties and the number of elements in the lens increase considerably. A lot of effort and expense are required to design a single-lens system with a channel capacity of more than about 1000×1000 fulfilling all necessary requirements.

One possible idea to increase the field to arbitrary sizes is to image with lenslet arrays instead of single lenses, as is done, for example, in some photocopiers [24]. However, for a nearly diffraction-limited imaging system all path lengths (through different lenses!) have to be correct to a fraction of the wavelength, otherwise different parts of the aperture do not combine coherently and we may lose resolving power and synchronicity. A good lenslet system for erect 1:1 imaging, with reasonable resolution, requires careful optical design, poses tight limits on optical and mechanical tolerances, and is difficult to make. In any case, lenslet arrays presently offer the possibility of arbitrarily wide interconnects, even though the resolution is compromised (which translates into unreasonable power requirements for devices that are large compared with the wavelength; see also Section 5).

If imaging is incorporated into a hologram (holographic lenses), then the same considerations apply regarding the focusing action of the hologram.

4.3. Space-invariant interconnects

It has been shown that simple space-invariant interconnects are sufficient to implement any digital circuits [6, 10]. Space-invariant interconnects can be implemented by spatial filtering, including holographic filters, or by direct split, shift and combine using beamsplitters. For each branch of our interconnect we need to couple into an extra mode on our non-linear device to prevent combining losses and interference problems. At first glance it looks possible to combine several beams onto the same mode coherently with rigid spatial filters, since the filter has, in principle, full control over the phase. However, any phase error coming from a coherent power supply laser or originating somewhere in the optical system reintroduces the interference problem. It is therefore safer (especially for global interconnections) always to combine onto different modes.

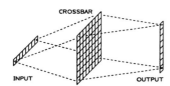

Figure 10 Optical matrix multipliers can be used to implement crossbars.

4.5. Some space-variant systems

In this section we describe some approaches to optical random interconnects. In principle, a completely random interconnect can be implemented using optical fibres or waveguides. However, basic problems that need to be solved are:

coupling losses between the fibres and the devices: note that the output of at least fanin fibres has to be coupled into the device, or we need Y-couplers for waveguides, or we need devices with multiple output ports;

'weaving' waveguides into random interconnects is difficult;

preserving synchronicity within a system is a problem: synchronicity is mandatory if we intend to use pulsed logic in long pipelines.

A second approach to optical random interconnects are crossbars (Fig. 10), implemented by matrix multipliers [22, 23]. The basic problems here are the following.

To interconnect a one-dimensional string of N inputs to N outputs, a crossbar uses N^2 crossing points. If the 'channel capacity' of the imaging system is 10^6, such a crossbar interconnects only 1000 devices. Assuming that the size of each device is near the diffraction limit, the crossbar uses all degrees of freedom offered by optics. It is, therefore, not easily expanded.

To illuminate all crossing-points, we need an enormous fanout of N. The fanin depends on the interconnection in the crossbar.

A third approach to random interconnects is branching (Fig. 11). Interconnects with a low degree of space variance M can be built from several space-invariant interconnects. This can be done either in separate imaging systems (split, convolve, mask and recombine) or in the same imaging system by multiplexing [21, K.-H. Brenner and A. Huang, unpublished results]. Using separate imaging systems we are faced with several problems.

The fanout has to be sufficient to drive all the M branches of the system (unless we use intermediate regeneration, which means partitioning the system at a lower level). In each of the branches we remove the unwanted pixels by masking, which leads to power losses.

Each branch is an optical system capable of handling the full channel capacity; besides increased cost, the larger volume requirements increase the overall latency.

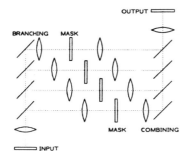

Figure 11 A space-variant interconnect can be built by branching and masking.

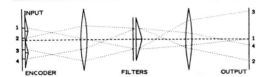

Figure 12 A space-variant interconnect by pupil division.

In some cases, the recombination can be done by normal beamsplitters, if there is no danger of interference in the resulting combined image. However, there are losses of intensity associated with normal beamsplitters [18]. If we use beam combination into different modes (polarization optical beamsplitters, different directions of light, different locations, different wavelengths or different arrival times) we can theoretically avoid power losses due to the beam combination. However, the number of accessible modes of the devices has to be M times higher.

This means that, for an interconnect with a degree of space variance M, we lose in branching the light absorbed in the masks; we also lose, in recombination, either a factor M in power or channel capacity, or both.

Another problem with branching is that if we use masks, we block the OFF signal from the previous arrays. This will be problematic if we use low-contrast devices. The consequences of this, for device operation and system tolerances, can be calculated by using the device parameters and the expressions given in Section 2 and in the Appendix. One way to get around this is to use renormalizing arrays in each branch. This adds to the complexity of the system, but may be necessary.

In pupil division (for example [21] or Fig. 12) the array is imaged through a single-lens system and spatial filtering operations are performed in different parts of the pupil. An array of holograms or prisms on the input array (encoder) directs the light from each individual device to the appropriate filtering operation. The difference to branching (as it was described in the previous paragraph), is that the light loss due to the masks is avoided. However, since the resolving power of the sub-pupils is reduced by a factor of M, the channel capacity is again diminished by the same factor.

Pupil division can be considered as multiplexing in direction. In principle, we can think of multiplexing schemes relying on polarization ($M = 2$) in wavelength or even in time. However, the number of accessible modes of the non-linear optical devices has to be, again, M times larger than the space-invariant case.

Faceted holograms [25, 26] (see, for example, Fig. 13) can be considered as multiplexing in space, where each device has its own optical system (fact of a hologram incorporating imaging and interconnection). However, for small devices we have to deal with large numerical apertures, which dictate a wide device-to-device separation and lead to unreasonably large systems.

In all free-space random interconnects discussed so far, we lose a factor of M (the degree of space variance) either in power or in channel capacity. Additionally, it is often difficult to maintain synchronicity of the signals (equal path lengths) and to image telecentric (for mode matching).

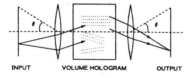

Figure 13 A space-variant interconnect built of two faceted holograms.

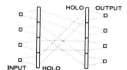

Figure 14 A space-variant interconnect based on a direction selective volume hologram.

4.6. Random interconnects based on volume holograms

Fig. 14 shows an optical interconnection set-up containing a volume hologram. The idea is that the first Fourier lens transforms each light source (device) in the object plane into a plane wave hitting the hologram from a different angle. Due to Bragg diffraction, a volume hologram can distinguish between different input directions and, it is hoped, that each incoming plane wave can 'read out' a different interconnection pattern for a random interconnect. Other set-ups, where the hologram is not located in the Fourier domain (with respect to the object), are possible, but are equivalent in terms of the following analysis of the interconnection power. An interconnection hologram could be recorded as a multiplex hologram. Besides practical problems such as limitations in the dynamic range of the holographic material [27], or the reciprocity law failure for multiplex exposures [28], there are more fundamental points we want to make about cross-talk.

We will prove in first-order Born approximation, that a holographically recorded three-dimensional structure is insufficient for arbitrary random interconnects within reasonably large field sizes (independent from material problems). Using the first-order Born approximation means taking into account only single scattering events; higher-order approximations are needed to explain multiple scattering. A full analysis is extremely complicated, as can be seen from the difficulty in completely analysing 'simple' gratings [29]. Although we conjecture that inclusion of higher-order terms will introduce more crosstalk, a closer investigation of multiple scattering events is necessary.

Our proof relies on the Ewald construction, which was introduced to explain X-ray diffraction in three-dimensional crystal structures [30]. Any three-dimensional structure can be Fourier decomposed. Any hologram which is recorded optically with a wavelength λ has a minimum period of $\lambda/2$ (the finest possible interference patterns with period $\lambda/2$ are created by counter-propagating waves). This means a hologram can only have frequency components within a sphere of radius $4\pi/\lambda$ in three-dimensional Fourier space. To create a connection between two elements we must diffract a wave with wavevector k_{in} into a wave with wavevector k_{out}. This means we need a grating with the difference frequency $k_{out} - k_{in}$. In Fourier space this occupies two dots which can be located anywhere in the allowed volume. The relative position of these two dots is fixed by the Ewald construction. That is, we can draw a sphere of radius $2\pi/\lambda$, and vectors k_{in} and k_{out} starting at the centre and ending at the dots. This situation is depicted in Fig. 15. For implementing a second interconnection we need at least a third dot in the Fourier space. In normal multiplex holography we would add for each interconnect a pair of Fourier components. Each time we add another dot to an existing pattern of p dots we are able to draw $2p$ new Ewald spheres. It is easy to see that p dots allow of the order of p^2 Ewald spheres. The physical meaning of these newly constructed Ewald spheres is that, in a multiple-exposure hologram, diffraction may occur not only on the gratings intentionally recorded during exposure, but also at 'virtual gratings' caused by beating between the recorded gratings. This means that, if there are already p interconnects, for each additional interconnect, we involuntarily add $2p - 1$ unwanted cross-terms.

These cross-terms will not cause unwanted interconnects if the devices are spaced sufficiently widely apart that the cross-terms can 'get lost between devices'. However, in

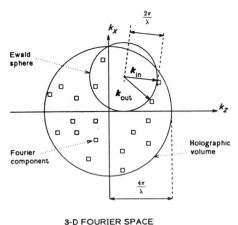

Figure 15 Illustration of the Ewald construction in three-dimensional Fourier space for determining diffraction on volume structures.

this case they may still cause power loss. Alternatively, if we work with a small field size, the cross-terms can be arranged to fall outside the field in which we are interested. In both cases we compromise the channel capacity of the interconnect, since less devices are interconnected than in a space-invariant system.

To estimate this effect, we will calculate the volume that each interconnect requires in three-dimensional Fourier space for a system with a given field angle θ on the input and the output side as shown in Fig. 14. This interconnect volume is a 'forbidden zone' around each dot in Fourier space. As soon as another dot enters this forbidden zone we can construct a Ewald sphere corresponding to a scattering from the object field into the image field, which would correspond to cross-talk. The maximum number of interconnects that can be realized is given by the ratio between the volume accessible by holography and the individual interconnect volume. This estimation is an upper limit, since it is not entirely clear how actually to produce such a structure holographically. In order that a scattering event from the input field into the output field can take place, the following relationships between the input/output wavevectors and the interconnecting vector between two dots Δk_{holo} have to be fulfilled:

$$\boldsymbol{k}_{\text{out}} = \boldsymbol{k}_{\text{in}} + \Delta \boldsymbol{k}_{\text{holo}} \tag{19}$$

$$|\Delta k_{\text{holo}}^{(x,y)}| = |k_{\text{out}}^{(x,y)} - k_{\text{in}}^{(x,y)}| < (4\pi/\lambda) \sin \theta \tag{20}$$

$$\Delta k_{\text{holo}}^{(z)} = (k^2 - k_{\text{out}}^{(x)2} - k_{\text{out}}^{(y)2})^{1/2} - (k^2 - k_{\text{in}}^{(x)2} - k_{\text{in}}^{(y)2})^{1/2} \tag{21}$$

These equations describe a volume which looks, for circular fields, somewhat similar to a doughnut (a cross-section is given in Fig. 16), and which is estimated by a rectangular box having the dimensions $(8\pi/\lambda) \sin \theta$ in the $k^{(x)}$- and the $k^{(y)}$-directions, and $(4\pi/\lambda)(1 - \cos \theta)$

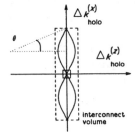

Figure 16 Cross-section through the 'forbidden volume' in Fourier space. If two frequency components of the volume structure (hologram) are located within this volume, then diffraction will occur from the input to the output (both limited by the maximum field angle).

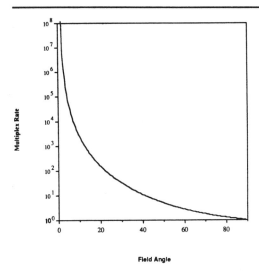

Figure 17 Maximum multiplex rate in a volume hologram as a function of the field angle, accessible without cross-talk.

in the $k^{(z)}$-direction. The volume accessible to holography is given by a sphere with radius $4\pi/\lambda$ and, therefore, the maximum number of interconnects which can be multiplexed into a volume hologram without cross-talk is estimated by:

$$M \sim \frac{\pi}{3} \frac{1}{(\sin^2\theta)(1 - \cos\theta)} \qquad (22)$$

A plot of this maximum multiplex rate versus the field angle θ is shown in Fig. 17. The performance increases drastically for very small field angles. Therefore we want a system with a very small field and a large numerical aperture to resolve many small devices and to collect all of the emitted light. Eventually, for a large number of small devices this leads to volume holograms of unreasonably large dimensions. Even if we had a system with appropriately high resolution and long working distance, the multiplex holographic recording on present materials is, as we stated previously, still burdened by problems with the dynamic range [27] and reciprocity failure [28].

5. Power requirements
5.1. Power dissipation

The power dissipation of optical logic gates has been discussed ever since optical devices were proposed for computing [31]. Nowadays, the total energy (optical plus electrical) required to perform optical logic operations is of the same order as that is required for electronic operations [4, 14]. Let us assume a 'chip' containing $N \times N$ devices, each having an active area A_{dev}. The packing constant α_{pack} is defined as the quotient of N^2 times the active device area and the chip area; it tells what portion of the available space is actually used by devices. The dissipated energy of a device is the product of its active area A_{dev} and the absorbed energy density ϱ_{sw} necessary to switch the device plus, in the case of optoelectronic devices, the dissipated electrical energy. For an energy-efficient device, it is essential that the excitation be confined to within the active area of the device during the switch. Any excitation which does not stay within the active area (that is, which escapes by thermal or carrier diffusion) is wasted. After the switching takes place we want to get rid of the excitation as quickly as possible. Surface recomination is an example of a fast way to remove excited carriers [32].

If, in a given system, we can remove a power density I_{cool} from the chip, without unacceptable thermal gradients or temperature rise, then the dissipated energy imposes an upper limit on the cycle time of a computer:

$$\tau_{cycle} > \frac{\varrho_{sw}}{I_{cool}} \alpha_{pack} \qquad (23)$$

In other words, given a certain type of non-linear device and a maximum cooling power, we find that the maximum possible number of bit-operations per unit time is directly proportional to the chip area. Since the smallest possible devices have an area of roughly λ^2/n^2, where λ is the wavelength and n the refractive index of the material, the single mode dissipation is $\lambda^2/n^2 \varrho_{sw}$. Since an imaging system can handle only a certain field size with the resolution needed for imaging the devices onto each other, reductions in packing density translate directly into a reduction of the number of addressable devices (constant channel capacity). If we potentially could address N^2 devices with our imaging system, then the achievable bit rate β is the number of usable devices divided by the cycle time:

$$\beta < \frac{I_{cool}}{\varrho_{sw}} N^2 \qquad (24)$$

Since we are assuming all the devices to be identical, the characteristics of each device in the array will change with temperature at the same rate. Ideally we would like this change to be minimal (a temperature-independent device). Under some circumstances a uniform temperature change can be compensated for. For instance, in a SEED [14], a temperature rise causes the bandgap of GaAs to move and hence the position of the excitonic absorption to change. In GaAs the band gap is 1.4 eV, and the temperature shift is -4×10^{-4} eV K^{-1}. This can probably be compensated for over a limited range by adjusting our power supply wavelength. In these cases we can tolerate a temperature shift if no temperature gradient exists across the array, which depends on how the devices are operated. For instance, if we use dual-rail logic [12], we can spread the power uniformly across our array. The efficiency of the heat sinking used is also very important. The details have to be very carefully thought out.

Present GaAs [14] devices have a ϱ_{sw} of about 10^{-14} J μm^{-2}. If we assume we can dissipate about 10 W cm^{-2} and we have a densely packed array, then we are limited in system cycle time to $\tau \simeq 100$ ns. If we use an imaging system with a channel capacity of 10^6, we end up with a limit for the bit rate of 10^{13} s^{-1}. To obtain a more reasonable cycle time, we have to reduce the packing density. However, reduction of the packing density reduces the number of devices within the field of our imaging system (we assume the channel capacity to be constant and maximal). Attempts to increase the field size will result in increased aberrations or worse resolution; this enforces devices with larger active areas. Thus, changes in packing density reduce the cycle time, yet the bit rate, which is proportional to the number of available devices, remains the same. If the devices are heat-transfer limited, then the issue is not how fast a single device will run, but how large ϱ_{sw}/I_{cool} is.

How severe is this limitation? Proven cooling systems used in cooling electronic systems [33, 34] can dissipate about 20 W cm^{-2}, but this is with a temperature rise of about 60 K. Detailed measurements of the temperature dependencies of existing devices are not available. For the SEED, the temperature dependence of the exciton peak means that variations in temperature across the array of devices of more than a few degrees are unacceptable. This would make the dissipation of about 10 W cm^{-2} extremely difficult. Using silicon microchannel plates, water-cooling thermal resistances of $0.083\, \text{K}\, \text{W}^{-1}$ have been measured [35]; the limits do not appear to have been reached [35]. So, with careful thermal engineering, it may be possible to do much better. There is also no inherent reason for optical devices based on resonant transitions being extremely temperature dependent. The quantum-well envelope state transition (QWEST) is almost completely independent of temperature over a range of more than 100 K [2]. More work has to be put into this area both in terms of thermal engineering and in terms of better (less temperature-sensitive) devices. However, as we can see, we are not yet near any fundamental limitations of cooling.

5.2. Engineering at the quantum limit

Considering these thermal problems, reductions of the dissipated energy of the device down to the lowest possible value is desirable. Note the difference between switching energy and dissipated energy. It has been shown that computers need not dissipate energy at all [36, 37]. One can use gates based on reversible effects, and even build reversible computers. It remains to be proven, however, that such computers are practicable and efficient. For example, most current non-linear optical devices absorb the light provided at the input ports and switch by introducing changes in refractive index or absorption of the non-linear material. It is theoretically conceivable to recover the absorbed light by stimulated emission and dump it elsewhere instead of heating the devices. Other schemes might rely on electronic non-linearities off resonance which require unreasonable powers. In both cases, practical realizations may fail, due to background absorption and other 'real-world effects'.

Therefore, it currently seems safe to assume that any optical logic gate will dissipate an energy corresponding to the absorption of a few photons. In fact, very small photon numbers ($N_{photon} = 25$) are sufficient to encode one bit of information safely; that is, with a low bit-error rate. However, as we have previously shown, a device with non-zero differential transmissions may significantly suffer from the photon shot noise $\Delta N_{photon} \simeq N_{photon}^{1/2}$. If we assume that we need 1000 photons for reliable switching, our lower limit on energy is

$$E_{min} = 1000 hc/\lambda \qquad (25)$$

For densely packed devices ($\alpha_{pack} = 1$) the limit for the cycle time imposed by power dissipation and shot noise may be approximated by

$$\tau_{cycle} > 1000 \frac{1}{I_{cool}} \frac{hcn^2}{\lambda^3} \qquad (26)$$

For GaAs near the band edge, the material constant hcn^2/λ^3 has a value of $4 \times 10^{-18}\, \text{J}\, \mu\text{m}^{-2}$. If we can dissipate $10\, \text{W}\, \text{cm}^{-2}$ without messing up the switching characteristic, then our minimum cycle time is 40 ns for the most densely packed system. In any real system α would be less than unity (to allow coupling in of our inputs and power supply beams), so our minimum cycle time would also be less. It is interesting to compare this with devices working farther in the infrared [2]: at $10.6\, \mu\text{m}$, under the same conditions, the minimum allowable cycle time is reduced by a factor of 2000, which is quite significant.

5.3. Power supply

One of the most difficult questions digital optical computing is facing right now is the power supply. We need a light source powerful enough to drive some 10^6 non-linear optical devices as quickly as they can run. For a small GaAs-device (active area of about λ^2) at 1 ns cycle time, this translates into approximately 10 W. Most optical non-linear devices so far discovered, or proposed, will only work with light in a small wavelength range. A laser, with its spatially and temporally coherent output, is probably the only way of generating sufficiently high powers, in the correct wavelength range and mode, for coupling into the devices. With larger devices or shorter cycle times which are physically possible [32], we soon run into laser limitations. Although this power requirement is of the same order as that of electronic computers, we suffer from the fact that it is, at present, expensive to convert electrical power into optical power in a laser. Large carbon dioxide lasers cost about US$50 per watt and are about 20% efficient. Cheap (and small) laser diodes are currently about $1000 per watt and up to 50% efficient. With the development of more-powerful laser diodes this cost may be lowered in the future.

One possible way around this is to use opto-electronic devices where some of the power is delivered to the device electrically. Furthermore, if the system has to be powered by a pulsed source, then mode-locking techniques for powerful lasers at our operating wavelengths with repetition rates of gigahertz have to be developed. In this aspect, lasers do have one considerable advantage over electronics: it is already possible to generate extremely short optical pulses optically. To test ultrahigh-speed electronics, it is common to generate short pulses optically, using a mode-locked laser, and then convert back to an electronic pulse.

We also need methods of homogeneously splitting our power supply; probably a single powerful laser beam into many (10^6) beams for illuminating the non-linear optical devices. At least for devices using critical biasing, the required accuracy of this array generator poses severe problems.

6. Conclusion

The conclusion of this paper is a list of requirements and problems for the different areas involved in building a digital optical computer. Architecturally there seem to be two approaches: fast sequential computers having architectures similar to present computers, and parallel synchronous pipelined computers with novel architectures. The sequential approach requires research on ultrafast, ultralow switching energy devices and on the problems associated with waveguide random interconnects.

The parallel architecture has to deal with the following issues.

For not inherently parallel problems, we have to learn how to map then on a very wide pipeline and how to avoid registering (breaking of the pipeline).
We want an architecture using devices with low fanin and fanout. Furthermore, all devices across an array should be used with the same constant fanin and fanout.
With current devices, architectures relying on large fanin or fanout are unsuitable: the accuracy requirements for the system become so tight that they defeat the idea of digital computing. Also, devices suffering from critical slowing down would require long switching times in such an environment.
Since the number of devices we can interconnect scales with the regularity of the interconnect, the interconnects will have to be as regular as possible.
An architecture which relies on synchronous interconnects allows the use of pulsed logic.
An efficient architecture can probably compete with electronics if we have devices with similar power dissipation and switching times.

Not all of the following requirements for non-linear optical devices are mandatory, but they should serve to judge the computational merits of a proposed device.

The devices must be infinitely cascadable and they must provide a fanin and fanout of at least two (including system losses). Further, they must allow construction of any Boolean function (complete logical cover).

The devices must have multiple input ports (or support multiple modes), to allow isolation and lossless beam combination, and to avoid interference problems.

The energy dissipation must be low (at least comparable with electronics).

The device must be small. If we get near to the quantum limit, it pays to work farther out in the infrared.

The device must be integratable into relatively densely packed two-dimensional arrays. Reliable semiconductor technology is highly developed for electronics and opto-electronics, and can be modified for many non-linear optical devices. The homogeneity of the devices on a two-dimensional array, yield, reliability and other practical questions have to be addressed.

The device should not drift with temperature or other environmental disturbances. Thermal engineering will be extremely important if we want to have high-speed, high-bandwidth systems. This favours temperature-tolerant devices. In any case, efficient cooling systems have to be devised.

Devices without internal feedback (transmission independent from transmitted power supply beam) do not rely on critical biasing, nor do they exhibit critical slowing down. They are, therefore, highly desirable, since they reduce the system accuracy required.

Devices with absolute gain ease system accuracy requirements. For a fanin and fanout of 2 only a small gain is required, depending on the contrast and system losses. High absolute gain is, in many cases, not an advantage.

As high a contrast as possible is desirable, but not at the expense of transmission.

The development of devices which threshold with respect to a local reference would alleviate the requirements on system accuracy. Similarly, local input renormalization before each threshold operation is desirable.

Generally, NOR-gates look better than AND-gates, since they have minimum threshold. Furthermore, a computer built out of NOR-gates does not need to use dual-rail logic.

At the current state of development opto-electronic devices, where some of the energy is provided electrically, may be favoured due to a lack of optical power supplies (lasers) for powering large arrays. This may change in the future [2].

Although there are huge amounts of literature on optical computing, the optical power supply is surprisingly seldom discussed.

We need powerful lasers at the wavelength of a strong optical non-linearity [2]. For a reasonably large computer the cost per watt of laser light deserves serious consideration.

For pulsed logic the laser should produce pulses with gigahertz rates.

For critically biased devices (and to a smaller degree for all other devices as well), we need an accurate array generator to illuminate large arrays of the devices. Alternatively, we need large two-dimensional arrays of equal lasers.

Of course, we have again also to consider questions like reliability, drift, accuracies, etc.

The optical interconnects for a parallel optical computer have the following properties:

For small devices (power consumption!) accurate model imaging over a large field is required to minimize coupling losses. Therefore, we need telecentric systems — preferably diffraction limited, with a large field and low geometrical distortion.

Optical considerations in the design of digital optical computers

To have the means for preventing back-propagation (isolation), and to combine all of the different inputs losslessly, we have to couple into different modes of a device. Also, we do not need interferometric precision in alignment of such a system.

If temporal stacking is to be used, then the path lengths must be controllable to within one pulse length. This is relatively easy with single-lens imaging systems.

For ultrafast pulses it is preferable to use refractive and reflective components (lenses, mirrors, etc.) instead of diffractive components (holograms). The chromatic aberration is easier to compensate. Fast pulses, incidentally, may be used even in systems with comparatively long cycle times.

Since small devices are desirable for low power dissipation, some optical tricks, such as pupil division become difficult to implement (for multiplex rates greater than about 4). Similarly, operations such as magnification and demagnification become difficult, since they screw up the mode patterns.

Although the authors have no conclusive proof, we believe that optical free-space random interconnects can be used for very small numbers of devices. In all the schemes for achieving space variance by multiplexing discussed, the number of devices is reduced by a factor of M, where M denotes the degree of space variance. Since the computing power scales with the number of possible parallel operations, a very low degree of space variance is desirable. There seems to be a very basic trade-off between the degree of space variance and the degree of parallelism.

Appendix
A non-binary non-inverting device

For the switching condition of ideal binary devices it makes no difference whether the characteristic (Fig. 2) is inverting or not. In this section we analyse a non-inverting device with positive differential transmissions (Fig. 2a). The aim is to find fanout and fanin conditions analogous to Equations 7 and 10, but corrected for the differential transmissions. First we have to work out the minimal and the maximal output powers in the LO and HI states, and perform worst-case estimations in Equations 6 and 8. The bias power is chosen such that threshold $-$ 1 HI inputs leave us a half 'switching event' $P_{OUT}^{HI} - P_{OUT}^{LO}$ apart from the switching point P_{sw}. Therefore, the minimal and maximal output powers are given by

$$P_{OUT}^{LO}(\max) = P_{OFF} - T_{LO}\left(\frac{1}{2}\frac{T_{SYS}}{\text{fanout}}(P_{OUT}^{HI} - P_{OUT}^{LO}) - \frac{\Delta P_{sw}}{2}\right) \quad (A1)$$

$$P_{OUT}^{LO}(\min) = P_{OFF} - T_{LO}\left(\frac{\text{threshold} - \frac{1}{2}}{\text{fanout}}T_{SYS}(P_{OUT}^{HI} - P_{OUT}^{LO}) - \frac{\Delta P_{sw}}{2}\right) \quad (A2)$$

$$P_{OUT}^{HI}(\min) = P_{ON} + T_{HI}\left(\frac{1}{2}\frac{T_{SYS}}{\text{fanout}}(P_{OUT}^{HI} - P_{OUT}^{LO}) - \frac{\Delta P_{sw}}{2}\right) \quad (A3)$$

$$P_{OUT}^{HI}(\max) = P_{ON} + T_{HI}\left(\frac{\text{fanin} - (\text{threshold} - \frac{1}{2})}{\text{fanout}}T_{SYS}(P_{OUT}^{HI} - P_{OUT}^{LO}) - \frac{\Delta P_{sw}}{2}\right) \quad (A4)$$

Since we assume the device to be fairly binary (otherwise it is useless), we approximate on the right-hand side of all these equations $P_{OUT}^{HI} - P_{OUT}^{LO} \simeq P_{ON} - P_{OFF}$ and we express everything in terms of the switching transmission, contrast, etc. Assuming that in the worst case the 'switching events' are as small as $P_{\text{OUT}}^{\text{HI}}(\min) - P_{\text{OUT}}^{\text{LO}}(\max)$ we find a modified fanout condition

$$\text{fanout} < \frac{T_{\text{SYS}} T_{\text{SW}} C_{\text{SW}}}{\sigma_{\text{eff}}} \left(1 + \frac{T_{\text{HI}} + T_{\text{LO}}}{2} T_{\text{SYS}} \frac{1}{\text{fanout}}\right) - \frac{T_{\text{HI}} + T_{\text{LO}}}{2} T_{\text{SYS}} \quad \text{(A5)}$$

Similarly, we obtain in the worst case the fanin condition

$$\text{fanin} \left(1 - C_{\text{SW}} - T_{\text{LO}} T_{\text{SYS}} C_{\text{SW}} \frac{1}{2\,\text{fanout}} + T_{\text{HI}} T_{\text{SYS}} C_{\text{SW}} \frac{\text{threshold} - \frac{1}{2}}{\text{fanout}} + \frac{T_{\text{LO}}}{T_{\text{SW}}} \frac{\sigma_{\text{eff}}}{2}\right)$$

$$+ (\text{threshold} - \tfrac{1}{2}) \left(C_{\text{SW}} + (T_{\text{LO}} - T_{\text{HI}}) T_{\text{SYS}} C_{\text{SW}} \frac{\text{threshold} - \frac{1}{2}}{\text{fanout}} - \frac{T_{\text{HI}} + T_{\text{LO}}}{T_{\text{SW}}} \frac{\sigma_{\text{eff}}}{2}\right)$$

$$< \frac{\text{fanout}}{T_{\text{SYS}} T_{\text{SW}}} \quad \text{(A6)}$$

A non-binary inverting device

Completely analogous to the previous section, we can derive the fanout and fanin conditions for a device having a characteristic such as Fig. 2c (inverting with positive differential transmissions). The fanout condition becomes

$$\text{fanout} < \frac{T_{\text{SYS}} T_{\text{SW}} C_{\text{SW}}}{\sigma_{\text{eff}}} \left(1 - (T_{\text{HI}} - T_{\text{LO}}) T_{\text{SYS}} \frac{\text{threshold} - \frac{1}{2}}{\text{fanout}} - T_{\text{LO}} T_{\text{SYS}} \frac{\text{fanin}}{\text{fanout}}\right)$$

$$+ \frac{T_{\text{HI}} + T_{\text{LO}}}{2} T_{\text{SYS}} \quad \text{(A7)}$$

The fanin condition is given by

$$\text{fanin} \left(1 - C_{\text{SW}} + T_{\text{LO}} T_{\text{SYS}} C_{\text{SW}} \frac{\text{fanin}}{\text{fanout}} - T_{\text{LO}} T_{\text{SYS}} C_{\text{SW}} \frac{\text{threshold} - \frac{1}{2}}{\text{fanout}} - \frac{T_{\text{LO}}}{T_{\text{SW}}} \frac{\sigma_{\text{eff}}}{2}\right)$$

$$+ (\text{threshold} - \tfrac{1}{2}) \left(C_{\text{SW}} - (T_{\text{LO}} + T_{\text{HI}}) T_{\text{SYS}} C_{\text{SW}} \frac{1}{2\,\text{fanout}} + \frac{T_{\text{HI}} + T_{\text{LO}}}{T_{\text{SW}}} \frac{\sigma_{\text{eff}}}{2}\right)$$

$$< \frac{\text{fanout}}{T_{\text{SYS}} T_{\text{SW}}} \quad \text{(A8)}$$

Similar (but different) conditions exist for devices where one or both differential transmissions are negative. We did not derive them.

References
1. A. HUANG, in 'IEEE 1983 10th International Computing Conference' (1983) p. 13.
2. L. C. WEST, Ph.D. Thesis (1985) Stanford University, California.
3. J. VON NEUMANN, in 'J. von Neumann: Collected Works', Volume 5 (Macmillan, New York, 1963).
4. H. M. GIBBS, P. MANDEL, N. PEYGHAMBAIAN and S. D. SMITH (eds), 'Optical Bistability III' (1985).
5. A. HUANG, Proc. IEEE 72 (1984) 780.
6. M. J. MURDOCCA, Masters Thesis (1984) Rutgers University, New Jersey.
7. T. YATAGAI, Appl. Opt. 25 (1986) 1571.
8. J. TANIDA and Y. ICHIOKA, ibid. 25 (1986) 1565.
9. K. H. BRENNER, A. HUANG and N. STREIBL, ibid. 25 (1986) 3054.
10. M. J. MURDOCCA, ibid. 26 (1987) 3365.

11. J. L. JEWELL, Y. H. LEE, J. F. DUFFY, A. C. GOSSARD, W. WIEGMANN and J. H. ENGLISH, in 'Proceedings of the Topical Meeting on Optical Bistability' (Tucson, Arizona, December 1985).
12. J. VON NEUMANN, 'J. von Neumann: Collected Works', Volume 5 (Macmillan, New York, 1963) p. 337.
13. B. S. WHERRETT, *IEEE J. Quantum Electron.* **QE-20** (1984) 646.
14. D. A. B. MILLER, D. S. CHEMLA, T. C. DAMEN, T. H. WOOD, C. A. BURRUS, A. C. GOSSARD and W. WIEGMANN, *ibid.* **QE-21** (1985) 1462.
15. D. A. B. MILLER, J. E. HENRY, A. C. GOSSARD and J. H. ENGLISH, *Appl. Phys. Lett.* **49** (1986).
16. E. GARMIRE, J. H. MARBURGER, S. D. ALLEN and H. G. WINFUL, *ibid.* **34** (1979) 374.
17. J. L. JEWELL, M. C. RUSHFORD and H. M. GIBBS, *ibid.* **44** (1984) 172.
18. J. W. GOODMAN, *Opt. Acta* **32** (1985) 1489.
19. R. W. KEYES, *ibid.* **32** (1985) 525.
20. R. W. KEYES, *Science* **230** (1985) 138.
21. A. W. LOHMANN, W. STORCK and G. STUCKE, *Appl. Opt.* **25** (1986) 1530.
22. A. A. SAWCHUK and B. K. JENKINS, in 'Proceedings of SPIE on Optical Computing', Volume 625 (Los Angeles, California, January 1986).
23. J. W. GOODMAN, A. R. DIAS and L. M. WOODY, *Opt. Lett.* **2** (1978) 1.
24. N. F. BORELLI, D. L. MORSE, R. H. BELLMAN and W. L. MORGAN, *Appl. Opt.* **24** (1985) 2520.
25. P. R. HAUGEN, H. BARTELT and S. K. CASE, *ibid.* **22** (1983) 2822.
26. J. W. GOODMAN, F. J. LEONBERGER, S. Y. KUNG and R. A. ATHALE, *Proc. IEEE* **72** (1984) 850.
27. S. A. COLLINS, S. F. HABIBY and P. C. GRIFFITH, in 'Proceedings of SPIE on Optical Computing', Volume 625 (Los Angeles, California, January 1986).
28. K. M. JOHNSON, M. ARMSTRONG, L. HESSELINK and J. W. GOODMAN, *Appl. Opt.* **24** (1985) 4467.
29. T. K. GAYLORD and M. G. MOHARAM, *Proc. IEEE* **73** (1985) 894.
30. P. P. EWALD, *Ann. Physik* **49** (1916) 1.
31. R. W. KEYES and J. A. ARMSTRONG, *Appl. Opt.* **8** (1969) 2549.
32. J. JEWELL, *CLEO* (1986).
33. S. OKTAY and H. C. KAMMERER, *IBM J. Res. Dev.* **26** (1982) 55.
34. E. A. WILSON, 'AFIPS Conference Proceedings, 1977 National Computer Conference' (Dallas, Texas, June 1977).
35. D. B. TUCKERMAN, Ph.D. Thesis (1984) Stanford University.
36. T. TOFFOLI, *Math. Systems Theory* **14** (1981) 13.
37. R. P. FEYNMAN, *Opt. News* (February 1984) 11.

Author Index

A

Alferness, R. C., 68, 71
Anderson, R. H., 191
Arthurs, E., 326, 392

B

Baran, J. E., 108
Bergvall, K., 86
Bogart, G. A., 89
Brackett, C. A., 318
Bradley, P. J., 155
Brenner, K.-H., 244
Bulley, R. M., 318, 392
Burrus, C. A., Jr., 132, 135, 321

C

Chemla, D. S., 132, 135
Chirovsky, L. M. F., 152
Combemale, Y., 59
Cooper, J. M., 326, 328, 392
Cunningham, J. E., 152
Curtis, L., 318

D

Damen, T. C., 132, 135
De Bernardi, C., 389
de Bosio, A., 389
Divino, M. D., 96
Döldissen, W., 86
Donuma, K.-I., 261
Downs, M. M., 283, 410
Dragone, C., 271

E

Eichmann, G., 241
Eisenstein, G., 321
El-Akkari, F. R., 103
English, J. H., 129, 149
Erikson, J. R., 89
Esener, S. C., 229
Evenkow, J. D., Jr., 354

F

Fan, T. R., 307
Feldman, M. R., 229
Fisher, A. D., 204
Friberg, S. R., 225
Fujii, Y., 261
Fujiwara, M., 381, 385

G

Gibbs, H. M., 126
Gimlett, J. L., 318
Giordano, P. P., 369
Glance, B. S., 321
Goldstein, E. L., 392
Goodman, M. S., 318, 326, 328, 392
Gossard, A. C., 129, 135, 149
Granestrand, P., 86
Guest, C. C., 229

H

Hamaguchi, H., 184
Heinrich, H., 86
Henry, J. E., 149, 152
Heritage, J. P., 315
Himeno, A., 176
Hinton, H. S., 38, 152, 405
Hoffmann, D., 86
Horimatsu, T., 184
Hornbeck, L. J., 197
Huang, A., 244
Huisman, R. F., 89

I

Iannone, P. P., 332
Ito, T., 261
Itoh, M., 385
Iwama, T., 184

J

Jahns, J., 247
Jewell, J. L., 126, 129

K

Kaede, K., 381
Kaminow, I. P., 332
Kasper, B. L., 321
Kim, B., 103
Kobayashi, M., 176
Kobrinski, H., 318, 326, 328, 392
Kogelnik, H., 62, 265
Komatsu, K., 363
Kondo, M., 363

L

Lam, D. K. W., 181
Lee, J. N., 204
Lee, S. H., 229
Lentine, A. L., 152
Li, Y., 241
Lohmann, A. W., 239, 287

M

MacDonald, R. I., 181
Makiuchi, M., 184
Marhic, M. E., 253, 256
Mathieu, X., 59
McCall, S. L., 129
McCaughan, L., 81
McCormick, F. B., 283
Melindo, F., 389
Midwinter, J. E., 24, 155, 397
Miller, D. A. B., 132, 135, 149, 152
Mistry, P., 155
Murata, S., 381, 385
Murdocca, M. J., 247

N

Nagashima, K., 363
Netravali, A. N., 345

Neyer, A., 101
Nishio, M., 381, 385
Numai, T., 385

O

Oikawa, Y., 184
Olsson, N. A., 378
O'Mahony, M. J., 159
Oshima, S., 261
Ostrowsky, D. B., 59

P

Padmanabhan, K., 345
Pape, D. R., 197
Papuchon, M., 59, 99
Parry, G., 155
Pate, M. A., 155
Payne, W. A., 405
Perlmutter, P., 108
Pollock, K., 321
Prise, M. E., 283, 410
Prucnal, P. R., 297, 307
Psaltis, D., 191

R

Ramaswamy, V., 96
Rieber, L., 59
Roberts, J. S., 155
Ross, W. E., 191
Roy, A. M., 59, 99
Rushford, M. C., 126

S

Saleh, A. A. M., 265
Salehi, J. A., 315
Santoro, M. A., 297, 307
Sasaki, M., 184
Scherer, A., 129
Schmidt, R. V., 62, 71
Schwider, J., 287
Sehgal, S. K., 297
Sejourne, B., 59
Sfez, B. G., 225
Shimosaka, N., 381
Silberberg, Y., 108, 173, 225

Smith, P. S., 225
Smith, P. W., 5
Smith, S. D., 111
Spanke, R. A., 89, 341
Standley, R. D., 96
Stoltz, B., 86
Stone, J., 332
Stork, W., 239
Streibl, N., 283, 287, 410
Stucke, G., 239
Stulz, L. W., 321, 332
Sugiyama, H., 261
Suzuki, A., 363
Suzuki, S., 363, 381, 385
Syrett, B. A., 181

T

Terakado, T., 363
Terui, H., 176
Thomas, J., 287
Thompson, R. A., 354, 369
Thylen, L., 86
Touge, T., 184
Tsai, C. S., 103
Tsang, W. T., 378
Tur, M., 326

V

Vecchi, M. P., 318, 326, 328, 392

W

Wada, O., 184
Walker, S. J., 283
Weiner, A. M., 225, 315
Werner, M., 59
Wheatley, P., 155
Whitehead, M., 155
Wiegmann, W., 132, 135
Wood, T. H., 132, 135

Y

Yamaguchi, K., 184
Yamazaki, S., 381

Subject Index

A

Active path optical switches, with electrical control, 400-401
All-optical nonlinear logic switches, 111-123
   device design/fabrication, 115-116
   figure of merit for optical circuit elements, 116-117
   optical bistability of, 112, 113, 115-116, 118-121
   optical computation requirements and, 112, 117-118, 121-123
   origin of giant nonlinearities, 113-115
All-optical repeater, 173-175
   advantages/disadvantages of, 173
   device design/fabrication, 173, 175
   optical bistability of, 173
   pulse propagation simulations, 174-175
   saturable characteristics, 173
Alternating delta/beta techniques
   directional couplers using, 71-80
   four sections of alternating delta/beta, 65
   N sections of alternating delta/beta, 65-67
   single step delta/beta reversal, 63-64
   switched directional couplers with, 62-67
   three sections of alternating delta/beta, 64-65
Architectural considerations
   device size, 40
   device-to-device interconnection, 50-52
   digital optical logic devices, 44-50
   directional couplers, 41-44
   for optical logic devices, 52
   parallelism, 39-40
   for photonic switching networks, 38-54
   power/speed/bandwidth, 38-39
   for smart pixel devices, 52-54
   spatial light modulators, 41
   wavelength-division switching systems, 41
   See also Optical switching elements
Array illuminator, based on phase contrast
   experimental verification, 290-292
   limitations of, 288-290
   phase contrast setup, 287-288
   quantitative approach, 288-290
   tolerances, 288-290
Asynchronous time multiplexing, 1
Attenuation characteristics, of large nonblocking optical space switches, 342

B

Balanced bridge modulator switch
   device design/fabrication, 96-98
   Mach-Zehnder interferometer and, 96, 98
   using Ti-diffused $LiNbO_3$ strip waveguides, 96-98
Bar state, polarization-independent optical directional coupler switch, 69-70
Beam-combination, optical digital computer design, 283-286
Beam propagation method (BPM), electro-optic x-switch and, 102
Bias direction, digital optical switch and, 108
Bistable devices, 6-7
   digital optical computers, 412-413
   See also Optical bistability
Bistable switching, all-optical nonlinear logic switches, 112, 113, 115-116, 118-121
Bit-error-rate performance, 4×4 passive splitter/active combiner switch (PSAC), 92-94
Block switching, 1
BOA-coupler. *See* Electrically active optical bifurcation (BOA)
Broadband networks, coherent photonic wavelength-division switching system for, 381-384

C

Circuit switching, 1
Cobra configuration
   electrically switched optical directional coupler and, 59, 60, 61
   switched directional couplers and, 62, 63
Co-channel interference measurement, wavelength division multiplexing coherent optical star network, 324
Code division multiple access (CDMA)
   conventional CDMA sequences, 307-309
   encoding/decoding femtosecond pulses, 315-317
   optical CDMA sequence design, 309-312
   spread spectrum fiber-optic local area network, 307-314
   synchronous CDMA, 300-301
   ultrafast all-optical synchronous multiple access fiber networks, 297-306
Coherent photonic wavelength-division switching system
   broadband Metropolitan-Area-Network and, 382
   for broadband networks, 381-384
   device design/fabrication, 381-382
   experimental results, 382-383
Combinational star couplers
   device design/fabrication, 254-255, 256-257
   distributed stars, 258
   efficiency of, 257
   implementation of, 257-258
   output uniformity, 256-257
   performance tradeoffs, 257
   for single-mode optical fibers, 256-258
   stability, 257
Compact optical generalized perfect shuffle, 241-243
   device design/fabrication, 242
   stretch-mask-add approach and, 242
Conjugate-shift invariant, Gaussian brackets and, 240
Coupling coefficient equalization, low-loss polarization-independent electro-optical switches, 82-83
Coupling strengths, polarization-independent optical directional coupler switch and, 68
Crossover networks
   device design/fabrication, 248
   limitations of, 249-251
   for optical digital computers, 247
   optical implementation of, 247-252
   perfect shuffle/Banyan compatibility, 251-252

D

Deformable mirror device
   device design/fabrication, 197-198
   DMD optical Fourier transform results, 201-203
   experimental setup, 201
   Fourier transform model, 198-201
   future development, 203
   membrane deflection model, 198
   for optical information processing, 197-203
   step grating optical Fourier transform results, 201
Demultiplexing and detection, frequency division multiple access/frequency shift-keyed (FDMA-FSK) star network, 334-336
Deterministic/statistic circuit assignment architectures
   Bitonic Sorter (BS) network characteristics, 390
   for optical switching systems, 389-391
   permutation network characteristics, 389-390
   single path network characteristics, 389, 390
Device design/fabrication
   all-optical nonlinear logic switches, 115-116
   all-optical repeater, 173, 175

Device design/fabrication *(continued)*
  balanced bridge modulator switch, 96-98
  coherent photonic wavelength-division switching system for, 381-382
  combinational star couplers, 254-255, 256-257
  compact optical generalized perfect shuffle, 242
  crossover networks, 248
  deformable mirror device, 197-198
  digital optical switch, 108-110
  dilated networks, 345-346
  directional couplers, 71-73, 78-80
  dual-core-fiber nonlinear coupler, 225-226
  eight-channel wavelength-division switching, 385-386
  electrically active optical bifurcation (BOA), 99-100
  electrically switched optical directional coupler, 59-60
  electro-optic x-switch, 101-102
  4×4 crossbar switch, 89
  4×4 opto-electric integrated circuit switch module, 184-188
  4×4 passive splitter/active combiner switch (PSAC), 90-91, 94
  frequency division multiple access/frequency shift-keyed (FDMA-FSK) star network, 332-333
  GaAs-AlAs monolithic microresonator arrays, 129-131
  guided-wave optical gate matrix switch, 176-179
  hard limiting opto-electronic logic devices, 155-156, 158
  hierarchic star couplers, 253-254
  high performance packet switching system (HYPASS), 328, 330
  high-speed optical time-division switching, 365-367
  hybrid opto-electric integrated circuit, 182-183
  integrated quantum-well self-electro-optic effect device, 149-150
  lithium niobate based photonic switching systems, 405
  low-loss polarization-independent electro-optical switches, 82, 83-84
  optical channel waveguide switch/coupler, 103-105
  optical perfect shuffle, 239
  optical switching/routing system, 378-379
  photonic switching modules, 354-355
  photonic time-slot interchanger, 369-375
  polarization-independent optical directional coupler switch, 69, 70
  quantum-well self-electro-optic effect device, 132-133
  reflective single-mode fiber-optic passive star couplers, 265-270
  semiconductor laser optical amplifiers, 160-161
  strictly nonblocking 8×8 integrated optical switch matrix, 86-87
  switched directional couplers, 62-63
  symmetric self-electro-optic effect device (S-SEED), 152-153
  two-dimensional magneto-optic spatial light modulator, 191-193
  two-dimensional spatial light modulators, 206-212
  wavelength division multiplexing coherent optical star network, 322-323
Device-to-device interconnection, 50-52
Digital optical computers
  accessible othogonal modes of device, 423-425
  beam combination and coherence, 421-422
  bistable devices, 412-413
  computational properties of, 413-414
  computer architecture, 410-412
  coupling, 423
  critical slowing down, 420-421
  error propagation, 416-420
  idealized binary devices, 416
  isolation, 422-423
  limitations of imaging, 426
  nonbinary inverting device, 437
  nonbinary noninverting device, 436-437
  optical digital computing, 410-412
  optical interconnects for, 425
  optical properties of, 414-416
  power dissipation, 431-433
  power supply, 433-434
  quantum limit engineering, 433
  random interconnects, based on volume holograms, 429-431
  space-invariant interconnects, 426
  space-variant systems, 427
  thresholding devices, 412-413

Digital optical logic devices, 44-50
Digital optical switch
  bias direction and, 108
  device design/fabrication, 108-110
  light propagation in, 109
  steplike response of, 109
Digital optics
  device performance, 30-31
  geometrical layouts, 33-34
  neural networks, 35-37
  optical computer development and, 34-35
  optical connections vs. electronics, 25
  optical interconnects, 25-26
  optically activated logic, 27-29
  quantum-well devices, 29-30
  thick holograms, 27
Dilated networks
  device design/fabrication, 345-346
  dilated Benes network, 346-348
  dilation of other multistage networks, 351-352
  network decomposition for compact layout, 350
  partially dilated Benes network, 348-349
  performance analysis, 349-350
  for photonic switching, 345-352
Directional couplers, 41-44
  device design/fabrication, 71-73, 78-80
  directional coupler filter, 78
  directional coupler switch/modulator, 74-78
  electrically switched optical directional coupler, 59-61
  4×4 guided-wave photonic switching systems, 89-94
  low-loss polarization-independent electro-optical switches, 81-84
  optical directional couplers, 73-74
  polarization-independent optical directional coupler switch, 68-70
  strictly nonblocking 8×8 integrated optical switch matrix, 86-88
  switched directional couplers, 62-67
  using alternating delta/beta techniques, 71-80
Directional coupler switch/modulator, 74-78
Dual-core-fiber nonlinear coupler
  device design/fabrication, 225-226
  femtosecond switching in, 225-227
  high-intensity illumination and, 227
  optical glass characteristics, 225
  pulse propagation and, 226-227
Dual-polarization operation, of polarization-independent optical directional coupler switch, 69

E

Efficient $N \times N$ star couplers
  nonuniform arrays, 276-277
  power transfer between circular arrays, 277-278
  using Fourier optics, 271-280
Eight-channel wavelength-division switching
  device design/fabrication, 385-386
  experimental results, 386-387
  using wide-tuning-range DFB LD filters, 385-388
Electrical bistability, of quantum-well self-electro-optic effect device, 142-143
Electrical interconnect model, 230-231
Electrically active optical bifurcation (BOA), 99-100
  device design/fabrication, 99-100
Electrically switched optical directional coupler, 59-61
  Cobra configuration and, 59, 60, 61
  fabrication technique, 59-60
Electronically-addressed devices (E-SLMs), two-dimensional spatial light modulators, 210-212
Electro-optic devices
  balanced bridge modulator switch, 96-98
  digital optical switch, 108-110
  directional coupler filter, 78
  directional coupler switch/modulator, 74-78
  electrically active optical bifurcation (BOA), 99-100

electrically switched optical directional coupler, 59-61
electro-optic x-switch, 101-102
  4 × 4 crossbar switch, 89-90
  4 × 4 guided-wave systems, 89-94
  4 × 4 passive splitter/active combiner switch (PSAC), 90-94
  low-loss polarization-independent electro-optical switches, 81-84
  optical channel waveguide switch/coupler, 103-106
  optical directional couplers, 73-74
  polarization-independent optical directional coupler switch, 68-70
  strictly nonblocking 8 × 8 integrated optical switch matrix, 86-88
  switched directional couplers, 62-67
Electro-optic x-switch
  beam propagation method (BPM) and, 102
  device design/fabrication, 101-102
  experimental results, 102
  using single-mode Ti: $LiNbO_3$ channel waveguides, 101-102
Encoding/decoding femtosecond pulses, 315-317
  CDMA optical telecommunications network, 315-317

F

Fan-out considerations, optical vs. electrical interconnects, 233-235
Fast wavelength tuning, opto-electronic packet switch using, 392-395
Feedback behavior, of quantum-well self-electro-optic effect device, 137-139
Femtosecond switching, in dual-core-fiber nonlinear coupler, 225-227
Fiber-optic local-area networks (FOLANs), star couplers and, 253-255
Fiber-to-fiber insertion loss measurements, 4 × 4 passive splitter/active combiner switch (PSAC), 91
4 × 4 crossbar switch, 89-90
  device design/fabrication, 89
  long term system performance, 89-90
4 × 4 guided-wave photonic switching systems
  4 × 4 crossbar switch, 89-90
  4 × 4 passive splitter/active combiner switch (PSAC), 90-94
4 × 4 optical switching network, 17-18
4 × 4 opto-electric integrated circuit switch module
  device design/fabrication, 184-188
  multichannel optical coupling, 185-187
  package scheme for, 185-187
  receiver OEIC chip, 185
  transmission experiment results, 187-188
  transmitter OEIC chip, 184-185
  using GaAs substrate, 184-188
4 × 4 passive splitter/active combiner switch (PSAC)
  applications of, 91
  bit-error-rate performance, 92-94
  contrast ratio drift under constant voltage, 91-92
  device design/fabrication, 90-91, 94
  fiber-to-fiber insertion loss measurements, 91
  initial contrast ratios and voltages, 91
  output power drift over time, 92
Fourier transform model, deformable mirror device, 198-201
Frequency division multiple access/frequency shift-keyed (FDMA-FSK) star network, 332-339
  demultiplexing and detection, 334-336
  device design/fabrication, 332-333
  experimental results, 337-339
  laser phase noise effects, 336-337
  single/tandem Fabry-Perot filters, 333-334
  with tunable optical filter demultiplexer, 332-339
Frequency tunable cleaved-coupled-cavity semiconductor lasers, optical switching/routing system using, 378-379

G

GaAs-AlAs monolithic microresonator arrays, 129-131
  device design/fabrication, 129-131
  pixel size and, 130
Gain dependence on polarization/temperature, semiconductor laser optical amplifiers, 163-164
Geometrical layouts, digital optics, 33-34

Gray scale considerations, two-dimensional magneto-optic spatial light modulator, 194-196
Guided-wave optical gate matrix switch
  device design/fabrication, 176-179
  high-silica guided-wave optical circuits, 176-178
  LD-gate integration, 179
  optical gates, 178
  switch configuration, 176
  switching characteristics, 179-180

H

Hard limiting opto-electronic logic devices, 155-158
  device design/fabrication, 155-156, 158
  inverting characteristics of, 157
  operational characteristics of, 156-157
  quantum efficiency of, 156, 157
Hierarchic star couplers, device design/fabrication, 253-254
High-intensity illumination, dual-core-fiber nonlinear coupler and, 227
High performance packet switching system (HYPASS)
  architecture for opto-electronic packet switch, 392-393
  device design/fabrication, 328, 330
  optical device requirements for, 328-329
High-silica guided-wave optical circuits, guided-wave optical gate matrix switch, 176-178
High-speed optical switch, 17
High-speed optical time-division switching
  bistable laser diode as optical memory, 364-365
  device design/fabrication, 365-367
  experimental results, 367
  optical switch as write/read gates, 365
  switching networks classification, 363-364
Holographic perfect shuffles, 245-246
Hybrid optically bistable switch, quantum-well self-electro-optic effect device as, 132-134
Hybrid opto-electric integrated circuit
  device design/fabrication, 182-183
  photoconductive detector arrays (PCDAs), 181
  planar optical waveguide, 181-182
Hybrid switching devices
  4 × 4 opto-electric integrated circuit switch module, 184-188
  hybrid opto-electric integrated circuit, 181-183

I

IF channel selection, wavelength division multiplexing coherent optical star network, 323
Image amplifier, 18-19
Influence of cavity (FP amplifier), semiconductor laser optical amplifiers, 162-163
Integrated quantum-well self-electro-optic effect device
  device design/fabrication, 149-150
  quantum-confined Stark effect (QCSE) and, 149
  2 × 2 array of optically bistable switches, 149-151
Interferometric perfect shuffles, 246
Inverting characteristic, of hard limiting opto-electronic logic devices, 157

L

LAMBDANET architecture
  device design/fabrication, 319
  high capacity of, 318-320
  as multiwavelength optical network, 318-320
  star configuration of, 318-320
Large nonblocking optical space switches
  architectures for, 341-344
  attenuation characteristics, 342
  optical component requirements, 341
  SNR characteristics, 342-343
  switch dimension limits, 343-344
Laser phase noise effects, frequency division multiple access/frequency shift-keyed (FDMA-FSK) star network, 336-337

LD-gate interaction, guided-wave optical gate matrix switch, 179
Light propagation, in digital optical switch, 109
Linear polarization, polarization-independent optical directional coupler switch and, 68
Linear repeater, semiconductor laser optical amplifiers, 165-167
Lithium niobate based photonic switching systems
   applications for, 408
   architectural considerations, 407-408
   component development for, 405
   device design/fabrication, 405
   optical amplifiers, 406-407
   polarization maintaining fiber, 405-406
   switch architectures and techniques, 407
   synchronization, 407-408
   transmission formats, 408
Local-area networks (LANs), star couplers for, 261-264
Logical switching elements
   all-optical nonlinear logic switches, 111-123
   GaAs-AlAs monolithic microresonator arrays, 129-131
   hard limiting opto-electronic logic devices, 155-158
   hybrid optically bistable switch, 132-134
   integrated quantum-well self-electro-optic effect device, 149-151
   nonlinear Fabry-Perot etalon, 126-128
   quantum-well SEED, 132-134
   quantum-well self-electro-optic effect device, 135-147
   symmetric self-electro-optic effect device (S-SEED), 152-154
Low-energy optical switch, 17
Low-loss polarization-independent electro-optical switches
   coupling coefficient equalization, 82-83
   device design/fabrication, 82, 83-84
   waveguide loss minimization, 82

M

Mach-Zehnder interferometer, balanced bridge modulator switch and, 96, 98
Membrane deflection model, deformable mirror device, 198
Multichannel optical coupling, 4×4 optoelectric integrated circuit switch module, 185-187
Multistage 4×4 elements, photonic switching modules, 358, 359
Multiwavelength optical crossconnect
   FOX architecture for, 326, 327
   for parallel-processing computers, 326-327

N

Negative resistance optoelectric oscillation, quantum-well self-electro-optic effect device, 143-144
Neural networks, digital optics, 35-37
Noise, semiconductor laser optical amplifiers and, 164
Nonlinear Fabry-Perot etalon
   relaxation characteristics of, 126-127
   transmission vs. time characteristics, 126
   used as optical logic gates, 126-128
Nonlinear Fabry-Perot structures
   all-optical nonlinear logic switches, 111-123
   GaAs-AlAs monolithic microresonator arrays, 129-131
   used as optical logic gates, 126-128

O

One-stage 4×4 element, photonic switching modules, 357-358
Optical amplifiers
   all-optical repeater, 173-175
   guided-wave optical gate matrix switch, 176-180
   lithium niobate based photonic switching systems, 406-407
   semiconductor laser optical amplifiers, 159-161
Optical bistability
   of all-optical repeater, 173
   integrated quantum-well self-electro-optic effect device and, 149-151
   quantum-well self-electro-optic effect device, 132-134, 139-142
   of symmetric self-electro-optic effect device (S-SEED), 152, 154
   of two-dimensional spatial light modulators, 209, 215
   See also Bistable devices
Optical channel waveguide switch/coupler
   device design/fabrication, 103-105
   using total internal reflection, 103-106
Optical computation requirements, all-optical nonlinear logic switches and, 112, 117-118, 121-123
Optical digital computer design
   beam combination using patterned reflectors, 283-284
   device characteristics, 283
   digital optics and, 34-35
   input/output module for, 284-286
   optical requirements, 283
   split and shift implementation, 283
Optical digital computers, crossover networks for, 247
Optical directional couplers, 73-74
Optical vs. electrical interconnects
   electrical interconnect model, 230-231
   fan-out considerations, 233-235
   improved optical link parameters, 236
   light modulators as optical signal transmitters, 236-237
   optical interconnect model, 229-230
   scaling electronic circuit dimensions, 235-236
   switching energy comparisons, 231-233
Optical fiber reentrant delay-line memories, photonic time-slot interchanger using, 369-376
Optical gain block, semiconductor laser optical amplifiers, 165
Optical gates, guided-wave optical gate matrix switch, 178
Optical glass characteristics, for dual-core-fiber nonlinear coupler, 225
Optical information processing, deformable mirror device and, 197-203
Optical interconnects, 25-26
   combinational star couplers, 254-255, 256-258
   compact optical generalized perfect shuffle, 241-243
   conjugate-shift invariant, 240
   crossover networks, 247-252
   efficient $N \times N$ star couplers, 271-280
   vs. electrical interconnects, 229-238
   hierarchic star couplers, 253-254
   model for, 229-230
   optical perfect shuffle, 239-240
   perfect shuffles, 244-246
   reflective single-mode fiber-optic passive star couplers, 265-270
   small loss-deviation tapered fiber star coupler, 261-264
Optical logic devices, 52
Optical logic gates, nonlinear Fabry-Perot etalon used as, 126-128
Optically activated logic, 27-29
Optically addressed devices (O-SLMs), two-dimensional spatial light modulators, 206-210
Optically controlled electronic logic, photonic switching, 401
Optically controlled optical logic, photonic switching, 401-402
Optical perfect shuffle, 239-240
   device design/fabrication, 239
Optical processing, spread spectrum fiber-optic local area network, 307-314
Optical receiver preamplifier, semiconductor laser optical amplifiers, 167-168
Optical set-reset latch, symmetric self-electro-optic effect device (S-SEED), 152-154
Optical switching elements
   bistable optical devices, 6-7
   compared to other switching technologies, 15-16
   4×4 optical switching network, 17-18
   high-speed optical switch, 17
   image amplifier, 18-19
   low-energy optical switch, 17
   optical time-division multiplexer/demultiplexer, 19-22
   size limitations, 12-15
   speed/power limitations, 7-12
Optical switching/routing system
   device design/fabrication, 378-379
   using frequency tunable cleaved-coupled-cavity semiconductor lasers, 378-379
Optical time-division multiplexer/demultiplexer, 19-22
Optimized multistage 4×4 Clos network, 360

Optimized multistage 4 × 4 elements, photonic switching modules, 358-360
Opto-electronic packet switch
  HYPASS architecture for, 392-393
  rapidly tunable receiver experiment, 393-395
  rapidly tunable transmitter experiment, 393, 394
  using fast wavelength tuning, 392-395

P

Packet switching, 1
  deterministic/statistic circuit assignment architectures for, 389-391
  opto-electronic packet switch, 392-395
Parallelism, architectural consideration, 39-40
Parallel-processing computers, multiwavelength optical crossconnect, 326-327
Passive pathway switches with electrical control, photonic switching, 398-400
Perfect shuffles
  compact optical generalized perfect shuffle, 241-243
  conjugate-shift invariant, 240
  crossover networks and, 247-252
  definition of, 244-245
  holographic perfect shuffles, 245-246
  interferometric perfect shuffles, 246
  optical interconnection implementations of, 244-246
  optical perfect shuffle, 239-240
Photoconductive detector arrays (PCDAs), hybrid opto-electric integrated circuit, 181
Photonic switching
  active path optical switches with electrical control, 400-401
  architectural considerations for networks, 38-54
  asynchronous time multiplexing, 1
  block switching, 1
  circuit switching, 1
  coherent photonic wavelength-division switching system, 381-384
  component characteristics vs. network requirements, 397-403
  device size, 40
  device-to-device interconnection, 50-52
  digital optical logic devices, 44-50
  dilated networks for, 345-352
  directional couplers, 41-44
  eight-channel wavelength-division switching, 385-388
  4 × 4 guided-wave systems, 89-94
  lithium niobate based photonic switching systems, 405-408
  modules with laser diode amplifier, 354-361
  for optical logic devices, 52
  optically controlled electronic logic, 401
  optically controlled optical logic, 401-402
  optimum component/technology choices, 403
  packet switching, 1
  parallelism, 39-40
  passive pathway switches with electrical control, 398-400
  photonic time-slot interchanger, 369-376
  power/speed/bandwidth, 38-39
  for smart pixel devices, 52-54
  spatial light modulators, 41
  for video applications, 89-94
  wavelength-division switching systems, 41
  wavelength routing technologies, 402-403
Photonic switching modules
  device design/fabrication, 354-355
  with laser diode amplifiers, 354-361
  multistage 4 × 4 elements, 358, 359
  one-stage 4 × 4 element, 357-358
  optimized multistage 4 × 4 Clos network, 360
  optimized multistage 4 × 4 elements, 358-360
  2 × 2 element, 355-357
Photonic time-slot interchanger
  device design/fabrication, 369-375
  future research in, 375-376
  using optical fiber reentrant delay-line memories, 369-376

Pixel size
  GaAs-AlAs monolithic microresonator arrays and, 130
  two-dimensional magneto-optic spatial light modulator and, 194-196
Planar optical waveguide, hybrid opto-electric integrated circuit, 181-182
Polarization-independent optical directional coupler switch
  bar state, 69-70
  coupling strengths and, 68
  device design/fabrication, 69, 70
  dual-polarization operation of, 69
  linear polarization and, 68
  switching efficiency and, 68
  using weighted coupling, 68-70
Polarization-maintaining fiber, lithium niobate based photonic switching systems, 405-406
Pulse propagation, dual-core-fiber nonlinear coupler and, 226-227

Q

Quantum-confined Stark effect (QCSE)
  integrated quantum-well self-electro-optic effect device and, 149
  quantum-well self-electro-optic effect device and, 135-136
Quantum-well devices, 29-30
Quantum-well self-electro-optic effect device
  absorption characteristics, 132, 133
  device design/fabrication, 132-133
  electrical bistability of, 142-143
  experimental procedures for, 136-137
  feedback behavior of, 137-139
  as hybrid optically bistable switch, 132-134
  negative resistance opto-electric oscillation, 143-144
  optical bistability of, 139-142
  optical stability of, 132-133
  performance scaling, 146-147
  quantum-confined Stark effect (QCSE) and, 135-136
  self-linearized modulation and optical level shifting, 144-146

R

Receiver OEIC chip, 4 × 4 opto-electric integrated circuit switch module, 185
Receiver sensitivity measurement, wavelength division multiplexing coherent optical star network, 323-324
Reflective single-mode fiber-optic passive star couplers, 265-270
  device design/fabrication, 265-270
  reflective n-stars, 267-269
  transmissive $n \times n$ star couplers, 266

S

Self-electro-optic effect device (SEED)
  hard limiting opto-electronic logic devices, 155-158
  integrated quantum-well self-electro-optic effect device, 149-151
  quantum-well self-electro-optic effect device, 132-134, 135-147
  symmetric self-electro-optic effect device (S-SEED), 152-154
Self-linearized modulation, quantum-well self-electro-optic effect device, 144-146
Semiconductor laser optical amplifiers
  amplifier equations, 161-162
  device design/fabrication, 160-161
  gain dependence on polarization/temperature, 163-164
  influence of cavity (FP amplifier), 162-163
  linear applications, 164-165
  linear repeater, 165-167
  noise and, 164
  nonlinear applications, 169-170
  optical gain block, 165
  optical receiver preamplifier, 167-168
  use in future fiber systems, 159-171
Signal processing, two-dimensional magneto-optic spatial light modulator and, 193-194
Signal-to-noise (SNR) characteristics, of large nonblocking optical space switches, 342-343
Single-mode optical fibers, for combinational star couplers, 256-258

Single/tandem Fabry-Perot filters, frequency division multiple access/frequency shift-keyed (FDMA-FSK) star network, 333-334
Size limitations, optical switching elements, 12-15
Small loss-deviation tapered fiber star coupler
   device design/fabrication, 261-263
   for LAN applications, 261-264
   redundancy optimization of number of parts, 263
   taper ratio of, 262-263
   transmission coefficient of, 261-262
Smart pixel devices, 52-54
Space-division switching
   dilated networks, 345-352
   large nonblocking optical space switches, 341-344
   photonic switching modules, 354-361
Spatial light modulators, 41
   deformable mirror device, 197-203
   two-dimensional magneto-optic spatial light modulator, 193-194
   two-dimensional spatial light modulators, 204-219
Speed/power limitations, optical switching elements, 7-12
Spot-array-generation, array illuminator based on phase contrast, 287-293
Spread spectrum fiber-optic local area network
   conventional CDMA sequences, 307-309
   experimental results, 312-313
   optical CDMA sequence design, 309-312
   for optical processing, 307-314
Star couplers
   combinational star couplers, 254-255, 256-258
   distributed stars, 258
   efficient $N \times N$ star couplers, 271-280
   fiber-optic local-area networks (FOLANs), 253-255
   hierarchic star couplers, 253-254
   reflective n-stars, 267-269
   reflective single-mode fiber-optic passive star couplers, 265-270
   small loss-deviation tapered fiber star coupler, 261-264
   transmissive $n \times n$ star couplers, 266
Strictly nonblocking 8×8 integrated optical switch matrix, 86-88
   device design/fabrication, 86-87
Switch dimension limits, of large nonblocking optical space switches, 343-344
Switched directional couplers
   Cobra configuration and, 62, 63
   device design/fabrication, 62-63
   four sections of alternating delta/beta, 65
   N sections of alternating delta/beta, 65-67
   single step delta/beta reversal, 63-64
   three sections of alternating delta/beta, 64-65
Switches
   coherent photonic wavelength-division switching system, 381-384
   dilated networks, 345-352
   eight-channel wavelength-division switching, 385-388
   high-speed optical time-division switching, 363-368
   large nonblocking optical space switches, 341-344
   optical switching/routing system, 378-379
   photonic switching modules, 354-361
   photonic time-slot interchanger, 369-376
Switching efficiency, polarization-independent optical directional coupler switch and, 68
Switching networks
   encoding/decoding femtosecond pulses, 315-317
   frequency division multiple access/frequency shift-keyed (FDMA-FSK) star network, 332-339
   high performance packet switching system (HYPASS), 328-331
   LAMBDANET architecture, 318-320
   multiwavelength optical crossconnect, 326-327
   spread spectrum fiber-optic local area network, 307-314
   ultrafast all-optical synchronous multiple access fiber networks, 297-306
   wavelength division multiplexing coherent optical star network, 321-325
Symmetric self-electro-optic effect device (S-SEED)
   device design/fabrication, 152-153
   vs. D-SEED operation, 152
   input/output characteristics of, 153-154
   optical bistability of, 152, 154
   optical set-reset latch, 152-154

T

Thick holograms, 27
Ti-diffused $LiNbO_3$ strip waveguides, balanced bridge modulator switch using, 96-98
Time division multiple access (TDMA)
   fixed assignment TDMA, 297-300
   ultrafast all-optical synchronous multiple access fiber networks, 297-306
Time-division switching
   high-speed optical time-division switching, 363-368
   photonic time-slot interchanger, 369-376
Total internal reflection switch, optical channel waveguide switch/coupler, 103-106
Transmission formats, for lithium niobate based photonic switching systems, 408
Transmitter OEIC chip, 4×4 opto-electric integrated circuit switch module, 184-185
Tunable optical filter demultiplexer, frequency division multiple access/frequency shift-keyed (FDMA-FSK) star network with, 332-339
2×2 element, photonic switching modules, 355-357
Two-dimensional magneto-optic spatial light modulator
   device design/fabrication, 191-193
   gray scale considerations, 194-196
   magneto-optic device and, 191-193
   for signal processing, 193-194
Two-dimensional spatial light modulators, 204-219
   device design/fabrication, 206-212
   early applications of, 204
   electronically addressed devices (E-SLMs), 210-212
   functional capabilities, 212-216
   future applications of, 218-219
   optical bistability of, 209, 215
   optically addressed devices (O-SLMs), 206-210
   performance characteristics, 216-218

U

Ultrafast all-optical synchronous multiple access fiber networks
   feasibility analysis, 304-305
   fixed assignment TDMA, 297-300
   power requirements, 301-304
   synchronous CDMA, 300-301
Ultrafast devices (subpicosecond), dual-core-fiber nonlinear coupler, 225-227

W

Wave guide loss minimization, low-loss polarization-independent electro-optical switches, 82
Wavelength division multiple access (WDMA)
   high performance packet switching system (HYPASS), 328-331
   multiwavelength optical crossconnect, 326-327
   wavelength division multiplexing coherent optical star network, 321-325
Wavelength division multiplexing coherent optical star network
   circuit description, 322-323
   co-channel interference measurement, 324
   IF channel selection, 323
   receiver sensitivity measurement, 323-324
   system throughput, 324-325
Wavelength-division switching, 41
   coherent photonic wavelength-division switching system, 381-384
   eight-channel wavelength-division switching, 385-388
   optical switching/routing system, 378-379
Weighted coupling, polarization-independent optical directional coupler switch using, 68-70
Wide-tuning-range DFB LD filters, eight-channel wavelength-division switching and, 385-388

X

X-switches, electro-optic x-switch, 101-102

Editors' Biographies

H. Scott Hinton received the BSEE degree from Brigham Young University in 1981 and the MSEE degree from Purdue University in 1982. He is currently the Department Head of the Photonic Switching Department at AT&T Bell Laboratories in Naperville, Illinois, where he has worked since 1981. He has been active in the field of photonic switching, having published numerous articles and co-edited two books, and having been granted 10 U.S. patents.

He is a member of both the Switching Committee and the Optical Communications Committee of the IEEE Communications Society, and served as the chairman of the Joint Subcommittee on Photonic Switching from 1987-1989. Mr. Hinton has organized several workshops on photonic switching in addition to serving on the program committee for the 1987 Topical Meeting on photonic switching, and as co-program chairman for the 1989 and 1990 Topical Meetings on photonic switching, and chairman of the 1991 Topical Meeting on photonic switching. Mr. Hinton is also serving as the 1989-1990 Chairman of the Chicago Chapter of the IEEE LEOS. He is a member of the IEEE and OSA.

John E. Midwinter (M'69, SM'81, F'83) received the BSc in physics in 1961 from King's College London. After working at the Royal Radar Establishment on Lasers and Non-Linear Optics, he also earned the PhD degree from London University in 1968. He subsequently spent several years in the research laboratories of the Perkin Elmer and Allied Chemical corporations in the United States. He returned to the UK in 1971 to head the development of optical fibers at the British Post Office (later to become British Telecom Research Labs [BTRL]). In 1977 he was assigned responsibility for the Optical Communications Technology program encompassing the advanced optical systems work in BTRL, where he saw the introduction of the first fiber systems and the development of the single-mode fiber and device technology. In 1984 he left BTRL and joined the Electronics Department at University College London, initially as British Telecom Professor of Optoelectronics and becoming Head of Department in 1988. While there, he has assembled a group to study digital optics and photonic switching.

Professor Midwinter is author of approximately 120 papers and several books, and editor of several books on photonic switching. He as worked in an editorial capacity for several IEEE (*JLT, LSM, JSAC,* etc.) and other journals. He was co-chairman of the 1989 IEEE/OSA Topical Meeting on photonic switching and program co-chairman of the 1987 meeting. He has served on a number of IEEE-COMSOC committees and is active in the UK in the Institution of Electrical Engineers, the Fellowship of Engineering, and The Royal Society.